Number Systems

Rational numbers e.g., $-\frac{2}{3}$, 1.62	
Integers e.g., $-2, 7, -3$ Whole numbers e.g., 0, 1, 3 Natural numbers e.g., 1, 5, 8	Irrational numbers e.g., $\sqrt{3}$, π, $\sqrt{35}$

Real numbers

Some Important Properties and Formulas

Fundamental Principle of Fractions

$$\frac{a}{b} = \frac{a \cdot k}{b \cdot k} \qquad b, k \neq 0$$

Zero-Product Rule

If $a \cdot b = 0$, then either $a = 0$ or $b = 0$ (or both).

Properties of Radicals for Nonnegative a and b

1. $\sqrt{ab} = \sqrt{a}\sqrt{b}$

2. $\sqrt{\dfrac{a}{b}} = \dfrac{\sqrt{a}}{\sqrt{b}} \qquad b \neq 0$

Equations of a Line

Point-slope form: $\quad y - y_1 = m(x - x_1)$

Slope-intercept form: $\quad y = mx + b$

The Quadratic Formula

The solutions to the equation $ax^2 + bx + c = 0$ are

$$x = \frac{-b \pm \sqrt{b^2 - 4ac}}{2a} \qquad \text{(where } a \neq 0\text{)}$$

Understanding Elementary Algebra with Geometry

A Course for College Students

Fifth Edition

Lewis Hirsch
Rutgers University

Arthur Goodman
Queens College of the City
University of New York

BROOKS/COLE

THOMSON LEARNING

Australia ■ Canada ■ Mexico ■ Singapore ■ Spain ■ United Kingdom ■ United States

BROOKS/COLE

THOMSON LEARNING ™

Sponsoring Editor: *Jennifer Huber*
Marketing Team: *Leah Thomson,*
  Samantha Cabaluna
Editorial Assistant: *Jonathan Wegner*
Project Editor: *Keith Faivre*
Production Service: *Susan L. Reiland*
Interior and Cover Design: *Andrew Ogus*
  Book Design

Cover Illustration: *Judith Harkness*
Print Buyer: *Vena Dyer*
Typesetting: *TSI Graphics*
Printing and Binding: *R.R. Donnelley &*
  Sons–Willard

For more information about this or any other Brooks/Cole product, contact:
BROOKS/COLE
511 Forest Lodge Road
Pacific Grove, CA 93950 USA
www.brookscole.com
1-800-423-0563 (Thomson Learning Academic Resource Center)

Library of Congress Cataloging-in-Publication Data

Hirsch, Lewis.
  Understanding elementary algebra with geometry: a course for college students/Lewis
  Hirsch, Arthur Goodman.--5th ed.
      p.   cm.
  Goodman's name appears first on the previous eds.
  Includes index.
  ISBN 0-534-38124-3
  1. Algebra. 2. Geometry. I. Goodman, Arthur. II. Title.

QA152.3 .H36 2002   2001035388
512′.12--dc21

CONTENTS

CHAPTER 4 Rational Expressions 145

CHAPTER 5 Graphing Straight Lines 203

CHAPTER 6 Interpreting Graphs and Systems of Linear Equations 255

C H A P T E R 7   Exponents and Polynomials 297

C H A P T E R 8   Factoring 345

C H A P T E R 9   More Rational Expressions 381

CHAPTER 10 Radical Expressions and Equations 429

CHAPTER 11 Quadratic Equations 465

CHAPTER G Geometry 513

APPENDIX A A Review of Arithmetic 595

APPENDIX B Using a Scientific Calculator 617

APPENDIX C Using a Graphing Calculator 623

APPENDIX D Introduction to Functions 629

PREFACE TO THE INSTRUCTOR

This fifth edition of *Understanding Elementary Algebra with Geometry* retains the same basic structure and philosophy as the previous editions.

Purpose

Understanding Elementary Algebra with Geometry, 5th edition, is an attempt on our part to offer a textbook that reflects our philosophy—that students can *understand* what they are doing in algebra and why.

We offer a view of algebra that takes every opportunity to explain why things are done in a certain way, and to show how supposedly "new" topics are actually just new applications of concepts already learned.

This book assumes only a basic knowledge of arithmetic. Appendix A includes a brief review of the arithmetic of decimals and percents.

Pedagogy

We believe that a student can successfully learn elementary algebra by mastering a few basic concepts and being able to apply them to a wide variety of situations. Thus, each section begins by relating the topic to be discussed to material previously learned. In this way the students can see algebra as a whole rather than as a series of isolated ideas.

Basic concepts, rules, and definitions are motivated and explained via numerical and algebraic examples. Formal proofs have been avoided except for those occasions when they illuminate the discussion.

Concepts are developed in a series of carefully constructed illustrative examples. Through the course of these examples we compare and contrast related ideas, helping the student to understand the sometimes subtle distinctions among various situations. In addition, these examples strengthen a student's understanding of how this "new" idea fits into the overall picture.

Every opportunity has been taken to point out common errors often made by students and to explain the misconception that often leads to a particular error.

Basic rules and/or procedures are highlighted so that students can find important ideas quickly and easily.

A spiral approach has been used for the presentation of some more difficult topics. That is, a topic is first presented at an elementary level and then returned to at increasing levels of complexity.

For example,

Simple rational expressions are covered in Chapter 4, whereas more complex rational expressions are dealt with in Chapter 9.

Factoring is covered in Section 2.3, and in Chapters 8 and 9.

Applications are covered repeatedly throughout the text.

A number of topics have been motivated by describing an application in which a particular type of equation is used to model a situation. We then return to this very same application after the necessary algebraic techniques have been developed. For example,

See the introductions to Chapter 4, Chapter 5, Chapter 8, Chapter 9, Section 10.5, and Chapter 11.

Additionally, special attention has been paid to using pattern recognition and numerical examples to illustrate the mathematical model being used in solving various applications.

Features

Examples The various steps in the solutions to examples are explained in detail. Many steps appear with annotations (highlighted in color) that involve the student in the solution. These comments explain how and why a solution is proceeding in a certain way.

Exercises There are over 3,000 homework exercises. Not only have the exercises been matched odd/even, but they have also been designed so that, in many situations, successive odd-numbered exercises compare and contrast subtle differences in applying the concepts covered in the section. Additionally, variety has been added to the exercise sets so that the student must be alert as to what the problem is asking.

For example, the exercise sets in Sections 4.3 and 9.3, which deal primarily with adding rational expressions, also contain some exercises on multiplying and dividing rational expressions. The exercise set in Section 4.4 on solving fractional equations also asks the student to combine rational expressions. The exercise sets in Sections 8.4 and 11.1, which deal primarily with quadratic equations, contain some linear equations as well.

Study Skill One of the main sources of students' difficulties is that they do not know how to study algebra. In this regard we offer a totally unique feature. Each section in the first four chapters concludes with a Study Skill. This is a brief paragraph discussing some aspect of studying algebra, doing homework, or preparing for or taking exams. Our students who have used the earlier editions of this book indicated that they found the Study Skills very helpful. The Algebra Study Skills sections in this text are based on ideas in the book *Studying Mathematics* by Mary Catherine Hudspeth and Lewis R. Hirsch (1982, Kendall/Hunt Publishing Company, Dubuque, Iowa). For more information and ideas on improving mathematics learning, we direct you to that book.

Questions for Thought Almost every exercise set contains Questions for Thought, which offer the student an opportunity to *think* about various algebraic ideas. They may be asked to compare and contrast related ideas, or examine an incorrect solution and explain why the solution is wrong. The Questions for Thought are intended to be answered in complete sentences and in grammatically correct English. The Questions for Thought can also be used by instructors as a vehicle for having students write across the curriculum.

Margin Comments Margin Comments have been added where appropriate in order to involve the students more actively in the learning process as they read the text. A margin comment will usually seek to emphasize a point made in the text presentation or ask a question requiring the student to focus attention on a particularly crucial aspect of the discussion.

Different Perspectives Different Perspectives boxes appear wherever there is an opportunity to highlight the connection between algebraic and geometric aspects of the same concept. In this way the student is encouraged to think about mathematical ideas from more than one point of view.

Thinking Out Loud Thinking Out Loud is a feature in which the solution to certain examples is presented in a question-and-answer format so that students can see examples of the thought processes involved in approaching and solving new or unfamiliar problems. This helps students develop more appropriate problem-solving strategies.

Mini-Review   Most sections contain a Mini-Review, which consists of exercises that allow students to periodically review important topics as well as help them prepare for the material to come. These Mini-Reviews afford the student additional opportunity to see new topics within the framework of what they have already learned.

Chapter Summary   Each chapter contains a chapter summary describing the basic concepts in the chapter. Each point listed in the summary is accompanied by an example illustrating the concept or procedure.

Review Exercises   There are over 750 review exercises. Each chapter contains a set of chapter review exercises and a chapter practice test. Additionally, there are four cumulative review exercise sets and four cumulative practice tests following Chapters 3, 6, 9, and 11. These offer the student more opportunities to practice choosing the appropriate procedure in a variety of situations.

Answer Section   The answer section contains answers to all the odd-numbered exercises, as well as to *all* the mini-reviews, chapter and cumulative review exercises, and practice test problems. The answer to each verbal problem contains a description of what the variable(s) represent and the equation (or system of equations) used to solve it. In addition, the answers to the cumulative review exercises and cumulative practice tests contain a reference to the section in which the relevant material is covered.

Using a Calculator   As in previous editions, exercises that require the use of a calculator have not been specially designated. We assume that the calculator is at the student's side for all problems, and part of the learning process is determining when a calculator is appropriate and/or necessary. Appendix B discusses the basic use of a scientific calculator.

New to the Fifth Edition

- A number of the exercise sets have been completely rewritten. Many of the exercise sets have been expanded to include exercise types suggested by reviewers.
- Many of the applications have been updated. Examples and exercises using real-world data (frequently in tabular form) have been added.
- Chapter G on geometry has been expanded to include a more detailed discussion of congruence and similarity.
- A new Appendix C discussing some graphing calculator issues relevant to this text has been added.
- A new Appendix D, An Introduction to Functions, has been added.
- The Thinking Out Loud feature in which an example is analyzed in a question-and-answer format has been applied to additional examples.
- A new Study Skill on Seeking Help has been added.
- Icons point students to material contained on the Brooks/Cole Website and on the Interactive Video Skillbuilder CD-ROM that accompanies the text.

Teaching Tools for the Instructor

Annotated Instructor's Edition (ISBN 0-534-38896-5)   This special version of the complete student text has answers printed next to the respective exercises. Graphs, tables, and other answers too long to appear next to their exercises are in a special answer section in the back of the text.

Test Bank (ISBN 0-534-38362-9)   The test bank includes 8 tests per chapter as well as 3 final exams. The tests are made up of a combination of multiple-choice, free-response, true/false, and fill-in-the-blank questions.

Complete Solutions Manual (ISBN 0-534-38359-9) The complete solutions manual provides worked-out solutions to all of the problems in the text.

BCA Testing (ISBN 0-534-38358-0) With a balance of efficiency and high-performance, simplicity and versatility, *Brooks/Cole Assessment* gives you the power to transform the learning and teaching experience. This revolutionary, internet-ready testing suite is text-specific and allows instructors to customize exams and track student progress in an accessible, browser-based format. BCA offers full algorithmic generation of problems and free-response mathematics. No longer are you limited to multiple-choice or true/false test questions. The complete integration of the testing and course management components simplifies your routine tasks. Test results flow automatically to your gradebook and you can easily communicate to individuals, sections, or entire courses.

Text-Specific Videotapes (ISBN 0-534-38363-7) This set of videotapes is available free upon adoption of the text. Each tape offers one chapter of the text and is broken down into ten 20-minute problem-solving lessons that cover each section of the chapter.

Learning Tools for the Student

Student Solutions Manual (ISBN 0-534-38361-0) The student solutions manual provides worked-out solutions to the odd-numbered problems in the text.

BCA Tutorial Instructor (ISBN 0-534-38841-8) and Student Versions (ISBN 0-534-38840-X) This text-specific, interactive tutorial software is delivered via the web (at http://bca.brookscole.com) and is offered in both student and instructor versions. Like *BCA Testing*, it is browser-based, making it an intuitive mathematical guide even for students with little technological proficiency. So sophisticated, it's simple, *BCA Tutorial* allows students to work with real math notation in real time, providing instant analysis and feedback. The tracking program built into the instructor version of the software enables instructors to carefully monitor student progress.

Interactive Video Skillbuilder CD (ISBN 0-534-38360-2) Packaged with each book, this is a single CD-ROM containing over eight hours of video instruction. There is at least one video lesson for each section of the book. The problems worked during each video lesson are listed next to the viewing screen, so students can work them ahead of time, if they choose. In order to help students evaluate their progress, each section contains a 10-question web quiz and each chapter contains a chapter test, with answers to each problem on each test.

Acknowledgments

The authors would like to thank the following reviewers for their helpful comments and suggestions: Theresa A. Barrie, Texas Southern University; Robert Begin, Blinn College; Laurette Blakey Foster, Prairie View A&M University; John Burghduff, Kingwood College; Mary Jane Ferguson, Houston Community College; Francis Foster, Oklahoma City Community College; Bob Harbison, Blinn College; Robert M. Kaufmann, University of Alabama, Birmingham; Terri Seirer, San Jacinto College; Charles Wheeler, Montgomery College.

The production of a textbook is a collaborative effort. We must thank our editor, Jennifer Huber, for her support and encouragement; project editor Keith Faivre, for his coordination of the entire production; assistant editor Julie Foster, for her help in preparing the ancillaries package that accompanies the text; and Sudhir Goel of Valdosta State University for his assistance in checking solutions.

Finally, we would like to thank our wives, Cindy and Sora, and our families for their constant support and encouragement.

PREFACE TO THE STUDENT

This text is designed to help you understand algebra. We are convinced that if you understand what you are doing and why, you will be a much better algebra student. (Our students who have used this book in its previous editions seem to agree with us.) This does not mean that after reading each section you will understand all the concepts clearly. Much of what you learn comes through the course of doing lots and lots of exercises and seeing for yourself exactly what is involved in completing an exercise. However, if you read the textbook carefully and take good notes in class, you will find algebra not quite so menacing.

Here are a few suggestions for using this textbook:

- Always read the textbook with a pencil and paper in hand. Reading mathematics is not like reading other subjects. *You* must be involved in the learning process. Work out the examples along with the textbook and *think* about what you are reading. Make sure you understand what is being done and why.

- You must work homework exercises on a daily basis. While attending class and listening to your instructor are important, do not mistake understanding someone else's work for the ability to do the work yourself. (Think about watching someone else driving a car, as opposed to driving yourself.) Make sure *you* know how to do the exercises.

- Read the Study Skills that appear at the end of each section in the first four chapters. They discuss the best ways to use the textbook and your notes. They also offer a variety of suggestions on how to study, do homework, and prepare for and take tests. If you want more information on improving your algebra study skills, we direct you to the book *Studying Mathematics,* by Mary Catherine Hudspeth and Lewis R. Hirsch (1982, Kendall/Hunt Publishing Company, Dubuque, Iowa). There is a complete list of the Study Skills (with page references) on the inside front cover of the book.

- This text assumes that you have (and know how to use) a scientific calculator. Appendix B discusses and illustrates some of the basic ways you will be using a scientific calculator in this course.

- Do not get discouraged if you have difficulty with some topics. Certain topics may not be absolutely clear the first time you see them. Be persistent. We all need time to absorb new ideas and become familiar with them. What was initially difficult will become less so as you spend more time with a subject. Keep at it and you will see that you are making steady progress.

CHAPTER 1

The Integers

STUDY SKILLS

1.1 What Is Algebra? Introduction and Basic Notation

Algebra allows us to make statements about mumbers in general without specifying in advance which numbers we are talking about.

When asked to explain what algebra is, most students have difficulty formulating a response. It is not a very difficult question, and it has a very straightforward answer. Algebra is a language. It happens to be the language of mathematics.

We are going to learn this language in the same way we would learn any new language. We will begin by learning the alphabet (that is, the symbols) we will be using. For the most part our alphabet consists of letters and symbols with which we are already familiar, such as the letters of the English alphabet, Arabic numerals, and the basic symbols of arithmetic. Next, we will learn the "grammar" of our new language, that is, the rules for putting the symbols together and manipulating them. After we learn the structure of the language, we can begin to actually use algebra to solve problems.

In some sense we can say that algebra is the generalization of arithmetic. We are going to let letters represent numbers and state our rules and our conclusions using letters, so that they will be valid for many or all numbers.

Sets

One concept that is used frequently is the idea of a set. The word *set* is used in mathematics in much the same way it is used in everyday life. A *set* is simply a well-defined collection of objects. The phrase *well-defined* means that there are clearly determined criteria for membership in the set. The criteria can be a list of those objects in the set, called the **elements** or **members** of the set, or they can describe those objects in the set.

For example, it is not sufficient to say "the set of all tall people in the class." *Tall* is a subjective criterion. It is possible to make the set well defined by saying "the set of all people in the class more than 6 feet tall."

One way to represent a set is to list the elements of the set and enclose the list in "set braces," which look like { }.

We often designate sets by using capital letters such as A, B, C. For example,

$$A = \{3, 4, 5, 8\}$$

$$B = \{a, e, i, o, u\}$$

$$C = \{red, white, blue\}$$

are three sets.

The symbol we use to indicate that an object is a member of a particular set is $\in$. Thus, $x \in S$ is a symbolic way of writing that x is a member or an *element* of S. We use the symbol $\notin$ to indicate that an object is *not* an element of a set. [In general, when we put a "/" through a mathematical symbol it means *not*. Thus, "$\neq$" means *not equal*.] For example, using the sets A, B, and C listed above, we have:

$5 \in A$ 5 is an element of A.

$p \notin B$ p is not an element of B.

In order to exhibit a set that contains many elements or a set that contains an infinite number of elements, we use a variation on the listing method. For example, the set $\{2, 4, 6, \ldots, 100\}$ represents the set of even numbers greater than 0 and less than or equal to 100. The three dots mean that the set continues according to the same pattern. The set $O = \{1, 3, 5, \ldots\}$ is the set of all odd numbers greater than 0. There is no number after the dots because this set is infinite; it has no last element.

Of course, this method of listing a set can be used only when the first few elements clearly show the pattern for *all* the elements in the set.

Certain frequently used sets of numbers are given special names.

The set of numbers we use for counting is called the set of **natural numbers** and is usually denoted with the letter N:

$$N = \{1, 2, 3, \ldots\}$$

If we add the number 0 to this set it is called the set of **whole numbers** and is denoted with the letter W:

$$W = \{0, 1, 2, 3, \ldots\}$$

Often when we describe a set we use the word "between," which can be ambiguous. When we say "the numbers between 5 and 10" do we mean to include or exclude 5 and 10? Let's agree that when we say "between" we mean "in between" and we do *not* include the first and last numbers.

EXAMPLE 1

List the elements of the following sets:

(a) The set A of whole numbers between 6 and 30

(b) The set B of odd numbers greater than 17

Solution

(a) The whole numbers are the same as the natural numbers, except that the whole numbers include 0. Note that 6 and 30 are not included.

$$A = \{7, 8, 9, \ldots, 29\} \quad {}^{*}$$

(b) Since no upper limit to this set is given, the answer is

$$B = \{19, 21, 23, 25, \ldots\} \qquad \textit{Note that } 17 \textit{ is not included.} \quad \blacksquare$$

Sometimes we cannot *list* the elements of a set, but rather we must describe the set. When this is the case we use what is called *set-builder notation*. **Set-builder notation** consists of the set braces, a **variable** that acts as a placeholder, a vertical bar (|) read "such that," and a sentence that describes what the variable can be. This last part is called the **condition** on the variable. For example:

Set-builder notation specifies a set by describing the elements of the set rather than listing them.

$$\{ \quad x \quad | \quad x \text{ is an even number greater than 0 and less than 10}\}$$

$$\uparrow \qquad \uparrow \qquad\qquad\qquad\qquad \uparrow$$

$$\textit{Variable} \quad \textit{Such that} \qquad\qquad \textit{Condition on the variable}$$

This is read "the set of all x such that x is an even number greater than 0 and less than 10," which is the set $\{2, 4, 6, 8\}$.

EXAMPLE 2

List the elements of the following set:

$$\{x \mid x \text{ is a whole number divisible by 3}\}$$

Solution

The number 0 is included because $0 \div 3 = 0$. Thus, our answer is

$$\{0, 3, 6, 9, \ldots\} \qquad\qquad\qquad \blacksquare$$

EXAMPLE 3

For each of the following sets, find another description using set-builder notation and also list the members of the set.

(a) $A = \{x \mid x \text{ is an even natural number less than 12}\}$

(b) $B = \{t \mid t \text{ is a natural number between 1 and 65 that ends in a zero}\}$

Solution

(a) If x is an even natural number less than 12, then x must be one of the numbers 2, 4, 6, 8, 10. Therefore we have $A = \{2, 4, 6, 8, 10\}$.

Another way to describe set A is

$$A = \{x \mid x \text{ is a natural number less than 12 that is divisible by 2}\}.$$

(b) If t is a natural number between 1 and 65 that ends in a zero, then t must be one of the numbers 10, 20, 30, 40, 50, 60. Therefore we have $A = \{10, 20, 30, 40, 50, 60\}$.

---

*Throughout the text we will use color boxes to indicate the final answer to an example.

Another way to describe set A is

$$A = \{t \mid t \text{ is one of the first six natural-number multiples of } 10\}$$

It is possible to place a condition on a set that no elements satisfy, as, for example,

$$F = \{x \mid x \text{ is an odd number divisible by } 2\}$$

Since it is impossible for an odd number to be divisible by 2, the set F has no members. It is called the **empty set** or the **null set** and it is symbolized by $\varnothing$. Thus, we have $F = \varnothing$.

Before we can continue we must introduce some terminology and notation.

Sums, Terms, Products, and Factors

Most of the errors made by students in algebra are the result of confusing terms with factors and factors with terms. We can do some things with factors that we cannot do with terms, and vice versa. We will define them now for arithmetic expressions and point out the differences throughout the book.

Sum is the word we use for addition. In an expression involving a sum, the numbers to be added in the sum are called the **terms.** The symbol used to indicate a sum is the familiar "+" sign.

Product is the word we use for multiplication. In an expression involving a product, the numbers being multiplied are called the **factors.** Saying that "a is a **multiple** of b" is equivalent to saying that "b is a **factor** of a." For example, we can say

20 is a multiple of 5   or   5 is a factor of 20.

48 is a multiple of 8   or   8 is a factor of 48.

Thus, a factor of n is a number that divides into n exactly, whereas a multiple of n is a number that is exactly divisible by n.

In algebra, we generally use the symbol "·" to indicate multiplication. We do not use the "×" to indicate multiplication because we very often use x in our work as a variable. Frequently we will also indicate a product simply by writing numbers or expressions next to each other with the appropriate "punctuation." For example, if we let x represent a number, then

Sums can be written as:	*Products can be written as:*
$3 + 4$ (the sum of 3 and 4)	$3 \cdot 4$ (the product of 3 and 4)
$7 + x$ (the sum of 7 and x)	$7 \cdot x$ (the product of 7 and x)
Note that in the sum $7 + x$, 7 and x are the terms.	$7x$ (also the product of 7 and x)
	Note that in the product $7x$, 7 and x are the factors.

If there is no operation symbol between two variables or between a number and a variable, such as in xy or $7a$, multiplication is understood. However, we cannot write "3 times 4" as 34. If we want to indicate multiplication of *numbers,* there must be some punctuation between the two numbers. We have already said that we can use the "·" and write $3 \cdot 4$. Alternatively, we can write "3 times 4" as (3)(4) or 3(4) or (3)4; in this way, the 3 and 4 are next to each other to indicate multiplication but the parentheses show us that 3 and 4 are two separate numbers.

EXAMPLE 4

List the elements of the following sets:

(a) $\{x \mid x \text{ is a whole number less than 30 and a multiple of } 5\}$

(b) $\{y \mid y \text{ is a natural number multiple of } 7\}$

(c) $\{f \mid f \text{ is a natural number factor of } 36\}$

Solution

(a) The answer is

$$\{0, 5, 10, 15, 20, 25\}$$

Note that 0 is included because 0 is a multiple of 5, since $0 = 0 \cdot 5$.

(b) The answer is

$$\{7, 14, 21, \ldots\}$$

(c) The answer is

$$\{1, 2, 3, 4, 6, 9, 12, 18, 36\}$$

 ■

If each element of set A is contained in set B, then we say that set A is a **subset** of set B. Thus, the set of even numbers greater than 0 is a subset of the natural numbers.

There is another subset of the natural numbers that is very important. It is called the set of *prime numbers*.

DEFINITION

A **prime number** is a natural number (excluding 1) that is divisible only by itself and 1. A natural number greater than 1 that is not prime is called **composite.**

In other words, a prime number is a natural number greater than 1 whose only factors are 1 and itself. For example, the numbers 5 and 13 are prime numbers because they are not divisible by any number other than themselves and 1. The number 12 is composite (not prime) because it is divisible by other natural numbers, such as 3 and 4.

The set of prime numbers is a perfect example of the necessity for set-builder notation. If we simply start listing the set of prime numbers, we would write

$$P = \{2, 3, 5, 7, 11, 13, 17, 19, 23, \ldots\}$$

Since the prime numbers have no pattern, unless someone knows this set is the set of prime numbers, they cannot tell what the next number in the set will be. On the other hand, having the definition of a prime number, we can write

$$P = \{m \mid m \text{ is a prime number}\}$$

and then we can determine whether or not a number is in the set.

Every composite number can be broken down (the word that is usually used is *factored*) into its prime factors. For example, we can break down 30 as follows:

$$30 = 2 \cdot 15 = \boxed{2 \cdot 3 \cdot 5}$$

Basically, we can pick *any* two factors we recognize and start with them. Then we continue breaking the factors down until *all* the factors are prime numbers.

EXAMPLE 5

Factor the number 48 into its prime factors.

Solution

We have many choices for the first two factors. We will illustrate just two of the possible paths to the answer:

$$48 = 2 \cdot 24 \qquad \textit{Factor 24.}$$
$$= 2 \cdot 2 \cdot 12 \qquad \textit{Factor 12.}$$
$$= 2 \cdot 2 \cdot 2 \cdot 6 \qquad \textit{Factor 6.}$$
$$= \boxed{2 \cdot 2 \cdot 2 \cdot 2 \cdot 3}$$

or

$$48 = 8 \cdot 6 \qquad \textit{Factor 8 and 6.}$$
$$= 2 \cdot 4 \cdot 2 \cdot 3 \qquad \textit{Factor 4.}$$
$$= \boxed{2 \cdot 2 \cdot 2 \cdot 2 \cdot 3}$$

No matter which factors you decide to start with, the final answer (since it involves *prime* factors only) will be the same. ■

The Number Line

The number line gives us a very useful geometric representation of the various sets of numbers with which we will be working.

Since our basic set of numbers so far is the set of whole numbers, we will, for the time being, associate the set of whole numbers with points on the number line in the following way. First, we draw a horizontal line. We mark off some point on the line and label it 0. Then we mark off another point to the right of 0 and label it 1. The distance between 0 and 1 is called the **unit length.**

We continue to mark off points 1 unit length apart moving toward the right. The arrow at the end of the number line indicates that the numbers are getting *larger* as we move in the direction of the arrow—to the right:

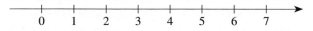

Thus, we have associated each whole number with the point on the number line *that many units to the right of 0.*

If we are asked to "graph" a set on the number line, it means that we want to indicate those points on the number line that are in the set. We usually indicate this by putting a heavy dot at those points that are in the given set.

EXAMPLE 6 Graph the set {3, 4, 6, 9} on the number line.

Solution

The number line gives us a very simple way of defining the idea of "order." For example, 3 is less than 7 because 3 is to the left of 7 on the number line. In general, we define a to be **less than** b if a is to the left of b on the number line. The symbol we use for "less than" is "<".

$a < b$ means that a is to the left of b on the number line.

$4 < 9$ is the symbolic statement for "4 is less than 9,"
which means that 4 is to the left of 9 on the number line.

Similarly, the symbol ">" is used for the expression **greater than.**

$a > b$ is the symbolic statement for "a is greater than b,"
which means that a is to the right of b on the number line.

The symbols "<" and ">" are called **inequality symbols.**

$$\text{lesser} < \text{greater} \qquad \text{greater} > \text{lesser}$$

Note that the inequality symbol always points toward the *smaller* number.

The accompanying box contains a list of all the equality and inequality symbols we will use, what each means, and one or two *true* statements using each symbol.

Equality and Inequality Symbols			
	$a = b$	a "equals" b	$7 + 3 = 12 - 2$
	$a \neq b$	a "is not equal to" b	$4 + 5 \neq 10$
	$a < b$	a "is less than" b	$3 < 8$
	$a \leq b$	a "is less than *or* equal to" b	$3 \leq 8; \ 3 \leq 3$
	$a > b$	a "is greater than" b	$7 - 2 > 4$
	$a \geq b$	a "is greater than *or* equal to" b	$6 \geq 6; \ 6 \geq 2$

Note that $a \leq b$ means that *either* $a < b$ **or** $a = b$, and similarly for $a \geq b$.

EXAMPLE 7   List the elements of each of the following sets and represent them on the number line.

(a) $\{x \mid x \in N \text{ and } x \leq 5\}$

(b) $\{a \mid a \in W \text{ and } a \geq 3 \text{ and } a < 9\}$

Solution   (a) The letter N stands for the set of natural numbers; therefore, the answer is

$\{1, 2, 3, 4, 5\}$ *Note that 5 is included, but 0 is excluded. Why?*

(b) The letter W stands for the set of whole numbers. Therefore, the answer is

$\{3, 4, 5, 6, 7, 8\}$ *Note that 3 is included but 9 is excluded.*   ■

Study Skills 1.1

Studying Algebra—How Often?

In most college courses, you are typically expected to spend 2 to 4 hours studying outside of class for every hour spent in class.

It is especially important that you spend this amount of time studying algebra since you must both *acquire* and *perfect* skills. As most of you who play a musical instrument or participate seriously in athletics already should know, it takes time and lots of practice to develop and perfect a skill.

It is also important that you distribute your studying over time. That is, do not try to do all your studying in 1, 2, or even 3 days, and then skip studying the other days. You will find that understanding algebra and acquiring the necessary skills are much easier if you spread your studying out over the week, doing a little each day. If you study in this way, you will need less time to study just before exams.

In addition, if your study sessions are more than an hour long, it is a good idea to take a 10-minute break within every hour you spend reading math or working exercises. This "break" helps to clear your mind, and allows you to think more clearly.

www EXERCISES 1.1

In Exercises 1–10, indicate whether the given statement is true or false.

1. $3 \in \{1, 3, 5, 8\}$

2. $5 \notin \{1, 3, 5, 8\}$

3. $8 \notin \{2, 3, 5, 7, 9\}$

4. $7 \in \{2, 4, 6, 9, 17\}$

5. $a \in \{b, c, d, a\}$

6. $g \notin \{r, g, c, b\}$

7. $17 \in N$

8. $12 \notin N$

9. $0 \in N$

10. $0 \in W$

In Exercises 11–34, list the elements of each of the following sets. Unless otherwise specified, assume that all numbers are whole numbers.

11. $\{x \mid x$ is a natural number less than 8$\}$

12. $\{x \mid x \in N$ and $x < 8\}$

13. $\{y \mid y$ is an even number less than 20$\}$

14. $\{m \mid m$ is an odd number greater than 7$\}$

15. $\{x \mid x \leq 6\}$

16. $\{n \mid n < 9\}$

17. $\{x \mid x < 6\}$

18. $\{n \mid n \geq 9\}$

19. $\{x \mid x \geq 6\}$

20. $\{n \mid n > 9\}$

21. $\{x \mid x > 6\}$

22. $\{n \mid n \leq 9\}$

23. $\{a \mid a$ is greater than 2 and less than 6$\}$

24. $\{a \mid a$ is greater than 6 and less than 2$\}$

25. $\{t \mid t$ is a factor of 24$\}$

26. $\{t \mid t$ is a factor of 30$\}$

27. $\{f \mid f$ is a factor of 12 or a factor of 15$\}$

28. $\{f \mid f$ is a factor of 12 and a factor of 15$\}$

29. $\{m \mid m$ is a multiple of 4$\}$

30. $\{n \mid n$ is a multiple of 5$\}$

31. $\{m \mid m$ is a multiple of 3 and a multiple of 4$\}$

32. $\{n \mid n$ is a multiple of 2 and a multiple of 5$\}$

33. $\{x \mid x$ is a multiple of 10 and not divisible by 5$\}$

34. $\{y \mid y$ is a multiple of 5 and not divisible by 10$\}$

In Exercises 35–40, find another description of the set using set-builder notation and also list the set using the roster method.

35. $A = \{x \mid x$ is an odd natural number less than 20$\}$

36. $B = \{x \mid x$ is an even natural number less than 20$\}$

37. $C = \{t \mid t$ is a natural number less than 50 that ends in a 5$\}$

38. $D = \{w \mid w$ is a natural number less than 60 that ends in a 0$\}$

39. $S = \{t \mid t$ is a natural number greater than 20 that ends in a double zero$\}$

40. $T = \{w \mid w$ is a natural number greater than 40 that ends in a triple zero$\}$

In Exercises 41–50, fill in the appropriate ordering symbol: either <, >, or =.

41. 4 _____ 2

42. 8 _____ 20

43. 7_____ 7

44. 5 _____ 0

45. 19 _____ 24 − 10

46. 18 − 5 _____ 10 + 4

47. 19 + 53 _____ 72

48. 16 _____ 12 + 4

49. 8 − 8_____ 4 · 0

50. 8 + 0 _____ 4 · 2

In Exercises 51–60, fill in as many of the following ordering symbols as make the statement true. Choose from <, ≤, >, ≥, =, ≠.

51. 5_____ 8

52. 17 _____ 12

53. 43 _____ 43

54. 5_____ 2 + 4

55. 4 · 2 _____ 4 + 2

56. 1 · 2 _____ 1 + 2

57. 6 · 0 _____ 6 + 0

58. 25 − 5 _____ 14 + 3

59. 16 _____ 12 − 5

60. 4 − 4 _____ 17 · 0

In Exercises 61–84, factor the given number into its prime factors. If the number is prime, say so.

61. 14	62. 26
63. 33	64. 35
65. 30	66. 50
67. 37	68. 80
69. 64	70. 41
71. 96	72. 120
73. 87	74. 91
75. 360	76. 420
77. 126	78. 165
79. 858	80. 561
81. 912	82. 1,332
83. 1,904	84. 3,024

In Exercises 85–88, graph the given set on a number line.

85. $\{0, 3, 7\}$

86. $\{x \mid x \text{ is an even natural number less than } 10\}$

87. $\{a \mid a \text{ is a natural number less than } 12 \text{ and not divisible by } 3\}$

88. $\{1, 4, 5, 8\}$

QUESTIONS FOR THOUGHT

89. What is the difference between the set N (the set of natural numbers) and the set W (the set of whole numbers)?

90. What is the difference between a *term* and a *factor*? Give an example to illustrate the difference.

91. What is the difference between a *factor* and a *multiple*? Give an example to illustrate the difference.

92. Let set A be the first ten multiples of 2 and let set B be the first ten multiples of 3. Let C be the set of numbers both sets have in common. How would you describe set C?

93. Repeat Exercise 92 with multiples of 4 and 6. Is the description of set C similar? Why or why not?

94. Describe each of the following expressions as a sum or a product. If the expression is a sum, list the terms. If the expression is a product, list the factors.

(a) $x + y + z$ (b) xyz (c) $xy + z$ (d) $x(y + z)$

1.2   Integers

Our everyday usage of numbers often requires us to indicate direction as well as size. For example, a checking account in which we have \$180 will show a balance of \$180,

TRANSACTION	PAYMENT		BALANCE	
Dan Smith	240	00	180	00
Car repair	110	00	70	00
Registration fee	250	00	−180	00

while a checking account in which we are overdrawn by $180 will show a balance of –$180.

Signed numbers tell us not only quantity or size, but also direction. Perhaps the most common everyday use of signed numbers is in describing temperatures: −10°F is 10 degrees *below* 0, while +20°F means 20 degrees *above* 0.

Mathematically, we want to extend our number system beyond the whole numbers to include these signed numbers as well. We do this as follows. We begin with our number line for the whole numbers:

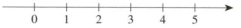

Now we extend the line to the left of 0 and mark off unit lengths going off to the *left*:

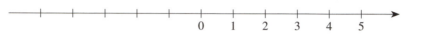

What notation shall we use to designate the units to the left of 0? How shall we name these points?

Rather than trying to memorize a new set of symbols, we can use the same symbols that we use to name units to the right of 0, except we place the symbol "–" before these numbers to indicate that they are to the *left* of 0. Our number line now looks like this:

We use the minus sign both for subtraction and to indicate a negative number.

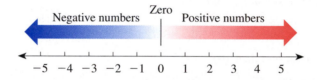

Thus, +2 is two units to the *right* of 0, while −2, which is read "negative 2," is 2 units to the *left* of 0. The set of numbers $\{\ldots, -3, -2, -1, 0, 1, 2, 3, \ldots\}$ is called the set of **integers** and is usually denoted with the letter Z (you will see later in this chapter that we are saving the letter I for a different set):

$$Z = \{\ldots, -3, -2, -1, 0, 1, 2, 3, \ldots\}$$

The set $\{1, 2, 3, \ldots\}$, which we have called the natural numbers, is also often called the set of **positive integers.** The set $\{-1, -2, -3, \ldots\}$ is called the set of **negative integers.** Note that the number 0, although it is an integer, is neither positive nor negative.

Very frequently when we write positive integers we do not write the "+" sign. That is, 5 is understood to mean +5. However, in this chapter, to make things clearer, we will usually write the "extra" + sign.

You might wonder about choosing the minus sign to indicate a negative number, when the minus sign already means subtraction. Will this lead to confusion? As we shall soon see, our various uses and interpretations of the minus sign will be consistent, and we will be free to think of the minus sign in any of several ways.

Opposites

Notice on the number line that −4 is the same distance from 0 as +4, but −4 is in the opposite direction from 0 than is +4 (see Figure 1.1). Similarly for +2 and −2, and +14 and −14. For that reason we call the pair of numbers x and $-x$ **opposites.**

Figure 1.1
Opposites

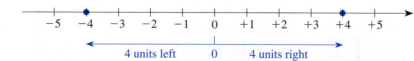

The number +4 is 4 units to the right of 0, and −4 is 4 units from 0 in the *opposite* direction. We can generalize this and say that if x is *any* number then $-x$ will be the same distance from 0 in the opposite direction.

Thus, in addition to naming the points on the left side of the number line, the minus (or negative) sign is also used to change the direction of a number.

> Putting a minus sign in front of a number changes a number into its opposite.

What we have just said is true for both positive and negative integers. That is,

$-(+5)$ can be read "the opposite of $+5$," which is -5

and in exactly the same way

$-(-5)$ can be read "the opposite of -5," which is $+5$

In general, we have "the opposite of $-x$ is x," regardless of whether x *itself* is positive or negative. That is,

$$-(-x) = x$$

regardless of whether x is positive or negative.

Our ideas about order on the number line continue to be true. That is, a is smaller than b means a is to the left of b, regardless of whether a and b are positive or negative. Thus, the following are true statements:

$2 < 7$ (2 is less than 7)   because 2 is to the left of 7 on the number line.

$-2 > -7$ (−2 is greater than −7)   because −2 is to the right of −7 on the number line.

$$\begin{array}{ccccccccccccccc} \mid & \mid & \mid & \mid & \mid & \mid & \mid & \mid & \mid & \mid & \mid & \mid & \mid & \mid \\ -7 & -6 & -5 & -4 & -3 & -2 & -1 & 0 & 1 & 2 & 3 & 4 & 5 & 6 & 7 \end{array}$$

Even though −7 is further away from 0 than 2 is, as integers −7 is smaller because it is further to the left. To use more familiar examples:

A temperature of $-9°$ is *lower* than a temperature of $-5°$.

40 feet *below* sea level (written -40 feet) is *higher* than 80 feet below sea level (written -80 feet).

As we shall see in the next section, there are times when we are not interested in the direction of a number, but only in its distance from 0.

DEFINITION | The *absolute value* of a number is its distance from 0 on the number line. The absolute value of x is written $|x|$.

Thus, for example, we have:

$$|5| = 5 \qquad |-5| = 5 \qquad |234| = 234 \qquad |-399| = 399 \qquad |0| = 0$$

EXAMPLE 1 | Compute each of the following:

(a) $|7 - 3|$ (b) $-|-6|$ (c) $-(-6)$

Solution | (a) Do not interpret the absolute value sign to mean "erase all negative signs." That is not what it means. We must first see what number is *inside* the absolute value sign, and then we take its distance from 0.

$$|7 - 3| = |4| = \boxed{4}$$

(b) Watch the minus signs very carefully. We compute the absolute value first.

$$-|-6| = -(6) \qquad \textit{Note that the first minus sign gets copied.}$$
$$= \boxed{-6}$$

$-|-6|$ is read "the opposite of the absolute value of −6."

(c) $-(-6)$ can be read as "the opposite of -6" and so the answer is $\boxed{6}$.

Note the difference between the meaning of this expression, $-(-6)$, and the expression $-|-6|$ in part **(b)**. ∎

Properties of the Integers

Although we are discussing the properties of the integers, all of the properties mentioned here apply as well to the sets of rational and real numbers, which we will discuss later in the text.

The foundation for much of the work that we do in algebra is the properties of the integers. For example, we are familiar with the fact that we can add or multiply two whole numbers in any order and get the same answer. That is,

$$6 + 2 = 2 + 6 \quad \text{and} \quad 6 \cdot 2 = 2 \cdot 6$$

These facts, which are true for integers (and in fact for all the numbers with which we will work), are called the **commutative properties.** We can state them algebraically as follows:

Commutative Properties		
	Commutative Property of Addition	$a + b = b + a$
	Commutative Property of Multiplication	$ab = ba$

The emphasis here should be placed on the fact that this property applies to addition and multiplication *only*. A similar property is *not* true for subtraction and division. That is, we cannot change the order of the numbers in a subtraction or division example and expect to get the same answer. For example,

$$6 - 2 \neq 2 - 6 \quad \text{and} \quad 6 \div 2 \neq 2 \div 6$$

Frequently we want to indicate that certain numbers or expressions are to be grouped together. We usually indicate these groupings by using parentheses like ( ) or brackets like [ ] or braces like { }. For example,

$4(3 \cdot 7)$ means multiply 3 by 7 and then multiply the result by 4

while

$(4 \cdot 3)7$ means multiply 4 by 3 and then multiply the result by 7

It is no surprise that in both cases we get the same answer of 84, because we know that we can group multiplications in any way we please. The same holds true for addition. These facts are called the **associative properties.** We can state them algebraically as follows:

Associative Properties		
	Associative Property of Addition	$a + (b + c) = (a + b) + c$
	Associative Property of Multiplication	$a(bc) = (ab)c$

As with the commutative properties, the associative properties do not pertain to subtraction or division. For example,

$$10 - (5 - 2) \neq (10 - 5) - 2 \qquad 36 \div (6 \div 3) \neq (36 \div 6) \div 3$$
$$10 - 3 \neq 5 - 2 \qquad\qquad\qquad 36 \div 2 \neq 6 \div 3$$
$$7 \neq 3 \qquad\qquad\qquad\qquad 18 \neq 2$$

In words, the commutative property says that we can *reorder* within a sum or a product, while the associative property says that we can *regroup* within a sum or a product.

The commutative and associative properties allow us to change expressions from one form into another. For example, if we begin with $4 + (x + 8)$ we can obtain the following:

$$
\begin{aligned}
4 + (x + 8) &= (4 + x) + 8 &&\textit{By using the associative property of addition} \\
&= (x + 4) + 8 &&\textit{By using the commutative property of addition} \\
&= x + (4 + 8) &&\textit{By using the associative property of addition} \\
&= x + 12 &&\textit{By addition}
\end{aligned}
$$

Procedures of this kind serve as the basis for much of the work we do in algebra.

Order of Operations

If we look at the expression $5 + 4 \cdot 3$, we can interpret the problem in two ways: Do we mean

$$5 + 4 \cdot 3 \overset{?}{=} 9 \cdot 3 \overset{?}{=} 27 \qquad \text{or} \qquad 5 + 4 \cdot 3 \overset{?}{=} 5 + 12 \overset{?}{=} 17$$

Both are potentially valid ways of working out the expression. Thus, we must come to some agreement as to what our *order of operations* is going to be, so that we will all understand a given expression to mean the same thing.

The accepted order of priority for our operations is given in the accompanying box.

Order of Operations	1. Evaluate expressions within grouping symbols first. If there are grouping symbols within grouping symbols then work from the innermost grouping symbol outward.
	2. Next, perform multiplications and divisions, working from *left to right*.
	3. Next, perform additions and subtractions, working from *left to right*.

EXAMPLE 2

Evaluate the following expressions:

(a) $8 - 3(6 - 4)$ (b) $(8 - 3)(6 - 4)$ (c) $19 - 2 \cdot 3 + 4$

Solution

(a) We begin by following the order of operations and starting inside the parentheses:

$$
\begin{aligned}
8 - 3(6 - 4) &= 8 - 3(2) \qquad \textit{Next we must do the multiplication.} \\
&= 8 - 6 \\
&= \boxed{2}
\end{aligned}
$$

(b) Again we follow the order of operations. Since we have two sets of parentheses, our first step is to evaluate within each set:

$$
\begin{aligned}
(8 - 3)(6 - 4) &= (5)(2) \\
&= \boxed{10}
\end{aligned}
$$

(c) Following the order of operations, we begin with the multiplication:

$$
\begin{aligned}
19 - 2 \cdot 3 + 4 &= 19 - 6 + 4 \\
&= \boxed{17}
\end{aligned}
$$

■

EXAMPLE 3

Evaluate. $5 + 2[15 - 4(2 + 1)]$

Solution

Following the order of operations, we work from the innermost grouping symbol outward. In this case, we first evaluate the expression within the parentheses and then within the brackets:

$$5 + 2[15 - 4(2 + 1)] = 5 + 2[15 - 4(3)] \qquad \textit{Next we continue inside the}$$
$$\textit{[ ] and do multiplication}$$
$$\textit{before subtraction.}$$

$$= 5 + 2[15 - 12] \qquad \textit{Now we subtract within the [ ].}$$
$$= 5 + 2[3] \qquad \textit{Again, we perform}$$
$$= 5 + 6 \qquad \textit{multiplication before addition.}$$
$$= \boxed{11} \qquad \blacksquare$$

Most of the time, when we want to indicate division we will use the fraction bar. That is, we will usually write

$$a \div b \quad \text{as} \quad \frac{a}{b}$$

EXAMPLE 4   Evaluate. $\dfrac{20 - 4(3)}{10 - 2(3)}$

Solution   A fraction bar acts as an understood grouping symbol. We must evaluate the numerator (top) and the denominator (bottom), and then divide the results. In both numerator and denominator remember to do the multiplication before the subtraction.

$$\frac{20 - 4(3)}{10 - 2(3)} = \frac{20 - 12}{10 - 6}$$
$$= \frac{8}{4}$$
$$= \boxed{2} \qquad \blacksquare$$

In the next section we will begin discussing how we perform the various arithmetic operations with integers.

Study Skills 1.2

Previewing Material

Before you attend your next class, preview the material to be covered beforehand. First, skim the section to be covered, look at the headings, and try to guess what the sections will be about. Then read the material carefully.

You will find that when you read the material before you go to class, you will be able to follow the instructor more easily, things will make more sense, and you will learn the material more quickly. If there was something you did not understand when you previewed the material, the teacher will be able to answer your questions *before* you work your assignment at home.

EXERCISES 1.2

In Exercises 1–26, determine whether the given statement is always true. If the statement is true, indicate which property of the integers it illustrates.

1. $9 + 3 = 3 + 9$

2. $10 + 5 = 5 + 10$

3. $x + 3 = 3 + x$

4. $a + 5 = 5 + a$

5. $9 - 3 = 3 - 9$

6. $10 - 5 = 5 - 10$

7. $x - 3 = 3 - x$

8. $a - 5 = 5 - a$

9. $9 \cdot 3 = 3 \cdot 9$

10. $10 \cdot 5 = 5 \cdot 10$

11. $9 \cdot x = x \cdot 9$

12. $5 \cdot x = x \cdot 5$

13. $9 \div 3 = 3 \div 9$

14. $10 \div 5 = 5 \div 10$

15. $6 + (3 + 5) = (6 + 3) + 5$

16. $4 + (7 + 2) = (4 + 7) + 2$

17. $6 + (x + y) = (6 + x) + y$

18. $4 + (a + b) = (4 + a) + b$

19. $6(3 \cdot 5) = (6 \cdot 3)5$

20. $(4 \cdot 7)2 = 4(7 \cdot 2)$

21. $6(xy) = (6x)y$

22. $(4a)b = 4(ab)$

23. $10 - (6 - 3) = (10 - 6) - 3$

24. $12 - (7 + 3) = (12 - 7) + 3$

25. $a - (b - c) = (a - b) - c$

26. $a - (b + c) = (a - b) + c$

In Exercises 27–34, name the property used in each step.

27. $(x + 7) + 5 = x + (7 + 5)$
$$= x + 12$$

28. $(a + 9) + 4 = a + (9 + 4)$
$$= a + 13$$

29. $3(4y) = (3 \cdot 4)y$
$$= 12y$$

30. $5(7s) = (5 \cdot 7)s$
$$= 35s$$

31. $3 + (a + 7) = 3 + (7 + a)$
$$= (3 + 7) + a$$
$$= 10 + a$$

32. $2 + (z + 6) = (2 + z) + 6$
$$= (z + 2) + 6$$
$$= z + (2 + 6)$$
$$= z + 8$$

33. $3(a \cdot 7) = 3(7a)$
$$= (3 \cdot 7)a$$
$$= 21a$$

34. $2(z \cdot 6) = 2(6z)$
$$= (2 \cdot 6)z$$
$$= 12z$$

In Exercises 35–56, evaluate each expression.

35. $-(+4)$

36. $-(+7)$

37. $-(-4)$

38. $-(-7)$

39. $|4|$

40. $|7|$

41. $|-4|$

42. $|-7|$

43. $-|4|$

44. $-|7|$

45. $-|-4|$

46. $-|-7|$

47. $|10 - 5|$

48. $|20 - 12|$

49. $|10| - |-5|$

50. $|20| - |-12|$

51. $|8 - 4 - 2|$

52. $|8| - |4 - 2|$

53. $|15 - 7 + 3|$

54. $|15| - |7 + 3|$

55. $|2| - |-2| + |-5|$

56. $|12| - |-2| - |-5|$

In Exercises 57–98, evaluate the given expression. Remember to follow the order of operations.

57. $7 + 3 \cdot 4$

58. $5 + 4 \cdot 6$

59. $4 \cdot 3 + 6$

60. $2 \cdot 8 + 3$

61. $11 - 4 \cdot 2$

62. $20 - 5 \cdot 3$

63. $15 - 5 \cdot 3$

64. $12 - 2 \cdot 5$

65. $6 + 3 \cdot 5 - 2$

66. $8 - 2 \cdot 3 - 1$

67. $12 - 4 \cdot 3 + 6$

68. $18 + 2 \cdot 9 + 5$

69. $6 + 4(3 + 2)$

70. $30 - 5(4 - 2)$

71. $6 + (4 \cdot 3 + 2)$

72. $30 - (5 \cdot 4 - 2)$

73. $6 + (4 \cdot 3) + 2$

74. $30 - (5 \cdot 4) - 2$

75. $(6 + 4)(3 + 2)$

76. $(30 - 5)(4 - 2)$

77. $30 \div 5(2)$

78. $48 \div 12 \div 4$

79. $30 \div (5 \cdot 2)$

80. $48 \div (12 \div 4)$

81. $15 + 12 \div 6 - 3$

82. $32 + 16 \div 8 - 6$

83. $\dfrac{15 + 2 \cdot 3}{3 + 2 \cdot 2}$

84. $\dfrac{12 + 6 \cdot 3}{6 + 3 \cdot 3}$

85. $\dfrac{18 - 4 \cdot 2}{13 - 4 \cdot 2}$

86. $\dfrac{40 - 5 \cdot 4}{20 - 5 \cdot 2}$

87. $\dfrac{24 - 4 \cdot 5 + 8}{2 + 3 \cdot 6 - 4 \cdot 2}$

88. $\dfrac{22 + 3 \cdot 5 - 1}{25 - 10 \cdot 2 + 4}$

89. $\dfrac{5 \cdot 2 + 3 \cdot 4 + 6 \cdot 3}{4 \cdot 5 + 2 \cdot 6 + 4 \cdot 2}$

90. $\dfrac{7 \cdot 2 + 4 \cdot 5 - 2 \cdot 2}{9 \cdot 3 - 2 \cdot 6 - 3 \cdot 3}$

91. $3 + 2[3 + 2(3 + 2)]$

92. $5 + 4[5 + 4(5 + 4)]$

93. $9 - 4[6 - 2(3 - 1)]$

94. $8 - 2[9 - 3(5 - 3)]$

95. $4 + \{3 + 5[2 + 2(3 + 1)]\}$

96. $5 + \{2 + 4[3 + 3(4 + 2)]\}$

97. $\dfrac{10 + 2(5 + 3)}{2 \cdot 5 + 3}$

98. $\dfrac{18 - 3(4 - 1)}{2 \cdot 3 + 3}$

Q QUESTIONS FOR THOUGHT

99. Make up a "real-life" example involving the use of negative numbers.

100. Draw a number line and label two points A and B where $A < B$.

1.3 Adding Integers

We would like to extend our arithmetic procedures to the set of integers. In this section we will focus our attention on determining how we are going to add integers.

One important thing to keep in mind is that whatever rule we develop for adding integers, we will insist that it allow us to add positive integers just as we always have. In other words, $5 + 2$ must still be equal to 7.

In fact, this is exactly the approach we are going to take. We are going to analyze the addition process we already know, and that will help us see how to reasonably define addition for integers.

Let's consider two similar examples:

$$5 + 2 = ? \quad \text{and} \quad 5 + (-2) = ?$$

To make our work even clearer, let's write these examples as

$$+5 + (+2) = ? \quad \text{and} \quad +5 + (-2) = ?$$

The rules of algebra do not allow us to write two operation symbols next to each other without some punctuation between them. Therefore, we cannot write $5 + -2$; instead we must write $5 + (-2)$.

Of course, we all know that the answer to the first example is 7, but let's try to analyze what is happening in +5 + (+2) on the number line.

We can visualize that +5 + (+2) means "start at +5 on the number line and move 2 units to the right." The following picture represents +5 + (+2) = 7:

In other words, *adding* +2 means "move 2 units to the right."

In view of this, how should we visualize 5 + (−2)? We are still starting at +5, but this time we are *adding* −2 instead of +2. It seems reasonable that if adding +2 means "move 2 units to the right," then adding −2 should mean "move 2 units to the *left*." (Remember, −2 is the opposite of +2.)

The following picture represents +5 + (−2):

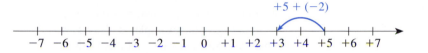

Therefore, we have +5 + (−2) = +3.

Let's now consider two more closely related examples:

$$-5 + (+2) = ? \qquad \text{and} \qquad -5 + (-2) = ?$$

Accepting our understanding of the first two examples, we really do not have much choice as to how we interpret these two. After all, the only difference in these two is that we are starting at −5 instead of at +5.

If adding +2 means moving 2 units to the right, then it should not matter whether we start at +5 or at −5, or at any other number. Similarly, if adding −2 means moving 2 units to the left, then again it should not matter whether we start at +5 or at −5 or anywhere else.

Let's draw the picture for each of these examples:

The picture for −5 + (+2) is: The picture for −5 + (−2) is:

Therefore, we have −5 + (+2) = −3.   Therefore, we have −5 + (−2) = −7.

Look carefully at the pictures for each of the four examples to see that they illustrate that adding a positive number means "move to the right," while adding a negative number means "move to the left."

Since it would be time-consuming to draw a number line every time we want to add integers, let's summarize these four examples and see if we can extract from them a general rule for adding integers. We have

$$+5 + (+2) = +7$$
$$+5 + (-2) = +3$$
$$-5 + (+2) = -3$$
$$-5 + (-2) = -7$$

In the first and last of these examples we are either adding two positive numbers or adding two negative numbers. In both cases, we are starting in a certain direction and then moving in the *same* direction; therefore, the two numbers are reinforcing each other. We are, in effect, *adding* the "number parts without regard to the sign," and just keeping the common sign of the two numbers. This "number part without regard to the sign" is just the number of units the number is away from 0, which is exactly what we called the absolute value of the number.

Rule for Adding Integers Part 1	When adding two integers with the *same sign,* add their absolute values and keep their common sign.

On the other hand, in the middle two examples we are adding numbers that have opposite signs. In these examples, we are starting in one direction and then moving in the *opposite* direction. Therefore, the two numbers are opposing each other. We are in effect *subtracting* the absolute values. However, in such a case we cannot, in general, tell whether the answer will be positive or negative until we know which number is "stronger." Again we make use of the idea of the absolute value of a number.

In the example $+5 + (-2)$ we get an answer of $+3$ because we have more positive strength $(+5)$ than negative strength (-2). In other words, since $+5$ has a larger absolute value than -2, the final answer is positive.

Similarly, in the example $-5 + (+2)$ we get an answer of -3 because -5 has a larger absolute value than $+2$, and therefore we have more negative strength than positive strength. Be careful here. We are *not* saying that -5 is greater than $+2$. In fact, the contrary is true; because $+2$ is to the right of -5, $+2$ is greater than -5. However, when we are adding integers with opposite signs, it is the larger *absolute value* with which we are concerned, not the larger number.

Rule for Adding Integers Part 2	When adding two integers with *opposite signs,* subtract the smaller absolute value from the larger and keep the sign of the number with the larger absolute value.

Our formal rule for adding integers is actually one rule with two parts. Do not be intimidated by the formal wording of this rule. In the next several examples we will see that using this rule is quite straightforward.

EXAMPLE 1

Compute each of the following:

(a) $-15 + (-8)$ (b) $-15 + (+8)$
(c) $+15 + (-8)$ (d) $+15 + (+8)$

Solution

(a) Since -15 and -8 have the same sign, we add their absolute values and keep the common sign, which is negative. Thus, we have

$$-15 + (-8) = -(15 + 8) = \boxed{-23} \;^*$$

(b) Since -15 and $+8$ have opposite signs, we subtract their absolute values as we would whole numbers (larger absolute value minus smaller absolute value) and keep the sign of the number with the larger absolute value (in this case it is -15), which makes the answer negative. Thus, we have

$$-15 + (+8) = -(15 - 8) = \boxed{-7}$$

(c) Since $+15$ and -8 have opposite signs, we proceed exactly as we did in part (b). This time, however, the answer will be positive since the number with the larger absolute value, $+15$, is positive. Thus, we have

$$+15 + (-8) = +(15 - 8) = \boxed{+7}$$

---

*Throughout the text we will use a color panel to indicate steps that are normally done *mentally,* but that we include to help clarify the procedure.

(d) Do not get carried away with the rule. This example is just $15 + 8$ and so we can get the answer without need of the rule, although we could use it if we wanted to. Thus, we have

$$+15 + (+8) = \boxed{23} \qquad \textit{Remember that 23 means +23.} \qquad \blacksquare$$

Once we know how to add two integers we can add more than two as well.

EXAMPLE 2

Compute each of the following:

(a) $-5 + (-9) + (-6)$ (b) $6 + (-4) + 7 + (-10)$

Solution

(a) Since we are adding three numbers whose signs are all the same (all negative), we add their absolute values and keep the negative sign:

$$-5 + (-9) + (-6) = -(5 + 9 + 6) = \boxed{-20}$$

(b) We have two choices. The first is to work from left to right as we have been. The second is to reorder and regroup the numbers so that we add all the positive and negative numbers separately first, and then add the results. We illustrate both methods.

Solution 1.   Working from left to right, we have

$$6 + (-4) + 7 + (-10) \qquad \textit{We begin by adding 6 and }-4\textit{ to get 2.}$$
$$= 2 + 7 + (-10) \qquad \textit{Next we add 2 and 7.}$$
$$= 9 + (-10) \qquad \textit{Now we add 9 and }-10.$$
$$= \boxed{-1}$$

Solution 2.   This time we compute the result by grouping positives and negatives first. Since we are *adding* the numbers, the commutative and associative properties allow us to rearrange and regroup the numbers any way we please.

$$6 + (-4) + 7 + (-10) \qquad \textit{We first group the positive and negative}$$
$$\textit{numbers separately.}$$
$$= (6 + 7) + [-4 + (-10)] \qquad \textit{Next we add the positives and negatives}$$
$$\textit{separately.}$$
$$= 13 + [-14] \qquad \textit{Now we add 13 and }-14.$$
$$= \boxed{-1} \qquad \blacksquare$$

EXAMPLE 3

Evaluate. $-|8 - 3| + |3 + (-8)|$

Solution

We begin by evaluating the expressions *within* the absolute value signs:

$$-|8 - 3| + |3 + (-8)| = -|5| + |-5|$$
$$= -(5) + 5$$

-5 and 5 have equal absolute values so that when we add we get 0.
Remember that 0 is neither positive nor negative so we do not write +0 or −0.

$$= \boxed{0} \qquad \blacksquare$$

We continue to work within the framework of the same order of operations we outlined in the last section.

EXAMPLE 4   Evaluate. $-8 + 3[5 + (-12) + 9]$

Solution   We begin inside the brackets by adding 5 and 9:

$$-8 + 3[5 + (-12) + 9] = -8 + 3[5 + 9 + (-12)]$$
$$= -8 + 3[14 + (-12)]$$
$$= -8 + 3[2]$$
$$= -8 + 6$$
$$= \boxed{-2} \quad \blacksquare$$

It is most important that you be able to add integers as automatically as you do whole numbers. The only way to acquire this skill is to *practice*. That is exactly what the exercise sets are designed for—to give you lots and lots of practice.

Study Skills 1.3

What to Do First

Before you attempt any exercises, either for homework or for practicing your skills, it is important to review the relevant portions of your notes and text.

As we mentioned in the introduction, memorizing a bunch of seemingly unrelated algebraic steps to follow in an example may serve you initially, but in the long run (most likely before Chapter 3), your memory will be overburdened—you will tend to confuse examples and/or forget steps.

Reviewing the material before doing exercises makes each solution you go through more meaningful. The better you understand the concepts underlying the exercise, the easier the material becomes, and the less likely you are to confuse examples or forget steps.

When reviewing the material, take the time to *think about what you are reading*. Try not to get frustrated if it takes you an hour to read and understand a few pages of a math text—that time will be well spent. As you read your text and your notes, think about the concepts being discussed: **(a)** how they relate to previous concepts covered and **(b)** how the examples illustrate the concepts being discussed. More than likely, worked-out examples will follow verbal material, so look carefully at these examples and try to understand why each step in the solution is taken. When you finish reading, take a few minutes and think about what you have just read.

EXERCISES 1.3

In Exercises 1–66, compute the value of each expression.

1. $+6 + (-8)$	2. $+8 + (-10)$
3. $-7 + (-5)$	4. $-11 + (-3)$
5. $-5 + (+12)$	6. $-8 + (+13)$
7. $+9 + (-4)$	8. $+7 + (-6)$
9. $+7 + (-3)$	10. $+9 + (-8)$
11. $-7 + (-3)$	12. $-9 + (-8)$
13. $-7 + (+3)$	14. $-9 + (+8)$
15. $+7 + (+3)$	16. $+9 + (+8)$
17. $-4 + (-11)$	18. $-8 + (-4)$
19. $12 + (-16)$	20. $16 + (-12)$
21. $9 + (-9)$	22. $-11 + 11$
23. $8 + (-15)$	24. $-19 + (-6)$

25. $-20 + (-24)$

26. $44 + (-29)$

27. $-4 + 10$

28. $-12 + 20$

29. $-7 + (-3) + (-5)$

30. $-10 + (-3) + (-1)$

31. $7 + (-3) + (-5)$

32. $10 + (-6) + (-1)$

33. $-7 + 3 + (-5)$

34. $10 + (-6) + 1$

35. $-7 + (-3) + 5$

36. $-10 + 6 + (-1)$

37. $15 + (-12) + (-8)$

38. $12 + (-14) + (-4)$

39. $-6 + 13 + (-6)$

40. $-2 + 15 + (-10)$

41. $-1 + (-9) + (-5)$

42. $-6 + (-7) + (-2)$

43. $18 + (-10) + (-2)$

44. $16 + (-1) + (-11)$

45. $-25 + (-5) + 26$

46. $-31 + (-9) + 37$

47. $22 + (-3) + (-14) + 1$

48. $-21 + 15 + (-2) + (-3)$

49. $-9 + 12 + (-14) + 5$

50. $-18 + (-3) + (-4) + 11$

51. $6 + (-3) + (-10) + (-15) + 27$

52. $-8 + 4 + 16 + 1 + (-8)$

53. $32 + (-61)$

54. $-48 + (-46)$

55. $-48 + (-19)$

56. $-63 + (-28)$

57. $-29 + (-31) + 51$

58. $34 + (-67) + 10$

59. $124 + (-237) + (-102)$

60. $-217 + (-86) + 300$

61. $-86 + 112 + (-78) + 201$

62. $123 + (-77) + (-325) + 115$

63. $-187 + 455$

64. $520 + (-366)$

65. $-84 + 127 + (-111)$

66. $156 + (-92) + (-181)$

In Exercises 67–74, evaluate the given expression.

67. $-6 + [9 + (-4)]$

68. $-6 + [5 + (-3)]$

69. $-7 + 7 \cdot 4 + (-20)$

70. $-10 + 6 \cdot 3 - 14$

71. $|1 + (-6)| + |1| + |-6|$

72. $|5 + (-9)| + |5| + |-9|$

73. $|-8| + (-8)$

74. $-|-8| + (-8)$

In Exercises 75–86, solve the problem by writing a sum of signed numbers and adding.

75. Niki has a balance of $548 in her checking account. She writes checks for $56, $16, and $71; then she deposits $145; then she writes checks for $180, $67, and $205; then she deposits $75, and finally she writes checks for $115 and $45. What is the final balance of her checking account?

76. Repeat Exercise 75 if Niki starts with a balance of –$82.

77. A certain location experienced a temperature variation of 80 degrees on a particular day. If the low temperature was –47°F, what was the high temperature that day?

78. A certain city experienced an annual temperature variation of 107 degrees during a particular year. If the high temperature was 102°F, what was the low temperature that year?

79. Carla has a balance of $48 in her checking account. She writes checks for $18, $22, and $15; then she deposits $50; then she writes checks for $28 and $12; then she deposits $20; then she writes another check for $17; and finally she deposits $27. What is the final balance of her checking account?

80. Repeat Exercise 79 if Carla begins with a balance of −$48.

81. A football team takes possession of the ball on their own 25-yard line. On first down they gain 8 yards; on second down they lose 14 yards; and on third down they lose 5 more yards. What is their location on fourth down?

82. On a certain morning the temperature at 6:00 A.M. is 4°C. Two hours later the temperature has fallen 9 degrees, and 3 hours after that it has risen 2 degrees. What is the temperature at 11:00 A.M.?

83. During the current fiscal year the Soles 'R' Us shoe store recorded the following quarterly earnings (positive numbers represent a quarterly profit, negative numbers represent a quarterly loss): $10,432, −$1,678, −$2,046, and $7,488. What was the total profit (or loss) for this year?

84. A record company begins a month with an inventory of 16,430 compact discs (CDs). Suppose that we represent CDs that are shipped as negative numbers and those received as positive numbers. The company records the following transactions during the month: −1,800, −465, 73, 92, 128, −166, and −2,750. What is the company's inventory at the end of the month?

85. Suppose that a plane is at an altitude of 10,000 feet and experiences the following altitude changes: +380 ft, +540 ft, −275 ft, −600 ft, and −72 ft. What is the final altitude of the plane?

86. An elevator begins its ascent with a passenger load of 1,538 pounds. At the first stop 220 pounds get off and 165 pounds get on. At the second stop 187 pounds get off. At the third stop 126 pounds get off. At the fourth stop 471 pounds get off and 204 pounds get on. How many pounds is the elevator carrying at this point?

❓ QUESTIONS FOR THOUGHT

87. What result should we get when we add an integer to its opposite? Why?

88. If the sum of two integers is 0, what can we conclude about the two integers? Why?

89. $7 - 4 = 3$ because $3 + 4 = 7$. Subtraction is defined in terms of addition. What should the answer to $4 - 7$ be and why?

90. What should the answer to $4 - (-7)$ be and why?

1.4 Subtracting Integers

We recall from our experience in arithmetic that subtraction and addition are two ways of looking at the same situation. In fact, subtraction is often defined in terms of addition. That is, if we want to answer $7 - 4 = ?$, we ask "What must be added to 4 to get 7?" In other words, we can translate $7 - 4 = ?$ into $4 + ? = 7$.

We continue the approach we took in the last section with addition: We begin with a familiar subtraction problem, analyze it, and see what it tells us about how to subtract integers in general.

As we did with addition, let's consider two examples:

$$5 - 2 = ? \quad \text{and} \quad 5 - (-2) = ?$$

Again, to make our work even clearer let's write these examples as

$$+5 - (+2) = ? \quad \text{and} \quad +5 - (-2) = ?$$

Of course, we all know the answer to the first example is 3, but let's try to analyze what is happening in $+5 - (+2)$ on the number line. We can visualize this as follows: $+5 - (+2)$ means "start at $+5$ on the number line and move 2 units to the *left*." This gives us the following picture:

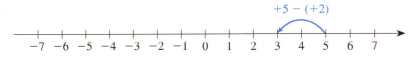

In other words, *subtracting* $+2$ means "move 2 units to the left."

In view of this, how should we view $+5 - (-2)$? We are still starting at $+5$, but this time we are *subtracting* -2. If *subtracting* $+2$ means "move 2 units to the left," then *subtracting* -2 must mean do the opposite—that is, "move 2 units to the right." (What other choice is there?) We visualize this on the number line as follows:

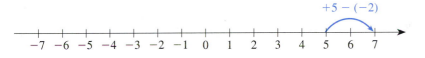

Therefore, $+5 - (-2) = 7$.

As with addition, if subtracting $+2$ means move 2 units to the left and subtracting -2 means move 2 units to the right, then it should not make any difference where we start. Thus, for the two closely related examples

$$-5 - (+2) = ? \quad \text{and} \quad -5 - (-2) = ?$$

we have the following pictures and resulting answers:

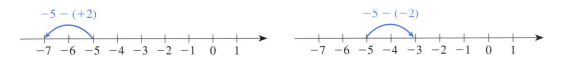

Therefore, $-5 - (+2) = -7$.   Therefore, $-5 - (-2) = -3$.

Keeping in mind the work we did in the last section on adding integers, we are familiar with the phrases "moving left" and "moving right." We have just seen that *subtracting* $+2$ means moving 2 units to the left, while in the last section we saw that *adding* -2 means exactly the same thing. Similarly, we have just seen that *subtracting* -2 means moving 2 units to the right, while in the last section we saw that *adding* $+2$ means exactly the same thing.

Thus, whenever we are faced with subtracting a number we can accomplish the same result by adding the opposite number, and then following the rules for addition. We state this fact algebraically (symbolically) as follows:

Rule for Subtracting Integers	$a - b = a + (-b)$

In words, this rule says that in order to subtract an integer we add its opposite. That is, when we are subtracting two integers, we change the subtraction to addition *and* change the sign of the number being subtracted. Note that the sign of the first number is unchanged.

EXAMPLE 1 | Compute. $-9 - (+6)$

Solution | We follow the rule for subtraction.

$$-9 - (+6) = -9 + (-6) \qquad \textit{We have changed the subtraction to addition, and}$$
$$\textit{we have changed the sign of the second number.}$$

$$= \boxed{-15} \qquad \textit{Now we follow the rule for addition.} \qquad ■$$

EXAMPLE 2 | Compute. $7 - 12$

Solution | This example means "$+7$ minus $+12$." At first, you may find it helpful to put in the "understood" positive signs in such an example.

$$7 - 12 = +7 - (+12) \qquad \textit{We follow the rule for subtraction. Change the sign}$$
$$\qquad\quad \downarrow \;\; \downarrow \qquad\qquad \textit{of the number being subtracted, and add.}$$

$$= +7 + (-12) \qquad \textit{Now we follow the rule for addition.}$$

$$= \boxed{-5} \qquad\qquad\qquad\qquad\qquad\qquad ■$$

We can apply this rule to examples involving more than one subtraction, or involving several subtractions or additions.

EXAMPLE 3 | Compute each of the following:
(a) $7 - 9 - (-4)$ (b) $-10 + (-3) - 5 + 11$

Solution | (a) We begin by inserting the understood positive signs.

$$7 - 9 - (-4) = +7 - (+9) - (-4) \qquad \textit{Change each subtraction to addition, and}$$
$$\qquad\qquad\quad \downarrow \;\; \downarrow \quad\;\; \downarrow \;\; \downarrow \qquad \textit{change the sign of the number following}$$
$$\textit{each subtraction.}$$

$$= +7 + (-9) + (+4) \quad \textit{Now we follow the rule for addition.}$$
$$= -2 + (+4)$$

$$= \boxed{2}$$

(b) We begin by inserting the understood positive signs. (If *you* feel this step is not necessary then skip it.)

$$-10 + (-3) - 5 + 11 = -10 + (-3) - (+5) + (+11) \qquad \textit{We follow the}$$
$$\qquad\qquad\qquad\qquad\qquad\qquad \downarrow \;\; \downarrow \qquad\qquad\qquad \textit{subtraction rule.}$$

$$= -10 + (-3) + (-5) + (+11) \qquad \textit{We add the negative}$$
$$\textit{numbers.}$$

$$= -18 + (+11) \qquad\qquad\qquad\qquad \textit{Now we use the}$$
$$\textit{addition rule.}$$

$$= \boxed{-7} \qquad\qquad\qquad\qquad\qquad\qquad\qquad ■$$

As the examples get more complicated we must keep the order of operations in mind.

EXAMPLE 4 | Evaluate each of the following:
(a) $4 - 5 - 7$ (b) $4 - (5 - 7)$

Solution (a) $4 - 5 - 7 = 4 - (+5) - (+7)$ *Following the rule for subtraction, we get*

$$= 4 + (-5) + (-7)$$
$$= 4 + (-12)$$
$$= \boxed{-8}$$

(b) The order of operations requires us to work inside the parentheses first.

$$4 - (5 - 7) = 4 - (5 - (+7))$$ *Note the use of the subtraction rule.*

$$= 4 - (5 + (-7))$$
$$= 4 - (-2)$$
$$= 4 + (+2)$$
$$= \boxed{6}$$

Note that the numbers in parts **(a)** and **(b)** are the same, but that the parentheses in part **(b)** change the meaning of the example a great deal. ■

We are all familiar with the use of signed numbers to indicate temperatures above and below zero. Similarly, positive numbers are used to indicate a height above sea level and negative numbers are used to indicate a depth below sea level.

EXAMPLE 5

The highest point in North America is the top of Mt. McKinley in Alaska, which is 20,320 ft above sea level. The lowest point in North America is Death Valley in California, which is 282 ft below sea level. Find the difference in height between these points.

Solution If we wanted to find the difference in height between a point 200 ft above sea level and a point 75 ft above sea level, we would simply compute "upper height − lower height." That is, 200 − 75 would give us a difference of 125 ft.

Similarly, the height of Mt. McKinley is 20,320 ft and the depth of Death Valley is −282 ft. Therefore the *difference* in height is

$$\text{Upper height} - \text{Lower height} = 20,320 - (-282) = 20,320 + 282$$
$$= \boxed{20,602 \text{ feet}}$$ ■

EXAMPLE 6

On a certain day the high temperature in Antarctica was −4°F and the low temperature was −39°F. What was the difference between the high and low temperatures that day?

Solution If we were finding the difference between two positive temperatures, such as 72° and 49°, we would compute 72 − 49 = 23 to get a difference of 23°. In other words, we computed Higher temperature − Lower temperature. We follow the same procedure in this example.

$$\text{Higher temperature} - \text{Lower temperature} =$$
$$-4° \quad - \quad (-39°) \quad = -4° + 39° = \boxed{35°F}$$ ■

EXAMPLE 7 | What are the terms in the expression $-6 - 4 + 3$?

Solution | Since terms are members of a sum, we rewrite the given expression as a sum. That is, we can rewrite $-6 - 4 + 3$ as $-6 + (-4) + (+3)$ and so we can more easily see that the terms are -6, -4, and 3. ∎

Up to this point we have been very careful to write each step in both our addition and subtraction examples. However, virtually every example and exercise we do from now on involves adding and subtracting integers in some way or another, and it is impractical for us to continue writing each step. In fact, as quickly as possible you must try to reach the point where you can subtract integers as easily and naturally as you subtract whole numbers. That is, in the same way you "just know" that $7 - 4 = 3$, so too you need to "just know" that $4 - 7 = -3$. This means that you must develop the ability to use the addition and subtraction rules mentally (perhaps not at first, of course, but eventually). The only way to acquire this ability is to do many exercises.

Study Skills 1.4

Doing Exercises

After you have finished reviewing the appropriate material, as discussed previously, you should be ready to do the relevant exercises. Although your ultimate goal is to be able to work out the exercises accurately *and* quickly, when you are working out exercises on a topic that is new to you it is a good idea to take your time and think about what you are doing while you are doing it.

Think about how the exercises you are doing illustrate the concepts you have reviewed. Think about the steps you are taking and ask yourself why you are proceeding in this particular way and not some other: Why this technique or step and not a different one?

Do not worry about speed now. If you take the time at home to think about what you are doing, the material becomes more understandable and easier to remember. You will then be less likely to "do the wrong thing" in an exercise. The more complex-looking exercises are less likely to throw you. In addition, if you think about these things in advance, you will need less time to think about them during an exam, and so you will have more time to work out the problems.

Once you believe you thoroughly understand what you are doing and why, you may work on increasing your speed.

EXERCISES 1.4

In Exercises 1–72, compute the value of each expression.

1. $6 - (+10)$

2. $4 - (+8)$

3. $-7 - (+4)$

4. $-9 - (+3)$

5. $3 - (-6)$

6. $5 - (-7)$

7. $-8 - (-2)$

8. $-6 - (-4)$

9. $-5 + (+8)$

10. $-7 + (+5)$

11. $5 - (+8)$

12. $4 - (+9)$

13. $5 - (-8)$

14. $4 - (-9)$

15. $-5 - (+8)$

16. $-4 - (+9)$

17. $-5 - (-8)$

18. $-4 - (-9)$

19. $5 + (-8)$

20. $4 + (-9)$

21. $-5 + (-8)$

22. $-4 + (-9)$

23. $6 - (-7)$

24. $2 - (-3)$

25. $-6 - (-7)$

26. $-2 - (-3)$

27. $2 + (-6) - (+7)$

28. $5 + (-8) - (+3)$

29. $2 - 6 - 7$

30. $5 - 8 - 3$

31. $2 - (6 - 7)$

32. $5 - (8 - 3)$

33. $7 - 9 - 3 + 2$

34. $2 - 3 + 1 - 8$

35. $11 - 5 + 4 - 7$

36. $4 - 7 - 2 + 3$

37. $2 - 3 - 6 - 2$

38. $-6 + 2 - 3 - 5$

39. $-10 + 4 - 9 - (-3)$

40. $-8 + 5 - 6 + 9$

41. $3 - 6 + 1 - (-4)$

42. $10 - (-3) + 1 - 4$

43. $-1 - 4 - 2 - (-5)$

44. $4 - 12 - (-2) - 3$

45. $4 - 8 - 6 + 3$

46. $9 - 10 - 2 + 5$

47. $4 - 8 - (6 + 3)$

48. $9 - (10 - 2) + 5$

49. $4 - (8 - 6) + 3$

50. $9 - 10 - (2 + 5)$

51. $4 - (8 - 6 + 3)$

52. $9 - (10 - 2 + 5)$

53. $-8 + 8$

54. $-6 + 6$

55. $-8 - 8$

56. $-6 - 6$

57. $-8 - (-8)$

58. $-6 - (-6)$

59. $18 - 35$

60. $-16 - 23$

61. $-31 + 17 - 12$

62. $-43 + 32 - 17$

63. $26 - (-41) - 52$

64. $57 - 82 - 31$

65. $-23 - 41 - 62$

66. $-49 - 26 - 33$

67. $100 - 83 - 45 - 24$

68. $225 - 56 - 97 - 115$

69. $52 - (-38) - 67$

70. $89 - (-27) - 62$

71. $-246 - 327 - (-542)$

72. $-561 - 412 - (-678)$

In Exercises 73–84, evaluate the given expression.

73. $9 - 5 \cdot 4 - 2$

74. $7 - 3 \cdot 5 - 1$

75. $9 - 5(4 - 2)$

76. $7 - 3(5 - 1)$

77. $9 - (5 \cdot 4 - 2)$

78. $7 - (3 \cdot 5 - 1)$

79. $|6 - 2| - |2 - 6|$

80. $|10 - 7| - |7 - 10|$

81. $|2 - 6| - (2 - 6)$

82. $|7 - 10| - (7 - 10)$

83. $|-4 - 3 + 2| - 4 - 3 + 2$

84. $|-8 + 6 - 5| - 8 + 6 - 5$

85. On a certain day the high temperature at the North Pole was $-6°C$ and the low temperature was $-19°C$. What was the difference between the high and low temperatures that day?

86. On a certain day the maximum temperature was $12°C$ and the minimum temperature was $-5°C$. By how many degrees did the temperature vary on this day?

87. At the beginning of the month Carla's checking account has a balance of $643.47, while at the end of the month the balance is −$82.94. If she made no deposits during this month, what was the total of the checks that Carla wrote during this month?

88. A helicopter is hovering at an altitude of 600 ft above sea level directly above a submarine that is submerged at a depth of 128 ft below sea level. What is the distance between the helicopter and the submarine?

In Exercises 89–96, use the accompanying diagram, which indicates the height and depth of various places on earth.

89. Find the difference in height between the top of Mt. Everest (the highest point on earth) and the Dead Sea (the lowest point on earth).

90. Find the difference in height between the top of Mt. Everest and the bottom of the Marianas Trench (the deepest known part of the ocean).

91. Find the difference in depth between the Dead Sea and the bottom of the Marianas Trench.

92. An undersea exploration vessel is traveling through the Marianas Trench at a depth of −487 ft. Find the difference in depth between the vessel and the bottom of the trench.

93. What are the terms in the expression −8 − 2 − 7?

94. What are the terms in the expression $x − y − 1$?

95. What are the terms in the expression −3 − m − n?

96. What are the terms in the expression $a − b − ab$?

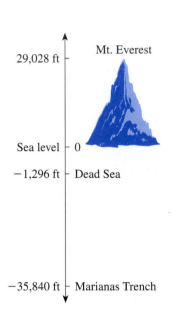

Ⓠ QUESTIONS FOR THOUGHT

97. If we denote a debt of $15,000 as −$15,000, explain how you could use *subtraction* of negative numbers to represent and calculate the following: Reduce a debt of $15,000 by $6,000.

98. What is *wrong* with |4 − 8| = |4| − |8|?

99. We know that multiplication means repeated addition. That is, 4 times 3 means add 3 four times. In view of this and the rule for adding negative integers, what should 4 times −3 be equal to?

100. When we multiply 4 times 3 we are multiplying *positive* 4 and so we interpret the multiplication as repetitive *addition*. The 4 tells us how many times to *add* 3. How might we interpret −4 times 3 and −4 times −3? Can we interpret multiplying by a negative as repetitive *subtraction?* If so, what should we get as answers?

1.5   Multiplying and Dividing Integers

When we credit an account we add a positive value; when we debit an account we add a negative value.

In order to develop rules for multiplying and dividing integers, we must keep in mind that multiplication is just a shorthand way of writing addition. For example, if we were to *credit* an account 4 times with $30, we could represent this as

$$+\$30 + (+\$30) + (+\$30) + (+\$30) = +\$120$$

or we could use multiplication to write

$$4(+\$30) = +\$120$$

Similarly, if we were to *debit* an account 4 times with $30, we could represent this as

$$-\$30 + (-\$30) + (-\$30) + (-\$30) = -\$120$$

or we could use multiplication to write

$$4(-\$30) = -\$120$$

Recognizing that multiplication is repetitive addition, whenever we multiply a positive times a negative, we will be repeatedly adding the *same* negative number and, according to our rule for addition, the answer will be negative. It does not matter whether the product is a positive times a negative or a negative times a positive. For example:

Positive times negative: $\quad 5(-2) = (-2) + (-2) + (-2) + (-2) + (-2)$
$$= -10$$

Negative times positive: $\quad (-2)5 = 5(-2) \qquad$ *By the commutative property*
$$= -10 \qquad \textit{As we just saw}$$

Thus, we see that the *product* of two numbers that have *opposite* signs should be *negative*.

Deciding what we want a negative times a negative to be requires a slightly more delicate analysis. When we first began talking about the integers in Section 1.2, we pointed out that a negative sign in front of a number can be thought of as meaning the "opposite of." Thus, -4 can be thought of as the opposite of $+4$. On the other hand, now that we have seen that a negative times a positive is negative, we can also think of -4 as -1 times 4 [because $-1(4) = -4$]. Thus, multiplying by -1 changes a number into its opposite. Thus, we can analyze -4 times -3 as follows:

$$(-4)(-3) = -1(4)(-3) \qquad \textit{Because } -4 = -1(4)$$
$$= -1(-12) \qquad \textit{Because 4 times } -3 \textit{ is equal to } -12$$
$$= +12 \qquad \textit{Because multiplying by } -1 \textit{ changes a number into its opposite}$$

We have come to the conclusion that the product of two negative numbers is positive. Since the product of two positives is also a positive, we can say that the *product of two numbers that have the *same* sign is *positive*.

Now that we see how multiplication of signed numbers works, what about division? Keep in mind that division is always defined in terms of multiplication. That is,

$$\frac{12}{3} = 4 \quad \text{because} \quad (+4) \cdot (+3) = +12 \qquad \textit{Remember that } \frac{a}{b} \textit{ means } a \div b.$$

Therefore, if we want to compute 12 divided by -3, we know that the answer must be -4 because -4 times -3 equals 12:

$$\frac{+12}{-3} = -4 \quad \text{because} \quad (-4)(-3) = +12$$

Similarly, we have

$$\frac{-12}{+3} = -4 \quad \text{because} \quad (-4)(+3) = -12$$

$$\frac{-12}{-3} = +4 \quad \text{because} \quad (+4)(-3) = -12$$

We can see that division behaves in the same way that multiplication does. The *quotient* of two numbers will be *positive* if the numbers have the *same* sign and it will be *negative* if the numbers have *opposite* signs.

Basically, we have seen that when we multiply or divide two signed numbers, we simply multiply or divide the numbers, ignoring their signs (that is, we use just their absolute values). Then we put down a positive sign if the signs are the same or a negative sign if the signs are opposite.

Let's formalize this into a rule.

Rule for Multiplying and Dividing Integers	When multiplying (or dividing) two integers, multiply (or divide) their absolute values. The sign of the answer is: *Positive* if both numbers have the *same* sign. *Negative* if the numbers have *opposite* signs.

EXAMPLE 1

Compute each of the following:

(a) $(-3)(-4)$ (b) $5(-6)$ (c) $\dfrac{12}{-6}$

Solution

(a) $(-3)(-4) = \boxed{12}$ *The product of two negatives is a positive.*

(b) $5(-6) = \boxed{-30}$ *The product of a positive and a negative is a negative.*

(c) $\dfrac{12}{-6} = \boxed{-2}$ *The quotient of a positive and a negative is a negative.* ∎

EXAMPLE 2

Compute. $(-5)(-8)(-2)$

Solution

$(-5)(-8)(-2) = (+40)(-2)$ *Because -5 times -8 is equal to $+40$*

$= \boxed{-80}$ *Because $+40$ times -2 is equal to -80* ∎

We must be extremely careful reading examples, particularly when they involve parentheses. For example, both

$$-4(-6) \qquad \text{and} \qquad -(4-6)$$

contain parentheses, but they serve different functions in the two expressions. In $-4(-6)$ the parentheses serve to indicate that -4 is multiplying -6, and therefore the answer is

$$-4(-6) = \boxed{24}$$

However, in $-(4-6)$, the parentheses indicate that we are to take the negative of the *result* of $4-6$, and therefore the answer is

$$-(4-6) = -(-2) = \boxed{2}$$

We mentioned earlier that even though we have chosen the same symbol (the minus sign) to indicate both subtraction and a negative number, there should not be any confusion because all our uses of the minus sign are consistent. We have seen the minus sign used as subtraction, to indicate a negative number, and to indicate the opposite (which we saw is the same as thinking of it as multiplying by -1). For example:

If we think of $-(-2)$ as the *opposite of negative* 2, we get $+2$ as our answer.

If we think of $-(-2)$ as *subtracting negative* 2, then the rule for subtraction gives $+(+2)$, and our answer is again $+2$.

If we think of $-(-2)$ as $-1(-2)$, then following our multiplication rule we also get $+2$.

We are free to think of the minus sign in the way we find most convenient provided we do not change the meaning of the expression.

EXAMPLE 3

Evaluate each of the following:

(a) $3 - 8(2)$ (b) $3(-8)(2)$

Solution

(a) Following the order of operations, we cannot do the subtraction $3 - 8$ first. We must do the multiplication first:

$$3 - 8(2) = 3 - 16$$
$$= \boxed{-13}$$

(b) This example involves multiplication only:

$$3(-8)(2) = -24(2)$$
$$= \boxed{-48}$$ ■

EXAMPLE 4

Compute. $8 - 5(-4)$

Solution

We will use this example to illustrate a slight shortcut, which also happens to offer the advantage of being less prone to making a sign error.

One way of evaluating this expression is as follows: Following the order of operations, we multiply 5 times -4 first.

$$8 - 5(-4) = 8 - (-20) \qquad \textit{Now we follow the rule for subtraction.}$$
$$= 8 + (+20)$$
$$= \boxed{28}$$

Instead of having to follow the subtraction rule, let's anticipate it by *thinking* addition at the outset. In order to do that we think of the minus sign in front of the 5 as part of the 5, so we say "-5 is multiplying -4." Keep in mind that we are thinking addition so we are going to *add* the result of the product. Thus, the alternate solution is as follows:

If we focus on $-5(-4)$ only, we would certainly read it as -5 times -4.

$$8 - 5(-4) = \;\boxed{8 + (-5)(-4)}\; = 8 + 20 \qquad \textit{We added the result of multiplying}$$
$$\qquad\qquad\qquad\qquad\qquad\qquad\qquad \textit{-5 times -4.}$$
$$= \boxed{28}$$

This approach not only saves a step in the solution, but it also often removes the necessity for "changing signs." The fewer times we have to change signs, the fewer opportunities for making a careless error.

Overall, this approach offers significant advantages, so we will use it from now on wherever it applies. ■

EXAMPLE 5

Evaluate. $6 - 4[3 - 7(2)]$

Solution

Following the order of operations, we begin by working inside the brackets first.

$$6 - 4[3 - 7(2)] = 6 - 4[3 - 14]$$
$$= 6 - 4[-11] \qquad \textit{We think of this as } -4 \textit{ times } -11, \textit{ which}$$
$$\textit{is then added to } 6.$$
$$= 6 + 44$$
$$= \boxed{50} \qquad\qquad\qquad \blacksquare$$

EXAMPLE 6   Evaluate. $10 - 2[8 - 3(4 - 9)]$

Solution   Again following the order of operations, we begin with the innermost grouping symbol. Within each grouping symbol we do multiplications before additions and subtractions.

$$10 - 2[8 - 3(4 - 9)] = 10 - 2[8 - 3(-5)] \qquad \textit{We computed } 4 - 9 = -5.$$
$$= 10 - 2[8 + 15] \qquad \textit{We computed } -3(-5) = 15.$$
$$= 10 - 2[23]$$
$$= 10 - 46$$
$$= \boxed{-36} \qquad\qquad\qquad \blacksquare$$

EXAMPLE 7   Evaluate. $\dfrac{-3 + 9}{1 - 3}$

Solution   Remember that a fraction bar is treated as if the numerator and denominator were each in parentheses. Therefore, we must compute the numerator and denominator first and then the resulting quotient.

$$\frac{-3 + 9}{1 - 3} = \frac{6}{-2} \qquad \textit{Since the signs are opposite, the answer is negative.}$$
$$= \boxed{-3} \qquad\qquad\qquad \blacksquare$$

EXAMPLE 8

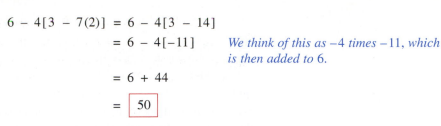

Solution

Evaluate. $\dfrac{-2(5)(-10)}{-2(5) - 10}$

Look at the example carefully! The numerator involves multiplication only, while the denominator involves multiplication and subtraction.

If you use a calculator to compute expressions like this, be sure to use the $\boxed{+/-}$ *key (not the subtraction key) to enter a negative number.*

$$\frac{-2(5)(-10)}{-2(5) - 10} = \frac{-10(-10)}{-10 - 10} \qquad \textit{In the denominator, } -10 - 10 \textit{ means } -10 + (-10).$$
$$= \frac{100}{-20}$$
$$= \boxed{-5} \qquad\qquad\qquad \blacksquare$$

As the example shows, the location of parentheses can make a great difference in what an example *means*. We will frequently emphasize the importance of reading an example carefully so that you clearly understand what it is saying, and what it is asking.

Properties of 0

Finally, we conclude this section with some special properties of the number 0.

We are familiar with the numerical facts that $5 \cdot 0 = 0$ and $11 \cdot 0 = 0$. In fact, 0 times any whole number is equal to 0, and this fact extends to the negative integers as well. We know that the product of any integer and 0 is equal to 0:

$$n \cdot 0 = 0 \quad \text{for all integers } n$$

What about dividing *into* 0? As long as we are not dividing *by* 0 (we will see why in a moment), dividing into 0 is perfectly legitimate. That is,

Keep in mind that

$$\frac{20}{4} = 5$$

because 5 · 4 = 20.

and

$$\frac{0}{8} = 0 \quad \text{because } 0 \cdot 8 = 0$$

$$\frac{0}{-4} = 0 \quad \text{because } 0 \cdot (-4) = 0$$

In general, we have

$$\frac{0}{n} = 0 \quad \text{for all integers } n \text{ not equal to } 0$$

What happens when we try to divide *by* 0? There are two cases to consider. If we try to divide 0 into a nonzero number—say, for example, 5—we cannot get an answer. If $\frac{5}{0}$ is going to be equal to some number, say *m*, then it must follow that $m \cdot 0 = 5$. But we know that *any* number times 0 is equal to 0, and so it is impossible to get 5. Therefore, dividing 0 into a nonzero number is *undefined*. That is, we have no available answer. (You might want to try dividing by 0 on your calculator.)

If we try to divide 0 into 0, we have a different problem. If $\frac{0}{0}$ is going to be equal to some number, say *r*, then it must follow that $r \cdot 0 = 0$. But this is true of *all* numbers *r*. This means that *any* number will work and so $\frac{0}{0}$ is not a unique number. Dividing 0 into 0 is said to be *indeterminate*. That is, we cannot determine a unique answer.

The terminology here is not particularly important for our purposes. What is important is that we realize that division by 0 does not make any sense, and therefore *is not allowed.*

Remember	If *a* is any nonzero number, then
	$\dfrac{0}{a} = 0$
	$\dfrac{a}{0}$ is undefined
	$\dfrac{0}{0}$ is indeterminate

Study Skills 1.5

Reading Directions

One important, frequently overlooked aspect of an algebraic problem is the verbal instructions. Sometimes the instructions are given in a single word, such as "simplify" or "solve" (occasionally it takes more time to understand the instructions than it takes to do the exercise). The verbal instructions tell us what we are expected to do, so make sure you read the instructions carefully and understand what is being asked.

Two examples may look the same, but the instructions may be asking you to do two different things. For example:

Identify the following property:

$$a + (b + c) = (a + b) + c$$

vs.

Verify the following property by replacing the variables with numbers:

$$a + (b + c) = (a + b) + c$$

(continued)

Study Skills 1.5 (continued)

Reading Directions

On the other hand, two different examples may have the same instructions but require you to do different things. For example:

Evaluate $2(3 - 8)$

vs.

Evaluate $2 + (3 - 8)$

You are asked to evaluate both expressions, but the solutions require different steps.

It is a good idea to familiarize yourself with the various ways the same basic instructions can be worded. In any case, always look at an example carefully and ask yourself what is being asked and what needs to be done, *before you do it.*

EXERCISES 1.5

Evaluate each of the following expressions, if possible.

1. $(+7)(-4)$

2. $(+6)(-3)$

3. $(-7)(+4)$

4. $(-6)(+3)$

5. $(-7)(-4)$

6. $(-6)(-3)$

7. $-7 - 4$

8. $-6 - 3$

9. $-(-7 - 4)$

10. $-(-6 - 3)$

11. $\dfrac{30}{-6}$

12. $\dfrac{24}{-8}$

13. $\dfrac{-30}{6}$

14. $\dfrac{-24}{8}$

15. $\dfrac{-30}{-6}$

16. $\dfrac{-24}{-8}$

17. $\dfrac{0}{6}$

18. $\dfrac{0}{-10}$

19. $\dfrac{6}{0}$

20. $\dfrac{-10}{0}$

21. $5(-3) - 7$

22. $6(-1) - 10$

23. $5 - 3 - 7$

24. $6 - 1 - 10$

25. $5 - (3 - 7)$

26. $6 - (1 - 10)$

27. $5(-3 - 7)$

28. $6(-1 - 10)$

29. $45 - 72 - 18$

30. $-63 - 19 + 41$

31. $105 - (28 - 81)$

32. $200 - (55 - 107)$

33. $-4(-2)(6)(-5)$

34. $9(-3)(2)(-4)$

35. $10 - 6(2 - 5)$

36. $7 - 3(5 - 9)$

37. $12 - 4(3 - 8)$

38. $11 - 6(2 - 7)$

39. $12 - 4 \cdot 3 - 8$

40. $11 - 6 \cdot 2 - 7$

41. $12 - (4 \cdot 3 - 8)$

42. $11 - (6 \cdot 2 - 7)$

43. $(12 - 4)(3 - 8)$

44. $(11 - 6)(2 - 7)$

45. $12(-8) - (-15)(-9)$

46. $-20(-7) - 13(6)$

47. $\dfrac{-15 - 6 + 3}{-9}$

48. $\dfrac{-14 - 6 + 5}{-3}$

49. $\dfrac{-20 - 8 - 12}{-4}$

50. $\dfrac{-10 - 25 - 15}{-5}$

51. $\dfrac{-20 - (8 - 12)}{-4}$

52. $\dfrac{-10 - (25 - 15)}{-5}$

53. $\dfrac{-20 - 8(-12)}{-4}$

54. $\dfrac{-10 - 25(-15)}{-5}$

55. $\dfrac{18}{-9} - \dfrac{20}{-5}$

56. $\dfrac{-28}{-4} - \dfrac{-32}{-8}$

57. $\dfrac{-44 - 16 + 80}{-5}$

58. $\dfrac{72 - 118 - 42}{-4}$

59. $\dfrac{9(-6)}{9 - 6}$

60. $\dfrac{12(-9)}{12 - 9}$

61. $\dfrac{6(4)(-8)}{6(4) - 8}$

62. $\dfrac{10(2)(-4)}{10(2) - 4}$

63. $\dfrac{8(-6) - 2}{-3 - 2}$

64. $\dfrac{-9(5) - 3}{-7 - 1}$

65. $\dfrac{9 - 5}{5 - 5}$

66. $\dfrac{-4 + 10}{-7 + 7}$

67. $\dfrac{4(-10)}{-3 - 1} - \dfrac{4 - 10}{-3(-1)}$

68. $\dfrac{6(-21)}{-5 - 1} + \dfrac{6 - 21}{-5(-1)}$

69. $\dfrac{28(-12)}{36 - 42}$

70. $\dfrac{64(-25)}{64 - 80}$

71. $-8[-4 - 6(4 - 7)]$

72. $-9[-2 - 5(1 - 8)]$

73. $12 - 4[7 - 3(6 - 2)]$

74. $18 - 10[9 - 5(4 - 1)]$

75. $56 - \dfrac{26 - 60}{2 - 4}$

76. $-72 - \dfrac{88 - 120}{8 - 12}$

77. $70 - \dfrac{40 + 5(-3)}{-5}$

78. $\dfrac{85 - 36 - 8(-4)}{-2}$

❓ QUESTIONS FOR THOUGHT

79. Discuss the similarities and differences between $5 - 2$ and $5(-2)$.

80. Discuss the similarities and differences between $6(-3) - 2$ and $6 - 3(-2)$.

81. Explain what is *wrong* (if anything) with each of the following:

(a) $-7 - 8 \overset{?}{=} -56$

(b) $-7 - 8 \overset{?}{=} +56$

(c) $-3 - (2 - 4) \overset{?}{=} -3 - 2 = -5$

(d) $-4 - 2(5 - 6) \overset{?}{=} -6(-1) = +6$

82. Taking into account the operations of addition, subtraction, multiplication, and division, is it accurate to say "two negatives make a positive"?

83. Consider the following sequence of computations:

$$4(-2) = -8$$
$$3(-2) = -6$$
$$2(-2) = -4$$
$$1(-2) = -2$$
$$0(-2) = 0$$
$$-1(-2) = ?$$
$$-2(-2) = ?$$

What does the pattern suggest the answers to the last two computations should be?
What does this suggest as to the result of multiplying two negative numbers?

1.6 The Real Number System

Throughout this chapter we have been working within the framework of the set of integers. However, we all recognize that the set of integers is not sufficient to supply us with all the numbers we need to describe various situations. For example, if we have to divide a 3-foot piece of wood into two *equal* pieces, then the length of each piece is 1.5 ft. The number 1.5 (or, if you prefer, the number $\frac{3}{2}$) is not an integer.

The set of "fractions" is called the set of **rational numbers** and is usually designated by the letter Q. The name "rational" comes from the word *ratio,* which means a quotient of two numbers. The set of rational numbers is difficult to list, primarily because no matter where you start, there is no *next* rational number. (Think about what the "first" fraction after 0 is.) Instead, we use set-builder notation to describe the set of rational numbers:

$$Q = \left\{ \frac{p}{q} \,\middle|\, p, q \in Z \text{ and } q \neq 0 \right\}$$

In words, this says that the set of rational numbers, which we are calling Q, is the set of all fractions whose numerators and denominators are integers (Z), provided the denominator is not equal to 0. Thus, the following are all rational numbers:

$$8\left(= \frac{8}{1}\right) \qquad \frac{-3}{7} \qquad 0\left(= \frac{0}{3}\right) \qquad 0.25\left(= \frac{1}{4}\right)$$

We can easily locate a rational number on the number line. For example, if we want to locate $\frac{7}{4} = 1\frac{3}{4}$, we would simply divide the unit interval into four equal parts, as shown here.

In order to convert a fraction into its decimal form (often called a **decimal fraction**), we divide the numerator by the denominator. (See Appendix A.) However, converting from the decimal form to a fraction is not quite so straightforward.

In the example above, we recognized that the decimal 0.25 is equal to $\frac{1}{4}$. Similarly, we recognize that the decimal $0.3333\overline{3}$ (where the dash above the last 3 indicates that the 3 repeats forever) is equal to the fraction $\frac{1}{3}$. On the other hand, it is highly unlikely that we would recognize the decimal $0.481481\overline{481}$ as being equal to the fraction $\frac{13}{27}$. (Divide 27 into 13 and verify that you get $0.481481\overline{481}$.)

In fact, not all decimals represent rational numbers. It turns out that if a decimal is nonterminating (it does not stop and give 0's after a while) and nonrepeating, then this decimal is *not* a rational number. In other words, such a decimal cannot be represented as the quotient of two integers.

The set of numbers on the number line that are not rational numbers is called the set of *irrational numbers* and is usually designated with the letter *I*. It is necessary for us to consider irrational numbers because just as the integers were insufficient to fill all our needs, so too the rational numbers do not quite do the job, either.

If we look for a number that when multiplied by itself gives a product of 9, we will fairly quickly come up with two answers:

$$3 \cdot 3 = 9 \quad \text{and} \quad (-3)(-3) = 9$$

A number x such that $x \cdot x = 9$ is called a *square root* of 9. (Square roots will be discussed in detail in Chapter 10.) Thus, we see that both 3 and -3 are square roots of 9.

Similarly, we can try to find a number that when multiplied by itself gives a product of 2 (such a number is called a *square root* of 2). It turns out that, if we try to find the answer by trial and error (using a calculator to do the multiplication would help), we can get closer and closer to 2 but we will *never* get 2 exactly. For example, if we try $(1.4)(1.4)$, we get 1.96, so we see that 1.4 is too small. If we try $(1.5)(1.5)$, we get 2.25 and we see that 1.5 is too big. If we continue in this way we can get better and better approximations to a square root of 2. We might reach the approximate answer 1.414214, but $(1.414214)(1.414214) = 2.0000012$ (rounded off to seven places). The point is that no matter how many places we get to, the decimal will never stop (because we never hit 2 exactly), and it never repeats. Thus, the square root of 2 (written $\sqrt{2}$) is an irrational number.

Another irrational number with which you may be familiar in π, which is discussed in Section G.7.

The important thing for us to recognize is that the irrational numbers also represent points on the number line. If we take all the rational numbers together with all the irrational numbers (both positive, negative, and zero), we get *all* the points on the number line. This set is called the set of *real numbers,* and is usually designated by the letter *R*:

$$R = \{x \mid x \text{ corresponds to a point on the number line}\}$$

We discuss the set of rational numbers in detail in Chapter 4, and irrational numbers in Chapter 10. Nevertheless, from now on unless we are told otherwise, we will assume that the set of real numbers serves as our basic frame of reference.

We should point out here that very often we describe a set of real numbers by using the *double inequality* notation. For instance, we may write

$$\{x \mid -1 < x \le 4\}$$

A double inequality of this sort is a "between" statement, and it means the real numbers *between* -1 and 4, *but* including 4 and excluding -1; that is, the double inequality represents the set of numbers x such that $-1 < x$ *and* $x \le 4$. This set is illustrated on the number line as follows:

Notice that we put an empty circle around the point -1 to indicate that it is excluded, and a solid circle around 4 to indicate that it is included.

EXAMPLE 1

Sketch the following sets on a number line:

(a) $\{x \mid x > -3\}$ (b) $\{x \mid 1 \le x < 5\}$

Solution

(a) The set $\{x \mid x > -3\}$ is the set of numbers to the right of -3:

(b) The set $\{x \mid 1 \le x < 5\}$ is the set of numbers that are at or to the right of 1 *and* to the left of 5:

It is important to mention that the properties of the integers that we have discussed, such as the commutative and associative properties, as well as the rules we have developed for adding, subtracting, multiplying, and dividing integers, carry over to the real number system.

EXAMPLE 2

Compute. $\dfrac{-5.8688}{1.4}$

Solution

If we want to compute $\dfrac{-5.8688}{1.4}$ we divide 5.8688 by 1.4, getting 4.192, and then make the answer negative because we are dividing opposite signs. Thus the answer is $\boxed{-4.192}$

This type of calculation is often done with a calculator. In order to use a calculator to do computations with signed numbers it is important to distinguish between the $\boxed{-}$ (subtraction key) and the $\boxed{+/-}$ (change of sign key), which changes the sign of the number on the display. If the number is positive, the $\boxed{+/-}$ changes it to negative and vice versa.

In order to do this computation with the calculator the following keystroke sequence (or something much like it) would be used.

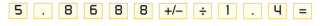

and the calculator display will read

$$\boxed{-4.192}$$

In the next chapter, we begin to build and manipulate algebraic expressions. The material on working with signed numbers is absolutely essential for all our work ahead, so be sure you can add, subtract, multiply, and divide signed numbers before you go ahead to the next chapter.

Study Skills 1.6

Estimation

As you work out exercises and solve problems, it is very important to be constantly aware of the reasonableness of your answer so that you do not propose impossible solutions to problems. This is particularly true when solving verbal problems, While it is obvious that if in a certain problem, x represents the number of 25-pound boxes, then an answer of $x = -7$ is ridiculous, it is also unreasonable to get an answer of $x = 8.3$. Why?

Sometimes recognizing an impossible answer is more subtle. For example, if a total of $5,000 is split into two investments and x represents one of the investments, then $x = \$6,500$ is impossible. Why?

Sometimes students are lulled into a false sense of security when they use a calculator to do computations. While it is true that the calculator does not make computational mistakes, you need to be sure that you have chosen to do the correct computations. Additionally, you may inadvertently hit the wrong operations key or put a decimal point in the wrong place. Having an estimate of the correct answer will make it much easier to recognize these types of errors.

EXERCISES 1.6

In Exercises 1–30, compute the given expression. Round off your answer to two decimal places where necessary.

1. $4.2 + 5.9$

2. $3.4 + 8.8$

3. $4(5.1)$

4. $6(3.9)$

5. $\dfrac{12.8}{3.2}$

6. $\dfrac{21.9}{7.3}$

7. $\dfrac{36.8}{1.5}$

8. $\dfrac{42.6}{2.4}$

9. $\dfrac{8}{0.2} + \dfrac{12}{0.4}$

10. $\dfrac{10}{0.25} + \dfrac{5}{0.1}$

11. $-2 - 5.3$

12. $-6 - 3.7$

13. $-2(5.3)$

14. $-6(3.7)$

15. $-2(-5.3)$

16. $-6(-3.7)$

17. $\dfrac{-8.4}{1.2}$

18. $\dfrac{-12.5}{2.5}$

19. $\dfrac{-6}{1.5} + \dfrac{21.6}{-1.2}$

20. $\dfrac{-12}{-1.2} - \dfrac{-10}{2.5}$

21. $0.831 - 0.746 - 0.294$

22. $28.7 - 32.56 - 18.61$

23. $0.53(21) - 0.42(85)$

24. $0.28(56) - 0.36(63)$

25. $12.4 - 20(0.8) + 4.7$

26. $15.7 - 35(0.6) - 4.6$

27. $0.02(28.6 - 13.5)$

28. $0.05(120.6 - 73.8)$

29. $5.2 - 1.4(2.8 - 0.7)$

30. $12.8 - 9.2(3.4 - 1.6)$

In Exercises 31–38, locate the given number between two successive integers on the number line. For example, if the given number is 2.6, it is located between 2 and 3 on the number line.

31. 4.8

32. -4.8

33. $-2\dfrac{1}{3}$

34. $2\dfrac{1}{3}$

35. $\dfrac{15}{2}$

36. $\dfrac{-15}{2}$

37. -0.24

38. 0.24

In Exercises 39–48, sketch the given set on a number line.

39. $\{x \mid x < 2\}$

40. $\{x \mid x > -3\}$

41. $\{x \mid x > 4\}$

42. $\{x \mid x < -2\}$

43. $\{x \mid -3 \le x \le 2\}$

44. $\{x \mid 2 < x < 7\}$

45. $\{x \mid 3 < x < 8\}$

46. $\{x \mid -2 \le x \le 2\}$

47. $\{x \mid 1 \le x < 3\}$

48. $\{x \mid -1 < x \le 3\}$

In Exercises 49–54, if possible list three numbers that are members and three numbers that are not members of the given set. If it is not possible, explain why.

49. $\{x \mid x$ is a real number but not an integer$\}$

50. $\{x \mid x$ is an integer but not a real number$\}$

51. $\{t \mid t$ is a rational number but not a real number$\}$

52. $\{t \mid t$ is a real number but not a rational number$\}$

53. $\{n \mid n$ is a rational number but not an integer$\}$

54. $\{w \mid w$ is an integer but not a rational number$\}$

(?) QUESTIONS FOR THOUGHT

55. Consider the following statements. If the statement is true, explain why. If the statement is false, explain why and/or give an example illustrating why.

(a) Every rational number is an integer.

(b) Every integer is a rational number.

(c) -15 is greater than -10.

(d) The absolute value of -15 is greater than the absolute value of -10.

(e) A rational number must be positive.

(f) An integer must be positive.

(g) Every real number is a rational number.

(h) Every rational number is a real number.

56. Explain how you could use the commutative and/or associative properties to make the following computations easier to do:

(a) $-1.7 + 12.9 - 8.3$

(b) $10(7)(-2.3)$

(c) $\dfrac{-10(3.2)}{-1.6}$

(d) $-30\left(\dfrac{1.6}{3}\right)(-10)$

57. Try to find an approximate value for a square root of 3. That is, try to find a number that when multiplied by itself gives a product of 3. Give answers correct to the nearest whole number, nearest tenth, nearest hundredth, and nearest thousandth. Use a calculator to help with the multiplication, but not to compute the square root itself.

CHAPTER 1 SUMMARY

After having completed this chapter you should be able to:

1. Use the listing method or set-builder notation to recognize sets (Section 1.1).

   For example:

   If we have the sets

$$A = \{3, 6, 9, 12, 15\}$$
$$B = \{x \mid x \text{ is an even integer between 8 and 30}\}$$
$$C = \{17, 19, 21, \ldots, 29\}$$

   Then

   (a) $6 \in A, \quad 10 \notin A$

   (b) $8 \notin B, \quad 28 \in B$

   (c) $\{x \mid x \in A \text{ and } x \in C\} = \varnothing$ (the empty set)

2. Evaluate a numerical expression using the order of operations (Section 1.2).

   For example:

$$8 + 4(7 - 2) = 8 + 4(5)$$
$$= 8 + 20$$
$$= \boxed{28}$$

3. Use the rules developed in this chapter to add, subtract, multiply, and divide signed numbers (Sections 1.3–1.6).

For example:

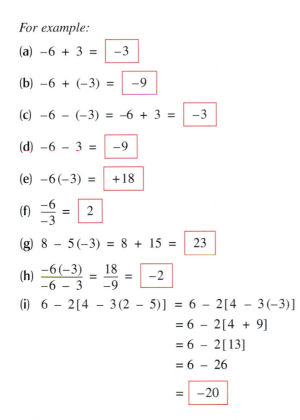

(a) $-6 + 3 =$ $\boxed{-3}$

(b) $-6 + (-3) =$ $\boxed{-9}$

(c) $-6 - (-3) = -6 + 3 =$ $\boxed{-3}$

(d) $-6 - 3 =$ $\boxed{-9}$

(e) $-6(-3) =$ $\boxed{+18}$

(f) $\dfrac{-6}{-3} =$ $\boxed{2}$

(g) $8 - 5(-3) = 8 + 15 =$ $\boxed{23}$

(h) $\dfrac{-6(-3)}{-6 - 3} = \dfrac{18}{-9} =$ $\boxed{-2}$

(i) $6 - 2[4 - 3(2 - 5)] = 6 - 2[4 - 3(-3)]$
$$= 6 - 2[4 + 9]$$
$$= 6 - 2[13]$$
$$= 6 - 26$$
$$= \boxed{-20}$$

CHAPTER 1   REVIEW EXERCISES

In Exercises 1–6, list the members of each set.

1. $\{x \mid x$ is a positive even integer less than 20$\}$

2. $\{y \mid y$ is a positive odd integer less than 20$\}$

3. $\{p \mid p$ is a prime number less than 20$\}$

4. $\{c \mid c$ is a positive composite number less than 20$\}$

5. $\{p \mid p$ is a prime number between 20 and 30$\}$

6. $\{x \mid x$ is a prime number and x is a composite number$\}$

In Exercises 7–12, factor the given number into its prime factors. If the number is prime, say so.

7. 30

8. 28

9. 47

10. 72

11. 100

12. 57

In Exercises 13–60, evaluate the given expression.

13. $3 - 7$

14. $4 - 9$

15. $-7 - 5$

16. $-8 - 2$

17. $-2 - (-6)$

18. $-6 - (-2)$

19. $-4 - 5 - 6$

20. $-2 - 3 - 4$

21. $-7 + 12 - 5$

22. $-6 + 8 - 12$

23. $7 - 4 + 3 - 9$

24. $2 - 5 - 3 + 1$

25. $8 - 5 - 6$

26. $4 - 9 - 3$

27. $8 - (5 - 6)$

28. $4 - (9 - 3)$

29. $8(5 - 6)$

30. $4(9 - 3)$

31. $8 - 3 - 6$

32. $4 - 9 - 2$

33. $8 - 3(-6)$

34. $4 - 9(-2)$

35. $8 - (3 - 6)$

36. $4 - (9 - 2)$

37. $8(-3)(-6)$

38. $4(-9)(-2)$

39. $9 - 4(3 - 7)$

40. $7 - 5(2 - 8)$

41. $9 - 4 \cdot 3 - 7$

42. $7 - 5 \cdot 2 - 8$

43. $9 - (4 \cdot 3 - 7)$

44. $7 - (5 \cdot 2 - 8)$

45. $(9 - 4)(3 - 7)$

46. $(7 - 5)(2 - 8)$

47. $|4 - 9| - |3 - 7|$

48. $|2 - 5| - |1 - 6|$

49. $\dfrac{-7 - 3}{-2(-5)}$

50. $\dfrac{-4 - 15 - 2}{1 - 4}$

51. $\dfrac{-4(-2)(-8)}{-4(2) - 8}$

52. $\dfrac{-3(2)(-6)}{3(-2) - 6}$

53. $\dfrac{6 - 4(3 - 1)}{-2(-3) - 4}$

54. $\dfrac{7 + 2(5 - 10)}{-3(-3) - 6}$

55. $8 + 2[3 - 4(1 - 6)]$

56. $6 - 4[7 + 2(1 - 5)]$

57. $28(-65) - 30(-8 - 7)$

58. $\dfrac{-108}{9} - \dfrac{768}{-24}$

59. $\dfrac{-48 - 5(36)}{47 - 53}$

60. $\dfrac{21(-106) - 35(-40)}{2(51 - 58)}$

In Exercises 61–64, evaluate the given expression. Round off to the nearest hundredth where necessary.

61. $-3.4(6.85) - 2.1$

62. $-3.4(6.85)(-2.1)$

63. $\dfrac{8.57 - 12.63}{-4.2}$

64. $\dfrac{-2.8}{0.4} - \dfrac{5.6}{-2.8}$

 CHAPTER 1 PRACTICE TEST

1. Let $A = \{0, 4, 8, 12, \ldots, 28\}$ and $B = \{x \mid x$ is an odd integer between 3 and 20$\}$.
   Answer parts (a)–(d) True or False.

   (a) $20 \in A$

   (b) $20 \in B$

   (c) $3 \in B$

   (d) Both A and B have the same number of elements.

   (e) List the elements of the set $C = \{x \mid x \in A$ and $x \in B\}$.

Evaluate each of the following.

2. $-9 - 4 + 3 - 6 + 5$

3. $|3 - 8| - |1 - 6|$

4. $4 - 7 - 3 - (-2)$

5. $-4(-3)(-2)$

6. $-3(-5)(-2)(-1)$

7. $-3(-5) - 2(-1)$

8. $\dfrac{10}{-2} - \dfrac{-18}{3}$

9. $\dfrac{4(-8)}{4-8} - \dfrac{3-6}{-2-1}$

10. $\dfrac{5(-4)(-3)}{1-7}$

11. $\dfrac{(-2)(-3)(-4)}{(-2)(-3)-4}$

12. $8 - 5(4-7)$

13. $8 - 5 \cdot 4 - 7$

14. $-7[5 + 4(3-7) - 2]$

15. $8 - 3[8 - 3(8-3)]$

16. $\dfrac{72 - 15(-45)}{3(17) - 60}$

17. $\dfrac{105(-42) - (-36)(-80)}{20(25) - 17(30)}$

18. Factor each of the following into its prime factors.

   (a) 84

   (b) 1,872

   (c) 79

CHAPTER 2

Algebraic Expressions

STUDY SKILLS

2.1 Variables and Exponents

In the previous chapter we discussed various number systems and some of their basic properties. When we wanted to make a general statement about numbers, we used letters to represent numbers. This is in fact the essence of algebra—it is the generalization of arithmetic.

When we want to describe even fairly simple properties of numbers in words, it can be quite cumbersome. It is usually much more convenient to use letters when we want to talk about "any numbers." The key idea here is the **variable,** a symbol that stands for a number (or numbers). In this book a variable will usually be represented as an italic Latin letter such as x, y, A, M, etc.

A variable is sometimes called an unknown.

In algebra, there are two ways variables are primarily used. One use of a variable is as a placeholder. That is, the variable is holding the place of a particular number (or numbers) that has not yet been identified but which needs to be found. The equation

$$x + 7 = 5$$

is an example of this type of use of a variable. Here we would like to figure out the number that when added to 7 gives 5. Variables are used in this way when we are solving equations.

A second use of a variable is to describe a general relationship between numbers and/or arithmetic operations. In the statement of the associative property, when we write

$$a + (b + c) = (a + b) + c$$

we mean that a, b, and c can be any real numbers.

Variables allow us to express mathematical ideas, concepts, and relationships in a shorthand way. Generally speaking, the mathematical notation we introduce as we proceed through this and subsequent chapters is a special mathematical shorthand that allows us to describe fairly complex ideas concisely.

While variables represent unknown quantities, a **constant,** on the other hand, is a symbol whose value is fixed. The numbers 8, -5, 2.43, and π are examples of constants.

Our goal in this chapter is to learn how to take algebraic expressions and, given a basic set of guidelines and properties, change them into simpler expressions. This process is called **simplifying algebraic expressions.**

We have used the phrase *algebraic expression* without having actually defined it. For the time being, we will accept the following definition of an algebraic expression. (We will extend this definition a bit further in Chapter 10.)

DEFINITION	An *algebraic expression* is made up of a finite number of additions, subtractions, multiplications, and divisions of constants and/or variables.

Note that, according to this definition, an algebraic expression *does not* contain an equals sign. That is, $3x + 7$ is an expression but $3x + 7 = 4$ is an equation, *not* an expression.

Do not be intimidated by this formal definition. It merely makes our terminology precise. For example, each of the following is an algebraic expression:

$$5 \qquad -8xy \qquad 3x + 4y - \frac{2}{w}$$

As we proceed through this course we will develop more specific terminology for different types of algebraic expressions.

To a great extent, the meaning of the word *simplify* will depend on the type of algebraic expression with which we are working.

In this chapter we deal with a certain type of expression that involves some new notation. We are frequently going to encounter expressions that involve a product with the same factor repeated numerous times. For example, Figure 2.1 illustrates a square of side 8 in. and a cube of side 20 cm. The area of the square is the product of the sides, which is 8 · 8. The volume of the cube is the product of its length, width, and height, which is 20 · 20 · 20.

Figure 2.1

The area of the square is 8 · 8   The volume of the cube is 20 · 20 · 20

The next example illustrates another situation that gives rise to repetitive multiplication.

EXAMPLE 1   Suppose that the price of a newly issued Internet stock is $3 per share, and that a stock broker projects that the price will double each month for 6 months. If the broker is correct, what will the price of the stock be at the end of the 6-month period?

Solution   We are told that the price of the stock when it is issued is $3. After 1 month the projected price would double to 2 · $3 = $6. After 2 months the price would double again, so we again multiply by 2, and the price would be 2 · 2 · $3 = $12. We continue this process for 6 months, as calculated in the following table.

Month	Price
1	2 · $3 = $6
2	2 · 2 · $3 = $12
3	2 · 2 · 2 · $3 = $24
4	2 · 2 · 2 · 2 · $3 = $48
5	2 · 2 · 2 · 2 · 2 · $3 = $96
6	2 · 2 · 2 · 2 · 2 · 2 · $3 = $192

■

The type of repeated multiplication we saw in Example 1 can be tedious to write and difficult to read. Consequently, we introduce a special notation for such situations. Consider the following expressions.

$3 \cdot 3 \cdot 3 \cdot 3 \cdot 3$ *Remember that we use · to indicate a product.*

$x \cdot x \cdot x \cdot x \cdot x$

If we multiply out the expression 3 · 3 · 3 · 3 · 3, we see that it has a numerical value of 243. On the other hand, the expression $x \cdot x \cdot x \cdot x \cdot x$ has no numerical value until we know the value of x. However, our focus now is not so much on the value of such expressions, but rather on their *form*. Consequently, we introduce a shorthand way of writing such products called **exponent notation.** We write

$$3 \cdot 3 \cdot 3 \cdot 3 \cdot 3 = 3^5$$

In the expression 3^5,

    3 is called the **base**

    5 is called the **exponent**

We read 3^5 as "3 raised to the fifth power." Similarly, $5^4 = 5 \cdot 5 \cdot 5 \cdot 5$. The exponent 4 tells us how many factors of 5 to *multiply.*

If, in addition, we want to *evaluate* or compute the value of 5^4, we would proceed as follows:

$$
\begin{aligned}
5^4 &= 5 \cdot 5 \cdot 5 \cdot 5 \\
&= 25 \cdot 5 \cdot 5 \\
&= 125 \cdot 5 \\
&= \boxed{625}
\end{aligned}
$$

DEFINITION	*Exponent notation:*
	$$x^n = \underbrace{x \cdot x \cdot x \cdot \cdots \cdot x}_{n\ times}$$
	where x appears as a factor n times. Of course, n is a positive integer.

The terminology x-squared *and* x-cubed *comes from the area of the following square and the volume of the following cube.*

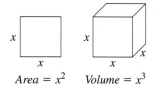

Area = x^2 Volume = x^3

In general, x^n is read "x to the nth power." However, exponents 2 and 3 are given special names:

x^2 is read "x squared"

x^3 is read "x cubed"

Note: $x = x^1$ but we usually do not write the exponent 1, even though we do often think of it.

EXAMPLE 2

Evaluate each of the following:

(a) 4^3 (b) $(-2)^4$ (c) -2^4 (d) $5 \cdot 3^2$

Solution

(a) $4^3 = 4 \cdot 4 \cdot 4$ *4 taken as a factor 3 times*

$ = 16 \cdot 4$

$ = \boxed{64}$

(b) $(-2)^4 = (-2)(-2)(-2)(-2)$

$ = 4(-2)(-2)$

$ = -8(-2)$

$ = \boxed{16}$

(c) $-2^4 = -(2 \cdot 2 \cdot 2 \cdot 2)$

$ = -(4 \cdot 2 \cdot 2)$

$ = -(8 \cdot 2)$

$ = \boxed{-16}$

Note the difference between parts **(b)** and **(c)**. In $(-2)^4$ the exponent 4 refers to the base -2; therefore, -2 appears as a factor 4 times. But in -2^4 the exponent 4 refers only to the base 2; therefore, 2 appears as a factor 4 times with only *one* negative sign in front. In other words, -2^4 is read as "the negative of 2^4." An exponent applies only to what is to its immediate left. If the intention is for the exponent 4 to apply to the -2 then parentheses are necessary, as in part **(b)** above.

(d) $5 \cdot 3^2 = 5 \cdot 3 \cdot 3$ or $5 \cdot 3^2 = 5 \cdot 3 \cdot 3$

$\qquad\qquad = 5 \cdot 9 \qquad\qquad\qquad\qquad = 15 \cdot 3$

$\qquad\qquad = \boxed{45} \qquad\qquad\qquad\qquad\quad = \boxed{45}$

Here again, do not make the mistake of first multiplying 5 times 3 and then squaring the result. The exponent 2 applies *only* to the 3. As long as we clearly keep this in mind we will carry out the order of operations correctly.   ■

EXAMPLE 3   Write each of the following with exponents.

(a) $xxyzzzz$ **(b)** $3xyyy$ **(c)** $xxx \cdot xxxx$ **(d)** $xxx + xxxx$

Solution   **(a)** $xxyzzzz = \boxed{x^2yz^4}$

(b) $3xyyy = \boxed{3xy^3}$

(c) $xxx \cdot xxxx = \boxed{x^7}$

(d) $xxx + xxxx = \boxed{x^3 + x^4}$ ■

EXAMPLE 4   Write each of the following without exponents.

(a) xy^3 **(b)** $(xy)^3$ **(c)** $x^2 \cdot x^3$ **(d)** $x^2 + x^3$

Solution   **(a)** $xy^3 = \boxed{xyyy}$

(b) $(xy)^3 = \boxed{(xy)(xy)(xy)}$

(c) $x^2 \cdot x^3 = \boxed{xx \cdot xxx}$

(d) $x^2 + x^3 = \boxed{xx + xxx}$ ■

Let's look at Example 4, parts **(c)** and **(d)**, more carefully:

$$x^2 \cdot x^3 = xx \cdot xxx = x^5 \qquad \text{but} \qquad x^2 + x^3 = xx + xxx \neq x^5$$

There is sometimes a tendency to look at $x^2 + x^3$ and to think something like

In order to understand how exponents work, we must be able to distinguish factors from terms.

"There are 5 x's so it must be x^5."

Algebraic notation is very precise. Thus, x^5 means x taken as a *factor* 5 times, which is not at all the same as x taken as a factor 2 times *added to* x taken as a factor 3 times. If you are still not sure of the difference, let's evaluate both expressions for a particular value of x, say $x = 4$:

$$4^5 = 4 \cdot 4 \cdot 4 \cdot 4 \cdot 4 = 1{,}024$$

$$\text{while} \qquad 4^2 + 4^3 = 4 \cdot 4 + 4 \cdot 4 \cdot 4 = 16 + 64 = 80$$

We get very different values because the expressions are really very different even though they may superficially look the same.

> When asked to simplify an expression involving exponents, we are expected to write an equivalent expression with bases and exponents occurring as few times as possible.

EXAMPLE 5

Simplify each of the following as completely as possible.

(a) x^2x^5 (b) a^4a^5 (c) yy^7 (d) $7^2 \cdot 7^4$

Solution

We can compute each expression by writing out all factors and simply counting them up.

(a) $x^2x^5 = xx \cdot xxxxx = \boxed{x^7}$ *Note the trend in exponents:* $2 + 5 = 7$

(b) $a^4a^5 = aaaa \cdot aaaaa = \boxed{a^9}$ *Note:* $4 + 5 = 9$

(c) $yy^7 = y \cdot yyyyyyy = \boxed{y^8}$ *Note:* $y = y^1$ *and* $1 + 7 = 8$

(d) $7^2 \cdot 7^4 = 7 \cdot 7 \cdot 7 \cdot 7 \cdot 7 \cdot 7 = \boxed{7^6}$ ■

From these examples we see that if we are multiplying powers of the same base, we keep the base and add the exponents. This follows from simply understanding what an exponent means, and being able to count. This observation generalizes as the first rule for exponents.

Exponent Rule 1	$a^m \cdot a^n = a^{m+n}$

This same rule extends to a product of more than two powers of the same base.

EXAMPLE 6

Simplify each of the following as completely as possible.

(a) $r^2r^4r^5$ (b) a^6aa (c) 3^43^6 (d) 2^35^2 (e) x^5y^4

Solution

We use exponent rule 1:

(a) $r^2r^4r^5 = r^{2+4+5} = \boxed{r^{11}}$ *Remember that the shaded steps are usually done mentally.*

(b) $a^6aa = a^{6+1+1} = \boxed{a^8}$

(c) $3^43^6 = 3^{4+6} = \boxed{3^{10}}$ *Note that the answer is **not** 9^{10}.*

We are not multiplying the 3's. Rule 1 says that we count up the factors of 3. There are 10 factors of 3, not 10 factors of 9!

(d) We cannot simplify this expression in the way we just did part **(c)**, because the bases are not the same; we can, however, compute its value.

$$2^35^2 = (2 \cdot 2 \cdot 2)(5 \cdot 5) = (8)(25) = \boxed{200}$$

(e) x^5y^4 remains unchanged, since the bases, x and y, are not identical.   ■

EXAMPLE 7

Simplify as completely as possible. $x^3x^5 + xx^2x^4$

Solution

$$x^3x^5 + xx^2x^4 = x^{3+5} + x^{1+2+4}$$

$$= \boxed{x^8 + x^7}$$

This is the complete answer. You do not get x^{15} as an answer. Rule 1 says that you add the exponents when you are *multiplying* the powers, not when you are *adding* them. ■

The commutative and associative properties of multiplication that we discussed in the last chapter allow us to multiply simple algebraic expressions by rearranging and regrouping the factors. For example:

$$(3x^2)(5x^4) = 3(x^2 \cdot 5)x^4 \qquad \textit{Associative property of multiplication}$$
$$= 3(5 \cdot x^2)x^4 \qquad \textit{Commutative property of multiplication}$$
$$= (3 \cdot 5)(x^2 \cdot x^4) \qquad \textit{Associative property of multiplication}$$
$$= \boxed{15x^6} \qquad \textit{Exponent rule 1}$$

Go back and reread this sequence of steps (which is usually done mentally), making sure you *understand* each step completely. Essentially, the commutative and associative properties allow us to ignore the original order and grouping of the factors so that we may multiply all constants together, and multiply identical variables together using exponent rule 1.

EXAMPLE 8

Simplify each of the following as completely as possible.

(a) $(4x^3)(7x^6)$ (b) $(-2x)(5x^4)(3x^3)$ (c) $(-2x^3y)(-6x^4y^5)$ (d) $(2x^2)^3$

Solution

Since these expressions all consist entirely of multiplication, we can use the commutative and associative properties to rearrange and regroup the factors.

(a) $(4x^3)(7x^6) = 4 \cdot x^3 \cdot 7 \cdot x^6 = (4 \cdot 7)(x^3 \cdot x^6) = \boxed{28x^9}$

Notice how we use the commutative and associative properties to rearrange and regroup the factors.

(b) $(-2x)(5x^4)(3x^3) = (-2)(5)(3)(xx^4x^3) = \boxed{-30x^8}$

(c) $(-2x^3y)(-6x^4y^5) = (-2)(-6)(x^3x^4yy^5) = \boxed{12x^7y^6}$

(d) $(2x^2)^3 = (2x^2)(2x^2)(2x^2) = (2 \cdot 2 \cdot 2)(x^2x^2x^2) = \boxed{8x^6}$

Notice how the coefficient (8) and exponent (6) of the final answer were computed. ∎

While we have only scratched the surface as far as simplifying expressions is concerned, understanding the examples we have just worked out is basic to being able to do the more complex problems in the sections ahead.

Study Skills 2.1

Comparing and Contrasting Examples

When learning most things for the first time, it is very easy to get confused and to treat things that are different as though they were the same because they "look" similar. Algebraic notation can be especially confusing because of the detail involved. Move or change one symbol in an expression and the entire example is different; change one word in a verbal problem and the whole problem may have a new meaning.

It is important that you be capable of making these distinctions. The best way to do this is by comparing and contrasting examples and concepts that look almost identical, but are not. It is also important that you ask

yourself in what ways these things are similar and in what ways they differ. For example, the associative property of addition is similar in some respects to the associative property of multiplication, but different from it in others. Also, the expressions $3 + 2 \cdot 4$ and $3 \cdot 2 + 4$ look similar, but are actually very different.

When you are working out exercises (or reading a concept), ask yourself, "What examples or concepts are similar to those which I am now doing? In what ways are they similar? How do I recognize the differences?" Doing this while you are working the exercises will help prevent you from making careless errors later on.

In Exercises 1–12, write the given expression without exponents.

1. x^6

2. y^5

3. $(-x)^4$

4. $(-y)^5$

5. $-x^4$

6. $-y^6$

7. $x^2 y^3$

8. $x^4 y^6$

9. $x^2 + y^3$

10. $x^4 + y^6$

11. xy^3

12. $(xy)^3$

In Exercises 13–22, write the given expression with exponents.

13. $aaaa$

14. $sssss$

15. $xxyyy$

16. $xx + yyy$

17. $-rrsss$

18. $(-r)(-r)sss$

19. $-xx(-y)(-y)(-y)$

20. $-xx - yyy$

21. $(xxx)(xxxxx)$

22. $(xxx) + (xxxxx)$

In Exercises 23–38, evaluate the given expression.

23. 3^5

24. 5^3

25. -2^3

26. $(-2)^3$

27. $(-2)^4$

28. -2^4

29. $3^2 + 3^3 - 3^4$

30. $4^2 + 4^3 - 4^4$

31. $4 \cdot 3^2 - 2 \cdot 5^2$

32. $7 \cdot 2^3 - 5 \cdot 3^2$

33. $(3 - 7)^2 - (4 - 5)^3$

34. $(6 - 8)^2 - (1 - 3)^3$

35. $3^2 4^3$

36. $2^3 2^4$

37. $3^2 3^3$

38. $2^3 3^4$

In Exercises 39–60, simplify the given expression as completely as possible.

39. $x^3 x^5$

40. $y^4 y^6$

41. $aa^2 a^4$

42. $m^3 mm^5$

43. $(3x)(5x)(4x)$

44. $(2a)(3a)(6a)$

45. $(3r^2)(2r^3)$

46. $(4w^5)(5w^4)$

47. $(-3x^3)(5x^2)$

48. $(-4x)(6x^5)$

49. $(-c^4)(2^3)(-5c)$

50. $(-3p)(-2p^5)(p^2)$

51. $(8x^3 y^2)(4xy^5)$

52. $(7x^2 y^3)(5xy^4)$

53. $(-2xy)(x^2 y^2)(-3xy)$

54. $(-8x)(-3xy^2)(5x^3 y^2)$

55. $(3a^4)^2$

56. $(4a^3)^2$

57. $(-4n^2)^3$

58. $(-5n^4)^3$

59. $(x^2)^3 (x^4)^2$

60. $(x^3)^2 (x^2)^4$

In Exercises 61–68, evaluate each expression. Round off your answer to the nearest thousandth where necessary.

61. $(0.52)^4$

62. $(1.83)^5$

63. $(1.4)^8$

64. $(-2.7)^9$

65. $5.1(4.6)^2$

66. $2.3(7.1)^3$

67. $(3.81)^3 - (2.64)^2$

68. $(-5.4)^2 - (4.1)^2$

⑦ QUESTIONS FOR THOUGHT

69. State in *words* the difference between $3 \cdot 2^4$ and $(3 \cdot 2)^4$.

70. Explain what is ***wrong*** with each of the following:

    (a) $3 \cdot 2^2 \stackrel{?}{=} 6^2 \stackrel{?}{=} 36$

    (b) $3x^4 \stackrel{?}{=} 3x \cdot 3x \cdot 3x \cdot 3x$

    (c) $5^4 \cdot 2 \stackrel{?}{=} 20 \cdot 2 \stackrel{?}{=} 40$

    (d) $(3x^3)^2 \stackrel{?}{=} 6x^6$

    (e) $-3^4 \stackrel{?}{=} (-3) \cdot (-3) \cdot (-3) \cdot (-3) \stackrel{?}{=} 81$

    (f) $x^4 + x^5 \stackrel{?}{=} x^9$

    (g) $x^2 \cdot x^7 \stackrel{?}{=} x^{14}$

    (h) $(x^2)^4 \stackrel{?}{=} x^6$

71. Compare and contrast the expressions $3 \cdot 5$ and 5^3.

72. Compare and contrast the expressions 3^5 and 5^3.

73. Compare and contrast the expressions 2^4 and 4^2. Does the fact that both of these expressions have a numerical value of 16 affect your answer?

2.2 Algebraic Substitution

Understanding the meaning of algebraic expressions allows us to use them to compute quantities encountered in everyday life.

EXAMPLE 1   Charlene wants to put a fence around a rectangular garden. The garden has a width of 9.5 meters and a length of 23.4 meters. The formula for the perimeter P of a rectangle with length L and width W is

$$P = 2L + 2W$$

Use this formula to compute the amount of fence Charlene needs.

Solution   While it is not always necessary to understand the real-life significance of the variables used in a formula, it certainly makes the computation more meaningful if we do. A rectangle is a four-sided figure with opposite sides equal and the sides meet at right angles. See Figure 2.2(a). The perimeter of a figure means the length around it.

Rectangles are discussed in Section G.3.

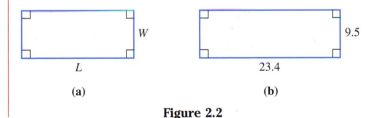

             (a)   **(b)**

Figure 2.2

Thus, computing the amount of fence Charlene needs requires us to compute the perimeter of the rectangle. In other words, we are finding the perimeter of the rectangle in Figure 2.2(b). To compute the perimeter we substitute the given values for the length and width into the formula:

$$P = 2L + 2W \qquad \textit{We substitute } L = 23.4 \textit{ and } W = 9.5.$$
$$P = 2(23.4) + 2(9.5)$$
$$P = 46.8 + 19 = 65.8$$

Therefore Charlene needs $\boxed{65.8}$ meters of fence to enclose the garden. ■

EXAMPLE 2

If a ball is thrown straight up into the air with an initial velocity of 40 feet per second, then its height h (in feet above the ground) t seconds after it is released is given by the formula

$$h = 40t - 16t^2$$

Use this formula to compute the height of the ball 2 seconds after it is released.

Solution

First of all, it is important to understand what the formula "says." The formula tells us that we can compute the height (h) of the ball above the ground after it is thrown by substituting the number of seconds (t) into the formula. So, for example, if $t = 1$ we get

$$h = 40(1) - 16(1^2) = 40 - 16 = 24$$

which means that 1 second after the ball is thrown it is 24 ft above the ground.

Frequently, the critical step in solving a mathematical problem is to completely understand the question and restate it in simpler or more familiar language. In this example we are given a formula relating the height of a ball to the number of seconds after it has been thrown upward. In order to find the height of the ball after 2 seconds we in effect reformulate the question to "find the value of h when $t = 2$." Therefore, we need to substitute the value $t = 2$ into the formula for h.

$$h = 40t - 16t^2 \qquad \textit{Substitute } t = 2.$$
$$h = 40(2) - 16(2)^2 = 80 - 16(4) = 80 - 64 = \boxed{16 \text{ feet}}$$ ■

EXAMPLE 3

A manufacturing firm determines that its monthly profit P (in dollars) is related to the number of items n it sells according to the equation

$$P = 15{,}800 + 8.12n - 0.002n^2$$

Use this equation to determine the monthly profit if the company sells 2,175 items.

Solution

Again let's first make sure we understand the given relationship. The formula tells us that we can compute the monthly profit (P) by substituting the number of items sold (n) into the formula.

In order to compute the monthly profit on the sale of 2,175 items we need to substitute $n = 2{,}175$ into the formula for P. We get

$$15{,}800 + 8.12(2{,}175) - 0.002(2{,}175)^2$$

It is important that you know how to use your calculator to compute powers.

A computation such as this is usually done with a calculator. Exactly how you carry out this computation using a calculator depends on the type of calculator you use. If your calculator has an $\boxed{x^2}$ key (as all scientific calculators do), you can evaluate this expression by using the following keystroke sequence (or something very much like it):

$\boxed{15800}$ $\boxed{+}$ $\boxed{8.12}$ $\boxed{\times}$ $\boxed{2175}$ $\boxed{-}$ $\boxed{.002}$ $\boxed{\times}$ $\boxed{2175}$ $\boxed{x^2}$ $\boxed{=}$

The display will then read $\boxed{\texttt{23999.75}}$. Thus, the profit on the sale of 2,175 items would be

$$\boxed{\$23{,}999.75}$$

EXAMPLE 4

An environmental consultant for a large factory finds that the cost C (in thousands of dollars) to remove p percent of the pollutants released by the factory is given by the formula

$$C = \frac{30p}{100 - p}$$

Determine how much it will cost to remove 75% of the pollutants.

Solution

Since p represents the percent of the pollutants to be removed, we substitute $p = 75$ in the given formula.

Example 7 in Section 9.4 discusses how we would use this formula to find how much it would cost to remove a given percent of the pollutants.

$$C = \frac{30p}{100 - p} \qquad \textit{Substitute } p = 75.$$

$$C = \frac{30(75)}{100 - 75} = \frac{30(75)}{25} = 90$$

The fact that $C = 90$ means that it would cost $90,000 to remove 75% of the pollutants.

Examples 2, 3, and 4 illustrate a common situation in which two quantities are related to each other in such a way that one of them *depends* on the other. In Example 2, the height h of the ball depends on the number t of seconds it has been in the air. Sometimes we say that the height h is a *function* of t, the number of seconds after the ball is thrown. Similarly, in Example 3, the profit P depends on the number n of items sold so we can say that P is a function of n; in Example 4, the cost C of removing pollutants depends on the percentage p to be removed so we can say that C is a function of p.

We will discuss functions a bit more in Section 5.5.

The following example illustrates how a formula derived from real data can be compared to the data itself.

EXAMPLE 5

The following data collected by the U.S. Department of Labor give the unemployment rate as a percentage of the total civilian labor force for January of each of the years listed.

Year	1993	1994	1995	1996	1997	1998
Unemp. rate	7.3%	6.6%	5.6%	5.7%	5.3%	4.7%

Based on these data, a statistician suggests that the following formula can be used to approximate the unemployment rate R in January of the years 1993 through 1998:

$$R = 0.05x^2 - 1.05x + 9.92$$

where $x = 3$ corresponds to 1993, $x = 4$ corresponds to 1994, etc.

(a) Use this formula to compute R for each of the years 1993 through 1998.

(b) Compare the values collected by the Department of Labor with the values obtained from the formula and comment about their accuracy.

Solution

(a) This part of the example is asking us to use the given formula to compute R for the values $x = 3, 4, 5, 6, 7,$ and 8. Thus if we substitute $x = 3$ into the formula for R we get

$$R = 0.05(3)^2 - 1.05(3) + 9.92$$

Computations such as these are usually carried out using a calculator. Exactly how you carry out the computation using a calculator depends on the type of calculator you use. If you are using a calculator that has an x^2 (as all scientific calculators do), then the following keystroke sequence (or something very much like it) will compute the value of R for $x = 3$ (that is, 1993).

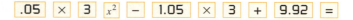

and the display will read 7.22 .

We repeat this computation for $x = 4, 5, 6, 7,$ and 8. The results are summarized in the following table. (We include the original data values for comparison.)

Year	1993	1994	1995	1996	1997	1998
Unemp. rate	7.3%	6.6%	5.6%	5.7%	5.3%	4.7%
x	3	4	5	6	7	8
R	7.22%	6.52%	5.92%	5.42%	5.02%	4.72%

(b) Comparing the original data values with those computed using R, we can see that the formula seems to give a reasonably accurate approximation to the unemployment rate. The formula is most accurate for 1998 and least accurate for 1995. ■

The process of replacing the variables in an expression with specific values and evaluating the result, as illustrated in this section, is called *algebraic substitution.*

A complaint often heard from students is that the expressions encountered in algebra all "look alike." If you do not happen to play the piano you may think the piano presents a similar problem—all the keys look alike. It is only through practice and study that you can learn to distinguish the keys. The same practice and study are required in algebra.

As we have begun to see in the last section, algebraic notation is very precise. It means exactly what it says. One of the most basic skills necessary for success in algebra is the ability to read carefully so that you can see what a particular problem says. For example, two expressions

$$(-5)(-3) \quad \text{and} \quad (-5) - (3)$$

may look quite similar at first glance, but they are in fact quite different.

$$(-5)(-3) = 15 \quad \text{while} \quad (-5) - (3) = -8$$

The first expression involves multiplication; the second involves subtraction. You must sensitize yourself to recognize these differences.

As you will see in the examples that follow, we must be careful to follow the order of operations and *not* change the arithmetic steps that appear in the example.

EXAMPLE 6

Evaluate each of the following for $x = 2$ and $y = -3$:

(a) $5x - 3$ **(b)** $-y$ **(c)** $x - y$ **(d)** $|x - y|$ **(e)** $|x| - |y|$

Solution

In each case we first replace every occurrence of a variable with its assigned value, and *then* we perform the indicated operations. In order to make this process even clearer you may find it is a good idea to put parentheses in wherever you see a variable, and then substitute the assigned value into the parentheses. However, be careful, because putting in too many parentheses can clutter up an example.

(a) Evaluate $5x - 3$ for $x = 2$:

$$5x - 3 = 5(\ \ ) - 3 \qquad \textit{We rewrite with parentheses.}$$
$$= 5(2) - 3 \qquad \textit{We put the value for x into the parentheses.}$$
$$= \boxed{7}$$

(b) Evaluate $-y$ for $y = -3$: $\quad -y = -(\ \ )$
$$= -(-3)$$
$$= \boxed{3}$$

(c) Evaluate $x - y$ for $x = 2$ and $y = -3$:

$$x - y = (\ \ ) - (\ \ )$$
$$= (2) - (-3) \qquad \textit{Now we follow the rule for subtraction.}$$
$$= (2) + (+3)$$
$$= \boxed{5}$$

(d) Evaluate $|x - y|$ for $x = 2$ and $y = -3$:

$$|x - y| = |(\ \ ) - (\ \ )|$$
$$= |(2) - (-3)|$$
$$= |2 + 3|$$
$$= |5|$$
$$= \boxed{5}$$

(e) Evaluate $|x| - |y|$ for $x = 2$ and $y = -3$:

$$|x| - |y| = |(\ \ )| - |(\ \ )|$$
$$= |(2)| - |(-3)| \qquad \textit{Note the difference here as compared to part}$$
$$= 2 - (+3) \qquad \textbf{(d)}\ \textit{above.}$$
$$= \boxed{-1}$$

EXAMPLE 7   Evaluate each of the following expressions for the given value of the variable.

(a) x^2 for $x = -3$ $\qquad\qquad$ (b) $-x^2$ for $x = -3$

(c) $(3t + 2)(2t + 1)$ for $t = 6$ $\qquad$ (d) $3w^2 + 4w - 2$ for $w = 5$

Solution   Again we begin by replacing each occurrence of the variable with parentheses and inserting the given value of the variable into the parentheses.

(a) For $x = -3$:

$$x^2 = (-3)^2 = (-3)(-3) = \boxed{9}$$

(b) For $x = -3$:

$$-x^2 = -(-3)^2 = -(-3)(-3) = \boxed{-9}$$

Note that the minus sign in front of the x^2 does not alter the fact that each x is replaced by -3.

(c) For $t = 6$:

$$(3t + 2)(2t + 1) = [3(6) + 2][2(6) + 1] \qquad \textit{We could have written}$$
$$= [18 + 2][12 + 1] \qquad \textit{(3(6) + 2)(2(6) + 1). How-}$$
$$= [20][13] \qquad \textit{ever, brackets can be used}$$
$$\textit{instead and using brackets}$$
$$= \boxed{260} \qquad \textit{makes it easier to read.}$$

(d) For $w = 5$:

$$3w^2 + 4w - 2 = 3(\quad)^2 + 4(\quad) - 2$$

$$= 3(5)^2 + 4(5) - 2$$ *Be sure to square the 5 first;*
the exponent 2 is only on the 5.

$$= 3(25) + 4(5) - 2$$

$$= 75 + 20 - 2$$

$$= \boxed{93}$$ ■

As we mentioned earlier, be careful not to change the arithmetic operations in the example when you substitute values. We will continue to insert parentheses whenever we think it makes the substitution clearer.

EXAMPLE 8

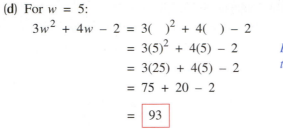

Evaluate each of the following for $x = 3$, $y = -4$, and $z = 5$:

(a) $x + yz$ **(b)** $(x + y)z$ **(c)** $x - y - z$ **(d)** $x - (y - z)$

(e) $x - 7(y - 2)$ **(f)** $(x - 7)(y - 2)$

(g) $(x + y)^2 - (xy)^2$ **(h)** $\dfrac{y - z}{x + y}$

Solution

(a) $x + yz = (\quad) + (\quad)(\quad)$

$$= (3) + (-4)(5)$$ *Watch your order of operations.*

$$= 3 + (-20)$$

$$= \boxed{-17}$$

(b) $(x + y)z = [3 + (-4)]5$ *Note the difference the given parentheses make*
*as compared with part **(a)**.*

$$= (-1)5$$

$$= \boxed{-5}$$

(c) $x - y - z = 3 - (-4) - 5$

$$= 3 + 4 - 5$$

$$= \boxed{2}$$

(d) $x - (y - z) = 3 - (-4 - 5)$ *Note the difference the given parentheses*
*make as compared with part **(c)**.*

$$= 3 - (-9)$$

$$= \boxed{12}$$

(e) $x - 7(y - 2) = 3 - 7(-4 - 2)$

$$= 3 - 7(-6)$$ *Watch your order of operations.*

$$= 3 + 42$$

$$= \boxed{45}$$

(f) $(x - 7)(y - 2) = (3 - 7)(-4 - 2)$ *Note the difference the extra parentheses*
*make as compared with part **(e)**.*

$$= (-4)(-6)$$

$$= \boxed{24}$$

(g) $(x + y)^2 - (xy)^2 = [3 + (-4)]^2 - [3(-4)]^2$ *Note that the first bracket has*
addition while the second
bracket has multiplication.

$$= [-1]^2 - [-12]^2$$

$$= 1 - 144$$

$$= \boxed{-143}$$

(h) $\dfrac{y - z}{x + y} = \dfrac{-4 - 5}{3 + (-4)}$ *We evaluate the numerator and denominator separately and then divide.*

$\qquad\qquad = \dfrac{-9}{-1}$

$\qquad\qquad = \boxed{9}$ ■

This seems to be an opportune place to review some terminology that we will find both very important and extremely useful.

Informally we have said that a *term* is a member of a sum while a *factor* is a member of a product. It would be a bit more precise to say that a **term** is an algebraic expression all of whose parts are connected by multiplication (and/or division). For example, the expression $2x^3y^2$ is a term because

$$2x^3y^2 = 2xxxyy$$

is a *product* of a constant and variables.

Similarly,

$$-4 \qquad \text{and} \qquad x \qquad \text{and} \qquad -7x^3y^4z^8$$

are examples of terms.

As we shall soon see, there are many situations in algebra where it is crucial to be able to distinguish between terms and factors.

Remember $3x + y - z$ is actually $3x + y + (-z)$.

$3x + y - z$ — consists of three terms, the first of which, $(3x)$, has two factors (3 and x).

$3(x + y)$ — is one term that is made up of two factors, 3 and $(x + y)$.

$3xy$ — is one term made up of three factors (3, x, and y).

In other words, a term is all *connected* by multiplication, while terms are *separated* by addition (or subtraction). (Those terms involving division of constants and variables will be discussed in Chapter 4.)

Here is some more terminology we will often use.

DEFINITION	The *constant* multiplier of a term is called the **numerical coefficient,** or usually just the **coefficient.** The *variable* part of a term (including its exponents) is often called the **literal** part of the term.

We always include the sign as part of the coefficient:

The coefficient of $5x$ is 5; its literal or variable part is x.

The coefficient of $-2y^3$ is -2; its literal part is y^3.

The expression $4x^2 - 3y$ is thought of as $4x^2 + (-3y)$ and consists of two terms. The first term, $4x^2$, has coefficient 4, while the second term, $-3y$, has coefficient -3.

If no constant appears, then the coefficient is understood to be 1. Thus, the term x^2yz has a coefficient of 1. Similarly, the term $-z^4$ has a coefficient of -1.

Thus, the process of algebraic substitution that we have described in this section can also be called *evaluating literal expressions.*

One final point: If we look at the expression $(3x)(-4x^3)$ carefully, we can see that it is one term. What is its coefficient? Normally we simplify an expression before we determine the coefficient:

$$(3x)(-4x^3) = -12x^4$$

and so the coefficient is -12.

EXAMPLE 9 — In each of the following expressions, determine the number of terms and the coefficient of each term.

(a) $8m - 3n$ **(b)** $8m(-3n)$

Solution

(a) $8m - 3n$ is made up of two terms. The coefficient of the first term is 8 and the coefficient of the second term is -3.

(b) $8m(-3n)$ is one term since it is all connected by multiplication. First we multiply:

$$8m(-3n) = -24mn$$

and we see that the coefficient is -24. ∎

Study Skills 2.2

Coping with Getting Stuck

All of us have had the frustrating experience of getting stuck on a problem; sometimes even simple problems can give us difficulty.

Perhaps you do not know how to begin; or, you are stuck halfway through an exercise and are at a loss as to how to continue; or, your answer and the book's answer do not seem to match. (Do not assume the book's solutions are 100% correct—we are only human even if we are math teachers. But do be sure to check that you have copied the problem accurately.)

Assuming you have reviewed all the relevant material beforehand, be sure you have spent enough time on the problem. Some people take one look at a problem and simply give up without giving the problem much thought. This is not what we regard as "getting stuck," since it is giving up before having even gotten started.

If you find after a reasonable amount of time, effort, and *thought,* that you are still not getting anywhere, if you have looked back through your notes and textbook and still have no clue as to what to do, try to find exercises similar to the one on which you are stuck (with answers in the back) that you can do. Analyze what you did to arrive at the solution and try to apply those principles to the problem you are finding difficult. If you have difficulty with those similar problems as well, you may have missed something in your notes or in the textbook. Reread the material and try again. If you are still not successful, go on to different problems or take a break and come back to it later.

If you are still stuck, wait until the next day. Sometimes a good night's rest is helpful. Finally, if you are still stuck after rereading the material, see your teacher (or tutor) as soon as possible.

EXERCISES 2.2

1. A car rental company uses the formula $C = 22.75 + 0.12m$ to compute the cost C (in dollars) of a 1-day rental for a car driven m miles. Compute the cost of a 1-day rental during which the car was driven 86 miles.

2. An office supply company uses the formula $C = 89.95 + 0.025n$ to compute the cost C (in dollars) of a 1-month rental of a copying machine making n copies. Compute the cost of a 1-month rental during which the copier made 682 copies.

3. A car rental company uses the formula $C = 18.95d + 0.14m$ to compute the cost C (in dollars) of renting a car for d days during which the car is driven m miles. Compute the cost of a 5-day rental during which the car was driven 264 miles.

4. An office supply company uses the formula $C = 76.50m + 0.021n$ to compute the cost C (in dollars) of renting a copying machine for m months during which the machine makes n copies. Compute the cost of renting a copying machine for 8 months during which 5,728 copies were made.

5. Suppose that the formula $h = 80 + 40t - 16t^2$ represents the height h (in feet) of an object above the ground t seconds after it is thrown. Find the height of the object 3.6 seconds after it is thrown. Round to the nearest tenth.

6. Suppose that the formula $h = 395 - 16t^2$ represents the height h (in feet) of an object above the ground t seconds after it is released. Find the height of the object 4.3 seconds after it is thrown. Round to the nearest tenth.

7. A company finds that its monthly cost C (in dollars) is related to the number n of items produced according to the equation $P = 12{,}600 + 6.35n - 0.002n^2$. Use this equation to determine the monthly cost if the company produces 4,280 items.

8. A retail firm finds that its yearly sales S (in dollars) are related to the number m of television advertising minutes it purchases during the course of the year according to the equation $S = 5{,}750 - 10.6m + 0.32m^2$. Find the yearly sales if the firm purchases 1,582 minutes of advertising time.

9. Niki earns a weekly income I (in dollars) given by $I = 415 + 17.5h$, where h is the number of overtime hours she works. Compute Niki's weekly income if she works 12 overtime hours.

10. Ivan earns a weekly salary S (in dollars) given by $S = 225 + 1.12n$, where n is the number of items he sells. Compute Ivan's weekly salary if he sells 285 items.

11. The telephone company computes Wanda's monthly charge C (in dollars) using the equation $C = 18.95 + 0.065n$, where n is the number of local calls she makes. Compute her monthly charge if she makes 82 local calls.

12. The gas company computes Antonio's monthly charge C (in dollars) using the equation $C = 12.80 + 0.122f$, where f is the number of cubic feet of gas he uses. Compute his monthly charge if he uses 115 cubic feet of gas.

13. The area A of a square of side s is $A = s^2$. Use this formula to find the area of a square of side 6.4 in. Round off the answer to the nearest tenth.

14. The volume V of a cube of edge e is $V = e^3$. Find the volume of a cube whose edge is 3.1 cm. Round off the answer to the nearest tenth.

15. The volume of a rectangular box of length L, width W, and height H is $V = LWH$. Find the volume V of a rectangular box of length 1.8 ft, width 2.3 ft, and height 1.1 ft. Round off the answer to the nearest tenth.

16. The surface area S of a closed rectangular box of length L, width W, and height H is $S = 2LW + 2WH + 2LH$. Find the surface area of a rectangular box of length 0.85 meter, width 1.60 m, and height 2.25 m. Round off the answer to the nearest hundredth.

17. Suppose a company purchases a piece of equipment for C dollars and expects the machinery to depreciate to zero dollars in N years. The Internal Revenue Service allows machinery to be depreciated according to the formula

$$V = C\left(1 - \frac{n}{N}\right)$$

where V represents the dollar value of the equipment in year n. Use this formula to compute the value of a piece of equipment 8 years after it was purchased for $18,000 and it is expected to depreciate to zero dollars in 15 years.

18. A publishing company determines that when printing n copies of a particular book, the average cost C in dollars per book is given by the formula

$$C = \frac{21.5n + 42{,}700}{n}$$

Use this formula to compute the average cost per book if 5,000, 10,000, and 20,000 books are printed.

19. According to an economist, the number N of computers that a small company can sell is related to the price x according to the formula

$$N = \frac{857{,}000}{0.01x^2 + 0.2x}$$

Determine the approximate number of computers that can be sold for $550.

20. According to an educational psychologist, a person can learn to type w words per minute after x weeks of a typing course according to the formula

$$w = \frac{72(x + 1)}{x + 4}$$

Determine the number of words per minute a student should expect to type after 3, 4, and 5 weeks of a typing course.

The data in Exercises 21–24 are from the Bureau of Labor Statistics.

21. The following table gives the percentage of females over the age of 16 in the civilian population that are employed in January of each of the years listed.

Year	1993	1994	1995	1996	1997	1998
Employment rate	53.7%	54.9%	55.5%	55.5%	56.5%	57.1%

Based on these data, a statistician suggests that the following formula can be used to approximate the female employment rate R in January of the years 1993 through 1998:

$$R = -0.03x^2 + 0.9x + 51.4$$

where $x = 3$ corresponds to 1993, $x = 4$ corresponds to 1994, etc.

(a) Use this formula to compute R for each of the years 1993 through 1998.

(b) Compare the values collected by the Department of Labor with the values obtained from the formula, and comment about the accuracy of the formula.

22. The following table gives the percentage of the civilian population between the ages of 16 and 19 years that are employed in April of each of the years listed.

Year	1990	1991	1992	1993	1994	1995
Employment rate	46.6%	42.8%	40.6%	41.2%	43.2%	44.6%

Based on these data, a statistician suggests that the following formula can be used to approximate the employment rate R for 16-to-19-year-olds in April of the years 1990 through 1995:

$$R = -0.14x^3 + 1.82x^2 - 5.98x + 46.72$$

where $x = 0$ corresponds to 1990, $x = 1$ corresponds to 1991, etc.

(a) Use this formula to compute R for each of the years 1990 through 1995.

(b) Compare the values collected by the Department of Labor with the values obtained from the formula and comment about the accuracy of the formula.

23. The following table gives the number of people (in millions) over the age of 16 who were unemployed for more than 15 weeks during September of each of the years listed.

Year	1990	1991	1992	1993	1994	1995
Number unemployed	1.7	2.5	3.5	3.0	2.9	2.3

Based on these data, a statistician suggests that the following formula can be used to approximate the number N of people unemployed given in the table:

$$N = 0.02x^3 - 0.36x^2 + 1.40x + 1.65$$

where $x = 0$ corresponds to 1990, $x = 1$ corresponds to 1991, etc.

(a) Use this formula to compute N for each of the years 1990 through 1995.

(b) Compare the values collected by the Department of Labor with the values obtained from the formula, and comment about the accuracy of the formula.

24. For people over the age of 16 who were unemployed during November of the years listed, the table gives the average number of weeks those people were unemployed as of November.

Year	1992	1993	1994	1995	1996	1997
Weeks unemployed	17.9	18.6	18.0	16.4	15.9	15.4

Based on these data, a statistician suggests that the following formula can be used to approximate the average number W of weeks given in the table:

$$W = -0.1x^2 + 0.27x + 18.15$$

where $x = 2$ corresponds to 1992, $x = 3$ corresponds to 1993, etc.

(a) Use this formula to compute W for each of the years 1992 through 1997.

(b) Compare the values collected by the Department of Labor with the values obtained from the formula, and comment about the accuracy of the formula.

In Exercises 25–80, evaluate the given expression for $x = 2$, $y = -3$, and $z = -4$.

25. $-y$

26. $-z$

27. $-|y|$

28. $-|z|$

29. $x + y$

30. $x + z$

31. $x - y$

32. $x - z$

33. $|x - y|$

34. $|x - z|$

35. $|x| - y$

36. $|y| - z$

37. $|x| - |y|$

38. $|y| - |z|$

39. $x + y + z$

40. $x - y - z$

41. $x + y - z$

42. $x - y + z$

43. $xy - z$

44. $x - yz$

45. $x(y - z)$

46. $(x - y)z$

47. $x - (y - z)$

48. $x - (y + z)$

49. xy^2

50. $(xy)^2$

51. $x + y^2$

52. $(x + y)^2$

53. $x^2 + y^2$

54. $x^2 + 2xy + y^2$

55. $(x + y + z)^2$

56. $x^2 + y^2 + z^2$

57. $-y^2$

58. $(-y)^2$

59. $-x^2(-z)$

60. $(-x)^2 - z$

61. xyz^2

62. $(xyz)^2$

63. $x(y - z)^2$

64. $y(x - z)^2$

65. $xy^2 - (xy)^2$

66. $(xy)^2 - (x - y)^2$

67. $(z - 3x)^2$

68. $z - 3x^2$

69. $(5x + y)(3x - y)$

70. $(y - 2z)(y + 3z)$

71. $y^2 - 3y + 2$

72. $2z^2 - 5z - 4$

73. $3x^2 + 4x + 1$

74. $5x^2 - 8x + 5$

75. $x^2 + 3x^2y - 3xy^2 + y^3$

76. $5x^2 - 2xy^2 + zx^2y$

77. $\dfrac{xy}{x + y}$

78. $\dfrac{z}{x} - y$

79. $\dfrac{2}{x} - \dfrac{y}{3} + \dfrac{z}{2}$

80. $\dfrac{xyz}{z - x}$

In Exercises 81–92, examine each of the following expressions as given and determine the number of terms. Also find the coefficient and the literal part of each term.

81. $3x - 4y$

82. $4a - 5b$

83. $3x(-4y)$

84. $4a(-5b)$

85. $3x(z - y)$

86. $4a(c - b)$

87. $4x^2 - 3x + 2$

88. $2a^2 + 7a - 3$

89. $-x^2 + y - 13$

90. $-a^2 - a - 1$

91. $3x(2x) + 4y(5y)$

92. $7a(2a) - 3c(6c)$

In Exercises 93–100, evaluate the given expression for $x = 0.24$ and $y = -0.5$. Round off to the nearest thousandth where necessary.

93. $8.4x - 0.03y$

94. $12.7x + 8y$

95. $10x^3 + 20y$

96. $0.25x^2 - 0.43y$

97. $xy^2 - (xy)^2$

98. $(x + y)^2$

99. $\dfrac{10}{x} + \dfrac{8}{y}$

100. $\dfrac{5}{xy}$

In Exercises 101–104, use a calculator to evaluate the expression for the given value in two ways: First, enter the given value as a fraction and then round off your answer to the nearest hundredth; second, round off the given fraction to the nearest hundredth, enter this value, and then round off your answer to the nearest hundredth. Compare the two answers. Which answer do you think is more accurate and why?

101. $x^2 - 8x + 5$ for $x = \dfrac{1}{3}$

102. $t^3 - 6t - 2$ for $t = -\dfrac{2}{7}$

103. $(2a - 1)(3a + 5)$ for $a = -\dfrac{2}{9}$

104. $\dfrac{s^2 - 5s + 1}{s^4 - s}$ for $s = \dfrac{5}{8}$

 QUESTIONS FOR THOUGHT

105. What is the difference between a *term* and a *factor*?

106. Make up an example of an expression that consists of three terms, one of which has one factor, one of which has two factors, and one of which has three factors.

2.3   The Distributive Property and Combining Like Terms

During the course of our earlier discussion of the real number system in Chapter 1, we talked about the special properties that the real numbers exhibit under addition and multiplication. Namely, we can reorder the terms of a sum and the factors of a product by the commutative property, and we can regroup those terms or factors by the associative property.

A third and essential property is one that describes how the operations of multiplication and addition interact. Let's begin by considering the numerical example $3(5 + 2)$. Following the order of operations, we would evaluate this as

$$3(5 + 2) = 3(7) = 21$$

However, we know that multiplication is really just a shorthand way of writing repeated addition. That is,

$$3(5 + 2) \quad \text{means} \quad \text{add } (5 + 2) \text{ three times.}$$

We can write this as

$$3(5 + 2) = (5 + 2) + (5 + 2) + (5 + 2) \qquad \textit{Let's regroup and reorder the addition.}$$

$$= (5 + 5 + 5) + (2 + 2 + 2) \qquad \textit{Now let's rewrite the addition}$$
$$= 3(5) + 3(2) \qquad\qquad\qquad \textit{back to multiplication.}$$
$$= 15 + 6$$
$$= 21$$

Thus, at the third step we see that $3(5 + 2) = 3(5) + 3(2)$.

Now you may be asking yourself why anyone would bother doing the computation the second way. The first way was easier and also followed our agreement on the order of operations. Of course you are right! The point of this example was not to suggest another method for doing this particular problem, but rather to illustrate a property of the real numbers. This property is called the ***distributive property of*** *multiplication over addition* (usually just the **distributive property** for short). It can be expressed algebraically as indicated in the box.

Distributive Property	$$a(b + c) = ab + ac$$ In words, this says that the *factor* outside the parentheses multiplies each *term* inside the parentheses.

Before we proceed to look at several examples, a few comments are in order. First, if you look back at our analysis of how $3(5 + 2)$ became $3 \cdot 5 + 3 \cdot 2$, you will see that it does not make any difference whether we are adding two numbers or 20 numbers in the parentheses. Each term inside still gets multiplied by the factor outside.

Second, since we have defined subtraction in terms of addition, the distributive property holds equally well for multiplication over subtraction or over addition *and* subtraction. That is,

$$a(b - c) = a \cdot b - a \cdot c$$

and

$$a(b + c - d) = a \cdot b + a \cdot c - a \cdot d$$

Third, because multiplication is commutative, we can also write the distributive property as

$$(b + c)a = b \cdot a + c \cdot a$$

When we use the distributive property in this way to remove a set of parentheses, we say that we are *multiplying out* the expression.

EXAMPLE 1

Use the distributive property to multiply out each of the following, and simplify as completely as possible.

(a) $5(x + 3)$ (b) $4(x^2 - y + 7)$ (c) $x(x^3 + 3x)$ (d) $(3 + 2a)a$

Solution

(a) $5(x + 3) = 5 \cdot x + 5 \cdot 3 = \boxed{5x + 15}$

(b) $4(x^2 - y + 7) = 4 \cdot x^2 - 4 \cdot y + 4 \cdot 7 = \boxed{4x^2 - 4y + 28}$

(c) $x(x^3 + 3x) = x \cdot x^3 + x \cdot 3x = \boxed{x^4 + 3x^2}$

(d) $(3 + 2a)a = 3 \cdot a + 2a \cdot a = \boxed{3a + 2a^2}$ ■

Recall that when we first described the distributive property as applied to the example $3(5 + 2) = 3 \cdot 5 + 3 \cdot 2$, we said that normally we would add $5 + 2$ and then multiply by 3 rather than use the distributive property. If you look at the example we have just completed, you will notice that we could not work inside the parentheses first. We had no choice here; we had to use the distributive property.

EXAMPLE 2

Multiply out each expression.

(a) $3(x + y)$ (b) $3(xy)$

Solution

It is very important to see that the distributive property applies to part (a) but *not* to part (b).

(a) $3(x + y) = 3 \cdot x + 3 \cdot y = \boxed{3x + 3y}$

(b) $3(xy)$ is not an expression to which we can apply the distributive property because there is no addition or subtraction within the parentheses. Do not make the *mistake* of thinking that $3(xy)$ is the same as $3x \cdot 3y$, which is $9xy$. This is *wrong*.

$$3(xy) = \boxed{3xy} \qquad \text{\textit{We simply "erase" the parentheses.}} \qquad \blacksquare$$

Since we will often be simplifying expressions by removing parentheses, knowing when a set of parentheses makes a difference is obviously important.

Let's agree to call parentheses **essential** if their erasure changes the arithmetic meaning of the problem. In other words, if we mentally erase the parentheses and that changes the meaning of the problem, then those parentheses are essential. If erasing parentheses does not make any difference, then those parentheses are called **nonessential.**

In Example 2 we have both types of parentheses. In part **(a)**, the parentheses in $3(x + y)$ are essential because *if* we erase them, we get $3x + y$ and the 3 will be multiplying only the x instead of the sum of x and y, which is not the same thing. On the other hand, in part **(b)** if we erase the parentheses in $3(xy)$ we get $3xy$, which is exactly the same thing because multiplication is associative. Thus, the parentheses in part **(b)** are nonessential.

EXAMPLE 3

In each of the following, determine whether each set of parentheses is essential or nonessential.

(a) $3x + (2y + z)$ **(b)** $3x + 2(y + z)$ **(c)** $x(y + z) + x(yz)$

Solution

(a) In $3x + (2y + z)$ the parentheses are nonessential since if we erase them we will still be adding the same three quantities. In other words, since addition is associative, the parentheses make no difference.

(b) In $3x + 2(y + z)$ the parentheses are essential because if we erase them we get $3x + 2y + z$ and then the 2 will be multiplying only the y instead of $y + z$. The parentheses do make a difference.

(c) In $x(y + z) + x(yz)$ the first parentheses are essential, the second are not. If we erase the first parentheses we get $xy + z$ and so x will be multiplying only y instead of $y + z$. If we erase the second parentheses, x will still be multiplying the product yz. $\blacksquare$

While the vocabulary of essential and nonessential parentheses is convenient, it is not—pardon the pun—essential. However, being able to recognize when parentheses make a difference in an expression is an extremely important skill.

As with any equality, the distributive property can be read two ways. If we read it from left to right, we are removing the parentheses and this is called **multiplying out.** If we read it from right to left, we are creating parentheses and this is called **factoring.**

$$a(b + c) = \boxed{a}\,(b + c) = \boxed{a} \cdot b + \boxed{a} \cdot c \qquad \textit{Multiplying out}$$

$$a \cdot b + a \cdot c = \boxed{a} \cdot b + \boxed{a} \cdot c = \boxed{a}\,(b + c) \qquad \textit{Factoring}$$

Notice that when we multiply out and remove parentheses, we are changing a product into a sum. When we factor and create parentheses, we are changing a sum into a product.

Thus far we have been using the distributive property to remove parentheses. Now let's turn our attention to the other side of the coin—using the distributive property to create parentheses, the process called **factoring.**

Looking at the distributive property as saying

$$ab + ac = a(b + c)$$

we see that both terms on the left-hand side have a *common factor* of a. The distributive property says that we can "factor out" the common factor of a by inserting the parentheses.

EXAMPLE 4   Use the distributive property to factor each of the following:

(a) $3x + 3y$ (b) $4x - 4y + 4z$ (c) $2x + 6$ (d) $yz + xz$

Solution   (a) Both $3x$ and $3y$ contain a common factor of 3. So we have

$$3x + 3y = \boxed{3}\,x + \boxed{3}\,y = \boxed{3}\,(x + y) = \boxed{3(x + y)}$$

(b) $4x - 4y + 4z = \boxed{4}\,x - \boxed{4}\,y + \boxed{4}\,z$

$$= \boxed{4}\,(x - y + z) = \boxed{4(x - y + z)}$$

(c) In $2x + 6$ it helps to think of 6 as $2 \cdot 3$ so we can "see" the common factor of 2.

$$2x + 6 = \boxed{2}\,x + \boxed{2}\cdot 3 = \boxed{2}\,(x + 3) = \boxed{2(x + 3)}$$

(d) $yz + xz = y \cdot \boxed{z} + x \cdot \boxed{z} = (y + x)\,z = \boxed{(y + x)z}$

Note that in each case we can immediately check our answer by multiplying it out. ■

While there is much more to say about factoring (which we will do in Chapter 8), we are already in a position to put this idea to use.

Many students can look at the expression $5x + 3x$ and "see" that the answer is $8x$. We can now justify this statement mathematically as follows:

$$5x + 3x = 5x + 3x \qquad \textit{The common factor is x.}$$
$$= (5 + 3)x \qquad \textit{We factor out the common factor of x.}$$
$$= 8 \cdot x$$

We are able to combine $5x$ and $3x$ because they are **like terms.**

DEFINITION   **Like terms** are terms whose variable parts are identical.

Thus, we similarly have

$$4a^2 + 5a^2 - 2a^2 = 4\,\boxed{a^2} + 5\,\boxed{a^2} - 2\,\boxed{a^2}$$

$$= (4 + 5 - 2)\,\boxed{a^2}$$

$$= (7)a^2$$

$$= \boxed{7a^2}$$

These are again like terms (they are all a^2 terms) and so we combine them by simply adding their coefficients. Note we said *add* the coefficients. Recall that we have defined the coefficient of a term to include the sign. Thus, in $4a^2 + 5a^2 - 2a^2$ the coefficients are 4, 5, and -2, and we are adding them.

This point of view makes it easy to state the following rule.

Rule for Combining Like Terms	To combine like terms simply add their coefficients.

We must take care to combine like terms properly.

EXAMPLE 5   Separate the following list into groups of like terms and name the coefficient of each term.

$$3x \qquad x^3 \qquad -2x \qquad -x^2y \qquad 5x \qquad xy^2 \qquad 2y^2x \qquad 5x^3$$

Solution   As our definition of like terms says, like terms must have variable parts that are *identical*. Do not think that just because all the above terms contain the variable x, they are all like terms. Remember that exponents are just shorthand for repetitive multiplication.

$$x^2 = x \cdot x \quad \text{while} \quad x^3 = x \cdot x \cdot x$$

and therefore they are not like terms.

The groups of like terms are:

Group 1: $3x, -2x, 5x$ with coefficients 3, -2, and 5, respectively

Group 2: x^3 and $5x^3$ with coefficients 1 and 5, respectively
(Remember that the coefficient 1 is usually not written.)

Group 3: $-x^2y$ with coefficient -1

Group 4: xy^2 and $2y^2x$ with coefficients 1 and 2, respectively
Note: By the commutative property, y^2x is the same as xy^2.   ∎

Important: From now on we are automatically expected to combine like terms whenever possible. The process of combining like terms is a basic step in simplifying an expression.

EXAMPLE 6   Simplify as completely as possible. $3x + 5y - 7x + 2y$

Solution   Keep in mind that we are "thinking addition" with the sign as part of the coefficient.

$3x + 5y - 7x + 2y = 3x - 7x + 5y + 2y$ *We have reordered by the commutative property.*

$$= 3 \; x \; - 7 \; x \; + 5 \; y \; + 2 \; y \qquad \textit{Combine like terms.}$$

$$= (3 - 7)x + (5 + 2)y$$

$$= \boxed{-4x + 7y} \quad \text{or} \quad \boxed{7y - 4x}$$ ∎

EXAMPLE 7   Simplify as completely as possible. $5x^2 + 9x - 7 - x^2 + x - 2$

Solution $$5x^2 + 9x - 7 - x^2 + x - 2 = (5 - 1)x^2 + (9 + 1)x - 7 - 2$$

$$= \boxed{4x^2 + 10x - 9}$$

Do not forget that the understood coefficient of x is 1 and of $-x^2$ is -1.   ∎

EXAMPLE 8

Multiply and simplify.

(a) $2(x^2 + 4y) + 3(x^2 - 2y)$ (b) $3x(x^3 + 2y) + 5(2 - xy)$

Solution

(a) $2(x^2 + 4y) + 3(x^2 - 2y) = 2x^2 + 8y + 3x^2 - 6y$ *We multiplied out by the distributive property.*

$$= (2 + 3)x^2 + (8 - 6)y \quad \textit{Combine like terms.}$$

$$= 5x^2 + 2y$$

(b) $3x(x^3 + 2y) + 5(2 - xy) = 3x \cdot x^3 + 3x \cdot 2y + 5 \cdot 2 - 5xy$

By the distributive property

$$= 3x^4 + 6xy + 10 - 5xy \quad \textit{Combine like terms.}$$

$$= 3x^4 + xy + 10$$

You probably noticed that in this example we used the distributive property both ways—multiplying out to remove the parentheses, and factoring in order to combine like terms. ■

In the next section we will look at expressions that require several steps before we reach their simplest form.

Study Skills 2.3

Seeking Help

In the last Study Skill we discussed some ideas for how to cope if you get stuck on a particular problem. However, sometimes you may feel that the difficulty you are having is more substantial than just the inability to do a problem. If you find yourself in such a situation it is important to recognize that one of the keys to success in a mathematics class is knowing when and where to get the help you need.

Your first step should always be to make an honest effort to review your class notes and the relevant text material. If this does not answer your question, then you should try to speak to your instructor after class. Sometimes just a word or two from your instructor can clarify a point that was interfering with your understanding of the material.

If this does not work, then you need to get additional help outside the class. Your instructor has most likely informed you of his or her office hours. These are the times that your instructor is available to give you individualized attention. Going to office hours affords you the opportunity to explain your difficulty to your instructor in much greater detail than is normally possible by asking questions during class. This enables your instructor to understand your difficulty much more clearly, and therefore to give you a more complete and meaningful response.

In order to make your visit as productive as possible, it is best to formulate your questions clearly beforehand. (It is a good idea to write them down.) Just telling your instructor "I don't understand" does not help your instructor in assisting you.

One additional point: Some students feel that they are a burden on their instructor when they come to office hours. This is not true. Your instructor is there to help you and is pleased that you are making the extra effort to learn the material.

EXERCISES 2.3

In Exercises 1–14, identify each set of parentheses as essential or nonessential.

1. $2x + 3(y + z)$

2. $5x + 4(y + z)$

3. $2x + (3y + z)$

4. $5x + (4y + z)$

5. $2(x + 3y + z)$

6. $5(x + 4y + z)$

7. $(2x + 3y + z)$

8. $(5x + 4y + z)$

9. $(2x + 3)y + z$

10. $(5x + 4)y + z$

11. $(2x + 3)(y + z)$

12. $(5x + 4)(y + z)$

13. $(2x + y) + 2(x + y)$

14. $5(x + y) + (5x + y)$

In Exercises 15–24, separate each list into groups of like terms, and name the coefficient and literal part of each term.

15. $x, \quad y, \quad 2x, \quad 3y$

16. $3u, \quad v, \quad 5v, \quad 7u$

17. $2x^2, \quad -3x, \quad 4x^3, \quad -x, \quad -x^2$

18. $5u^3, \quad 4u^2, \quad -u, \quad u^2, \quad u^3$

19. $4, \quad 4u, \quad 4u^2, \quad 5, \quad 5u$

20. $-3, \quad -3x, \quad -3x^2, \quad -3x^3, \quad x$

21. $5x^2, \quad 5x^2y, \quad 5y^2, \quad 5xy^2$

22. $7s^2, \quad 7st^2, \quad 7s^2t, \quad 7t^2$

23. $-x^2y, \quad 2xy^2, \quad x^2y^2, \quad 3xy^2, \quad -2x^2y$

24. $-4s^2t, \quad 2st^2, \quad 3s^2t, \quad -s^2t^2, \quad s^2t^2$

In Exercises 25–44, multiply out and simplify as completely as possible.

25. $3(x + 4)$

26. $5(a + 3)$

27. $5(y - 2)$

28. $7(b - 4)$

29. $-2(x + 7)$

30. $-3(y + 2)$

31. $3(5x + 2)$

32. $5(3x + 4)$

33. $-4(3x + 1)$

34. $-6(2x + 3)$

35. $x(x + 3)$

36. $a(a + 4)$

37. $x(x^2 + 3x)$

38. $a(a^2 + 4a)$

39. $5x(2x - 4)$

40. $2y(5y - 4)$

41. $0.42(40 - x)$

42. $1.56(2x + 80)$

43. $1.28(3x + 60)$

44. $0.08(60 - 5x)$

In Exercises 45–52, factor the given expression by taking out the common factor.

45. $2x + 10$

46. $3x + 12$

47. $5y - 20$

48. $7x - 14$

49. $9x + 3y - 6$

50. $10m - 2n - 8$

51. $x^2 + xy$

52. $3x^3 + x$

In Exercises 53–96, simplify by combining like terms whenever possible.

53. $2x + 5x$

54. $3a + 7a$

55. $2x^2 + 5x^2$

56. $3a^2 + 7a^2$

57. $3a - 8a + 2a$

58. $4m - 9m + 3m$

59. $-3y + y - 2y$

60. $-2z - 5z + z$

61. $-x - 2x - 3x$

62. $-a - 4a - 2a$

63. $2x - 3y - 7x + 5y$

64. $3s - 4t - 8s + t$

65. $3x + 5y + 2z$

66. $4a - 2b - 5c$

67. $3x^2 + 7x + x^2 + 3x$

68. $5m^2 + 7m + m^2 + 2m$

69. $x^2 - 2x + x^2 - x$

70. $y^2 - 6y + y^2 - y$

71. $5x^2y - 3x^2 + x^2y - x^2$

72. $8s^2t - 4s^2 - s^2 - s^2t$

73. $2x + 5x - 3y + y - 7x$

74. $m + 9m - 3n - n - 10m$

75. $-5s^2 + 3st - s^2 + 6s^2$

76. $-3x^2 + 5xy - xy - x^2 - 2x^2$

77. $3a^2b + ab^2 - ab^2 - 2a^2b - ab^2$

78. $2x^2y + xy^2 - x^2y - x^2y - xy^2$

79. $0.8x - 5 - x + 2.7$

80. $0.45x + 1.8 - 0.62x$

81. $0.28x + 5.48 + 0.54x$

82. $0.085x + 28.75 + 0.072x + 54.8$

83. $3(x + y) + 4x - y$

84. $4(a + b) + 4a - b$

85. $3(x + y) + 4(x - y)$

86. $4(a + b) + 4(a - b)$

87. $3(x + y) + 4x(-y)$

88. $4(a + b) + 4a(-b)$

89. $5x(x^2 + 3) + 2x(3 + x^2)$

90. $7y^2(y + 2) + 2y(5 + y^2)$

91. $5x(x^2 + 3) + 2x(3x^2)$

92. $7y^2(y + 2) + 2y(5y^2)$

93. $0.06x + 0.09(x + 5,000)$

94. $0.18x + 0.25(72 - x)$

95. $0.35x + 0.42(2x) + 0.54(80 - 3x)$

96. $0.08x + 0.09(x + 1,500) + 0.12(4,500 - 3x)$

ⓠ QUESTIONS FOR THOUGHT

97. Which property of the real numbers allows us to combine like terms?

98. Discuss what is ***wrong*** with each of the following (if anything):

(a) $5 + 3(x - 4) \overset{?}{=} 8(x - 4) \overset{?}{=} 8x - 32$

(b) $5 + 3(x - 4) \overset{?}{=} 5 + 3x - 4 \overset{?}{=} 3x + 1$

(c) $5 + 3(x - 4) \overset{?}{=} 5 + 3x - 12 \overset{?}{=} 8x - 12$

(d) $5 + 3(x - 4) \overset{?}{=} 5 + 3x - 12 \overset{?}{=} 3x - 7$

99. How can the distributive property be used to make it easier to compute the following expressions?

(a) $45(116) - 45(16)$ (b) $60(278) - 60(28)$

2.4 Simplifying Algebraic Expressions

Up to this point we have been developing a variety of concepts and procedures that we shall now integrate into one unified approach to simplifying algebraic expressions. Let's begin by looking at a few examples that offer additional practice and will help us appraise the type of examples we can currently handle.

EXAMPLE 1   Multiply and simplify. $2(3x - 4) + 5(2x - 1)$

Solution   We remove each set of parentheses by using the distributive property.

$$2(3x - 4) + 5(2x - 1) = 2 \cdot 3x - 2 \cdot 4 + 5 \cdot 2x + 5(-1)$$

$$= 6x - 8 + 10x - 5$$

$$= (6 + 10)x - 8 - 5 \qquad \textit{Combine like terms.}$$

$$= \boxed{16x - 13} \qquad \blacksquare$$

EXAMPLE 2 Multiply and simplify. $3(4a - 7) + (2a - 3) + 5(a + 2)$

Solution We note that the first and third sets of parentheses are essential while the second is not. We remove the first and third sets by the distributive property.

$$3(4a - 7) + (2a - 3) + 5(a + 2) = 12a - 21 + 2a - 3 + 5a + 10$$

$$= (12 + 2 + 5)a - 21 - 3 + 10$$

Combine like terms.

$$= \boxed{19a - 14} \qquad \blacksquare$$

EXAMPLE 3 Multiply and simplify. $2x^2(x - 3) + x(3x^2 - 5x)$

Solution We multiply out using the distributive property.

$$2x^2(x - 3) + x(3x^2 - 5x) = 2x^2 \cdot x - 2x^2 \cdot 3 + x \cdot 3x^2 - x \cdot 5x$$

$$= 2x^3 - 6x^2 + 3x^3 - 5x^2$$

$$= (2 + 3)x^3 + (-6 - 5)x^2 \qquad \textit{Combine like terms.}$$

$$= \boxed{5x^3 - 11x^2} \qquad \blacksquare$$

EXAMPLE 4 Multiply and simplify. $2x^2y(x + 3y) + 4y(x^3 + 2x^2y)$

Solution While our basic outline remains the same, we do have to exercise more care as the examples become more complex.

$$2x^2y(x + 3y) + 4y(x^3 + 2x^2y) = 2x^2y \cdot x + 2x^2y \cdot 3y + 4y \cdot x^3 + 4y \cdot 2x^2y$$

$$= 2x^3y + 6x^2y^2 + 4yx^3 + 8x^2y^2$$

Notice that $2x^3y$ and $4yx^3$ are like terms. The order of the factors does not matter.

$$= \boxed{6x^3y + 14x^2y^2} \qquad \blacksquare$$

In order to make it easier to recognize like terms, it helps to have the variables in the same order in all the terms. From now on, as we obtain our products when we multiply out, we will use the same order for the variables throughout a particular example. In this text we will normally use alphabetical order.

Next let us examine how our understanding of a minus sign, our definition of a numerical coefficient, and the distributive property work together.

If we use the distributive property on the next two expressions and write out *all* the steps, we get the following:

$$-3(x + 4) = -3x + (-3)(4) = -3x + (-12) = \boxed{-3x - 12}$$

$$-3(x - 4) = -3x - (-3)(4) = -3x - (-12) = \boxed{-3x + 12}$$

Since any subtraction can be converted to addition, it is much easier to do these examples by performing the addition mentally as we do with integers. We simply multiply the coefficients according to the distributive property. In other words, we would do the above examples as follows:

$$-3(x + 4) = -3x - 12 \qquad \textit{Keep in mind that the addition is understood.}$$
$$\textit{−3x − 12 means −3x + (−12).}$$

$$-3(x - 4) = -3x + 12 \qquad \textit{−3 times −4 equals +12.}$$

This way of looking at these examples becomes even more useful as the examples become more complex.

EXAMPLE 5 Simplify. $2x - 5(x - 2)$

Solution We think of -5 as the coefficient of $(x - 2)$, and so we distribute -5 into the parentheses.

$$2x - 5(x - 2) = 2x - 5x + 10 \qquad \textit{−5 times x and −5 times −2}$$

$$= \boxed{-3x + 10} \qquad \textit{After combining like terms} \qquad ∎$$

EXAMPLE 6 Simplify as completely as possible. $2(3x - y) - (10x - 3y)$

Solution In using the distributive property, we multiply out each set of parentheses by the coefficient that precedes it. The coefficient of the first set of parentheses is 2. What is the coefficient of the second set? Anytime we have a negative sign we can think of it as -1. For example, $-5 = -1 \cdot 5$. In applying the distributive property it is helpful to have some number to use as the coefficient of the parentheses. Thus, the answer to the question is that the coefficient of the second set of parentheses is -1.

$$2(3x - y) - (10x - 3y) = 2(3x - y) \; -1 \,(10x - 3y) \qquad \textit{We distribute the −1.}$$

$$= 6x - 2y - 10x + 3y$$

$$= \boxed{-4x + y}$$

Note that this answer can also be written as $y - 4x$. Since we are "thinking" addition, the terms in an expression can always be rearranged as long as the coefficient of each term is unchanged.   ∎

EXAMPLE 7 Simplify as completely as possible. $8 - 3[x - 4(x - 3)]$

Solution Remember that both square brackets and parentheses are types of grouping symbols. Whenever one grouping symbol occurs within another, it is usually easiest and most efficient to work from the innermost grouping symbol outward. We begin by removing the parentheses around $x - 3$ by distributing the -4.

$$8 - 3[x - 4(x - 3)] = 8 - 3[x - 4x + 12] \qquad \textit{Watch the order of operations. We}$$
$$\textit{do not perform the subtraction}$$
$$\textit{8 − 3 = 5. Remember that multi-}$$
$$\textit{plication precedes addition and}$$
$$\textit{subtraction. Next combine like}$$
$$\textit{terms within the brackets.}$$

$$= 8 - 3[-3x + 12] \qquad \textit{Now remove the brackets by dis-}$$
$$= 8 + 9x - 36 \qquad \textit{tributing the −3.}$$

$$= \boxed{9x - 28}$$

Note that in the second step we could have distributed the -3 to remove the bracket, but we chose to combine like terms within the bracket first. This is usually the better procedure to follow. As soon as like terms present themselves, combine them.   ∎

Having looked at these examples, let's pause to consolidate our ideas into a procedural outline.

Procedure for Simplifying Algebraic Expressions	1. Remove any essential grouping symbols. If necessary, use the distributive property to multiply each term inside the grouping symbol by its coefficient. Remember that the coefficient includes the sign that precedes it.
	2. If there are grouping symbols within grouping symbols, always work from the innermost one outward, following the order of operations.
	3. When multiplying terms, keep the same order of the variables to make it easier to recognize like terms. The best method is to write terms with their numerical coefficients first, followed by the variables in alphabetical order.
	4. In each step of the solution look for any like terms that can be combined.

In each of the remaining examples in this section, simplify the given expression as completely as possible.

EXAMPLE 8

$4xy(x - 3) - (xy^2 - 12xy) - 3y(xy + x^2)$

Solution   Following the outline given in the box, we proceed as follows:

$$4xy(x - 3) - (xy^2 - 12xy) - 3y(xy + x^2)$$

Use the distributive property to remove each set of parentheses.
$x - 3$ *gets multiplied by* $4xy$.
$xy^2 - 12xy$ *gets multiplied by* -1.
$xy + x^2$ *gets multiplied by* $-3y$.

$$= 4x^2y - 12xy - xy^2 + 12xy - 3xy^2 - 3x^2y$$

Note that since the very first term came out with x appearing before y, we kept that same order throughout. Now we are ready to combine like terms.

$$= \boxed{x^2y - 4xy^2}$$ ■

EXAMPLE 9

$4x(3x^2 - y) + (x^3 - 4xy) + 2x(3x)(-4y)$

Solution   First of all, we should recognize that the second set of parentheses in this example is nonessential, because its coefficient is 1. Second, it is important to see that the last term in the example does *not* call for the distributive property. Why not?

$$4x(3x^2 - y) + (x^3 - 4xy) + 2x(3x)(-4y) = 12x^3 - 4xy + x^3 - 4xy - 24x^2y$$

Combine like terms.

$$= \boxed{13x^3 - 8xy - 24x^2y}$$ ■

EXAMPLE 10

$4 - [4 - 4(4 - a)]$

Solution   We begin by working on the innermost parentheses first by distributing the -4.

$$4 - [4 - 4(4 - a)] = 4 - [4 - 16 + 4a]$$ *Combine like terms within [   ].*

$$= 4 - [-12 + 4a]$$ *Remove [   ] by distributing -1.*

$$= 4 + 12 - 4a$$

$$= \boxed{16 - 4a}$$

EXAMPLE 11 $x\{2x^2 + x[x - 3(x - 1)]\}$

Solution   We begin by distributing the -3 to remove the parentheses.

$$x\{2x^2 + x[x - 3(x - 1)]\} = x\{2x^2 + x[x - 3x + 3]\} \quad \text{\textit{Combine like terms within [ ].}}$$

$$= x\{2x^2 + x[-2x + 3]\} \quad \text{\textit{Remove [ ] by distributing x.}}$$

$$= x\{2x^2 - 2x^2 + 3x\} \quad \text{\textit{Combine like terms within \{   \}.}}$$

$$= x\{3x\}$$

$$= \boxed{3x^2}$$

Study Skills 2.4

Reviewing Old Material

One of the most difficult aspects of learning algebra is that each skill and concept is dependent on those previously learned. If you have not acquired a certain skill or learned a particular concept well enough, this will, more than likely, affect your ability to learn the next skill or concept.

Thus, even though you have finished a topic that was particularly difficult for you, you should not breathe too big a sigh of relief. Eventually you will have to learn that topic well in order to understand subsequent topics. It is important that you try to master all skills and understand all concepts.

Whether or not you have had difficulty with a topic, you should be constantly reviewing previous material as you continue to learn new subject matter. Reviewing helps to give you a perspective of the material you have covered. It helps you tie the different topics together and makes them *all* more meaningful.

Some statement you read 3 weeks ago, and which may have seemed very abstract then, is suddenly simple and obvious in the light of all you know now.

Since many problems require you to draw on many of the skills you have developed previously, it is important for you to review so that you will not forget or confuse them. You will be surprised to find how much constant reviewing aids in the learning of new material.

When working the exercises, always try to work out some exercises from earlier chapters or sections. Try to include some review exercises at every study session, or at least every other session. Take the time to reread the text material in previous chapters. When you review, think about how the material you are reviewing relates to the topic you are presently learning.

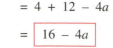 **EXERCISES 2.4**

Simplify each of the following expressions as completely as possible.

1. $4x + y + 4(x + y)$

2. $3r + s + 3(r + s)$

3. $5(m + 2n) + 3(m - n)$

4. $5(a - 2b) + 4(2a + b)$

5. $-2(x - 3y) + 5(y - x)$

6. $-4(2r - 3s) + 6(s - r)$

7. $5 + 3(x - 2)$

8. $7 + 2(x - 5)$

9. $5 - 3(x - 2)$

10. $7 - 2(x - 5)$

11. $(5 - 3)(x - 2)$

12. $(7 - 2)(x - 5)$

13. $4(m - 3n) + 2(5m + 6n)$

14. $6(3u + 4v) + 4(7u - 6v)$

15. $2(a - 2b) - 4(b - 2a)$

16. $5(r + 4s) - 3(3s - 2r)$

17. $3(2x^2 - 4y) + 4(5y - 3x^2)$

18. $5(a^2 - 4b^2) + 2(b^2 - 4a^2)$

19. $8 - (3x - 4)$

20. $10 - (6x - 5)$

21. $5y - (1 - 2y)$

22. $7a - (2 - 3a)$

23. $5(x - 3y) - x - 3y$

24. $7(a - 5b) - (a - 5b)$

25. $5(x - 3y) - (x - 3y)$

26. $7(a - 5b) - a - 5b$

27. $5(x - 3y) - x(-3y)$

28. $7(a - 5b) - a(-5b)$

29. $5x(-3y) - x(-3y)$

30. $7a(-5b) - a(-5b)$

31. $5x(-3y)(-x)(-3y)$

32. $7a(-5b)(-a)(-5b)$

33. $2x^2(x - 2) + x(3x^2 - 4x)$

34. $5y(y^2 - 3) + y^2(y - 3)$

35. $3a(4a - 1) - a(4 - a)$

36. $6z(z - 2) - z(4z - 12)$

37. $4(x^2 + 7x) - (x^2 + 7x)$

38. $9(m^3 - 4m) - (m^3 - 4m)$

39. $3a(a^2 + 3b) + 4b^2(a^2 - b)$

40. $7z(z^2 - 4y) + 3z^2(z - 2)$

41. $3x^2 - 7x + 4 - 8x^2 - 3 - x$

42. $4y^3 - y^2 + 7 - 2y - y^2 + y$

43. $x^2y(xy - x) - 5xy(x^2y - x^2)$

44. $2a^2b(ab - a) - 3ab(a^2b - a^2)$

45. $4u^2v(u - v) - (uv^3 + u^2v^2)$

46. $6st^2(s - t) + (s^2t^2 + 3s^3t)$

47. $4(x + 3y) + (4x + 3y) + 4x(3y)$

48. $5(a + 2b) + (5a + 2b) + 5a(2b)$

49. $6(m - 2n) + (6m - 2n) + 6m(-2n)$

50. $3(u - 4v) + (3u - 4v) + 3u(-4v)$

51. $3t^5(t^4 - 4) - (t^5 + t^4) - 2t^3(3t)(-t^5)$

52. $u^3(u^3 - 5) - (u^6 + u^3) - 3u^2(u^3)(-u)$

53. $-3(-x + 2) + (8 - 5x) - (2 - 2x)$

54. $-5(-t + 3) - (3t - 6) + (9 - 2t)$

55. $a - 2[a - 2(a - 2)]$

56. $x - 3[x - 3(x - 3)]$

57. $x\{x - 4[x - (x - 4)]\}$

58. $m^2\{2m - 5[m - (m - 5)]\}$

59. $3(x + 2) + 4[x - 3(2 - x)]$

60. $5(y - 3) + 2[y - 5(3 - y)]$

61. $4(y - 3) - 2[3y - 5(y - 1)]$

62. $7(a - 1) - 6[2a - 4(a - 3)]$

63. $t[3t - 4(t + 5)] - 2(t^2 - 4t)$

64. $8z^2[z^2 - z(z - 6)] - 7z(z^2 - 3z)$

65. $-a^2(3a - 7) - 2a[a^2 - 4(a - 2)]$

66. $-2d^3[3d - (6 - 5d)] - 4d(8 - 7d)$

? QUESTIONS FOR THOUGHT

67. In words, describe the error (or errors) in each of the following "solutions" and explain how to correct the errors.

(a) $3 + 2[x + 4(x + 3)] \stackrel{?}{=} 3 + 2[x + 4x + 12]$

$\stackrel{?}{=} 3 + 2[5x + 12]$

$\stackrel{?}{=} 5[5x + 12]$

$\stackrel{?}{=} 25x + 60$

(b) $3x + 5[x + (x + 3)] \stackrel{?}{=} 3x + 5[x^2 + 3x]$

$\stackrel{?}{=} 3x + 5x^2 + 15x$

$\stackrel{?}{=} 5x^2 + 18x$

68. Discuss what is *wrong* (if anything) with each of the following examples:

(a) $5 - 3(x - 4) \overset{?}{=} 2(x - 4) \overset{?}{=} 2x - 8$

(b) $5 - 3(x - 4) \overset{?}{=} 5 - 3x - 4 \overset{?}{=} 1 - 3x$

(c) $5 - 3(x - 4) \overset{?}{=} 5 - 3x - 12 \overset{?}{=} -7 - 3x$

(d) $5 - 3(x - 4) \overset{?}{=} 5 - 3x + 12 \overset{?}{=} 17 - 3x$

2.5 Translating Phrases and Sentences Algebraically

Even though it is still quite early in our study of algebra, we have already developed *algebraic* skills adequate enough for us to begin solving "real-life problems." This we will do in the next chapter. However, in order to be able to apply these skills to such problems, we need to be able to formulate and translate a problem stated in words into its equivalent algebraic form.

In this section we will focus our attention on translating phrases and sentences into their algebraic form, and leave the formulation and solution of entire problems to the next chapter.

In order to express a relationship algebraically, we often need to understand the underlying structure of the relationship. One effective strategy for uncovering the structure is to examine one or more numerical examples of the relationship in search of a *pattern* that we can generalize.

EXAMPLE 1

Translate each of the following phrases into an algebraic expression.

(a) The value of q quarters in cents

(b) The total weight of c cartons that weigh 44 pounds each

(c) The distance covered if you travel at an average rate of 52 miles per hour for h hours

Solution

(a) Suppose we have 14 quarters. Since each quarter is worth 25 cents, the 14 quarters are worth $14 \cdot 25 = 350$ cents. If we had 29 quarters, they would be worth $29 \cdot 25 = 725$ cents. We can see that to compute the value of a number of quarters we multiply

$$(\textit{The number of quarters}) \cdot (\textit{The value of 1 quarter})$$

By exactly the same logic, q quarters are worth

$$q \cdot 25 = \boxed{25q \text{ cents}}$$

(b) Suppose we had 17 of these cartons. Since each carton weighs 44 pounds, the 17 cartons weigh $17 \cdot 44 = 748$ pounds in total. If we had 32 cartons, they would weigh $32 \cdot 44 = 1,408$ pounds. We can see that to compute the total weight of a number of cartons, we multiply

$$(\textit{The number of cartons}) \cdot (\textit{The weight of 1 carton})$$

By exactly the same logic, c cartons weigh

$$c \cdot 44 = \boxed{44c \text{ pounds}}$$

(c) Suppose we travel for 6 hours at this rate. Since we travel 52 miles each hour, in 6 hours we would travel $6 \cdot 52 = 312$ miles. If we travel 11.5 hours at this rate, we would travel $11.5(52) = 598$ miles. We can see that to compute the total distance traveled in a certain number of hours, we multiply

$$(The\ number\ of\ hours\ traveled) \cdot (Rate\ of\ travel)$$

Similarly, if we travel for h hours at an average speed of 52 mph we would travel

$$h \cdot 52 = \boxed{52h\ miles}$$

Even though the three parts of this example may seem to be unrelated, in fact, their structure is actually very similar. Notice the similarities in the various parts of this example. In each case we have a number of items (quarters, cartons, hours) and we know how much each one of the items is "worth" (how many cents it's worth, how much it weighs, how many miles you travel). In each case we multiply the *number* of items we have by the "value" of one item. This type of relationship has very wide application, so be on the lookout for it. ∎

EXAMPLE 2   Translate each of the following phrases into an algebraic expression.

(a) The cost per gallon of heating oil that costs $240 for x gallons

(b) The monthly salary for an employee who earns D dollars per year

(c) The amount of acid in x ounces of a solution marked as 12%

Solution   As we did in the previous example, let's begin by examining a numerical example and generalizing the procedure.

(a) Suppose we paid $240 for 200 gallons of heating oil. We would then compute the per gallon cost by dividing the total cost by the number of gallons. Thus

$$Per\ gallon\ cost = \frac{Total\ cost}{Number\ of\ gallons} = \frac{\$240}{200\ gallons} = \$1.20\ per\ gallon$$

Similarly, if we pay $240 for x gallons the cost per gallon is $\boxed{\dfrac{240}{x}\ dollars\ per\ gallon}$.

(b) In order to compute monthly income, we divide the annual income by 12. Thus if the annual income is D dollars

$$Monthly\ income = \frac{Annual\ income}{12} = \boxed{\frac{D}{12}\ dollars\ per\ month}$$

(c) Suppose we have 20 oz of a 50% acid solution. This means that we have 20 ounces of liquid but only 50% of the liquid is actually acid. Thus we know that we have 10 ounces of actual acid (50% of the 20 ounces). Thus, in general, to compute the amount of actual substance in a solution we multiply:

$$(Concentration\ of\ the\ solution\ as\ a\ decimal) \cdot (Amount\ of\ solution)$$

If we have 20 oz of a 50% acid solution we get $0.50(20) = 10$ ounces of actual acid.

In this example we are told that we have x ounces of a 12% solution and so the amount of acid is $(0.12)x = \boxed{0.12x\ oz}$. ∎

EXAMPLE 3   Suppose that the price of an item is $\$x$ and that the price is discounted by 15%. Express the discounted price in terms of x.

Solution   In order to compute a given percent of a number we multiply the number by the percent written as a decimal. 15% written as a decimal is 0.15, so, for example, to compute 15% of 280 we multiply $0.15(280) = 42$. Similarly, to get 15% of the price x we multiply $0.15x$.

You can find a review of percents in Appendix A.

A 15% discount means that we are to subtract 15% of the original price from the original price to get the discounted price.

Discounted price = Original price − 15% of the original price

If the original price is x, we have

$$= x - 15\% \text{ of } x$$
$$= x - 0.15x \quad \textit{Combine like terms. Remember that x means 1x.}$$
$$= \boxed{0.85x}$$

It is always a good idea to examine the reasonableness of an answer. The answer $0.85x$ means that to get the discounted price we multiply the original price by 0.85, which is the same as 85% of x. Thus, the answer is saying that after a 15% discount, the price is 85% of the original price. ■

Next let's consider some common phrases and how they are translated from English into algebra.

EXAMPLE 4

In each of the following, let x represent the *number,* and write an algebraic expression that translates the given phrase.

(a) 5 more than a number   (b) 5 less than a number
(c) 5 times a number   (d) the product of 5 and a number

Solution

Translating a phrase algebraically does not mean simply reading the phrase from left to right and replacing each word with an equivalent algebraic expression. We must translate the *meaning* of the phrase as well as the words. Keep in mind that we are letting x represent the number.

(a) "5 more than a number" is translated as

$$\boxed{x + 5}$$

We recognize that the phrase "5 more than" conveys addition. We are starting with something (a number, which in this example we are calling x) and adding 5 to it. This becomes $x + 5$. While $5 + x$ is also correct, $x + 5$ is usually considered preferable. The next part of this example will clarify why.

(b) "5 less than a number" is translated as

$$\boxed{x - 5}$$

We recognize that the phrase "5 less than" conveys subtraction. We are starting with something (again, a number that we are calling x) and subtracting 5 *from it.* This becomes $x - 5$, *not* $5 - x$, which is what we would get if we ignored the *meaning* of the statement and simply translated from left to right.

Remember: $5 - x$ is not the same as $x - 5$. For example, if we translate "5 less than 8" correctly we should get $8 - 5$, which is 3 (not $5 - 8$, which is −3).

(c) "5 times a number" is translated as

$$\boxed{5x}$$

The word "times" conveys multiplication.

(d) "The product of 5 and a number" is also translated as

$$\boxed{5x}$$

The word "product" also conveys multiplication. ■

Many students look for *key words* to help them quickly determine which algebraic symbols to use in translating an English phrase or sentence. For instance, in Example 4 we recognized the key words "more than" to indicate addition, "less than" to indicate subtraction, and "product" to indicate multiplication.

However, the key words do not usually indicate *how* the symbols should be put together to yield an accurate translation of the words. In order to be able to put the symbols together in a meaningful way we must understand how the key words are being used in the context of the given verbal statement.

Let's look at two verbal expressions that use the same key words and yet do not have the same meaning:

1. Three times the sum of five and four

2. The sum of three times five and four

Both expressions contain the following key words:

"three"	the number 3	(3)
"times"	meaning multiply	($\cdot$)
"sum"	meaning addition	(+)
"five"	the number 5	(5)
"four"	the number 4	(4)

How should the symbols be put together?

Expression 1 says: Three times . . . .

Three times what? . . . The sum.

Thus, we must *first* find the sum of 5 and 4, before multiplying by 3.

Expression 2 says: The sum of . . . .

The sum of what *and* what? . . . The sum of 3 times 5 *and* 4.

Thus, we must *first* compute 3 times 5 before we can determine what the sum is.

Translating the two verbal expressions algebraically, we get:

Expression 1 translates to $3(5 + 4) = 3 \cdot 9 = 27$.

Expression 2 translates to $3 \cdot 5 + 4 = 15 + 4 = 19$.

Thus, in comparing expressions 1 and 2, we can see that the simple change of word positions in a phrase can change the meaning quite a bit.

EXAMPLE 5

Translate each of the following algebraically:

(a) Eight more than three times a number

(b) Two less than five times a number is 18.

(c) The sum of two numbers is four less than their product.

Solution

(a) "Eight more than" means we are going to *add* 8 to something. To what? To "three times a number." If we represent the number by n (you are free to choose any letter you like), we get

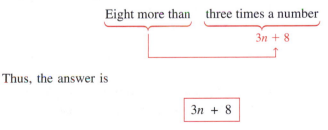

Thus, the answer is

$$3n + 8$$

Of course, $8 + 3n$ is also correct.

(b) "Two less than" means we are going to *subtract* 2 from something. From what? From "5 times a number."

The word "is" translates to "=." If we represent the number by s, we get

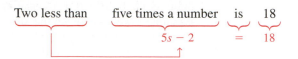

$$5s - 2 = 18$$

Thus, the final answer is

$$\boxed{5s - 2 = 18}$$

It is very important to note here that we must subtract 2 *from* $5s$ and therefore $2 - 5s = 18$ is *not* a correct translation.

(c) The sentence mentions two numbers, so let's call them x and y.

The sum of two numbers   is   four less than   their product

$$x + y \qquad = \qquad\qquad\qquad x \cdot y - 4$$

Thus, our translation is

$$\boxed{x + y = xy - 4}$$ ■

EXAMPLE 6

Translate each of the following algebraically:

(a) The sum of two consecutive integers

(b) The sum of two consecutive even integers

(c) The sum of two consecutive odd integers

Solution

This example illustrates how important it is to state clearly what the variable you are using represents. Let's begin by looking at some numerical examples for guidance.

Examples of two consecutive integers	*Examples of two consecutive even integers*	*Examples of two consecutive odd integers*
2 and 3	4 and 6	7 and 9
19 and 20	10 and 12	15 and 17
25 and 26	28 and 30	31 and 33
x and $x + 1$	x and $x + 2$	x and $x + 2$

(a) The case of two consecutive integers is fairly straightforward. If we let x represent the first integer, then the next consecutive integer is $x + 1$. Therefore, the sum of two consecutive integers can be represented as

$$x + (x + 1) \quad \text{or} \quad \boxed{2x + 1}$$

(b) Based on the numerical examples given above, we notice that the two cases of consecutive even or odd integers are basically the same. In order to get from one even integer to the next we have to add 2, and to get from one odd integer to the next we have to add 2. It is not the adding of 2 that makes the numbers even or odd, but rather whether the *first* number is even or odd. If x is even, then so is $x + 2$. If x is odd, then so is $x + 2$.

If we let x represent the first *even* integer, then the next even integer is represented as $x + 2$. Thus, the sum of two consecutive even integers can be represented as

$$x + (x + 2) \quad \text{or} \quad \boxed{2x + 2}$$

(c) If we let x represent the first *odd* integer, then the next odd integer is represented as $x + 2$. Thus, the sum of two consecutive odd integers can be represented as

$$x + (x + 2) \quad \text{or} \quad \boxed{2x + 2}$$

It all depends on how we designate x. ∎

The next example again illustrates the need to understand clearly the meaning of the words in an example in order to be able to put them together properly.

While it is certainly necessary to recognize that particular words and phrases imply specific arithmetic operations, and to understand what the problem means taken as a whole, there is still one important factor necessary to be able to successfully translate from verbal expressions to algebraic ones. This additional factor is the ability to take your basic general knowledge and common sense and apply them to a particular problem.

Let's look at how these various components blend together in formulating the solution to a problem.

EXAMPLE 7

The Rent-A-Relic car rental company charges a daily rental fee of $21.95 plus a mileage charge of $0.12 per mile. Write an expression that represents the total charge for a 1-day rental in which the car is driven m miles, and use this expression to compute the total charge for a 1-day rental in which the car is driven 180 miles.

Solution

The total daily charge is equal to the daily rental fee plus the mileage charge. To compute the mileage charge we multiply the number of miles driven (which is represented by m) by the charge per mile, which is $0.12. Thus, if we let C represent the total daily charge, we have

$$\boxed{C = 21.95 + 0.12m}$$

We often say that this equation is a *formula* for the total daily charge in terms of the number of miles driven.

In order to compute the total charge for a 1-day rental in which the car is driven 180 miles, we can use the formula we just obtained for $m = 180$. In other words, we simply substitute $m = 180$ into the above formula:

$$C = 21.95 + 0.12m \qquad \textit{We substitute } m = 180.$$
$$= 21.95 + 0.12(180)$$
$$= 21.95 + 21.60 = \boxed{\$43.55}$$

∎

EXAMPLE 8

Tanya exercises for 20 minutes per day for D days and exercises M minutes per day for 10 days.

(a) Express the total number of minutes T that Tanya exercises in terms of D and M.

(b) Use the result of part (a) to compute the total number of minutes Tanya exercises if $D = 6$ days and $M = 35$ minutes.

Solution

(a) If Tanya exercises 20 minutes each day for 4 days, she would have exercised $20(4) = 80$ minutes. We can see that to compute the number of minutes she exercises over a period of days, we multiply the number of minutes she exercises per day by the number of days. Therefore, using the information given in the example, the total number of minutes she exercises is

$$T = 20 \cdot D + M \cdot 10 = \boxed{20D + 10M \text{ minutes}}$$

(b) We use the formula for T obtained in part **(a)** and substitute the given values.

$$T = 20D + 10M \qquad \textit{Substitute } D = 6 \textit{ and } M = 35.$$
$$= 20(6) + 10(35)$$
$$= 120 + 350 = \boxed{470 \text{ minutes}}$$ ■

EXAMPLE 9   Suppose film costs \$3 per roll and you buy two multiroll packs of film. The first pack contains a certain number of rolls of film (let's say n rolls), and the second pack contains five less than the number of rolls in the first pack.

(a) How many rolls of film are there in the first pack?

(b) How much does the first pack cost?

(c) How many rolls are there in the second pack?

(d) How much does the second pack cost?

(e) What is the total cost of the two packs of film?

Solution   **(a)** The example tells us that the number of rolls of film in the first pack is to be represented by

$$\boxed{n}$$

(b) How much do n rolls of film cost? Let's think numerically for a moment.

If n were equal to 4 (that is, if there were 4 rolls of film in the first pack), then the cost of the four rolls would be $4 \cdot \$3 = \$12$.

If n were equal to 7, the cost would be $7 \cdot \$3 = \$21$.

If n were equal to 11, the cost would be $11 \cdot \$3 = \$33$.

Whatever the number of rolls in the pack, we multiply by 3 to get the cost of the entire pack. Therefore, the cost of the first pack of film is

$$n \cdot \$3 = \boxed{3n \text{ dollars}}$$

(c) In part **(a)** we followed the suggestion in the example and let n represent the number of rolls in the first pack; how then do we represent the number of rolls in the second pack?

We are told that the number of rolls in the second pack is "five less than the number in the first pack." But we have already represented the number of rolls in the first pack as n. So the number in the second pack is

5 less than n

We know this translates to

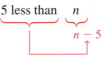

Thus, the answer to part **(c)** is

$$\boxed{n - 5}$$

(d) To represent the cost of the second pack of film, we reason as we did in part **(b)**. Since the number of rolls in the second pack is $n - 5$, the second pack costs

$$(n - 5) \cdot \$3 = \boxed{3(n - 5) \text{ dollars}}$$

(e) To get the total cost of the two packs of film we add the results obtained in parts (b) and (d):

$$3n + 3(n - 5) \text{ dollars}$$

Simplifying this, we get

$$3n + 3n - 15 \text{ dollars} \quad \text{or} \quad 6n - 15 \text{ dollars}$$

Notice that in order to successfully answer the various parts of this question we must be able to distinguish clearly between the *number* of rolls of film in each pack, and the *cost* of the film in each pack. These are two distinct ideas and this is where our own basic knowledge came into play.

We have been quite detailed in our discussion of this example in order to emphasize that you must understand what the problem is asking as well as what it is saying. ∎

Study Skills 2.5

Reflecting

When you have finished reading or doing examples, it is always a good idea to take a few minutes to think about what you have just covered. Think about how the examples relate to the verbal material, and how the material just covered relates to what you have learned previously. How are the examples and concepts you have just covered similar to or different from those you have already learned?

EXERCISES 2.5

*In Exercises 1–18, let n represent the **number** and translate each phrase or sentence algebraically.*

1. Four more than a number

2. Four times a number

3. Four less than a number

4. A number increased by four

5. A number decreased by four

6. The product of four and a number

7. Six more than five times a number

8. Six less than five times a number

9. Nine less than twice a number

10. Five more than twice a number

11. The product of a number and seven more than the number

12. The product of a number and seven less than the number

13. The product of two more than a number and six less than the number

14. The product of three less than a number and four more than the number

15. Eight less than twice a number is fourteen.

16. One less than three times a number is seven.

17. Four more than five times a number is two less than the number.

18. Ten less than a number is three more than six times the number.

In Exercises 19–22, let r and s represent two numbers and translate the sentence algebraically.

19. The sum of two numbers is equal to their product.

20. Five more than the sum of two numbers is equal to their product.

21. Twice the sum of two numbers is three less than their product.

22. Five times the product of two numbers is eight more than their sum.

In Exercises 23–32, use any letter you choose to translate the given phrase or sentence algebraically. Be sure to identify clearly what your variable represents.

23. The sum of two consecutive integers

24. The product of two consecutive integers

25. The sum of two consecutive even integers

26. The sum of two consecutive odd integers

27. The product of three consecutive odd integers

28. The product of four consecutive even integers

29. Eight times an even integer is four less than seven times the next even integer.

30. Five less than twice an odd integer is four more than the next odd integer.

31. The sum of the squares of three consecutive integers is five.

32. The cube of the sum of two consecutive odd integers is sixty-four.

33. Shawna is working at a part-time job that pays her $8.65 per hour. Write an expression for the amount A of money she earns in h hours. Use this expression to compute the amount she earns if she works 14 hours.

34. Lamont is driving at 58 kilometers per hour. Write an expression for the distance d he travels in h hours. Use this expression to compute the number of miles he travels in 6.5 hours.

35. A printer charges a set-up fee of $225 plus a per-page fee of $0.045. Write an expression for the total cost C for a job that contains p pages. Use this expression to compute the total cost of a job that contains 4,800 pages.

36. In addition to a weekly salary of $480, a salesperson earns $2.15 for each item that he sells. Write an expression for the total amount A earned in a week in which i items are sold. Use this expression to compute the total amount earned in a week in which 115 items are sold.

37. A campsite charges a flat fee of $42 per night plus $9 per night for each person. Write an expression for the cost C of the campsite per night for p people. Use this expression to compute the cost of the campsite for 6 people for one night.

38. A campsite charges a flat fee of $42 per night plus $10 per night for each adult and $4 per night for each child. Write an expression for the cost T of the campsite per night for a adults and c children. Use this expression to compute the cost of the campsite for 5 adults and 6 children.

39. A long-distance phone company charges $0.65 per minute for the first minute and $0.42 for each minute thereafter. Write an expression for the cost C of a long-distance call that lasts m minutes. Use this expression to compute the cost of a long-distance call that lasts 12 minutes.

40. An electric company charges a flat fee of $42.50 for the first 300 kilowatt-hours of usage, and $0.15 for each kilowatt-hour above 300. Write an expression for the cost C of using k kilowatt-hours (where k is more than 300). Use this expression to compute the cost of using 752 kilowatt-hours.

41. Marty runs 8 km per day for d days, and he runs k km per day for 10 days. Express the total number T of kilometers that Marty runs in terms of d and k. Use this expression to compute T when $d = 12$ days and $k = 15$ km.

42. Gene packs B boxes per hour for 9 hours and then packs 7 boxes per hour for H hours. Express the total number T of boxes that Gene packs in terms of B and H. Use this expression to compute T when $B = 20$ boxes per hour and $H = 15$ hours.

43. John mows L lawns per day for 6 days, and he mows 9 lawns per day for D days. Express the total number T of lawns that John mows in terms of L and D. Use this expression to compute T when $L = 5$ lawns per day and $D = 8$ days.

44. Jessica assembles M machines per week for 6 weeks and then assembles 8 machines per week for W weeks. Express the total number T of machines that

Jessica assembles in terms of M and W. Use this expression to compute T when $M = 12$ machines per week and $W = 16$ weeks.

45. Express the amount you save on a book that is priced at d dollars with a 30% discount.

46. Express the amount of tax you pay on a car that costs c dollars at a tax rate of 7%.

47. Express the total cost of a coat that sells for d dollars plus 4% tax.

48. Calculate the total cost of dinner if the meal costs $53.25 and you leave an 18% tip.

49. Express the sale price of a computer priced at p dollars with a 20% discount.

50. Express the sale price of a stereo priced at p dollars with a 35% discount.

51. Express the selling price of a TV that a store buys for $380 and marks up 28%.

52. Express the selling price of a computer system that a store buys for $635 and marks up 30%.

53. A furniture store is having a 20%-off sale. The store then advertises that buyers can take an additional 10% off the sale price.
    (a) Determine the price of a lamp that originally sold for $120.
    (b) Are the consecutive 20% and 10% discounts equivalent to a single 30% discount off the original price? Explain.

54. A home heating oil company raises its prices by 10%. Then due to a shortage, the new prices are raised an additional 15%.
    (a) Determine the price of a gallon of oil that originally sold for $1.20.
    (b) Are the consecutive 10% and 15% raises equivalent to a single 25% raise of the original price? Explain.

55. Express the weekly income of a salesperson who earns a weekly salary of $285 plus a commission of $0.52 for each of x items sold.

56. A contractor charges a flat fee of $1,650 plus $27 per square foot of construction. Find the total cost of a construction job of s square feet.

57. A student earns $6 per hour as a tutor in the Math Lab and $14 per hour for private tutoring. During a certain month he works h hours in the lab and 7 fewer hours as a private tutor. Express his tutoring income for the month in terms of h.

58. A part-time employee earns $10.25 per hour during the week and $18 per hour on weekends. During a certain month, she worked h weekend hours and 36 more than that number of hours during the week. Express her weekly income in terms of h.

59. Shawna is driving to her vacation. For t hours she averages 48 mph and for 2 fewer hours she averages 54 mph. Express the total distance she travels in terms of t.

60. Daryl made two truck deliveries. One trip took 4 hours at an average speed of r kph and the second trip took 3.5 hours at an average speed that was 8 kph faster. Express the total distance Daryl traveled in terms of r.

61. A box contains 8 nickels, 12 dimes, and 9 quarters.
    (a) How many nickels are in the box?
    (b) What is the *value* of the nickels in the box?
    (c) How many dimes are in the box?
    (d) What is the *value* of the dimes in the box?
    (e) How many quarters are in the box?
    (f) What is the *value* of the quarters in the box?
    (g) All together, how many coins are there in the box?
    (h) What is the *total value* of all the coins in the box?

62. Repeat Exercise 61 for *n* nickels, *d* dimes, and *q* quarters.

63. Electronic mark sense reader A can grade 200 exams per minute and grades for 15 minutes, while electronic mark sense reader B can grade 160 exams per minute and grades for 20 minutes.

(a) How many exams per minute does A grade?

(b) How many minutes does A grade?

(c) How many exams does A grade?

(d) How many exams per minute does B grade?

(e) How many minutes does B grade?

(f) How many exams does B grade?

(g) How many exams do A and B grade all together?

64. Repeat Exercise 63 if machine A grades *x* exams per minute for 25 minutes, and machine B grades 250 exams per minute for *t* minutes.

65. Ruth walks at the rate of 100 meters per minute and jogs at the rate of 220 meters per minute. She walks for 25 minutes and then jogs for 35 minutes.

(a) How fast does she walk?

(b) For how long does she walk?

(c) How far does she walk?

(d) How fast does she jog?

(e) For how long does she jog?

(f) How far does she jog?

(g) How much distance has Ruth covered all together?

66. Repeat Exercise 65 if Ruth walks at the rate of *r* meters per minute, jogs at twice that rate, walks for *t* minutes, and jogs for 20 minutes longer than she walks.

CHAPTER 2   SUMMARY

After having completed this chapter you should be able to:

1. Understand and evaluate algebraic expressions involving exponents (Section 2.1).

For example:

(a) *Write without exponents.* $-x^2y^3 = \boxed{-xxyyy}$

(b) *Write with exponents.* $2xxx + 3xxxx = \boxed{2x^3 + 3x^4}$

(c) *Evaluate.* $-6^2 = -6 \cdot 6 = \boxed{-36}$

$$(-6)^2 = (-6)(-6) = \boxed{36}$$

2. Substitute numerical values into algebraic expressions and evaluate the results (Section 2.2).

For example:

Evaluate. $xy^2 - (x - y)^3$ for $x = 2$ and $y = -1$

$$xy^2 - (x - y)^3 = (2)(-1)^2 - [2 - (-1)]^3$$
$$= (2)(1) - [3]^3$$
$$= 2 - 27$$
$$= \boxed{-25}$$

3. Distinguish between *terms* and *factors* (Section 2.2).

    For example:

    $x + y + z$ consists of three terms, each term consisting of one factor.

    xyz is one term consisting of three factors.

4. Use the distributive property to *multiply out, factor,* and *combine like terms* (Section 2.3).

    For example:

    (a) *Multiply out.* $3x(x^2 - 5x) = \boxed{3x^3 - 15x^2}$

    (b) *Factor.* $14x + 28y - 7 = \boxed{7(2x + 4y - 1)}$ *Factor out the common factor of 7.*

    (c) *Combine like terms.* $3x^2 - 5x - x^2 + 2x = \boxed{2x^2 - 3x}$

5. Simplify algebraic expressions by removing grouping symbols and combining like terms (Section 2.4).

    For example:

    $$3x^2(x - 4) - 2x(x^2 + 3) - (x - 12x^2)$$

    Remove each set of parentheses by distributing its coefficient.

    $= 3x^3 - 12x^2 - 2x^3 - 6x - x + 12x^2$ *Now combine like terms.*

    $= \boxed{x^3 - 7x}$

6. Translate phrases and sentences from their English language form into their algebraic form (Section 2.5).

    For example:

    Four less than three times a number is 26.

    If we let n represent the number, then the translation becomes:

    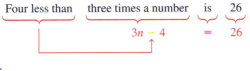

    Thus, the answer is

    $$\boxed{3n - 4 = 26}$$

CHAPTER 2   REVIEW EXERCISES

In Exercises 1–6, write the given expression without exponents.

1. xy^3

2. $(xy)^3$

3. $-x^4$

4. $(-x)^4$

5. $3x^2$

6. $(3x)^2$

In Exercises 7–10, write the given expression with exponents.

7. $xxyyy$

8. $xx + xxx$

9. $aa - bbb$

10. $-aa(bbb)$

In Exercises 11–20, evaluate the given expression for $x = -3$, $y = 4$, and $z = -2$.

11. $-z^4$

12. $(-z)^4$

13. xy^2

14. $(xy)^2$

15. $xyz - (x + y + z)$

16. $xyz - x + y + z$

17. $|xy| + z - |z|$

18. $|x + y + z| + |x| - |y| - |z|$

19. $2x^2 - (x + y)^2$

20. $3z^3 + (x - z)^2$

In Exercises 21–46, simplify each expression as completely as possible.

21. $x^3 x^4 x$

22. $r^5 r r^3$

23. $a^7 a^2 + a^3 a^6$

24. $2y^4 y^2 + 3y^3 y^3$

25. $4x^3 x^2 + 3x^2 x^4$

26. $5a^7 a + 2a^4 a^3$

27. $3x^2 - 7x + 7 - 5x^2 - x - 3$

28. $4m^3 - 8 - m^2 - 3m^2 - m^3 - 1$

29. $2a^2 b(3ab^4)$

30. $-5ab(2a^2 b^3)$

31. $2a^2 b(3a + b^4)$

32. $-5ab(2a^2 + b^3)$

33. $3x(2x + 4) + 5(x^2 - 3)$

34. $2z(3z - 5) + 4(z^2 + 1)$

35. $4y^2(y - 2) - y(y^2 - 5y)$

36. $c^3(2c - 3) - c^2(c^2 - 8c)$

37. $3xy(x^2 - 2y) + 4xy^2(y - x)$

38. $2r^2 s(s - rs) - 5s(r^2 s + r^2 s^2)$

39. $3(2x - 4y) - (x + 2y) - (x - 2y)$

40. $4(3a - 2b) - (a - 2b) - (a + 2b)$

41. $3x^4(x^3 - 2y^2) - 4x(x^2)(x^4)$

42. $6y^5(x^3 - 5x^4) + 2x^2(3xy^2)(5xy^3)$

43. $3 - [x - 3 - (x - 3)]$

44. $3 - x[3 - x(3 - x)]$

45. $0.02x + 0.07(2x + 1{,}250)$

46. $0.06x + 0.25(280 - 3x)$

In Exercises 47–50, use the distributive property to factor the given expression.

47. $5x^2 + 10$

48. $2a^5 + 16$

49. $3y - 6z + 9$

50. $22x - 33y + 11$

In Exercises 51–54, translate the given statement algebraically. Let n represent the number.

51. The sum of a number and seven is four less than three times the number.

52. Five less than twice a number is four more than three times the number.

53. The sum of two consecutive odd integers is five less than the smaller one.

54. We have three consecutive integers such that the sum of the first integer and four times the second integer is eight more than three times the third integer.

55. What is the price of a television priced at d dollars after taking a 35% discount?

56. A beeper company offers a plan that charges a flat fee of $4.50 a month plus $0.02 per call. Express the monthly charge M for c calls.

57. John earns a commission of $2 for each newspaper subscription he sells, and $5 for each magazine subscription he sells. He sells 12 newspaper subscriptions and 9 magazine subscriptions.

(a) How many newspaper subscriptions does he sell?

(b) How much does he earn for each newspaper subscription?

(c) How many magazine subscriptions does he sell?

(d) How much does he earn for each magazine subscription?

(e) How much does he earn for the newspaper subscriptions?

(f) How much does he earn for the magazine subscriptions?

(g) How much does he earn all together?

58. Repeat Exercise 57 if John sells n newspaper subscriptions and four less than twice that many magazine subscriptions.

CHAPTER 2   PRACTICE TEST

In Problems 1–4, evaluate each of the given expressions.

1. -3^4

2. $(-3)^4$

3. $(-3 - 4 + 6)^5$

4. $7 - 3(2 - 6)^2$

In Problems 5–8, evaluate the given expression for $x = -2$, $y = 3$, and $z = -4$.

5. $x - y - z$

6. $xy - z$

7. $x^3 - z^2$

8. $2yz^2 - |x|$

In Problems 9–13, perform the indicated operations and simplify as completely as possible.

9. $4x^2y - 5xy + y^2 - 3xy - 2y^2 - x^2y$

10. $-3xy^2(-4x^2y)(-2x^3)$

11. $2x(x^2 - y) - 3(x - xy) - (2x^3 - 3x)$

12. $3x^3(x^2 - 4xy) - 2x(xy)(-6x^2)$

13. $4 - [x - 4(x - 4)]$

14. $0.06x + 0.12(x + 500)$

15. If we let n stand for the *number,* translate each of the following phrases:

(a) Four more than twice a number

(b) Twenty less than five times a number is equal to the number.

16. A blank 60-minute audio cassette costs \$2, and a blank 90-minute audio cassette costs \$3. Someone buys a certain number of 60-minute cassettes (let's say x) and five less than twice that many 90-minute cassettes.

(a) How much does a 90-minute cassette cost?

(b) How many 60-minute cassettes were bought?

(c) How much does a 60-minute cassette cost?

(d) How many 90-minute cassettes were bought?

(e) How much was spent on the 90-minute cassettes?

(f) How much was spent on the 60-minute cassettes?

(g) How much was spent all together on the cassettes?

17. A business pays \$99 per month to lease a copying machine plus \$0.15 per copy. Express the monthly cost M if the business makes C copies.

Study Skills 2.6

Checking Your Work

We develop confidence in what we do by knowing that we are right. One way to check to see if we are right is to look at the answers usually provided in the back of the book. However, few algebra texts provide *all* the answers. And of course, answers are not provided during exams, when we need the confidence most.

Isn't it frustrating to find out that you incorrectly worked a problem on an exam, and then to discover that you would easily have seen your error had you just taken the time to check over your work? Therefore, you should know how to check your answers.

The method of checking your work should be different from the method used in the solution. In this way you are more likely to discover any errors you might have made. If you simply rework the problem the same way, you cannot be sure you did not make the same mistake twice.

Ideally, the checking method should be quicker than the method for solving the problem (although this is not always possible).

Learn how to check your answers, and practice checking your homework exercises as you do them.

CHAPTER 3

First-Degree Equations and Inequalities

STUDY SKILLS

One of our main goals in algebra is to develop the ability to solve equations. As we discuss various types of expressions, we will also attempt to describe methods for solving equations involving these expressions. Once we have the ability to solve equations we can then apply algebraic techniques to solving real-life problems, which we will begin to do in Section 3.2.

In Chapter 2 we mentioned that variables are used in algebra primarily in two ways: **(a)** To describe a general relationship between numbers and/or arithmetic operations, as for example in the statement of the distributive property; and **(b)** as a placeholder where the variable represents a number that is as yet not identified, but which needs to be found. It is this second use of variables that we are going to discuss in this chapter.

3.1 Types of Equations

When we read the mathematical sentences $5 + 2 = 7$ and $4 + 7 = 12$, we see clearly that the first is true and the second is false. The first is true because the left-hand side of the equation, $5 + 2$, represents the same number as the right-hand side of the equation, 7. The second is false because the two sides of the equation do not represent the same number.

On the other hand, the sentence $x + 3 = 5$ is neither true nor false. Since the letter x represents some number, we cannot tell whether the sentence is true until we know the value of x. Such an equation, whose truth or falsehood depends on the value of the variable, is called a ***conditional equation.***

The value or values of the variable that make a conditional equation true are called the ***solutions*** of the equation. They are the values that, when substituted for the variable into the equation, make both sides of the equation equal. Any such value is said to ***satisfy*** the equation.

In the equation mentioned above, $x + 3 = 5$, we can see that $x = 2$ is a solution to the equation because $2 + 3 = 5$ is true, and so $x = 2$ satisfies the equation. Another number such as $x = 8$ does not satisfy the equation because $8 + 3 \neq 5$. Therefore, $x + 3 = 5$ is a conditional equation.

Not every equation that involves a variable is necessarily conditional. For example, the equation

$$5(x + 2) - 4 = 5x + 6$$

is not conditional. If we simplify the left-hand side of this equation we get

$$5(x + 2) - 4 = 5x + 6$$
$$5x + 10 - 4 = 5x + 6$$
$$5x + 6 = 5x + 6$$

which is *always true,* no matter what the value of the variable is. Such an equation, which is always true regardless of the value of the variable, is called an ***identity.***

On the other hand, the equation

$$5 = 2(x + 3) - 2x$$

is also not conditional. If we simplify the right-hand side of this equation we get

$$5 = 2x + 6 - 2x$$
$$5 = 6$$

which is *always false,* no matter what the value of the variable is. Such an equation, which is always false regardless of the value of the variable, is called a ***contradiction.*** It has no solution.

Often, if we simplify both sides of an equation as completely as possible, we can recognize it as an identity or a contradiction. (Here is where we begin to use the skills we developed in the last chapter.)

EXAMPLE 1

Determine the type of each of the following equations:

(a) $4x - 3(x - 2) = x + 6$

(b) $5y(y - 4) - y(5y - 20) = 3(2y + 1) - (6y - 7)$

Solution

Looking at each equation in the form that it is given, it is not clear what type of equation we have. We will simplify both sides of each equation as completely as possible according to the methods we learned in Chapter 2. Hopefully, we will then be able to recognize what type of equation we have.

(a) $4x - 3(x - 2) = x + 6$ *Remove the parentheses by distributing the −3.*

     $4x - 3x + 6 = x + 6$ *Combine like terms.*

         $x + 6 = x + 6$

Since $x + 6 = x + 6$ is always true, we see that the original equation is an identity.

(b) $5y(y - 4) - y(5y - 20) = 3(2y + 1) - (6y - 7)$

Remove the parentheses by applying the distributive property.
Remember: $-(6y - 7) = -1(6y - 7) = -6y + 7$

$5y^2 - 20y - 5y^2 + 20y = 6y + 3 - 6y + 7$ *Combine like terms.*

                 $0 = 10$

Since $0 = 10$ is always false, we see that the original equation is a contradiction. ∎

As you might expect, the most interesting situation is that of a conditional equation. Since a conditional equation is neither always true nor always false, we would like to be able to determine exactly which values satisfy the equation. The process of finding the values that satisfy an equation is called ***solving the equation.***

Once we find the values that we think are the solutions to an equation we would, of course, like to be able to check our answers. This consists of substituting the proposed value into the original equation and verifying that it satisfies the equation.

Several examples will illustrate this idea.

EXAMPLE 2

In each of the following, determine which of the listed values satisfies the given equation.

(a) $2x + 11 = 2 - x$;   for $x = 4, -3$

(b) $5 - 2(4 - a) = 8a + 4(2 - a) - 12$;   for $a = 0, \dfrac{1}{2}$

(c) $3z^2 = 1 - 2z$;   for $z = -1, 2, \dfrac{1}{3}$

Solution

We replace each occurrence of the variable with the value we are checking and evaluate each side separately according to the order of operations.

(a) ***Check*** $x = 4$: $2x + 11 = 2 - x$

                   $2(4) + 11 \overset{?}{=} 2 - (4)$

                      $8 + 11 \overset{?}{=} -2$

                         $19 \ne -2$

Therefore, $x = 4$ does *not* satisfy the equation.

Check $x = -3$: $2x + 11 = 2 - x$

                   $2(-3) + 11 \overset{?}{=} 2 - (-3)$

                     $-6 + 11 \overset{?}{=} 2 + 3$

                          $5 \overset{\checkmark}{=} 5$

Therefore, $x = -3$ *does* satisfy the equation.

(b) ***Check*** $a = 0$: $5 - 2(4 - a) = 8a + 4(2 - a) - 12$

$$5 - 2(4 - 0) \stackrel{?}{=} 8(0) + 4(2 - 0) - 12$$

$$5 - 2(4) \stackrel{?}{=} 0 + 4(2) - 12$$

$$5 - 8 \stackrel{?}{=} 8 - 12$$

$$-3 \neq -4$$

Therefore, $a = 0$ does *not* satisfy the equation.

Check $a = \frac{1}{2}$: $5 - 2(4 - a) = 8(a) + 4(2 - a) - 12$

$$5 - 2\left(4 - \frac{1}{2}\right) \stackrel{?}{=} 8\left(\frac{1}{2}\right) + 4\left(2 - \frac{1}{2}\right) - 12$$

$$5 - 2\left(\frac{8}{2} - \frac{1}{2}\right) \stackrel{?}{=} \frac{8}{1} \cdot \frac{1}{2} + 4\left(\frac{4}{2} - \frac{1}{2}\right) - 12$$

$$5 - \frac{2}{1} \cdot \frac{7}{2} \stackrel{?}{=} 4 + \frac{4}{1} \cdot \frac{3}{2} - 12$$

$$5 - 7 \stackrel{?}{=} 4 + 6 - 12$$

$$-2 \stackrel{\checkmark}{=} -2$$

Therefore, $a = \frac{1}{2}$ *does* satisfy the equation.

(c) ***Check*** $z = -1$: $3z^2 = 1 - 2z$

$$3(-1)^2 \stackrel{?}{=} 1 - 2(-1)$$

$$3 \cdot 1 \stackrel{?}{=} 1 + 2$$

$$3 \stackrel{\checkmark}{=} 3$$

Therefore, $z = -1$ *does* satisfy the equation.

Check $z = 2$: $3z^2 = 1 - 2z$

$$3(2)^2 = 1 - 2(2)$$

$$3 \cdot 4 \stackrel{?}{=} 1 - 4$$

$$12 \neq -3$$

Therefore, $z = 2$ does *not* satisfy the equation.

Check $z = \frac{1}{3}$: $3z^2 = 1 - 2z$

$$3\left(\frac{1}{3}\right)^2 \stackrel{?}{=} 1 - 2\left(\frac{1}{3}\right)$$

$$\frac{3}{1} \cdot \frac{1}{9} \stackrel{?}{=} 1 - \frac{2}{3}$$

$$\frac{1}{3} \stackrel{\checkmark}{=} \frac{1}{3}$$

Therefore, $z = \frac{1}{3}$ *does* satisfy the equation.

Note that all the equations in the example are conditional. They are neither always true nor always false. ∎

EXERCISES 3.1

In Exercises 1–16, determine whether the given equation is an identity or a contradiction.

1. $2(x - 3) = 2x - 6$

2. $3(x + 4) = 3x + 12$

3. $5(x + 2) = 5x + 2$

4. $4x - 1 = 4(x - 1)$

5. $2a + 4 + 3a = 6a + 4 - a$

6. $3a - 5 + a = 6a + 5 - 2a$

7. $2z^2 + 3z - 2z^2 - z = z + z + 1$

8. $2z + z - 5 = 3z^2 + z - 3z^2 + 2z$

9. $5u - 4(u - 1) - u = u - 4 - (u - 2)$

10. $7u - 5(u - 1) - 2u = 2u - 1 - (2u - 3)$

11. $7 - 3(y - 2) = y + 4(5 - y) - 7$

12. $8 - 2(y - 4) = y + 3(4 - y) + 4$

13. $w(w - 2) - w^2 + 2w = 3(w + 1) - (3w - 1)$

14. $2w(w + 1) - (2w^2 + 2w) = 4(w - 3) - 4w + 12$

15. $2(x^2 - 3) - x(2x - 1) + x = 2 - x - (x + 8) + 4x$

16. $6(x^2 - x) + 2x(3 - 3x) + 1 = 4 + x - (x + 4) - 1$

In Exercises 17–34, determine whether the given equation is satisfied by the values listed following it.

17. $x + 5 = -2; \ x = -3, -7$

18. $x - 5 = 2; \ x = -3, 3$

19. $2 - a = 3; \ a = -5, 5, 1$

20. $4 - a = -2; \ a = -6, 2, 6$

21. $5y - 6 = y - 3; \ y = 2, \dfrac{3}{4}$

22. $5y - 2 = 2y + 2; \ y = -2, \dfrac{4}{3}$

23. $6 - 2w = 10 - 3w; \ w = -4, 1$

24. $11 - 5w = -1 - w; \ w = 3, -3$

25. $4(x - 7) - (x + 1) = 15 - x; \ x = \dfrac{2}{5}, 7$

26. $3(x - 6) - (x - 2) = 10 - x; \ x = \dfrac{1}{2}, 6$

27. $3z + 2(z - 1) = 4(z + 2) - (z + 5)$; $z = -2, 1$

28. $5z + 4(z - 2) = 2(z + 6) - (z + 4)$; $z = -3, 2$

29. $x^2 - 3x = 2x - 6$; $x = -2, 2$

30. $x^2 - 5x = 15 - 3x$; $x = 0, -3$

31. $a^2 - 4a = 4 - a$; $a = -1, 4$

32. $a^2 - 5a = 10 - 2a$; $a = -2, 5$

33. $y(y + 6) = (y + 2)^2$; $y = -2, 2$

34. $y(y + 5) = (y + 2)^2$; $y = -4, 4$

ⓠ QUESTIONS FOR THOUGHT

35. What does it mean when we say that a value satisfies or is a solution to an equation?

36. What is the difference between a conditional equation, an identity, and a contradiction?

◇ MINI-REVIEW

Evaluate each of the following:

37. $-5 + 7$

38. $4 - 9$

39. $8 - 11$

40. $-6 + 10$

41. $-8 - 8$

42. $-8(-8)$

43. $9 - 12$

44. $5 - 16$

45. $8 - 5(-3)$

46. $10 + 4(-6)$

47. $-6 + 3(-9)$

48. $-5 - 2(-8)$

3.2 Solving First-Degree Equations in One Variable and Applications

In the previous section we discussed how to check whether a particular number is a solution to a specific equation. The next order of business is to develop a method to actually *find* the solution(s) to conditional equations. As we continue through the text, we will see that the methods we develop depend on the type of equation we are trying to solve. However, in order to develop a systematic method for solving any type of equation we must begin by discussing the basic properties of equalities.

If we think of the "=" sign as indicating that the two sides of the equation are "perfectly balanced," then we will not upset this balance by changing each side of the equation in exactly the same way. If we start with an equation of the form

$$a = b$$

and we

   add the same quantity to both sides of the equation,

or   subtract the same quantity from both sides of the equation,

or   multiply both sides of the equation by the same quantity,

or   divide both sides of the equation by the same (nonzero) quantity,

then the two sides of the equation remain equal. Algebraically we can write this as shown in the box at the top of page 99.

When we write $a = b$ we mean that in any expression where a and b occur they are interchangeable. This idea is known as the **substitution principle.**

Looking carefully at property 1 of equality and applying the substitution principle, we can see that what property 1 actually says is

If $a = b$ then $a + c = b + c$ *Now substitute a for b.*

giving us $a + c = a + c$

Properties of Equality	Addition Property of Equality	1. If $a = b$ then $a + c = b + c$
	Subtraction Property of Equality	2. If $a = b$ then $a - c = b - c$
	Multiplication Property of Equality	3. If $a = b$ then $a \cdot c = b \cdot c$
	Division Property of Equality	4. If $a = b$ then $\dfrac{a}{c} = \dfrac{b}{c}; c \neq 0$

Sometimes this idea is expressed as "If equals are added to equals, the results are equal."

Consider the following four equations:

$$3x - (x - 1) = x + 3$$
$$2x + 1 = x + 3$$
$$x + 1 = 3$$
$$x = 2$$

All of these equations are conditional. Even the equation $x = 2$ is conditional. We are not saying that x is 2 (even though we often read it that way). It is just that it is obvious that the only solution to the equation $x = 2$ is 2.

We can easily check that $x = 2$ is a solution to each of them (verify this). As we shall soon discover, $x = 2$ is the only solution to these four equations. Equations that have exactly the same set of solutions are called **equivalent equations.** Of the four equations, the solution to the last one was the most obvious. An equation of this kind (that is, $x = 2$) is often called an **obvious equation.**

When we are faced with the task of solving an equation, perhaps the most important thing to do first is to ask, "What kind of equation am I trying to solve?" As we shall see time and again, the method of solution will very much depend on the kind of equation we are trying to solve.

For the time being, we are going to restrict ourselves to solving what we refer to as **first-degree equations in one variable,** that is, equations that involve only one variable and in which that variable appears to the first power only. For example, the equation $3x - 5 = x + 3$ is a first-degree equation, while $3x^2 = x + 5$ is called a **second-degree equation.** (The degree of an equation is not always obvious and we will discuss this in a bit more detail later in this section.)

Given an equation to solve, we can apply the four properties of equality that were listed in the previous box to transform the given equation into one for which the solution may be more obvious. Since we are interested in the solution to a given equation, we want to make sure that each time we apply one of the properties of equality we obtain an equivalent equation—that is, an equation with the same solution set.

Applying the four properties of equality appropriately does yield an equivalent equation with one important exception. Property 3 says "If $a = b$ then $ac = bc$." While this *is* always true, if $c = 0$ we do not necessarily obtain an equivalent equation. For example, the equation $x = 3$ is a conditional equation whose only solution is 3. If we multiply both sides of this equation by 0, we get $0 \cdot x = 0 \cdot 3$ or $0 = 0$, which is an identity, and so all numbers are solutions. Consequently, if we want to be sure that we will always obtain an equivalent equation, we must restrict property 3 to have $c \neq 0$.

We can summarize this as shown in the box.

> An equation can be transformed into an equivalent equation by adding or subtracting the same quantity to both sides of the equation, or by multiplying or dividing both sides of the equation by the same *nonzero* quantity.

Recalling that the simplest possible equation is one of the form $x =$ "a number" (which we called the *obvious* equation), we will solve a first-degree equation in one variable by applying the properties of equality to transform the original equation into progressively simpler *equivalent* equations until we eventually end up with an obvious equation. Let's illustrate the method with several examples.

EXAMPLE 1

Solve for x. $x - 5 = 2$

Solution

Since the obvious equation is always of the form "$x =$ number" (or "number $= x$"), our goal in solving an equation is to "isolate x" (or whatever the variable happens to be) on one side of the equation. Since 5 is being subtracted from x on the left-hand side of the equation, we can "undo" the subtraction by adding 5 to both sides of the equation:

$$\begin{array}{rl} x - 5 &= 2 \\ +5 \quad &+5 \\ \hline \boxed{x} &\boxed{= 7} \end{array}$$ *We add 5 to both sides of the equation using the addition property of equality.*

Check $x = 7$: $x - 5 = 2$
$$(7) - 5 \overset{\checkmark}{=} 2$$ ∎

EXAMPLE 2

Solve for x. $x + 7 = 2$

Solution

Before we proceed to the solution a comment is in order. Solving equations involves not only using the properties of equality, but also using the properties in a way that leads to the solution. For instance, if we choose, we may add 3 to both sides of the equation:

$$\begin{array}{rl} x + 7 &= 2 \\ +3 \quad &+3 \\ \hline x + 10 &= 5 \end{array}$$ *By the addition property of equality*

The equation we have obtained, $x + 10 = 5$, is equivalent to the given equation. Adding 3 to both sides of the equation is mathematically correct, but it is *not* particularly useful, since we are no closer to a solution.

Learning to solve equations means learning how to use the properties of equality in a way that leads to a solution. The proper way to solve the equation is

$$\begin{array}{rl} x + 7 &= 2 \\ -7 \quad &-7 \\ \hline \boxed{x = -5} \end{array}$$ *We can think of this as the subtraction property (subtract 7 from both sides) or as the addition property (add −7 to both sides).*

Check $x = -5$: $x + 7 = 2$
$$-5 + 7 \overset{\checkmark}{=} 2$$ ∎

EXAMPLE 3

Solve for x. $24 = 3x$

Solution

In order to isolate x we want to eliminate the 3 that is multiplying the x, so we *divide both sides* of the equation by 3.

$$24 = 3x$$
$$\frac{24}{3} = \frac{3x}{3}$$ *By the division property of equality*

$$\boxed{8 = x}$$ *Note that it does not matter on which side we isolate x.*

Check $x = 8$: $24 = 3x$

$$24 \overset{\checkmark}{=} 3(8)$$

■

Let's analyze the three examples given above to see if we can determine why a particular property was needed in each example in order to achieve our goal of isolating x (the variable) and obtaining the obvious equation.

If we examine	*what is being done to x,*	*then we can determine how to isolate x.*
↓	↓	↓
In $x - 5 = 2$	5 is *subtracted* from x,	so *add* 5 to both sides in order to isolate x.
In $x + 7 = 2$	7 is *added* to x,	so *subtract* 7 from both sides in order to isolate x.
In $24 = 3x$	3 is *multiplying* x,	so *divide* both sides by 3 in order to isolate x.

Note the pattern: We use the *inverse* operation to isolate x. Recall that addition and subtraction are inverse operations, as are multiplication and division.

What if the equation involves more than one operation?

EXAMPLE 4   Solve for a. $3a + 7 = 8$

Solution   In order to isolate a we must deal with the 3 multiplying a and the 7 being subtracted. There is nothing wrong with using the division property first to divide both sides of the equation by 3. However, this means we must divide the *entire* left-hand side by 3 and the *entire* right-hand side by 3. Thus, the entire left-hand side, $3a + 7$, *not* just the $3a$, must be divided by 3. Hence, the equation would look like this:

$$3a + 7 = 8$$

$$\frac{3a + 7}{3} = \frac{8}{3} \qquad \textcolor{blue}{\textit{Dividing both entire sides of the equation by 3}}$$

This equation is equivalent to the given equation but, because it involves fractions, it is a bit more difficult to solve (even if done properly). Whenever we use the division property we run the risk of introducing fractions into the equation. Therefore, while fractions in an equation may be inevitable, as a general procedure it is best to use the addition and subtraction properties first whenever possible.

We proceed as follows. We first isolate the term containing the variable, then we isolate the variable itself.

$$3a + 7 = 8$$

$$\underline{\quad -7 \quad -7 \quad} \qquad \textcolor{blue}{\textit{Subtract 7 from both sides (subtraction property).}}$$

$$3a = 1$$

$$\frac{3a}{3} = \frac{1}{3} \qquad \textcolor{blue}{\textit{Divide both sides by 3 (division property).}}$$

$$\frac{\cancel{3}a}{\cancel{3}} = \frac{1}{3}$$

$$\boxed{a = \frac{1}{3}}$$

Check $a = \frac{1}{3}$: $3a + 7 = 8$

$$3\left(\frac{1}{3}\right) + 7 \overset{?}{=} 8$$

$$1 + 7 \overset{\checkmark}{=} 8 \qquad\blacksquare$$

Sometimes we have to simplify the equation before we proceed to solve it.

EXAMPLE 5

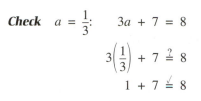

Solve for y. $3(12 - y) - 16 = 2$

Solution

The first two steps in this solution illustrate why we learned how to simplify expressions before solving equations.

Before we actually isolate the variable we want to make sure that each side of the equation is simplified as completely as possible. In this example, the left-hand side is not.

$3(12 - y) - 16 =$	2		*Multiply out using the distributive property.*
$36 - 3y - 16 =$	2		*Combine like terms.*
$20 - 3y =$	2		*Now we proceed to isolate y.*
-20	-20		*Subtract 20 from both sides.*

$$-3y = -18 \qquad \text{\textit{In order to isolate y, we divide both sides by the}}$$
$$\text{\textit{coefficient of y, which is }} -3.$$

$$\frac{-3y}{-3} = \frac{-18}{-3}$$

$$\boxed{y = 6}$$

Check $y = 6$: $3(12 - y) - 16 = 2$

$$3(12 - 6) - 16 \overset{?}{=} 2$$

$$3(6) - 16 \overset{?}{=} 2$$

$$18 - 16 \overset{\checkmark}{=} 2 \qquad\blacksquare$$

One additional comment: It is important that we do not confuse an "expression" with an "equation."

$$\underbrace{3(12 - y) - 16}_{} \qquad \underbrace{3(12 - y) - 16 = 2}_{}$$

This is an expression.   *This is an equation.*

We can simplify the expression $3(12 - y) - 16$ to $20 - 3y$ and that would be our final answer; we are not looking for a value for y. On the other hand, in Example 5 we had an equation that we solved to get $y = 6$. Remember that where possible we "simplify" expressions whereas we "solve" equations. If you recognize and use the appropriate language you are more likely to follow the appropriate procedures.

One of our main goals is to develop the ability to apply our algebraic skills to real-life situations, often presented as verbal problems. Now that we have learned how to solve some equations, we can combine this skill with our experience in translating phrases and sentences algebraically to illustrate some applications. (You may want to review Section 2.5 before proceeding.)

Before we proceed any further, it is worth mentioning that you may be able to solve some of the problems that follow by "playing around with the numbers." This may be a valid way of getting the answer to a specific problem, *but* it is a very risky strategy for problem solving. What if the numbers are complicated, the trial-and-error procedure is very long, or the problem has no answer? Consequently, we will not consider the trial-and-error method as an acceptable method of solution for our purposes. Most applications of mathematics in the "real" world involve translating a particular problem into mathematical language, getting an equation out of this translation, and solving it.

Let's begin by looking at several examples.

EXAMPLE 6

The length of a rectangular garden is five less than twice its width. If the perimeter of the garden is 83 ft, find the dimensions of the rectangle.

Solution

In order to solve this problem we need to know exactly what a rectangle looks like, what perimeter means, and how to compute it.

A diagram is particularly useful in a problem of this kind. A rectangle is a four-sided figure in which the opposite sides are equal and the sides meet at right angles (see Figure 3.1). We call two sides the length, labeled L, and two sides the width, labeled W. The perimeter, P, of a geometric figure means the distance around it. Thus, for a rectangle,

$$P = L + W + L + W, \quad \text{or in simplified form,} \quad P = 2L + 2W$$

Figure 3.1

Rectangle

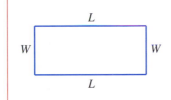

In this example we are told that the length is five less than twice the width. Since the length is described in terms of the width, we can let

$$W = \text{Width of the rectangle}$$

Then

$$2W - 5 = \text{Length of the rectangle} \quad \text{(length is 5 less than twice the width)}$$

Alternatively, we could simply draw a diagram and label it as shown in Figure 3.2.

Figure 3.2

Length described in terms of width

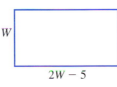

The perimeter being 83 ft gives us the equation

$$2W \quad + \quad 2(2W - 5) \quad = \quad 83 \quad *$$

Twice the width + Twice the length = Perimeter

Now we proceed to solve this equation:

$$2W + 2(2W - 5) = 83$$
$$2W + 4W - 10 = 83$$
$$6W - 10 = 83$$
$$\underline{+10 \quad +10}$$
$$6W = 93$$
$$\frac{6W}{6} = \frac{93}{6} \qquad \textit{Reduce to lowest terms.}$$
$$W = \frac{31}{2} = 15.5$$

If the width is 15.5 ft, then the length is

$$2W - 5 = 2(15.5) - 5 = 31 - 5 = 26 \text{ feet}$$

---

*We will indicate the equation we obtain from each problem by enclosing it in a rectangular shaded panel in this way.

Check: Is the length 5 less than twice the width? Is 26 equal to 5 less than twice 15.5? Yes. Is the perimeter 83 ft? Is $2(15.5) + 2(26) \overset{?}{=} 83$? Yes.

Thus, the answer to the problem is: The dimensions of the rectangle are

$$W = 15.5 \text{ feet}; \quad L = 26 \text{ feet}$$ ■

EXAMPLE 7   The sum of two numbers is 36. If the second number is four more than three times the first, find the numbers.

Solution   Since the algebraic skills we have learned thus far limit us to solving equations involving only one variable, we shall try to translate all the relevant information in the problem in terms of one variable.

We are seeking two numbers with the properties described in the example. To make it easier to talk about the two numbers, let's refer to them as the "first number" and the "second number."

Let x = the first number.

The problem tells us that the "second number" is "four more than three times" the first number. Therefore, we can represent the second number as

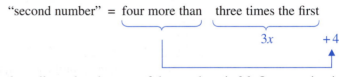

"second number" = four more than   three times the first
$3x$ $+4$

The problem also tells us that the sum of the numbers is 36. Our equation is

"first number" + "second number"   is equal to   36

or written algebraically,

$$x \quad + \quad 3x + 4 \quad = \quad 36$$

Now we proceed to solve the equation.

$$
\begin{aligned}
x + 3x + 4 &= 36 \qquad \textit{Combine like terms.} \\
4x + 4 &= 36 \\
\underline{-4 \quad\; -4}& \\
4x &= 32 \\
\frac{4x}{4} &= \frac{32}{4} \\
x &= 8
\end{aligned}
$$

Thus, the first number is 8. The second number is

$$3x + 4 = 3(8) + 4 = 24 + 4 = 28$$

The *answer* to the problem is the pair of numbers

$$\boxed{8 \text{ and } 28}$$

In order to check this answer it is not sufficient to check the answer in the equation since we may have formulated our equation incorrectly. In other words, we may have the right answer to a wrong equation. Instead, we must check our answer in the original words of the problem.

Check: Is 28 equal to 4 more than three times 8? Yes. Is the sum of 8 and 28 equal to 36? Yes. Our answer checks. ■

EXAMPLE 8

Monica works as a salesperson. She earns a base salary of $125 plus a commission of $2.40 for each item she sells each week.

(a) Determine her weekly salary if she sells 200 items.

(b) Determine the number of items she sold during a week in which her income was $528.20.

Solution   In order to answer part **(a)** we can simply compute her commission on 200 items and add her base salary to obtain her weekly income. However, the two parts of this example highlight the difference between substituting into an equation versus solving an equation. Consequently we postpone the solutions to both parts of the example until after we obtain an equation relating Monica's weekly income to the number of items she sells.

Let's begin with our description of how we compute Monica's salary. We can write this relationship as follows:

$$\text{Commission} + \text{Salary} = \text{Weekly income}$$

$$(\text{Number of items sold})(2.40) + \quad 125 \quad = \text{Weekly income}$$

Let's calculate Monica's salary for different numbers of items sold and record our information in a table with the goal of finding a general formula for the weekly income, which we will call I. See Table 3.1.

Again we are looking for a pattern.

Table 3.1

Number of Items Sold	Commission	Salary	Weekly Income
50	$50(2.40) = \$120$	$125	$I = 120 + 125 = \$245$
92	$92(2.40) = \$220.80$	$125	$I = 220.80 + 125 = \$345.80$
150	$150(2.40) = \$360$	$125	$I = 360 + 125 = \$485$
n	$n(2.40) = \$2.40n$	$125	$I = 2.40n + 125$

The last entry in the table, $I = 2.40n + 125$, is exactly what we are looking for. It is an equation that relates Monica's weekly income (I) to the number of items sold (n).

Such an equation that expresses how two quantities are related is often called a *mathematical model* of the relationship.

Let's now return to the questions asked in this example.

(a) As we mentioned previously, in order to compute the weekly income if she sells 200 items, we can simply compute her commission on the 200 items and then add her base salary. This is exactly equivalent to substituting $n = 200$ into the equation we just derived for I.

$$I = 2.40n + 125 \qquad \textit{Substitute } n = 200.$$

$$I = 2.40(200) + 125 = \boxed{\$605}$$

(b) Here we are asked to find the number of items she must sell to earn a particular weekly income. One possible way we could approach the problem is to compute Monica's weekly income if she sells 100 items, 125 items, 140 items, and so on until we (hopefully) find a weekly income of $528.20 as required. This trial-and-error process can be quite time-consuming.

Think about how the trial-and-error process would work if there were no number of items that yield a particular weekly income.

However, in terms of the equation (formula) we have for I, we are looking for the value of n that makes $I = 528.20$. We do this by substituting for I and *solving* the equation for n.

$$I = 2.40n + 125 \qquad \textit{Substitute } 528.20 \textit{ for } I.$$

$$528.20 = 2.40n + 125 \qquad \textit{We solve for n. Subtract } 125 \textit{ from both sides.}$$

$$\underline{-125 \qquad\qquad\qquad -125}$$

$$403.20 = 2.40n \qquad \textit{Divide both sides by } 2.40.$$

$$\frac{403.20}{2.40} = \frac{2.40n}{2.40}$$

$$\boxed{168 = n}$$

Thus, Monica sold 168 items in order to earn a weekly income of $528.20. Note that Table 3.1 tells us that the sale of 150 items yields a weekly income of $485, so that our answer of 168 items to produce a weekly income of approximately $528 seems reasonable.

It is always a good idea to think about the reasonableness of an answer. If an answer seems to be unreasonable, that is a signal to check over your work carefully and, one hopes, to find the error.

Finally, notice the difference in how the equation we derived for I is used in parts **(a)** and **(b)**. In part **(a)** we simply substitute and compute, whereas in part **(b)** we still need to solve the resulting equation after we substitute. We will have more to say about this distinction as we proceed to other types of equations. ∎

In the next section we will continue our discussion of solving equations and look at more applications with the goal of generalizing our solutions into a strategy outline for solving verbal problems.

Study Skills 3.2

Preparing for Exams: When to Study

When to start studying and how to distribute your time studying for exams is as important as how to study. To begin with, "pulling all-nighters" (staying up all night to study just prior to an exam) seldom works. As with athletic or musical skills, algebraic skills cannot be developed overnight. In addition, without an adequate amount of rest, you will not have the clear head you need to work on an algebra exam. It is usually best to start studying early—from $1\frac{1}{2}$ to 2 weeks before the exam. In this way you have the time to perfect your skills and, if you run into a problem, you can consult your teacher to get an answer in time to include it as part of your studying.

It is also a good idea to distribute your study sessions over a period of time. That is, instead of putting in 6 hours in one day and none the next 2 days, put in 2 hours each day over the 3 days. You will find that not only will your studying be less boring, but also you will retain more with less effort.

As we mentioned before, your study activity should be varied during a study session. It is also a good idea to take short breaks and relax. A study "hour" could consist of about 50 minutes of studying and a 10-minute break.

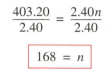 **EXERCISES 3.2**

Solve each of the following equations. If the equation is always true or has no solutions, indicate so.

1. $x + 3 = 8$
2. $x + 5 = 11$
3. $y - 4 = 7$
4. $y - 6 = 3$
5. $a + 3 = 1$
6. $a + 5 = 2$
7. $a - 5 = -8$
8. $a - 3 = -7$
9. $3x = 21$
10. $8x = 48$
11. $4x = 15$
12. $2x = 37$
13. $-4x = -12$
14. $-9x = -36$
15. $3x + x = 8 - 12$
16. $x + 2x = 1 - 10$
17. $4x - x = 2 - 8$
18. $6x - x = 4 - 14$

19. $2z - 3z - 11z = -4(6)$

20. $4z - z - 9z = -8(3)$

21. $3w - 7w + 4w = 8 - 3$

22. $9w - 3w - 6w = 4 - 9$

23. $2x + 1 = 7$

24. $5x - 2 = 8$

25. $10x - 4 = 22$

26. $6x + 3 = 35$

27. $2t = 3t + 5$

28. $3z = 4z - 6$

29. $9 = 6 - 3a$

30. $15 = 7 - 4b$

31. $20 = 3w - 1$

32. $37 = 13r - 2$

In Exercises 33–38, if the exercise is an expression, simplify it; if it is an equation, solve it.

33. $8 - 3(x - 4)$

34. $8 - 3(x - 4) = 14$

35. $4t - 2(t - 3) = 12$

36. $4t - 2(t - 3)$

37. $3a + 5(2a + 4) + a$

38. $34 = 3a + 5(2a + 4) + a$

In Exercises 39–58, solve each of the problems algebraically. That is, set up an equation and solve it. Be sure to clearly label what the variable represents. Round your answer to the nearest tenth where necessary.

39. Find the width of a rectangle whose length is 6 in. and whose area is 40 sq in.

40. Find the length of a rectangle whose width is 4 ft and whose area is 22 sq ft.

41. Find the length of a rectangle whose width is 3.5 yd and whose area is 73.5 sq yd.

42. Find the width of a rectangle whose length is 15 cm and whose area is 85.5 sq cm.

43. How long will it take someone driving at 52 mph to travel 234 miles?

44. What average rate of speed must a car maintain if it is to travel 729 miles in 15 hours?

45. A car rental company charges $29.95 per day and 17¢ per mile. If a 3-day car rental cost $123.17, how many miles were driven?

46. A car rental company has a daily charge plus a mileage charge of 14¢ per mile. During a 5-day car rental, 340 miles were driven and the total bill was $172.55. What is the daily rental charge?

47. A trucking company determines that the cost C (in dollars per mile) of operating a truck is given by $C = 0.003s + 0.21$ where s is the average speed of the truck.

    (a) Find the cost per mile if the truck averages 55 miles per hour.

    (b) Find the average speed that yields a cost per mile of $0.35.

48. Anthropologists often extrapolate the appearance of a human being from the parts of a skeleton they uncover. The following mathematical model relates the height h (in centimeters) of the individual to the length f (in centimeters) of the femur (thigh bone):

$$h = 3.68f + 21.44$$

    (a) Use this model to find the height of an individual whose femur measures 35 cm.

    (b) Use this model to find the length of a femur corresponding to an individual who is 180 cm tall.

49. Nature experts claim that the number C of times that a cricket chirps per minute is related to the Fahrenheit temperature T according to the model

$$C = 4T - 160$$

(a) Use this model to determine the temperature at which crickets will chirp 150 times per minute.

(b) Use this model to determine how many times per minute a cricket will chirp at a temperature of 90°F.

50. Metals expand when they are heated. Suppose that the length L (in centimeters) of a particular metal bar varies with the Celsius temperature T according to the model

$$L = 0.009T + 5.82$$

(a) Use this model to determine the temperature at which the bar will be 6.5 cm long.

(b) Use this model to determine the length of the bar at a temperature of 120°C.

51. As dry air rises, it expands due to the lower atmospheric pressure, and as a result the air cools at the rate of approximately 5.4°F for each 1,000 feet of increase in altitude (up to an altitude of approximately 40,000 ft). Thus, if the ground-level temperature is 46°F, the temperature at an altitude of A feet above the ground is given by the equation

$$T = 46 - 0.0054A$$

(a) Determine the temperature at an altitude of 5,000 feet.

(b) Determine the altitude at which the temperature will be 32°F.

52. One of the various models that is proposed for the proper weight W (in pounds) of a man h inches tall is

$$W = 5.7h - 228$$

(a) According to this model, what is the proper weight for a person 6 feet tall?

(b) According to this model, how tall is a person whose proper weight is 200 pounds?

53. A jogger's heart rate N (in beats per minute) increases as his or her speed s (in feet per second) increases. A physiologist collects data on a particular jogger and finds that the equation $N = 1.67s + 55$ seems to fit the data quite well. According to this model, what speed would produce a heart rate of 85 beats per minute?

54. A business buys a piece of machinery for $32,800. For tax purposes, the business depreciates the value V of the machine after y years according to the formula $V = 32,800 - 1,850y$. After how many years is the machine worth $20,000?

55. The length of a rectangle is 7 more than twice its width. If the perimeter of the rectangle is 50 cm, find the dimensions of the rectangle.

56. The width of a rectangle is 8 less than three times its length. If the perimeter of the rectangle is 24 in., find the dimensions of the rectangle.

57. The width of a rectangle is 10 less than five times the length. If the perimeter of the rectangle is 100 yd, find the dimensions of the rectangle.

58. The length of a rectangle is 12 more than four times its width. If the perimeter of the rectangle is 134 meters, find the dimensions for the rectangle.

? QUESTIONS FOR THOUGHT

59. Summarize the properties of equality in one sentence.

60. When are two equations equivalent?

61. Separate the following equations into groups of equivalent equations.

$x + 1 = 16$	$2x = 18$	$2x = 10$	$x = 15$
$x = -1$	$x - 3 = 12$	$x + 2 = 11$	$2x = 30$
$x + 7 = 6$	$x = 9$	$5x = 25$	$x - 2 = 7$

62. How do you check your answer after you have solved an equation?

63. What is **wrong** (if anything) with each of the following and why?
Solve for x.

(a)
$$2x + 4 = 8$$
$$\underline{-4 \quad -4}$$
$$2x = 12$$
$$\frac{2x}{2} = \frac{12}{2}$$
$$x = 6$$

(b)
$$3x - 2 = 19$$
$$\underline{-3 \qquad -3}$$
$$x - 2 = 16$$
$$x - 2 = 16$$
$$\underline{+2 \quad +2}$$
$$x = 18$$

(c)
$$3x - 6 = 12$$
$$\frac{3x - 6}{3} = \frac{12}{3}$$
$$x - 6 = 4$$
$$\underline{+6 \quad +6}$$
$$x = 10$$

(d)
$$5x = 2x$$
$$\frac{5x}{x} = \frac{2x}{x}$$
$$5 = 2$$
contradiction

(e)
$$2x = 5x - 6$$
$$\underline{-5x \quad -5x}$$
$$-3x = -6$$
$$\frac{-3x}{3} = \frac{-6}{3}$$
$$x = -2$$

(f)
$$2x = 5x + 6$$
$$\underline{-5x \quad -5x}$$
$$-3x = 6$$
$$\frac{-3x}{3} = \frac{6}{3}$$
$$-x = 2$$

(g)
$$7x = 4x$$
$$\underline{-4x \quad -4x}$$
$$3x = 0$$
$$\frac{3x}{3} = \frac{0}{3}$$
$$x = 0$$

◇ **MINI-REVIEW**

In Exercises 64–69, evaluate the given expression.

64. $|5 - 9| - |-3 - 8|$

65. $\dfrac{-15 - 5}{-2(-5)(-2)}$

66. $\dfrac{-4(-2)(-6)}{-4(-2) - 6}$

67. $7 - 5[7 - 5(7 - 5)]$

68. Write 312 as a product of prime factors.

69. Write 83 as a product of prime factors.

3.3 More First-Degree Equations and Applications

In this section we continue to develop our ability to solve first-degree equations and verbal problems. We will examine several somewhat more complicated equations and applications with the goal of establishing procedural outlines that we can use to solve such problems.

In the previous section all the equations we solved had the variable appearing on one side of the equation only. However, frequently the variable may appear on both sides of the equation. The next example illustrates how we handle this type of situation.

EXAMPLE 1 | Solve for x. $18 - 7x = 3 - x$

Solution | We want to isolate x on one side of the equation. As illustrated in the two solutions that follow, which side we decide to isolate x on does not matter.

Solution 1:

$$18 - 7x = 3 - x$$
$$\underline{+7x \qquad +7x}$$
$$18 = 3 + 6x$$
$$\underline{-3 \quad -3}$$
$$15 = 6x$$

If we decide to isolate x on the right-hand side, we add 7x to both sides of the equation, which eliminates x from the left-hand side.

$$\frac{15}{6} = \frac{6x}{6}$$

$$\frac{15}{6} = x \qquad \textit{We reduce the fraction.}$$

$$\boxed{\frac{5}{2} = x}$$

*In algebra, the equations
$x = 2$ and $2 = x$ say
exactly the same thing. In
most cases, since we usually
want to focus our attention
on the value of the variable,
we read both equations as
x equals 2.*

Solution 2:

$$18 - 7x = 3 - x \qquad \textit{If we decide to isolate x on the left-hand side, we}$$
$$\underline{\quad +x \qquad\quad +x} \qquad \textit{add x to both sides of the equation, which eliminates}$$
$$18 - 6x = \quad 3 \qquad \textit{x from the right-hand side.}$$
$$\underline{-18 \qquad\qquad -18}$$
$$-6x = -15$$

$$\frac{-6x}{-6} = \frac{-15}{-6} \qquad \textit{A negative divided by a negative is positive.}$$

$$x = \frac{15}{6} \qquad \textit{We reduce the fraction.}$$

$$\boxed{x = \frac{5}{2}} \qquad \textit{Thus, both approaches yield the same solution.}$$

Check $x = \frac{5}{2}$: $\qquad 18 - 7x = 3 - x$

$$18 - 7\left(\frac{5}{2}\right) \overset{?}{=} 3 - \frac{5}{2}$$

$$\frac{18}{1} - \frac{7}{1} \cdot \frac{5}{2} \overset{?}{=} \frac{3}{1} - \frac{5}{2}$$

$$\frac{36}{2} - \frac{35}{2} \overset{?}{=} \frac{6}{2} - \frac{5}{2}$$

$$\frac{1}{2} \overset{\checkmark}{=} \frac{1}{2} \qquad\qquad\qquad\qquad\qquad ■$$

Sometimes we need to simplify the equation before we actually solve it, as illustrated in the next example.

EXAMPLE 2 | Solve for a. $6a - 22 - 3(a - 4) = 20$

Solution | In order to isolate a on one side of the equation we must first simplify the left-hand side of the equation.

$$6a - 22 - 3(a - 4) = 20 \qquad \textit{We first remove the parentheses by distributing }-3.$$
$$6a - 22 - 3a + 12 = 20 \qquad \textit{Combine like terms.}$$
$$3a - 10 = 20 \qquad \textit{Now solve; add 10 to both sides.}$$
$$\underline{\quad +10 \quad +10}$$
$$3a = 30$$
$$\frac{3a}{3} = \frac{30}{3}$$
$$\boxed{a = 10}$$

Check $\qquad 6a - 22 - 3(a - 4) = 20 \qquad \textit{We substitute } a = 10.$

$$6(10) - 22 - 3(10 - 4) \overset{?}{=} 20$$

$$60 - 22 - 3(6) \stackrel{?}{=} 20$$
$$60 - 22 - 18 \stackrel{\checkmark}{=} 20 \qquad \blacksquare$$

Before proceeding to several more examples, let's pause to organize an outline of the strategy we have used to solve first-degree equations.

Strategy for Solving First-Degree Equations	1. Remove any grouping symbols by using the distributive property.
	2. Simplify both sides of the equation as completely as possible.
	3. By using the addition and/or subtraction properties of equality, isolate the variable term on one side of the equation and the numerical term on the other side.
	4. By using the division property of equality, make the coefficient of the variable equal to 1 and, in so doing, obtain an obvious equation.
	5. Check the solution in the original equation.

EXAMPLE 3

Solve for t. $2t - 5(t - 2) = 20 - (2t + 6)$

Solution

Following the outline given in the box, we first remove the parentheses in the equation by using the distributive property.

$$2t - 5(t - 2) = 20 - (2t + 6) \qquad \textit{Remove parentheses.}$$
$$2t - 5t + 10 = 20 - 2t - 6 \qquad \textit{Combine like terms on both sides.}$$
$$-3t + 10 = 14 - 2t$$
$$\underline{+2t \qquad\qquad +2t} \qquad \textit{Eliminate t from the right-hand side.}$$
$$-t + 10 = 14$$
$$\underline{-10 \quad -10}$$
$$-t = 4 \qquad \textit{We are not finished yet. We want t alone,}$$
$$\frac{-t}{-1} = \frac{4}{-1} \qquad \textit{so we divide both sides of the equation by the coefficient of t, which is } -1.$$
$$\boxed{t = -4}$$

Check $t = -4$:
$$2t - 5(t - 2) = 20 - (2t + 6)$$
$$2(-4) - 5(-4 - 2) \stackrel{?}{=} 20 - (2(-4) + 6)$$
$$-8 - 5(-6) \stackrel{?}{=} 20 - (-8 + 6)$$
$$-8 + 30 \stackrel{?}{=} 20 - (-2)$$
$$22 \stackrel{\checkmark}{=} 20 + 2 \qquad \blacksquare$$

As we mentioned earlier, the technique we are using here of isolating x applies to first-degree equations—that is, to equations involving the variable to the first power only. The fact that an equation is of the first degree is not always obvious, as we see in the next example.

EXAMPLE 4

Solve for x. $x(x + 5) - 2(x - 1) = x^2 + 2$

Solution

This equation may not appear to be a first-degree equation (because of the x^2 term), but in fact we shall see that it is. This example also highlights why it is important to simplify equations first.

$$x(x + 5) - 2(x - 1) = x^2 + 2 \qquad \textit{Remove parentheses.}$$

$$x^2 + 5x - 2x + 2 = x^2 + 2 \qquad \textit{Combine like terms on the left-hand side.}$$

$$x^2 + 3x + 2 = x^2 + 2 \qquad \textit{Subtract } x^2 \textit{ from both sides.}$$

$$\underline{-x^2 \qquad\qquad -x^2}$$

$$3x + 2 = 2 \qquad \textit{Now we see that, in fact, we do have}$$
$$\textit{a first-degree equation.}$$

$$\underline{-2 \quad -2}$$

$$3x = 0 \qquad \textit{Do not treat 0 any differently.}$$

$$\frac{3x}{3} = \frac{0}{3}$$

$$\boxed{x = 0}$$

Check $x = 0$: $x(x + 5) - 2(x - 1) = x^2 + 2$

$$0(0 + 5) - 2(0 - 1) \stackrel{?}{=} 0^2 + 2$$

$$0(5) - 2(-1) \stackrel{?}{=} 0 + 2$$

$$0 + 2 \stackrel{\checkmark}{=} 2 \qquad\qquad\blacksquare$$

To determine the degree of an equation, isolate all the variable terms on one side of the equation and simplify; the highest exponent of the variable is the degree of the equation. If no variable appears at all, then the original equation is equivalent to either an *identity* or a *contradiction*. Examples 5 and 6 illustrate these possibilities.

EXAMPLE 5   Solve for a. $2(a + 1) - 3(a - 1) = 5 - a$

Solution
$$2(a + 1) - 3(a - 1) = 5 - a \qquad \textit{Remove parentheses.}$$
$$2a + 2 - 3a + 3 = 5 - a \qquad \textit{Combine like terms.}$$
$$-a + 5 = 5 - a$$
$$\underline{+a \qquad\qquad +a}$$
$$5 = 5$$

Since $5 = 5$ is *always* true and is equivalent to the original equation, the original equation is an identity. In fact, you might have already noticed that we have an identity in the third line of our solution because the expressions $-a + 5$ and $5 - a$ are always equal regardless of the value of a.

In solving an equation that is in fact an identity, the variable drops out entirely, and we get an equation of the form "a number is equal to itself," which is always true.

$$\boxed{\text{All values of } a \text{ make the original equation true.}} \qquad\qquad\blacksquare$$

EXAMPLE 6   Solve for y. $2 - (y - 4) + 3y = 2y + 1$

Solution
$$2 - (y - 4) + 3y = 2y + 1 \qquad \textit{Remove the parentheses.}$$
$$2 - y + 4 + 3y = 2y + 1 \qquad \textit{Combine like terms.}$$
$$6 + 2y = 2y + 1$$
$$\underline{-2y \quad -2y}$$
$$6 = 1$$

Since $6 = 1$ is always false and is equivalent to the original equation, the original equation is a contradiction. In solving an equation that is in fact a contradiction, the variable drops out entirely, and we get an equation that tries to assert that two unequal numbers are equal, which is always false.

$$\boxed{\text{There are no solutions to the original equation.}} \qquad\qquad\blacksquare$$

Based on the examples we have done we can state that if a first-degree equation is conditional, then it has a unique solution, which we find by the methods outlined above. Otherwise, it is an identity or a contradiction.

Now that we have developed the ability to solve more complex equations, we can handle somewhat more complicated applications.

EXAMPLE 7

A wooden board, 20 meters long, is cut into three pieces. The medium piece is twice as long as the shortest piece, and the longest piece is 5 meters longer than the medium piece. Find the lengths of the three pieces.

Solution

Recall that an algebraic solution requires us to write and solve an equation to answer the question. Let's analyze this problem carefully.

THINKING OUT LOUD

What do we need to find?

The length of each piece

Where do we start?

When we write an equation, we are saying that two quantities are equal. In order to identify such quantities, it is frequently helpful to draw a simple diagram to represent the given information. See Figure 3.3(a).

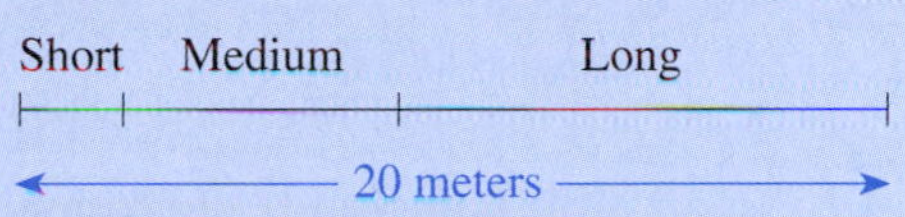

Figure 3.3(a)
Diagram for Example 7

The diagram helps us see that when we add the lengths of the three pieces the total length is 20 meters.

What is our equation going to say?

Based on the diagram we have

Length of short piece + Length of medium piece + Length of longest piece = 20

What information are we given?

The example tells us that the medium piece is twice the length of the shortest piece, and the longest piece is 5 meters longer than the medium piece.

How do we label the pieces?

Since the length of the medium piece is described in terms of the length of the shortest piece, it seems reasonable to begin as follows:

Let x = Length of the shortest piece.

$2x$ = Length of the medium piece. (The example tells us that the medium piece is twice the length of the shortest piece.)

$2x + 5$ = Length of the longest piece. (The example tells us that the longest piece is 5 meters longer than the medium piece.)

We can now redraw the figure as shown in Figure 3.3(b).

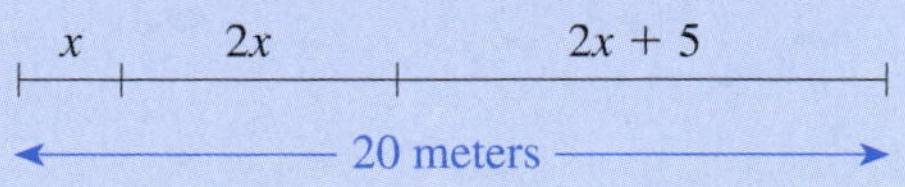

Figure 3.3(b)

Therefore, since the lengths of the three pieces of the board add up to 20 meters, the equation becomes

$$x + 2x + 2x + 5 = 20$$

Now we solve this equation.

$x + 2x + 2x + 5 = 20$	*Combine like terms.*
$5x + 5 = 20$	*Subtract 5 from both sides.*
$5x = 15$	*Divide both sides by 5.*
$x = 3$	

Thus,   The length of the shortest piece is $x =$ $\boxed{3 \text{ meters}}$.

The length of the medium piece is $2x =$ $\boxed{6 \text{ meters}}$.

The length of the longest piece is $2x + 5 = 2(3) + 5 =$ $\boxed{11 \text{ meters}}$.

The check is left to the student.   ■

EXAMPLE 8

A student collected $116 on the sale of 40 tickets to a school play. If the ticket prices were $5 for adults and $2 for students, how many of each type were purchased?

Solution

In formulating a solution to this problem, it is important to recognize the difference between how many tickets were sold and how much they are worth, or their *value*. For example, if 30 student tickets and 10 adult tickets were sold, then we have 40 tickets. However, to compute how much money was collected for these tickets (that is, the total value of the tickets) we must consider the value of each type of ticket.

The value of 30 student tickets is $30(2) = \$60$

The value of 10 adult tickets is $10(5) = \$50$

The total *number* of tickets is 40   **but**   the total *value* of the tickets is $110.

We multiply the number of student tickets by the value of one student ticket ($2) to get the value of all the student tickets. Similarly, we multiply the number of adult tickets by the value of one adult ticket ($5) to get the value of all the adult tickets. The total value of all the tickets is the *sum* of the value of the student tickets plus the value of the adult tickets.

In exactly the same way, the value of 40 tickets consisting of 15 student tickets and 25 adult tickets is

$$15(2) + 25(5) = 30 + 125 = \$155$$

In order to solve this problem algebraically we need to write an equation. An equation is a statement of equality. Therefore, we must identify some relationship in this example that allows us to write an equation.

Based on our numerical analysis of several possible scenarios of 40 student and adult tickets, we recognize the following relationship.

Value of the student tickets	+	Value of the adult tickets	= Total value

$$\left(\begin{array}{c}\text{Number of}\\\text{student tickets}\end{array}\right) \cdot \left(\begin{array}{c}\text{Value of 1}\\\text{student ticket}\end{array}\right) + \left(\begin{array}{c}\text{Number of}\\\text{adult tickets}\end{array}\right) \cdot \left(\begin{array}{c}\text{Value of 1}\\\text{adult ticket}\end{array}\right) = \text{Total value}$$

$$\left(\begin{array}{c}\text{Number of}\\\text{student tickets}\end{array}\right) \cdot (2) + \left(\begin{array}{c}\text{Number of}\\\text{adult tickets}\end{array}\right) \cdot (5) = 116 \qquad (*)$$

Looking at this last "equation," we see that all that remains to be done is to label the number of student tickets and the number of adult tickets using *one* variable. It may seem that there are two unknown quantities, but in reality there is only one. As we saw previously, if it turns out that there are 30 student tickets then, since there are 40 tickets in all, there *would have to be* 40 – 30 = 10 adult tickets. Similarly, if there are 15 student tickets then there must be 40 – 15 = 25 adult tickets. Consequently, if we

$$\text{Let } x = \text{Number of student tickets sold}$$

then

$40 - x = $ Number of adult tickets sold   *Since there are 40 tickets all together*

Now that we have expressed the number of student tickets and adult tickets in terms of one variable, we can substitute into the equation we marked (∗), giving us the equation

(Number of student tickets) · (2) + (Number of adult tickets) · (5) = 116   (∗)

$\quad\quad (x) \cdot (2) \quad\quad\quad + \quad\quad (40 - x) \cdot (5) \quad\quad = 116$

Thus, the equation we will use to solve this example is

$$2x + 5(40 - x) = 116$$

Now we solve this equation.

$$2x + 5(40 - x) = 116$$
$$2x + 200 - 5x = 116$$
$$-3x + 200 = 116$$
$$-200 = -200$$
$$-3x = -84$$
$$\frac{-3x}{-3} = \frac{-84}{-3}$$
$$x = 28$$

Thus, there were $\boxed{28 \text{ student tickets}}$ and 40 – 28, or $\boxed{12 \text{ adult tickets}}$.

Check: 28 student tickets and 12 adult tickets gives us 40 tickets all together, as required.

$$28 \text{ student tickets are worth } 28(2) = \$56$$
$$12 \text{ adult tickets are worth } 12(5) = \$60$$
$$\text{Total value} = \$116$$

Two points are worth mentioning. First, we did not have to let x be the number of student tickets. We could just as well have let x be the number of adult tickets. Solving the equation would then have yielded $x = 12$ instead of 28, but the answer to the problem would have still turned out to be 28 student tickets and 12 adult tickets. In some cases the way a problem is worded suggests the choice of what the variable should represent (see Example 7), and the "proper" choice may make the problem a bit easier. Only experience will teach you how to make the best choice.

Second, since we knew that there were 40 tickets all together, once we labeled the number of one type of ticket as x, then the number of the other type had to be $40 - x$. This idea comes up fairly frequently. If we have two quantities and we know their total and we represent the number of one of the unknown quantities by x, then the other quantity must be *total* – x. ∎

Analyzing the solutions to the verbal problems we have solved thus far leads us to the following general outline for solving verbal problems.

Outline of Strategy for Solving Verbal Problems	1. Read the problem carefully, as many times as is necessary to understand what the problem is saying and what it is asking.
	2. Use diagrams whenever you think it will make the given information clearer.
	3. Ask whether there is some underlying relationship or formula you need to know. If not, then the words of the problem themselves give the required relationship.
	4. Clearly identify the unknown quantity (or quantities) in the problem, and label it (them) using one variable.
	Step 4 is very important, and not always easy.
	5. By using the underlying formula or relationship in the problem, write an equation involving the unknown quantity (or quantities).
	Step 5 is the *crucial step.*
	6. Solve the equation.
	7. Make sure you have answered the question that was asked.
	8. Check the answer(s) in the original words of the problem.
	Keep the reasonableness of your answer in mind. Did you obtain a negative weight or length? Is your answer for the time or distance much too large or much too small?

Being asked to solve a problem *algebraically* means to apply this outline (or some variation of it) to the problem, as opposed to a trial-and-error procedure. Do not be surprised if it takes a while to solve a problem, especially if it is a type you have not seen before.

One other word of advice: Take the time to look over your work and think about what you have done. The more time you spend thinking about the problem and the relationships you have uncovered, the easier subsequent problems will be. If you can learn to apply this outline to relatively simple problems, it is more likely that you will be able to apply it to more complicated ones as well.

EXAMPLE 9   A carpenter charges $36 per hour for her time and $22 per hour for her assistant's time. On a certain kitchen remodeling job, the assistant works by himself doing the preparatory work and is then joined by the carpenter so that they finish the job working together. If the assistant worked 11 hours more than the carpenter and the total labor charge for the job was $1,228, how many hours did each of them work?

Solution   Having analyzed the previous example very carefully, we can apply our experience to this example. As before, we begin with what we think is the underlying equality relationship in this example:

Wages paid to carpenter + Wages paid to assistant = Total labor charge

We recognize that in order to compute each person's wages we multiply the number of hours each person worked by his or her hourly pay rate. Therefore, the last equation becomes

(Number of hours carpenter works) · *(Hourly rate for carpenter)*
    + *(Number of hours assistant works)* · *(Hourly rate for assistant)*
                                                        = *Total labor charge*

(Number of hours carpenter works) · (36)
    + *(Number of hours assistant works)* · (22)   = $1,228

Looking at this last "equation," we see that all that remains to be done is to label the number of hours that each person worked.

Let *x* = Number of hours the carpenter works

then

$x + 11$ = Number of hours the assistant works

The assistant worked for 11 *hours before being joined by the carpenter. Therefore, the assistant worked* 11 *hours more than the carpenter.*

Therefore, the equation becomes

$$\left(\begin{array}{c} \textit{Number of hours} \\ \textit{carpenter works} \end{array}\right) \cdot (36) + \left(\begin{array}{c} \textit{Number of hours} \\ \textit{assistant works} \end{array}\right) \cdot (22) = \$1{,}228$$

$$(x) \qquad \cdot (36) + \qquad (x + 11) \qquad \cdot (22) = \$1{,}228$$

We now solve this equation.

$$36x + 22(x + 11) = 1{,}228$$
$$36x + 22x + 242 = 1{,}228$$
$$58x + 242 = 1{,}228$$
$$\underline{ -242 \qquad -242}$$
$$58x = 986$$
$$\frac{58x}{58} = \frac{986}{58}$$
$$x = 17$$

Therefore,

the carpenter worked for 17 hours

and since $17 + 11 = 28$,

her assistant worked for 28 hours

The check is left to the student. ■

EXAMPLE 10

Susan jogs to the post office at the rate of 12 kilometers per hour (kph) and walks home at the rate of 6 kph. If her total time, jogging and walking, is 3 hours, how far is the post office from her home?

Solution

Let's begin by drawing a little diagram to help us visualize the problem.

HOME ⇄ POST OFFICE
Jogging →
← Walking

The diagram emphasizes for us the fact that the jogging distance is the same as the walking distance.

The underlying relationship or formula needed in this problem is that if you are traveling at a constant rate of speed then

Distance covered = [Rate (or speed) at which you travel] · [Time traveling]

In short, we write

$$d = rt$$

For example, if you travel at 50 miles per hour for 3 hours, you have covered 150 miles:

$$d = rt$$

$$d = 50\,\frac{\text{miles}}{\text{hr}} \cdot 3\,\text{hr}$$

$$d = 150\,\text{miles}$$

We have already noted the fact that the distances covered jogging to and walking from the post office are the same. That is, we have

$$d_{\text{jogging}} = d_{\text{walking}}$$ *d_{jogging} is called a **subscripted variable** and is used to indicate the distance covered jogging. d_{jogging} and d_{walking} are two different variables, just like x and y.*

Using the formula $d = rt$, we can rewrite this as

$$r_{\text{jogging}} \cdot t_{\text{jogging}} = r_{\text{walking}} \cdot t_{\text{walking}}$$

We are given the information that $r_{\text{jogging}} = 12$ kph and that $r_{\text{walking}} = 6$ kph. So our equation now looks like

$$12 \cdot t_{\text{jogging}} = 6 \cdot t_{\text{walking}}$$

Thus, we need to figure out how the different times are related.

Since we are told that Susan's *total* time is 3 hours, let

$$t = t_{\text{jogging}}, \quad \text{the amount of time for Susan to jog to the post office}$$

Then

$$3 - t = t_{\text{walking}}, \quad \text{the amount of time for Susan to walk home}$$

(The total time was 3 hours, so if $t =$ time jogging, then the time walking is what is left over, or $3 - t$.) We can now fill in the information and our equation becomes

$$12 \cdot t = 6 \cdot (3 - t)$$

Now we solve this equation.

$$12t = 6(3 - t)$$
$$12t = 18 - 6t$$
$$18t = 18$$
$$t = 1$$

Therefore, the time jogging was 1 hour, but we are not finished yet. The example asks for the distance of the post office from her home:

$$\text{Distance jogging} = \left(12\,\frac{\text{km}}{\text{hr}}\right) \cdot (1\,\text{hr}) = \boxed{12\text{ kilometers}}$$

Check: If Susan jogged for 1 hour then she walked for $3 - 1 = 2$ hours.

$$\text{Distance walking} = \left(6\,\frac{\text{km}}{\text{hr}}\right) \cdot (2\,\text{hr}) \overset{\checkmark}{=} 12\text{ kilometers}$$

Note that this problem was a bit different in that the variable t did not represent the quantity that was asked for (distance). However, once we found t we were easily able to compute the distance we were asked to find. ■

In doing the exercises, do not be discouraged if you do not get a complete solution to every problem. Make an honest effort to solve the problems and keep a written record of your work so that when you go over the problems in class you can see how far you got and exactly where you may have encountered difficulties.

Study Skills 3.3

Preparing for Exams: Study Activities

If you are going to learn algebra well enough to be able to demonstrate high levels of performance on exams, then you must concern yourself with both developing your skills in algebraic manipulation and understanding what you are doing and why you are doing it.

Many students concentrate only on skills and resort to memorizing the procedures for algebraic manipulations. This may work for quizzes or a test covering just a few topics. For exams covering a chapter's worth of material or more, this can be quite a burden on the memory. Eventually interference occurs and problems and procedures get confused. If you find yourself doing well on quizzes but not on longer exams, this may be your problem.

Concentrating on understanding what a method is and why it works is important. Neither the teacher nor the textbook can cover every possible way in which a particular concept may present itself in a problem. If you understand the concept, you should be able to recognize it in a problem. But again, if you concentrate only on understanding concepts and not on developing skills, you may find yourself prone to making careless and costly errors under the pressure of an exam.

In order to achieve the goal of both skill development and understanding, your studying should include four activities: (**1**) practicing problems, (**2**) reviewing your notes and textbook, (**3**) drilling with study cards (to be discussed in the next section), and (**4**) reflecting on the material being reviewed and the exercises being done.

Rather than doing any one of these activities over a long period of time, it is best to do a little of the first three activities during a study session and save some time for reflection at the end of the session.

EXERCISES 3.3

In Exercises 1–44, solve the given equation. If the equation is always true or has no solution, say so.

1. $8y + 4 = 5y + 19$

2. $7y + 5 = 3y + 17$

3. $2a + 5 = 4a + 12$

4. $3a + 7 = 10a - 1$

5. $5r - 8 = 3r - 20$

6. $9r - 7 = 6r - 19$

7. $10 - x = 4 - 3x$

8. $9 - 5x = 1 - x$

9. $-4 - 3u = -2 - u$

10. $-13 - u = -5u - 1$

11. $x + 7 = 7 - x$

12. $x + 5 = 5 - x$

13. $x + 7 = 7 + x$

14. $x + 5 = 5 + x$

15. $x - 7 = 7 + x$

16. $x - 5 = 5 + x$

17. $x - 7 = 7 - x$

18. $x - 5 = 5 - x$

19. $2(t + 1) + 4t = 29$

20. $3(t + 2) + 5t = 36$

21. $2(y + 3) + 4(y - 2) = 22$

22. $6(y - 3) + 3(y + 5) = 51$

23. $4 + 3(3y - 5) = 2y - 11 + y$

24. $6 + 4(y - 2) = 5y - 8 - y$

25. $3(a - 2) + 4(2 - a) = a + 2(a + 1)$

26. $5(a - 4) + 3(5 - a) = a + 4(a - 1) - 1$

27. $8z - 3(z - 3) = -9$

28. $12z - 8(z - 1) = -5$

29. $20t - 5(t - 1) = 0$

30. $8t + 4(t - 2) = 3$

31. $20 - 5(t - 1) = 0$

32. $8 + 4(t - 2) = 3$

33. $2(y - 3) - 3(y - 5) = 5y - 5(y - 2)$

34. $4(y - 2) - 5(y - 3) = 7y - 7(y - 1)$

35. $4x - 3(x + 8) = 5x - 2(x - 12) - 2x$

36. $7a - 5(a - 2) - a = 4a - 2(a - 5) - a$

37. $a - (5 - 3a) = 7a - (a - 3) - 8$

38. $2y - (7 - 4y) = 10y - (y - 2)$

39. $x^2 + 3x - 7 = x^2 - 5x + 1$

40. $x^2 - 5x + 2 = x^2 + x + 8$

41. $x(x + 2) + 3x = x(x - 1) - 12$

42. $a(a - 2) - a = a(a + 1) - 8$

43. $2z(z + 1) + 3(z + 2) = 3z(z + 2) - z^2$

44. $3y(y - 1) = 2y(y - 2) - (3 - y^2)$

In Exercises 45–50, solve the given equation. Round off your answers to the nearest hundredth where necessary.

45. $0.3x - 0.82 = 1.13$

46. $6.7a - 13.4 = 2.8a + 110.23$

47. $2.3t - 1.6(t + 0.1) = -0.139$

48. $1.4 + 0.5(8 - t) = 0.7t$

49. $3.4(t - 8) = 10.6(t + 3)$

50. $0.03w - 0.8(w - 0.62) = 40$

51. A company finds that the revenue R (in dollars) earned on the sale of x items is given by $R = 2.8x$ and the cost C (in dollars) of producing x items is given by $C = 1.60x + 2{,}100$. How many items must the company sell in order to break even, that is, when are the revenue and cost equal?

52. A carpenter finds that the revenue R (in dollars) she earns on the sale of p picture frames is given by $R = 15.75p$ and her cost C (in dollars) is given by $C = 4.25p + 460$. How many picture frames must she sell in order to break even, that is, when are the revenue and cost equal?

53. A wooden board 30 ft long is cut into two pieces so that the longer piece is 8 ft longer than the shorter piece. Find the lengths of the two pieces.

54. A wooden board 27 ft long is cut into two pieces so that the longer piece is 8 times as long as the shorter piece. Find the lengths of the two pieces.

55. One number is 4 more than 3 times another. If the sum of the two numbers is 24, find the numbers.

56. One number is 8 more than twice another. If the sum of the two numbers is 38, find the numbers.

57. One number is 5 less than 4 times another. If the sum of the two numbers is 11, find the numbers.

58. One number is 9 less than twice another. If the sum of the two numbers is 8, find the numbers.

59. If a number is added to 4 less than 5 times itself, the result is 27. Find the number.

60. If a number is added to 3 more than 5 times itself, the result is 30. Find the number.

61. If a number is subtracted from 4 more than twice itself, the result is 12. Find the number.

62. If a number is subtracted from 6 less than 3 times itself, the result is 18. Find the number.

63. The sum of three numbers is 80. The largest number is 10 more than twice the smallest, and the middle number is 5 less than twice the smallest. Find the three numbers.

64. The sum of three numbers is 68. The largest number is 6 less than twice the smallest, and the middle number is 10 less than the largest. Find the three numbers.

65. The sum of three consecutive integers is 45. What are they?

66. The sum of four consecutive integers is 14 less than 5 times the smallest integer. Find the four integers.

67. The sum of three consecutive odd integers is 29 more than twice the largest. Find them.

68. The sum of four consecutive even integers is 172. Find them.

69. The length of a rectangle is one more than twice the width. If the perimeter is 26 cm, find the dimensions of the rectangle.

70. If the length of a rectangle is 4 more than 5 times the width, and the perimeter is 32 meters, what are its dimensions?

71. The lengths of the sides of a triangle are three consecutive integers. If the perimeter of the triangle is 24 cm, what are the lengths of the three sides?

72. The first side of a triangle is 10 in. more than the second side. If the third side is 3 times as long as the second side and the perimeter is 45 in., find the lengths of the three sides.

73. The length of a rectangle is 6 more than the width. If the width is increased by 10 while the length is tripled, the new rectangle has a perimeter that is 56 more than the original perimeter. Find the original dimensions of the rectangle.

74. The length of a rectangle is 2 less than 3 times the width. If the width is tripled while the length is decreased by 2, the new perimeter is 12 more than the original perimeter. Find the original dimensions.

75. A truck rental costs $45 per day plus $0.40 per mile. If a 2-day rental costs $170, how many miles were driven?

76. A TV repairman charges $60 for a service call plus $32 per hour for repairs. If the total bill for a customer comes to $108, how many hours did the repair take?

77. A computer rental company has an installation charge of $125 plus a daily rental charge. If a 5-day rental costs $275, what is the daily rental charge?

78. An employment agency charges clients a $250 placement fee plus a daily fee. If it places an employee for 12 days and the total fee is $742, what is the daily fee?

79. A collection of 20 coins consisting of dimes and quarters has a total value of $4.25. How many of each type of coin are there?

80. Jack bought 34 stamps at the post office. Some were 33¢ stamps and the rest were 50¢ stamps. If the total cost of the stamps was $12.75, how many of each type did he buy?

81. Advanced-purchase tickets to an art exhibition cost $10, while tickets purchased at the door cost $12. If a total of 150 tickets were sold and $1,580 was collected, how many advanced-purchase tickets were sold?

82. In a collection of nickels, dimes, and quarters, there are twice as many dimes as nickels, and 3 fewer quarters than dimes. If the total value of the coins is $4.50, how many of each type of coin are there?

83. In a collection of dimes, quarters, and half-dollars, there are 45 coins in all. There are 11 more quarters than half-dollars and the remaining coins are dimes. If the total value of the coins is $11.10, how many of each type of coin are there?

84. Serena works part-time for a publisher and earns a commission of $2.20 for each book she sells and $3.50 for each magazine subscription she sells. During a certain week she made a total of 100 sales and earned $313.60. How many books did she sell?

85. An electrician charges $45 per hour for her time and $24 per hour for her assistant's time. On a certain job the assistant worked alone for 4 hours preparing the site, and then the electrician and her assistant completed the job together. If the total labor bill for the job was $464, how many hours did the electrician work?

86. A carpenter charges $42 per hour for his time and $20 per hour for his apprentice's time. On a certain job the apprentice does some preparatory work alone and then the carpenter finishes the job alone. If the job took a total of 11 hours and the total bill was $324.50, how long did each work?

87. Two trains leave cities 300 miles apart at 10:00 A.M., traveling toward each other. One train travels at 60 mph and the other train travels at 90 mph. At what time do they pass by each other?

88. Repeat Exercise 87, if the 90-mph train leaves at 8:00 A.M. and the 60-mph train leaves at 10:00 A.M.

89. Two people leave by car from the same location, traveling in opposite directions. One leaves at 2:00 P.M. driving at 55 kph, while the other leaves at 3:00 P.M. driving at 45 kph. At what time will they be 280 kilometers apart?

90. How long would it take someone driving at 80 kph to overtake someone driving at 50 kph with a 1-hour head start?

91. A relay race requires each team of two contestants to complete a 172-kilometer course. The first person runs part of the course, then the second person completes the remainder of the course by bicycle. One team covers the running section at 18 kph and the bicycle section at 50 kph in a total of 6 hours. How long did it take to complete the running section? How long is the running section of the course?

92. A bike race consists of two segments whose total length is 90 km. The first segment is covered at 10 kph and takes 2 hours longer to complete than the second segment, which is covered at 25 kph. How long is each segment?

93. A person can drive from town A to town B at a certain rate of speed in 5 hours. If he increases his speed by 15 kph, he can make the trip in 4 hours. How far is it from town A to town B?

94. A person can drive from city A to city B at a certain rate of speed in 6 hours. If she decreases her speed by 20 mph, she can make the trip in 8 hours. How far is it from city A to city B?

95. A secretary and a trainee are processing a pile of 124 forms. The secretary processes 15 forms per hour while the trainee processes 7 forms per hour. If the trainee begins working at 9:00 A.M. and is then joined by the secretary 2 hours later, at what time will they finish the pile of forms?

96. An optical mark reader is being used to grade 9,500 grade sheets. A machine that reads 200 sheets per minute begins the job but breaks down before completing all the grade sheets. A new machine, which reads 300 sheets per minute, then completes the job. If it took a total of 40 minutes to read all the grade sheets, how long did the first machine work before it broke down?

97. Margaret wants to buy 200 shares of stock in an Internet company. She buys some shares at $8.125 per share, and the rest when the price has gone up to $9.375 per share. If she spent a total of $1,725 for all the shares, how many shares did she buy at each price?

98. A store manager orders carpet for her showroom and office space. The showroom area requires 225 sq yd more carpet than the office space. She chooses carpet for the showroom costing $15.39 per sq yd, and carpet for the office costing $9.99 per sq yd. If the total cost of the carpet ordered is $7,650.45, how many square yards were ordered for the office?

99. A truck is loaded with boxes of two different weights. Some of the boxes weigh 6.58 kg each and the rest of the boxes weigh 9.32 kg each. There were 89 more heavier boxes than lighter boxes, and the total weight of all the boxes was 1,974.28 kg. How many boxes were there altogether?

100. A manufacturer orders a batch of 5,000 screws for $79.60. The batch contained two types of screws: regular screws costing $0.0125 each and magnetic screws costing $0.0185 each. How many magnetic screws were ordered?

101. A commercial Web site company charges $675 for designing and setting up a basic Web site and $1,850 for a deluxe Web site, which includes 2 months of technical support. During a certain month the company bills $17,350 for 17 Web sites. How many Web sites of each type were designed and set up?

102. A Web site manager notices that on a certain day her Web site experiences 12,468 "hits." She separates the hits into three categories:

    Category 1:   Hits that spend less than 1 minute on the Web site.

    Category 2:   Hits the spend between 1 minute and 5 minutes on the Web site.

    Category 3:   Hits that spend more than 5 minutes on the Web site.

    There were twice as many hits in category 2 as in category 3, and the total number in category 1 is equal to the numbers in categories 2 and 3 together. How many hits were in each category?

3.4 Types of Inequalities and Basic Properties of Inequalities

In Section 1.1 we discussed the meaning of the inequality symbols $<, >, \leq, \geq$. In this section we are going to discuss inequalities in much greater detail. Much of our discussion will follow along the same lines as our previous one on equalities earlier in this chapter.

Much of the vocabulary we use to describe inequalities is the same as that used for equations.

Inequalities such as $-2 < 5$ and $3 > 8$ can be categorized as being true or false, while an inequality such as $x + 3 < 5$ is neither true nor false since its truth depends on the value of x. An inequality whose truth depends on the value of the variable is called a ***conditional inequality.***

As we saw with equations, the presence of a variable does not necessarily mean that we have a conditional inequality. If an inequality is always true regardless of the value of the variable it is called an ***identity,*** while if it is always false it is called a ***contradiction.*** We will examine these possibilities and learn how to recognize them in the next section.

The values of the variable that make a conditional inequality true are called the ***solutions*** of the inequality.

EXAMPLE 1   In each of the following inequalities determine whether the given value of the variable is a solution.

(a) $x < 5$; for $x = 3$ (b) $x > -4$; for $x = -3$

(c) $a \leq 7$; for $a = 7$ (d) $4x - 10 < -20$; for $x = -2$

Solution   (a) $x < 5$ *Substitute 3 for x.*

$3 \overset{?}{<} 5$

$3 \overset{\checkmark}{<} 5$ *Therefore, 3 is a solution.*

(b) $x > -4$ *Substitute -3 for x.*

$-3 \overset{?}{>} -4$

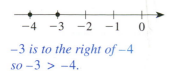

-3 is to the right of -4 so $-3 > -4$.

$-3 \overset{\checkmark}{>} -4$ *Remember that $a > b$ means that a is to the right of b on the number line. Since -3 is to the right of -4, we have $-3 > -4$. Therefore, -3 is a solution.*

(c) $a \leq 7$ *Substitute 7 for a.*

$7 \overset{?}{\leq} 7$

$7 \overset{\checkmark}{\leq} 7$ *Therefore, a = 7 is a solution.*

Remember that $a \leq b$ means $a < b$ or $a = b$ so that $7 \leq 7$ means $7 < 7$ or $7 = 7$, which is true.

(d) $4x - 10 < -20$ *Substitute −2 for x.*

$4(-2) - 10 \overset{?}{<} -20$

$-8 - 10 \overset{?}{<} -20$

$-18 \not< -20$ *Therefore, −2 is not a solution. On the number line, −18 is not to the left of −20.* ■

EXAMPLE 2   Determine whether the specified value satisfies the given inequality.

(a) $8 - 3(y - 5) \geq 17$;   for $y = 1$ **(b)** $2 < 3x + 8 < 7$;   for $x = 0$

Solution   **(a)** $8 - 3(y - 5) \geq 17$ *Substitute $y = 1$.*

$8 - 3(1 - 5) \overset{?}{\geq} 17$ *Remember the order of operations.*

$8 - 3(-4) \overset{?}{\geq} 17$

$8 + 12 \overset{?}{\geq} 17$

$20 \overset{?}{\geq} 17$

$20 \overset{\checkmark}{\geq} 17$ *Therefore, $y = 1$ does satisfy this inequality.*

(b) Remember that a double inequality like $2 < 3x + 8 < 7$ means $2 < 3x + 8$ *and* $3x + 8 < 7$. In other words, $3x + 8$ must be *between* 2 and 7.

$2 < 3x + 8 < 7$ *Substitute $x = 0$.*

$2 \overset{?}{<} 3(0) + 8 \overset{?}{<} 7$

$2 \overset{?}{<} 0 + 8 \overset{?}{<} 7$

$2 \overset{?}{<} 8 \overset{?}{<} 7$

$2 < 8 \not< 7$ *Therefore, $x = 0$ does not satisfy the inequality.* ■

Recall from our discussion in Section 1.6 that first-degree inequalities in one variable can be illustrated on a number line. Table 3.2 summarizes the variety of inequalities we will encounter, what each one means, and its graph on the number line.

Table 3.2 Graphs of Inequalities

Inequality	Description	Graph of the Solution Set
$x < 3$	x is less than 3.	
$x \leq 3$	x is less than or equal to 3.	*The 3 is filled in because 3 is included in the solution set.*
$x > -1$	x is greater than −1.	
$x \geq -1$	x is greater than or equal to −1.	
$-1 < x < 3$	x is greater than −1 *and* is less than 3.	

(continued)

Table 3.2 Graphs of Inequalities (continued)

Inequality	Description	Graph of the Solution Set
$-1 \leq x \leq 3$	x is greater than or equal to -1 *and* is less than or equal to 3.	
$-1 < x \leq 3$	x is greater than -1 *and* is less than or equal to 3.	
$-1 \leq x < 3$	x is greater than or equal to -1 *and* is less than 3.	

In solving first-degree equations our goal was to isolate x. In order to do this we used the properties of equality to transform the original equation into simpler equivalent equations until we ended up with an obvious equation. ***First-degree inequalities,*** meaning those that involve the variable to the first power only, are going to be handled in a very similar way.

Keep in mind, however, that while a first-degree equality such as $x = 4$ has only one solution, a first-degree inequality such as $x > 4$ can have infinitely many solutions. Any number greater than 4 satisfies this inequality.

In discussing inequalities, those of the form

$$x < 3 \qquad a \geq -2 \qquad 1 < y < 5$$

(that is, those in which the variable is isolated) are called ***obvious inequalities.*** Given our experience with solving equations, it seems natural to ask what are the properties of inequalities that we can use to obtain equivalent inequalities (that is, inequalities with the same solution set).

Let's start with a true inequality such as $-6 < 8$, perform the same kinds of operations on this inequality as we did when we were solving equations, and see what happens.

$-6 < 8$	Add 2 to both sides.	$-6 + 2 \ ? \ 8 + 2$	
		$-4 \ ? \ 10$	
		$-4 < 10$	*The inequality symbol remains the same.*
$-6 < 8$	Subtract 2 from both sides.	$-6 - 2 \ ? \ 8 - 2$	
		$-8 \ ? \ 6$	
		$-8 < 6$	*The inequality symbol remains the same.*
$-6 < 8$	Multiply both sides by 2.	$2(-6) \ ? \ 2(8)$	
		$-12 \ ? \ 16$	
		$-12 < 16$	*The inequality symbol remains the same.*
$-6 < 8$	Multiply both sides by -2.	$-2(-6) \ ? \ -2(8)$	
		$12 \ ? \ -16$	
		$12 > -16$	*The inequality symbol is reversed.*

$-6 < 8$ Divide both sides by 2.

$$\frac{-6}{2} \; ? \; \frac{8}{2}$$
$$-3 \; ? \; 4$$
$$-3 \; < \; 4$$ *The inequality symbol remains the same.*

$-6 < 8$ Divide both sides by -2.

$$\frac{-6}{-2} \; ? \; \frac{8}{-2}$$
$$3 \; ? \; -4$$
$$3 \; > \; -4$$ *The inequality symbol is reversed.*

If the reversal of the inequality symbol were simply a random occurrence, we would not be able to formulate the properties of inequalities. Fortunately, the examples we have just looked at illustrate what happens in general. If we illustrate a few of the above examples on the number line, the mechanics of what is going on should become clear.

On the number line, the original inequality $-6 < 8$ is true because -6 is to the left of 8:

Adding or subtracting the same quantity from both sides of the inequality simply shifts both numbers the same number of units to either the left or right. In either case, the number further to the left remains further to the left, and hence *the inequality symbol remains the same.*

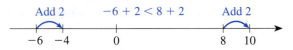

Multiplying both sides by 2 looks like this:

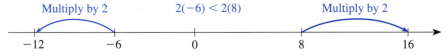

and again *the inequality symbol remains the same.*

However, multiplying both sides by -2 looks like this:

Multiply by -2

We can see that after multiplying by -2, the positions of the resulting numbers are reversed, and therefore *the inequality symbol must be reversed.*

(You might find it interesting to consider some examples starting with an inequality $a < b$ where a and b are both positive or both negative and see what happens when you multiply or divide the inequality first by a positive and then by a negative number.)

While we have only analyzed some numerical evidence, inequalities behave this way all the time. We formulate these properties as indicated in the box.

Properties of Inequalities	1. If we add or subtract the same quantity to each side of an inequality, the inequality symbol remains the *same.*
	2. If we multiply or divide each side of an inequality by a *positive* quantity, the inequality symbol remains the *same.*
	3. If we multiply or divide each side of an inequality by a *negative* quantity, the inequality symbol is *reversed.*

These properties tell us what operations we can perform on an inequality and still obtain an equivalent inequality. Algebraically, we can write these properties as shown in the next box. (In each case, the $<$ symbol can be replaced by $>$, $\leq$, or $\geq$.)

Properties of Inequalities	1. If $a < b$ then $a + c < b + c$.
	2. If $a < b$ then $a - c < b - c$.
	3. If $a < b$ then $\begin{cases} ac < bc & \text{when } c \text{ is positive.} \\ ac > bc & \text{when } c \text{ is negative.} \end{cases}$
	4. If $a < b$ then $\begin{cases} \dfrac{a}{c} < \dfrac{b}{c} & \text{when } c \text{ is positive.} \\ \dfrac{a}{c} > \dfrac{b}{c} & \text{when } c \text{ is negative.} \end{cases}$

EXAMPLE 3

In each of the following, use the properties of inequalities to perform the indicated operation on the given inequality. Sketch the resulting equivalent inequality on a number line.

(a) $x - 5 < 4$; add 5 to each side

(b) $a + 3 \geq -2$; subtract 3 from each side

(c) $12 < 4y$; divide each side by 4

(d) $-3x \geq 6$; divide each side by -3

(e) $3 < x + 2 \leq 7$; subtract 2 from each member

Solution

(a) $x - 5 < 4$
$\underline{+5 +5}$

$\boxed{x < 9}$ *The inequality symbol remains the same under addition (see property 1).*

The open circle at 9 means that 9 is *excluded.*

(b) $a + 3 \geq -2$
$\underline{-3 -3}$

$\boxed{a \geq -5}$ *The inequality symbol remains the same under subtraction (see property 2).*

The solid circle at -5 means that -5 is *included.*

(c) $12 < 4y$

$\dfrac{12}{4} < \dfrac{4y}{4}$ *The inequality symbol remains the same when we divide by a **positive** number (see property 4).*

$\boxed{3 < y}$ *This can be read as "3 is less than y" or "y is greater than 3."*

(d) $-3x \geq 6$

$$\frac{-3x}{-3} \leq \frac{6}{-3}$$ *The inequality symbol **reverses** when we divide by a **negative number** (see property 4).*

$\boxed{x \leq -2}$

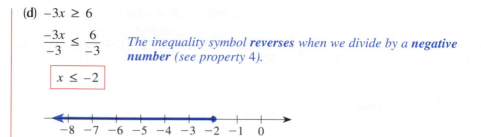

(e) The double inequality $3 < x + 2 \leq 7$ has three *members* to it. These members are 3, $x + 2$, and 7. In order to produce an equivalent inequality we must perform our operations on each member of the inequality. In this example we are asked to subtract 2 from each member.

$$3 < x + 2 \leq 7$$
$$\underline{-2 \qquad -2 \ -2}$$
$$\boxed{1 < x \leq 5}$$

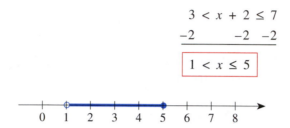

Study Skills 3.4

Preparing for Exams: Making Study Cards

Study cards are 3″ × 5″ or 5″ × 8″ index cards that contain summary information needed for convenient review. We will discuss three types of cards: the definition/principle card, the warning card, and the quiz card.

The **definition/principle (D/P) cards** are cards that contain a single definition, concept, or rule for a particular topic.

Here is an example of a D/P card.

The front of each D/P card should contain

1. A heading of a few words

2. The definition, concept, or rule accurately recorded

3. If possible, a restatement of the definition, concept, or rule in your own words

The back of the card should contain examples illustrating the idea on the front of the card.

FRONT

> The Distributive Property
> $a(b + c) = ab + ac$
>   or
> $(b + c)a = ba + ca$
> Multiply each <u>term</u> by a.

BACK

> (1) $3x(x + 2y) = 3x(x) + 3x(2y)$
> $= 3x^2 + 6xy$
> (2) $-2x(3x - y) = -2x(3x) - 2x(-y)$
> $= -6x^2 + 2xy$
> (3) $(2x + y)(x + y)$
> $= 2x(x + y) + y(x + y)$

Warning (W) cards are cards that contain errors that you may be consistently making on homework, quizzes, or exams, or those common errors pointed out by your teacher or your text. The front of the warning card should contain the word WARNING; the back of the card should contain an example of both the correct and incorrect way an example should be done. Be sure to label clearly which solution is correct and which is not. For example:

(continued)

Study Skills 3.4 (continued)

Preparing for Exams: Making Study Cards

```
        FRONT                          BACK

  ┌──────────────────────┐   ┌──────────────────────────────┐
  │ WARNING              │   │ EXAMPLES                     │
  │ EXPONENTS            │   │ 2·3² = 2·3·3 NOT (2·3)(2·3)   │
  │   An exponent refers │   │ (−3)² = (−3)(−3) = 9          │
  │ only to the          │   │   ↑Parentheses mean −3 is the │
  │ factor immediately   │   │    factor to be squared.     │
  │ to the left of       │   │ BUT                          │
  │ the exponent.        │   │ −3² = −3·3 = −9               │
  │                      │   │   ↑The factor being squared  │
  │                      │   │    here is 3, not −3.        │
  └──────────────────────┘   └──────────────────────────────┘
```

    Quiz cards are another type of study card. We will discuss how to use them later. For now, go through your text and pick out a few of the odd-numbered exercises (just the problem) from each section, putting one or two problems on one side of each card. Make sure you copy the *instructions* as well as the problem accurately. On the back of the card write down the exercise number and section of the book where the problem was found. For example:

```
        FRONT                          BACK

  ┌──────────────────────┐   ┌──────────────────────────────┐
  │ Translate the        │   │ Exercise 17                  │
  │ following            │   │ Section 2.5                  │
  │ algebraically using  │   │                              │
  │ n as the             │   │                              │
  │ number.              │   │                              │
  │ Four more than five  │   │                              │
  │ times a              │   │                              │
  │ number is two less   │   │                              │
  │ than the             │   │                              │
  │ number.              │   │                              │
  └──────────────────────┘   └──────────────────────────────┘
```

EXERCISES 3.4

In Exercises 1–20, determine whether the given value of the variable satisfies the inequality.

1. $x + 4 < 3$; $x = -2$

2. $x + 7 < 5$; $x = -1$

3. $a - 2 > -1$; $a = -3$

4. $a - 4 > 2$; $a = -4$

5. $-y + 3 \leq 5$; $y = -2$

6. $-y + 1 \leq 7$; $y = -6$

7. $-8 \leq 2 - x$; $x = 6$

8. $4 - x \geq 3$; $x = 1$

9. $2z - 5 < -3$; $z = 1$

10. $3z - 7 \leq 5$; $z = 4$

11. $12 < 5 + 2u$; $u = 3$

12. $15 < 6 + 4u$; $u = 2$

13. $7 - 4x < 8$; $x = -4$

14. $8 - 3x > -5$; $x = 5$

15. $-2 < 8 - x < 3$; $x = 6$

16. $-4 \leq 9 - 2x < 7$; $x = -1$

17. $6 + 2(a - 3) < 1$; $a = -2$

18. $5 + 3(a - 7) > -12$; $a = 2$

19. $-12 < 9 - 5(x + 1) < -5$; $x = 3$

20. $-5 < 8 - 2(x + 3) \leq 0$; $x = 4$

In Exercises 21–54, perform the indicated operations on the given inequality. Sketch the resulting inequality on a number line.

21. $x - 3 < 2$; add 3 to each side

22. $x - 5 < 1$; add 5 to each side

23. $a + 7 > 4$; subtract 7 from each side

24. $3 \leq a + 6$; subtract 6 from each side

25. $-4 > w + 2$;   subtract 2 from each side

26. $w + 4 < -1$;   subtract 4 from each side

27. $z + 3 > 0$;   subtract 3 from each side

28. $z + 7 > 0$;   subtract 7 from each side

29. $3x \leq 12$;   divide each side by 3

30. $10 \leq 5x$;   divide each side by 5

31. $4y > -8$;   divide each side by 4

32. $2y < -10$;   divide each side by 2

33. $-3x < 6$;   divide each side by -3

34. $-6x > -12$;   divide each side by -6

35. $-3x < -6$;   divide each side by -3

36. $-6x > 12$;   divide each side by -6

37. $7a > 0$;   divide each side by 7

38. $2a < 0$;   divide each side by 2

39. $-7a \geq 0$;   divide each side by -7

40. $-2a < 0$;   divide each side by -2

41. $-x < 3$;   multiply each side by -1

42. $-x < 3$;   divide each side by -1

43. $-x < -3$;   multiply each side by -1

44. $-x < -3$;   divide each side by -1

45. $-5 < a - 4 \leq 2$;   add 4 to each member

46. $-2 < a - 5 < 1$;   add 5 to each member

47. $1 \leq x + 3 \leq 5$;   subtract 3 from each member

48. $2 < x + 4 \leq 6$;   subtract 4 from each member

49. $-6 < 3y < 3$;   divide each member by 3

50. $-4 \leq 4y \leq 12$;   divide each member by 4

51. $0 \leq -2x < 2$;   divide each member by -2

52. $0 < -5x < 10$;   divide each member by -5

53. $-5 < -x < -1$;   multiply each member by -1

54. $-5 < -x < -1$;   divide each member by -1

(?) QUESTIONS FOR THOUGHT

55. What does it mean for one number to be less than another?

56. Summarize the properties of inequalities in two sentences.

57. What is the basic difference between the properties of equalities and those of inequalities?

58. Explain in words what the inequality $2 < x \leq 5$ means.

◇ MINI-REVIEW

Perform the indicated operations and simplify as completely as possible.

59. $(2x^2)(3x^3)$

60. $(-4x^3y)(-2x^3y^4)$

61. $4(x - 3y) + 2(3x - y)$

62. $x(x - 2y) + y(x + y)$

63. $x^2(x^3 - 4x) - (2x^5 - 4x^3)$

64. $3ab(a^2 - 5b) - 3b(a^3 + 4ab)$

3.5   Solving First-Degree Inequalities in One Variable and Applications

If we put the method we developed in Section 3.2 for solving first-degree equations together with the properties of inequalities discussed in the previous section, we can formulate the procedure described in the accompanying box.

To Solve a First-Degree Inequality	Use the same procedure as in solving a first-degree equation, *except* that when multiplying or dividing an inequality by a *negative* quantity the inequality symbol must be *reversed*.

Let's apply this outline to several examples.

EXAMPLE 1   Solve for x. $x + 4 < 6$

Solution   Our goal is to isolate x and obtain an obvious inequality. If the example were an equation, that is, $x + 4 = 6$, we would subtract 4 from each side. Therefore, according to our outline we proceed here in exactly the same way:

$$\begin{aligned} x + 4 &< 6 \\ -4 \quad &\phantom{<} -4 \\ \hline \end{aligned}$$

$\boxed{x < 2}$ *The inequality symbol remains the same under subtraction.*

The solution $x < 2$ means that *any number* less than 2 makes the original inequality true. It is impossible to check every number less than 2 in the inequality. Instead, we can check a number less than 2 to see that it does satisfy the inequality, *and* a number greater than or equal to 2 to see that it does not satisfy the inequality.

Check:

We can choose any number less than 2, say $x = 1$ because $1 < 2$. This *should* satisfy the inequality.

$$\begin{aligned} x + 4 &< 6 \\ 1 + 4 &\overset{?}{<} 6 \\ 5 &\overset{\checkmark}{<} 6 \end{aligned}$$

We can choose any number greater than or equal to 2, say $x = 3$ because $3 > 2$. This *should not* satisfy the inequality.

$$\begin{aligned} x + 4 &< 6 \\ 3 + 4 &\overset{?}{<} 6 \\ 7 &\not< 6 \end{aligned}$$

While this is not a conclusive check, as in the case of an equation, it does make us feel much more confident about our solution. ■

EXAMPLE 2   Solve for t. $-4t < 8$

Solution $-4t < 8$ *We divide both sides by -4 to isolate t.*

$\dfrac{-4t}{-4} > \dfrac{8}{-4}$ *Since we are dividing by a **negative** number, we **reverse** the inequality symbol.*

$\boxed{t > -2}$

Check:

We can choose any number greater than -2, say $t = -1$. This should satisfy the inequality.

$$\begin{aligned} -4t &< 8 \\ -4(-1) &\overset{?}{<} 8 \\ 4 &\overset{\checkmark}{<} 8 \end{aligned}$$

We can choose any number less than or equal to -2, say $t = -3$. This should not satisfy the inequality.

$$\begin{aligned} -4t &< 8 \\ -4(-3) &\overset{?}{<} 8 \\ 12 &\not< 8 \end{aligned}$$
■

EXAMPLE 3   Solve for a. $3a + 5 \geq 2$

Solution $$\begin{aligned} 3a + 5 &\geq 2 \\ -5 \quad & -5 \\ \hline 3a &\geq -3 \end{aligned}$$ *The inequality symbol remains the same under subtraction.*

$\dfrac{3a}{3} \geq \dfrac{-3}{3}$ *The inequality symbol remains the same when we divide by a positive number. The fact that we are dividing **into** -3 is irrelevant.*

$\boxed{a \geq -1}$

Check:

We can choose any number greater than or equal to –1, say $a = 0$. This should satisfy the inequality.

$$3a + 5 \geq 2$$
$$3(0) + 5 \overset{?}{\geq} 2$$
$$5 \overset{\checkmark}{\geq} 2$$

We can choose any number less than –1, say $a = -2$. This should not satisfy the inequality.

$$3a + 5 \geq 2$$
$$3(-2) + 5 \overset{?}{\geq} 2$$
$$-6 + 5 \overset{?}{\geq} 2$$
$$-1 \ngeq 2 \qquad \blacksquare$$

EXAMPLE 4

Solution

Solve for x and sketch the solution set on a number line. $5x - 1 < 7x + 9$

Since x appears on both sides of the inequality, we can choose to isolate x on either side. Let's do it both ways and compare the results.

Solution 1

$$5x - 1 < 7x + 9$$
$$\underline{-7x \qquad -7x}$$
$$-2x - 1 < 9$$
$$\underline{\qquad +1 \quad +1}$$
$$-2x < 10$$
$$\frac{-2x}{-2} > \frac{10}{-2}$$
$$\boxed{x > -5}$$

Since we are dividing by a negative number, the inequality symbol is reversed.

Solution 2

$$5x - 1 < 7x + 9$$
$$\underline{-5x \qquad -5x}$$
$$-1 < 2x + 9$$
$$\underline{-9 \qquad \quad -9}$$
$$-10 < 2x$$
$$\frac{-10}{2} < \frac{2x}{2}$$
$$\boxed{-5 < x}$$

Since we are dividing by a positive number, the inequality symbol remains the same.

It is important to look at these two answers and see that they say the same thing. They are both saying that x is greater than -5. Both methods of solution are correct. The sketch of the solution set is shown in Figure 3.4.

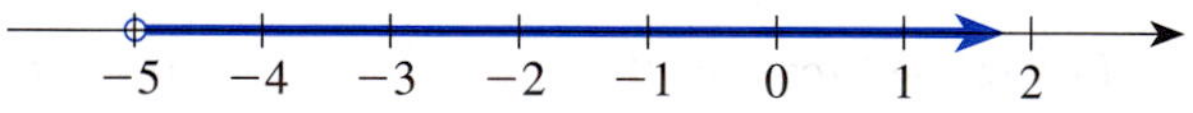

Figure 3.4
Solution set for Example 4

[number line from –5 to 2 with open circle at –5 and arrow to the right]

$\blacksquare$

EXAMPLE 5

Solve for x and sketch the graph of the solution set. $2 < x + 4 < 5$

Solution

First let's review what $2 < x + 4 < 5$ means. Essentially we are trying to find numbers which when added to 4 will yield a result that is both greater than 2 *and* less than 5. In other words, we are really solving two inequalities:

$$2 < x + 4 \qquad \text{and} \qquad x + 4 < 5$$

Both inequalities can be solved by applying the same properties.

$$\begin{array}{ccc} 2 < x + 4 & & x + 4 < 5 \\ \underline{-4 \qquad -4} & & \underline{-4 \quad -4} \\ -2 < x & \text{and} & x < 1 \end{array}$$

Combining these two answers and using the double inequality ("between") notation, we get

$$-2 < x < 1$$

Since we are doing the same thing to both inequalities we can think of the solution in the following simpler way. Solving a double inequality means that we want to isolate x in the middle.

$$2 < x + 4 < 5$$
$$\underline{-4 \qquad -4 \quad -4}$$

We subtract 4 from each member of the inequality. The inequality symbol remains the same.

$$\boxed{-2 < \quad x \qquad < 1}$$

The sketch of the solution set is shown in Figure 3.5.

Figure 3.5

Solution set for Example 5

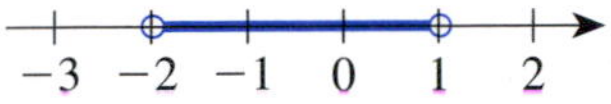

The check here requires us to choose three numbers: one number less than or equal to −2, which should not satisfy the inequality; one number between −2 and 1, which should satisfy the inequality; and one number greater than or equal to 1, which should not satisfy the inequality. It is left to the student to carry out this check. ■

EXAMPLE 6

Solve for a and sketch the graph of the solution set. $5 < 5 - 2a \le 11$

Solution

We want to isolate a in the middle.

$$5 < \quad 5 - 2a \le 11$$
$$\underline{-5 \quad -5 \qquad \quad -5}$$
$$0 < \qquad -2a \le 6$$
$$\frac{0}{-2} > \quad \frac{-2a}{-2} \ge \frac{6}{-2}$$

We divide all three members of the inequality by −2. Both inequality symbols must be reversed.

$$\boxed{0 > \quad a \qquad \ge -3}$$

This same answer could be written as

$$\boxed{-3 \le a < 0}$$

This second form of the inequality is much easier to visualize, although both forms are correct. The graph of the solution set is shown in Figure 3.6.

Figure 3.6

Solution set for Example 6

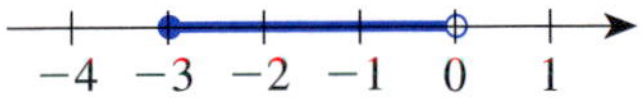

The check is left to the student. ■

As was mentioned in the last section, not all inequalities involving variables are conditional. The next two examples in this section illustrate these possibilities.

EXAMPLE 7

Solve for x. $5x - 3(x - 2) < 2x + 9$

Solution

$$5x - 3(x - 2) < 2x + 9$$
$$5x - 3x + 6 < 2x + 9$$
$$2x + 6 < 2x + 9$$
$$\underline{-2x \qquad -2x}$$
$$6 < 9 \qquad \textit{This statement is always true.}$$

The variable has been eliminated entirely and the resulting inequality is always true. Therefore, the original inequality is an *identity*.

> All values of x are solutions to this inequality.

■

EXAMPLE 8 Solve for z. $4z - (z - 7) < 3z + 4$

Solution

$$4z - (z - 7) < 3z + 4$$
$$4z - z + 7 < 3z + 4$$
$$3z + 7 < 3z + 4$$
$$\underline{-3z \qquad -3z}$$
$$7 < 4 \qquad \textit{This statement is always false.}$$

Therefore, the original inequality is a *contradiction*.

> No values of z are solutions to this inequality.

■

Many problems involving real-life situations can be modeled and solved using first-degree inequalities. The outline suggested in Section 3.3 for solving verbal problems applies equally well to problems that give rise to inequalities.

EXAMPLE 9 What numbers satisfy the condition that "3 less than 4 times the number is less than 29"?

Solution

In translating this problem algebraically, we must be careful to distinguish the phrase "3 less than" from the phrase "3 *is* less than." The phrase "3 less than" indicates that we are subtracting 3 from some quantity—it is *not* a statement of inequality. An example of this would be the phrase "3 less than 8," which is translated as $8 - 3$. On the other hand, the phrase "3 *is* less than" is a statement of inequality. An example of this would be the phrase "3 is less than 8," which is translated as $3 < 8$.

Let $x =$ a number satisfying the condition given in the example. The condition given in the problem can be translated as follows:

3 less than 4 times the number	is less than	29
$4x - 3$	$<$	29

Thus, the inequality is

$$4x - 3 < 29$$

Now we solve this inequality.

$$4x - 3 < 29$$
$$\underline{+3 \qquad +3}$$
$$4x < 32$$
$$\frac{4x}{4} < \frac{32}{4}$$
$$\boxed{x < 8}$$

Thus, any number less than 8 satisfies the conditions of this problem. The check is left to the student.

■

EXAMPLE 10 Joan earns a quarterly bonus commission of 4.8% on all sales exceeding $15,000. Determine what Joan's quarterly sales must be in order to earn a commission of at least $750.

Solution

We can translate the example using an inequality statement as follows:

Joan's commission $\geq$ $750

Note that we have used an inequality symbol because the example tells us that Joan must earn *at least* $750 in commission, which means her commission must be greater than or equal to $750.

If she makes sales worth $18,000, then she earns commission on $18,000 − $15,000 = $3,000. In order to compute 4.8% of $3,000, we multiply 3,000 by 0.048.

In order for Joan to earn any commission at all, she must sell at least $15,000 worth of merchandise. For example, if she has sales worth $18,000, she will earn the 4.8% commission on $3,000, which is 0.048(3,000) = $144 in commission. In general, we need to compute 4.8% of the excess *above* $15,000. We compute this excess by subtracting 15,000 from the amount of her sales.

Let x = Joan's monthly sales in dollars

The analysis we have just made translates to

$$0.048(x - 15,000) \geq 750$$

We can now solve this inequality.

$0.048(x - 15,000) \geq 750$	*Multiply out the left-hand side.*
$0.048x - 0.048(15,000) \geq 750$	.
$0.048x - 720 \geq 750$	*Add 720 to both sides.*
$0.048x \geq 1470$	*Divide both sides by 0.048.*
$\dfrac{0.048x}{0.048} \geq \dfrac{1470}{0.048}$	

$$x \geq 30,625$$

Thus, Joan must sell at least $30,625 worth of merchandise if she wants to earn at least $750 in commission.

Check: If Joan makes sales worth $30,625, she will earn the 4.8% commission on $30,625 − $15,000 = $15,625. The commission would be 0.048(15,625) = $750. If she makes sales worth more than $30,625 then she will earn more than $750 in commission, as required. ■

Study Skills 3.5

Preparing for Exams: Using Study Cards

The very process of making up study cards is a learning experience in itself. Study cards are convenient to use—you can carry them along with you and use them for review in between classes or as you wait for a bus.

Use the (D/P and W) cards as follows:

1. Look at the heading of a card and, covering the rest of the card, see if you can remember what the rest of the card says.

2. Continue this process with the remaining cards. Pull out those cards you know well and put them aside, but do review them from time to time. Study those cards you do not know.

3. Shuffle the cards so that they are in random order and repeat the process again from the beginning.

4. As you go through the cards, ask yourself the following questions (where appropriate):

   (a) When do I use this rule, method, or principle?

   (b) What are the differences and similarities between problems?

   (c) What are some examples of the definitions or concepts?

   (d) What concept is illustrated by the problem?

   (e) Why does this process work?

   (f) Is there a way to check this problem?

Solve each of the following inequalities:

1. $x + 5 < 3$

2. $x + 7 < 4$

3. $a - 2 > -3$

4. $a - 5 > -2$

5. $2y < 7$

6. $4y < 13$

7. $2y > -7$

8. $4y > -13$

9. $-2y < 7$

10. $-4y > 13$

11. $-2y > -7$

12. $-4y < -13$

13. $-x < 4$

14. $-t > 1$

15. $-1 > -y$

16. $-w > -6$

17. $5x + 3 \leq 8$

18. $3x + 7 \geq 13$

19. $2x - 9 \geq 16$

20. $4x - 5 \geq -8$

21. $2(z - 3) + 4 \geq -6$

22. $4(z - 1) + 3z < -4$

23. $3(x + 4) + 2(x - 1) < 20$

24. $2(x - 5) + 5(x + 3) \geq 19$

25. $5(w + 3) - 7w \leq 7$

26. $2(w + 4) - 5w < 2$

27. $3(a + 4) - 4(a - 1) < 10$

28. $5(a - 2) - 6(a + 1) > -5$

29. $4(y - 3) - (3y - 12) \geq 2$

30. $7(y + 1) - (6y + 7) \leq 4$

31. $2(u + 2) - 2(u - 1) < 5$

32. $3(u - 1) - 3(u + 1) < 2$

33. $4(x - 2) - (4x - 3) < 6$

34. $5(x + 1) - (5x - 1) > 12$

35. $x + 3 < 2x + 7$

36. $2x + 5 < 3x + 8$

37. $5t - 3 \geq 3t + 10$

38. $3t - 8 \leq t - 7$

39. $2(a - 5) + 3a > 6a - 6$

40. $3(a - 4) + 5a \leq 9a - 8$

41. $4(w + 2) - 3(w - 1) > 5(w - 1) - 5w$

42. $5(r + 3) - 4(r - 2) > 7(r - 4) - 7r$

43. $2y - 4(y + 1) \leq 8 - (y + 2)$

44. $4y - 6(y - 2) \geq 10(y - 6)$

45. $2 < x + 7 < 10$

46. $1 < x + 5 < 9$

47. $3 < 2a + 5 < 7$

48. $2 < 3a + 2 \leq 8$

49. $-5 \leq -4y + 3 < 7$

50. $-1 < -2y + 5 \leq 13$

51. $1 \leq 6 - x < 3$

52. $0 < 9 - x \leq 5$

In Exercises 53–64, solve the inequality and sketch the solution set on a number line.

53. $x + 4 < 2x - 1$

54. $x + 5 < 2x - 3$

55. $3(a + 2) - 5a \geq 2 - a$

56. $5(a - 1) - 8a \geq 3 - a$

57. $-1 < x + 3 < 2$

58. $-3 < x + 5 < 4$

59. $-1 \leq y - 3 \leq 2$

60. $-3 \leq y - 5 \leq 4$

61. $-3 \leq 4t + 5 < 9$

62. $-6 < 5t - 1 \leq -1$

63. $-5 < 3 - 2x \leq 9$

64. $-1 \leq 2 - 3x < 11$

In Exercises 65–68, solve the given inequality. Round off your answers to the nearest hundredth where necessary.

65. $0.8 - 0.45(x - 2) \leq 0.26$

66. $0.5(x - 0.3) > 0.25(x + 3)$

67. $5.468 < 2.9t - 12.86 < 20.519$

68. $-15.45 \leq 53.67 - 0.45t < 36.93$

In Exercises 69–76, translate the given phrase or sentence algebraically.

69. 7 is greater than x.

70. 7 greater than x

71. 10 less than t

72. 10 is less than t.

73. x is at least 5.

74. a is at most 5.

75. y is no more than 2.

76. h is no less than 8.

In Exercises 77–88, set up an inequality and solve it. Be sure to clearly label what the variable represents.

77. What numbers satisfy the condition "four less than three times the number is less than seventeen"?

78. What numbers satisfy the condition "five less than four times the number is greater than 19"?

79. If 12 more than 6 times a number is greater than 3 times the number, how large must the number be?

80. If 9 less than 5 times a number is less than twice the number, how small must the number be?

81. The width of a rectangle is 8 cm. If the perimeter is to be at least 80 cm, how large must the length be?

82. If the width of a rectangle is 10 meters and the perimeter is not to exceed 120 meters, how large can the length be?

83. The length of a rectangle is 18 in. If the perimeter is to be at least 50 in. but not greater than 70 in., what is the range of values for the width?

84. The medium side of a triangle is 2 cm longer than the shortest side, and the longest side is twice as long as the shortest side. If the perimeter of the triangle is to be at least 30 cm and no more than 50 cm, what is the range of values for the shortest side?

85. An organization wants to sell tickets to a concert. They plan on selling 300 reserved-seat tickets and 150 tickets at the door. The price of a reserved-seat ticket is to be $2 more than a ticket at the door. If they want to collect at least $3,750, what is the minimum price they can charge for a reserved-seat ticket?

86. An aide in the mathematics department office gets paid $3 per hour for clerical work and $8 per hour for tutoring. If she wants to work a total of 20 hours and earn at least $135, what is the maximum number of hours she can spend on clerical work?

87. Marc earns a yearly bonus of 1.6% of all sales he makes during the year in excess of $82,000. Determine what Marc's annual sales must be in order to earn a year-end bonus of at least $1,800.

88. Celene receives a yearly commission of 2.1% of her total annual sales above $125,000. During a 10-year period her minimum annual commission was $1,470 and her maximum annual commission was $3,927. Find the range of her total annual sales during this period.

QUESTIONS FOR THOUGHT

89. What is *wrong* with each of the following "solutions"?

(a)
$$
\begin{array}{rcl}
2 + 7x & \leq & 5x \\
-7x & & -7x \\
\hline
2 & \leq & -2x \\
-2 & \leq & x
\end{array}
$$

(b)
$$
\begin{array}{rcl}
2x + 4 & < & 2 \\
-4 & & -4 \\
\hline
2x & > & -2 \\
x & > & -1
\end{array}
$$

(c)
$$
\begin{array}{rcl}
3x - 9 & > & 6x \\
-3x & & -3x \\
\hline
-9 & > & 3x \\
-3 & < & x
\end{array}
$$

90. Look at each of the following inequality statements and determine whether they make sense. Explain your answers.

(a) $-3 < x < 2$

(b) $-5 < x < -8$

(c) $7 < x < 4$

(d) $6 > x < 3$

(e) $3 < x < -2$

(f) $-5 > x > 4$

CHAPTER 3   SUMMARY

After having completed this chapter you should be able to:

1. Understand and recognize the basic types of equations (*conditional, identity,* and *contradiction*) and the properties of equality (Section 3.1).

2. Determine whether or not a particular value is a solution to a given equation or inequality (Sections 3.1, 3.4).

   For example:

   Does $x = -3$ satisfy the equation $3x - 4(2 - x) = -5$?

   $$
   \begin{array}{rcl}
   3x - 4(2 - x) & = & -5 \qquad \textit{Substitute } x = -3. \\
   3(-3) - 4(2 - (-3)) & \overset{?}{=} & -5 \\
   -9 - 4(5) & \overset{?}{=} & -5 \\
   -9 - 20 & \overset{?}{=} & -5 \\
   -29 & \neq & -5 \qquad \textit{Therefore, } x = -3 \textit{ does not satisfy}
   \end{array}
   $$
   the equation.

3. Use the properties of equations and inequalities to solve first-degree equations and inequalities (Sections 3.2, 3.3, 3.5).

   For example:

   (a) *Solve for t.* $\quad 3t - 4(2 - t) = 48$

   Solution:
   $$
   \begin{array}{rcl}
   3t - 4(2 - t) & = & 48 \qquad \textit{Simplify.} \\
   3t - 8 + 4t & = & 48 \\
   7t - 8 & = & 48 \\
   +8 & & +8 \qquad \textit{Add 8 to both sides.} \\
   \hline
   7t & = & 56 \qquad \textit{Divide both sides by 7.} \\
   \dfrac{7t}{7} & = & \dfrac{56}{7} \\
   \boxed{t = 8}
   \end{array}
   $$

   (b) *Solve for y.* $\quad 5 - 3(y + 1) < 8$

   Solution:
   $$
   \begin{array}{rcl}
   5 - 3y - 3 & < & 8 \qquad \textit{Simplify.} \\
   2 - 3y & < & 8 \\
   -2 & & -2 \\
   \hline
   -3y & < & 6 \qquad \textit{Dividing both sides by } -3 \textit{ reverses the inequality sign.}
   \end{array}
   $$

$$\frac{-3y}{-3} > \frac{6}{-3}$$

$$\boxed{y > -2}$$

4. Solve verbal problems that give rise to first-degree equations or inequalities in one variable (Sections 3.2, 3.3, 3.5).

For example:

Six less than 3 times a number is 12 more than the number. Find the number.

Solution: Let $n =$ the number. We translate the given relationship as follows:

Six less than	3 times a number	is	12 more than	the number
	$3n \quad - 6$	$=$		$n \quad + 12$

Thus, the equation is

$$\boxed{3n - 6 = n + 12}$$

$$
\begin{array}{rcl}
3n - 6 &=& n + 12 \\
-n & & -n \\
\hline
2n - 6 &=& 12 \\
+6 & & +6 \\
\hline
2n &=& 18 \\
\dfrac{2n}{2} &=& \dfrac{18}{2}
\end{array}
$$

$$\boxed{n = 9}$$

Study Skills 3.6

Preparing for Exams: Reviewing Your Notes and Text; Reflecting

Another activity we suggested as an important facet of studying for exams is to review your notes and text. Your notes are a summary of the information you feel is important at the time you write it down. In the process of reviewing your notes and text you may turn up something you missed: some gap in your understanding may get filled that may give more meaning to (and make it easier to remember) some of the definitions, rules, and concepts on your study cards. Perhaps you will understand a shortcut that you missed the first time around.

Reviewing the explanations or problems in the text *and* your notes gives you a better perspective and helps tie the material together. Concepts will begin to make more sense when you review and think about how they are interrelated. It is also important to practice review problems so that you will not forget those skills you have already learned. Do not forget to review old homework exercises, quizzes, and exams—especially those problems that were incorrectly done. Review problems also offer an excellent opportunity to work on your speed as well as your accuracy.

We discussed reflecting on the material you are reading and the exercises you are doing. Your thinking time is usually limited during an exam, and you want to anticipate variations in problems and make sure that your careless errors will be minimized at that time. For this reason it is a good idea to try to think about possible problems ahead of time. Make as clear as possible the distinctions that exist in those areas where you tend to get confused.

As you review material, ask yourself the study questions given in Study Skills 2.1. Also look at the Questions for Thought at the end of most of the exercise sets and ask yourself those questions as well.

CHAPTER 3   REVIEW EXERCISES

In Exercises 1–4, determine whether the given equation or inequality is conditional, an identity, or a contradiction.

1. $5(x - 4) - 3(x - 3) = 3 - (14 - 2x)$

2. $3x - 4(x - 3) < 3 - (x - 4)$

3. $3a(a + 3) - a(2a + 4) = a^2 + 10$

4. $a(a - 5) - 2(a - 5) = a^2 - 7a + 10$

In Exercises 5–14, determine whether the given values of the variable satisfy the equation or inequality.

5. $2x - 5 = -7$; $x = -6, -1$

6. $3x - 7 = 5$; $x = -4, 4$

7. $4y + 3 \leq 10 + 2y$; $y = 4, \frac{5}{2}$

8. $6w + 11 \geq 31 - 9w$; $w = -2, \frac{4}{3}$

9. $3t + 2(t + 1) = 3t + 3$; $t = -2, \frac{1}{2}$

10. $5z + 3(z - 1) = 10z - 12$; $z = -3, \frac{6}{5}$

11. $8 - 3(x - 2) > x - 4$; $x = -5, 5$

12. $7 - 5(x - 4) \leq x + 2$; $x = -4, 4$

13. $a^2 + (a - 2)^2 = 20$; $a = -2, 2$

14. $u^2 - (u - 5)^2 = 4$; $u = -3, 3$

In Exercises 15–26, solve the equation or inequality.

15. $5x + 8 = 2x - 7$

16. $3x - 11 = 7x + 5$

17. $2(y + 4) - 2y = 8$

18. $3r + 2(r - 4) = r + 4(r - 1)$

19. $2(3a + 4) + 8 = 5(3a - 1)$

20. $5(2t - 3) + 2(t - 2) = 5$

21. $8x - 3(x - 4) = 4(x + 3) + 28$

22. $4(x - 2) - 7x = 2x - 3(x + 2)$

23. $a(a + 3) - 2(a - 1) = a(a - 1) + 7$

24. $2w(w + 1) - 3(w - 2) = w^2 - 1 + w(w - 8)$

25. $8 - 3(x - 1) < 2$

26. $9 - 5(x - 2) \geq 4$

In Exercises 27–30, solve the inequality and sketch the solution set on a number line.

27. $2(x - 3) - 4(x - 1) \geq 7 - x$

28. $3(2x + 1) - 4(3x - 1) < 17 - 4x$

29. $2 \leq 3a + 8 < 20$

30. $-2 < 4 - 2t < 12$

*Solve each of the following problems **algebraically.***

31. One number is 3 less than twice another. If their sum is 18, find the numbers.

32. The larger of two numbers is 7 less than 3 times the smaller. If their sum is 8 more than the smaller number, find the numbers.

33. The length of a rectangle is 4 more than 5 times the width. If the perimeter is 80 cm, find the dimensions of the rectangle.

34. The width of a rectangle is 8 less than twice the length. If the perimeter is 1 less than 4 times the width, find the dimensions of the rectangle.

35. A manufacturer sells first-quality skirts for $15 each, and irregulars for $9 each. If a wholesaler spends $2,010 on 150 skirts, how many of each type did she buy?

36. A round-trip by car takes 7 hours. If the rate going was 45 kph and the rate returning was 60 kph, how far was the round-trip?

37. A laborer earns $12/hour regular wages and $18/hour for overtime (overtime being computed for working more than 40 hours per week). If a laborer wants to earn at least $570 during a certain week, what is the minimum number of overtime hours he must work?

38. A store advertises a 20% discount on all shirts. If the sale prices ranged from $18.36 to $27.96, what was the original price range of the shirts?

1. Determine whether the given equation is conditional, an identity, or a contradiction.

   (a) $3x - 5(x - 2) = -2x + 8$

   (b) $3x - 5(x - 2) = -2x + 10$

   (c) $3x - 5(x - 2) = 2x - 10$

2. Determine whether the given value is a solution to the equation or inequality.

   (a) $2(x + 3) - (x - 1) = 9 - x; \quad x = 1$

   (b) $a^2 - 3a = a(a + 1) + 5; \quad a = -2$

   (c) $8 - 3(t - 4) < 10; \quad t = 3$

3. Solve each of the following equations or inequalities:

   (a) $6 - 3x = 3x - 10$

   (b) $2(3y - 5) - 4y = 2 - (y + 12)$

   (c) $2a^2 - 3(a - 4) = 2a(a - 6) + 3a$

   (d) Solve and sketch the solution set on a number line: $9 - 5(x - 2) \geq 4$.

   (e) Solve and sketch the solution set on a number line: $1 < 3 - x \leq 5$.

4. The length of a rectangle is 5 less than 4 times its width. If the perimeter is 11 more than 7 times the width, find the dimensions of the rectangle.

5. At a flea market used audio cassettes sell for $1 each while new ones sell for $3 each. If a person spends $46 on 20 cassettes, how many of each type did he buy?

6. If 8 less than four times a number is 1 more than twice the number, what is the number?

7. Jan earns a 1.25% commission on all sales above $10,000. How much must she sell to earn a commission of at least $1,500?

Study Skills 3.7

Preparing for Exams: Using Quiz Cards

A few days before the exam, select an appropriate number of problems from the quiz cards or old exams or quizzes, and make up a practice test for yourself. You may need the advice of your teacher as to the number of problems and the amount of time to allow yourself for the test. If available, old quizzes and exams may help guide you.

Now find a quiet, well-lit place with no distractions, set your clock for the appropriate time limit (the same as your class exam will be), and take the test. Pretend it is a real test; that is, do not leave your seat or look at your notes, books, or answers until your time is up. (Before giving yourself a test, you may want to refer to next chapter's discussion on taking exams.)

When your time is up, stop; you may now look up the answers and grade yourself. If you are making errors, check over what you are doing wrong. Find the section where those problem types are covered, review the material, and try more problems of that type.

If you do not finish your practice test on time, you should definitely work on your speed. Remember that speed as well as accuracy is important on most exams.

Think about what you were doing as you took your test. You may want to change your test-taking strategy or reread the next chapter's discussion on taking exams. If you were not satisfied with your performance and you have the time after the review, make up and give yourself another practice test.

CHAPTERS 1–3   CUMULATIVE REVIEW

In Exercises 1–24, *perform the indicated operations and simplify as completely as possible.*

1. $-8 - 5 - 7$

2. $-8 - 5(-7)$

3. $12 - 4(3 - 5)$

4. $12 - (4 \cdot 3 - 5)$

5. -5^2

6. $(-5)^2$

7. xx^2x^3

8. $x^2x^3x^4 + x^5x^3x$

9. $x^2y - 2xy^2 - xy^2 - 3x^2y$

10. $3z^2 - 5z - 7 - 2z^2 - z - 1$

11. $2x(3x^2 - 4y)$

12. $2x(3x^2)(-4y)$

13. $-3u^2(u^3)(-5v)$

14. $-3u^2(u^3 - 5v)$

15. $4(m - 3n) + 3(2m - n)$

16. $7(2t^2 - 3r^3) + 5(r^3 - 3t^2)$

17. $2ab(a^2 - ab) - 4a^2(ab - b^2)$

18. $3x^2yz(xz - 6y^2) - 9y^2z(x^2 - 2x^2y)$

19. $x^2y - xy^2 - (xy^2 - x^2y)$

20. $8(x - y) - (8x - y)$

21. $x - 3[x - 4(x - 5)]$

22. $a - [b - a(b - a)]$

23. $3xy(4x^3y - 2y) - 2x(3y^2)(2x^3)$

24. $x(x - y) - y(y - x) - (x^2 - y^2)$

In Exercises 25–32, evaluate the given expression for $x = -2$, $y = -3$, and $z = 5$.

25. x^2

26. $-x^2$

27. $xy^2 - (xy)^2$

28. $(x - z)^2$

29. $|x - y - z|$

30. $xyz + xy - z$

31. $2x - 4y^2$

32. $|3x - z|$

In Exercises 33–44, solve the equation or inequality. In the case of an inequality, sketch the solution set on a number line.

33. $2x + 11 = 5x + 10$

34. $9 - 5t = 13 - 4t$

35. $9(a + 1) - 3(2a - 2) = 12$

36. $3(4w - 5) - (8w + 1) = -1$

37. $4(5 - x) - 2(6 - 2x) = 8$

38. $8 - 3(a - 7) \leq 2$

39. $2(s + 4) + 3(2s + 2) > 6s$

40. $6(t - 3) + 4(6 - t) = 8 + 2(t - 5)$

41. $1 < 2y - 5 \leq 3$

42. $1 \leq 4 - z \leq 6$

43. $4(3d - 2) + 6(8 - d) = 9d - 4(d - 10)$

44. $2[x + 2(x + 2)] = 8$

Solve each of the following problems algebraically. Be sure to label clearly what the variable represents.

45. Eight less than twice a number is greater than or equal to 4 more than the number. Find the number.

46. The length of a rectangle is 8 more than 3 times its width. If the perimeter is 44 cm, find the dimensions of the rectangle.

47. A bakery charges 40¢ for "danishes" and 55¢ for "pastries." Louise pays $8.25 for an assortment of 18 danishes and pastries. How many of each type were in the assortment?

48. An office has an old copier that makes 20 copies per minute and a newer model that makes 25 copies per minute. If the older machine begins making copies at 10:00 A.M. and is joined by the newer machine at 10:15 A.M., at what time will they have made a total of 885 copies?

CHAPTERS 1–3   CUMULATIVE PRACTICE TEST

1. Evaluate each of the following:

   (a) $-3^4 + 3(-2)^3$

   (b) $(3 - 7 - 2)^2$

2. Evaluate each of the following for $u = -4$, $v = -1$.

    (a) $(u - v)^2 - uv^2$ (b) $|2u - 3v|$

3. Perform the indicated operations and simplify as completely as possible.

    (a) $5x^2 - 4x - 8 - 7x^2 - x + 11$ (b) $2(x - 3y) + 5(y - x)$

    (c) $3a(a^2 - 2b) - 5(a^3 - ab)$ (d) $2(u - 4v) - 3(v - 2u) - (8u - 11v)$

    (e) $4x^2y^3(x - 5y) - 2xy(5y^2)(-2xy)$ (f) $6 - a[6 - a(6 - a)]$

4. Solve each of the following equations and inequalities.

    (a) $8 - 8x = 4 - 3x$ (b) $2(w - 4) - 3(w - 2) = 7$

    (c) $9 - 5(x - 2) \leq 4$ (d) $3(a + 5) - 2(1 - a) = 1 - (8 - a)$

    (e) $6(2 - z) - 5(3 - z) = 4 - z$ (f) $5t + 2(t - 7) - (t - 3) = 6t - 11$

5. Solve the following inequalities and sketch the solution set on a number line.

    (a) $7 - 3a \geq 13$ (b) $1 \leq 4x - 3 < 17$

6. Solve each of the following problems algebraically.

    (a) One number is 5 less than 3 times another. If their sum is 27, find the numbers.

    (b) A tire retailer pays $1,124 for 40 tires. Some were new tires costing $32 each, while the rest were retreads costing $19 each. How many of each type were there?

    (c) Judy completes a bike race at an average speed of 15 kph. If she had averaged 25 kph, she would have finished the race 1 hour faster. What was the distance that the race covered?

    (d) A store advertises sweaters at an 18% discount. If the least expensive sale price is $36.49, what was the least expensive original price?

CHAPTER 4

Rational Expressions

STUDY SKILLS

In the first three chapters we have focused primarily on algebraic expressions involving the addition, subtraction, and multiplication of variables. However, just as we need numerical fractions to describe many real-life situations, we also need *algebraic fractions* to describe many real-life relationships. An algebraic fraction is a fractional expression such as $\frac{x + 1}{3x}$, in which variables appear in the *numerator* (top) and/or the *denominator* (bottom).

For example, in photography the following equation is quite useful:

$$\frac{1}{f} = \frac{1}{d_s} + \frac{1}{d_i}$$

This equation is discussed further in Example 7 of Section 9.6.

This is called the *lens equation* and relates the focal length of the lens f, the distance d_s from the subject to the lens, and the distance d_i from the image to the lens. See the figure below.

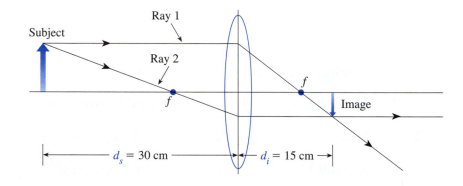

In the sections that follow, we will begin by reviewing the various arithmetic operations for arithmetic fractions and then see how the same procedure carries over to algebraic fractions.

4.1   Fundamental Principle of Fractions

Recall that an arithmetic fraction (rational number) is simply the quotient of two integers, where the denominator is not equal to 0. In Chapter 1 we defined the set of rational numbers as those numbers that *can be* represented as the quotient of integers.

For example, $\frac{2}{3}$, $-\frac{3}{5}$, and 4 are all arithmetic fractions. Remember that every integer is also a fraction since it has an understood denominator of 1 that, while it is seldom written, is very useful to think of in many situations. Thus, $4 = \frac{4}{1}$.

We can locate any rational number, $\frac{p}{q}$, on the number line by recalling that $\frac{p}{q}$ means "p of q equal parts." For example, $\frac{5}{8}$ is 5 of 8 equal parts. In order to locate the point $\frac{5}{8}$, we divide the interval between 0 and 1 on the number line into 8 equal parts and count out 5 of those parts to the right of 0:

Similarly, $-\frac{5}{4}$ would be 5 of 4 equal parts counted out to the left of 0.

The basic starting point in dealing with and understanding fractions is simply recognizing the fact that two fractions can look different and yet be equal (represent the same amount). It does not make any difference if a certain whole is divided into 3 equal parts and you have 2, or if that same whole is divided into 6 equal parts and you have 4. In both cases you have the same amount, as indicated in the next figure. Thus, the fractions $\frac{2}{3}$ and $\frac{4}{6}$ are equal—they each represent the same amount.

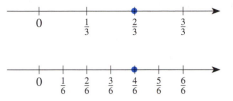

What we have just illustrated is called the **_Fundamental Principle of Fractions._** It says that the value of a fraction is unchanged if the numerator and denominator are both multiplied or divided by the same nonzero quantity. (We must specify nonzero, for if not, we would be dividing by 0, which, as we have already seen, does not make any sense.) Algebraically, we have the result given in the box.

Fundamental Principle of Fractions	$\dfrac{a}{b} = \dfrac{a \cdot k}{b \cdot k}$ where $b, k \neq 0$

If we read this from left to right, we are multiplying both the numerator and the denominator by k—this is called **_building fractions to higher terms._** If we read from right to left, we are dividing both the numerator and the denominator by k—this is called **_reducing the fraction to lower terms._**

$$\frac{12}{15} = \frac{12 \cdot 4}{15 \cdot 4} = \frac{48}{60} \qquad \textit{This is an example of building fractions.}$$

$$\frac{12}{15} = \frac{4 \cdot 3}{5 \cdot 3} = \frac{4 \cdot \cancel{3}}{5 \cdot \cancel{3}} = \frac{4}{5} \qquad \textit{This is an example of reducing fractions.}$$

The slashes indicate that we have divided both numerator and denominator by 3. When we use the Fundamental Principle in this way, we frequently say that we have "canceled" the common factor of 3. Crossing out the 3's is just a convenient way of indicating that we are dividing the numerator and denominator by 3. The key thing to keep in mind is that we are allowed to cancel only *common factors* and not common terms. Thus, it is very important that we be able to distinguish terms from factors. Recall from Chapter 2 that a **_term_** is an algebraic expression that is connected by multiplication and/or division. For example, the expression $3at + xyz$ has two terms, each of which consists of three factors, while the expression $3xyz$ is one term consisting of four factors.

The following very common error should be noted:

$$\frac{3 + 2}{1 + 2} \neq \frac{3 + \cancel{2}}{1 + \cancel{2}} \neq 3 \quad \text{because} \quad \frac{3 + 2}{1 + 2} = \frac{5}{3} \neq 3$$

The 2's *cannot* be canceled because 2 is not a common factor, but a common term.

Remember	$\dfrac{a \cdot k}{b \cdot k} = \dfrac{a}{b}$ but $\dfrac{a + k}{b + k} \neq \dfrac{a}{b}$ where $k \neq 0$

In words, this says **_common terms cannot be canceled._**

We will further discuss the concept of building fractions when we get to addition and subtraction of fractions in Section 4.3. For the time being, we will concentrate on reducing fractions.

For the sake of simplicity and uniformity, we require that all final answers be reduced completely. Such an answer is said to be reduced to *lowest terms.*

DEFINITION

A fraction is said to be reduced to *lowest terms* if the numerator and the denominator have no common factor other than 1 or −1.

EXAMPLE 1

Reduce to lowest terms.

(a) $\dfrac{42}{70}$ (b) $\dfrac{18x^5}{12x^2}$

Solution

(a) Using the Fundamental Principle of Fractions, we get

$$\frac{42}{70} = \frac{3 \cdot 14}{5 \cdot 14} = \frac{3 \cdot \cancel{14}}{5 \cdot \cancel{14}} = \boxed{\frac{3}{5}}$$

This entire process is often written in a shorthand fashion as follows:

$$\frac{\overset{3}{\cancel{42}}}{\underset{5}{\cancel{70}}} = \boxed{\frac{3}{5}}$$

Writing the solution this way "hides" the fact that we used a common factor of 14 to reduce the fraction. Additionally, it is quite possible that we might not see that 14 is a common factor of both 42 and 70. Consequently, we could also have proceeded as follows:

$$\frac{42}{70} = \frac{6 \cdot 7}{10 \cdot 7} = \frac{6 \cdot \cancel{7}}{10 \cdot \cancel{7}} = \frac{6}{10} = \frac{3 \cdot 2}{5 \cdot 2} = \frac{3 \cdot \cancel{2}}{5 \cdot \cancel{2}} = \boxed{\frac{3}{5}}$$

In shorthand, this would be written

$$\frac{\overset{\overset{3}{\cancel{6}}}{\cancel{42}}}{\underset{\underset{5}{\cancel{10}}}{\cancel{70}}} = \boxed{\frac{3}{5}}$$

While $\frac{6}{10}$ is a *partially* reduced form of $\frac{42}{70}$, it is not the final answer because 6 and 10 still have a common factor of 2. We must use the common factor of 2 to reduce the fraction in order for the fraction to be in the *lowest terms.*

To make the reducing process as efficient as possible, start reducing with the *largest* common factor that you see.

(b) This same example could have been written as $(18x^5) \div (12x^2)$ because the fraction bar is an alternative way of writing division.

In this example, as in all those that follow, we automatically exclude any value of the variable that makes the denominator equal to 0. Thus, for this expression, $x \neq 0$.

$$\frac{18x^5}{12x^2} = \frac{2 \cdot 3 \cdot 3xxxxx}{2 \cdot 2 \cdot 3xx} = \frac{2 \cdot 3 \cdot 3x\cancel{xxxx}}{2 \cdot 2 \cdot 3\cancel{xx}} = \boxed{\frac{3x^3}{2}}$$

Or, we could write this process as follows:

$$\frac{18x^5}{12x^2} = \frac{3 \cdot 6 \cdot x^3 \cdot x^2}{2 \cdot 6 \cdot x^2} = \frac{3 \cdot \cancel{6} \cdot x^3 \cdot \cancel{x^2}}{2 \cdot \cancel{6} \cdot \cancel{x^2}} = \boxed{\frac{3x^3}{2}}$$

This reduction process is usually written as

$$\frac{18x^5}{12x^2} = \frac{\overset{\overset{3}{x^3}}{\cancel{18x^5}}}{\underset{2}{\cancel{12x^2}}} = \boxed{\frac{3x^3}{2}}$$

■

One of the basic steps in simplifying a fraction is reducing it to lowest terms. When a fraction is reduced to lowest terms it is usually easier to comprehend the size of a fraction, as illustrated in the following example.

EXAMPLE 2

We will discuss ratio further in Section 4.5.

On a particular day, the New York Stock Exchange reported that 1,176 stocks went up in value (called *gainers*) and 420 stocks went down in value (called *losers*). Express the ratio of gainers to losers in simplest form.

Solution

A **ratio** is another name for a fraction. The ratio of gainers to losers means a fraction with the number of gainers in the numerator and the number of losers in the denominator.

$$\frac{\text{Number of gainers}}{\text{Number of losers}} = \frac{1{,}176}{420}$$ *Let's use the prime factorizations to reduce the fraction.*

$$= \frac{2 \cdot 2 \cdot 2 \cdot 3 \cdot 7 \cdot 7}{2 \cdot 2 \cdot 3 \cdot 5 \cdot 7}$$

$$= \boxed{\frac{14}{5}}$$

Therefore, the ratio of gainers to losers is 14 to 5. This simplified ratio makes it easier to see that the ratio of gainers to losers was almost 3 to 1. In other words, there were nearly 3 gainers for each loser. ■

EXAMPLE 3

Reduce to lowest terms.

(a) $\dfrac{-12x^3y^4}{2x^5y^3}$ (b) $\dfrac{x^2}{x+2}$

Solution   (a) $\dfrac{-12x^3y^4}{2x^5y^3} = \dfrac{-6 \cdot 2x^3 \cdot y^3 \cdot y}{2x^3 \cdot x^2 \cdot y^3} = \dfrac{-6 \cdot \cancel{2x^3} \cdot \cancel{y^3} \cdot y}{\cancel{2x^3} \cdot x^2 \cdot \cancel{y^3}} = \boxed{\dfrac{-6y}{x^2}}$

As we indicated in the last example, much of this work is often done mentally, and the solution would look like

$$\frac{-12x^3y^4}{2x^5y^3} = \frac{\overset{-6}{\cancel{-12}}x^3\overset{y}{\cancel{y^4}}}{\underset{x^2}{\cancel{2x^5}}y^3} = \boxed{\frac{-6y}{x^2}}$$

In the shorthand solution we have reduced 2 with -12, x^3 with x^5, and y^3 with y^4.

(b) We cannot apply the Fundamental Principle of Fractions to $\dfrac{x^2}{x+2}$.

$\dfrac{x^2}{x+2}$ *cannot* be reduced because there is no common *factor.*

The x in the numerator is a factor but the x in the denominator is *not* a factor—it is a term. The fraction as given is already in lowest terms. ■

EXAMPLE 4

Reduce to lowest terms.

(a) $\dfrac{(4x^3)(-3xy)}{(6xy)(x^2)}$ (b) $\dfrac{(3x^2y)^2}{(2xy^2)^3}$ (c) $\dfrac{8x - 7x}{4x^2 - 7x^2}$

Solution   (a) As with many problems of this type, there is more than one correct method. We illustrate two possible approaches.

First approach

We first multiply out and then reduce.

$$\frac{(4x^3)(-3xy)}{(6xy)(x^2)} = \frac{-12x^4y}{6x^3y} \qquad \textit{Now reduce.}$$

$$= \frac{-\overset{-2}{\cancel{12}}x^{\overset{x}{\cancel{4}}}y}{\cancel{6}x^{\cancel{3}}\cancel{y}}$$

$$= \frac{-2x}{1} = \boxed{-2x}$$

Second approach

We first reduce and then multiply out.

$$\frac{(4x^3)(-3xy)}{(6xy)(x^2)} = \frac{(4x^3)(-3x\cancel{y})}{(6x\cancel{y})(x^2)}$$

We can reduce further.

$$= \frac{-12x^3}{6x^2}$$

$$= \frac{-\overset{-2}{\cancel{12}}x^{\overset{x}{\cancel{3}}}}{\cancel{6}x^{\cancel{2}}} = \frac{-2x}{1} = \boxed{-2x}$$

(b) We *cannot* cancel within the parentheses because of the outside exponents. Instead, we can proceed as follows.

$$\frac{(3x^2y)^2}{(2xy^2)^3} = \frac{(3x^2y)(3x^2y)}{(2xy^2)(2xy^2)(2xy^2)} \qquad \textit{Multiply.}$$

$$= \frac{9x^4y^2}{8x^3y^6} \qquad \textit{Now reduce.}$$

$$= \frac{9x^{\overset{x}{\cancel{4}}}y^2}{8x^{\cancel{3}}y^{\underset{y^4}{\cancel{6}}}} = \boxed{\frac{9x}{8y^4}}$$

(c) Our first step must be to simplify the numerator and denominator by combining like *terms*. Then we can look further to see if we have any common *factors* that we can use to reduce the fraction.

$$\frac{8x - 7x}{4x^2 - 7x^2} = \frac{x}{-3x^2} \qquad \textit{Now we can reduce.}$$

$$= \frac{\cancel{x}}{-3x^{\underset{x}{\cancel{2}}}}$$

$$= \frac{1}{-3x} = \boxed{-\frac{1}{3x}} \qquad \textit{Don't forget the 1 in the numerator.} \quad ■$$

Signs of Fractions

A comment is in order about the location of a minus sign in a fraction. Since dividing quantities with opposite signs yields an answer with a negative sign, it makes no difference whether the minus sign appears in the numerator, denominator, or in front of the fraction. However, the preferred form is to have the minus sign either in front or in the numerator.

$$-\frac{6}{7} = \frac{-6}{7} = \frac{6}{-7} \qquad \textit{These fractions are all equal, but the first and second are the preferred forms.}$$

Notice that in part **(c)** of Example 4 we wrote our final answer in the preferred form.

Study Skills 4.1

Taking an Algebra Exam: Just Before the Exam

You will need to concentrate and think clearly during the exam. For this reason it is important that you get plenty of rest the night before the exam and that you have adequate nourishment.

(continued)

Study Skills 4.1 (continued)

Taking an Algebra Exam: Just Before the Exam

It is *not* a good idea to study up until the last possible moment. You may find something that you missed and become anxious because there is not enough time to learn it. Then rather than simply missing a problem or two on the exam, the anxiety may affect your performance on the entire exam. It is better to stop studying some time before the exam and do something else. You could, however, review formulas you need to remember and warnings (common errors you want to avoid) just before the exam.

Also, be sure to give yourself plenty of time to get to the exam.

EXERCISES 4.1

Reduce each of the following rational expressions to lowest terms.

1. $\dfrac{18}{30}$

2. $\dfrac{12}{28}$

3. $\dfrac{-9}{21}$

4. $\dfrac{-10}{24}$

5. $\dfrac{-15}{-6}$

6. $\dfrac{-14}{-8}$

7. $\dfrac{-5 + 8}{10 - 4}$

8. $\dfrac{4 - 6}{4 - 2}$

9. $\dfrac{8 - 5(2)}{6 - 4(3)}$

10. $\dfrac{3 - 6(4)}{-2 - 3(4)}$

11. $\dfrac{x^3}{x}$

12. $\dfrac{y^5}{y^2}$

13. $\dfrac{x}{x^3}$

14. $\dfrac{y^2}{y^5}$

15. $\dfrac{10x}{4x^2}$

16. $\dfrac{8y^2}{6y^3}$

17. $\dfrac{-3z^6}{5z^2}$

18. $\dfrac{-7w^9}{9w^3}$

19. $\dfrac{12t^5}{30t^{10}}$

20. $\dfrac{-28u^8}{16u^4}$

21. $\dfrac{6ab^5}{-2a^3b^2}$

22. $\dfrac{10a^2b^6}{-5a^4b^4}$

23. $\dfrac{(2x^3)(6x^2)}{(4x)(3x^4)}$

24. $\dfrac{(4y^4)(5y^5)}{(8y^3)(y^6)}$

25. $\dfrac{(r^3t^2)(-rt^3)}{2r^2t^7}$

26. $\dfrac{(-a^4b^3)(a^2b^5)}{3a^3b^6}$

27. $\dfrac{3a(5b)(-4ab^3)}{6ab(2a^2b^2)}$

28. $\dfrac{-2s^2(8t^2)(5st)}{6s^3t^2(5st^3)}$

29. $\dfrac{(2x)^5}{(4x)^3}$

30. $\dfrac{(3y)^4}{(6y)^2}$

31. $\dfrac{(-4x^3)^2}{(-2x^4)^3}$

32. $\dfrac{(-6y^2)^2}{(-2y^4)^3}$

33. $\dfrac{(2xy^2)^3}{(4x^2y^3)^3}$

34. $\dfrac{(3x^2y)^4}{(9xy^2)^2}$

35. $\dfrac{5x - 2x}{10x - 4x}$

36. $\dfrac{6y - 3y}{12y - 6y}$

37. $\dfrac{5a(2x)}{15a(8x)}$

38. $\dfrac{5a - 2x}{15a - 8x}$

39. $\dfrac{4s - 3t}{8s - 9t}$

40. $\dfrac{4s(3t)}{8s(9t)}$

41. $\dfrac{7a^2 - 5a^2 - 6a^2}{4a - 8a}$

42. $\dfrac{10z^2 - 8z^2 - 4z^2}{5z^3 - 7z^3}$

43. $\dfrac{5x^2 - 3x - x^2 + 3x}{6x^2 - 5x - 2x^2 + 5x}$

44. $\dfrac{2y^2 + 2y - y^2 - 2y}{5y^2 + y - 4y^2 - y}$

❓ QUESTIONS FOR THOUGHT

45. In your own words, state the Fundamental Principle of Fractions.

46. If each step of the following were correct, what would you conclude?

$$2 = \frac{6}{3} = \frac{4 + 2}{1 + 2} \overset{?}{=} \frac{4 + 2}{1 + 2} \overset{?}{=} \frac{4}{1} = 4$$

Which step was *incorrect*? Explain why.

47. Discuss what is **wrong** with each of the following:

(a) $\dfrac{3x^2}{2x} \overset{?}{=} \dfrac{3}{2x}$

(b) $\dfrac{5x}{25x^2} \overset{?}{=} \dfrac{5x}{5 \cdot 5xx} \overset{?}{=} \dfrac{0}{5x} \overset{?}{=} 0$

(c) $\dfrac{2xy}{6x^3y^2} \overset{?}{=} \dfrac{2xy}{2 \cdot 3xyx^2y} \overset{?}{=} 3x^2y$

(d) $\dfrac{3x + 2x^2}{x} \overset{?}{=} \dfrac{3x + 2x^2}{x} \overset{?}{=} 3 + 2x^2$

48. Group the equivalent fractions together:

(a) $-\dfrac{3}{4}$ (b) $\dfrac{-3}{-4}$ (c) $\dfrac{3}{-4}$ (d) $-\dfrac{-3}{-4}$

(e) $\dfrac{-3}{4}$ (f) $-\dfrac{-3}{4}$ (g) $-\dfrac{3}{-4}$

◇ MINI-REVIEW

Solve each of the following problems algebraically. Be sure to label what the variable represents.

49. The length of a rectangle is 5 cm more than twice the width. If the perimeter of the rectangle is 34 cm, find its dimensions.

50. A collection of nickels and dimes is worth $1.30. If there are 5 fewer dimes than nickels, how many of each are there?

4.2 Multiplying and Dividing Rational Expressions

Having begun to handle algebraic fractions in the last section, we now turn our attention to formulating methods for performing the arithmetic operations with them. We will begin with the procedures for multiplication and division, which are simpler than those for addition and subtraction.

Multiplication

We take as our starting point the multiplication of ordinary arithmetic fractions. For example,

Similarly,
$3 \cdot \frac{1}{2} = \frac{3}{1} \cdot \frac{1}{2} = \frac{3}{2}$

$$\frac{2}{3} \cdot \frac{5}{7} = \frac{2 \cdot 5}{3 \cdot 7} = \boxed{\frac{10}{21}}$$

The multiplication is accomplished by multiplying the numerators and dividing by the product of the denominators. It is quite natural, then, to extend this rule to the multiplication of any two algebraic fractions.

Multiplication of Rational Expressions	$\dfrac{a}{b} \cdot \dfrac{c}{d} = \dfrac{a \cdot c}{b \cdot d}$ where $b, d \neq 0$

Before we look at an example involving variables, let's look at one more numerical example.

EXAMPLE 1 Multiply. $\dfrac{8}{9} \cdot \dfrac{3}{10}$

Solution Following the rule for multiplying fractions as stated in the box, we get

$$\frac{8}{9} \cdot \frac{3}{10} = \frac{8 \cdot 3}{9 \cdot 10}$$

$$= \frac{24}{90} \qquad \textit{Now we reduce the fraction.}$$

$$= \frac{4 \cdot 6}{15 \cdot 6}$$

$$= \frac{4 \cdot \cancel{6}}{15 \cdot \cancel{6}}$$

$$= \boxed{\frac{4}{15}} \qquad\qquad ■$$

In getting the final answer in Example 1, we followed the ground rules laid down in the last section requiring that our final answer be reduced to lowest terms.

When the multiplication is carried out, any factor in a numerator ends up in the numerator of the product, and any factor in a denominator ends up in the denominator of the product. Therefore, it is much more efficient to reduce *before* we actually carry out the multiplication.

In other words, it is much easier to do Example 1 as follows:

$$\frac{8}{9} \cdot \frac{3}{10} = \frac{\overset{4}{\cancel{8}}}{\underset{3}{\cancel{9}}} \cdot \frac{\overset{1}{\cancel{3}}}{\underset{5}{\cancel{10}}} \qquad \textit{We reduce the 3 with the 9 and the 8 with the 10.}$$

$$= \frac{4 \cdot 1}{3 \cdot 5} = \boxed{\frac{4}{15}}$$

In the examples that follow we will adopt the approach of trying to use any common factors to reduce before we multiply.

EXAMPLE 2 | Multiply. $\dfrac{-14}{x^6} \cdot \dfrac{x^3}{4}$

Solution $\dfrac{-14}{x^6} \cdot \dfrac{x^3}{4} = \dfrac{-14}{x^3 x^3} \cdot \dfrac{x^3}{4}$ *We illustrate the process of reducing the fractions.*

$= \dfrac{-\overset{-7}{\cancel{14}}}{x^3 x^3} \cdot \dfrac{\cancel{x^3}}{\underset{2}{\cancel{4}}}$

Verify that you get the same result if you multiply the fractions first and then reduce.

$= \boxed{\dfrac{-7}{2x^3}}$ ∎

EXAMPLE 3 | Multiply. $\dfrac{5x^3}{4y^2} \cdot \dfrac{-6y^8}{25x^4}$

Solution An example such as this one, which has many common factors that can be used to reduce the fraction, can be difficult to follow. Consequently, we will show each step in the reduction process separately. (The order in which we choose to use the common factors is arbitrary.) When you do such an exercise, however, you will most likely do the reducing all in one step.

$\dfrac{5x^3}{4y^2} \cdot \dfrac{-6y^8}{25x^4}$ *Reduce the 5 with the 25 and the 4 with the −6.*

$= \dfrac{\cancel{5}x^3}{\underset{2}{\cancel{4}}y^2} \cdot \dfrac{-\overset{-3}{\cancel{6}}y^8}{\underset{5}{\cancel{25}}x^4}$ *Next reduce x^3 with x^4.*

$= \dfrac{\cancel{x^3}}{2y^2} \cdot \dfrac{-3y^8}{5\underset{x}{\cancel{x^4}}}$ *Finally we reduce y^2 with y^8.*

$= \dfrac{1}{2y^2} \cdot \dfrac{-3\overset{y^6}{\cancel{y^8}}}{5x}$

$= \dfrac{-3y^6}{2 \cdot 5x} = \boxed{\dfrac{-3y^6}{10x}}$ ∎

Now that we have worked a bit with multiplication, let's look at the Fundamental Principle of Fractions again and see how the two ideas are related. We know that

$$\frac{k}{k} = 1, \qquad k \neq 0$$

Since multiplying by 1 does not change the value of an expression, we have the following:

$$\frac{a}{b} \cdot 1 = \frac{a}{b} \qquad \textit{Multiply by 1.}$$

$$\frac{a}{b} \cdot \frac{k}{k} = \frac{a}{b} \qquad \textit{Since } \frac{k}{k} = 1$$

$$\frac{a \cdot k}{b \cdot k} = \frac{a}{b} \qquad \textit{From the definition of multiplication}$$

This last line is exactly what the Fundamental Principle asserts. In other words, the Fundamental Principle simply says that multiplying by 1 does not change the value of a fraction.

When more than two fractions are to be multiplied together, we proceed in exactly the same way.

EXAMPLE 4

Multiply. $\dfrac{2}{3y} \cdot \dfrac{x}{5a} \cdot \dfrac{5}{8x^2}$

Solution

We reduce the 2 with the 8; the 5 with the 5; the x with the x^2.

$$\frac{2}{3y} \cdot \frac{x}{5a} \cdot \frac{5}{8x^2} = \frac{\cancel{2}}{3y} \cdot \frac{\cancel{x}}{\cancel{5}a} \cdot \frac{\cancel{5}}{\underset{4x}{\cancel{8x^2}}}$$

$$= \boxed{\frac{1}{12axy}} \qquad \textit{Note that the final answer is } \frac{1}{12axy} \textit{ and not } 12axy.$$

Keep in mind that when we reduce here we are left with a factor of 1 in each numerator.

■

EXAMPLE 5

Multiply. $5 \cdot \dfrac{3}{x}$

Solution

When doing examples involving expressions, some of which have denominators and some of which do not, it is a very good idea to put in the "understood" denominator of 1 for those expressions without a denominator. Thus, in this example we think of (and write) 5 in the fractional form $5 = \frac{5}{1}$.

$$5 \cdot \frac{3}{x} = \frac{5}{1} \cdot \frac{3}{x} = \boxed{\frac{15}{x}}$$

Do not make the mistake of multiplying both the numerator 3 and the denominator x by 5.

■

It is important to recognize the different forms fractional expressions can take. For example,

$$\frac{3}{5} = 3 \cdot \frac{1}{5} \qquad \text{because} \qquad 3 \cdot \frac{1}{5} = \frac{3}{1} \cdot \frac{1}{5} = \frac{3}{5}$$

Thus, in general

Similarly, $\dfrac{3}{4}x = \dfrac{3x}{4}$.

$$\boxed{\frac{a}{b} = a \cdot \frac{1}{b}}$$

Division

You probably remember the "rule" for dividing fractions as "invert and multiply." Before we state the rule explicitly and explain why the division of fractions is carried out in this way, let's keep in mind that division is defined to be the inverse of multiplication.

If someone asks "Why is $35 \div 7 = 5$?" a reasonable reply would be "Because $5 \cdot 7 = 35$." In other words, the answer to a division problem is checked by multiplication. So let's state the rule for dividing fractions, and then verify it.

Division of Rational Expressions	$\dfrac{a}{b} \div \dfrac{c}{d} = \dfrac{a}{b} \cdot \dfrac{d}{c} \qquad b, c, d \neq 0$

To see why division is defined in this way, remember that we verified

$$35 \div 7 = \quad 5 \qquad \text{because} \qquad 5 \quad \cdot \, 7 = 35 \quad \text{is true.}$$

$$\frac{a}{b} \div \frac{c}{d} = \frac{a}{b} \cdot \frac{d}{c} \quad \text{because} \quad \left(\frac{a}{b} \cdot \frac{d}{c}\right) \cdot \frac{c}{d} = \frac{a}{b} \quad \text{is true!} \quad \left(\text{since } \frac{a}{b} \cdot \frac{d}{c} \cdot \frac{c}{d} = \frac{a}{b}\right)$$

In other words, the rule for division was formulated in such a way as to make sure that it works out correctly.

EXAMPLE 6

A kilometer is approximately $\frac{5}{8}$ of a mile. Determine the number of kilometers in 20 miles.

Solution

Since each $\frac{5}{8}$ of a mile is approximately a kilometer, determining the number of kilometers in 20 miles is the same as determining how many times $\frac{5}{8}$ goes into 20. To find out how many times $\frac{5}{8}$ goes into 20, we need to divide 20 by $\frac{5}{8}$.

If we want to know how many times 6 goes into 30, we divide 30 by 6.

$$20 \div \frac{5}{8} = \frac{20}{1} \div \frac{5}{8} = \frac{20}{1} \cdot \frac{8}{5} = \boxed{32}$$

Therefore, there are approximately 32 kilometers in 20 miles. ∎

Another definition would be appropriate here.

DEFINITION

The *reciprocal* of a nonzero number x is defined to be $\dfrac{1}{x}$.

Thus, the reciprocal of 3 is $\dfrac{1}{3}$. The reciprocal of $\dfrac{3}{4}$ is $\dfrac{1}{\frac{3}{4}}$. Remember that this means $1 \div \dfrac{3}{4}$. Therefore,

$$\frac{1}{\frac{3}{4}} = 1 \div \frac{3}{4} = 1 \cdot \frac{4}{3} = \boxed{\frac{4}{3}}$$

Thus, we can see that in general,

The reciprocal of $\dfrac{a}{b}$ is $\dfrac{b}{a}$ where $a, b \neq 0$.

In light of this definition, the rule for dividing fractions can be restated as

"To divide by a fraction, multiply by its reciprocal."

EXAMPLE 7

Divide. $\dfrac{4x^2y^3}{5a^4} \div \dfrac{20ax}{y^3}$

Solution

Following the rule for division, we get

$$\frac{4x^2y^3}{5a^4} \div \frac{20ax}{y^3} = \frac{4x^2y^3}{5a^4} \cdot \frac{y^3}{20ax} \qquad \textit{Reduce the 4 with the 20, the x with the } x^2.$$

$$= \frac{4\overset{x}{x^2}y^3}{5a^4} \cdot \frac{y^3}{\underset{5}{20ax}}$$

$$= \frac{xy^3 \cdot y^3}{5a^4 \cdot 5a}$$

$$= \boxed{\dfrac{xy^6}{25a^5}}$$ ■

EXAMPLE 8   Divide. $\dfrac{x}{y} \div (xy)$

Solution   As we mentioned previously, it is very helpful to think of xy as $\dfrac{xy}{1}$.

$$\dfrac{x}{y} \div (xy) = \dfrac{x}{y} \div \dfrac{xy}{1} \qquad \textit{Use the rule for division.}$$

$$= \dfrac{x}{y} \cdot \dfrac{1}{xy}$$

$$= \dfrac{\not{x}}{y} \cdot \dfrac{1}{\not{xy}}$$

$$= \boxed{\dfrac{1}{y^2}}$$ ■

EXAMPLE 9   Perform the indicated operations and simplify. $\dfrac{a^2}{b}\left(\dfrac{ab}{c} \div \dfrac{c^2}{b^2}\right)$

Solution   Following the order of operations, we begin by working within the parentheses:

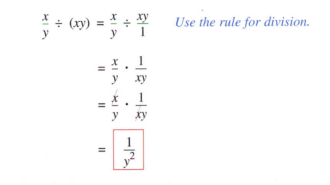

$$\dfrac{a^2}{b}\left(\dfrac{ab}{c} \div \dfrac{c^2}{b^2}\right) \qquad \textit{We follow the rule for dividing fractions.}$$

$$= \dfrac{a^2}{b}\left(\dfrac{ab}{c} \cdot \dfrac{b^2}{c^2}\right) \qquad \textit{Now we multiply within the parentheses.}$$

$$= \dfrac{a^2}{b} \cdot \dfrac{ab^3}{c^3} \qquad \textit{Next we reduce.}$$

$$= \dfrac{a^2}{\not{b}} \cdot \dfrac{ab^{\overset{b^2}{\not{3}}}}{c^3}$$

$$= \boxed{\dfrac{a^3b^2}{c^3}}$$ ■

Since a quotient *is* a fraction, another way of expressing a quotient of fractions is by using a large fraction bar rather than the division sign "÷". Thus, we can write

$$\dfrac{a}{b} \div \dfrac{c}{d} \quad \text{as} \quad \dfrac{\dfrac{a}{b}}{\dfrac{c}{d}}$$

A fractional expression that contains fractions within it, such as the preceding one, is called a ***complex fraction.***

EXAMPLE 10   Change the following to a simple fraction reduced to lowest terms. $\dfrac{\dfrac{3x^2}{2y}}{\dfrac{9xy^2}{4x}}$

Solution   The given complex fraction means

$$\frac{3x^2}{2y} \div \frac{9xy^2}{4x}$$

We therefore apply the rule for division of fractions to change the division to multiplication and to invert the divisor.

$$\frac{\dfrac{3x^2}{2y}}{\dfrac{9xy^2}{4x}} = \frac{3x^2}{2y} \div \frac{9xy^2}{4x}$$

$$= \frac{3x^2}{2y} \cdot \frac{4x}{9xy^2} \qquad \textcolor{blue}{\text{Reduce.}}$$

$$= \frac{3x^2}{2y} \cdot \frac{\overset{2}{4x}}{\underset{3}{9xy^2}}$$

$$= \boxed{\frac{2x^2}{3y^3}}$$

Let's take a look at how fractional computations are done with a calculator.

EXAMPLE 11

Use a calculator to compute $\dfrac{15}{32} \div \dfrac{5}{4}$.

Solution

In order to do this calculation correctly we must use parentheses so that the calculator will know exactly what is being divided by what. In other words, we think of the given computation as

$$\left(\frac{15}{32}\right) \div \left(\frac{5}{4}\right)$$

The following keystrokes will carry out this computation:

$$\boxed{(}\ \boxed{15}\ \boxed{\div}\ \boxed{32}\ \boxed{)}\ \boxed{\div}\ \boxed{(}\ \boxed{5}\ \boxed{\div}\ \boxed{4}\ \boxed{)}\ \boxed{=}$$

and the display will read

$$\boxed{.375}$$

Let's check this computation by doing the division manually.

$$\frac{15}{32} \div \frac{5}{4} = \frac{15}{32} \cdot \frac{4}{5} = \frac{3}{8}$$

If we use a calculator to compute $\dfrac{3}{8}$, the display reads $\boxed{.375}$, which agrees with our previous answer.

Study Skills 4.2

Taking an Algebra Exam: Beginning the Exam

At the exam, make sure that you listen carefully to the instructions given by your instructor or the proctor.

As soon as you are allowed to begin, jot down the formulas you think you might need, and write some key words (warnings) to remind you to avoid common errors or errors you have previously made. Writing down the formulas first will relieve you of the burden of worrying about whether you will remember them when you need to, thus allowing you to concentrate better.

You should refer back to the relevant warnings as you go through the exam to make sure you avoid those errors.

Remember to read the directions carefully.

EXERCISES 4.2

Perform the indicated operations. Final answers should be reduced to lowest terms.

1. $\dfrac{-4}{9} \cdot \dfrac{-2}{3}$

2. $\dfrac{6}{-25} \cdot \dfrac{-3}{5}$

3. $\dfrac{-6}{10} \cdot \dfrac{15}{9}$

4. $\dfrac{12}{20} \cdot \dfrac{-15}{8}$

5. $\dfrac{2}{3y} \cdot \dfrac{x}{5}$

6. $\dfrac{4}{5m} \cdot \dfrac{n}{7}$

7. $\dfrac{x^2}{4y} \cdot \dfrac{5x}{3y}$

8. $\dfrac{a^3}{2b} \cdot \dfrac{3a^2}{4b}$

9. $\dfrac{4}{5} \div \dfrac{5}{4}$

10. $\dfrac{-7}{10} \div \dfrac{10}{-7}$

11. $\dfrac{6x}{y} \div \dfrac{y^2}{2x^2}$

12. $\dfrac{8a}{3b} \div \dfrac{b^3}{4a^2}$

13. $\dfrac{3}{2t} \cdot \dfrac{tw}{6}$

14. $\dfrac{5}{4} \cdot \dfrac{mn}{20}$

15. $4 \cdot \dfrac{x}{12}$

16. $8 \cdot \dfrac{y}{4}$

17. $4 \div \dfrac{x}{12}$

18. $8 \div \dfrac{y}{4}$

19. $\dfrac{x}{12} \div 4$

20. $\dfrac{y}{4} \div 8$

21. $\dfrac{-2x}{3y^2} \cdot \dfrac{-9y}{4x}$

22. $\dfrac{3x^2}{-4y} \cdot \dfrac{-16y}{12x^3}$

23. $\dfrac{m^3n^2}{2m} \cdot \dfrac{6}{n^3}$

24. $\dfrac{3t}{r^2t^3} \cdot \dfrac{r^3}{9}$

25. $\dfrac{3uv^2}{5w} \div \dfrac{6u^2v}{15w}$

26. $\dfrac{21y^2z^2}{12uv} \div \dfrac{14yz}{3v}$

27. $6xy \cdot \dfrac{2x}{3y}$

28. $(10a^2b) \cdot \dfrac{2a}{5b}$

29. $6xy \div \dfrac{2x}{3y}$

30. $(10a^2b) \div \dfrac{2a}{5b}$

31. $\dfrac{2x}{3y} \div (6xy)$

32. $\dfrac{2a}{5b} \div (10a^2b)$

33. $\dfrac{-4x}{9y} \cdot \dfrac{x^2}{y^2} \cdot \dfrac{3y}{2x}$

34. $\dfrac{5m}{4n} \cdot \dfrac{m^3}{n^2} \cdot \dfrac{-2}{10m^2n}$

35. $\dfrac{9}{a^2}\left(\dfrac{a}{3} \div \dfrac{3}{a}\right)$

36. $\dfrac{x^2}{10}\left(\dfrac{2}{x} \div \dfrac{x}{5}\right)$

37. $\dfrac{9}{a^2} \div \left(\dfrac{a}{3} \cdot \dfrac{3}{a}\right)$

38. $\dfrac{x^2}{10} \div \left(\dfrac{2}{x} \cdot \dfrac{x}{5}\right)$

39. $\dfrac{9}{a^2} \div \left(\dfrac{a}{3} \div \dfrac{3}{a}\right)$

40. $\dfrac{x^2}{10} \div \left(\dfrac{2}{x} \div \dfrac{x}{5}\right)$

41. $\left(\dfrac{9}{a^2} \div \dfrac{a}{3}\right) \div \dfrac{a}{3}$

42. $\left(\dfrac{x^2}{10} \div \dfrac{x}{2}\right) \div \dfrac{x}{2}$

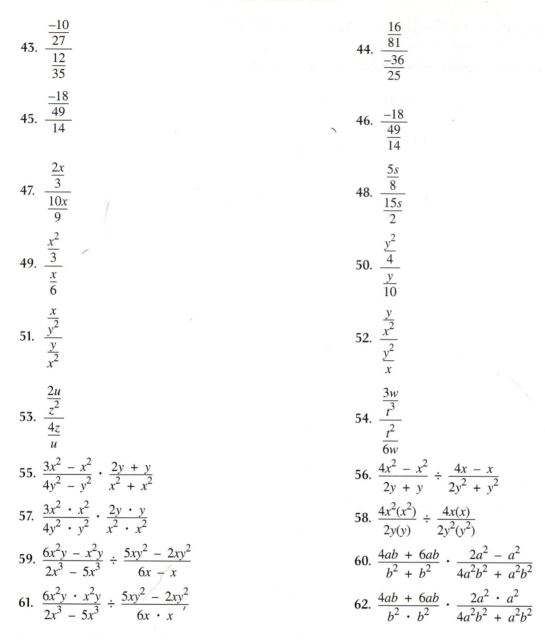

43. $\dfrac{\dfrac{-10}{27}}{\dfrac{12}{35}}$

44. $\dfrac{\dfrac{16}{81}}{\dfrac{-36}{25}}$

45. $\dfrac{\dfrac{-18}{49}}{14}$

46. $\dfrac{-18}{\dfrac{49}{14}}$

47. $\dfrac{\dfrac{2x}{3}}{\dfrac{10x}{9}}$

48. $\dfrac{\dfrac{5s}{8}}{\dfrac{15s}{2}}$

49. $\dfrac{\dfrac{x^2}{3}}{\dfrac{x}{6}}$

50. $\dfrac{\dfrac{y^2}{4}}{\dfrac{y}{10}}$

51. $\dfrac{\dfrac{x}{y^2}}{\dfrac{y}{x^2}}$

52. $\dfrac{\dfrac{y}{x^2}}{\dfrac{y^2}{x}}$

53. $\dfrac{\dfrac{2u}{z^2}}{\dfrac{4z}{u}}$

54. $\dfrac{\dfrac{3w}{t^3}}{\dfrac{t^2}{6w}}$

55. $\dfrac{3x^2 - x^2}{4y^2 - y^2} \cdot \dfrac{2y + y}{x^2 + x^2}$

56. $\dfrac{4x^2 - x^2}{2y + y} \div \dfrac{4x - x}{2y^2 + y^2}$

57. $\dfrac{3x^2 \cdot x^2}{4y^2 \cdot y^2} \cdot \dfrac{2y \cdot y}{x^2 \cdot x^2}$

58. $\dfrac{4x^2(x^2)}{2y(y)} \div \dfrac{4x(x)}{2y^2(y^2)}$

59. $\dfrac{6x^2y - x^2y}{2x^3 - 5x^3} \div \dfrac{5xy^2 - 2xy^2}{6x - x}$

60. $\dfrac{4ab + 6ab}{b^2 + b^2} \cdot \dfrac{2a^2 - a^2}{4a^2b^2 + a^2b^2}$

61. $\dfrac{6x^2y \cdot x^2y}{2x^3 - 5x^3} \div \dfrac{5xy^2 - 2xy^2}{6x \cdot x}$

62. $\dfrac{4ab + 6ab}{b^2 \cdot b^2} \cdot \dfrac{2a^2 \cdot a^2}{4a^2b^2 + a^2b^2}$

In Exercises 63–74, express your answer in fractional from.

63. There are approximately $\frac{8}{5}$ kilometers in one mile. Determine the number of miles in 100 kilometers.

64. If a board is $8\frac{1}{2}$ ft long, find the length of one-third of the board.

65. If you need to equally divide $\frac{3}{8}$ of a cake among 4 people, how much of the cake will each person receive?

66. There are approximately $2\frac{1}{2}$ centimeters in one inch. Determine the number of inches in 38 centimeters.

67. If a bucket holds $7\frac{1}{2}$ gallons of water, how many gallons will $5\frac{1}{4}$ buckets hold?

68. How many $\frac{3}{4}$-gallon bottles can be filled from a bucket that holds $7\frac{1}{2}$ gallons of water?

69. How many $\frac{3}{4}$-ounce tuna hors d'oeuvres can be made from a $6\frac{1}{4}$-ounce can of tuna?

70. If a can of tuna contains $6\frac{1}{4}$ ounces, how much tuna is obtained from $3\frac{1}{2}$ cans?

71. How many miles can be driven with $8\frac{1}{3}$ gallons of gas in a car that gets $21\frac{4}{5}$ miles per gallon?

72. How many gallons of gas are needed to drive $198\frac{7}{8}$ miles in a car that gets $18\frac{1}{2}$ miles per gallon?

73. A plastic sheet covers $12\frac{2}{3}$ square meters. How many such sheets are needed to cover an area of 76 square meters?

74. A plastic sheet covers $12\frac{2}{3}$ square meters. How much area is covered by $18\frac{1}{6}$ such sheets?

In Exercises 75–80, use a calculator to do the computation. Round off to three decimal places where necessary.

75. $\dfrac{3}{10} \cdot \dfrac{8}{5}$

76. $\dfrac{5}{6} \div \dfrac{3}{4}$

77. $\dfrac{2}{5} \div \dfrac{8}{9}$

78. $\dfrac{12}{7} \cdot \dfrac{5}{8}$

79. $\left(\dfrac{5}{6} \div \dfrac{3}{4}\right) \cdot \dfrac{1}{2}$

80. $\dfrac{1}{8} \cdot \left(\dfrac{2}{5} \div \dfrac{1}{2}\right)$

QUESTIONS FOR THOUGHT

81. Explain the rule for dividing fractions. Discuss why it works.

82. Explain what is **wrong** with each of the following:

   (a) $\dfrac{3x}{2y} \div \dfrac{2y}{3x} \stackrel{?}{=} \dfrac{3x}{2y} \div \dfrac{2y}{3x} \stackrel{?}{=} 1$

   (b) $5 \cdot \dfrac{3x}{2} \stackrel{?}{=} \dfrac{5 \cdot 3x}{5 \cdot 2} \stackrel{?}{=} \dfrac{15x}{10}$

83. Make up a verbal problem involving money that would give rise to the following equation (be sure to indicate what the variable would represent):

$$25x + 10(x + 3) = 275$$

MINI-REVIEW

Solve each of the following problems algebraically. Be sure to label what the variable represents.

84. A car rental company charges $15 per day plus $0.20 per mile. How many miles were driven during a 2-day rental if the total cost was $65?

85. The sum of the angles of a triangle is 180°. If the largest angle is twice the smallest and the third angle is 20° more than the smallest, find the size of each angle.

4.3 Adding and Subtracting Rational Expressions

As we mentioned previously, many of the basic procedures in algebra have as their basis the distributive property, and the method for combining fractions is a prime illustration of this.

Recall that when we learned how to combine like terms, we reasoned by the distributive property that $5x + 3x = (5 + 3)x = 8x$. In light of our discussion of multiplication of fractions in the last section, we can see that any fraction $\frac{a}{c}$ can be written as $a \cdot \frac{1}{c}$ because

$$a \cdot \frac{1}{c} = \frac{a}{1} \cdot \frac{1}{c} = \frac{a}{c}$$

For example, $\frac{3}{7} = 3 \cdot \frac{1}{7}$. Thus, to add $\frac{3}{7} + \frac{2}{7}$ we can proceed as follows:

$$\frac{3}{7} + \frac{2}{7} = 3 \cdot \frac{1}{7} + 2 \cdot \frac{1}{7}$$

$$= (3 + 2) \cdot \frac{1}{7} \qquad \textit{We have factored out the common factor of } \frac{1}{7}.$$

$$= 5 \cdot \frac{1}{7} = \boxed{\frac{5}{7}}$$

This process can, of course, be used whenever the fractions have the same denominator. Such fractions are said to have a **common denominator**.

As long as the denominators are exactly the same, we can perform addition, subtraction, or a combination of addition and subtraction of several fractions. We can formulate this idea as follows:

Rule for Adding and Subtracting Rational Expressions	$\dfrac{a}{c} + \dfrac{b}{c} = \dfrac{a + b}{c} \qquad \dfrac{a}{c} - \dfrac{b}{c} = \dfrac{a - b}{c}, \qquad c \neq 0$

In words, this rule says that we can add or subtract fractions with common denominators by adding or subtracting the numerators and putting the result over the common denominator.

EXAMPLE 1 Add. $\dfrac{x}{6} + \dfrac{x + 3}{6}$

Solution Since the denominators are the same, we can apply the above rule to add the numerators and keep the common denominator.

$$\frac{x}{6} + \frac{x + 3}{6} = \frac{x + x + 3}{6}$$

$$= \boxed{\frac{2x + 3}{6}} \qquad \textit{Note that this answer } \textbf{cannot} \textit{ be reduced any further; the 2 and the 3 are } \textbf{not} \textit{ factors of the numerator.} \qquad \blacksquare$$

EXAMPLE 2 Combine. $\dfrac{3x + 5}{4x} - \dfrac{7}{4x} + \dfrac{2 - x}{4x}$

Solution Again, since the denominators are the same, we simply "combine the numerators and keep the denominator."

$$\frac{3x + 5}{4x} - \frac{7}{4x} + \frac{2 - x}{4x} = \frac{3x + 5 - 7 + 2 - x}{4x}$$ *Combine like terms.*

$$= \frac{2x}{4x}$$ *Reduce.*

$$= \frac{\overset{1}{\cancel{2x}}}{\underset{2}{\cancel{4x}}}$$ *Remember the understood 1 in the numerator.*

$$= \boxed{\frac{1}{2}}$$ ■

Even when the denominators are the same, subtraction problems often require extra care.

EXAMPLE 3

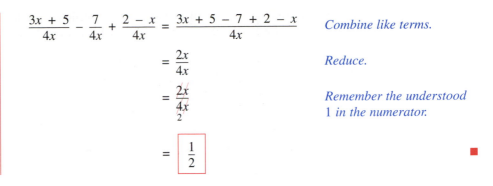

Subtract. $\dfrac{3a - 7}{10} - \dfrac{a - 7}{10}$

Solution

The rule we have stated tells us that we must subtract the *entire* second numerator from the first.

$$\frac{3a - 7}{10} - \frac{a - 7}{10} = \frac{3a - 7 - (a - 7)}{10}$$ *Notice that the parentheses are essential; we must subtract the entire second numerator.*

$$= \frac{3a - 7 - a + 7}{10}$$ *Because $-(a - 7) = -a + 7$*

$$= \frac{2a}{10}$$ *Reduce.*

$$= \frac{\overset{}{\cancel{2a}}}{\underset{5}{\cancel{10}}}$$

$$= \boxed{\frac{a}{5}}$$ ■

The next type of problem to consider is one in which the denominators are not the same. For example, if two people are working together to complete a task and one of them completes $\frac{2}{3}$ of the task and the other person completes $\frac{1}{5}$ of the task, how much of the task have they completed? In order to answer this question we must be able to add $\frac{2}{3}$ and $\frac{1}{5}$. In arithmetic we learned to add these fractions by changing each one into an equivalent fraction with the same denominator, as follows:

$$\frac{2}{3} + \frac{1}{5} = \frac{10}{15} + \frac{3}{15} = \frac{13}{15}$$

Therefore, the two people have completed $\frac{13}{15}$ of the job.

Let's examine this arithmetic process of adding fractions with unlike denominators more carefully in order to generalize it to algebraic fractions as well. If the denominators are not the same, we will apply the Fundamental Principle of Fractions to build each denominator into a common denominator. In order to keep the fractions as simple as possible, we generally try to use what is called the ***least common denominator*** (LCD for short), which is the least common multiple of the denominators—that is, the "smallest" expression that is exactly divisible by each of the denominators.

While the LCD can sometimes be found by simply looking at the denominators, it is very useful, particularly for algebraic fractions, to have a mechanical procedure for finding it. We will first state the procedure, and then explain its use with a numerical example.

Outline for Finding the LCD	**Step 1.** Factor each denominator as completely as possible and list the distinct factors. **Step 2.** The LCD consists of the product of each of these *distinct* factors the *maximum* number of times (that is, to the highest power) it appears in any one denominator.

Unless it is stated to the contrary, whenever we discuss factoring, we mean using integer factors.

A numerical example will help clarify the process.

EXAMPLE 4   Combine. $\dfrac{5}{18} + \dfrac{7}{24} - \dfrac{11}{30}$

Solution   Since the denominators are not the same, we seek the LCD, which we will find by using the outline given in the box. In dealing with a numerical example, to factor as completely as possible means to break each number into its prime factorization.

Step 1. $18 = 2 \cdot 3 \cdot 3$
$24 = 2 \cdot 2 \cdot 2 \cdot 3$ *Notice that the **distinct** factors are 2, 3, 5.*
$30 = 2 \cdot 3 \cdot 5$

Finding the LCD for 18, 24, and 30 means we are looking for the smallest number that is exactly divisible by 18, 24, and 30.

Step 2.   We must make a decision for each distinct factor.

How many factors of 2 do we need?

2 appears as a factor once in 18, three times in 24, and once in 30. Therefore, according to the outline, we need to take *three* factors of 2. We take the factor 2 the *maximum* number of times it appears in any *one* denominator, *not* the total number of times it appears all together.

How many factors of 3 do we need?

3 appears as a factor twice in 18, once in 24, and once in 30. Therefore, we need to take *two* factors of 3.

How many factors of 5 do we need?

5 appears as a factor zero times in 18, zero times in 24, and once in 30. Therefore, we need to take *one* factor of 5.

Following the instructions in step 2 of the outline, we get

$$\text{LCD} = 2 \cdot 2 \cdot 2 \cdot 3 \cdot 3 \cdot 5 = 360$$

If you look at 360 in its factored form, you can see each of the original denominators contained in it, and therefore it is a common denominator. For example, 360 "contains" 18 and 24 and 30.

$2 \cdot 2 \cdot \underbrace{2 \cdot 3 \cdot 3 \cdot 5}_{18}$ and $\underbrace{2 \cdot 2 \cdot 2 \cdot 3}_{24} \cdot 3 \cdot 5$ and $2 \cdot 2 \cdot 3 \cdot \underbrace{2 \cdot 3 \cdot 5}_{30}$

On the other hand, because we chose each factor the *maximum* number of times it appears as a factor in any of the denominators, we do not have any extra factors. If we try to delete any of the factors, the LCD will not be divisible by all three denominators. Thus, 360 is the *least common denominator*.

Now that we have decided on 360 as the LCD, we want to build each of our original denominators into 360 by applying the Fundamental Principle. The easiest way to see how to build a denominator into the LCD (360 in this example) is to look at the LCD in its factored form and fill in the missing factors. In other words, we look at 360 and each denominator in its factored form, and then use the Fundamental Principle to fill in the factors necessary to change the denominator into 360.

Looking at $18 = 2 \cdot 3 \cdot 3$ and $360 = 2 \cdot 2 \cdot 2 \cdot 3 \cdot 3 \cdot 5$, we can see that 18 is missing $2 \cdot 2 \cdot 5$. Thus, we multiply the numerator and denominator by $2 \cdot 2 \cdot 5$.

$$\frac{5}{18} = \frac{5}{2 \cdot 3 \cdot 3} = \frac{5}{2 \cdot 3 \cdot 3} \cdot \frac{(2 \cdot 2 \cdot 5)}{(2 \cdot 2 \cdot 5)} = \frac{100}{360}$$

Now we repeat this process for the other two fractions.

$$\frac{7}{24} = \frac{7}{2 \cdot 2 \cdot 2 \cdot 3} = \frac{7}{2 \cdot 2 \cdot 2 \cdot 3} \cdot \frac{(3 \cdot 5)}{(3 \cdot 5)} = \frac{105}{360}$$

$$\frac{11}{30} = \frac{11}{2 \cdot 3 \cdot 5} = \frac{11}{2 \cdot 3 \cdot 5} \cdot \frac{(2 \cdot 2 \cdot 3)}{(2 \cdot 2 \cdot 3)} = \frac{132}{360}$$

Even though we are focusing our attention on the denominator to tell us what the missing factors are in each case, the Fundamental Principle tells us nevertheless that in order to obtain an equivalent fraction we must multiply both the numerator and denominator by the same quantity.

Therefore, the original example has become

$$\frac{5}{18} + \frac{7}{24} - \frac{11}{30} = \frac{100}{360} + \frac{105}{360} - \frac{132}{360}$$
$$= \frac{100 + 105 - 132}{360}$$
$$= \boxed{\frac{73}{360}} \qquad \blacksquare$$

While it may seem to have taken a very long time to complete Example 4 due to all the explanations along the way, the effort was worthwhile because this same process can be used in every problem where we need to combine fractions with unlike denominators.

You may have previously learned to build arithmetic fractions by asking

$$\frac{5}{18} = \frac{?}{360}$$

dividing 18 into 360, which gives 20, and then multiplying 5 times 20 to get $\frac{100}{360}$. While this procedure for building fractions works well for arithmetic fractions, it can become very messy with algebraic fractions. On the other hand, the process of building fractions by analyzing and filling in the missing factors works well for both types of fractions.

EXAMPLE 5   Add. $\dfrac{5}{2x^3} + \dfrac{7}{3x^2}$

Solution   Since the denominators are not the same, we want to find the LCD. You may be able to "see" that the LCD is $6x^3$. If so, fine. If not, we can find it by following the outline.

Step 1.   We factor each denominator completely.

$$\left. \begin{array}{l} 2x^3 = 2 \cdot x \cdot x \cdot x \\ 3x^2 = 3 \cdot x \cdot x \end{array} \right\} \qquad \textit{The distinct factors are 2, 3, and x.}$$

While this step is not absolutely necessary here ($2x^3$ and $3x^2$ are already in factored form), it probably helps to make the distinct factors clearer.

Step 2. We take each distinct factor the maximum number of times it appears in any one denominator. Therefore, we take one factor of 2, one factor of 3, and three factors of x. We have the LCD $= 2 \cdot 3 \cdot x \cdot x \cdot x = 6x^3$.

Now we want to build each of our original fractions into an equivalent fraction having $6x^3$ as its denominator.

$$\frac{5}{2x^3} = \frac{5}{2 \cdot x \cdot x \cdot x} = \frac{5 \cdot (3)}{2 \cdot x \cdot x \cdot x \cdot (3)} = \frac{15}{6x^3}$$

$$\frac{7}{3x^2} = \frac{7}{3 \cdot x \cdot x} = \frac{7 \cdot (2x)}{3 \cdot x \cdot x \cdot (2x)} = \frac{14x}{6x^3}$$

Thus, the original example has become

$$\frac{5}{2x^3} + \frac{7}{3x^2} = \frac{15}{6x^3} + \frac{14x}{6x^3}$$

$$= \boxed{\frac{15 + 14x}{6x^3}}$$

It is important to note that this final answer cannot be reduced any further. The x that appears in the numerator is *not a factor* of the numerator. ■

EXAMPLE 6 Multiply. $\dfrac{5}{2x^3} \cdot \dfrac{7}{3x^2}$

Solution Read examples carefully! This is a multiplication problem, not an addition problem, so no common denominator is needed.

$$\frac{5}{2x^3} \cdot \frac{7}{3x^2} = \boxed{\frac{35}{6x^5}}$$ ■

EXAMPLE 7 Subtract. $\dfrac{7}{4xy^2} - \dfrac{1}{10y^4}$

Solution We begin by finding the LCD.

Step 1. $\left.\begin{array}{l} 4xy^2 = 2 \cdot 2 \cdot xyy \\ 10y^4 = 2 \cdot 5 \cdot yyyy \end{array}\right\}$ *The distinct factors are 2, 5, x, y.*

          Step 2.   LCD $= 2 \cdot 2 \cdot 5 \cdot xyyyy = 20xy^4$

Next we build the original fractions into equivalent fractions that have the LCD as denominator.

$$\frac{7}{4xy^2} = \frac{7 \cdot (5y^2)}{4xy^2 \cdot (5y^2)} = \frac{35y^2}{20xy^4}$$

$$\frac{1}{10y^4} = \frac{1 \cdot (2x)}{10y^4 \cdot (2x)} = \frac{2x}{20xy^4}$$

Thus, the original problem has become

$$\frac{7}{4xy^2} - \frac{1}{10y^4} = \frac{35y^2}{20xy^4} - \frac{2x}{20xy^4}$$

$$= \boxed{\frac{35y^2 - 2x}{20xy^4}}$$

  ■

The procedure used in Example 7 can be applied to examples involving more than two fractions.

EXAMPLE 8   Combine. $\dfrac{9}{10x^2y^3} + \dfrac{7}{6xy^2} - \dfrac{2}{15x^3y}$

Solution   Here is the solution without numbering all the steps and in a more concise form.

$$\left.\begin{array}{l} 10x^2y^3 = 2 \cdot 5 \cdot xxyyy \\ 6xy^2 = 2 \cdot 3 \cdot xyy \\ 15x^3y = 3 \cdot 5 \cdot xxxy \end{array}\right\} \quad \textit{The distinct factors are 2, 3, 5, x, and y.}$$

The LCD is $2 \cdot 3 \cdot 5 \cdot xxxyyy = 30x^3y^3$. For each fraction, look at the denominator to determine the missing factors, then apply the Fundamental Principle.

$$\frac{9}{10x^2y^3} + \frac{7}{6xy^2} - \frac{2}{15x^3y} = \frac{9 \cdot (3x)}{10x^2y^3 \cdot (3x)} + \frac{7 \cdot (5x^2y)}{6xy^2 \cdot (5x^2y)} - \frac{2 \cdot (2y^2)}{15x^3y \cdot (2y^2)}$$

$$= \frac{27x}{30x^3y^3} + \frac{35x^2y}{30x^3y^3} - \frac{4y^2}{30x^3y^3}$$

$$= \boxed{\frac{27x + 35x^2y - 4y^2}{30x^3y^3}}$$

  ■

EXAMPLE 9   Combine. $\dfrac{5}{4x} + \dfrac{1}{2x^2} + 3y$

Solution   It helps to think of $3y$ as $\dfrac{3y}{1}$.

$$\left.\begin{array}{l} 4x = 2 \cdot 2 \cdot x \\ 2x^2 = 2 \cdot x \cdot x \\ 1 = 1 \end{array}\right\} \quad \textit{The distinct factors are 2 and x.}$$

The LCD is $2 \cdot 2 \cdot x \cdot x = 4x^2$.

$$\frac{5}{4x} + \frac{1}{2x^2} + \frac{3y}{1} = \frac{5 \cdot (x)}{4x \cdot (x)} + \frac{1 \cdot (2)}{2x^2 \cdot (2)} + \frac{3y \cdot (4x^2)}{1 \cdot (4x^2)}$$

$$= \frac{5x}{4x^2} + \frac{2}{4x^2} + \frac{12x^2y}{4x^2}$$

$$= \boxed{\frac{5x + 2 + 12x^2y}{4x^2}}$$

  ■

Study Skills 4.3

Taking an Algebra Exam: What to Do First

Not all exams are arranged in ascending order of difficulty (from easiest to most difficult). Since time is usually an important factor, you do not want to spend so much time working on a few problems that you find difficult and then realize that you do not have enough time to solve the problems that are easier for you. Therefore, it is strongly recommended that you first look over the exam and then follow the order given below:

1. Start with the problems that you know how to solve quickly.

2. Then go back and work on problems that you know how to solve but that take longer.

3. Then work on those problems that you find more difficult, but for which you have a general idea of how to proceed.

4. Finally, divide the remaining time between the problems you find most difficult and checking

your solutions. Do not forget to check the warnings you wrote down at the beginning of the exam.

You probably should not be spending a lot of time on any single problem. To determine the average amount of time you should be spending on a problem, divide the amount of time given for the exam by the number of problems on the exam. For example, if the exam lasts for 50 minutes and there are 20 problems, you should spend an average of $\frac{50}{20} = 2\frac{1}{2}$ minutes per problem. Remember, this is just an estimate. You should spend less time on "quick" problems (or those worth fewer points), and more time on the more difficult problems (or those worth more points). As you work the problems, be aware of the time; if half the time is gone, you should have completed about half of the exam.

 **EXERCISES 4.3**

In each of the following exercises, perform the indicated operations. Express your answer as a single fraction reduced to lowest terms.

1. $\frac{5}{3} + \frac{4}{3}$

2. $\frac{6}{5} + \frac{4}{5}$

3. $\frac{3}{5} - \frac{7}{5}$

4. $\frac{5}{7} - \frac{8}{7}$

5. $\frac{7}{9} - \frac{5}{9} - \frac{8}{9}$

6. $\frac{4}{15} - \frac{11}{15} - \frac{2}{15}$

7. $\frac{2}{3} + \frac{4}{5}$

8. $\frac{3}{2} \cdot \frac{5}{7}$

9. $\frac{2}{3} \cdot \frac{4}{5}$

10. $\frac{3}{2} + \frac{5}{7}$

11. $\frac{2}{3} - \frac{5}{6}$

12. $\frac{1}{2} - \frac{7}{10}$

13. $3 + \frac{3}{4} - \frac{3}{8}$

14. $5 + \frac{5}{3} - \frac{5}{6}$

15. $\frac{3}{2} - \frac{4}{5} + \frac{7}{10}$

16. $\frac{7}{3} - \frac{4}{5} + \frac{2}{15}$

17. $\frac{5}{6} - \frac{3}{8} + \frac{1}{4}$

18. $\frac{3}{4} - \frac{5}{18} - \frac{1}{9}$

19. $2 + \frac{1}{3} - \frac{1}{2}$

20. $3 - \frac{1}{3} + \frac{1}{4}$

21. $\dfrac{8}{3x} + \dfrac{4}{3x}$

22. $\dfrac{5}{4y} + \dfrac{3}{4y}$

23. $\dfrac{8}{3x} \cdot \dfrac{4}{3x}$

24. $\dfrac{5}{4y} \cdot \dfrac{3}{4y}$

25. $\dfrac{7}{6x} + \dfrac{13}{6x} - \dfrac{11}{6x}$

26. $\dfrac{9}{10y} - \dfrac{7}{10y} + \dfrac{13}{10y}$

27. $\dfrac{3y}{7x} - \dfrac{5y}{7x} + \dfrac{4y}{7x}$

28. $\dfrac{4a}{5b} - \dfrac{7a}{5b} + \dfrac{6a}{5b}$

29. $\dfrac{w}{9z} - \dfrac{5w}{9z} + \dfrac{4w}{9z}$

30. $\dfrac{3m}{5n} - \dfrac{7m}{5n} + \dfrac{4m}{5n}$

31. $\dfrac{-5a}{7b} + \dfrac{3a}{7b} - \dfrac{12a}{7b}$

32. $\dfrac{-7a}{15b} + \dfrac{3a}{15b} - \dfrac{6a}{15b}$

33. $\dfrac{x + 3}{3x} + \dfrac{x - 6}{3x}$

34. $\dfrac{x + 5}{2x} + \dfrac{2x - 3}{2x}$

35. $\dfrac{3y^2 - 5}{4y} + \dfrac{5 - 4y^2}{4y}$

36. $\dfrac{2y^2 - 1}{6y} + \dfrac{y^2 + 1}{6y}$

37. $\dfrac{5x + 2}{10x} - \dfrac{x + 2}{10x}$

38. $\dfrac{3x + 7}{4x} - \dfrac{x + 7}{4x}$

39. $\dfrac{2a - 1}{3a} - \dfrac{5a - 1}{3a}$

40. $\dfrac{5a + 3}{4a} - \dfrac{13a + 3}{4a}$

41. $\dfrac{w - 4}{6w} - \dfrac{w - 3}{6w} + \dfrac{5}{6w}$

42. $\dfrac{w - 3}{8w} - \dfrac{2w - 3}{8w} + \dfrac{w + 2}{8w}$

43. $\dfrac{t^2 - 3t + 2}{5t^2} - \dfrac{7t + t^2}{5t^2}$

44. $\dfrac{t^2 + 2t + 5}{7t^2} - \dfrac{5 - 8t + t^2}{7t^2}$

45. $\dfrac{3}{x} + \dfrac{2}{y}$

46. $\dfrac{4}{x} \cdot \dfrac{7}{y}$

47. $\dfrac{3}{x} \cdot \dfrac{2}{y}$

48. $\dfrac{4}{x} + \dfrac{7}{y}$

49. $\dfrac{5}{3x} - \dfrac{7}{2}$

50. $\dfrac{4}{5x} - \dfrac{5}{3}$

51. $\dfrac{5}{48x} + \dfrac{3}{24y}$

52. $\dfrac{7}{90y} + \dfrac{2}{45x}$

53. $\dfrac{4}{x^2} - \dfrac{3}{2x}$

54. $\dfrac{2}{x^3} - \dfrac{4}{3x}$

55. $\dfrac{4}{x^2} \cdot \dfrac{3}{2x}$

56. $\dfrac{2}{x^3} \cdot \dfrac{4}{3x}$

57. $\dfrac{2}{3x^2} + \dfrac{3}{2x^2}$

58. $\dfrac{4}{5x^3} + \dfrac{2}{3x^4}$

59. $\dfrac{7}{12a^2} - \dfrac{9}{84a}$

60. $\dfrac{5}{18a^3} - \dfrac{1}{90a^2}$

61. $\dfrac{1}{x} + 2$

62. $\dfrac{2}{y} + 1$

63. $\dfrac{5}{3xy} + \dfrac{1}{6y^2}$

64. $\dfrac{2}{5xy} + \dfrac{1}{10x^2}$

65. $\dfrac{7}{6a^2b} + \dfrac{3}{4ab^3}$

66. $\dfrac{5}{4ab^2} + \dfrac{9}{10a^3b}$

67. $\dfrac{7}{6a^2b} \cdot \dfrac{3}{4ab^3}$

68. $\dfrac{5}{4ab^2} \cdot \dfrac{9}{10a^3b}$

69. $\dfrac{7}{24x^2} + \dfrac{1}{60x}$

70. $\dfrac{6}{35a^2} + \dfrac{1}{30a^3}$

71. $\dfrac{15}{48rt^2} - \dfrac{7}{72t^3}$

72. $\dfrac{5}{28c^2d^3} + \dfrac{1}{18cd^4}$

73. $\dfrac{3y}{20x^2} + \dfrac{9x}{50y}$

74. $\dfrac{11c}{60a^2b} - \dfrac{2c}{45ab^3}$

75. $\dfrac{4y}{3x^2} - \dfrac{3}{2x} + \dfrac{y}{x^2}$

76. $\dfrac{3y}{5x} - \dfrac{2}{x^2} - \dfrac{y}{2x}$

77. $\dfrac{3}{4m^2n} - \dfrac{5}{6mn^3} + \dfrac{1}{8n^2}$

78. $\dfrac{1}{6rt^2} - \dfrac{7}{4r^3t} + \dfrac{4}{9r^2}$

79. $\dfrac{x}{y} + \dfrac{y}{x} + \dfrac{3x}{2y}$

80. $\dfrac{3y}{x} - \dfrac{x}{4y} + \dfrac{2y}{5x}$

81. $t - \dfrac{3}{t}$

82. $u^2 + \dfrac{2}{u}$

83. $3x^2 + \dfrac{1}{x} - \dfrac{2}{x^2}$

84. $5y^3 - \dfrac{2}{y} + \dfrac{1}{2y}$

85. $\dfrac{2x+3}{24} + \dfrac{x}{18}$

86. $\dfrac{3x+2}{15} + \dfrac{x}{40}$

87. $\dfrac{2x+3}{x} + \dfrac{x}{2}$

88. $\dfrac{3x+2}{x} + \dfrac{x}{3}$

89. $\dfrac{a-5}{2} + \dfrac{3}{a}$

90. $\dfrac{a-4}{3} + \dfrac{5}{a}$

 QUESTIONS FOR THOUGHT

91. What is a least common denominator? Why is it needed?

92. Explain why the procedure outlined in this section produces the LCD for a given set of denominators.

93. What is the LCD for $\frac{3}{10}$ and $\frac{7}{9}$? What is the LCD for $\frac{5}{6}$ and $\frac{3}{4}$? Looking at these two LCDs, can you determine when the LCD of two fractions will simply be the product of the denominators?

94. Discuss in detail what is **wrong** with each of the following.

(a) $\dfrac{x+3}{x} - \dfrac{5-x}{x} \overset{?}{=} \dfrac{x+3-5-x}{x} \overset{?}{=} \dfrac{-2}{x}$

(b) $\dfrac{2x}{y} + \dfrac{3y}{x} \overset{?}{=} \dfrac{2\cancel{x}}{\cancel{y}} + \dfrac{3\cancel{y}}{\cancel{x}} \overset{?}{=} 2 + 3 \overset{?}{=} 5$

(c) $\dfrac{5x}{2y} + \dfrac{7y}{6x} \overset{?}{=} \dfrac{5x}{6xy} + \dfrac{7y}{6xy} \overset{?}{=} \dfrac{5x+7y}{6xy}$

(d) $\dfrac{5x}{2y} + \dfrac{7y}{6x} \overset{?}{=} \dfrac{15x^2}{6xy} + \dfrac{7y^2}{6xy} \overset{?}{=} \dfrac{\overset{5\ x}{\cancel{15x^2}}}{\underset{2}{\cancel{6xy}}} + \dfrac{\overset{y}{\cancel{7y^2}}}{\cancel{6xy}} \overset{?}{=} \dfrac{5x}{2y} + \dfrac{7y}{6x} \overset{?}{=} \dfrac{5x+7y}{2y+6x}$

(e) $\dfrac{5x}{2y} + \dfrac{7y}{6x} \overset{?}{=} \dfrac{15x^2}{6xy} + \dfrac{7y^2}{6xy} \overset{?}{=} \dfrac{15x^2 + 7y^2}{6xy} \overset{?}{=} \dfrac{\cancel{15x^2}^{\,5x^2}+7\cancel{y^2}^{\,y^2}}{\cancel{6xy}_{2}} \overset{?}{=} \dfrac{5x + 7y}{2}$

95. Look back at the solution to Example 8 of this section. The next to the last line of the solution contains three fractions that *can* be reduced. Why were they *not* reduced?

MINI-REVIEW

Solve each of the following problems algebraically. Be sure to label what the variable represents.

96. A collection of 28 coins consisting of nickels and dimes is worth $2. How many of each are there?

97. An air freight service charges a $10 pick-up charge plus a $4 per pound delivery charge. If a package is picked up and delivered for a total cost of $28, how many pounds did the package weigh?

4.4 Solving Fractional Equations and Inequalities

Solving equations that contain fractional expressions involves combining the ideas we learned in Chapter 3 with the material in the previous section. By using the idea of the least common denominator and the multiplication property of equality, we can convert an equation involving fractional expressions into an equivalent equation without fractions. We can then solve it by the methods we have already learned.

Several examples will illustrate the process.

EXAMPLE 1　Solve for x.　$\dfrac{x}{2} - 5 = \dfrac{1}{2}$

Solution　Rather than solve this equation as is, which would involve working with fractions, we can use the multiplication property of equality to multiply both sides of this equation by 2. We choose to multiply by 2 because this will eliminate the denominators and give us an equivalent equation without fractions. It will then be much easier to isolate x. We proceed as follows:

$$\frac{x}{2} - 5 = \frac{1}{2}$$　*Multiply both sides of the equation by 2. (Write 2 as $\frac{2}{1}$.)*

$$\frac{2}{1}\left(\frac{x}{2} - 5\right) = \frac{2}{1} \cdot \frac{1}{2}$$　*We must use the distributive property on the left side.*

$$\frac{2}{1} \cdot \frac{x}{2} - 2 \cdot 5 = \frac{2}{1} \cdot \frac{1}{2}$$　*Note that each **term** gets multiplied by 2. Now we cancel.*

$$x - 10 = 1$$　*This equation is equivalent to the original one, but much easier to solve.*

$$\underline{+10 \quad +10}$$

$$\boxed{x = 11}$$

Check $x = 11$: $\dfrac{x}{2} - 5 = \dfrac{1}{2}$

$$\dfrac{11}{2} - 5 \stackrel{?}{=} \dfrac{1}{2}$$

$$\dfrac{11}{2} - \dfrac{10}{2} \stackrel{\checkmark}{=} \dfrac{1}{2}$$

■

EXAMPLE 2   Solve for x. $\dfrac{x}{4} - \dfrac{2}{3} = \dfrac{7}{12}$

Solution   Before we attempt to isolate x on one side of the equation, we first ask if there is a way of eliminating *all* the denominators. We need to multiply both sides of the equation by a number that is exactly divisible by each of the denominators and so will cancel each of the denominators. The LCD is exactly the smallest number that will do the job, so we multiply both sides of the equation by 12, which is the LCD for 3, 4, and 12.

$$\dfrac{x}{4} - \dfrac{2}{3} = \dfrac{7}{12}$$ *Multiply both sides of the equation by* 12. *According to the multiplication property of equality, multiplying both sides by* 12 *yields an equivalent equation.*

$$\boxed{12}\left(\dfrac{x}{4} - \dfrac{2}{3}\right) = \boxed{12} \cdot \dfrac{7}{12}$$ *We must use the distributive property on the left side.*

$$\dfrac{12}{1} \cdot \dfrac{x}{4} - \dfrac{12}{1} \cdot \dfrac{2}{3} = \dfrac{12}{1} \cdot \dfrac{7}{12}$$ *Note that each* **term** *gets multiplied by* 12.

$$\dfrac{\overset{3}{\cancel{12}}}{1} \cdot \dfrac{x}{4} - \dfrac{\overset{4}{\cancel{12}}}{1} \cdot \dfrac{2}{3} = \dfrac{\cancel{12}}{1} \cdot \dfrac{7}{\cancel{12}}$$

In this way we have obtained the following *equivalent* equation *without* fractions:

$$3x - 8 = 7$$
$$\underline{+8 \quad +8}$$
$$3x = 15$$
$$\dfrac{\cancel{3}x}{\cancel{3}} = \dfrac{15}{3}$$
$$\boxed{x = 5}$$

Note that we did not use the LCD to convert each of the original fractions into an equivalent one with the LCD as denominator. Rather, we used the LCD to multiply both sides of the equation to "clear" the denominators.

Check $x = 5$: $\dfrac{x}{4} - \dfrac{2}{3} = \dfrac{7}{12}$

$$\dfrac{5}{4} - \dfrac{2}{3} \stackrel{?}{=} \dfrac{7}{12}$$

$$\dfrac{15}{12} - \dfrac{8}{12} \stackrel{?}{=} \dfrac{7}{12}$$

$$\dfrac{7}{12} \stackrel{\checkmark}{=} \dfrac{7}{12}$$

■

At this point it is very important to distinguish between the example we have just completed and the examples we did in the last section. We do this in Example 3.

EXAMPLE 3 | Combine. $\dfrac{x}{4} - \dfrac{2}{3} + \dfrac{7}{12}$

Solution | Note that this is an expression, *not* an equation, and so we are not solving for *x*. We are going to use the LCD of 12, but *not* to eliminate the denominators. We cannot eliminate the denominators because we do not have an equation where we can multiply *both* sides. This example has only "one side." However, as we discussed in great detail in the last section, we do use the LCD to convert each fraction into an equivalent one with a denominator of 12.

$$\dfrac{x}{4} - \dfrac{2}{3} + \dfrac{7}{12} = \dfrac{x \cdot (3)}{4 \cdot (3)} - \dfrac{2 \cdot (4)}{3 \cdot (4)} + \dfrac{7}{12}$$

$$= \dfrac{3x}{12} - \dfrac{8}{12} + \dfrac{7}{12}$$

$$= \dfrac{3x - 8 + 7}{12}$$

$$= \boxed{\dfrac{3x - 1}{12}}$$

Since this was not an equation, we did not get a solution for x. ■

Look over Examples 2 and 3 very carefully to make sure you see the difference between them.

EXAMPLE 4

Solve for *a*. $\dfrac{a + 10}{6} + a = \dfrac{1}{2}$

Solution | Since the LCD is 6, we multiply both sides of the equation by 6, to "clear" the denominators.

$$\dfrac{a + 10}{6} + a = \dfrac{1}{2}$$

$$6 \left(\dfrac{a + 10}{6} + a \right) = 6 \cdot \dfrac{1}{2}$$

$$\dfrac{6}{1} \cdot \left(\dfrac{a + 10}{6} \right) + 6 \cdot a = \dfrac{6}{1} \cdot \dfrac{1}{2}$$

$$\dfrac{\cancel{6}}{1} \cdot \left(\dfrac{a + 10}{\cancel{6}} \right) + 6a = \dfrac{\overset{3}{\cancel{6}}}{1} \cdot \dfrac{1}{2}$$

$$a + 10 + 6a = 3$$

$$7a + 10 = 3$$

$$7a = -7$$

$$\boxed{a = -1}$$

Check $a = -1$: $\dfrac{a + 10}{6} + a = \dfrac{1}{2}$

$$\dfrac{-1 + 10}{6} + (-1) \overset{?}{=} \dfrac{1}{2}$$

$$\dfrac{9}{6} - 1 \overset{?}{=} \dfrac{1}{2}$$

$$\dfrac{3}{2} - \dfrac{2}{2} \overset{?}{=} \dfrac{1}{2}$$

$$\dfrac{1}{2} \overset{\checkmark}{=} \dfrac{1}{2}$$ ■

EXAMPLE 5 | Solve for a. $\dfrac{1}{4}a + \dfrac{3a - 4}{10} = a - 4$

Solution | The LCD for 4 and 10 is 20. We multiply both sides of the equation by 20.

$$\frac{a}{4} + \frac{3a - 4}{10} = a - 4 \qquad \textit{Remember that } \frac{1}{4}a = \frac{1}{4} \cdot \frac{a}{1} = \frac{a}{4}.$$

$$20 \left(\frac{a}{4} + \frac{3a - 4}{10}\right) = 20\,(a - 4)$$

$$\frac{20}{1} \cdot \frac{a}{4} + \frac{20}{1} \cdot \left(\frac{3a - 4}{10}\right) = 20(a - 4)$$

$$\overset{5}{\cancel{\frac{20}{1}}} \cdot \frac{a}{\cancel{4}} + \overset{2}{\cancel{\frac{20}{1}}} \cdot \left(\frac{3a - 4}{\cancel{10}}\right) = 20a - 80$$

$$5a + 2(3a - 4) = 20a - 80$$

$$5a + 6a - 8 = 20a - 80$$

$$11a - 8 = 20a - 80$$

$$\underline{-11a \qquad\qquad -11a}$$

$$-8 = 9a - 80$$

$$\underline{+80 \qquad\quad +80}$$

$$72 = 9a$$

$$\frac{72}{9} = \frac{9a}{9}$$

$$\boxed{8 = a}$$

Check $a = 8$: $\qquad \dfrac{1}{4}a + \dfrac{3a - 4}{10} = a - 4$

$$\frac{1}{4}(8) + \frac{3(8) - 4}{10} \overset{?}{=} 8 - 4$$

$$2 + \frac{24 - 4}{10} \overset{?}{=} 4$$

$$2 + \frac{20}{10} \overset{?}{=} 4$$

$$2 + 2 \overset{\checkmark}{=} 4$$

EXAMPLE 6 | Solve for y. $\dfrac{8y}{9} - \dfrac{2y + 1}{12} = \dfrac{4y + 3}{18}$

Solution | Here the LCD is not quite so obvious.

$$\left.\begin{array}{l} 9 = 3 \cdot 3 \\ 12 = 2 \cdot 2 \cdot 3 \\ 18 = 2 \cdot 3 \cdot 3 \end{array}\right\} \quad \textit{The LCD is } 2 \cdot 2 \cdot 3 \cdot 3 = 36.$$

$$36 \left(\frac{8y}{9} - \frac{2y + 1}{12}\right) = 36 \left(\frac{4y + 3}{18}\right)$$

$$\frac{36}{1} \cdot \frac{8y}{9} - \frac{36}{1} \cdot \left(\frac{2y + 1}{12}\right) = \frac{36}{1} \cdot \left(\frac{4y + 3}{18}\right)$$

$$\overset{4}{\cancel{\frac{36}{1}}} \cdot \frac{8y}{\cancel{9}} - \overset{3}{\cancel{\frac{36}{1}}} \cdot \left(\frac{2y + 1}{\cancel{12}}\right) = \overset{2}{\cancel{\frac{36}{1}}} \cdot \left(\frac{4y + 3}{\cancel{18}}\right)$$

$$4(8y) - 3(2y + 1) = 2(4y + 3)$$ *Watch the signs.*

$$32y - 6y - 3 = 8y + 6$$
$$26y - 3 = 8y + 6$$
$$\underline{-8y \qquad\quad -8y}$$
$$18y - 3 = 6$$
$$\underline{\qquad +3 \quad +3}$$
$$18y = 9$$

$$\boxed{y = \frac{1}{2}}$$

Check $y = \frac{1}{2}$: $\qquad \dfrac{8y}{9} - \dfrac{2y + 1}{12} = \dfrac{4y + 3}{18}$

$$\dfrac{8\left(\frac{1}{2}\right)}{9} - \dfrac{2\left(\frac{1}{2}\right) + 1}{12} \stackrel{?}{=} \dfrac{4\left(\frac{1}{2}\right) + 3}{18}$$

$$\dfrac{4}{9} - \dfrac{1 + 1}{12} \stackrel{?}{=} \dfrac{2 + 3}{18}$$

$$\dfrac{4}{9} - \dfrac{2}{12} \stackrel{?}{=} \dfrac{5}{18}$$

$$\dfrac{4}{9} - \dfrac{1}{6} \stackrel{?}{=} \dfrac{5}{18}$$

$$\dfrac{8}{18} - \dfrac{3}{18} \stackrel{?}{=} \dfrac{5}{18}$$

$$\dfrac{5}{18} \stackrel{\checkmark}{=} \dfrac{5}{18}$$ ∎

As we saw in Chapter 3, our procedure for solving inequalities is basically the same as for solving equations, except that if we multiply or divide the inequality by a *negative* number we must *reverse* the inequality.

EXAMPLE 7 Solve for *x*. $\dfrac{3}{5} - \dfrac{x - 3}{15} \leq \dfrac{1}{3}$

Solution The LCD is 15.

$$\dfrac{3}{5} - \dfrac{x - 3}{15} \leq \dfrac{1}{3}$$ *Multiply both sides of the inequality by 15.*

$$15 \left(\dfrac{3}{5} - \dfrac{x - 3}{15}\right) \leq 15 \cdot \dfrac{1}{3}$$ *Use the distributive property on the left-hand side.*

$$\dfrac{15}{1} \cdot \dfrac{3}{5} - \dfrac{15}{1} \cdot \left(\dfrac{x - 3}{15}\right) \leq \dfrac{15}{1} \cdot \dfrac{1}{3}$$

$$\dfrac{\overset{3}{\cancel{15}}}{1} \cdot \dfrac{3}{\cancel{5}} - \dfrac{\cancel{15}}{1} \cdot \left(\dfrac{x - 3}{\cancel{15}}\right) \leq \dfrac{\overset{5}{\cancel{15}}}{1} \cdot \dfrac{1}{\cancel{3}}$$ ***Be careful.** This minus sign in front of the x − 3 belongs to both terms. It is incorrect to write −x − 3.*

$$9 - (x - 3) \leq 5$$ *The parentheses are necessary.*
$$9 - x + 3 \leq 5$$
$$12 - x \leq 5$$
$$\underline{-12 \qquad\quad -12}$$
$$-x \leq -7$$ *To get x alone, divide both sides by −1.*

Remember that dividing by a negative number reverses the inequality.

$$\frac{-x}{-1} \geq \frac{-7}{-1}$$

$$\boxed{x \geq 7}$$

Recall that we cannot check every value of $x \geq 7$, so we check one value of x that is greater than or equal to 7 to verify that it is a solution, and we check one value of $x < 7$ to verify that it is not.

Check $x = 7$: This should satisfy the inequality.

$$\frac{3}{5} - \frac{x - 3}{15} \leq \frac{1}{3}$$

$$\frac{3}{5} - \frac{7 - 3}{15} \overset{?}{\leq} \frac{1}{3}$$

$$\frac{3}{5} - \frac{4}{15} \overset{?}{\leq} \frac{1}{3}$$

$$\frac{9}{15} - \frac{4}{15} \overset{?}{\leq} \frac{5}{15}$$

$$\frac{5}{15} \overset{\checkmark}{=} \frac{5}{15}$$

Check $x = 6$: This should not satisfy the inequality.

$$\frac{3}{5} - \frac{x - 3}{15} \leq \frac{1}{3}$$

$$\frac{3}{5} - \frac{6 - 3}{15} \overset{?}{\leq} \frac{1}{3}$$

$$\frac{3}{5} - \frac{3}{15} \overset{?}{\leq} \frac{1}{3}$$

$$\frac{9}{15} - \frac{3}{15} \overset{?}{\leq} \frac{5}{15}$$

$$\frac{6}{15} \nleq \frac{5}{15}$$

■

EXAMPLE 8

Solve for t. $0.23t + 0.7(t - 20) = 172$

Solution

(If you would like to review decimal arithmetic, see Appendix A in the back of the book.)

While it is not necessary to rewrite this equation in fractional form, it will help us see that a problem involving decimals is handled in the same way as one involving fractions.

$$0.23t + 0.7(t - 20) = 172$$

rewritten in fractional form becomes

$$\frac{23}{100}t + \frac{7}{10}(t - 20) = 172$$

We can now see that the LCD is 100. In other words, if we look at the original equation in decimal form, wanting to "clear" the decimals is equivalent to "clearing" the fractions. If we want to move the decimal point one place to the right, we multiply by 10; to move the decimal point two places to the right, we multiply by 100; to move the decimal point three places to the right, we multiply by 1,000, etc. In this example we want to move the decimal point two places to the right (so that 0.23 will become 23); therefore, we multiply both sides of the equation by 100.

$$100[0.23t + 0.7(t - 20)] = 100(172) \qquad \textit{Use the distributive property}$$
$$100(0.23t) + 100(0.7)(t - 20) = 100(172) \qquad \textit{on the left-hand side.}$$
$$23t + 70(t - 20) = 17{,}200$$
$$23t + 70t - 1{,}400 = 17{,}200$$
$$93t - 1{,}400 = 17{,}200$$
$$\underline{ +1{,}400 \quad +1{,}400}$$
$$93t = 18{,}600$$

$$\frac{93t}{93} = \frac{18,600}{93}$$

$$\boxed{t = 200}$$

Check $t = 200$: $\qquad 0.23t + 0.7(t - 20) = 172$

$$0.23(200) + 0.7(200 - 20) \overset{?}{=} 172$$

$$46 + 0.7(180) \overset{?}{=} 172$$

$$46 + 126 \overset{?}{=} 172$$

$$172 \overset{\checkmark}{=} 172 \qquad \blacksquare$$

In the next two sections we will use our ability to solve fractional equations and inequalities to examine a wider variety of applications.

Study Skills 4.4

Taking an Algebra Exam: Dealing with Panic

In the first two chapters of this text we have given you advice on how to learn algebra. In the last chapter we discussed how to prepare for an algebra exam. If you followed this advice and put the proper amount of time to good use, you should feel fairly confident and less anxious about the exam. But you may still find during the course of the exam that you are suddenly stuck or you "draw a blank." This may lead you to panic and say irrational things like "I'm stuck. . . . I can't do this problem. . . . I can't do any of these problems. . . . I'm going to fail this test." Your heart may start to beat faster and your breath may quicken. You are entering a panic cycle.

These statements are irrational. Getting stuck on a few problems does not mean that you cannot do any algebra. These statements only serve to interfere with your concentrating on the exam itself. How can you think about solving a problem while you are telling yourself that you cannot? The increased heart and breath rate are part of this cycle.

What we would like to do is to break this cycle. What we recommend that you do is first put aside the exam and silently say to yourself **STOP!** Then try to relax, clear your mind, and encourage yourself by saying to yourself such things as "This is only one (or a few) problems, not the whole test" or "I've done problems like this before, so I'll get the solution soon." (Haven't you ever talked to yourself this way before?)

Now take some slow deep breaths and search for some problems that you know how to solve and start with those. Build your concentration and confidence up slowly with more problems. When you are through with the problems you can complete, go back to the ones on which you were stuck. If you have the time, take a few minutes and rest your head on your desk, and then try again. But make sure you have checked the problems you have completed.

EXERCISES 4.4

Solve each of the following equations and inequalities.

1. $\dfrac{x}{3} = 9$

2. $\dfrac{y}{5} = 4$

3. $\dfrac{a}{4} = -6$

4. $\dfrac{a}{2} = -9$

5. $\dfrac{y}{6} = \dfrac{5}{4}$

6. $\dfrac{z}{10} = \dfrac{7}{6}$

7. $\dfrac{w}{8} = -\dfrac{7}{6}$

8. $\dfrac{u}{9} = -\dfrac{5}{12}$

9. $\dfrac{3x}{2} = -18$

10. $30 = \dfrac{5}{6}x$

11. $16 = \dfrac{4}{5}x$

12. $\dfrac{2x}{7} = -14$

13. $\dfrac{x}{3} - 2 = \dfrac{2}{3}$

14. $\dfrac{x}{5} - 3 = \dfrac{4}{5}$

15. $\dfrac{3a}{4} + 2 = \dfrac{5}{4}$

16. $\dfrac{5a}{2} + 9 = \dfrac{3}{2}$

17. $\dfrac{u}{2} - \dfrac{u}{4} = 2$

18. $\dfrac{u}{3} - \dfrac{u}{9} = 4$

19. $\dfrac{y}{3} + \dfrac{y}{5} < \dfrac{6}{5}$

20. $\dfrac{y}{2} + \dfrac{y}{7} > \dfrac{5}{7}$

21. $3x - \dfrac{2}{3}x = \dfrac{4}{3}$

22. $\dfrac{3}{5}x - 3x = \dfrac{14}{5}$

23. $\dfrac{a}{4} - \dfrac{a}{3} \geq \dfrac{5}{2}$

24. $\dfrac{a}{6} - \dfrac{a}{5} \leq \dfrac{4}{3}$

25. $0.7x + 0.4x = 5.5$

26. $0.6x + 0.8x = 8.4$

27. $0.3x - 0.25x = 2$

28. $0.9x - 0.36x = 2.7$

29. $0.8m + 0.05m = 0.34$

30. $0.65m + 1.5m = 0.43$

31. $\dfrac{w + 3}{4} = \dfrac{w + 4}{3}$

32. $\dfrac{w - 2}{6} = \dfrac{w - 1}{9}$

33. $\dfrac{w + 3}{4} + 1 = \dfrac{w + 4}{3}$

34. $\dfrac{w - 2}{6} - 2 = \dfrac{w - 1}{9}$

35. $\dfrac{x + 1}{2} + x = 6$

36. $\dfrac{x + 3}{5} - x = 2$

37. $\dfrac{y}{6} - \dfrac{y - 2}{4} > 1$

38. $\dfrac{y}{5} - \dfrac{y - 1}{2} > 2$

39. $3 - \dfrac{a + 1}{4} = \dfrac{a + 4}{2}$

40. $4 - \dfrac{a + 3}{10} = \dfrac{a + 11}{5}$

41. $\dfrac{2y - 3}{2} - \dfrac{y - 5}{3} = \dfrac{1}{6}$

42. $\dfrac{y + 2}{14} - \dfrac{4y + 1}{7} = 1$

43. $\dfrac{x + 2}{3} - \dfrac{2x + 3}{4} = \dfrac{x + 4}{8}$

44. $\dfrac{5x - 1}{2} - \dfrac{x - 2}{5} = \dfrac{8x + 11}{6}$

45. $\dfrac{t}{2} + \dfrac{t - 1}{3} + \dfrac{t - 6}{4} = t - 2$

46. $\dfrac{t}{6} + \dfrac{t + 3}{5} + \dfrac{t - 2}{10} = t - 6$

47. $0.5(x + 2) - 0.3(x - 4) = 3$

48. $0.6(x - 4) - 0.4(x - 5) = 4.6$

49. $0.06x + 0.04(5,000 - x) = 320$

50. $0.05x + 0.08(x + 3,000) = 890$

51. $0.035x + 0.052(x + 5,000) = 1,130$

52. $0.155x + 0.075(80 - x) = 8.4$

53. $3(y + 2) + \dfrac{y + 3}{5} = \dfrac{9y + 8}{2}$

54. $5(y - 3) + \dfrac{2 - y}{3} = \dfrac{7y + 1}{4}$

55. $z + \dfrac{z + 5}{3} - \dfrac{z - 2}{6} = \dfrac{z + 4}{4} + 1$

56. $5z - \dfrac{3 - z}{2} + \dfrac{z + 4}{5} = 8 - \dfrac{z + 8}{3}$

57. $3 \leq \dfrac{x}{3} - \dfrac{x + 1}{2} \leq 6$

58. $\dfrac{2}{5} < \dfrac{2x + 3}{5} - \dfrac{3x + 1}{7} < \dfrac{6}{7}$

In the exercises that follow, if the exercise is an equation, solve it; if not, perform the indicated operations and express your answer as a single fraction.

59. $\frac{x}{3} + \frac{x}{2} + \frac{x}{5}$

60. $x + \frac{x}{3} - \frac{x}{4}$

61. $\frac{x}{3} + \frac{x}{2} + \frac{x}{5} = 62$

62. $x + \frac{x}{3} - \frac{x}{4} = 26$

63. $\frac{x + 5}{2} - \frac{x - 1}{4} = 2$

64. $\frac{2x - 1}{5} - \frac{x - 7}{3} = 2$

65. $\frac{x + 5}{2} - \frac{x - 1}{4}$

66. $\frac{2x - 1}{5} - \frac{x - 7}{3}$

? QUESTION FOR THOUGHT

67. Make up a verbal problem involving a purchase of some items that would give rise to the following equation (be sure to indicate what the variable would represent):

$$30x + 50(80 - x) = 3{,}600$$

◇ MINI-REVIEW

Solve each of the following problems algebraically. Be sure to label what the variable represents.

68. The width of a rectangle is 20 less than 3 times the length. If the perimeter is 32 ft, find its dimensions.

69. An express trucking company charges $7 for the first pound and $4 for each additional pound. If a package costs $43 to deliver, how much does it weigh?

4.5 Ratio and Proportion

One of the most common and useful applications of the techniques involving fractions we have developed so far is in dealing with the ideas of ratio and proportion.

A *ratio* is simply a fraction. If we say that the ratio of boys to girls in a certain class is 2 to 3, that means that there are 2 boys for every 3 girls. The ratio 2 to 3 is written as the fraction $\frac{2}{3}$ (sometimes it is also written 2 : 3). However, the ratio of girls to boys would be written $\frac{3}{2}$, since there are 3 girls for every 2 boys.

EXAMPLE 1 If a certain car dealer sold 350 domestic cars and 210 imported cars, find each of the following:

(a) The ratio of domestic cars sold to imported cars sold

(b) The ratio of imported cars sold to domestic cars sold

(c) The ratio of domestic cars sold to the total number of cars sold

Solution

(a) $\dfrac{\textit{Number of domestic cars sold}}{\textit{Number of imported cars sold}} = \dfrac{350}{210} = \dfrac{5 \cdot 70}{3 \cdot 70} = \boxed{\dfrac{5}{3}}$

(b) $\dfrac{\textit{Number of imported cars sold}}{\textit{Number of domestic cars sold}} = \dfrac{210}{350} = \dfrac{3 \cdot 70}{5 \cdot 70} = \boxed{\dfrac{3}{5}}$

(c) $\dfrac{\textit{Number of domestic cars sold}}{\textit{Total number of cars sold}} = \dfrac{350}{350 + 210} = \dfrac{350}{560} = \dfrac{5 \cdot 70}{8 \cdot 70} = \boxed{\dfrac{5}{8}}$ ∎

A *proportion* is an equation between two ratios. Frequently we can interpret given information in terms of a proportion, and the resulting equation is often easy to solve.

EXAMPLE 2

If the male to female ratio in a certain factory is 8 to 3 and there are 45 women in the factory, how many men are there?

Solution

If we let x = number of males in the factory, then the given information of the ratio of male to female being 8 to 3 allows us to write the following proportion:

$$\frac{\textit{Number of males}}{\textit{Number of females}} = \frac{8}{3}$$

$$\frac{x}{45} = \frac{8}{3} \qquad \textit{Since we want to solve for x, we multiply both sides of the equation by 45.}$$

$$45 \cdot \frac{x}{45} = 45 \cdot \frac{8}{3}$$

$$\frac{\cancel{45}}{1} \cdot \frac{x}{\cancel{45}} = \frac{\overset{15}{\cancel{45}}}{1} \cdot \frac{8}{3}$$

$$\boxed{x = 120}$$

There are 120 males in the factory. ∎

EXAMPLE 3

A recent survey found that of the total number of people surveyed, 192 preferred brand X. If the ratio of those who did not prefer brand X to those who did was 3 to 4, how many people were surveyed altogether?

Solution

Let n = number of people who did not prefer brand X. Then our proportion is

$$\frac{\textit{Number of people who did not prefer brand X}}{\textit{Number of people who did prefer brand X}} = \frac{3}{4}$$

$$\frac{n}{192} = \frac{3}{4} \qquad \textit{To solve for n we multiply both sides of the equation by 192.}$$

$$\frac{\cancel{192}}{1} \cdot \frac{n}{\cancel{192}} = \frac{\overset{48}{\cancel{192}}}{1} \cdot \frac{3}{4}$$

$$n = 144$$

We are not finished yet. The example asks for the *total* number of people surveyed. Thus the answer is

$$144 + 192 = \boxed{336}$$

An alternative solution is to consider the following proportion:

$$\frac{\textit{Total number of people surveyed}}{\textit{Number of people who prefer brand X}} = \frac{7}{4}$$

The ratio of $\frac{7}{4}$ reflects the fact that if the ratio of those who did not prefer brand X to those who did is 3 to 4, then out of 7 people surveyed, 4 preferred brand X.

If we let n = total number of people surveyed, this proportion becomes

$$\frac{n}{192} = \frac{7}{4}$$

$$\frac{\cancel{192}}{1} \cdot \frac{n}{\cancel{192}} = \frac{\overset{48}{\cancel{192}}}{1} \cdot \frac{7}{4}$$

$$\boxed{n = 336}$$ ■

EXAMPLE 4 If there are 2.54 centimeters to 1 inch, how wide, in inches, is a table that is 72 centimeters wide? (Round off your answer to the nearest tenth.)

Solution Let x = number of inches in 72 centimeters.
It is usually easier to solve a fractional equation if the variable is in the numerator. We therefore translate the given information into the following proportion:

If we set up the ratio with inches over centimeters on one side of the proportion, then we must have inches over centimeters on the other side as well.

$$\frac{\text{Length of the table in inches}}{72 \text{ cm}} = \frac{1 \text{ in.}}{2.54 \text{ cm}}$$ *Notice how the units in a proportion must agree.*

$$\frac{x}{72} = \frac{1}{2.54}$$

$$\frac{\cancel{72}}{1} \cdot \frac{x}{\cancel{72}} = \frac{72}{1} \cdot \frac{1}{2.54}$$

$$x = \frac{72}{2.54}$$ *Use a calculator to get*

$$\boxed{x = 28.3 \text{ inches}}$$ *Rounded to the nearest tenth* ■

EXAMPLE 5 If there is 0.946 liter in 1 quart, how many ounces are there in a 2-liter bottle?

Solution Let n = number of ounces in a 2-liter bottle. Our proportion is

$$\frac{n \text{ ounces}}{2 \text{ liters}} = \frac{1 \text{ quart}}{0.946 \text{ liter}}$$ *Since units must be consistent throughout the proportion, we convert 1 quart into 32 ounces.*

$$\frac{n \text{ ounces}}{2 \text{ liters}} = \frac{32 \text{ ounces}}{0.946 \text{ liter}}$$

$$\frac{\cancel{2}}{1} \cdot \frac{n}{\cancel{2}} = \frac{2}{1}\left(\frac{32}{0.946}\right)$$

$$n = \frac{64}{0.946}$$

$$\boxed{n = 67.7 \text{ ounces}}$$ *Rounded to the nearest tenth* ■

EXAMPLE 6  In a certain town the real-estate tax on a house with an assessed value of $112,500 is $830. Assuming the same tax rate, find the real-estate tax due on a house with an assessed value of $165,000.

Solution Since the real-estate tax is being computed at the same rate, we can set up a proportion comparing the tax due to the assessed value.

$$\frac{\text{Tax due on house assessed at } \$165,000}{\text{Assessed value of } \$165,000} = \frac{\text{Tax due on house assessed at } \$112,500}{\text{Assessed value of } \$112,500}$$

Let x = the tax due on a house with an assessed value of $165,000. Thus, the proportion becomes

$$\frac{x}{165,000} = \frac{830}{112,500}$$

Multiply both sides of the equation by 165,000 to solve for x.

$$\frac{165,000}{1} \cdot \frac{x}{165,000} = \frac{165,000}{1} \cdot \frac{830}{112,500}$$

$$x = \frac{165,000\,(830)}{112,500}$$

Using a calculator, we compute this to be

$$x = 1,217.33$$

Thus the real-estate tax on a house assessed at $165,000 is $\boxed{\$1,217.33}$.

The check is left to the student.   ■

EXAMPLE 7

Cherise and Jerome agree to share the profits in their business in the ratio of 3 to 2. How should they split up an annual profit of $85,600?

Solution

Exercise 65 discusses how this problem would be solved by writing a proportion directly from the given information.

Rather than solve this problem by translating the given information directly into a proportion, let's try a different approach.

The fact that the ratio of Cherise's share to Jerome's share is 3 to 2 means that we can represent their shares as follows:

Let $3x$ = Cherise's share   and $2x$ = Jerome's share

Note that

$$\frac{\textit{Cherise's share}}{\textit{Jerome's share}} = \frac{3x}{2x} = \frac{3x}{2x} = \frac{3}{2} \quad \text{as required}$$

Once we have represented the two shares, we can write the following equation:

$3x + 2x = 85,600$ *Since the total of the two shares is 85,600*

$$5x = 85,600$$

$$x = \frac{85,600}{5} = 17,120$$

Now that we know the value of x, we can compute the individual shares.

Cherise's share $= 3x = 3\,(17,120) = \boxed{\$51,360}$.

Jerome's share $= 2x = 2\,(17,120) = \boxed{\$34,240}$.

The check is left to the student.   ■

Chapter G at the end of the book covers the basic elements of geometry used in this book.

In geometry, if the three angles of one triangle are equal to the three angles of a second triangle, the triangles are called *similar triangles*. Similar triangles have the same shape but not necessarily the same size, as shown in Figure 4.1.

Figure 4.1

$\triangle ABC$ is similar to $\triangle DEF$

$\angle A = \angle D$

$\angle B = \angle E$

$\angle C = \angle F$

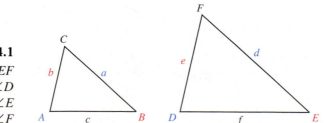

As indicated in Figure 4.1, we usually label each side of a triangle with the lower-case letter of the angle opposite it. By the corresponding sides of similar triangles we mean the sides of the two triangles that are opposite equal angles. Thus in Figure 4.1, a corresponds to d, b corresponds to e, and c corresponds to f. It is a fact, demonstrated in elementary geometry, that the corresponding sides of similar triangles are in proportion. This means that if, for example, one pair of corresponding sides is in the ratio of 3 to 2, then all the pairs of corresponding sides are in the ratio of 3 to 2. Thus in Figure 4.1, if $\dfrac{d}{a} = \dfrac{3}{2}$, then we also have $\dfrac{e}{b} = \dfrac{3}{2}$ and $\dfrac{f}{c} = \dfrac{3}{2}$.

EXAMPLE 8

Suppose the following two triangles are similar, with $\angle A = \angle D$, $\angle B = \angle E$, and $\angle C = \angle F$. Find the values of x and y.

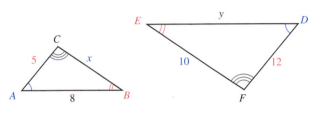

Solution

Given that these two triangles are similar, the side labeled x corresponds to the side labeled 10, the side labeled y corresponds to the side labeled 8, and the side labeled 12 corresponds to the side labeled 5. Thus we may set up the following proportions:

$$\frac{x}{10} = \frac{5}{12} \qquad\qquad \frac{y}{8} = \frac{12}{5}$$

$$x = \frac{50}{12} = \boxed{4\frac{1}{6}} \qquad\qquad y = \frac{96}{5} = \boxed{19\frac{1}{5}} \qquad \blacksquare$$

EXAMPLE 9

Suppose that at a certain time a 6-ft man casts a 3.4-ft shadow. Find the height of a tree (to the nearest tenth) that casts an 11.8-ft shadow at the same time.

Solution

The following diagram serves to illustrate the given information. We let x represent the height of the tree, $\overline{AC}$ represent the shadow cast by the man, and $\overline{AE}$ the shadow cast by the tree.

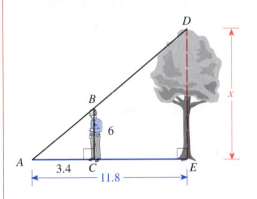

Since the angles of $\triangle ABC$ are the same as the angles of $\triangle ADE$, the two triangles are similar. We may therefore write the following proportion:

$$\frac{x}{6} = \frac{11.8}{3.4}$$

$$x = \frac{6(11.8)}{3.4} = \boxed{20.8 \text{ ft}} \qquad \textit{Rounded to the nearest tenth} \qquad \blacksquare$$

In Chapter 9 we will examine additional examples involving ratios and proportions.

Study Skills 4.5

Taking an Algebra Exam: A Few Other Comments About Exams

Do not forget to check over all your work as we have suggested on numerous occasions. Reread all directions and make sure that you have answered all the questions as directed.

If you are required to show your work (such as for partial credit), make sure that your work is neat. Do not forget to put your final answers where directed or at least indicate your answers clearly by putting a box or a circle around your answer. For multiple-choice tests be sure you have filled in the correct space.

One other bit of advice: Some students are unnerved when they see others finishing the exam early. They begin to believe that there may be something wrong with themselves because they are still working on the exam. They should not be concerned, for there are some students who can do the work quickly and others who leave the exam early because they give up, not because the exam was too easy for them.

In any case, do not be in a hurry to leave the exam. If you are given 1 hour for the exam, then take the entire hour. If you have followed the suggestions in this chapter such as checking your work, etc., and you still have time left over, relax for a few minutes and then go back and check over your work again.

EXERCISES 4.5

In Exercises 1–10, write each of the phrases as a ratio.

1. 7 red to 5 black

2. 9 short to 2 long

3. 5 black to 7 red

4. 2 long to 9 short

5. 11 with to 5 without

6. 5 with to 11 without

7. x to $3x$

8. $3x$ to x

9. a to the sum of b and c

10. the sum of b and c to a

In Exercises 11–18, solve each proportion.

11. $\dfrac{x}{5} = \dfrac{12}{3}$

12. $\dfrac{x}{6} = \dfrac{10}{2}$

13. $\dfrac{a}{6} = \dfrac{5}{3}$

14. $\dfrac{a}{8} = \dfrac{9}{2}$

15. $\dfrac{y}{15} = \dfrac{20}{6}$

16. $\dfrac{y}{10} = \dfrac{35}{14}$

17. $\dfrac{y}{9} = \dfrac{4}{3}$

18. $\dfrac{y}{18} = \dfrac{10}{9}$

In Exercises 19–24, solve each proportion. Round off your answers to the nearest hundredth where necessary.

19. $\dfrac{x}{0.43} = \dfrac{0.26}{0.9}$

20. $\dfrac{a}{2.61} = \dfrac{3.82}{7.41}$

21. $\dfrac{t}{61.95} = \dfrac{8.8}{47.02}$

22. $\dfrac{14.61}{218.36} = \dfrac{w}{97.03}$

23. $\dfrac{y + 0.3}{0.7} = \dfrac{y - 0.2}{0.9}$

24. $\dfrac{u - 2.6}{4.5} = \dfrac{u + 7.8}{6.6}$

Solve each of the following exercises by first setting up a proportion or an equation. Round off your answers to the nearest hundredth.

25. A jar contains marbles in the ratio of 7 red to 5 black. If there are 210 black marbles, how many red marbles are there?

26. Repeat Exercise 25 with the ratio of red to black reversed.

27. On a certain test a math teacher found that the ratio of grades 90 or above to those below 90 was $\frac{3}{8}$. If 24 students got below 90, how many got 90 or above?

28. Repeat Exercise 27 if the ratio of grades 90 or above to those below 90 was 5 to 12.

29. If the ratio of the length of a rectangle to its width is $\frac{9}{4}$, and the length is 18 cm, what is the width of the rectangle?

30. If the ratio of the width of a rectangle to its length is $\frac{3}{7}$, and the length is 35 mm, find the width of the rectangle.

31. If the sides of a rectangle are as shown in the accompanying diagram, what is the ratio of the shorter side to the longer side?

32. In Exercise 31, what is the ratio of the shorter side to the perimeter?

33. In a scale drawing, actual sizes are all diminished in the same proportion. If a 12-meter wall is represented by a 5-cm length, what length would represent a 20-meter wall?

34. Repeat Exercise 33 if the 12 meter wall is represented by a 7-cm length.

35. If there are 2.2 pounds in 1 kilogram, how many kilograms are there in 10 pounds?

36. If there are 1.06 quarts in 1 liter, how many liters are there in 2 gallons?

37. If there is 0.92 meter in 1 yard, how many yards are there in 100 meters?

38. There is approximately 0.625 mile in 1 kilometer. If a speedometer reads 55 mph, how would it read in kilometers per hour?

39. The scale on a map indicates that 5 cm is equivalent to 10 miles. If the distance between two cities on the map is 12.6 cm, how far apart are the two cities?

40. Using the same scale as in Exercise 39, how far apart on the map would two cities be if they are actually 176 miles from each other?

41. Two partners decide to share the profits of their business in the same ratio as their initial investments. Amanda invested $18,500 and her share of the profits was $2,800. How much was Rolfe's share if he invested $12,800?

42. A music store sells CDs and cassettes in the ratio of 13 to 6. During a certain week the store sold 498 cassettes. How many CDs were sold?

43. A recipe for 6 servings of a pasta sauce calls for 2 teaspoons of garlic and 5 tablespoons of olive oil. How much garlic and olive oil would be needed for 20 servings?

44. A recipe for 8 servings of salad dressing requires 50 grams of sugar and 12 grams of ginger. How much sugar and ginger would be needed for 15 servings?

45. In a certain town the annual real-estate tax on a property assessed at $126,400 is $788. Find the tax due on a property assessed at $98,200.

46. Lena owns a home in a town in which the property tax rate is $6.58 per $1,000 of assessed value. Lena's home is assessed at $187,500, and she receives a tax bill for $1,357.85. Is her tax bill correct? If not, what should it be?

47. A bronze alloy consists of copper and tin in the ratio of 11 to 2, respectively. If the bronze weighs 52 oz, how much copper and tin does it contain?

48. A brass alloy contains only copper and zinc in the ratio of 5 to 3, respectively. If the alloy contains 86 grams of copper, how much does the alloy weigh altogether?

49. Samantha and Greg agree to share their company's annual profits in the ratio of 7 to 5, respectively. If the annual profit is $128,650, how much did each receive?

50. A waiter and busboy agree to pool their tips and then divide them in the ratio of 11 to 7, respectively. If they collect a total of $368, how much did each receive?

51. On a certain day, the stock exchange reports that the ratio of stocks that went up (gainers) to those that went down (losers) is 8 to 3. If there were 976 gainers, how many losers were there?

52. On a certain day, the stock exchange reports that the ratio of stocks that went up (gainers) to those that went down (losers) is 5 to 9. If there were 756 losers, how many stocks were traded that day?

53. A will provides that Janice and Wanda should divide an estate in the ratio of 7 to 4. If Janice's share of the estate is $56,300, how much was the estate worth altogether?

54. In a legal settlement, a client and lawyer agree to split the settlement in the ratio of 8 to 3, respectively. If the lawyer's share was $36,500, how much was the settlement worth altogether?

55. A common method for estimating an animal population that is difficult to count is to capture, tag, and then release a certain number of animals. After allowing time for the tagged animals to mix together with the untagged, a second group of animals is captured. If we assume that the ratio of tagged to untagged animals is the same for the whole population as it is for the second group, we can set up a proportion to estimate the total population.

     If 48 deer are captured from a certain area and tagged, and a sample of 60 deer taken after mixing shows 18 are tagged, how many deer are estimated in the entire area?

56. Repeat Exercise 55 if 28 deer were tagged originally and a sample of 45 deer taken after mixing shows that 12 are tagged.

In Exercises 57–60, the given triangles are similar. The equal angles are indicated by the same number of arcs within the angles. Find the lengths of the missing sides in each of the triangles. Round off your answers to the nearest hundredth where necessary.

57.

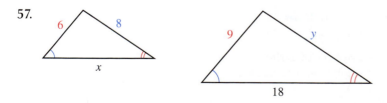

58.

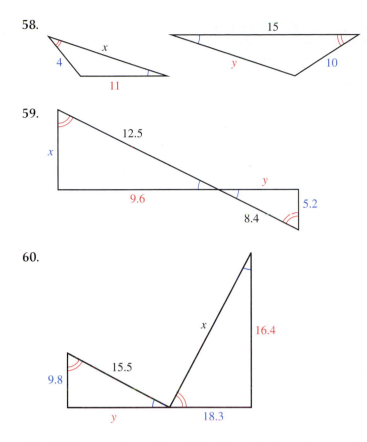

59.

60.

61. A 6-foot man is standing 15 feet from the base of a street light and casts an 8-foot shadow. How high above the ground is the street light? See the figure below.

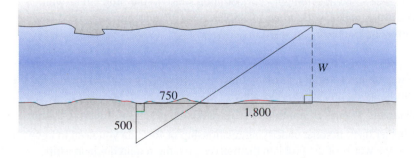

62. Suppose that a street light is 20 feet above the ground. How far from the street light is a 5-foot girl standing if she casts a 4-foot shadow? [*Hint:* Use a figure similar to the one for the previous exercise.]

63. A surveyor is attempting to measure the width of a river. He makes the measurements in the accompanying figure. Use the given measurements to compute the width *w* of the river. All measurements are given in meters.

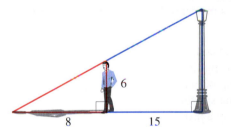

64. A surveyor is attempting to find the distance between points *A* and *B* across a lake, as illustrated in the accompanying figure. Use the given measurements she found to determine the distance between *A* and *B*. All measurements are given in feet.

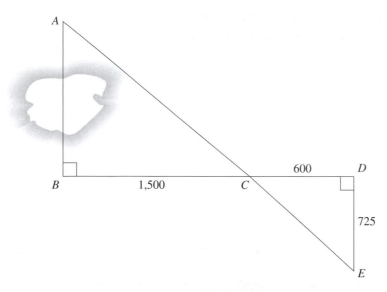

---

② QUESTION FOR THOUGHT

65. Solve Example 7 by writing a proportion directly from the given information. That is, write and solve the proportion

$$\frac{Cherise's\ share}{Jerome's\ share} = \frac{3}{2}$$

[*Hint:* Let *x* = Cherise's share; then $85,600 − *x* = Jerome's share.]

4.6 Applications

Now that we have developed the ability to handle a wider range of first-degree equations in one variable, we can apply our knowledge to a greater variety of applications.

It is worth repeating that no attempt is being made to make you an expert in any particular type of verbal problem. As we do more problems, we will see that while problems may seem at first to be very different, their solutions often show a similar structure.

As in our previous work with verbal problems, not only do we need to know how to translate certain phrases and sometimes certain mathematical formulas algebraically, but we also need to be able to apply our common sense to the problem as well.

Before we proceed with some examples, let's restate our suggested outline for solving verbal problems.

Outline of Strategy for Solving Verbal Problems	1. Read the problem carefully, as many times as is necessary to understand what the problem is saying and what it is asking.
	2. Use diagrams whenever you think it will make the given information clearer.
	3. Ask whether there is some underlying relationship or formula you need to know. If not, then the words of the problem themselves give the required relationship.

(continued)

Outline of Strategy for Solving Verbal Problems	*(continued)*

4. Clearly identify the unknown quantity (or quantities) in the problem, and label it (them) using one variable.

    Step 4 is very important and not always easy.

5. By using the underlying formula or relationship in the problem, write an equation involving the unknown quantity (or quantities).

    Step 5 is the *crucial step.*

6. Solve the equation.

7. Make sure you have answered the question that was asked.

8. Check the answer(s) in the original words of the problem.

    Keep the reasonableness of your answer in mind. Did you obtain a negative weight or length? Is your answer for the time or distance much too large or much too small?

EXAMPLE 1

If 3 more than three-fifths of a number is 1 less than that number, find the number.

Solution

Let x = the number. Next we translate the phrases in the problem.

$$\text{``three-fifths of the number''} = \frac{3}{5} \cdot x = \frac{3x}{5}$$

$$\text{``3 more than three-fifths of the number''} = \frac{3x}{5} + 3$$

$$\text{``1 less than the number''} = x - 1 \quad \textbf{Remember:} \quad Not\ 1 - x$$

The word "is" is translated as "equal to," so that the equation is

$$\frac{3x}{5} + 3 = x - 1 \qquad \textit{We multiply both sides of the equation by 5 to clear the fraction.}$$

$$5 \left(\frac{3x}{5} + 3 \right) = 5\ (x - 1)$$

$$\frac{5}{1} \cdot \frac{3x}{5} + 5 \cdot 3 = 5x - 5$$

$$3x + 15 = 5x - 5$$

$$20 = 2x$$

$$\boxed{10 = x}$$

Check: $\frac{3}{5}$ of the number $= \frac{3}{5}(10) = 6.$

3 more than $\frac{3}{5}$ of the number $= 6 + 3 = 9.$

One less than the number $= 10 - 1 = 9.$ The answer checks. ■

EXAMPLE 2

In a certain triangle the medium side is 2 less than twice the shortest side, while the longest side is $2\frac{1}{2}$ times the shortest side. If the perimeter of the triangle is 2 more than 5 times the shortest side, find the lengths of the three sides of the triangle.

Solution

Since the sides of the triangle are all described in terms of the shortest side, let

$$x = \text{Length of the shortest side}$$

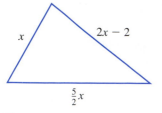

Figure 4.2
Triangle for Example 2

$2x - 2 = $ Length of the medium side

$\frac{5}{2}x = $ Length of the longest side   ***Remember:*** $2\frac{1}{2} = 2 + \frac{1}{2} = \frac{4}{2} + \frac{1}{2} = \frac{5}{2}$

(See Figure 4.2.) The problem also tells us that the perimeter is 2 more than 5 times the shortest side, so that the perimeter is $5x + 2$. The fact that the perimeter is the length around the triangle gives us the equation

$$x + 2x - 2 + \frac{5}{2}x = 5x + 2$$ *Combine like terms.*

$$3x - 2 + \frac{5}{2}x = 5x + 2$$ *Multiply both sides of the equation by 2 using the distributive property.*

$$2 \cdot 3x - 2 \cdot 2 + 2 \cdot \frac{5x}{2} = 2(5x + 2)$$

$$6x - 4 + 5x = 10x + 4$$
$$11x - 4 = 10x + 4$$
$$x = 8$$

Therefore, the shortest side $= \boxed{8}$.

The medium side $= 2x - 2 = 2(8) - 2 = \boxed{14}$.

The longest side $= \frac{5}{2}x = \frac{5}{2}(8) = \boxed{20}$.

The check is left to the student.   ■

EXAMPLE 3

A men's shop purchases a total of 60 shirts for $929.75. If some were dress shirts costing $17.50 each and the rest were sport shirts costing $14.25 each, how many of each were purchased?

Solution

As usual, we begin by trying to identify the relationship in the problem that will enable us to write an equation. For instance, we can begin with the following:

Cost of all the dress shirts + *Cost of all the sport shirts* = *Total cost*

$\left(\begin{matrix}Number\ of \\ dress\ shirts\end{matrix}\right) \cdot \left(\begin{matrix}Cost\ of\ 1 \\ dress\ shirt\end{matrix}\right) + \left(\begin{matrix}Number\ of \\ sport\ shirts\end{matrix}\right) \cdot \left(\begin{matrix}Cost\ of\ 1 \\ sport\ shirt\end{matrix}\right) = $ *Total cost*

$\left(\begin{matrix}Number\ of \\ dress\ shirts\end{matrix}\right) \cdot (\$17.50) + \left(\begin{matrix}Number\ of \\ sport\ shirts\end{matrix}\right) \cdot (\$14.25) = \$929.75$

We could also have let n = number of sport shirts, and 60 − n = number of dress shirts.

Now let $n = $ number of dress shirts. Then $60 - n = $ number of sport shirts. (Why?) The equation becomes

$$17.50n + 14.25(60 - n) = 929.75$$

$\underbrace{}_{\substack{Cost\ of \\ dress\ shirts}} \quad \underbrace{}_{\substack{Cost\ of \\ sport\ shirts}} \quad \underbrace{}_{Total\ cost}$

$$17.50n + 14.25(60 - n) = 929.75$$ *Multiply both sides of the equation by 100.*
$$1{,}750n + 1{,}425(60 - n) = 92{,}975$$
$$1{,}750n + 85{,}500 - 1{,}425n = 92{,}975$$ *Combine like terms.*
$$325n + 85{,}500 = 92{,}975$$
$$325n = 7{,}475$$

$$n = \frac{7{,}475}{325} = 23$$

They sold $\boxed{23 \text{ dress shirts}}$ and $60 - 23 = \boxed{37 \text{ sport shirts}}$

Check:

$$23 \text{ shirts } + 37 \text{ shirts } = 60 \quad \checkmark$$
$$23 \text{ shirts @ } \$17.50 = 23(17.50) = \$402.50$$
$$37 \text{ shirts @ } \$14.25 = 37(14.25) = \underline{\$527.25}$$
$$\$929.75 \quad \checkmark$$

EXAMPLE 4

A total of $6,000 was split into two investments. Part was invested in a bank account paying 7% interest per year, and the remainder was invested in stocks paying 11% interest per year. If the total yearly interest from the two investments is $492, how much was invested in the bank account?

Solution

In order to solve this problem we must know how to compute simple interest and how to write a percent as a decimal. (See Appendix A.)

To compute simple interest we use the formula

$$I = Prt$$

where I = interest; P = principal (the amount invested); r = rate of interest, which is the percentage written as a decimal; and t = time (the number of time periods for which the interest is being computed). For example, if $800 is invested at 6% per year for 2 years, then

$$P = 800, \quad r = 0.06 \ (6\% \text{ written as a decimal}), \quad \text{and} \quad t = 2$$

The formula would yield

$$I = (800)(0.06)(2) = \$96$$

In this example, as in most of the examples that we will do, $t = 1$ year.

In order to obtain an equation for this example we can begin by writing the following "verbal equation" summarizing the given information:

$$\left(\begin{array}{c} \textit{Amount of interest} \\ \textit{from 7\% investment} \end{array} \right) + \left(\begin{array}{c} \textit{Amount of interest} \\ \textit{from 11\% investment} \end{array} \right) = \$492$$

Based on what we have just said, we compute each amount of interest by multiplying the principal by the rate. Thus the verbal equation becomes

(Amount invested at 7%) · (0.07) + (Amount invested at 11%) · (0.11) = $492

Thus, if we let x = Amount invested at 7%   *Not the interest, but the actual amount invested*

then $6{,}000 - x$ = Amount invested at 11%   *Because there was $6,000 invested altogether*

Then our previous equation finally becomes

$$0.07x \quad + \quad 0.11(6{,}000 - x) \quad = \quad 492$$

Interest from the 7% investment   *Interest from the 11% investment*   *Total interest*

$$0.07x + 0.11(6{,}000 - x) = 492 \qquad \textit{Multiply both sides of the equation by}$$
$$7x + 11(6{,}000 - x) = 49{,}200 \qquad \textit{100 to clear the decimals.}$$

$$7x + 66,000 - 11x = 49,200$$

$$-4x + 66,000 = 49,200$$

$$\underline{-66,000 \quad -66,000}$$

$$-4x = -16,800$$

$$x = \frac{-16,800}{-4}$$

$$\boxed{x = \$4,200}$$ *This is the amount invested in the bank at 7%.*

Check: The amount invested at 11% is $6,000 - x = 6,000 - 4,200 = 1,800$.

$$4,200 + 1,800 \overset{\checkmark}{=} 6,000$$

$$0.07(4,200) + 0.11(1,800) = 294 + 198 \overset{\checkmark}{=} 492 \qquad ■$$

EXAMPLE 5 How many milliliters (ml) of a 30% alcohol solution must be mixed with 50 ml of a 70% alcohol solution to produce a 55% alcohol solution?

Solution It may help to visualize the problem as shown in Figure 4.3.

Figure 4.3
Alcohol solutions for
Example 5

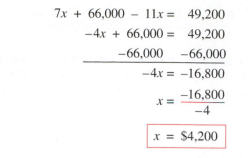

$$\begin{array}{ccccc} x \text{ ml} & + & 50 \text{ ml} & = & (x + 50) \text{ ml} \\ 30\% & & 70\% & & 55\% \end{array}$$

If you ask how much *alcohol* is in the second container in Figure 4.3, the answer is not 50 ml nor 70%: 50 ml is the amount of *solution,* not the amount of actual alcohol, and 70% is a *percent,* not an amount. To figure out how much actual alcohol is in the second container, you compute

$$70\% \text{ of } 50 \text{ ml} = 0.70(50) = 35 \text{ ml of alcohol}$$

Using the same idea, we can write the following equation:

$$0.30x \qquad + \qquad 0.70(50) \qquad = \qquad 0.55(x + 50)$$

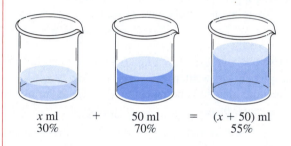

*Amount of alcohol   Amount of alcohol   Total amount of
in first container   in second container   alcohol in the mixture*

$$0.30x + 0.70(50) = 0.55(x + 50) \qquad \textit{Multiply both sides of the equation by } 100.$$

$$30x + 70(50) = 55(x + 50)$$

$$30x + 3,500 = 55x + 2,750$$

$$750 = 25x$$

$$\boxed{30 \text{ ml} = x}$$

The check is left to the student. ■

If we take a moment to look back at Examples 3, 4, and 5 (and, as you might also recognize, the coin problems from the last chapter), we can see that while these problems may have seemed to have nothing in common, once we analyzed them the resulting equations exhibited a very similar structure. The types of problems that result in such equations are often called *value problems* or *mixture problems.* Looking for similarities in different situations is often more illuminating than looking for differences.

EXAMPLE 6

At 10 A.M. Dana leaves for work riding her motorbike at 30 mph. Her husband Bill realizes that she left some important papers and so leaves 20 minutes later to deliver the papers. If he drives the same route at 50 mph, how long will it take him to catch up to her?

Solution

As we have mentioned previously, this text is interested primarily in algebraic solutions to verbal problems. This means we must write and solve an equation in order to answer the question posed in a given example or exercise.

Let's analyze this problem carefully.

THINKING OUT LOUD

What do we need to find?
How long (that is, how much time) it takes Bill to catch up to her

Where do we start?
An algebraic solution requires us to write an equation. In general, an equation says that two quantities are *equal.* In order to help us identify such quantities, it is frequently helpful to draw a simple diagram to represent the given information.

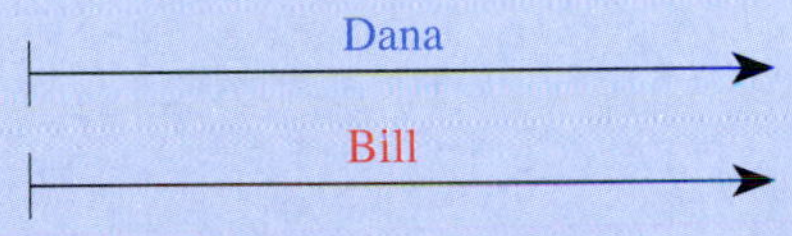

Figure 4.4
Diagram for Example 6

$$\text{Dana} \longrightarrow$$
$$\text{Bill} \longrightarrow$$

Figure 4.4 helps us see that when Bill catches up to her, they will have traveled the *same* distance.

What is our equation going to say?
Based on the diagram we have

Dana's distance = Bill's distance

$$d_{\text{Dana}} = d_{\text{Bill}}$$

How do we compute the distance traveled?
We do not know the actual distance traveled; however, we do know that the distance traveled is found by multiplying each person's rate (speed) by the time that person travels. Thus, we need to find each person's rate and each person's time.

What information are we given?
We know the rates since we are given Dana's speed and Bill's speed. We do not know their times, but we are told that Dana left 20 minutes before Bill (or, alternatively, that Bill left 20 minutes after she did).

How do we use this information?
We have already recognized that the equation will say that the distance traveled by Dana is equal to the distance traveled by Bill. Since distance is equal to rate times time, we must use the given information to identify both rates and both times.

In other words, we use the fact that $d = rt$ to compute each distance individually and set them equal to each other.

$$d_{\text{Dana}} = d_{\text{Bill}}$$

$$(\text{Dana's rate}) \cdot (\text{Dana's time}) = (\text{Bill's rate}) \cdot (\text{Bill's time})$$

$$r_{\text{Dana}} \cdot t_{\text{Dana}} = r_{\text{Bill}} \cdot t_{\text{Bill}}$$

If we now let

$$t = \text{Number of hours Bill drives until he catches up to Dana}$$

then

$$t + \frac{1}{3} = \text{Number of hours Dana drives until Bill catches up}$$

(*Note:* Since Dana left 20 minutes before Bill, she rides for 20 minutes *more* than Bill drives. Because the rate is given in miles per *hour*, we must convert the time of 20 minutes into one-third of an hour. We must use the same units throughout a problem.)

Since Dana's rate is 30 mph and Bill's rate is 50 mph, the equation is

$$30\left(t + \frac{1}{3}\right) = 50t \qquad \text{This equation is the algebraic translation of } r_{\text{Dana}} \cdot t_{\text{Dana}} = r_{\text{Bill}} \cdot t_{\text{Bill}}.$$

$$30t + 10 = 50t$$

$$10 = 20t$$

$$\frac{1}{2} = t$$

It will take Bill one-half hour to catch up to Dana.

Check: In one-half hour Bill travels $\frac{1}{2}(50) = 25$ miles.

In five-sixths of an hour Dana travels $\frac{5}{6}(30) = 25$ miles. ✓ ∎

Dana's time is $\frac{1}{3}$ of an hour more than Bill's time.

$$\frac{1}{2} + \frac{1}{3} = \frac{5}{6}$$

EXAMPLE 7 Harry can process 12 forms per hour, while Susan can process 16 forms per hour. If Harry starts processing forms at 1:00 P.M. and Susan starts at 2:20 P.M., at what time will Susan have processed as many forms as Harry?

Solution The basic idea in the problem is that

$$\text{Number of forms processed} = \begin{pmatrix}\text{Rate at which they}\\\text{are processed}\end{pmatrix} \cdot (\text{Time spent processing})$$

For example, if you process forms at the rate of 8 per hour for 3 hours, you could process $8 \cdot 3 = 24$ forms.

The problem is asking at what time will they both have processed the same number of forms. That is, at what time will

$$\text{Number of forms Harry processed} = \text{Number of forms Susan processed}$$

In other words, at what time will

$$\begin{pmatrix}Harry's\\processing\ rate\end{pmatrix} \cdot \begin{pmatrix}Harry's\\processing\ time\end{pmatrix} = \begin{pmatrix}Susan's\\processing\ rate\end{pmatrix} \cdot \begin{pmatrix}Susan's\\processing\ time\end{pmatrix}$$

Let

$$t = \text{Number of hours Susan works until she catches up to Harry}$$

Then

$$t + \frac{4}{3} = \text{Number of hours Harry works}$$

(because Harry works 1 hr 20 min more than Susan; 1 hr and 20 min $= 1\frac{1}{3}$ hr $= \frac{4}{3}$ hr).

Therefore, the equation is

$$12\left(t + \frac{4}{3}\right) = 16t$$

$$12t + 12 \cdot \frac{4}{3} = 16t$$

$$12t + 16 = 16t$$

$$16 = 4t$$

$$4 = t$$

It takes 4 hours for Susan to catch up with Harry. Since she started at 2:20, she will have processed as many forms as Harry at

6:20 P.M.

(Do not forget to answer the original question.)
The check is left to the student. ■

Looking back at Examples 6 and 7, we can again see basic similarities in the structure of our solutions to these problems. Problems of this type are often called *rate–time* problems.

One final reminder: Do not just read a problem and give up. Make an honest effort at a solution. Time spent in an unsuccessful attempt at a solution is not wasted. With continued effort you will find that each attempt will bring you closer to a complete solution.

EXERCISES 4.6

In each of the following exercises, set up an equation or inequality and solve the problem. Be sure to indicate clearly what quantity your variable represents. Round to the nearest tenth where necessary.

1. If 5 more than two-thirds of a number is 9, find the number.

2. If 3 less than five-sevenths of a number is 12, find the number.

3. If 2 less than three-fourths of a number is 7 less than one-eighth of the number, find the number.

4. If 2 less than three-fourths of a number is less than one-eighth of the number, how large can the number be?

5. If the width of a rectangle is $\frac{1}{2}$ of its length and the perimeter is 36 meters, find the dimensions of the rectangle.

6. If the width of a rectangle is 1 more than $\frac{2}{3}$ of its length and the perimeter is 32 cm, what are the dimensions of the rectangle?

7. The medium side of a triangle is $\frac{3}{4}$ of the longest side, and the shortest side is $\frac{1}{2}$ of the medium side. If the perimeter of the triangle is 17 in., find the lengths of the sides of the triangle.

8. The medium side of a triangle is 6 more than $\frac{1}{2}$ the shortest side, and the longest side is 4 times the shortest side. If the perimeter is 17 in., find the lengths of the sides of the triangle.

9. An amusement park sells regular admission tickets as well as combination tickets that cover admission and a number of rides. The price of a regular admission ticket is $15, while a combination ticket costs $22. If a total of $6,895 was collected on the sale of 350 tickets, how many of each type were sold?

10. A plumber and her assistant work on a certain installation job. First the assistant does some preparatory work alone, and then the plumber completes the job alone. The plumber gets paid $48 per hour, and the assistant gets paid $28 per hour. If the installation job took a total of 9 hours to complete and the total labor cost for the plumber and assistant was $327, how many hours did each work on the job?

11. A collection of coins consisting of dimes and quarters has a value of $2.55. If the number of dimes is 3 more than twice the number of quarters, how many of each type of coin are there?

12. A collection of 40 coins consisting of nickels and dimes has a value of $2.65. How many of each type of coin are there?

13. A certain machine can sort screws at the rate of 175 per minute. A newer, faster machine can do the sorting at the rate of 250 per minute. If the older machine begins sorting a batch of 13,675 screws at 10 A.M. and then 15 minutes later the newer machine joins in the sorting process, at what time will the sorting be completed?

14. At 8:00 A.M. a line of 650 cars is waiting for passage through a toll barrier. At 8:00 A.M. a toll machine begins admitting cars at the rate of 15 per minute. At 8:10 A.M. a toll-taker comes on duty in the next booth and begins admitting cars at the rate of 10 per minute, and they continue working together until all 650 cars have been admitted. At what time will the last car pass through the toll barrier?

15. A certain sum of money is invested at 6.35%, and $4,000 more than that amount is invested at 7.28%. If the annual interest from the two investments is $972.70, how much was invested at 6.35%?

16. A certain sum was invested at 5.68%, and $2,000 less than that amount was invested at 6.52%. If the annual interest from the two investments was $601.60, how much was invested at each rate?

17. A total of $800 was split into two investments. Part paid 9% and the remainder paid 6%. If the annual interest from the two investments was $67.50, how much was invested at each rate?

18. A total of $7,500 was invested as follows: a certain amount at 7%, twice that amount at 10%, and the remainder at 12%. If the annual interest from the three investments was $738, how much was invested at each rate?

19. A total of $6,000 is to be split into two investments, part at 8% and the remainder at 12%, in such a way that the annual interest is 9% of the amount invested. How should the $6,000 be split?

20. Repeat Exercise 19 if the total to be invested is $10,000 and the yearly interest is to be 11.8% of the amount invested.

21. How many milliliters of a 30% hydrochloric acid solution must be mixed with 30 milliliters of a 50% hydrochloric acid solution to produce a 45% solution of hydrochloric acid?

22. How many ounces of a 40% alcohol solution should be mixed with 60 oz of a 70% alcohol solution to produce a 60% alcohol solution?

23. How much of each of a 2.4% salt solution and a 4.6% salt solution must be mixed together to produce 90 liters of a 3% salt solution?

24. How many ounces of each of a 5.2% and a 6.1% iodine solution need to be mixed together to produce 40 oz of a 6% iodine solution?

25. How much pure antifreeze should be added to a radiator that contains 10 gallons of a 30% antifreeze solution to produce a 50% antifreeze solution? [*Hint:*   Pure antifreeze is 100% antifreeze.]

26. How much water should be added to a radiator that contains 10 gallons of an 80% antifreeze solution to dilute it to a 50% antifreeze solution? [*Hint:*   Pure water is 0% antifreeze.]

27. How many pounds of candy selling at $3.75/lb should be mixed with 35 pounds of candy selling at $5/lb to produce a mixture that should sell at $4.25/lb?

28. How many pounds of coffee beans selling at $6.65/lb should be mixed with how many pounds of coffee beans selling for $4.25/lb to produce a 24-lb mixture that should sell for $5.75/lb?

29. John and Susan leave their homes at 8:00 A.M., going toward each other along the same route, which is 9 miles long. John walks at 4 mph while Susan jogs at 8 mph. At what time will they meet?

30. Two trains leave two cities that are 400 km apart. They both leave at 11:00 A.M. traveling toward each other on parallel tracks. If one train travels at 70 kph and the other travels at 90 kph, at what time will they pass each other?

31. Repeat Exercise 29 if John leaves at 7:45 and Susan leaves at 8:00.

32. Repeat Exercise 30 if the slower train leaves at 9:40 A.M. and the faster train leaves at 11:00 A.M.

33. If David walks at the rate of 5 mph and jogs at the rate of 9 mph and it takes 2 hours to cover a distance of 16 miles, how much time was spent jogging?

34. A person drives from town A to town B at the rate of 50 mph and then flies back at the rate of 160 mph. If the total traveling time is 21 hours, how far is it from town A to town B?

35. Sal has $3,200 invested in an account paying 7.2% interest per year. At what rate must he invest an additional $2,800 so that the annual interest from the two investments will be $390?

36. Irma invests $2,600 in a corporate bond paying 8.4% interest per year. At what rate must she invest an additional $1,800 so that her annual interest will be $339?

37. A chemist wants to make 90 ml of a 40% hydrochloric acid solution. She also wants to use all of the 30 ml of 20% hydrochloric acid she has on hand. What percent solution must she use for the remaining 60 ml?

38. The owner of a gourmet coffee shop wants to create 200 pounds of a special blend of coffee to sell at $8.15 per pound. He also wants to use all 80 pounds of the $6.50 blend he has on hand. How expensive should the coffee be that he uses for the remaining 120 pounds?

39. The length of a rectangle is 1 less than twice the width. If the width is 1 more than one-seventh of the perimeter, find the dimensions of the rectangle.

40. The width of a rectangle is one-third the length. If the length is 4 less than one-half of the perimeter, find the dimensions of the rectangle.

41. If the temperature range on a certain day was between 25°C and 40°C, what was the temperature range in degrees Fahrenheit? Use the formula $C = \frac{5}{9}(F - 32)$.

42. On a certain day the minimum temperature was 14°F while the maximum temperature was 23°F. What was the temperature range in degrees Celsius? (Use $F = \frac{9}{5}C + 32$.)

43. A store advertises a 20% sale on skirts. If the reduced price range is between $12.60 and $20.76, what was the original range of prices?

44. A total of $5,000 is to be invested, part at 8% and the remainder at 10%. If the interest from the two investments is to be at least $460, what is the least that must be invested at 10%?

CHAPTER 4 SUMMARY

After having completed this chapter, you should be able to:

1. Use the Fundamental Principle of Fractions to reduce fractions (Section 4.1).

   For example:

   Reduce to lowest terms. $\dfrac{6x^2y^5}{8x^3y^2}$

   Solution. $\dfrac{6x^2y^5}{8x^3y^2} = \dfrac{2 \cdot 3 \cdot x^2 \cdot y^2 \cdot y^3}{2 \cdot 4 \cdot x^2 \cdot x \cdot y^2} = \boxed{\dfrac{3y^3}{4x}}$

2. Multiply and divide algebraic fractions (Section 4.2).

   For example:

   (a) $\dfrac{2a^2b}{9x} \cdot \dfrac{3x^2}{4b} = \dfrac{2a^2b}{9x} \cdot \dfrac{3x^2}{4b}$

   $\qquad = \boxed{\dfrac{a^2x}{6}}$

   (b) $\dfrac{8yz}{5x^3} \div (10xyz) = \dfrac{8yz}{5x^3} \cdot \dfrac{1}{10xyz}$

   $\qquad = \dfrac{8yz}{5x^3} \cdot \dfrac{1}{10xyz}$

   $\qquad = \boxed{\dfrac{4}{25x^4}}$

3. Find the LCD of several rational expressions (Section 4.3).

   For example: *Find the LCD of the fractions.* $\dfrac{2}{3xy^2}, \dfrac{1}{6x^2y}, \dfrac{3}{8x^3}$

   $\left.\begin{array}{l} 3xy^2 = 3xyy \\ 6x^2y = 2 \cdot 3xxy \\ 8x^3 = 2 \cdot 2 \cdot 2xxx \end{array}\right\}$ *The distinct factors are 2, 3, x, and y.*

   Following our outline we find that the LCD $= 24x^3y^2$.

4. Combine rational expressions, using the Fundamental Principle to build fractions where necessary (Section 4.3).

   For example:

   (a) $\dfrac{6x+5}{2x^2} + \dfrac{x-5}{2x^2} = \dfrac{6x+5+x-5}{2x^2}$

$$= \frac{7x}{2x^2} \qquad \textit{Reduce.}$$

$$= \boxed{\frac{7}{2x}}$$

(b) $\dfrac{2}{3xy^2} - \dfrac{1}{6x^2y} + \dfrac{3}{8x^3}$ $\qquad$ *In* **(3)** *above we saw that the LCD for these fractions is* $24x^3y^2$.

$$= \frac{2 \cdot (8x^2)}{3xy^2 \cdot (8x^2)} - \frac{1 \cdot (4xy)}{6x^2y \cdot (4xy)} + \frac{3 \cdot (3y^2)}{8x^3 \cdot (3y^2)}$$

$$= \boxed{\frac{16x^2 - 4xy + 9y^2}{24x^3y^2}}$$

5. Solve a first-degree equation or inequality in one variable with rational coefficients (Section 4.4).

   For example: Solve for t.

$$\frac{t + 1}{2} - \frac{t - 2}{3} = 2 \qquad \textit{Multiply both sides of the equation by the LCD, which is 6, to clear the fractions.}$$

$$\frac{\overset{3}{\cancel{6}}}{1} \cdot \frac{t + 1}{2} - \frac{\overset{2}{\cancel{6}}}{1} \cdot \frac{t - 2}{3} = 6 \cdot 2$$

$$3(t + 1) - 2(t - 2) = 12$$

$$3t + 3 - 2t + 4 = 12$$

$$t + 7 = 12$$

$$\boxed{t = 5}$$

6. Write and solve proportions (Section 4.5).

   For example:

   The ratio of those employees of a certain firm who have college degrees to those who do not is 6 to 5. If there are 180 employees who do not have college degrees, how many do?

   Let n = Number of employees who have a college degree.

$$\frac{n}{180} = \frac{6}{5} \qquad \textit{Multiply both sides of the equation by } 180.$$

$$n = 180 \cdot \frac{6}{5}$$

$$\boxed{n = 216}$$

7. Translate and solve a wide variety of verbal problems that give rise to first-degree equations and inequalities in one variable (Section 4.6).

   For example:

   The length of a rectangle is 4 more than $\frac{2}{3}$ the width. If the perimeter of the rectangle is 28 meters, find the dimensions of the rectangle.

   Solution: The underlying idea is that the perimeter of a rectangle is twice the width plus twice the length:

$$2L + 2W = P$$

This problem says that the length is 4 more than $\frac{2}{3}$ the width. If we let W = width, then

$$\frac{2}{3} \cdot W + 4 = \frac{2W}{3} + 4 = \text{length}$$

If we like, we can express the same information in a diagram, as shown here.

From the fact that the perimeter is equal to 28 we can write the equation

W ⬚

$\frac{2}{3}W + 4$

$$2\left(\frac{2W}{3} + 4\right) + 2W = 28$$

$$\frac{4W}{3} + 8 + 2W = 28 \qquad \textit{Multiply both sides of the equation by 3 to clear the fraction.}$$

$$\frac{3}{1} \cdot \frac{4W}{3} + 3 \cdot 8 + 3 \cdot 2W = 3 \cdot 28$$

$$4W + 24 + 6W = 84$$
$$10W + 24 = 84$$
$$10W = 60$$
$$W = 6$$

Since the width is 6, the length is $\frac{2}{3}(6) + 4 = 4 + 4 = 8$ meters. Thus, the solution is

> width = 6 meters; length = 8 meters

Check: The perimeter of the rectangle is $2(6) + 2(8) = 12 + 16 \overset{\checkmark}{=} 28$.

CHAPTER 4   REVIEW EXERCISES

In Exercises 1–8, reduce each fraction to lowest terms.

1. $\dfrac{-18}{42}$

2. $-\dfrac{14}{35}$

3. $\dfrac{15x^6}{6x^2}$

4. $\dfrac{8a^3}{28a^9}$

5. $\dfrac{-10x^3y^5}{4xy^{10}}$

6. $\dfrac{24x^8y}{-15x^4y^5}$

7. $\dfrac{3t - 7t - t}{-2t^2 - 3t^2}$

8. $\dfrac{2w^2z + 4w^2z}{8wz^2 - 2wz^2}$

In Exercises 9–26, perform the indicated operations; express your final answer reduced to lowest terms.

9. $\dfrac{a}{4} \cdot \dfrac{a}{4}$

10. $\dfrac{a}{4} + \dfrac{a}{4}$

11. $\dfrac{7a}{6} - \dfrac{5a}{6}$

12. $\dfrac{7a}{6} \div \dfrac{5a}{6}$

13. $\dfrac{4x - 3}{6x} - \dfrac{x - 1}{6x}$

14. $\dfrac{7x - 2}{5x} - \dfrac{2x - 3}{5x}$

15. $\dfrac{2y^2 - 3y}{4} - \dfrac{y^2 - 3y}{4} + \dfrac{y^2}{4}$

16. $\left(\dfrac{2y^2 - 3y}{4} - \dfrac{y^2 - 3y}{4}\right) \div \dfrac{y^2}{4}$

17. $\left(\dfrac{x^2}{4} \cdot \dfrac{6}{xy^2}\right) \div (2xy)$

18. $\dfrac{x^2}{4} \div \left(\dfrac{6}{xy^2} \cdot (2xy)\right)$

19. $\dfrac{a}{2} \cdot \dfrac{a}{4}$

20. $\dfrac{a}{2} + \dfrac{a}{4}$

21. $\dfrac{x^2}{2} - \dfrac{x^2}{6} + \dfrac{x^2}{3}$

22. $\dfrac{x^2}{2} \cdot \dfrac{x^2}{6} \div \dfrac{x^2}{3}$

23. $\dfrac{4}{x^2} + \dfrac{3}{2x}$

24. $\dfrac{5}{3y^2} + \dfrac{3}{2y}$

25. $\dfrac{3}{4a^2 b} - \dfrac{5}{6ab} + \dfrac{7}{8b^3}$

26. $\dfrac{3}{8rt^2} + \dfrac{7}{12rt} - \dfrac{5}{3t^3}$

In Exercises 27–36, solve the equation or inequality.

27. $\dfrac{x}{6} - \dfrac{1}{4} = \dfrac{7}{12}$

28. $\dfrac{x}{8} - \dfrac{5}{6} = \dfrac{1}{24}$

29. $\dfrac{t + 1}{2} + \dfrac{t + 2}{3} < \dfrac{t + 7}{6}$

30. $\dfrac{2a + 5}{4} + \dfrac{4a + 1}{6} = 2$

31. $\dfrac{y + 3}{5} - \dfrac{y - 2}{3} = 1$

32. $\dfrac{z + 4}{15} - \dfrac{z - 4}{5} > \dfrac{z - 8}{3}$

33. $\dfrac{x}{3} = \dfrac{x + 1}{6}$

34. $\dfrac{x}{3} = \dfrac{x}{6} + 1$

35. $2x + 0.2(x + 6) = 10$

36. $\dfrac{x}{2} + 0.3x = 16$

In Exercises 37–41, solve each problem by first writing an appropriate equation or inequality. Round to the nearest hundredth where necessary.

37. If there are 28.4 grams in 1 ounce, how many ounces are there in 1 kilogram (1,000 grams)?

38. If 1 less than three-fourths of a number is 4 less than the number, find the number.

39. A total of $7,000 is split into three investments. A certain amount is invested at 6%, twice that amount at 7%, and the remainder at 8%. If the annual interest from the three investments is to be at least $500, what is the most that can be invested at 6%?

40. A collection of 30 coins consists of nickels, dimes, and quarters. There are twice as many dimes as nickels, and the rest are quarters. If the value of the collection is $3.50, how many of each type are there?

41. After driving along at a certain rate of speed for 5 hours, Bill realizes that he could have covered the same distance in 3 hours if he had driven 20 mph faster. What is his present speed?

CHAPTER 4   PRACTICE TEST

1. Reduce each of the following to lowest terms:

   (a) $\dfrac{-10}{24}$

   (b) $\dfrac{x^{10}}{x^2}$

   (c) $\dfrac{-6a^6}{-3a^3}$

   (d) $\dfrac{25r^2t^3}{-15r^4t}$

2. Perform the indicated operations and express your answer in lowest terms.

   (a) $\dfrac{2y}{3x} \cdot \dfrac{9x^2}{4y^2}$

   (b) $\dfrac{2y}{3x^2} \div \dfrac{8y^2}{9x^4}$

   (c) $\dfrac{-3ab^2}{4a^2} \cdot \dfrac{2a^3}{3b} \cdot \dfrac{9a^2b}{-6a^2b^2}$

   (d) $\dfrac{4xy^3}{9x^2} \div 18xy$

   (e) $\dfrac{a}{5} + \dfrac{a}{5}$

   (f) $\dfrac{a}{5} \cdot \dfrac{a}{5}$

   (g) $\dfrac{9}{4x} + \dfrac{3}{4x}$

   (h) $\dfrac{a}{3} + \dfrac{a}{8}$

   (i) $\dfrac{5}{4x} + \dfrac{1}{6x}$

   (j) $\dfrac{3}{10x^2y} + \dfrac{7}{3x}$

   (k) $\dfrac{x^2 - 3x - 2}{8x} - \dfrac{6 - 3x + x^2}{8x}$

3. Solve each of the following equations and inequalities:

   (a) $\dfrac{x}{3} + \dfrac{x}{5} = 8$

   (b) $\dfrac{x - 5}{2} + \dfrac{x}{5} \geq 8$

   (c) $\dfrac{a + 3}{5} - \dfrac{a - 2}{4} = 1$

   (d) $0.03t + 0.5t = 10.6$

Solve each of the following problems algebraically. Round to the nearest hundredth where necessary.

4. If there are 1.61 kilometers in a mile, how many miles are there in 50 kilometers?

5. A number when added to $\frac{2}{3}$ of itself is 5 less than twice the number. Find the number.

6. Advance-sale tickets to a show cost $18.50 each, while tickets at the door cost $15 each. If $6,770 was collected on the sale of 400 tickets, how many tickets were sold at the door?

7. A total of $7,000 is split into two investments, one paying 8% interest and the other paying 13%. If the annual interest from the two investments is $750, how much is invested at each rate?

8. How many ounces of a 20% sulfuric acid solution must be added to 24 oz of a 65% sulfuric acid solution to produce a 50% solution?

9. Two people leave a town traveling in opposite directions. The first person leaves at 11:00 A.M. driving at 48 kph, while the second person leaves at 3:00 P.M. that afternoon and drives at 55 kph. At what time will they be 604 km apart?

CHAPTER 5

Graphing Straight Lines

STUDY SKILLS

For the most part we have handled applications and verbal problems by using one variable only. One of the main goals of this chapter and the next is to develop our ability to deal with equations involving two variables.

For example, suppose that Lenore keeps track of the number of hours she exercises each week (h) and the number of pounds she loses that week (p) and comes up with the following data for a 3-week period:

	Week 1	Week 2	Week 3
h (Number of hours exercising)	1	3	5
p (Number of pounds lost)	0	1	2

Having a "picture" of this relationship and/or an equation relating h and p would give Lenore a better understanding of how her exercise program and weight loss are related. (We will discuss Lenore and her exercise program further at the beginning of Section 5.3 and again in Example 6 of Section 5.4.)

5.1   The Rectangular (Cartesian) Coordinate System

Let's begin by considering the following equation in *two* variables:

$$2x + y = 6$$

What does it mean to have a solution to this equation? A moment's thought will make us realize that a *single* solution to this equation consists of *two* numbers—a value for x together with a value for y. In other words, *a solution* to this equation consists of *a pair of numbers*.

Thus, in the equation $2x + y = 6$, we can easily "see" some pairs of numbers that work:

$$x = 1, \quad y = 4 \quad \text{works because} \quad 2(1) + 4 = 6.$$
$$x = 3, \quad y = 0 \quad \text{works because} \quad 2(3) + 0 = 6.$$
$$x = -1, \quad y = 8 \quad \text{works because} \quad 2(-1) + 8 = 6.$$

On the other hand, we can generate solutions to this equation by simply picking a value for either x or y and solving for the other variable. For example, we may choose $x = 2$, and then substitute $x = 2$ into the equation and solve for y:

$$2x + y = 6$$
$$2(2) + y = 6$$
$$4 + y = 6$$
$$\boxed{y = 2} \qquad \textit{Thus, we know that } x = 2, y = 2 \textit{ is a solution.}$$

If instead we choose $y = 6$, then we substitute $y = 6$ into the equation and solve for x:

$$2x + y = 6$$
$$2x + 6 = 6$$
$$2x = 0$$
$$\boxed{x = 0} \qquad \textit{Thus, we know that } x = 0, y = 6 \textit{ is a solution.}$$

The important thing to realize is that in this way we can generate an unlimited number of solutions to this equation. Even though this equation is not always true (for example, $x = 4, y = -1$ does not work), it does have *infinitely* many solutions.

One way of keeping track of the solutions is in a table. We could list the solutions mentioned above as follows:

x	y
1	4
3	0
−1	8
2	2
0	6

These are some of the solutions to $2x + y = 6$.

EXAMPLE 1 Complete the following table for the equation $2x + y = 6$.

x	y
−2	
	−3
$\frac{1}{2}$	
	9

Solution To fill in the first row of the table, we need to use the given equation $2x + y = 6$ to find the value of y that corresponds to $x = -2$.

$$2x + y = 6 \qquad \text{\textit{Substitute }} x = -2.$$
$$2(-2) + y = 6 \qquad \text{\textit{Now solve for y.}}$$
$$y = 10$$

Similarly, to fill in the second row of the table, we need to use the given equation to find the value of x that corresponds to $y = -3$.

$$2x + y = 6 \qquad \text{\textit{Substitute }} y = -3.$$
$$2x + (-3) = 6 \qquad \text{\textit{Now solve for x.}}$$
$$2x = 9 \quad \text{and so} \quad x = \frac{9}{2}$$

It is left to the student to substitute $x = \frac{1}{2}$ to fill in the third row and $y = 9$ to fill in the fourth row. Thus the table becomes

x	y
−2	10
$\frac{9}{2}$	−3
$\frac{1}{2}$	5
$-\frac{3}{2}$	9

■

As these examples illustrate, a first-degree equation in two variables has infinitely many solutions. We can simply choose a value for x (or y) and then find the corresponding value of the other variable. Since there is no way to list infinitely many solutions, we are going to develop an alternative way of exhibiting all the solutions. We are going to draw a "picture" of the solution set. In order to do this, we leave these equations for a little while, but we will return to these ideas at the end of this section.

The Rectangular Coordinate System

The number line is often called a **one-dimensional coordinate system.** The word *coordinate* basically means location. *One-dimensional* means that we can move only right or left on the number line. We cannot move off the number line.

On the number line we determine the location of a point by using a *single* number. This number tells us the point's distance and direction from 0 on the number line. Thus, −6 is called the coordinate of the point that is 6 units to the left of 0.

In order to be able to describe locations on a flat surface, which is usually called a **two-dimensional plane,** we will need two coordinates. Using two coordinates to describe a location is a common practice with which we are all familiar. For example, to locate a seat in a theater, your ticket usually contains two "coordinates" such as A-4 or G-26. The first coordinate is the letter, which indicates the row. The row closest to the front is usually lettered A, and all remaining rows behind it are ordered alphabetically. The second coordinate is the number indicating your position in the row, usually starting from the left side of the theater (see Figure 5.1). Note that we need *two* coordinates to determine an exact seat in the theater.

Figure 5.1
Typical arrangement of
seats in a theater

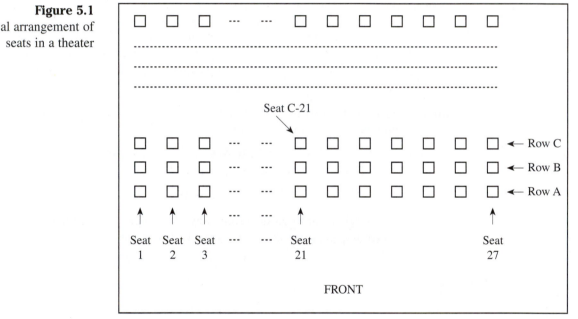

Naming points in the plane utilizes the same idea. Starting from some agreed-on point of reference, we use two coordinates: one coordinate indicates right or left, the other indicates up or down.

We build our system in the following way. We begin with a number line,

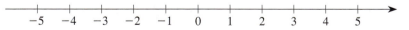

and then we draw another number line perpendicular to the first, so that their 0 points coincide (see Figure 5.2).

We call the horizontal number line the **x-axis** and the vertical number line the **y-axis.** Together they are called the **coordinate axes.** The point of intersection of the coordinate axes is called the **origin.**

The usual convention is to mark the positive values on the y-axis above the origin, and the negative values below the origin. Such a system of coordinate axes is called a **rectangular** (or **Cartesian,** named after the famous French philosopher and mathematician René Descartes) **coordinate system.**

To describe the location of a point in the plane, we need *two* numbers, which are called an **ordered pair.** An ordered pair of numbers is a pair of numbers enclosed in parentheses and separated by a comma in the following way: (x, y).

Figure 5.2

Coordinate axes and the location of (2, 3)

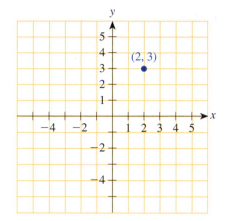

We imagine ourselves situated at the origin. The first number, called the **x-coordi-nate,** tells us how far to move right or left: Positive means right, and negative means left. The second number, called the **y-coordinate,** tells us how far to move up or down: Positive means up, and negative means down.

For example, Figure 5.2 shows the location of the point (2, 3). To locate the point (2, 3), we start at the origin and move 2 units right and then 3 units up. Of course, we could also move 3 units up and then 2 units right. What is important to remember is that the x-coordinate tells us "right–left," while the y-coordinate tells us "up–down."

Locating a point on a plane is called ***plotting the point.***

EXAMPLE 2

Plot each of the following points.

(a) (3, 4)   (b) (4, 3)   (c) (−3, 5)

(d) (−5, −4)   (e) (2, −2)   (f) (0, −5)

Solution

See Figure 5.3. Note that points (3, 4) and (4, 3) are *not* the same. The order in an ordered pair is important.

Figure 5.3

Coordinate axes for Example 1

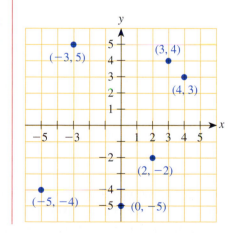

EXAMPLE 3

Plot the following points.

(a) (−6, 0)   (b) (0, 4)   (c) (0, 0)

Solution

(a) For the point (−6, 0), the x-coordinate, −6, tells us to move 6 units to the left, and the y-coordinate, 0, tells us to move 0 units in the vertical direction. Thus, the point (−6, 0) is on the x-axis, 6 units to the left of the origin (see Figure 5.4 on page 208). In fact, any time the y-coordinate of a point is 0, the point must lie on the x-axis.

Figure 5.4
Coordinate axes
for Example 2

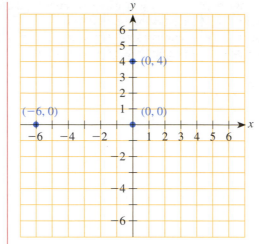

(b) For the point (0, 4), the *x*-coordinate, 0, tells us to move 0 units in the horizontal direction, and the *y*-coordinate, 4, tells us to move up 4 units from the origin. Thus, the point (0, 4) is on the *y*-axis, 4 units above the origin (see Figure 5.4). In fact, any time the *x*-coordinate of a point is 0, the point must lie on the *y*-axis.

(c) The point (0, 0) is the origin.   ■

As Figure 5.5 illustrates, the coordinate axes divide the plane into four **quadrants,** which are numbered in a counterclockwise direction starting in the upper right quadrant.

Figure 5.5
Division of plane
into quadrants

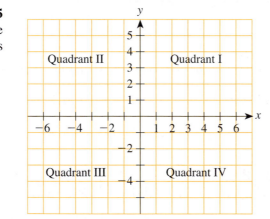

EXAMPLE 4   Indicate which quadrant contains each of the following points.

(a) (2, −5)   (b) (−3, 7)

Solution   (a) Since the point (2, −5) has a positive *x*-coordinate and a negative *y*-coordinate, we must move to the right and downward, which puts us into quadrant IV.

(b) Since the point (−3, 7) has a negative *x*-coordinate and a positive *y*-coordinate, we must move to the left and upward, which puts us into quadrant II.

Note that a point that lies on either the *x*-axis or the *y*-axis does *not* lie in any of the quadrants.   ■

Let us return now to the equation $2x + y = 6$, which we discussed at the beginning of this section. Using the notation of ordered pairs, we now have another way of recording those *x* and *y* values which *together* are a solution to the equation. Since we have agreed that in our ordered pair notation the first coordinate stands for *x* and the second for *y*, instead of saying that $x = 1$, $y = 4$ is a solution to the equation, we can say that the ordered pair $(1, 4)$ is a solution to the equation.

In our table of solutions to the equation $2x + y = 6$, we can add a column that records the same solutions in ordered pair notation:

x	y	(x, y)
1	4	$(1, 4)$
3	0	$(3, 0)$
−1	8	$(−1, 8)$
2	2	$(2, 2)$
0	6	$(0, 6)$

We plot these points in part **a** of Figure 5.6. This picture suggests very strongly that the points lie on a straight line and that we should "connect the dots." In fact, this turns out to be true, which means that the straight line is a "picture" of the solution set, as shown in part **b**.

Figure 5.6

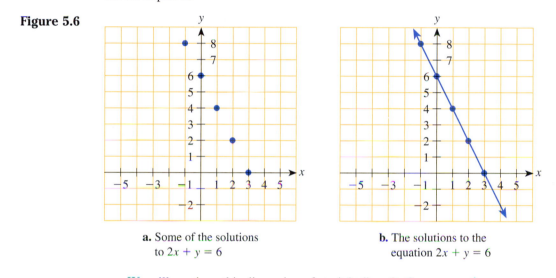

a. Some of the solutions to $2x + y = 6$

b. The solutions to the equation $2x + y = 6$

We will continue this discussion of straight lines in the next section.

Study Skills 5.1

Reviewing Your Exam: Diagnosing Your Strengths and Weaknesses

Your exam will be a useful tool in helping you to determine what topics, skills, or concepts you need to work on in preparation for the next topic or in preparation for future exams. After you get your exam back, you should review it carefully; examine what you did correctly and what problems you missed.

Do not quickly gloss over your errors and assume that any errors were minor or careless. Students often mistakenly label many of their errors as "careless," when in fact, they are a result of not clearly understanding a certain concept or procedure. Be honest with yourself. Do not delude yourself into thinking that all errors are careless. Ask yourself the following questions about your errors:

Did I understand the directions or perhaps misunderstand them?

Did I understand the topic to which the question relates?

Did I misuse a rule or property?

Did I make an arithmetic error?

Look over the entire exam. Did you consistently make the same type of error throughout the exam? Did you consistently miss problems covering a particular topic or concept? You should try to follow your work and see what you were doing on the exam. If you think you have a problem understanding a concept, topic, or approach to a problem, you should immediately seek help from your teacher or tutor, and reread relevant portions of your text.

In Exercises 1–20, plot the given point.

1. (2, 1)

2. (1, 2)

3. (−2, 1)

4. (1, −2)

5. (2, −1)

6. (−1, 2)

7. (−2, −1)

8. (−1, −2)

9. (4, 0)

10. (0, 4)

11. (−4, 0)

12. (0, −4)

13. (0, 3)

14. (6, 0)

15. (0, −3)

16. (−6, 0)

17. (4, 5)

18. (−3, −4)

19. (5, 5)

20. (−5, −5)

In Exercises 21–32, indicate which quadrant contains the given point. If a point lies on one of the coordinate axes, indicate which one.

21. (2, 6)

22. (−2, 6)

23. (−2, −6)

24. (2, −6)

25. (0, 3)

26. (3, 0)

27. (−2.6, 4.9)

28. (0, −1.6)

29. (0, 0)

30. (8, 47)

31. (489, −16)

32. (−586, 0)

In Exercises 33–44, determine which, if any, of the ordered pairs listed satisfy the given equation.

33. $2y - 4x = 10$;   (7, 1), (1, 7), (5, 0)

34. $5x - 3y = -2$;   (2, 4), (1, −1), (−1, 1)

35. $y = 4x - 1$;   (8, 7), (7, 8), (0, −1)

36. $x = 3y - 6$;   (5, 10), (−6, 0), (0, 2)

37. $6x - 4y - 8 = 0$;   (−2, −5), (6, 7), (−10, −17)

38. $-7x - 2y + 20 = 0$;   (−1, −3), (−3, −1), (2, 3)

39. $\frac{2}{3}x - \frac{1}{2}y = 1$;   (6, 2), (−6, −10), (3, −2)

40. $8y - 9x = 5$; $\left(\frac{1}{4}, \frac{1}{3}\right), \left(-\frac{1}{2}, -\frac{1}{9}\right), \left(\frac{3}{4}, \frac{2}{3}\right)$

41. $y = x^2 - 3x - 4$;   (−2, 8), (1, −6), (2, 8)

42. $y = -x^2 + 5x - 1$;   (−1, −5), (−3, −25), (0, −1)

43. $y = \frac{1}{x + 2}$;   (0, 2), $\left(1, \frac{1}{3}\right)$, (−1, 1)

44. $y = \frac{2x + 3}{x - 4}$;   (3, 9), $\left(0, -\frac{3}{4}\right)$, (5, 13)

In Exercises 45–48, use the given equation to fill in the missing values of the table.

45. $4y - 3x = 7$

x	y
−1	
	−5
	0
0	

46. $2x - 5y = 10$

x	y
	−4
−10	
−2	
	−5

47. $y = \dfrac{2}{3}x + 1$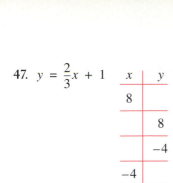

x	y
8	
	8
	−4
−4	

48. $y = -\dfrac{1}{2}x - 3$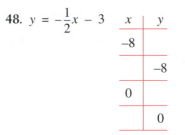

x	y
−8	
	−8
0	
	0

? QUESTIONS FOR THOUGHT

49. Plot the points in the following set: $\{(x, y) \mid y = x + 2, \text{ and } x = -3, 0, 4\}$.

50. Plot the points in the following set: $\{(x, y) \mid x = y + 2, \text{ and } x = -3, 0, 4\}$.

51. Given the equation $3x + 2y = 12$, complete the given ordered pairs:

$$(2, \quad) \quad (0, \quad) \quad (\quad, -3) \quad (\quad, 0)$$

52. Given the equation $4x - y = 8$, complete the given ordered pairs:

$$(-2, \quad) \quad (0, \quad) \quad (\quad, 4) \quad (\quad, 0)$$

53. Why is the point (x, y) called an *ordered* pair?

54. What is the difference between (x, y) and $\{x, y\}$?

55. Plot a few points that satisfy the equation $y = x^2$. Do you think the graph of this equation is a straight line? Explain.

56. Plot a few points that satisfy the equation $y = x^3$. Do you think the graph of this equation is a straight line? Explain.

◇ MINI-REVIEW

Solve each of the following problems algebraically. Be sure to label what the variable represents.

57. Marina does one-third of her homework problems when she comes home, 8 problems just before supper, and the remaining two-fifths of the problems after supper. How many homework problems did she have?

58. On a certain history exam, 46% of the questions were multiple-choice questions, 5 were essay questions, and the remaining 44% were fill-in-the-blank questions. How many multiple-choice questions were there?

5.2   Graphing a Linear Equation in Two Variables

In the last section we saw that a solution to the equation $2x + y = 6$ consists of a *pair* of numbers, and that there are infinitely many pairs of numbers that will work. We also saw that every ordered *pair* of real numbers can be thought of as a *point* in a rectangular coordinate system.

We are going to limit our discussion here to first-degree equations involving two variables.

If a pair of numbers (x, y) makes an equation true, we say that **the point (x, y) satisfies the equation.**

EXAMPLE 1

In each of the following, determine whether the given point satisfies the given equation.

(a) Point: $(10, 1)$; equation: $y - x = 9$

(b) Point: $(-4, -5)$; equation: $-3x = 7 - y$

Solution

To determine whether a particular point satisfies a given equation, we substitute the first coordinate (the x value) for x and the second coordinate (the y value) for y to see whether we obtain a true equation.

(a) In order to check the point $(10, 1)$, we substitute $x = 10$, $y = 1$ into the equation $y - x = 9$. Be careful not to confuse the order of x and y in the equation with the order of x and y in the ordered pair.

$$y - x = 9$$
$$1 - 10 \stackrel{?}{=} 9$$

$$-9 \neq 9 \qquad \boxed{\text{The point } (10, 1) \text{ does } \textit{not} \text{ satisfy the equation.}}$$

(b) To check the point $(-4, -5)$, we substitute $x = -4$, $y = -5$ into the equation $-3x = 7 - y$.

$$-3x = 7 - y$$
$$-3(-4) \stackrel{?}{=} 7 - (-5)$$

$$12 \stackrel{\checkmark}{=} 12 \qquad \boxed{\text{The point } (-4, -5) \textit{ does } \text{satisfy the equation.}} \qquad \blacksquare$$

We might also be given an equation and *one* of the two coordinates in an ordered pair and be asked to find the other coordinate that makes the point satisfy the equation.

EXAMPLE 2

Complete the following ordered pairs so that they satisfy the equation

$$3x - 2y = 12$$

(a) $(\ \ , 0)$ (b) $(-6, \ \ )$ (c) $(0, \ \ )$

Solution

(a) In order to complete the ordered pair $(\ \ , 0)$, we substitute the value $y = 0$ into the equation and then solve for x:

$$3x - 2y = 12$$
$$3x - 2(0) = 12$$
$$3x = 12$$
$$x = 4 \qquad \text{Thus, the ordered pair is } \boxed{(4, 0)}.$$

(b) In order to complete the ordered pair $(-6, \ \ )$, we substitute the value $x = -6$ into the equation and solve for y:

$$3x - 2y = 12$$
$$3(-6) - 2y = 12$$
$$-18 - 2y = 12$$
$$-2y = 30$$
$$y = -15 \qquad \text{Thus, the ordered pair is } \boxed{(-6, -15)}.$$

(c) We substitute $x = 0$ and solve for y:

$$3x - 2y = 12$$
$$3(0) - 2y = 12$$
$$-2y = 12$$
$$y = -6 \qquad \text{Thus, the ordered pair is } \boxed{(0, -6)}.$$

As is often the case, completing an ordered pair in which one of the coordinates is 0 [as in parts (**a**) and (**c**)] is particularly easy. We will bring this idea up again in a moment. ■

We will find the following terminology useful.

DEFINITION	The set of all points that satisfy an equation is called the *graph* of the equation.

As we saw in the last section when we plotted some solutions to the equation $2x + y = 6$, the points that satisfy the equation seem to lie in a straight line. While we are not going to prove this fact, we will accept its truth.

In mathematics, the statement of an important fact *that can be proven* is called a *theorem.*

THEOREM	The graph of an equation of the form $Ax + By = C$ is a straight line (provided A and B are not both equal to 0).

It is for this reason that a first-degree equation in two variables is called a *linear equation.* When a linear equation is written in the form $Ax + By = C$, we say that the equation is in *general form.*

Keep in mind that not every linear equation is given in general form. It is enough for us to know that we can put it into general form, if necessary.

When we say that "the solution set to a first-degree equation in two variables is a straight line," we mean to say *two* things. The first is that if we plot all the solutions to such an equation, they will all fall on a straight line. The second is that if we pick any point on this straight line, the point will satisfy the equation.

One suggestive way to think of an equation is to think of it as a condition. For example, the equation $2x + y = 6$ places the following condition on any point that is a member of the solution set:

Two times the x-coordinate plus the y-coordinate must be equal to 6.

Since we know that the graph of a linear equation is a straight line, we would like to be able to sketch its graph. We know that a straight line is determined by *two* points, and therefore we need to find two points that satisfy the equation. But which two points should we try to find in order to draw the line? We could find *any* two points (pairs of numbers) that satisfy the equation by arbitrarily choosing a number for one variable and solving for the other. However, we just saw in Example 2 that, given an equation, completing ordered pairs where one of the coordinates is 0 is quite easy. Why not use these points to draw the line?

Let's illustrate this approach with several examples.

Suppose we want to sketch the graph of the equation $2x - 3y = 12$. We begin by completing the ordered pairs $(0, \ )$ and $(\ , 0)$. To complete $(0, \ )$, we substitute $x = 0$ and solve for y:

$$2x - 3y = 12$$
$$2(0) - 3y = 12$$
$$-3y = 12$$

$$y = -4 \qquad \text{Thus, the first point is} \boxed{(0, -4)}.$$

To complete $(\ , 0)$, we substitute $y = 0$ and solve for x:

$$2x - 3y = 12$$
$$2x - 3(0) = 12$$

$$2x = 12$$

$$x = 6 \qquad \text{Thus, the second point is } \boxed{(6, 0)}.$$

Having these two points, we can draw our line (see Figure 5.7).

However, we must be careful here. Since any two points determine a line, how do we know we have the correct line? It is a very good idea to get one more point on the line just to check that we have not made an error. What additional point shall we find? We simply look at the equation and try to pick a "convenient" value for either x or y, and then solve for the other variable.

Let's pick $x = 3$; we substitute $x = 3$ into the equation and solve for y:

$$2x - 3y = 12$$
$$2(3) - 3y = 12$$
$$6 - 3y = 12$$
$$-3y = 6$$
$$y = -2 \qquad \text{Thus, the third point is } \boxed{(3, -2)}.$$

If we now plot this point, we see that it falls on the line we have already drawn (see Figure 5.8).

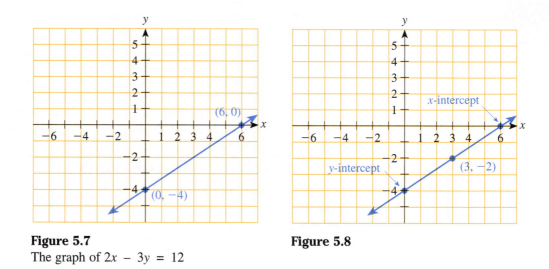

Figure 5.7
The graph of $2x - 3y = 12$

Figure 5.8

In the last section, we pointed out that if the x-coordinate of a point is 0, then that point must lie on the y-axis. Similarly, if the y-coordinate of a point is 0, then the point must lie on the x-axis.

DEFINITION

The y-coordinate of the point where a line intersects the y-axis is called the **y-intercept** of the line.

The x-coordinate of the point where a line intersects the x-axis is called the **x-intercept** of the line.

Thus for the equation $2x - 3y = 12$, the y-intercept of the line is -4, and the x-intercept of the line is 6. See Figure 5.8.

Keep in mind that we have drawn only a representative portion of the line. The line extends indefinitely in both directions.

The method we have just outlined is called the ***intercept method*** for graphing a straight line. In most cases it is the preferred method to follow because the points are the easiest to find and the simplest to plot.

DIFFERENT PERSPECTIVES: The Graph of an Equation

An equation and its graph offer two ways of looking at a relationship

ALGEBRAIC DESCRIPTION

An equation such as $y = 3x - 5$ describes a relationship in which the y value is 5 less than 3 times the x value. Ordered pairs such as $(1, -2)$, $(0, -5)$, and $(3, 4)$ satisfy this relationship and hence are solutions to the equation.

GEOMETRIC DESCRIPTION

The graph of an equation gives a pictorial representation of the relationship. The following is the graph of the equation $y = 3x - 5$.

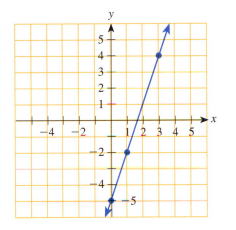

Every point (x, y) on the graph also gives a pair of numbers that satisfies the equation, and every pair of numbers that satisfies the equation corresponds to a point on the graph. We can see that the points $(1, -2)$, $(0, -5)$, and $(3, 4)$ are on the graph, and by substituting we can check that they satisfy the equation $y = 3x - 5$.

EXAMPLE 3

Sketch the graph of $5y = 2x + 10$.

Solution

Since this is a linear equation, its graph is a straight line. We will find the x- and y-intercepts as the two points we need to draw the line.

To find the x-intercept, we set $y = 0$ and solve for x.

$$5y = 2x + 10 \qquad \textit{Substitute } y = 0.$$
$$5(0) = 2x + 10$$
$$0 = 2x + 10$$
$$-10 = 2x$$
$$-5 = x$$

Therefore, the x-intercept is -5. We plot the point $(-5, 0)$.

To find the *y*-intercept, we set *x* = 0, and solve for *y*.

$$5y = 2x + 10 \qquad \textit{Substitute } x = 0.$$
$$5y = 2(0) + 10$$
$$5y = 10$$
$$y = 2$$

Therefore, the *y*-intercept is 2. We plot the point (0, 2).

For our "check" point, we choose *x* = 5. (We choose *x* = 5 so that we will hopefully get a nice *y* value. The reason we hope for a nice answer is that in order to solve for *y* we need to divide by 5. Therefore, we pick an *x* value that is divisible by 5.)

$$5y = 2x + 10 \qquad \textit{Substitute } x = 5.$$
$$5y = 2(5) + 10$$
$$5y = 20$$
$$y = 4$$

We have our third point (5, 4).

We can now draw the graph of the line $5y = 2x + 10$ (Figure 5.9).

Figure 5.9
The graph of $5y = 2x + 10$

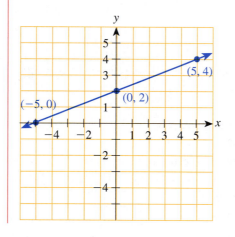

The straight line is the set of all points (and only those points) that satisfy the condition "5 times the y-coordinate is 10 more than twice its x-coordinate."

There is one difficulty with the intercept method. As we will see in the next example, if a line passes through the origin, (0, 0), both the *x*- and *y*-intercepts are the same. Rather than getting the necessary two points, we get only one.

EXAMPLE 4

Sketch the graph of *y* = 3*x*.

Solution

We proceed as before and try to find the *x*- and *y*-intercepts.
To find the *x*-intercept, we set *y* = 0 and solve for *x*:

$$y = 3x \qquad \textit{Substitute } y = 0.$$
$$0 = 3x$$
$$0 = x \qquad \text{Thus, the } x\text{-intercept is 0. We plot the point (0, 0).}$$

If a line crosses the *x*-axis at the origin, then it must cross the *y*-axis there as well. That is, the *x*- and *y*-intercepts of this line coincide. (Verify for yourself that if you substitute *x* = 0, you will get *y* = 0.) Since we get only one point from the intercept method, we must find another point. We are free to choose any convenient value for either *x* or *y* and then solve for the other variable.
This time we choose *y* = 3 and solve for *x*:

$$y = 3x \qquad \textit{Substitute } y = 3.$$
$$3 = 3x$$
$$1 = x \qquad \text{Therefore, our second point is } (1, 3).$$

For our check point, we choose $x = -1$ and solve for y:

$$y = 3x$$
$$y = 3(-1) = -3 \qquad \text{We have as our check point } (-1, -3).$$

The graph appears in Figure 5.10.

Figure 5.10
The graph of $y = 3x$

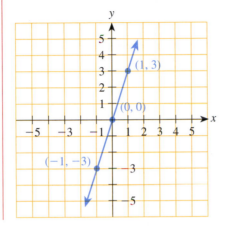

Let's summarize the intercept method for graphing a straight line.

The Intercept Method	

The Intercept Method

1. Locate the x-intercept, which is the x-coordinate of the point where the line crosses the x-axis.

   Do this by substituting $y = 0$ into the equation and solving for x.

2. Locate the y-intercept, which is the y-coordinate of the point where the line crosses the y-axis.

   Do this by substituting $x = 0$ into the equation and solving for y.

3. Locate a third "check" point by choosing a convenient value for either x or y and then solving for the other variable.

4. Draw the line passing through all three points.

There are two other special types of lines where the intercept method fails. We illustrate these in the next example.

EXAMPLE 5

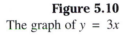

Sketch the graphs of the following equations in a rectangular coordinate system.

(a) $x = 3$ (b) $y = -2$

Solution

At first glance both the wording of the example and the appearance of the given equations may seem peculiar. Why the extra words "in a rectangular coordinate system"? Why do the equations not have two variables as all the other equations in this chapter have had?

The answers to these questions are closely related. If we are simply asked to graph the solution set of the equation $x = 3$, we might quite reasonably answer as follows:

The fact that the question specifies "in a rectangular coordinate system" means that we have to think of the equations $x = 3$ and $y = -2$ in the context of a first-degree equation in *two* variables. As we pointed out earlier, it is often helpful to think of an equation as a condition that x and y must satisfy. If only one of the variables appears, then there is *no* condition on the other variable.

(a) In the equation $x = 3$, y does not appear. Therefore, the equation $x = 3$ places a condition on x only. That condition is that the x-coordinate of the point be 3. In other words, all points that satisfy this equation must be located 3 units to the right. (If you think of the general form $Ax + By = C$, this is just a case where $B = 0$. That is, the equation $x = 3$ can be thought of as $1 \cdot x + 0 \cdot y = 3$. This equation implies that x must be equal to 3 but that y can be anything!)

Our graph is going to be a vertical line parallel to the y-axis and 3 units to the right. The graph appears in Figure 5.11.

Figure 5.11
The graph of $x = 3$

For any point on this vertical line, the x-coordinate must be 3 but the y-coordinate can be any number.

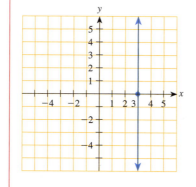

(b) Similarly, in the equation $y = -2$, x does not appear. (This is the special case of $Ax + By = C$ with $A = 0$.) This means that the condition on our point is that it always be located 2 units down. On the other hand, x can be anything!

The graph is going to be a horizontal line parallel to the x-axis and 2 units down. The graph appears in Figure 5.12.

Figure 5.12
The graph for $y = -2$

For any point on this horizontal line, the y-coordinate must be −2 but the x-coordinate can be any number.

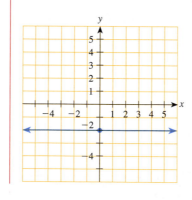

> In general, the graph of an equation of the form $x = $ constant is a vertical line parallel to the y-axis. The graph of an equation of the form $y = $ constant is a horizontal line parallel to the x-axis.

Sometimes we obtain fractional values for our points. Since we are just "sketching" the graph, we simply plot the points in a reasonably accurate location.

DIFFERENT PERSPECTIVES: Intercepts

Consider the geometric and algebraic interpretations of the x- and y-intercepts of the graph of $3x - 2y = 6$.

GEOMETRIC INTERPRETATION

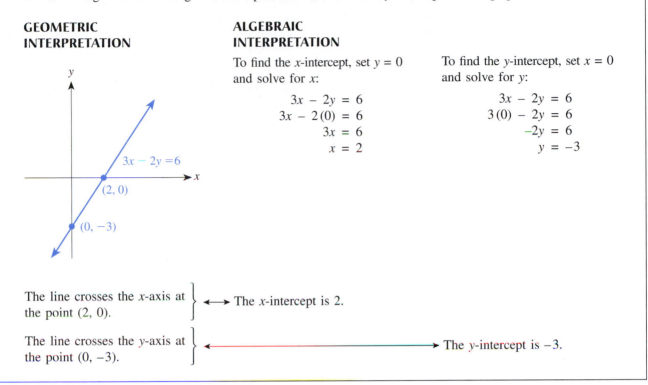

ALGEBRAIC INTERPRETATION

To find the x-intercept, set $y = 0$ and solve for x:

$$3x - 2y = 6$$
$$3x - 2(0) = 6$$
$$3x = 6$$
$$x = 2$$

To find the y-intercept, set $x = 0$ and solve for y:

$$3x - 2y = 6$$
$$3(0) - 2y = 6$$
$$-2y = 6$$
$$y = -3$$

The line crosses the x-axis at the point $(2, 0)$. $\longleftrightarrow$ The x-intercept is 2.

The line crosses the y-axis at the point $(0, -3)$. $\longleftrightarrow$ The y-intercept is -3.

EXAMPLE 6   Sketch the graph of $3x + 4y = 10$.

Solution   We sketch the graph using the intercept method.

To find the x-intercept, set $y = 0$ and solve for x:

$$3x + 4y = 10$$
$$3x + 4(0) = 10$$
$$3x = 10$$
$$x = \frac{10}{3} = 3\frac{1}{3}$$

The x-intercept is $3\frac{1}{3}$.

To find the y-intercept, set $x = 0$ and solve for y:

$$3x + 4y = 10$$
$$3(0) + 4y = 10$$
$$4y = 10$$
$$y = \frac{10}{4} = 2\frac{1}{2}$$

The y-intercept is $2\frac{1}{2}$.

The graph appears in Figure 5.13. We leave it to the student to check this graph by finding a third point and verifying that it lies on the line.

Figure 5.13
The graph of $3x + 4y = 10$

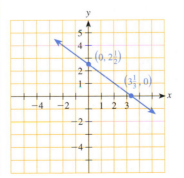

Thus far we have discussed first-degree equations in two variables and explained that the solutions to such equations can be viewed as ordered pairs in a rectangular coordinate system. We also stated that the graph of all the ordered pairs satisfying a first-degree equation in two variables is a straight line.

With this background we may use two variables (they do not have to be x and y) to represent quantities and graph the resulting relationship. (In Chapter 6 we will use two-variable equations to solve verbal problems.)

EXAMPLE 7

Linda has two part-time jobs. She works as a math tutor, for which she charges $12.50 per hour, and for a telemarketing firm, for which she earns $8 per hour. During a certain month she wants to earn a total of $400.

(a) Let m be the number of hours Linda tutors and t be the number of hours she works for the telemarketing firm. Write an equation representing the relationship between the number of hours she must work at each job in order to earn a total of $400.

(b) Sketch the graph of the relationship found in part (a).

(c) If she plans to work 25 hours for the telemarketing firm, how many hours must she work tutoring?

Solution

(a) Since Linda earns $12.50 per hour tutoring and she tutors for m hours, she earns $12.50m$ dollars tutoring. Similarly, she earns $8t$ dollars working for the telemarketing firm. Since she will earn $400 for the month, the equation is

$$12.50m + 8t = 400$$

(b) In order to sketch the graph of this equation, we must decide which quantity to represent along the horizontal axis and which along the vertical axis. We are free to label the axes either way; however, keep in mind that if we label the horizontal axis m and the vertical axis t, then we must record the ordered pairs as (m, t). On the other hand, if we choose to do the reverse, then we must also reverse the ordered pairs.

In both cases we have sketched the graph using the intercept method.

$$12.50m + 8t = 400 \qquad \textit{Find the m-intercept; set } t = 0.$$
$$12.50m + 8(0) = 400$$
$$12.50m = 400$$
$$m = \frac{400}{12.50} = 32$$

$$12.50m + 8t = 400 \qquad \textit{Find the t-intercept; set } m = 0.$$
$$12.50(0) + 8t = 400$$
$$8t = 400$$
$$t = \frac{400}{8} = 50$$

Thus, the m-intercept is 32 and the t-intercept is 50. The graphs appearing in Figure 5.14 show the results of the two choices of labeling the axes.

(c) The equation $12.50m + 8t = 400$ expresses the relationship between the number of hours tutoring (m) and the number of hours working for the telemarketing company (t). If we know one of these quantities we can substitute and solve for the other. We are told that she will work 25 hours for the telemarketing company ($t = 25$) and we want to find how many hours she must tutor ($m = ?$).

$$12.50m + 8t = 400 \qquad \textit{Substitute } t = 25 \textit{ and solve for m.}$$
$$12.50m + 8(25) = 400$$
$$12.50m + 200 = 400$$

$$12.50m = 200$$

$$m = \frac{200}{12.5} = 16 \qquad \text{Linda needs to tutor for} \boxed{\text{16 hours.}}$$

Figure 5.14
The graph of
$12.50m + 8t = 400$

Note that we have drawn the graph in the first quadrant only because neither m nor t can be negative.

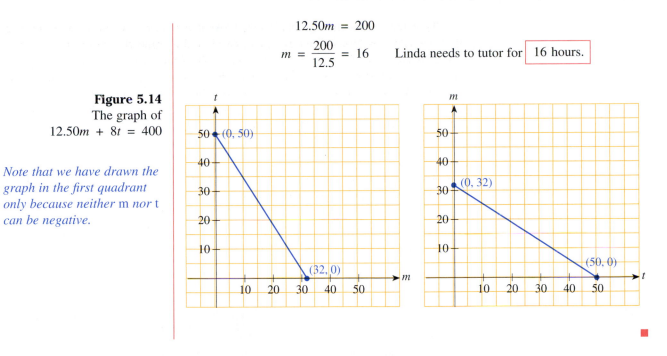

EXAMPLE 8

The following graph illustrates the relationship between the number n of inches of snowfall and the price P of snow shovels (in dollars).

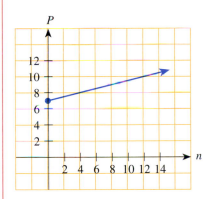

(a) Use the graph to determine the price of a snow shovel if there is a snowfall of 12 inches.

(b) Suppose that we know that the equation of the line is $P = 0.25n + 7$. Use this equation to find the price of a snow shovel when the snowfall is 12 inches.

Solution

(a) 12 inches of snow means that $n = 12$, and we want to read the corresponding P value (the price) from the graph.

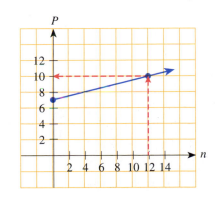

In other words, the point (12, 10) is on the line.

As we can see from the graph, if $n = 12$ we must go up vertically 10 units in order to intersect the graph. Therefore, the price of a snow shovel is $10 when the snowfall is 12 inches.

(b) Given the equation $P = 0.25n + 7$, in order to find the price when the snowfall is 12 inches we must substitute $n = 12$ and find P.

$$P = 0.25n + 7 \qquad \textit{Substitute } n = 12.$$

$$P = 0.25(12) + 7 = 3 + 7 = \boxed{10}$$

In other words, the ordered pair (12, 10) satisfies the equation.

Therefore, the price of a snow shovel is $10 when the snowfall is 12 inches, which agrees with our previous answer. ∎

In Chapter 6 we see how we can use first-degree equations in two variables to model relationships and solve verbal problems.

Study Skills 5.2

Reviewing Your Exam: Checking Your Understanding

If you carefully looked over your exam and you believe that you understand the material and what you did wrong, then do the following:

Copy the problems over on a clean sheet of paper and rework them without your text, notes, or exam. When you are finished, check to determine whether your answers are correct. If your answers are correct, try to find problems in the text similar to those problems and work the new problems on a clean sheet of paper (again without notes, text, or exam). If some of your new answers are incorrect, then you may have

learned how to solve your test problems, but you probably have not thoroughly learned the topic being tested. You may need to repeat these steps several times until you are confident you understand your errors.

In any case, you should keep your exam (with the correct answers) since it is a good source of information for future studying. You may want to record errors that you consistently made on the exam on your warning cards: exam problem types can be used when you make up your quiz cards (see Section 3.4).

EXERCISES 5.2

In Exercises 1–8, complete each ordered pair so that it satisfies the given equation.

1. $3x - 7y = 21$; (, 15), (14,), (−2,)

2. $5y + 6x = 30$; (−5,), (, −6), (, 4)

3. $2y + 9x = 36$; (6,), (0,), (, 0)

4. $4x + 7y = 56$; (, 2), (, 0), (0,)

5. $3x - 5y = 10$; $\left(-\dfrac{2}{3}, \ \right), \left(\ , -\dfrac{4}{5}\right), (5, \ )$

6. $7x + 9y = 20$; $\left(\ , \dfrac{1}{3}\right), \left(\dfrac{2}{7}, \ \right), (\ , 4)$

7. $y = -\dfrac{1}{2}x + 5$; (−6,), (, 4), (3,)

8. $y = \dfrac{2}{3}x - 1$; (, 7), (−6,), (, 5)

In Exercises 9–22, find the x- and y-intercepts of the equation.

9. $x - y = 7$

10. $x + y = 9$

11. $y - x = -4$

12. $y + 3x = -6$

13. $2x + 3y = 12$

14. $3x + 2y = 12$

15. $y = 4x$

16. $y = -3x$

17. $2x = 5y$

18. $3x = -7y$

19. $x = 5$

20. $y = 4$

21. $x - 4 = 3y$

22. $3x - 8 = 4y$

In Exercises 23–44, use the intercept method to sketch the graph of the given equation in a rectangular coordinate system.

23. $x + y = -5$

24. $x - y = 4$

25. $x - y = 8$

26. $x + y = -2$

27. $3x - 4y = 12$

28. $3x - 7y = 21$

29. $y = 2x - 10$

30. $y = 4x + 12$

31. $y = -4x$

32. $y = 3x$

33. $y + 7 = x - 5$

34. $x + 3 = y - 4$

35. $y = 5$

36. $x = 6$

37. $x = -4$

38. $y = -1$

39. $y = \dfrac{x - 3}{2}$

40. $x = \dfrac{y + 2}{3}$

41. $2(x - 3) = 4(y + 2)$

42. $6(y - 1) = 3(x + 2)$

43. $\dfrac{2x + 1}{2} = \dfrac{y - 3}{4}$

44. $\dfrac{y + 6}{3} = \dfrac{x - 4}{4}$

45. Sketch the graph of $d = 3t + 4$ using the horizontal axis for t values and the vertical axis for d values.

46. Sketch the graph of $P = 10c - 5$ using the horizontal axis for c values and the vertical axis for P values.

47. Sketch the graph of $h = -6t + 30$ using the horizontal axis for t values and the vertical axis for h values.

48. Sketch the graph of $T = -3d - 4$ using the horizontal axis for d values and the vertical axis for T values.

49. Sketch the graph of $u - 4v = 8$ using the horizontal axis for u values and the vertical axis for v values.

50. Sketch the graph of $u - 4v = 8$ using the horizontal axis for v values and the vertical axis for u values.

51. An electronics discount store wants to use up a credit of $9,110 with its supplier to order a shipment of VCRs and TVs. Each VCR costs $125 and each TV costs $165.

    (a) Let v represent the number of VCRs and t represent the number of TVs. Write an equation that reflects the given situation.

    (b) Sketch the graph of this relationship. Be sure to label the coordinate axes clearly.

    (c) If 28 VCRs are ordered, use the equation you obtained in part (a) to find the number of TVs.

52. A computer store budgets $12,000 to buy computers and laser printers. Each computer costs $650 and each printer costs $200.

    (a) Let C represent the number of computers and P represent the number of printers. Write an equation that reflects the given situation.

    (b) Sketch the graph of this relationship. Be sure to label the coordinate axes clearly.

(c) If the shipment contains 16 computers, use the equation you obtained in part **(a)** to find the number of printers.

53. Mary is working as a clerk in a bank where she inspects documents. It takes her 12 minutes to check a loan application and 8 minutes to check a credit report. On a certain day she is scheduled to spend 360 minutes inspecting documents.

   (a) Let a represent the number of loan applications and c represent the number of credit reports. Write an equation that reflects the given situation.

   (b) Sketch the graph of this relationship. Be sure to label the coordinate axes clearly.

   (c) If Mary inspected 25 loan applications, use the equation you obtained in part **(a)** to find the number of credit reports she checked.

54. A seamstress can hem a skirt in 10 minutes and make the cuffs on a pair of pants in 15 minutes. On a certain day she has allotted 300 minutes to doing these tasks.

   (a) Let s represent the number of skirts she hems and p represent the number of pairs of pants she cuffs. Write an equation that reflects the given situation.

   (b) Sketch the graph of this relationship. Be sure to label the coordinate axes clearly.

   (c) If she hems 18 skirts, use the equation you obtained in part **(a)** to find the number of pairs of pants she cuffs.

ⓠ QUESTIONS FOR THOUGHT

55. Sketch the graph of the line whose x-intercept is 3 and whose y-intercept is 5.

56. Sketch the graph of the line whose points have equal x- and y-coordinates. What would the equation of this line be?

57. Sketch the graph of the line whose points have x- and y-coordinates that are negatives of each other. What would the equation of this line be?

58. Define x- and y-intercepts in two ways:

   (a) In terms of the graph of an equation

   (b) In terms of an algebraic solution to the equation

59. Consider the equation $c = 3n + 2$.

   (a) Sketch the graph of this equation using the horizontal axis for the n values and the vertical axis for the c values.

   (b) Use the graph to describe what happens to the values of c as n changes.

   (c) Sketch the graph of this equation using the horizontal axis for the c values and the vertical axis for the n values.

   (d) Use the graph to describe what happens to the values n as c changes.

   (e) Does the way we choose to label the axes affect the appearance of the graph? Does the way we choose to label the axes affect the relationship between the variables?

◇ MINI-REVIEW

In Exercises 60–62, simplify the given expression.

60. $x^2(x^3)(x^2)^3$

61. $(2x^2)(5xy)(-y^2) + x^3y^3$

62. $(2x^2)(5xy - y^2) - x^2y^2$

Solve the following problem algebraically. Be sure to label what the variable represents.

63. Lamont has invested $1,300 in a savings account that pays 4% annual interest. At what interest rate must an additional $800 be invested to produce $100 per year in interest?

5.3 The Slope

In the last section we discussed the procedure for sketching the graph of a line whose equation is given. We now turn our attention to the reverse situation: How can we obtain the equation of a line given its graph?

Suppose Lenore knows that if she exercises for 3 hours per week she will lose 1 pound that week and if she exercises 5 hours per week she will lose 2 pounds that week. Additionally, let's assume that the number of hours she exercises per week and the resulting weight loss are related by a first-degree equation.

Let's draw a graph of this relationship using h for the horizontal axis, representing the number of hours Lenore exercises per week, and p for the vertical axis, representing the number of pounds she loses that week. This means that we will record ordered pairs as (h, p). Therefore, the given information concerning exercise and accompanying weight loss is recorded as $(3, 1)$ and $(5, 2)$. In Figure 5.15, we draw the straight line through these points.

Figure 5.15

The graph of the exercise/weight loss relationship

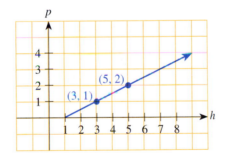

The point $(3, 1)$ means that 3 hours of exercise results in a weight loss of 1 pound. What does the point $(5, 2)$ mean?

Note that we have drawn the graph only in the first quadrant since both h (the number of hours exercising) and p (the number of pounds lost) cannot be negative.

A natural question Lenore might ask is "What should I expect my weight loss to be if I exercise 8 hours per week?" Looking at the graph, it is a bit difficult to determine exactly what p would be if $h = 8$. However, if we had the *equation* of the line drawn in Figure 5.15 we could simply substitute $h = 8$ into the equation and solve for p to determine the number of pounds lost. In effect, we would like to determine the equation of a line from its graph.

We will return to answer this question in the next section, but in order to do so we must first introduce the idea of the *slope* of a line. As we shall see in a moment, the slope of a line is a number that indicates the steepness of a line.

In looking for the equation of a line, we are looking for a condition that x and y must satisfy. For example, if the equation of a line is $y = x + 3$, this tells us that in order for a point to be on this line its y-coordinate must be 3 more than its x-coordinate. The equation $y = x + 3$ is the condition that x and y must satisfy. The slope of a line, which we will define in a moment, tells us how the *change* in y is related to the *change* in x. We will then be able to use the slope to derive an equation of a given line.

In Figure 5.16 on page 226, we have drawn a nonvertical line and labeled two points on the line as (x_1, y_1) and (x_2, y_2). We are using subscripted variables so that we can retain the letters x and y to indicate the first and second coordinates of an ordered pair. Thus we know that x_1 and x_2 are both *first* coordinates and that y_1 and y_2 are both *second* coordinates.

Figure 5.16

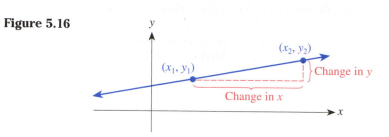

Why do we insist that the line not be vertical?

It is a basic geometric fact that as we move from any point on a nonvertical line to any other point on the line the ratio of the change in the y-coordinates to the change in the x-coordinates is *constant*. This fact is exactly what we are looking for—a condition that every point on the line must satisfy.

We define the following:

DEFINITION

Let $P_1(x_1, y_1)$ and $P_2(x_2, y_2)$ be any two points on a nonvertical line L. The **slope** of L, denoted by m, is given by

$$m = \frac{y_2 - y_1}{x_2 - x_1} \qquad (x_1 \neq x_2)$$

Be careful! Make sure to subtract the y-coordinates and the x-coordinates in the *same* order when computing the slope of a line.

The remainder of this section is devoted to further development of this idea. In the next section we will return to answer the question raised earlier about how to obtain an equation for a line when we have its graph.

EXAMPLE 1

Compute the slope of the line passing through each of the following pairs of points. Sketch the graph of each line.

(a) $(2, -1)$ and $(5, 3)$ (b) $(-1, 2)$ and $(1, -3)$

(c) $(4, 5)$ and $(-2, 5)$

Solution

(a) Using the definition of slope that we have just given, we can let the "first" point be $(x_1, y_1) = (2, -1)$ and the "second" point be $(x_2, y_2) = (5, 3)$. We then get

$$m = \frac{y_2 - y_1}{x_2 - x_1} = \frac{3 - (-1)}{5 - 2} = \boxed{\frac{4}{3}}$$

Notice that the order in which we choose the points is irrelevant. We could just as well have let $(x_1, y_1) = (5, 3)$ and $(x_2, y_2) = (2, -1)$. We then get

$$m = \frac{y_2 - y_1}{x_2 - x_1} = \frac{-1 - 3}{2 - 5} = \frac{-4}{-3} = \boxed{\frac{4}{3}}$$

What is important is that we subtract the x- and y-coordinates in the *same* order.

The graph appears in Figure 5.17. Note that to get from the point $(2, -1)$ to the point $(5, 3)$ we move right 3 units (the change in x) and up 4 units (the change in y).

Figure 5.17

Note that this line with positive slope goes up from left to right.

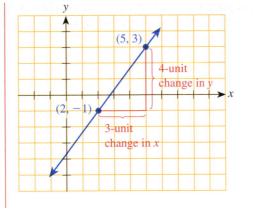

(b) The given points are $(-1, 2)$ and $(1, -3)$. We get

$$m = \frac{\text{Change in } y}{\text{Change in } x} = \frac{y_2 - y_1}{x_2 - x_1} = \frac{-3 - 2}{1 - (-1)} = \boxed{\frac{-5}{2}}$$

The graph appears in Figure 5.18. Note that to get from the point $(-1, 2)$ to the point $(1, -3)$ we move right 2 units (the change in x) and down 5 units (the change in y).

Figure 5.18

Note that this line with negative slope goes down from left to right.

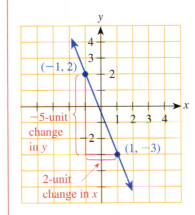

(c) The given points are $(4, 5)$ and $(-2, 5)$. We get

$$m = \frac{\text{Change in } y}{\text{Change in } x} = \frac{y_2 - y_1}{x_2 - x_1} = \frac{5 - 5}{4 - (-2)} = \frac{0}{6} = \boxed{0}$$

The graph appears in Figure 5.19. Note that this line, which has zero slope, is *horizontal*.

Figure 5.19

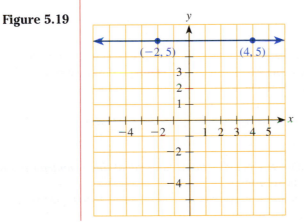

It is important to keep in mind that the slope of a line is constant. Basic geometry tells us that regardless of which two points we choose on a particular line, the slope determined by them is the same.

Before we continue, let's agree that whenever we describe a graph, we describe it for increasing values of x, that is, moving from left to right.

The line in Figure 5.20**a** is rising (as we move from left to right), while the line in Figure 5.20**b** is falling (as we move from left to right).

Figure 5.20

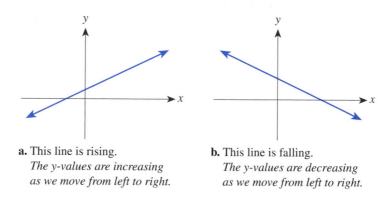

a. This line is rising.
The y-values are increasing as we move from left to right.

b. This line is falling.
The y-values are decreasing as we move from left to right.

Let us now examine what this number, the slope, tells us about a line. As mentioned earlier, the slope of a line is a number that indicates the steepness and direction of the line.

Figure 5.21 illustrates lines with various slopes through the point (3, 1). Note that lines with positive slope rise (go up) as we move from left to right, while lines with negative slope fall (go down) as we move from left to right. The larger the slope in absolute value, the steeper the line.

Figure 5.21

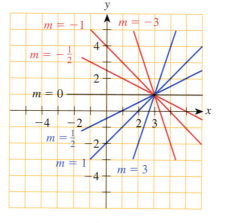

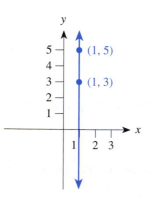

If you look back at our definition of slope, you will notice that we specified that the line be nonvertical. Why? If we try to compute the slope of a vertical line, such as the one passing through the points (1, 3) and (1, 5), we would get

$$m = \frac{5 - 3}{1 - 1} = \frac{2}{0}$$

which is *undefined.* Thus, *the slope of a vertical line is undefined.* (We may also say a vertical line has *no slope.*)

Do not confuse a horizontal line, which has slope equal to 0, with a vertical line, whose slope is undefined (see Figure 5.22).

Figure 5.22

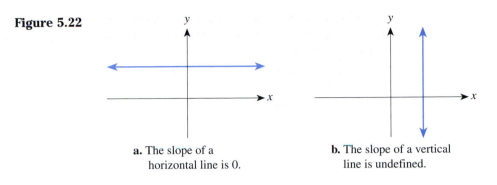

a. The slope of a horizontal line is 0.

b. The slope of a vertical line is undefined.

We summarize this discussion in the accompanying box.

Summary of Slopes	A line with positive slope is rising (goes up from left to right).
	A line with negative slope is falling (goes down from left to right).
	A line with slope equal to 0 is horizontal.
	A line with slope undefined (no slope) is vertical.

EXAMPLE 2   Sketch the graph of the line with slope $\frac{2}{5}$ that passes through the point $(-3, 1)$.

Solution   In order to sketch the line we need *two* points on the line. The slope being $\frac{2}{5}$ means that

$$\frac{\text{Change in } y}{\text{Change in } x} = \frac{2}{5}$$

In other words, for every 5-unit change in x, we get a 2-unit change in y. Starting from the point $(-3, 1)$ we can move 5 units to the right and 2 units up, which brings us to the point $(2, 3)$. We can now sketch the graph of the line through the points $(-3, 1)$ and $(2, 3)$ (see Figure 5.23).

Figure 5.23

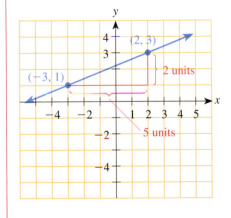

As this example illustrates, if we know the slope of the line and a single point on the line, the line is completely determined. (There is only one such line.)   ■

EXAMPLE 3   Sketch the graph of the line with slope $-\frac{1}{2}$ that passes through the point $(4, 3)$.

Solution   Again, in order to sketch the line we need *two* points on the line. The slope of $-\frac{1}{2}$ can be interpreted as

$$\frac{\text{Change in } y}{\text{Change in } x} = \frac{-1}{2} \quad \text{or} \quad \frac{\text{Change in } y}{\text{Change in } x} = \frac{1}{-2}$$

In other words, starting from the point (4, 3) we can move 2 units to the right and 1 unit *down,* which brings us to the point (6, 2); or we can move 2 units *left* and 1 unit up, which brings us to the point (2, 4). Figure 5.24 shows that either interpretation gives us the same line because all three points lie on the same line.

Figure 5.24

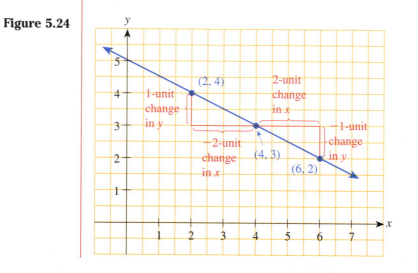

EXAMPLE 4

The straight line segment in Figure 5.25 illustrates the relationship between the demand *d* (in hundreds of items) for a certain item and the price *p* (in dollars) per item.

Figure 5.25

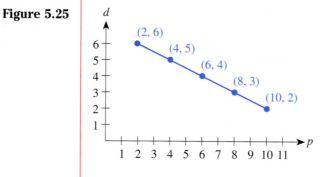

(a) Use the graph to determine the slope of the line.

(b) How many items are in demand at a price of $6 per item?

(c) Describe how the graph illustrates the relationship between the price of the item and demand for it.

Solution

(a) From the graph we can see that the points (8, 3) and (10, 2) are on the line. We can use these two points to compute the slope of the line as

$$m = \frac{2 - 3}{10 - 8} = \frac{-1}{2} = \boxed{-\frac{1}{2}}$$

Let's keep in mind that each unit along the *p*-axis represents a one-dollar change in the price and that each unit along the *d*-axis represents a 100-unit change in the demand. Therefore, when we compute the slope of this line segment, we actually have

$$m = \frac{\text{Change in demand}}{\text{Change in price}}$$

Consequently, the slope of $-\frac{1}{2}$ means that for every additional \$2 the price is increased, the demand will *decrease* by 100 items. Look carefully at Figure 5.25 to verify that the graph shows this same relationship.

(b) To find the demand when the price is \$6, we must find the d value corresponding to $p = 6$. Looking at the graph, we see that the point $(6, 4)$ is on the graph, which means that when the price p is equal to \$6, the demand d is equal to

$$\boxed{400 \text{ items}}$$

(c) We would expect that as the price per item increases, the demand for the item decreases. This is exactly what the graph shows. The line is falling, indicating that as p increases, d decreases. This also agrees with the result obtained in part **(a)** that the line has a negative slope. ∎

EXAMPLE 5

Suppose that the Futurecom Company had 3.4 million cable subscribers in 1990 and that during the next decade the number of subscribers increased by 1.2 million subscribers per year.

(a) Letting $x = 0, 1, 2$, etc. represent 1990, 1991, 1992, etc., and letting y represent the number of subscribers (in millions), write an equation relating x and y.

(b) Sketch the graph of the equation obtained in part **(a)**.

(c) Find the slope of the line graphed in part **(b)** and relate it to the equation obtained in part **(a)**.

Solution

(a) One method that often helps us find an equation relating two quantities is to examine some numerical data and hopefully expose a pattern. The example tells us that the company starts with 3.4 million subscribers in 1990 and adds 1.2 million subscribers each year for the next 10 years. Let's summarize this information in the following table.

Year (x)	Number of Subscribers (y) in millions
$x = 0$ (1990)	$y = 3.4$
$x = 1$ (1991)	$y = 3.4 + 1.2 = 4.6$
$x = 2$ (1992)	$y = 3.4 + 2(1.2) = 5.8$
$x = 3$ (1993)	$y = 3.4 + 3(1.2) = 7$
$x = 4$ (1994)	$y = 3.4 + 4(1.2) = 8.2$
x	$y = 3.4 + x(1.2) = 3.4 + 1.2x$

Thus the equation relating x and y is $\boxed{y = 1.2x + 3.4}$.

(b) We recognize that the equation found in part **(a)** is a first-degree equation and so its graph is a straight line. Thus, we need only find two points on the line. Based on the table in part **(a)**, we find that the points $(0, 3.4)$ and $1, 4.6)$ are on the line. The graph appears in Figure 5.26 on page 232.

Figure 5.26

The graph of
$y = 1.2x + 3.4$

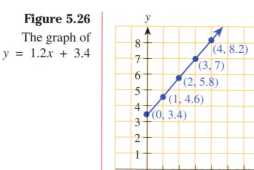

(c) We can use the same two points, (0, 3.4) and (1, 4.6), to compute the slope. In order to understand the significance of the slope let's include the meaning of the variables.

$$m = \frac{\text{Change in } y}{\text{Change in } x} = \frac{\text{Change in number of subscribers}}{\text{Change in years}} = \frac{4.6 - 3.4}{1 - 0}$$

$$= 1.2 \quad \text{million subscribers per year}$$

Note that the slope is exactly the same as the rate at which the number of subscribers is changing. This idea of the slope of a line being interpreted as a rate of change will be discussed further. ■

Study Skills 5.3

Preparing for a Comprehensive Final Exam

Many algebra courses require students to take a comprehensive final exam—an exam covering the entire course. Because of the amount of material covered in this type of exam, your preparation should necessarily differ from your preparation for other exams.

To succeed in a comprehensive final, even those students who have been doing well on exams all along will need to take the time to review the ideas and procedures covered earlier in the course. As more material is learned, it becomes easier to forget and/or confuse concepts learned previously.

For example, we often find that, on finals, students manage to correctly solve complex problems from the later chapters but have difficulty with some of the simpler problems covered in the earlier chapters. This is because the more complex material was most recently learned, whereas some of the "simple" topics, covered much earlier in the course, were forgotten or became obscured by the newer material. Even if you knew the material well earlier in the course, there is no substitute for timely review of the *entire* course syllabus.

Your studying should begin at least two weeks before the exam. Starting two weeks early means that you will probably be learning new material as you are studying for the final exam. You should consider studying for the final as separate from learning the new material. This means that your total math study time should be increased and divided between learning the new material and reviewing previous material (even the "simple" topics).

Your studying should include the following:

1. Review your class notes. Pay particular attention to those topics you may have found a bit difficult the first time you learned them.

2. If you have been writing out Study Cards, be sure to review them. Pay particular attention to the Warning Cards.

3. Review a selection of homework exercises. Be sure to choose a wide variety of exercises to cover all the topics in your syllabus and the various types of exercises within each topic.

4. Review all your class exams and quizzes. Be sure you understand how every single problem is solved and where you made your errors.

5. Find out whether copies of previous final exams are available for you to examine. Be sure you know what the format for the final exam is: Will the exam be multiple choice, fill in, or a combination? Will there be partial credit?

(continued)

Study Skills 5.3 *(continued)*

Preparing for a Comprehensive Final Exam

6. A few days before the final exam, make up a practice final exam, and follow the directions given in Study Skills 2.6: Using Quiz Cards.

Remember, going over all this material takes time, so be sure to start your studying well in advance of the final exam date.

EXERCISES 5.3

In Exercises 1–32, find the slope of the line passing through the given points. Round to the nearest hundredth where necessary.

1. (3, 5) and (6, 9)
2. (1, 4) and (7, 6)
3. (2, 6) and (1, 3)
4. (5, 1) and (1, 9)
5. (−1, 4) and (3, −2)
6. (2, −5) and (−4, 3)
7. (−1, −2) and (−3, −4)
8. (−4, −3) and (−2, −5)
9. (0, 2) and (3, 0)
10. (2, 0) and (0, 3)
11. (4, 7) and (−3, 7)
12. (−1, 5) and (3, 5)
13. (2, 6) and (2, 9)
14. (3, −1) and (3, 2)
15. $\left(1, \frac{1}{2}\right)$ and $\left(\frac{3}{4}, 2\right)$
16. $\left(4, \frac{1}{3}\right)$ and $\left(6, \frac{2}{3}\right)$
17. $\left(-\frac{1}{3}, \frac{1}{5}\right)$ and $\left(\frac{3}{2}, \frac{1}{4}\right)$
18. $\left(\frac{4}{5}, -\frac{1}{2}\right)$ and $\left(-\frac{1}{3}, \frac{3}{4}\right)$
19. (−1, 3) and (3, −1)
20. (4, −5) and (−2, 3)
21. (2, −5) and (−3, −5)
22. (4, 5) and (4, −3)
23. (0, a) and (a, 0), $a \neq 0$
24. (0, −b) and (−b, 0), $b \neq 0$
25. (a, b) and (2a, 2b), $a \neq 0$
26. (a, b) and (a + b, 2b), $b \neq 0$
27. (0.8, 2.65) and (1.3, 4.72)
28. (12.63, 10.44) and (9.48, 7.96)
29. (3.7, −1.05) and (−2.16, 4.9)
30. (−8.65, 2.8) and (12.5, −3.72)
31. (9.62, 8.77) and (−1.4, 8.77)
32. (3.45, 10.88) and (3.45, −4.69)

In Exercises 33–46, sketch the graph of the line satisfying the given conditions.

33. Passing through (1, 3) with slope 3
34. Passing through (1, 3) with slope $\frac{1}{3}$
35. Passing through (3, 2) with slope $\frac{1}{2}$
36. Passing through (3, 2) with slope 2
37. Passing through (2, 1) with slope $\frac{3}{2}$
38. Passing through (2, 1) with slope $\frac{2}{3}$

39. Passing through $(-1, 0)$ with slope 4

40. Passing through $(-1, 0)$ with slope -4

41. Passing through $(1, -2)$ with slope $\dfrac{-2}{3}$

42. Passing through $(1, -2)$ with slope $\dfrac{-3}{2}$

43. Passing through $(4, 3)$ with 0 slope

44. Passing through $(4, 3)$ and whose slope is undefined

45. Passing through $(-1, -5)$ and whose slope is undefined

46. Passing through $(-1, -5)$ with 0 slope

In Exercises 47–54, determine the slope of the line from its graph.

47.

48.

49.

50.

51.

52.

53.

54.

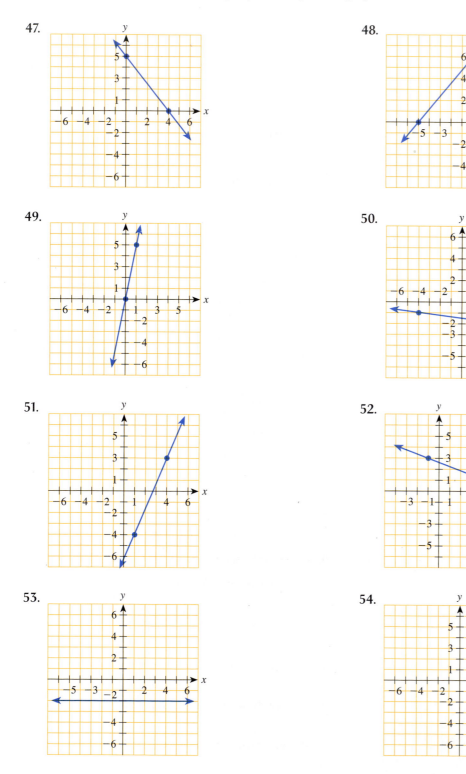

In Exercises 55–58, use the given graph to determine whether the line has positive slope, negative slope, zero slope, or its slope is undefined.

55.

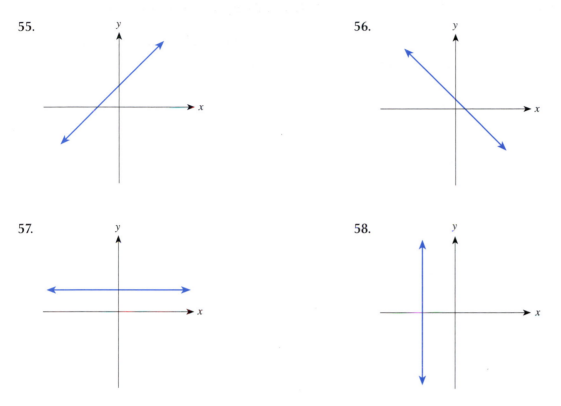

56.

57.

58.

59. The following figure illustrates three lines with slopes m_1, m_2, and m_3. List these slopes in increasing order.

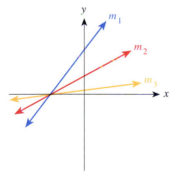

60. The following figure illustrates three lines with slopes m_1, m_2, and m_3. List these slopes in decreasing order.

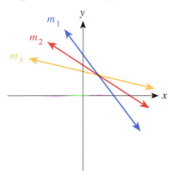

61. The straight line segment in the accompanying diagram represents the relationship between the number of calories burned *c* during a brisk walk that lasts *m* minutes.

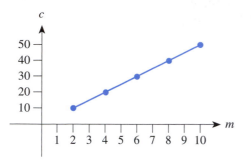

(a) How many calories are burned during a brisk walk that lasts 8 minutes?

(b) Use the graph to determine the slope of the line.

(c) How would you interpret the slope of this line?

62. The straight line segment in the accompanying diagram represents the relationship between a manufacturer's cost *per* radio, *c* (in dollars), and the number *r* of radios manufactured (in thousands).

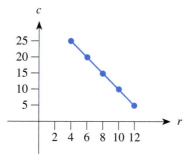

(a) What is the cost per radio if 10,000 radios are manufactured?

(b) Use the graph to determine the slope of the line.

(c) How would you interpret the slope of this line?

63. The straight line segment in the accompanying diagram represents the relationship between the number of gallons *g* that a car uses to travel *m* miles during a 450-mile trip.

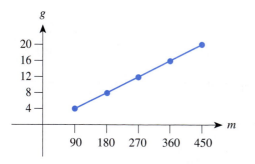

(a) How many gallons are used if the car travels 360 miles?

(b) How many miles can the car travel using 8 gallons?

(c) Compute the car's gas consumption in miles per gallon using the results of parts (a) and (b). Are your results the same? Should they be?

(d) Use the graph to determine the slope of the line.

(e) How would you interpret the slope of this line?

64. The straight line segment in the accompanying diagram represents the relationship between the number of miles m that a car travels in a period of h hours during a 260-mile trip.

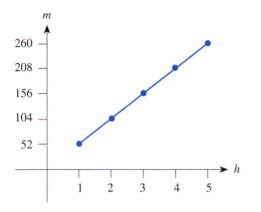

(a) How many miles does the car travel in 2 hours?

(b) How many hours does it take for the car to travel 260 miles?

(c) Compute the car's average speed using the results of parts **(a)** and **(b).** Are the results the same? Should they be?

(d) Use the graph to determine the slope of the line.

(e) How would you interpret the slope of this line?

65. Suppose that Jane works part-time making deliveries for a caterer. She gets paid a base salary of $80 per day plus $15 for each delivery she makes that day.

(a) Letting d represent the number of deliveries she makes and letting A represent the amount she earns for each day that she works, write an equation relating d and A.

(b) Sketch the graph of the equation obtained in part **(a),** representing d along the horizontal axis.

(c) Find the slope of the line graphed in part **(b)** and relate it to the equation obtained in part **(a).**

66. Suppose that a swimming pool has 500 gallons of water in it when a valve is turned on that feeds 150 gallons of water into the pool each minute.

(a) Letting m represent the number of minutes after the valve is opened and letting N represent the number of gallons of water in the pool, write an equation relating m and N.

(b) Sketch the graph of the equation obtained in part **(a),** representing m along the horizontal axis.

(c) Find the slope of the line graphed in part **(b)** and relate it to the equation obtained in part **(a).**

 QUESTION FOR THOUGHT

67. How could you use the idea of slope to show that the three points $(-1, -2)$, $(2, 0)$, and $(5, 2)$ all lie on a straight line?

◇ **MINI-REVIEW**

68. *Solve for x.* $\dfrac{x}{3} - \dfrac{x-1}{4} = \dfrac{x+1}{2}$ **69.** *Combine.* $\dfrac{x}{3} - \dfrac{x-1}{4} + \dfrac{x+1}{2}$

Solve the following problem algebraically. Be sure to label what the variable represents.

70. Tamika leaves point A at 10:00 A.M. traveling due east at 60 kph. One-half hour later, Ramon leaves the same location traveling due west at 70 kph. At what time will they be 257.5 km apart?

5.4 The Equation of a Line

We are now prepared to answer the question raised at the beginning of the last section: How can we obtain the equation of a line given its graph? We have already seen that if we are given the slope of a line and a point on the line we can draw its graph, but what is its equation?

Let the given slope be m and the given point be (x_1, y_1). What condition must a point (x, y) satisfy in order to be on the line? Keeping in mind that the slope of a line is a constant and looking at Figure 5.27, we can see that if (x, y) is any other point on the line when the slope determined by the two points (x_1, y_1) and (x, y) must be m. We must have

$$\frac{y - y_1}{x - x_1} = m \qquad \textit{Multiply both sides by } (x - x_1).$$

$$y - y_1 = m(x - x_1)$$

Figure 5.27

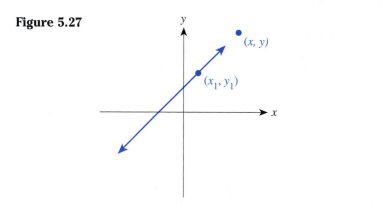

We have just derived the following:

Point–Slope Form of the Equation of a Line	An equation of the line with slope m that passes through the point (x_1, y_1) is $$y - y_1 = m(x - x_1)$$

EXAMPLE 1   Write an equation of the line with slope 4 passing through the point $(-2, 1)$.

Solution   We use $m = 4$ and $(x_1, y_1) = (-2, 1)$ in the point–slope form.

$$y - y_1 = m(x - x_1)$$

$$y - 1 = 4(x - (-2))$$

$$\boxed{y - 1 = 4(x + 2)}$$ ∎

Remember	x_1 and y_1 are the coordinates of the *given* point.
	x and y are the variables that appear in the equation.

EXAMPLE 2

Write an equation of the line with slope m and y-intercept b.

Solution

We are given the slope m. Remember that the y-intercept is the y-coordinate of the point where the line crosses the y-axis. Therefore, the fact that the y-intercept is b means that the line crosses the y-axis at the point $(0, b)$.

Applying the *point–slope form,* we get

$$y - y_1 = m(x - x_1) \qquad \textit{Substitute } (0, b) \textit{ for } (x_1, y_1).$$
$$y - b = m(x - 0)$$
$$y - b = mx$$

$$\boxed{y = mx + b}$$

This last equation is called the ***slope–intercept form*** of the equation of a line. ∎

Slope–Intercept Form of the Equation of a Line	An equation of the line with slope m and y-intercept b is given by $$y = mx + b$$

One of the most useful features of the slope–intercept form is that it makes it very easy to determine the slope of a line from its equation.

EXAMPLE 3

Determine the slopes of the lines whose equations are given.

(a) $y = 3x - 7$ (b) $2y + 3x = 9$

Solution

The slope–intercept form shows us that when an equation of a line is solved *explicitly* for y, the coefficient of x is the slope.

(a) We simply "read off" the slope from the equation.

$$y = mx + b$$
$$\updownarrow$$
$$y = 3x - 7$$

Thus, $\boxed{m = 3}$. We can also see that the y-intercept is -7.

(b) We must first solve the equation explicitly for y. That is, isolate y on one side of the equation.

$$2y + 3x = 9$$
$$2y = -3x + 9 \qquad \textit{Divide both sides of the equation by 2.}$$

$$y = \frac{-3}{2}x + \frac{9}{2}$$

$$\updownarrow$$

$$y = mx + b$$

Thus, $\boxed{m = -\frac{3}{2}}$.   Again we can also see that the y-intercept is $\frac{9}{2}$.

Keep in mind that if we were asked to find the y-intercept we could just as easily substitute $x = 0$ into the original equation and solve for y.   ■

EXAMPLE 4   Sketch the graphs of each of the following equations:

(a) $y = 2x + 4$ (b) $y = 2x - 6$

Solution   We can graph these equations using the intercept method outlined in Section 5.2.

For $y = 2x + 4$:   We set $x = 0$ and get $y = 4$; we set $y = 0$ and get $x = -2$. This gives us the points $(0, 4)$ and $(-2, 0)$.

For $y = 2x - 6$:   We set $x = 0$ and get $y = -6$; we set $y = 0$ and get $x = 3$. This gives us the points $(0, -6)$ and $(3, 0)$.

The graphs of both equations appear in Figure 5.28.

Figure 5.28

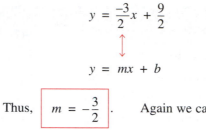

From the graphs shown in Figure 5.26, it appears that the two lines are parallel. We also note that both lines have the same slope. (Both equations are given in slope–intercept form and we can see that the slope of each line is 2.)

In fact, this is always the case.

> If two distinct lines have the same slope, then they are parallel.

It is also true that if two lines are parallel and have a slope then the slopes must be equal.

EXAMPLE 5   Write an equation of the line passing through the points $(1, -5)$ and $(-2, 2)$.

Solution   Let's first analyze this problem in order to develop a strategy for the solution.

THINKING OUT LOUD

What do we need to find?	The equation of a line
What is needed to find the equation of a line?	A point and the slope of the line
What information is given in the problem?	Two points on the line
How do we use the given information?	We can use the two points to compute the slope. Then we can use the slope with either point in the point–slope form.

Based on our analysis, we first compute the slope of the line.

$$m = \frac{y_2 - y_1}{x_2 - x_1} = \frac{2 - (-5)}{-2 - 1} = \frac{7}{-3} = -\frac{7}{3}$$

Now we can substitute this slope and one of the points into the point–slope form. First, using the point $(1, -5)$:

$$y - y_1 = m(x - x_1) \qquad \textit{Substitute } m = -\frac{7}{3} \textit{ and } x_1 = 1 \textit{ and } y_1 = -5.$$

$$y - (-5) = -\frac{7}{3}(x - 1)$$

$$\boxed{y + 5 = -\frac{7}{3}(x - 1)}$$

Second, using the point $(-2, 2)$:

$$y - y_1 = m(x - x_1) \qquad \textit{Substitute } m = -\frac{7}{3} \textit{ and } x_1 = -2 \textit{ and } y_1 = 2.$$

$$y - 2 = -\frac{7}{3}(x - (-2))$$

$$\boxed{y - 2 = -\frac{7}{3}(x + 2)}$$

While these two answers may look different, if we solve both of these answers for y we will see that they are actually equivalent. (Try it!)

In Example 7 we discuss how you can use the slope–intercept form to write an equation of a line in an example such as this. ■

Let's return to the situation described at the beginning of Section 5.3.

EXAMPLE 6

Suppose Lenore knows that if she exercises for 3 hours per week she will lose 1 pound that week and if she exercises 5 hours per week she will lose 2 pounds that week. Additionally, let's assume that the number of hours she exercises per week and the resulting weight loss are related by a first-degree equation. Determine the number of pounds she can expect to lose during a week in which she exercises for 8 hours.

Solution

As we did previously, let's draw a graph of this relationship using h for the horizontal axis, representing the number of hours Lenore exercises per week, and p for the vertical axis, representing the number of pounds she loses that week. This means that we will record ordered pairs as (h, p). Therefore, the given information concerning exercise and accompanying weight loss is recorded as $(3, 1)$ and $(5, 2)$. In Figure 5.29 on page 242, we draw the straight line through these points.

Figure 5.29

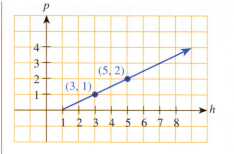

Asking for the weight loss accompanying a week in which she exercises for 8 hours can be restated as "Determine what p would be if $h = 8$." This is a bit difficult to read off the graph, but rather straightforward to compute if we first get the equation of this line.

We will use the point–slope form of the equation of a line, but we will use h in place of x and p in place of y. That is, we will use the equation

$$p - p_1 = m(h - h_1)$$

First we need to compute the slope. We use the points $(3, 1)$ and $(5, 2)$.

$$m = \frac{2 - 1}{5 - 3} = \frac{1}{2}$$

Now we can use the point–slope form with either of the given points. We will use $(3, 1)$ for (h_1, p_1).

$$p - p_1 = m(h - h_1) \qquad \textit{First substitute } m = \frac{1}{2}, h_1 = 3, \textit{ and } p_1 = 1.$$

$$p - 1 = \frac{1}{2}(h - 3) \qquad \textit{Now we can substitute } h = 8.$$

$$p - 1 = \frac{1}{2}(8 - 3) = \frac{1}{2}\left(\frac{5}{1}\right) = \frac{5}{2} \qquad \textit{So we have}$$

$$p - 1 = \frac{5}{2} \qquad \textit{Now solve for p.}$$

$$p = \frac{5}{2} + 1 = \frac{5}{2} + \frac{2}{2} = \frac{7}{2} = 3\frac{1}{2}$$

Therefore, if Lenore exercises for 8 hours she can expect to lose $\boxed{3\frac{1}{2} \text{ pounds}}$. ∎

The previous example illustrates how useful it is to have both the graph of a relationship and its equation.

EXAMPLE 7 Write an equation of the line with slope 3 that passes through the point $(2, -5)$.

Solution We may choose to use either the point–slope form or the slope–intercept form to write the equation.

Solution 1:   If we use the point–slope form, we get

$$y - (-5) = 3(x - 2)$$

which gives us

$$\boxed{y + 5 = 3(x - 2)}$$

Solution 2:   If we use the slope–intercept form, then we already know the slope is 3. Therefore, our equation is

$$y = 3x + b$$

We still need to determine b, the y-intercept. We can do this by recognizing that since the point $(2, -5)$ is on the line, this point must satisfy the equation $y = 3x + b$.

$y = 3x + b$ *We substitute the point $(2, -5)$. That is, we substitute $x = 2$ and $y = -5$.*

$-5 = 3(2) + b$ *Now we solve for b.*

$-5 = 6 + b$

$-11 = b$ *Now that we know the value of b, we can substitute into $y = 3x + b$.*

$\boxed{y = 3x - 11}$ *This is the slope–intercept form for the equation of the given line.*

Note that the answers obtained by the two methods appear to be different. It is left to the student to verify that both answers are equivalent.

In conclusion then, when asked to write an equation of a line, we are free to use either the point–slope form or the slope–intercept form. Our choice should be guided by which form makes better use of the given information.   ■

When two quantities (represented by two variables) are related by a first-degree equation, we say that the two quantities are related to each other linearly, or that the quantities vary linearly with respect to each other. Conversely, if we say that two quantities are linearly related, we mean that the equation relating them is a first-degree equation and therefore the graph describing the relationship is a straight line.

EXAMPLE 8   A telemarketing advertiser finds that after a commercial appears on television, the number of calls made to place orders within the next hour is linearly related to the number of times the phone number appears during the commercial. If the phone number appears 3 times, then 95 calls are made to place orders. If the phone number appears 5 times, then 145 calls are made to place orders.

(a) Write an equation expressing the number of calls n made to place orders in terms of the number of times p the phone number appears.

(b) Use this equation to predict the number of calls that would be made if the phone number appears 7 times.

(c) Using a coordinate system in which the horizontal axis is labeled p and the vertical axis is labeled n, sketch a graph of this relationship.

(d) Find the n-intercept and discuss its significance.

Solution   (a) We are told that when $p = 3$, $n = 95$, and when $p = 5$, $n = 145$. If we view the given information in the form of ordered pairs (p, n), we are told that $(3, 95)$ and $(5, 145)$ are two ordered pairs that satisfy a linear relationship. Therefore n and p are related by an equation of the form $n - n_1 = m(p - p_1)$. It is important to recognize that when we compute the slope of this line we have

$$m = \frac{\text{Change in } n}{\text{Change in } p}$$

Using the ordered pairs $(3, 95)$ and $(5, 145)$, we obtain

$$m = \frac{145 - 95}{5 - 3} = \frac{50}{2} = 25$$

Therefore, we can write the equation

$$n - 95 = 25(p - 3) \quad \text{which becomes} \quad \boxed{n = 25p + 20}$$

(b) In order to predict the number of calls that would be made if the phone number appeared 7 times, we substitute $p = 7$ into one of the equations we obtained in part (a). Since we are looking for n, we use the second form:

$$n = 25p + 20 \qquad \textit{Substitute } p = 7.$$

$$n = 25(7) + 20 = \boxed{195 \text{ phone calls}}$$

Thus we would expect 195 phone calls to place orders if the phone number is shown 7 times during the commercial.

(c) We can use the points $(3, 95)$ and $(5, 145)$ to sketch the graph, as shown in Figure 5.30.

Figure 5.30

The graph of $n = 25p + 20$

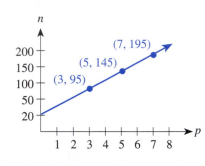

Note that we have sketched the graph only for $p \geq 0$ since it does not make any sense to consider a negative number of phone calls.

(d) From the equation $n = 25p + 20$ we can see that the n-intercept (that is, the value of n when $p = 0$) is 20. We can interpret this to mean that even if the phone number would not appear during the commercial, the advertiser would still receive 20 phone calls. Presumably this is the result of customers obtaining the phone number from some other source. ■

EXERCISES 5.4

In Exercises 1–32, write an equation of the line satisfying the given conditions.

1. Passing through $(1, 5)$ with slope 3

2. Passing through $(2, 7)$ with slope -3

3. Passing through $(-1, 4)$ with slope $\dfrac{1}{2}$

4. Passing through $(5, -3)$ with slope $\dfrac{3}{4}$

5. Passing through $(-3, -5)$ with slope $-\dfrac{2}{3}$

6. Passing through $(-1, -4)$ with slope $-\dfrac{4}{5}$

7. Passing through $(0, 6)$ with slope 5

8. Passing through $(0, 2)$ with slope 4

9. Passing through $(0, -2)$ with slope $\dfrac{1}{4}$

10. Passing through $(-4, 0)$ with slope $\dfrac{1}{5}$

11. Passing through $(-2, 0)$ with slope $-\dfrac{3}{4}$

12. Passing through $(0, -4)$ with slope $-\dfrac{2}{7}$

13. Passing through $(4, -2)$ with slope 1

14. Passing through $(-3, 2)$ with slope -1

15. Passing through $(5, 6)$ with slope 0

16. Passing through $(4, 7)$ whose slope is undefined

17. Passing through $(2, 3)$ and $(5, 9)$

18. Passing through $(1, 5)$ and $(3, 11)$

19. Passing through $(-1, 4)$ and $(2, -2)$

20. Passing through $(1, 5)$ and $(3, -4)$

21. Passing through $(0, 5)$ and $(5, 2)$

22. Passing through $(2, 3)$ and $(0, -2)$

23. Passing through $(-1, 0)$ and $(0, -1)$

24. Passing through $(0, -2)$ and $(-2, 0)$

25. Horizontal line passing through (2, 3)

26. Vertical line passing through (2, 0)

27. Vertical line passing through (4, −3)

28. Horizontal line passing through (4, −3)

29. Line has *x*-intercept 5 and *y*-intercept 2

30. Line has *y*-intercept 3 and *x*-intercept 7

31. Line has *x*-intercept −3 and *y*-intercept 4

32. Line has *x*-intercept −5 and *y*-intercept −1

In Exercises 33–36, two sets of values are given for variables having a linear relationship. In each case write the slope–intercept form for the equation of the line corresponding to the given set of values and answer the accompanying question.

33.

x (Number of hours spent studying)	2	3
y (Grade on math exam)	75	82

What would the grade be if a student studies for 5 hours?

34.

x (Number of hours practicing video game)	2	3
y (Grade on math exam)	75	70

What would the grade be if a student practices video games for 4 hours?

35.

x (Number of clerks working)	6	8
y (Number of minutes waiting time)	8	5

What would the waiting time be if 4 clerks are working?

36.

x (Number of people waiting in line)	6	8
y (Number of minutes waiting time)	10	15

What would the waiting time be if 15 people are waiting in line?

In Exercises 37–46, determine the slope of the line from its equation.

37. $y = 5x + 7$

38. $y = 2x - 11$

39. $y = -3x - 1$

40. $y = -4x + 2$

41. $x + y = 7$

42. $x - y = 5$

43. $2y + 3x = 6$

44. $3y - 5x = 12$

45. $2x - 5y + 7 = 0$

46. $5x - 6y - 7 = 0$

In Exercises 47–54, write an equation of the line whose graph is given.

47.

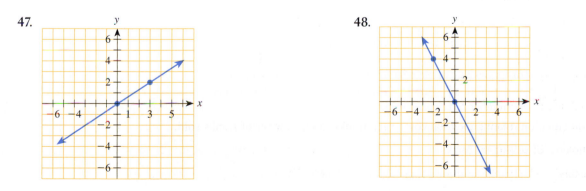

48.

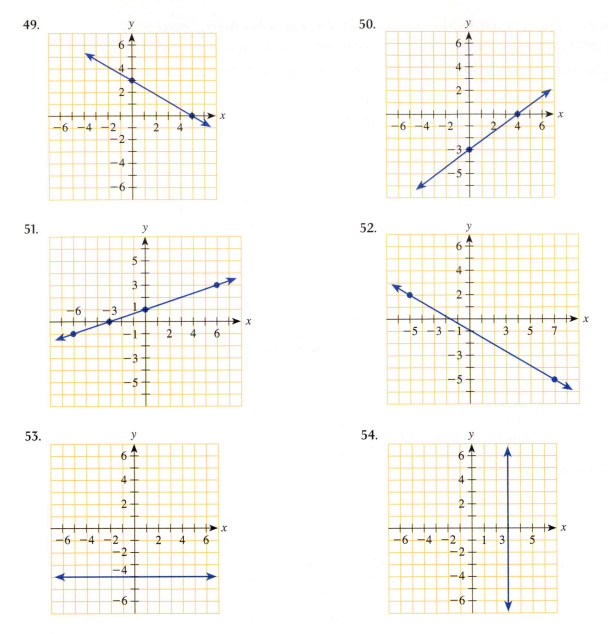

49.   50.   51.   52.   53.   54.

55. What is the slope of a line that is parallel to the line whose equation is $y = 3x - 8$?

56. What is the slope of a line that is parallel to the line whose equation is $4y - 5x = 12$?

57. Write an equation of the line that passes through the point $(-3, 0)$ and is parallel to the line whose equation is $y = -x + 4$.

58. Write an equation of the line that passes through the point $(2, -1)$ and is parallel to the line whose equation is $4x - 3y = 6$.

In Exercises 59–68, *round off to the nearest hundredth when necessary.*

59. A sidewalk food vendor knows that selling 80 franks costs a total of $73 and selling 100 franks costs a total of $80. Assume that the total cost c of the franks is linearly related to the number f of franks sold.

   (a) Write an equation relating the total cost of the franks to the number of franks sold.

   (b) Find the cost of selling 90 franks.

   (c) How many franks can be sold for a total cost of $90.50?

(d) The costs that exist even if no items are sold are called the *fixed costs*. Find the vendor's fixed costs. [*Hint:* If no franks are sold then $f = 0$.]

60. A carpenter knows that he can make 8 picture frames at a total cost of $46 and 10 picture frames at a total cost of $55. Assume that the total cost c of the picture frames is linearly related to the number n of picture frames made.

   (a) Write an equation relating the total cost of the picture frames to the number of picture frames made.

   (b) Find the cost of making 20 picture frames.

   (c) How many picture frames can be made for a total cost of $91?

   (d) The costs that exist even if no items are produced are called the *fixed costs*. Find the carpenter's fixed costs. [*Hint:* If no picture frames are made then $n = 0$.]

61. In physiology, a jogger's heart rate N, in beats per minute, is related linearly to the jogger's speed s. A certain jogger's heart rate is 80 beats per minute at a speed of 15 ft/sec and 82 beats per minute at a speed of 18 ft/sec.

   (a) Write an equation relating the jogger's speed and heart rate.

   (b) Predict this jogger's heart rate if she jogs at a speed of 20 ft/sec.

   (c) According to the equation obtained in part (a), what is the jogger's heart rate at rest? [*Hint:* At rest the jogger's speed is 0.]

62. A business buys a piece of machinery for $12,800. Suppose that the value of the machinery depreciates linearly to a value of $2,400 in 8 years.

   (a) Write an equation relating the value V of the machinery after n years. [*Hint:* The value of the machinery in year 0 is $12,800.]

   (b) Find the value of the machine in year 5.

   (c) How long will it take for the machine to depreciate to a value of zero?

63. A gourmet food shop sells custom blended coffee by the ounce. Suppose that 1 oz sells for $1.70 and 2 oz sells for $3.20. Assume that the cost c of the coffee is linearly related to the number of ounces n purchased (where $n \geq 1$).

   (a) Write an equation relating the cost of the coffee to the number of ounces purchased.

   (b) What would be the cost of 3.5 oz of coffee?

   (c) Suppose a package of coffee is marked at $6.50 but has no indication of how much coffee it contains. Determine the number of ounces of coffee this package contains.

   (d) Sketch a graph of this equation for $n \geq 1$ using the horizontal axis for n and the vertical axis for c.

64. Suppose that the monthly cost of maintaining a car is linearly related to the number of miles driven during that month, and that for a particular model, on average, it costs $160 per month to maintain a car driven 540 miles and $230 per month to maintain a car driven 900 miles.

   (a) Write an equation relating the cost c of monthly maintenance and the number n of miles driven that month.

   (b) What would be the average cost of maintaining a car driven 750 miles per month?

   (c) Based on this relationship, if your average monthly maintenance is $250, how many miles do you drive per month?

   (d) Sketch a graph of this equation using the horizontal axis for n and the vertical axis for c.

(e)   What is the c-intercept of this graph? How would you interpret the fact that even if no miles are driven there is still a monthly maintenance cost?

65.   Assume that the maximum speed your car can travel on a hill, whether going uphill or downhill, is linearly related to the steepness of the hill as measured by the angle the hill makes with a horizontal line. Suppose that your maximum speed on a 6° incline is 62 mph and your maximum speed on a 2° decline (that is, an incline of −2°) is 115 mph.

(a)   Write an equation relating the maximum speed s of the car and the number n of degrees of the incline.

(b)   What would be the maximum speed on a 4.5° incline?

(c)   What would be the maximum speed on a −2.8° decline?

(d)   Sketch a graph of this equation using the horizontal axis for n and the vertical axis for s.

(e)   What is the s-intercept of this graph? What is the significance of this s-intercept?

(f)   What is the n-intercept of this graph? What is the significance of this n-intercept?

66.   Nature experts tell us that crickets can act as an outdoor thermometer because the rate at which a cricket chirps is linearly related to the temperature. At 60°F they make an average of 80 chirps per minute, and at 68°F they make an average of 112 chirps per minute.

(a)   Write an equation relating the average number of chirps c and the temperature t.

(b)   What would be the average number of chirps at 90°F?

(c)   If you hear an average of 88 chirps per minute, what is the temperature?

(d)   Sketch a graph of this equation using the horizontal axis for t and the vertical axis for c.

(e)   What is the t-intercept of this graph? What is the significance of this t-intercept?

67.   Bridges (and many concrete highways) are constructed with "expansion joints," which are small gaps in the roadway between one section of the bridge and the next. These expansion joints allow room for the roadway to expand during hot weather. Suppose that a bridge has a gap of 1.5 cm when the air temperature is 24°C, that the gap narrows to 0.7 cm when the air temperature is 33°C, and that the width of the gap is linearly related to the temperature.

(a)   Write an equation relating the width of the gap w and the temperature t.

(b)   What would be the width of a gap in this roadway at 28°C?

(c)   At what temperature would the gap close completely?

(d)   If the temperature exceeds the value found in part (c) that causes the gap to close, it is possible that the roadway could buckle. Is this likely to occur? Explain.

68.   An efficiency expert finds that the average number of defective items produced in a factory is approximately linearly related to the average number of hours per week the employees work. If the employees average 34 hours of work per week, then an average of 678 defective items are produced. If the employees average 45 hours of work per week, then an average of 834 defective items are produced.

(a)   Write an equation relating the average number of defective items d and the average number of hours h the employees work.

(b)   What would be the average number of defective items if the employees average 40 hours per week?

(c) According to this relationship, how many hours per week would the employees need to average in order to reduce the average number of defective items to 500? Do you think this is a practical goal? Explain.

② QUESTIONS FOR THOUGHT

69. Suppose you are asked to write an equation of a line satisfying certain conditions. Explain how the given information influences your choice of whether to use the point–slope form or the slope–intercept form to write the equation.

70. A company has recently set up a Web site on the Internet and keeps track of the number of visits to the Web site (called *hits*) each day and the number of calls placed to the company's 800 number each day. The accompanying graph contains the data collected. The horizontal axis, labeled *h*, represents the number of daily hits to the Web site, and the vertical axis, labeled *n*, represents the number of phone calls each day to the company's 800 number.

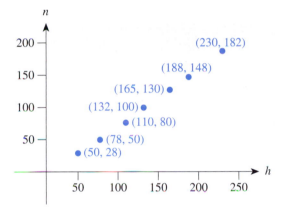

(a) Using the given data, write an equation of a line that you think best fits the given data.

(b) Check each data point in your equation by substituting *h* and seeing how close the equation's value of *n* compares with the actual value of *n*.

(c) Use the equation obtained in part (a) to predict the number of phone calls on a day on which there are 150 hits to the Web site. When we predict within the range of the given *h* values we say that we are *interpolating* from the given data.

(d) If it turns out that on the day on which there were 150 hits there were 130 phone calls, discuss how well this fits the equation. You may want to refer to the results you obtained in part (b).

(e) Use the equation obtained in part (a) to predict the number of phone calls on a day on which there are 225 hits to the Web site. When we predict outside the range of the given *h* values we say that we are *extrapolating* from the given data.

71. This exercise discusses the relationship between the slopes of perpendicular lines.

(a) Sketch the graphs of $y = 2x + 4$ and $y = -\frac{1}{2}x + 4$ on the same coordinate system.

(b) Based on your graph, does it appear that these lines are perpendicular?

(c) What is the relationship between the slopes of these two lines?

(d) It is a fact that the slopes of perpendicular lines are negative reciprocals of each other (provided that neither of the lines is vertical). What is the slope of a line perpendicular to the line whose equation is $y = \frac{2}{5}x + 7$?

◇ MINI-REVIEW

Solve each of the following problems algebraically. Be sure to label what the variable represents.

72. The width of a rectangle is 4 less than $\frac{2}{3}$ its length. If the perimeter of the rectangle is 3 times the length, find its dimensions.

73. A beaker contains 40 ml of a 60% alcohol solution. What percentage alcohol solution must be added to produce 70 ml of a 45% solution?

CHAPTER 5   SUMMARY

After having completed this chapter, you should be able to:

1. Determine whether an ordered pair satisfies a given equation (Sections 5.1, 5.2).

    For example: Does the point $(3, -6)$ satisfy the equation $2x - 10 = y - 3$?
    To check, we substitute $x = 3$ and $y = -6$ into the equation:

    $$2x - 10 = y - 3$$
    $$2(3) - 10 \stackrel{?}{=} -6 - 3$$
    $$-4 \neq -9 \qquad \text{Therefore, } (3, -6) \text{ does not satisfy the equation.}$$

2. Complete an ordered pair for an equation, given one of the coordinates (Sections 5.1, 5.2).

    For example: Complete the ordered pair $(-2, \quad)$ for the equation $x - 5y = 3$.
    We substitute the value $x = -2$ into the equation and solve for y:

    $$x - 5y = 3$$
    $$-2 - 5y = 3$$
    $$-5y = 5$$
    $$y = -1 \qquad \text{Therefore, the ordered pair is } \boxed{(-2, -1)} \, .$$

3. Sketch the graph of a linear equation using the intercept method (Section 5.2).

    For example: Sketch the graph of $3x - 2y = 6$.
    To find the x-intercept, we set $y = 0$ and solve for x:

    $$3x - 2y = 6$$
    $$3x - 2(0) = 6$$
    $$3x = 6$$
    $$x = 2 \qquad \text{Therefore, the } x\text{-intercept is 2. We plot the point } (2, 0).$$

    To find the y-intercept, we set $x = 0$ and solve for y:

    $$3x - 2y = 6$$
    $$3(0) - 2y = 6$$
    $$-2y = 6$$
    $$y = -3 \qquad \text{Therefore, the } y\text{-intercept is } -3. \text{ We plot the point } (0, -3).$$

For our "check point" we choose $x = 4$ and solve for y:

$$3x - 2y = 6$$
$$3(4) - 2y = 6$$
$$12 - 2y = 6$$
$$-2y = -6$$
$$y = 3 \qquad \text{Therefore, our check point is} \boxed{(4, 3)}.$$

We now plot the three points and draw the line passing through them (see Figure 5.31).

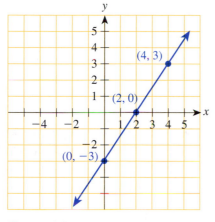

Figure 5.31
The graph of $3x - 2y = 6$

4. Use the definition of *slope* to find the slope of a line (Section 5.3).

   For example: Find the slope of the line passing through the points $(2, -3)$ and $(-1, 4)$.

   $$m = \frac{y_2 - y_1}{x_2 - x_1} = \frac{4 - (-3)}{-1 - 2} = \frac{7}{-3} = \boxed{-\frac{7}{3}}$$

5. Write an equation of a specified line (Section 5.4).

   For example: Write an equation of the line with slope $\frac{1}{2}$ that passes through the point $(-2, 4)$.

   Using the *point–slope form* for the equation of a straight line, we get

   $$y - y_1 = m(x - x_1)$$
   $$y - 4 = \frac{1}{2}(x - (-2))$$

   $$\boxed{y - 4 = \frac{1}{2}(x + 2)}$$

CHAPTER 5 REVIEW EXERCISES

In Exercises 1–6, plot the given points in a rectangular coordinate system.

1. $(-4, 3)$ 　　　　　　　　2. $(8, 0)$

3. $(6, -2)$ 　　　　　　　　4. $(-7, -9)$

5. $(0, 0)$ 　　　　　　　　6. $(0, -4)$

In Exercises 7–12, fill in the missing coordinate for the ordered pairs of the given equations.

7. $2x + 4y = 14$; (3,)

8. $2x - 3y = 8$; (0,)

9. $x + 2y = 4$; (, −3)

10. $2x - 3 = 4y$; (, 5)

11. $x - y = 0$; (0,)

12. $x - 2y = 7$; (, −1)

In Exercises 13–26, graph the given equation in a rectangular coordinate system, using the intercept method.

13. $x - y = 8$

14. $x + y = 8$

15. $3x + 7y = 21$

16. $2y - 6x = 18$

17. $y = 3x + 2$

18. $3x + 2y = 5$

19. $x - 2y = 4$

20. $x = y$

21. $x = 4$

22. $y = -3$

23. $3x + 2y = 12$

24. $y + x = 5$

25. $3y = 6$

26. $-2x = 5$

In Exercises 27–32, find the slope of the line passing through the given points.

27. (2, 5) and (1, 7)

28. (−1, 4) and (2, −3)

29. (−3, −2) and (3, 2)

30. (6, 2) and (6, 5)

31. (1, 8) and (−1, 8)

32. (2, −1) and (−3, −11)

In Exercises 33–38, write an equation of the line satisfying the given conditions.

33. Passing through (3, 2) with slope 4

34. Passing through (−5, −1) with slope $\frac{3}{4}$

35. Passing through the points (2, −6) and (1, 0)

36. Passing through the points (−7, 2) and (0, 3)

37. Horizontal line passing through (3, 8)

38. Vertical line passing through (3, 8)

39. What is the slope of a line whose equation is
$$y = -\frac{3}{4}x - 7?$$

40. What is the slope of a line whose equation is
$$3x - 8y + 2 = 0?$$

41. Write an equation of the line that passes through the point (−3, 5) and is parallel to the line whose equation is $y = 5x - 1$.

42. Write an equation of the line that passes through the point (3, 0) and is parallel to the line whose equation is $2y + 5x = 8$.

In Exercises 43–44, write an equation of the line whose graph is given.

43.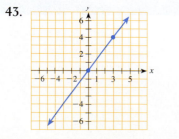

44.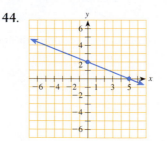

45. A quality-control inspector finds that the number of defective items that a worker produced is linearly related to the number of overtime hours that worker put in. If the worker put in 8 overtime hours, there were 12 defective items produced, and if the worker put in 10 overtime hours, there were 17 defective items produced. Write an equation relating the number of overtime hours and the number of defective items, and use this equation to predict the number of defective items you would expect to find if the worker puts in 20 overtime hours.

46. A student finds a linear relationship between the number of hours she studies for a math test and the grade she receives. If she studies for 2 hours she gets a grade of 72, and if she studies for 5 hours she gets a grade of 91. Write an equation relating the number of hours she studies and the exam grade she receives, and use this equation to predict her exam grade if she studies for 4 hours.

 CHAPTER 5   PRACTICE TEST

1. Determine whether the point $(-3, -4)$ satisfies the equation.

$$\frac{2y - x}{5} = x - y$$

2. Find the missing coordinates for the ordered pair of the given equation.

   (a) $3x - y = 12$; (5,)

   (b) $x - 3y = 10$; (, -2)

   (c) $4x + 5y = 15$; (0,)

3. Sketch the graphs of each of the following in a rectangular coordinate system using the intercept method.

   (a) $3x - 5y = 15$

   (b) $y = 3x - 7$

   (c) $y + 2x = 0$

   (d) $x = 4$

   (e) $y = -3$

4. Find the slope of the line passing through the points $(-1, -5)$ and $(2, -3)$.

5. Write an equation of the line with slope $\frac{4}{3}$ that passes through the point $(4, -1)$.

6. Write an equation of the line passing through the points $(2, 5)$ and $(4, 0)$.

7. Write the equations of the horizontal and vertical lines that pass through the point $(3, 5)$.

8. What is the slope of the line whose equation is $5x - 7y + 10 = 0$?

9. Write an equation of the line that passes through the point $(-4, 0)$ and is parallel to the line whose equation is $y = -3x + 4$.

10. Write an equation of the line whose graph appears in the following figure.

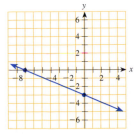

11. A charity solicitation firm finds that the amount of charity pledged is linearly related to the number of phone calls made. If they make 15 phone calls they get $80 in pledges, and if they make 21 phone calls they get $115 in pledges. Write an equation relating the number of phone calls made and the amount of charity pledged, and use this equation to predict the amount of charity pledged if they make 30 phone calls.

CHAPTER 6

Interpreting Graphs and Systems of Linear Equations

In Chapter 5 we discussed the equations and graphs of straight lines. We pointed out that the equation of a line describes the particular relationship that x- and y-coordinates must satisfy in order for a point to be on the line. With this background, this chapter begins with more general graphs and shows how they can be used to describe a relationship between two quantities. The chapter then continues with a discussion of systems of linear equations and their applications to verbal problems.

6.1   Interpreting Graphs

Most of us have had some experience with graphs outside a mathematics class—graphs often accompany newspaper and magazine articles or appear in textbooks from a variety of disciplines. A graph is often used to give us a convenient picture illustrating some relationship between two quantities. The graph gives us a visual impression of the relationship, and allows us to give more detail about the nature of the particular relationship under discussion. In mathematics, a graph is a legitimate tool that helps us to visualize a trend or relationship between two quantities. While an equation can be extremely useful in finding exact values that satisfy a relationship, general trends may be difficult to observe from the equation. On the other hand, a graph can be used not only to get reasonable estimates of these values, but also has the additional advantage of allowing us to "see" the nature of the relationship.

As we discussed in the previous chapter, an equation such as $y = 2x - 4$ describes a particular relationship between x and y, and the graph of all the ordered pairs satisfying this relationship is a straight line. We can describe this relationship verbally by saying that in order for a point to lie on this line its y-coordinate must be 4 less than twice its x-coordinate. The graph of this equation (which appears in Figure 6.1) describes this relationship pictorially.

Figure 6.1
The graph of $y = 2x - 4$

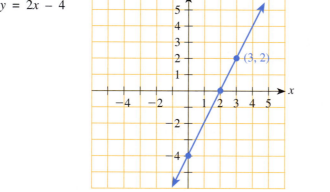

In particular, we can see that the point $(3, 2)$ is on the graph, so we know that when $x = 3$ we have $y = 2$. We can also verify this by substituting the values $x = 3$, $y = 2$ into the equation $y = 2x - 4$.

In general, we can see that the line is rising. (Remember that we always describe a graph as we move from left to right.) We can restate the fact that the line is rising by saying that "as x increases, y also increases." Thus the graph shows us what happens to y as x changes.

EXAMPLE 1   Maureen jogs on a regular basis. She finds that the number of miles she can jog is related to the temperature. The graph in Figure 6.2 illustrates this relationship as the temperature varies from 80°F to 90°F. The horizontal axis, labeled t, represents the temperature in degrees Fahrenheit, and the vertical axis, labeled m, represents the average number of miles she jogs. Use this graph to describe this relationship.

Figure 6.2

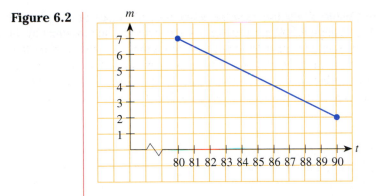

Solution   Since the line segment is falling, we can see that as the temperature increases, the number of miles Maureen can jog steadily decreases. In particular, whereas she can jog 7 miles when the temperature is 80°F, she can jog only 2 miles when the temperature is 90°F. ∎

EXAMPLE 2

The graph in Figure 6.3 illustrates how the profit a company expects on the sale of a certain item depends on its selling price. The horizontal axis is labeled s and represents the selling price of the item (in dollars). The vertical axis is labeled P and represents the profit (in thousands of dollars) that the company earns. Use the information given in this graph to describe how the profit relates to the selling price.

Figure 6.3

Solution   Examining the graph, we can draw the following conclusions:

1. The point $(0, -100)$ is on the graph. This means that when the selling price is 0 ($s = 0$), the profit is $-\$100,000$ ($P = -100$). We can interpret this negative profit to mean that if the item is given away free, the company would have a *loss* of $100,000. This loss may be due to fixed costs such as rent and taxes that must be paid regardless of what the selling price is or how many items are sold.

2. As the selling price starts to increase, the profit increases as well. The point $(10, 0)$ on the graph tells us that when the selling price is $10 ($s = 10$) the profit is 0 ($P = 0$). (This is often called the *break-even point.*)

3. As the selling price continues to increase up to $30, the profit continues to increase as well. The highest point on the graph is $(30, 80)$. This highest point corresponds to the *maximum profit* the company can earn, which is $80,000.

4. As the selling price increases beyond $30, the profit decreases until, when the selling price is $50, the profit is again 0. This can be explained by the fact that once the selling price gets too high, fewer people will buy the item, thus decreasing the profit.

Again, this graph gives a "snapshot" of the relationship between the selling price and the expected profit. A graph offers a visual way of presenting a great deal of information about a relationship. ∎

EXAMPLE 3

Jani is traveling on business in her car. Suppose that the graph in Figure 6.4 describes her day's travel. The horizontal axis is labeled t and represents the number of hours since Jani began her trip. The vertical axis is labeled d and represents the distance (in miles) that Jani is from her home. Use the information given in this graph to describe her day as best you can.

Figure 6.4

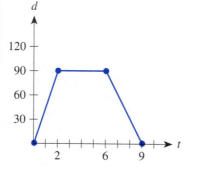

Solution

Examining the graph, we can draw the following conclusions:

1. The point $(0, 0)$ is on the graph. This means that when $t = 0$, $d = 0$. Keeping in mind that t is the number of hours since Jani began her trip, $t = 0$ means that Jani has not yet begun her trip, and since d is the number of miles that Jani is from home, $d = 0$ means that she is 0 miles from home. In other words, Jani is starting her trip from home.

2. The line is rising from $(0, 0)$ to $(2, 90)$, which means that during this 2-hour period Jani's distance from home is *increasing*.

3. The point $(2, 90)$ is on the graph, which means that when $t = 2$, $d = 90$. In other words, after 2 hours, Jani is 90 miles from home. Since Jani covers a distance of 90 miles in 2 hours, her average speed during her first 2 hours of travel is

$$\text{Average speed} = \frac{90 \text{ miles}}{2 \text{ hours}} = 45 \text{ mph}$$

4. The next portion of the graph is horizontal, which means that the distance is not changing. The distance from home remains 90 miles for the next 4 hours. Possibly Jani is at a business meeting that lasts for 4 hours.

5. The line is falling from $(6, 90)$ to $(9, 0)$, which means that during this 3-hour period Jani's distance from home is *decreasing*.

6. The point $(9, 0)$ means that after 9 hours Jani's distance from home is again 0. After 9 hours Jani has returned to her home.

7. During the 3-hour period from $t = 6$ to $t = 9$, Jani covers a distance of 90 miles in 3 hours, and so her average speed during her return trip home is 30 mph.

The main point here is that the graph allows us to see at a glance some important aspects of Jani's trip. ∎

EXAMPLE 4

The accompanying graph illustrates two possible cellular phone plans that a company may choose for its salespeople. The line labeled R represents the monthly cost C (in dollars) under the regular plan for m total minutes of air time. The line labeled S represents the monthly cost C (in dollars) under the special plan for m total minutes of air time.

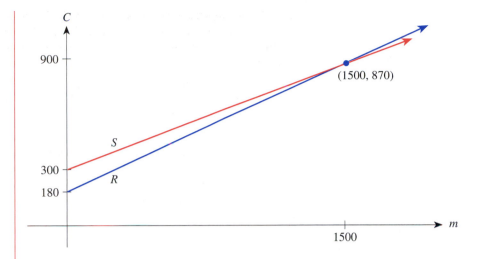

(a) Use the graph to describe how the monthly costs differ under the two plans.

(b) Explain the significance of the point of intersection of the two lines.

Solution

(a) We can see that if no minutes of air time are used (that is, $m = 0$, which corresponds to the C intercept), then the regular plan costs $180 and the special plan costs $300.

As m starts to increase we see that line S is above line R, which means that the C values on line S are larger than the C values on line R. This means that as salespeople start to use air time the special plan is more expensive than the regular plan.

Eventually, as m gets larger, line S falls below line R, which means that if enough minutes of air time are used the special plan costs less than the regular plan.

(b) The fact that the lines cross at the point (1500, 870) means that if a total of 1,500 minutes of air time is used, then both plans cost the same: $870.

This also tells us that if fewer than 1,500 minutes of air time are used, then the regular plan is cheaper, but if more than 1,500 minutes of air time are used, then the special plan is cheaper. ■

EXERCISES 6.1

1. The following two graphs describe certain characteristics of two aircraft, labeled A and B: their age, cruising speed, size, and cruising range.

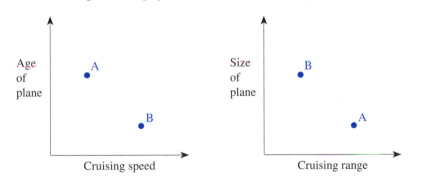

(a) Which plane is larger?

(b) Which plane has a faster cruising speed?

(c) Which plane has the more extended cruising range?

(d) Which plane is newer?

2. The two graphs at the top of page 260 describe certain characteristics of three people labeled A, B, and C: their height, weight, age, and income.

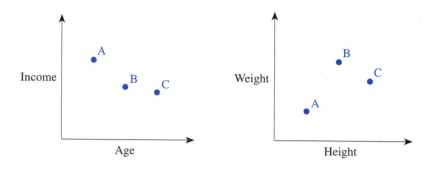

(a) List the people in increasing order of their weight.

(b) List the people in increasing order of their income.

(c) List the people in decreasing order of their height.

(d) List the people in increasing order of their age.

3. The following graph illustrates the level of telephone usage in a small town. The horizontal axis, which is labeled h, represents the hour of the day. That is, 0 = midnight, 1 = 1 A.M., 14 = 2 P.M., etc. The vertical axis, labeled n, represents the average number of phone calls (in thousands).

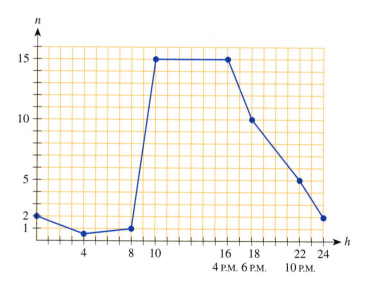

(a) What is the average number of phone calls made at midnight?

(b) At what time is the average number of phone calls a minimum?

(c) What is the average number of phone calls made at 6:00 P.M.?

(d) During what period of time is the average number of phone calls decreasing?

(e) At what time is the average number of phone calls a maximum?

(f) During what period of time does the average number of phone calls remain constant?

4. The accompanying graph illustrates the level of electrical power usage in a small town during a 1-year period. The horizontal axis, labeled m, represents the month of the year. That is, 1 to 2 = January, 2 to 3 = February, ..., 12 to 13 = December. The vertical axis, labeled K, represents the number of megawatts of electrical power being used.

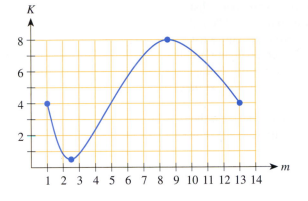

(a) During which month does the power usage reach a maximum?

(b) During which month does the power usage reach a minimum?

(c) During which month does the power usage increase?

(d) During which months does the power usage decrease?

5. The following graph illustrates the relationship between air temperature and altitude. The horizontal axis, labeled a, represents the altitude in miles. The vertical axis, labeled T, represents the temperature in degrees Celsius.

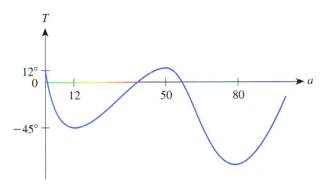

(a) It is commonly believed that the temperature drops steadily as the altitude increases. Does this graph confirm or contradict this belief?

(b) If a balloon with a temperature-measuring device is sent aloft, between what altitudes will the temperature be decreasing?

(c) As the balloon rises from an altitude of 45 miles to an altitude of 55 miles, how is the temperature changing?

6. The accompanying graph illustrates the level of a certain substance in the blood after a medication containing this substance is taken. The horizontal axis, labeled m, represents the number of minutes after the medication is taken. The vertical axis, labeled A, represents the amount of the substance (measured in milligrams) present in the blood.

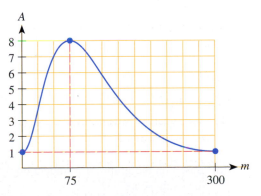

(a) How many milligrams of the substance are present in the blood before any medication is taken? This is the body's normal level of this substance. [*Hint:* Find A when $m = 0$.]

(b) How long does it take until the amount of this substance reaches its maximum level in the blood?

(c) How long after taking the medication does the level of the substance return to normal?

7. In a psychology experiment, a group of students was asked to memorize a list of 20 nonsense syllables (meaningless three-letter words such as "brg," "odu," etc.). After successfully demonstrating that they had memorized the entire list, all the students were retested at various time intervals. The accompanying graph (often called a *retention curve*) illustrates the relationship between the average number of words the group remembered and the number of hours later that they were retested.

   The horizontal axis, labeled t, represents the number of hours later that the students were retested. The vertical axis, labeled n, represents the average number of words the group remembered.

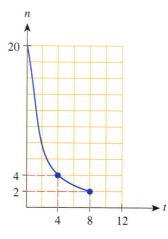

   (a) What does the point $(0, 20)$ on the graph represent?

   (b) Do the students remember more words after 4 hours or after 5 hours?

   (c) During which 2-hour period do the students forget the most words?

   (d) How many words are forgotten during the first 4 hours? During the next 4 hours?

8. The accompanying graph illustrates the relationship between the number of hours a student studies for a math exam and the score on the exam.

   The horizontal axis, labeled h, represents the number of hours the student studied. The vertical axis, labeled s, represents the score on the exam.

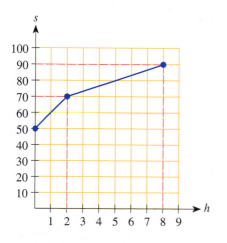

   (a) Based on the graph, if a student does not study for the exam, what would the exam score be?

   (b) If a student studies 2 hours for the exam, how much would the exam score increase?

(c) Does the score increase at the same rate if the study time is increased from 0 to 2 hours as it does if the study time is increased from 6 to 8 hours?

(d) Does an additional 2 hours of studying always increase the score by the same amount? Explain.

9. The deer population in a certain region over a 10-year period is illustrated in the accompanying graph.

    The horizontal axis, labeled t, represents the year: $t = 0$ represents 1990; $t = 1$ represents 1991, and so on. The vertical axis, labeled N, represents the number of deer (in hundreds).

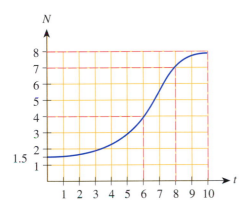

(a) How many deer are present in the region at the beginning of this 10-year period?

(b) During which 2-year period does the deer population appear to be increasing most rapidly?

(c) As we approach the end of this 10-year period, it appears that the number of deer is leveling off at a certain number. This number is called the environment's *carrying capacity* for deer. Based on this graph, what is the carrying capacity of this environment for deer?

10. Animals are being taught to press a bar to get food after a certain time interval. After being trained in this way for a prolonged period, the animals are then removed from this situation for a while and later returned to the bar. The following graph illustrates the relationship between the time delay in receiving the food and the number of times the animal presses the bar during a 5-minute period after being returned to the bar.

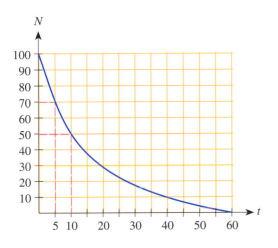

    The horizontal axis, labeled t, represents the number of seconds between the time the animal presses the bar and the food is received. The vertical axis, labeled N, represents the number of times the animal presses the bar during the 5-minute period.

(a) If the animal is trained by receiving food immediately, how many times will it press the bar?

(b) If the food is delayed 5 seconds, how many fewer times will the animal press the bar?

(c) If the food is delayed by 60 seconds, how many times will the animal press the bar?

(d) If the animal presses the bar 50 times, how many seconds was the food delayed?

11. A company has a choice of two machines it can purchase. The accompanying graph illustrates the value of the two machines in the years following their purchase. The line labeled A represents the value V (in dollars) of one machine n years after it is purchased. The line labeled B represents the value V (in dollars) of the other machine n years after it is purchased.

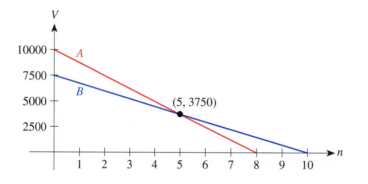

(a) What is the purchase price of each of the machines? [*Hint:* We are asking for the price in year 0.]

(b) How many years does it take each of the machines to depreciate to a value of zero dollars?

(c) Is there any time when the two machines have the same value? If so, when?

(d) Describe the comparative values of the two machines over the period of time indicated in the graph.

12. The accompanying graph illustrates two possible wage plans for an employee that include a weekly salary plus commission. The line labeled L represents the weekly wages W (in dollars) under the first plan if the employee sells n items. The line labeled M represents the weekly wages W (in dollars) under the second plan if the employee sells n items.

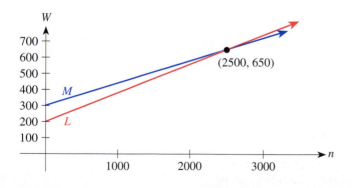

(a) If no items are sold, find the weekly salary under each plan. This is called the *base salary*.

(b) Under what conditions will the wages received under the two plans be the same?

(c) Describe how an employee should decide whether to choose plan L or plan M.

13. The following graph illustrates how the price of a stock fluctuated from the closing price on Monday through the closing price on Friday during a particular week. The horizontal axis represents the days and the vertical axis represents the price of the stock in dollars.

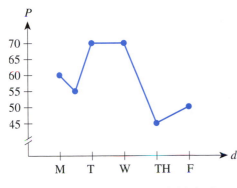

(a) On what day did the stock hit its lowest price? What was this lowest price?

(b) On what day did the price of the stock both rise and fall?

(c) On what day did the price of the stock change the most? By how much?

(d) How much did the price change from its closing price on Tuesday to its closing price on Friday?

14. The following graph illustrates how the price of a stock fluctuated from 9:00 A.M. to 4:00 P.M. during a particular day. The horizontal axis represents the hour of day and the vertical axis represents the price of the stock in dollars.

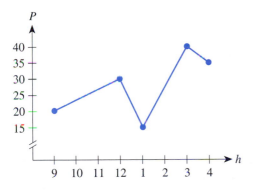

(a) During which hour did the price change the most? By how much?

(b) If you bought 100 shares of this stock at 9:00 A.M. and sold those 100 shares at 1:00 P.M., did you make money or lose money? How much?

(c) If you wanted to earn as much as possible on the purchase and sale of 100 shares, at what time should you have bought them and sold them? How much would the profit be?

In Exercises 15 and 16, sketch a graph that represents the scenario described in the exercise. Be sure to clearly label any variables and the coordinate axes. Keep in mind that various graphs may be drawn to represent each situation.

15. Suppose that the temperature of a metal bar starts at 40°C. Over the course of 30 minutes the temperature of the bar steadily increases to a temperature of 160°C. It remains at that temperature for 45 minutes and then steadily cools off to a temperature of 60°C over the next 20 minutes.

16. The price of a certain stock starts the day at $15 per share. Over the first 2 hours of trading, the price of the stock steadily declines to $13 per share. It remains at that price for 3 hours and then declines to $11.50 per share over the next hour.

MINI-REVIEW

17. *Simplify.* $\dfrac{6x^2}{25yz^3} \div (15xyz)$

18. *Solve for t.* $\dfrac{t+3}{5} - \dfrac{t-2}{4} \le 2$

19. A total of $1,185 was spent on 36 pairs of slacks. Some were dress slacks costing $40 each, and the rest were casual slacks costing $25 each. How many of each were bought?

6.2 Solving Systems of Linear Equations: Graphical Method

Let us begin by considering the following situation.

EXAMPLE 1

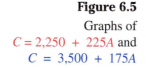

A homeowner wants to build a family room as an extension to her house. She receives bids from two competing contractors. The ILF (It Lasts Forever) Construction Company charges a flat fee of $2,250 for architectural plans and obtaining permits, plus $225 per square meter for the actual construction. The WBB (We Build Better) Construction Company charges a flat fee of $3,500 plus $175 per square meter. For what size room will the two companies charge the same price?

Solution

We can analyze the question as follows. The cost, C, of building the room can be expressed in terms of the area, A, of the room:

Cost = Flat fee + Actual building costs

= Flat fee + (Cost per square meter) (Number of square meters)

For ILF this would be: $C = 2,250 + 225A$

For WBB this would be: $C = 3,500 + 175A$

Each of these equations is a first-degree equation in two variables (a linear equation) and hence has as its graph a straight line. Asking for what area, A, the costs will be the same is equivalent to asking where the lines intersect.

At the point of intersection the C value on both lines is the same.

We sketch the graphs of each line using the methods discussed in Chapter 5, and then we estimate the point of intersection (see Figure 6.5). (Naturally, we draw the graph only for $A \ge 0$, to the right of the C-axis, since a negative area makes no sense.) It appears that the graphs intersect when $A = 25$.

Figure 6.5

Graphs of
$C = 2,250 + 225A$ and
$C = 3,500 + 175A$

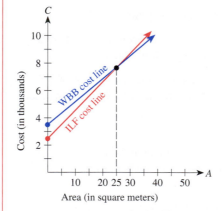

Check: For ILF: The cost of building a 25-square-meter room is

$$C = 2,250 + 25(225) = 2,250 + 5,625 = \$7,875$$

For WBB: The cost of building a 25-square-meter room is

$$C = 3,500 + 25(175) = 3,500 + 4,375 = \$7,875$$

Thus, the answer to the example is 25 square meters.

For A < 25, the ILF line is below the WBB line; for A > 25, the WBB line is below the ILF line.

A few comments are in order here about the "graphical solution" obtained in Example 1. What can be said in favor of this graphical solution is that not only did we answer the question that was asked, but in addition, by looking at the graphs in Figure 6.5, we can see that for a room whose area is less than 25 square meters the ILF Company is cheaper, whereas for a room whose area is greater than 25 square meters the WBB Company is cheaper.

While this graphical method gives us valuable information about the two equations, depending on our eyes to determine a point of intersection can be rather inaccurate. What if the answer to the last example had actually been $A = 24.6$ square meters?

The remainder of this section discusses the graphical method described in Example 1, and the next section describes algebraic methods that allow us to find the exact solutions to such problems.

In Sections 5.1 and 5.2 we saw that a linear equation in two variables has infinitely many solutions. If we have two such linear equations, a reasonable question to ask is "Are there and can we find solutions (that is, ordered pairs) that satisfy both equations?" Such an ordered pair is called a **simultaneous solution** to the two equations. For example, is there and can we find an ordered pair that satisfies the two equations:

$$\begin{cases} x + y = 6 \\ x - y = 2 \end{cases}$$

This is called a **system of linear equations.** The large brace on the left indicates that the two equations are to be solved simultaneously.

With a little bit of trial and error we might find that the ordered pair (4, 2) works in both equations. (Check it!) However, before we proceed to discuss a systematic method for finding this solution, let's see what we can expect from such a system of equations.

If we keep in mind that the graph of a linear equation is a straight line, then asking for a solution to a system of two linear equations in two variables is the same as asking "What point do the two lines have in common?" In other words, where do the two lines intersect? Basically, there are three possibilities.

The first possibility is that the two lines intersect in exactly one point. This point is the unique solution to the system of equations. Such a system is called **consistent** (see Figure 6.6).

The second possibility is that the two lines are parallel, and so they never intersect. In this case there are no solutions to the system of equations. Such a system is called **inconsistent** (see Figure 6.7).

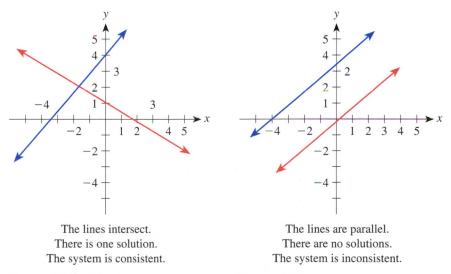

The lines intersect.	The lines are parallel.
There is one solution.	There are no solutions.
The system is consistent.	The system is inconsistent.

Figure 6.6   **Figure 6.7**

The third possibility is that the graphs of the two equations coincide (they are both the same line). In this case, there are infinitely many solutions. All the points that satisfy one equation also satisfy the other. Such a system is called **_dependent_** (see Figure 6.8).

Figure 6.8

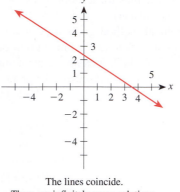

The lines coincide.
There are infinitely many solutions.
The system is dependent.

This analysis leads us to the method of solution we will discuss in this section. It is called the **_graphical method._** In order to solve a system of linear equations, we will graph the two equations on the same coordinate axes (by the method we learned in Section 5.2) and then "read off" the point of intersection from our picture. Once we have what we think is the solution, we will check it by substituting the values into both equations.

EXAMPLE 2

Solve the system of equations. $\begin{cases} x - y = 2 \\ x + y = 6 \end{cases}$

Solution

We graph both lines by the intercept method. (In order to shorten the solution a bit, we leave the finding of a check point to the student.)

$$x - y = 2$$

To get the x-intercept, we set $y = 0$ and solve for x. We get 2 as the x-intercept. Therefore, the line passes through (2, 0).

To get the y-intercept, we set $x = 0$ and solve for y. We get -2 as the y-intercept. Therefore, the line passes through (0, -2).

$$x + y = 6$$

Similarly, for the second equation we get 6 as the x-intercept and 6 as the y-intercept.

With this information we graph each of the straight lines (see Figure 6.9). We can see from the picture that the lines appear to cross at the point (4, 2). As we verified before, this point does satisfy both equations.

Figure 6.9

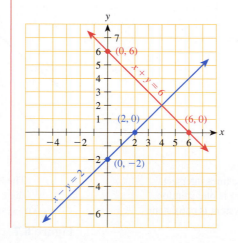

EXAMPLE 3   Solve the following system of equations. $\begin{cases} x + 2y = 2 \\ 2x + 4y = 0 \end{cases}$

Solution   Again we graph both lines by the intercept method.

$$x + 2y = 2$$

To get the x-intercept, we set $y = 0$ and solve for x. Thus, the x-intercept is 2.

To get the y-intercept, we set $x = 0$ and solve for y. Thus, the y-intercept is 1.

$$2x + 4y = 0$$

To get the x-intercept, we set $y = 0$ and solve for x. Thus, the x-intercept is 0. This is also the y-intercept since the line passes through the origin.

Consequently, we need to find a second point on the line. We choose $x = 4$ and solve for y. We get $y = -2$. So our second point is $(4, -2)$.

With this information we graph each of the straight lines, as shown in Figure 6.10. From the graphs we can see that the two lines seem to be parallel, which means that the system of equations has no solution and therefore is inconsistent.

Recall that in the last section we noted that if two distinct lines have the same slope they are parallel. If we take the two equations in this system and put them in the form $y = mx + b$, the first becomes $y = -\frac{1}{2}x + 1$ and the second becomes $y = -\frac{1}{2}x$. Thus, we can see that both lines have slope equal to $-\frac{1}{2}$; they are in fact parallel.

Figure 6.10

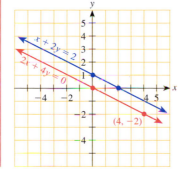

EXAMPLE 4   Solve the following system of equations. $\begin{cases} x - y = 5 \\ 3x = 3y + 15 \end{cases}$

Solution   If we find the intercepts for these two equations, we find that both lines have the same x-intercept of 5 and the same y-intercept of -5. Therefore, the lines pass through the points $(5, 0)$ and $(0, -5)$.

The graphs of these two equations appear in Figure 6.11. Since the graphs for the two equations are exactly the same line, there are infinitely many solutions to the system. Any ordered pair that makes one equation true also makes the other equation true. The system is therefore dependent.

Figure 6.11
The graph of $x - y = 5$ and
$3x = 3y + 15$

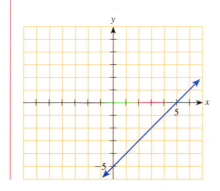

If we rewrite the first equation as $x = y + 5$, we can see that the second equation can be obtained from the first equation by multiplying both sides of the first equation by 3. Naturally, then, any ordered pair that satisfies one equation must also satisfy the other one. Thus, the solution set is

$$\{(x, y) \mid x - y = 5\}$$

Some questions concerning solving a system of equations by the graphical method may have occurred to you by now.

What happens if the actual solution to a system is the point $(2\frac{1}{5}, 4\frac{1}{3})$? Can we be expected to "read" such an answer from the graphs? These difficulties force us to conclude that the graphical method of solution can be imprecise.

Consequently, we need to find another method that will allow us to solve a system of linear equations algebraically—that is, by manipulating the two equations. We will do this in the next section.

EXERCISES 6.2

1. A potter has fixed costs of $80. It costs her $12 to produce each piece, and she sells each piece for $20. Therefore, her total cost C for producing n pieces is given by the equation $C = 80 + 12n$. Her total revenue R for producing n pieces is given by the equation $R = 20n$.

   (a) Sketch the graph of both equations on the same coordinate system, labeling the horizontal axis n.

   (b) The potter will break even when her costs and revenue are equal. Use the graph in part (a) to determine the point at which the two lines cross. This is called the *break-even* point.

   (c) How many pieces must she sell in order to break even?

2. A small software firm has a new computer game that it wants to market on a CD-ROM. The fixed costs are $2,400, and it costs $6 to produce each CD-ROM. Therefore, their total cost C for producing n CDs is given by the equation $C = 2,400 + 6n$. They plan on selling the CDs for $18 each. Their total revenue R for producing n CDs is given by the equation $R = 18n$.

   (a) Sketch the graph of both equations on the same coordinate system, labeling the horizontal axis n and the vertical axis C.

   (b) The company will break even when their costs and revenue are equal. Use the graph in part (a) to determine the point at which the two lines cross. This is called the *break-even* point.

   (c) How many CDs must they sell in order to break even?

In each of the following exercises, use the graphical method to solve the given system of equations for x and y.

3. $\begin{cases} x + y = 4 \\ x - y = 2 \end{cases}$

4. $\begin{cases} x - y = 5 \\ x + y = 3 \end{cases}$

5. $\begin{cases} 3x + y = 3 \\ x - y = 1 \end{cases}$

6. $\begin{cases} 4x - y = 4 \\ x - y = -2 \end{cases}$

7. $\begin{cases} 3x + y = 6 \\ 6x + 2y = 12 \end{cases}$

8. $\begin{cases} 5x + 2y = 10 \\ 10x + 4y = 40 \end{cases}$

9. $\begin{cases} x + y = 0 \\ x - y = 0 \end{cases}$

10. $\begin{cases} 2x - y = 0 \\ x - 2y = 0 \end{cases}$

11. $\begin{cases} 3x + 2y = 6 \\ x - y = 2 \end{cases}$

12. $\begin{cases} 5x + y = 10 \\ x + y = 6 \end{cases}$

13. $\begin{cases} 3x - y = 9 \\ 6x - 2y = 18 \end{cases}$

14. $\begin{cases} 8x - 6y = 24 \\ 4x - 3y = 12 \end{cases}$

15. $\begin{cases} 5x - 3y = 15 \\ 2x - y = 4 \end{cases}$

16. $\begin{cases} 3x - 2y = 6 \\ x - 2y = -2 \end{cases}$

17. $\begin{cases} y = 2x + 6 \\ y = x + 1 \end{cases}$

18. $\begin{cases} y = 3x - 9 \\ y = x - 5 \end{cases}$

19. $\begin{cases} y = x - 3 \\ y = x + 4 \end{cases}$

20. $\begin{cases} y = x + 5 \\ y = x - 2 \end{cases}$

21. $\begin{cases} 2x = y - 8 \\ 3x = y + 6 \end{cases}$

22. $\begin{cases} x = 3y - 12 \\ x = 4y - 8 \end{cases}$

23. $\begin{cases} 3x + y = 3 \\ y = x + 5 \end{cases}$

24. $\begin{cases} 2x - y = 2 \\ x = y + 3 \end{cases}$

25. $\begin{cases} 5x - 4y = 20 \\ y = x - 3 \end{cases}$

26. $\begin{cases} 5x - 4y = -3 \\ x = y - 1 \end{cases}$

27. $\begin{cases} 2x + y = 10 \\ 2y = 20 - 4x \end{cases}$

28. $\begin{cases} x + 3y = 6 \\ 6y = 15 - 3x \end{cases}$

29. $\begin{cases} x + y = 8 \\ y = x \end{cases}$

30. $\begin{cases} x - y = 4 \\ y = x \end{cases}$

31. $\begin{cases} x + y = 6 \\ y = -2 \end{cases}$

32. $\begin{cases} x - y = 5 \\ y = -3 \end{cases}$

33. $\begin{cases} 2x - y = 2 \\ x = -4 \end{cases}$

34. $\begin{cases} 3x + y = 3 \\ x = 2 \end{cases}$

35. $\begin{cases} y = 4 \\ x = -1 \end{cases}$

36. $\begin{cases} x = 3 \\ y = -2 \end{cases}$

 QUESTIONS FOR THOUGHT

37. Discuss what is meant by a solution to a system of equations.

38. Is it possible for a system of two linear equations to have *exactly* two solutions? Why or why not?

39. Match the lettered items in column II with the numbered items in column I.

Column I	Column II
(1) Simultaneous equations in two variables with exactly one solution	(a) Conditional
(2) An equation in one variable with no solutions	(b) Consistent
(3) An equation in one variable with exactly one solution	(c) Dependent
	(d) Inconsistent

(Columns continue on page 272.)

Column I	Column II
(4) Simultaneous equations in two variables with infinitely many solutions	(e) Contradiction
(5) Simultaneous equations in two variables with no solutions	(f) Identity
(6) An equation in one variable with infinitely many solutions	

◇ **MINI-REVIEW**

In Exercises 40–43, solve the given inequality and sketch the solution set on a number line.

40. $4x + 7 \leq 6x + 17$ **41.** $9 - 5(x + 4) > 9$

42. $-2 \leq 3x + 1 < 13$ **43.** $1 \leq 7 - 2x < 13$

44. An orchestral society put on a concert. The members sold 200 tickets in advance and 75 tickets at the door. They charged $1.50 more for tickets at the door than for advanced-sale tickets. If they collected a total of $1,075, how much did they charge for tickets at the door?

45. The U.S. House of Representatives has 435 members, all of whom voted on a certain bill. The ratio of those who voted in favor of the bill to those who voted against was 8 to 7. How many were in favor of the bill?

6.3 Solving Systems of Linear Equations: Algebraic Methods

As we saw in the last section, the graphical method for solving a system of linear equations has some severe limitations. Consequently, we would like to develop an algebraic method for solving such a system.

By an algebraic method, we mean a method in which we manipulate the equations according to the rules of algebra in order to arrive at the solution to the system. Since we know how to solve a first-degree equation in *one* variable, our strategy is going to be to transform our system of two equations in two variables into one equation in one variable.

The Elimination Method

Let's again consider the system of equations that we solved graphically in Example 2 of Section 6.2.

EXAMPLE 1 | Solve the system of equations. $\begin{cases} x - y = 2 \\ x + y = 6 \end{cases}$

Solution | We know that we can add the same quantity to both sides of an equation and obtain an equivalent equation (an equation with the same solution set). Suppose we add 6 to both sides of the first equation in this system. We obtain

$$\begin{array}{r} x - y = 2 \\ +6 \quad +6 \\ \hline x - y + 6 = 8 \end{array} \quad \textit{This does not seem to help much.}$$

Now for the slight twist. Since we are seeking a *simultaneous* solution to the system, any solution we find must satisfy *both* equations in the system. The second equation in the system says that the quantity $x + y$ is equal to 6. Therefore, instead of adding 6 to both sides of the first equation, we can add $x + y$ to the left side and 6 to the right side (because $x + y$ and 6 are equal!). This is called "adding the two equations." In this way we still obtain an equivalent equation.

Let's see what happens when we do this.

$$
\begin{array}{l}
x - y = 2 \\
\underline{x + y = 6} \\
\quad\quad 2x = 8 \\
\quad\quad\ \ x = 4
\end{array}
$$

*We get an equation in **one** variable only. We can solve this equation for x.*

Now we can substitute $x = 4$ into the second equation to solve for y:

$$
\begin{array}{l}
x + y = 6 \\
4 + y = 6 \\
\\
\quad\quad y = 2
\end{array}
$$

Thus, our solution is the point $\boxed{(4, 2)}$.

Finally, we should check this answer in both equations.

Check: $\quad x - y = 2 \qquad x + y = 6$

$\qquad\qquad 4 - 2 \overset{\checkmark}{=} 2 \qquad 4 + 2 \overset{\checkmark}{=} 6$ ■

The procedure we used in Example 1 *eliminated* one of the variables from our system of equations and thereby allowed us to solve for the two variables one at a time. This procedure is called the ***elimination method***.

In the system we solved in Example 1, what made the procedure work was that we were fortunate to have the variable y appearing with *opposite* coefficients. When we added the two equations, the y's were eliminated. Naturally, we cannot expect to have such a system in all cases.

EXAMPLE 2

Solve the system of equations.
$$
\begin{cases}
4x - y = 8 \\
3x - y = 9
\end{cases}
$$

Solution

While we could subtract the two equations in this system in order to eliminate y, subtracting equations often leads to sign errors; therefore, we will always use *addition* to eliminate one of the variables.

If we simply add the two equations in our system, we will get an equivalent equation, *but* we will not eliminate anything.

$$
\begin{array}{l}
4x - \ y = \ 8 \\
\underline{3x - \ y = \ 9} \\
7x - 2y = 17
\end{array}
$$

This is still an equation in two variables.

Before we add our equations, we need to have one of the variables appearing with *opposite* coefficients. Then, when we add the two equations, that variable will be eliminated. In order to accomplish this we will multiply the second equation by -1.

Keep in mind it is only the sign of the y coefficient we care about, but the multiplication property of equality requires us to multiply *both sides* of the equation by -1.

$$
\begin{cases}
4x - y = 8 \\
3x - y = 9
\end{cases}
\qquad
\begin{array}{c}
\xrightarrow{\textit{As is}} \\
\xrightarrow{\textit{Multiply by } -1}
\end{array}
\qquad
\begin{array}{l}
4x - y = \ \ 8 \\
\underline{-3x + y = -9} \\
\quad\quad\ x = -1
\end{array}
$$

Now add the resulting equations.

Now we can substitute $x = -1$ into either one of the *original* equations to solve for y. We will substitute $x = -1$ into the first equation.

$$4x - y = 8$$
$$4(-1) - y = 8$$
$$-4 - y = 8$$
$$-y = 12$$
$$y = -12 \qquad \text{Thus, our solution is the point } \boxed{(-1, -12)}.$$

Check: We check this solution by substituting $x = -1$, $y = -12$ into both equations:

$$4x - y = 8 \qquad\qquad 3x - y = 9$$
$$4(-1) - (-12) \stackrel{?}{=} 8 \qquad\qquad 3(-1) - (-12) \stackrel{?}{=} 9$$
$$-4 + 12 \stackrel{\checkmark}{=} 8 \qquad\qquad -3 + 12 \stackrel{\checkmark}{=} 9 \qquad\qquad ■$$

In order to avoid multiplying the equations by fractions, it is sometimes necessary to multiply *both* equations by an appropriate constant in order to eliminate one of the variables.

EXAMPLE 3

Solve the system of equations. $\begin{cases} 6x - 7y = 26 \\ -4x - 5y = 2 \end{cases}$

Solution

First we must decide which variable we want to eliminate. In order to convert either $6x$ into $4x$ or $-4x$ into $-6x$ so that the x's will drop out, we would need to multiply by fractions; similarly for the y coefficients.

Instead, we can convert the coefficients of one of the variables in both equations into opposite coefficients. The easiest approach is to convert both coefficients of one of the variables into their least common multiple (LCM). Recall that the least common multiple of two numbers is the smallest number exactly divisible by each of the numbers. For example, the LCM for 4 and 6 is 12; the LCM for 5 and 7 is 35. Since the arithmetic is a little easier, we choose to eliminate x.

We are going to convert the coefficients of x into 12 and -12 so that the x terms will be eliminated when we add. Remember that we must multiply both sides of the equation in each case.

$$\begin{cases} 6x - 7y = 26 \quad \xrightarrow{\textit{Multiply by 2}} \quad 12x - 14y = 52 \\ -4x - 5y = 2 \quad \xrightarrow{\textit{Multiply by 3}} \quad \underline{-12x - 15y = 6} \quad \textit{Add the resulting} \\ \phantom{-4x - 5y = 2 \quad \xrightarrow{\textit{Multiply by 3}} \quad} -29y = 58 \qquad \textit{equations.} \\ \phantom{-4x - 5y = 2 \quad \xrightarrow{\textit{Multiply by 3}} \quad} y = -2 \end{cases}$$

Now we substitute $y = -2$ into one of the original equations (we will use the first one) and solve for x.

$$6x - 7y = 26$$
$$6x - 7(-2) = 26$$
$$6x + 14 = 26$$
$$6x = 12$$
$$x = 2 \qquad \text{Thus, our solution is the point } \boxed{(2, -2)}.$$

Check: We check by substituting $x = 2$, $y = -2$ into both original equations:

$$6x - 7y = 26 \qquad\qquad -4x - 5y = 2$$
$$6(2) - 7(-2) \stackrel{?}{=} 26 \qquad\qquad -4(2) - 5(-2) \stackrel{?}{=} 2$$
$$12 + 14 \stackrel{\checkmark}{=} 26 \qquad\qquad -8 + 10 \stackrel{\checkmark}{=} 2 \qquad\qquad ■$$

The choice of which variable to eliminate is yours. In many instances, one of the variables offers some advantages. As you do more exercises, you will learn to recognize how to make the easier choice. However, even if you choose the "harder" variable to eliminate, the procedure is still the same.

Let's pause to summarize the elimination method.

The Elimination Method	1. Decide which variable you want to eliminate.
	2. Multiply one or both equations by appropriate constants so that the variable you have chosen to eliminate appears with opposite coefficients.
	3. Add the two resulting equations.
	4. Solve the equation in *one* variable obtained in step 3.
	5. Substitute the value of the variable obtained in the previous step into one of the original equations and solve for the other variable.
	6. Check your solution in both original equations.

EXAMPLE 4

Solve the system of equations. $\begin{cases} 6x - 5y = 2 \\ 9x - 2y = 3 \end{cases}$

Solution   Since the LCM for 6 and 9 is 18, while the LCM for 5 and 2 is 10, we choose to eliminate the y terms. Since the y coefficients have the same sign, we will also need to change the sign of one of them. We will change the sign in the first equation. We do this by multiplying by -2 instead of $+2$.

$$\begin{cases} 6x - 5y = 2 & \xrightarrow{\text{Multiply by } -2} & -12x + 10y = -4 \\ 9x - 2y = 3 & \xrightarrow{\text{Multiply by } 5} & \underline{45x - 10y = 15} \quad \text{\textit{Add the resulting}} \\ & & 33x = 11 \quad \text{\textit{equations.}} \end{cases}$$

$$x = \frac{11}{33} = \frac{1}{3}$$

Now we substitute $x = \frac{1}{3}$ into one of the original equations. This time we will use the second one and solve for y.

$$9x - 2y = 3$$
$$9\left(\frac{1}{3}\right) - 2y = 3$$
$$3 - 2y = 3$$
$$-2y = 0$$

$$y = 0 \qquad \text{Thus, the solution is the point } \boxed{\left(\frac{1}{3}, 0\right)}.$$

Check: We check by substituting $x = \frac{1}{3}$, $y = 0$ into both original equations:

$$6x - 5y = 2 \qquad\qquad 9x - 2y = 3$$
$$6\left(\frac{1}{3}\right) - 5(0) \overset{?}{=} 2 \qquad\qquad 9\left(\frac{1}{3}\right) - 2(0) \overset{?}{=} 3$$
$$2 - 0 \overset{\checkmark}{=} 2 \qquad\qquad 3 - 0 \overset{\checkmark}{=} 3 \qquad\qquad \blacksquare$$

In order for the elimination method to work most easily, it is preferable that the two equations have the like variables lined up in the same column, with the constant on the other side of the equation. Additionally, we would certainly prefer to work with equations without fractional expressions. Consequently, we may have preliminary steps to go through before we are actually ready for the elimination procedure.

EXAMPLE 5 Solve the system of equations.
$$\begin{cases} \dfrac{x}{2} + \dfrac{y}{3} = 5 \\ y = 2x + 1 \end{cases}$$

Solution In order to get the system of equations into a more familiar form, we begin by multiplying the first equation by 6 in order to clear the denominators, and subtracting $2x$ from both sides of the second equation:

$$\begin{cases} \dfrac{x}{2} + \dfrac{y}{3} = 5 \\ y = 2x + 1 \end{cases} \qquad \xrightarrow{\text{Multiply by 6}} \qquad 3x + 2y = 30$$

$$\xrightarrow[\text{from both sides}]{\text{Subtract } 2x} \qquad -2x + y = 1$$

Now we can proceed to the elimination process. We choose to eliminate y because it requires multiplying only one of the equations.

$$\begin{cases} 3x + 2y = 30 \\ -2x + y = 1 \end{cases} \qquad \xrightarrow{\text{As is}} \qquad 3x + 2y = 30$$

$$\xrightarrow{\text{Multiply by } -2} \qquad 4x - 2y = -2 \qquad \text{\textit{Now we add the}}$$

$$\overline{ 7x = 28} \qquad \text{\textit{resulting equations.}}$$

$$x = 4$$

Now we substitute $x = 4$ into one of the original equations. This time we use the second one (because it is simpler) and solve for y.

$$y = 2x + 1$$
$$y = 2(4) + 1$$
$$y = 9 \qquad \text{Thus, our solution is the point } \boxed{(4, 9)}.$$

Check: We check by substituting $x = 4$, $y = 9$ into both original equations:

$$\frac{x}{2} + \frac{y}{3} = 5 \qquad\qquad y = 2x + 1$$

$$\frac{4}{2} + \frac{9}{3} \overset{?}{=} 5 \qquad\qquad 9 \overset{?}{=} 2(4) + 1$$

$$2 + 3 \overset{\checkmark}{=} 5 \qquad\qquad 9 \overset{\checkmark}{=} 8 + 1 \qquad\qquad ■$$

We should mention that once we get the value of one of our variables we can substitute that value into any equation that contains both variables in order to solve for the missing variable. We illustrate this point in the next example.

EXAMPLE 6 Solve the system of equations.
$$\begin{cases} x + \dfrac{y}{2} = 3 \\ \dfrac{2x}{5} - y = -2 \end{cases}$$

Solution In order to clear the fractions, we multiply both sides of the first equation by 2 and both sides of the second equation by 5.

$$\begin{cases} x + \dfrac{y}{2} = 3 \\ \dfrac{2x}{5} - y = -2 \end{cases} \qquad \begin{array}{l} \xrightarrow{\text{Multiply by 2}} \quad 2x + y = 6 \\ \xrightarrow{\text{Multiply by 5}} \quad 2x - 5y = -10 \end{array}$$

Next we proceed to the elimination process. We choose to eliminate y.

$$\begin{cases} 2x + y = 6 \\ 2x - 5y = -10 \end{cases}$$

$\xrightarrow{\textit{Multiply by 5}}$ $10x + 5y = 30$

$\xrightarrow{\textit{As is}}$

$$\begin{array}{r} 10x + 5y = 30 \\ 2x - 5y = -10 \\ \hline 12x = 20 \end{array}$$ *Now we add the resulting equations.*

$$x = \frac{20}{12} = \frac{5}{3}$$

Instead of substituting $x = \frac{5}{3}$ into one of the original equations, we might prefer to substitute it into one of the equations with the denominators already cleared. If we substitute $x = \frac{5}{3}$ into the equation $2x + y = 6$, we can solve for y.

$$2x + y = 6$$

$$2\left(\frac{5}{3}\right) + y = 6$$

$$\frac{10}{3} + y = 6$$

$$y = 6 - \frac{10}{3}$$

$$y = \frac{18}{3} - \frac{10}{3}$$

$$y = \frac{8}{3}$$ Thus, our solution is the point $\boxed{\left(\frac{5}{3}, \frac{8}{3}\right)}$.

Now we check the solution in *both* original equations.

Check: $x + \dfrac{y}{2} = 3$ $\dfrac{2x}{5} - y = -2$

$$\frac{5}{3} + \frac{\frac{8}{3}}{2} \stackrel{?}{=} 3 \qquad\qquad \frac{2\left(\frac{5}{3}\right)}{5} - \frac{8}{3} \stackrel{?}{=} -2$$

$$\frac{5}{3} + \frac{4}{3} \stackrel{?}{=} 3 \qquad\qquad 2 \cdot \frac{5}{3} \cdot \frac{1}{5} - \frac{8}{3} \stackrel{?}{=} -2$$

$$\frac{9}{3} \stackrel{\checkmark}{=} 3 \qquad\qquad \frac{2}{3} - \frac{8}{3} \stackrel{\checkmark}{=} -2$$ ■

The Substitution Method

There is another method for solving a system of equations, called the **substitution method.** In the substitution method we solve for one of the variables in one of the equations and then substitute for that variable in the other equation. For example, to solve the system

$$\begin{cases} x + 2y = 1 \\ x - y = 4 \end{cases}$$

we proceed as follows. We solve the second equation for x, obtaining $x = y + 4$, and then substitute the quantity $x = y + 4$ into the first equation.

$$x + 2y = 1 \qquad \textit{Substitute } y + 4 \textit{ for } x.$$

$$\overbrace{y + 4} + 2y = 1$$

$$3y + 4 = 1$$

$$3y = -3$$

$$y = -1$$

Once we know the y value, we can proceed to find x as we would in the elimination method. Substituting $y = -1$ into the equation $x = y + 4$, we get

$$x = y + 4$$
$$x = -1 + 4$$
$$x = 3 \qquad \text{Thus, the solution to the system is } \boxed{(3, -1)}.$$

The check is left to the student.
Let's summarize the substitution method.

The Substitution Method	1. Use one of the equations to solve explicitly for one of the variables.
	2. Substitute the expression obtained in step 1 into the *other* equation.
	3. Solve the resulting equation in one variable.
	4. Substitute the value obtained into one of the equations containing both variables (usually the one solved explicitly that was found in step 1) and solve for the other variable.
	5. Check the solution in the original equations.

While both the elimination and substitution methods will work for all systems of equations, unless you can easily solve explicitly for one of the variables, you will probably find the elimination method simpler to use.

EXAMPLE 7

Solve the system of equations. $\begin{cases} 8x + 4y = 10 \\ 4x + 2y = 5 \end{cases}$

Solution

Looking at this system of equations, we can see that neither variable looks particularly easy to solve for. This suggests that we use the elimination method rather than the substitution method.

If we choose to use the elimination method, there really is no preference here as to which variable to eliminate. In fact, whichever variable we choose to eliminate, we will multiply the second equation by -2.

$$\begin{cases} 8x + 4y = 10 \\ 4x + 2y = 5 \end{cases} \quad \xrightarrow{\text{As is}} \quad \begin{array}{r} 8x + 4y = 10 \\ \underline{-8x - 4y = -10} \\ 0 = 0 \end{array} \quad \textit{Add the resulting equations.}$$

Both variables have been eliminated, and we are left with an identity. The implication is that the two equations are identical. In fact, if we look back at the two original equations, we can see that the first equation can be obtained from the second equation by multiplying both sides of the second equation by 2. Thus, both equations have exactly the same solution set. Their graphs are exactly the same line, and so there are infinitely many solutions to this system of equations. The system is dependent. We can use either equation to write the solution set. The solution set is

$$\boxed{\{(x, y) \mid 4x + 2y = 5\}}$$

■

EXAMPLE 8

Solve the system of equations. $\begin{cases} \dfrac{x}{2} - 2y = 3 \\ x - 4y = -8 \end{cases}$

Solution

We solve this system by the substitution method.
To solve the second equation for x, we add $4y$ to both sides.

$$x = 4y - 8 \qquad \textit{Now we substitute } 4y - 8 \textit{ for } x \textit{ in the first equation.}$$

$$\frac{x}{2} - 2y = 3$$

$$\frac{4y - 8}{2} - 2y = 3 \qquad \textit{Factor the numerator and reduce the fraction.}$$

$$\frac{\overset{2}{4}(y - 2)}{\underset{1}{2}} - 2y = 3$$

$$2y - 4 - 2y = 3 \qquad \textit{Simplify.}$$

$$-4 = 3 \qquad \textit{This is a contradiction.}$$

Both variables have been eliminated, and we are left with a contradiction.

Our assumption that there is a solution to the system of equations has led to a contradiction. Therefore, this assumption must be false, and there are no solutions to this system of equations. The system is inconsistent. If we graphed these two lines, they would be parallel.

This system of equations has no solutions . ∎

Examples 7 and 8 illustrate that if we eliminate both variables from our system, then there are two possibilities:

If the resulting equation is an identity, then the two equations have the same solution set and the same line for their graphs; the equations are dependent.

If the resulting equation is a contradiction, then the two equations have no simultaneous solution, and their graphs are parallel lines; the equations are inconsistent.

EXERCISES 6.3

In Exercises 1–14, *solve the system of equations using the elimination method.*

1. $\begin{cases} x + y = 1 \\ x - y = 3 \end{cases}$

2. $\begin{cases} x - y = 4 \\ x + y = 6 \end{cases}$

3. $\begin{cases} 2x + y = 5 \\ x - y = 4 \end{cases}$

4. $\begin{cases} -x + 3y = 8 \\ x - 2y = -6 \end{cases}$

5. $\begin{cases} 7x + 2y - 15 = 0 \\ 3x - 2y + 5 = 0 \end{cases}$

6. $\begin{cases} a - 5b - 30 = 0 \\ a + 5b + 40 = 0 \end{cases}$

7. $\begin{cases} 2x + y = 15 \\ x - 2y = 0 \end{cases}$

8. $\begin{cases} x - 3y = 1 \\ 2x + y = 9 \end{cases}$

9. $\begin{cases} 3x + 2y = -11 \\ x + 3y = 1 \end{cases}$

10. $\begin{cases} 5x - y = 18 \\ x + 2y = -3 \end{cases}$

11. $\begin{cases} 4s - 5t = 4 \\ 2s + 10t = 7 \end{cases}$

12. $\begin{cases} 6u - w = 2 \\ 2u - 3w = 2 \end{cases}$

13. $\begin{cases} 12c - 20d = 19 \\ 18c - 12d = 15 \end{cases}$

14. $\begin{cases} 14w + 9z = 6 \\ 21w + 15z = 11 \end{cases}$

In Exercises 15–28, *solve the system of equations using the substitution method.*

15. $\begin{cases} x + 2y = 9 \\ y = 3x + 1 \end{cases}$

16. $\begin{cases} x + 3y = 5 \\ y = 4x - 7 \end{cases}$

17. $\begin{cases} 2x - 3y = 8 \\ \quad\; x = 2y + 4 \end{cases}$

18. $\begin{cases} 3x + 4y = 12 \\ \quad\; x = 3y - 9 \end{cases}$

19. $\begin{cases} 5x + 4y = 6 \\ \quad x - y = 3 \end{cases}$

20. $\begin{cases} 3a + 5b = 1 \\ \quad b - a = 5 \end{cases}$

21. $\begin{cases} 2r - 5s = 9 \\ \qquad s = 1 - r \end{cases}$

22. $\begin{cases} 4u + 3v = 0 \\ \qquad u = -1 - v \end{cases}$

23. $\begin{cases} 7x + 2y = 9 \\ 2x + 3y = 5 \end{cases}$

24. $\begin{cases} 6x - 5y = 2 \\ 5x - 2y = 6 \end{cases}$

25. $\begin{cases} 6u - w = 2 \\ 2u - 3w = 2 \end{cases}$

26. $\begin{cases} a - 4b = 1 \\ 2a - 5b = 3 \end{cases}$

27. $\begin{cases} 4w - 3t = 8 \\ 6w - t = 5 \end{cases}$

28. $\begin{cases} 18p + 2r = 1 \\ 6p - r = 2 \end{cases}$

In Exercises 29–62, solve the system of equations using any method you choose.

29. $\begin{cases} r + 2t = 10 \\ 3r + t = -15 \end{cases}$

30. $\begin{cases} 5m + n = 5 \\ \qquad m = 2n + 12 \end{cases}$

31. $\begin{cases} 6x + y = 6 \\ 4x + 1 = y \end{cases}$

32. $\begin{cases} 3x + 6y = 2 \\ -3x - 3y = 1 \end{cases}$

33. $\begin{cases} 8a + 6b = -3 \\ 12a + 9b = -5 \end{cases}$

34. $\begin{cases} 8a + 6b = 6 \\ 12a - 9b = 3 \end{cases}$

35. $\begin{cases} 11a - 2b = 30 \\ 3a + 3b = -6 \end{cases}$

36. $\begin{cases} 7a - 5b = 17 \\ 3a - 2b = 7 \end{cases}$

37. $\begin{cases} 2x + 3 = 4y \\ 6x = 9 - 12y \end{cases}$

38. $\begin{cases} 3y = 24 - 9x \\ 3x + y = 8 \end{cases}$

39. $\begin{cases} 5x + 2y = 4y + 9 \\ \qquad y = x - 3 \end{cases}$

40. $\begin{cases} 4x + 3 = 2y - 5 \\ \quad\; x = y - 4 \end{cases}$

41. $\begin{cases} a - b = 1 \\ \dfrac{a}{3} + \dfrac{b}{5} = 1 \end{cases}$

42. $\begin{cases} \dfrac{a}{4} - 4b = 2 \\ \dfrac{a}{8} - 5b = 2 \end{cases}$

43. $\begin{cases} \dfrac{w}{2} - \dfrac{t}{5} = 11 \\ \dfrac{w}{8} - \dfrac{t}{9} = 0 \end{cases}$

44. $\begin{cases} \dfrac{x}{3} - \dfrac{y}{6} = 5 \\ \dfrac{x}{4} - \dfrac{y}{2} = 15 \end{cases}$

45. $\begin{cases} \dfrac{2r}{5} - \dfrac{3t}{8} = 14 \\ \dfrac{4r}{5} + \dfrac{3t}{4} = 4 \end{cases}$

46. $\begin{cases} \dfrac{3w}{5} + \dfrac{4z}{3} = 44 \\ \dfrac{5w}{8} + \dfrac{7z}{6} = 45 \end{cases}$

47. $\begin{cases} 5x - 3y = 20 \\ 7x + 2y = 28 \end{cases}$

48. $\begin{cases} 6x - 9y = 45 \\ 8x - 5y = 25 \end{cases}$

49. $\begin{cases} 7x + 4y = 5 \\ 4x + 3y = 0 \end{cases}$

50. $\begin{cases} 6x + 5y = 13 \\ 5x + 3y = 5 \end{cases}$

51. $\begin{cases} \dfrac{x}{2} + \dfrac{y}{3} = 1 \\ \dfrac{x}{4} - y = 11 \end{cases}$

52. $\begin{cases} \dfrac{x}{3} - \dfrac{y}{4} = 2 \\ x + \dfrac{y}{3} = -7 \end{cases}$

53. $\begin{cases} \dfrac{x+y}{2} = 4 \\ 3x = 5 - 3y \end{cases}$

54. $\begin{cases} \dfrac{x-y}{3} = 1 \\ x = y - 1 \end{cases}$

55. $\begin{cases} 0.4x + 0.2y = 8 \\ 0.7x - 0.3y = 1 \end{cases}$

56. $\begin{cases} 0.2x + 0.02y = 8 \\ 4x - y = 20 \end{cases}$

57. $\begin{cases} 5.2x + 3y = 14 \\ 0.3x - 2y = 9.5 \end{cases}$

58. $\begin{cases} 8x - 2.8y = 4 \\ 0.3x + y = 11.2 \end{cases}$

59. $\begin{cases} 3.6x + 2.9y = 23.71 \\ 1.7x - 4.5y = 14.64 \end{cases}$

60. $\begin{cases} 2.8x - 4.6y = 3.5 \\ 5.2x - 3.4y = 10.1 \end{cases}$

61. $\begin{cases} 11.2x - 2.6y = 22.84 \\ 6.7x + 15.3y = 3.55 \end{cases}$

62. $\begin{cases} -3.5x + 7.8y = 32.55 \\ 8.2x - 3.8y = -25.6 \end{cases}$

? QUESTIONS FOR THOUGHT

63. A student began his solution to a system of equations as follows:

$$\begin{cases} 3x + 2y = 19 & \xrightarrow{\text{Multiply by 3}} & 9x + 6y = 57 \\ 2x - 3y = 4 & \xrightarrow{\text{Multiply by 2}} & 4x - 6y = 4 \end{cases}$$

$$13x = 61$$

$$x = \frac{61}{13}$$

What error has the student already made?

64. A student began her solution to a system of equations as follows:

$$\begin{cases} 7x - 2y = 3 & \xrightarrow{\text{As is}} & 7x - 2y = 3 \\ 5x + y = 0 & \xrightarrow{\text{Multiply by 2}} & 10x + 2y = 2 \end{cases}$$

$$17x = 5$$

$$x = \frac{5}{17}$$

What error has the student already made?

6.4 Applications

Many of the applications we discuss in this section are very similar to those we have discussed previously. In fact, we will continue to use most of the outline we first gave in Chapter 3 (page 116) for solving verbal problems. (You might find it useful to review that outline before you proceed with this section.)

However, since we now know how to solve a system of two linear equations in two variables, we are no longer restricted to solutions that involve one variable only. As we shall soon see, there are situations where using two variables makes it easier to formulate the solution to a problem.

Let's review an example of the type we have done previously and then illustrate how we would solve the same problem using a two-variable approach.

EXAMPLE 1

An office supply company shipped a total of 24 printers and charged the client $6,795. The shipment consisted of bubble-jet printers costing $225 each and laser printers costing $380 each. How many of each were shipped?

Solution

We recognize that there are two unknown quantities: the number of bubble-jet printers and the number of laser printers.

First, we will solve this problem using the approach outlined in Chapter 3. From now on we will call this the *one-variable approach*.

One-variable approach:
We begin by recognizing the fundamental relationship.

$$\text{Cost of the bubble-jet printers} \quad + \quad \text{Cost of the laser printers} \quad = \text{Total cost}$$

$$\left(\begin{array}{c}\text{Number of bubble-jet}\\\text{printers}\end{array}\right) \cdot (\$225) + \left(\begin{array}{c}\text{Number of laser}\\\text{printers}\end{array}\right) \cdot (\$380) = \$6{,}795 \quad (*)$$

Looking at this last "equation," we see that all that remains to be done is label the number of bubble-jet printers and the number of laser printers using *one* variable. Therefore

$$\text{Let } x = \text{Number of bubble-jet printers sold}$$

Then

$$24 - x = \text{Number of laser printers sold} \qquad \textit{Since there are 24 printers sold altogether}$$

Now that we have expressed the number of bubble-jet printers and laser printers in terms of one variable, we can substitute into the equation we marked $(*)$, giving us the equation

$$\begin{aligned}225x + 380(24 - x) &= 6{,}795 \qquad \textit{Now we solve this equation.}\\225x + 9{,}120 - 380x &= 6{,}795\\-155x + 9{,}120 &= 6{,}795\\\underline{-9{,}120 \quad -9{,}120}\\-155x &= -2{,}325\\x &= \frac{-2{,}325}{-155}\\x &= 15\end{aligned}$$

Thus, there were $\boxed{15 \text{ bubble-jet printers}}$ and $24 - 15$, or $\boxed{9 \text{ laser printers}}$. The check is left to the student.

The second approach is to label the two unknown quantities with two different variables.

Two-variable approach:

$$\text{Let } x = \text{Number of bubble-jet printers sold}$$

and

$$\text{Let } y = \text{Number of laser printers sold}$$

Since we are using two variables, we must use the information given in the example to write *two* equations. The two relationships we have are

The total number of printers is 24.

The total cost of the printers is $6,795.

If we rewrite these relationships using x and y, we get the following system of equations.

$$\begin{cases} x + y = 24 & \textit{This equation says: Number of bubble-jet printers} \\ & \qquad\qquad\textit{+ Number of laser printers } = 24 \\ 225x + 380y = 6{,}795 & \textit{This equation says: Cost of the bubble-jet printers} \\ & \qquad\qquad\textit{+ Cost of the laser printers } = \$6{,}795 \end{cases}$$

We will solve this system of equations by the substitution method. Solving the first equation for x, we get $x = 24 - y$. Now we substitute for x in the second equation.

Solve the first equation for y and substitute into the second equation. Compare the resulting equation to the one used in the one-variable approach.

$$225x + 380y = 6{,}795 \qquad \textit{Substitute } x = 24 - y.$$
$$225(24 - y) + 380y = 6{,}795$$
$$5{,}400 - 225y + 380y = 6{,}795$$
$$5{,}400 + 155y = 6{,}795$$
$$155y = 1{,}395$$
$$y = \frac{1{,}395}{155} = 9$$

Now we substitute $y = 9$ into the first equation, $x + y = 24$, and we solve for x to get $x = 15$. Thus the solution is the same as in the one-variable approach: 15 bubble-jet printers and 9 laser printers. ■

In some examples the one-variable approach may be easier, while in others the two-variable approach may offer some advantages. The two procedures are similar in that they both require two relationships between the variables. They differ in the way these relationships are used. In general, you should use the method you can adapt most easily to the particular problem you are trying to solve.

In the next few examples we will illustrate the two-variable approach in several different types of situations.

EXAMPLE 2 The ratio of two numbers is 5 to 2. If the sum of the two numbers is 50, what are the two numbers?

Solution

$$\text{Let } x = \text{the larger number.}$$
$$\text{Let } y = \text{the smaller number.}$$

The first relationship is that the ratio of the two numbers is 5 to 2. We can translate this as

$$\frac{x}{y} = \frac{5}{2}$$

The second relationship is that the sum of the two numbers is 50. We translate this as

$$x + y = 50$$

We now proceed to solve this system of equations. Notice that our first step is to clear the fractions in the first equation (by multiplying both sides by $2y$) and to get the y term on the left-hand side.

$$\begin{cases} \dfrac{x}{y} = \dfrac{5}{2} & \xrightarrow[\textit{sides by } 2y]{\textit{Multiply both}} & 2x = 5y & \xrightarrow[\textit{from both sides}]{\textit{Subtract } 5y} & 2x - 5y = 0 \\[2mm] x + y = 50 & \xrightarrow{\textit{As is}} & x + y = 50 & \xrightarrow{\textit{As is}} & x + y = 50 \end{cases}$$

Now we are ready to eliminate one of the variables. We choose to eliminate x.

$$\begin{array}{r} 2x - 5y = 0 \\ x + y = 50 \end{array} \quad \begin{array}{c} \xrightarrow{\textit{As is}} \\ \xrightarrow{\textit{Multiply by } -2} \end{array} \quad \begin{array}{r} 2x - 5y = 0 \\ -2x - 2y = -100 \\ \hline -7y = -100 \end{array} \quad \begin{array}{l} \textit{Add the two} \\ \textit{equations.} \end{array}$$

$$y = \frac{-100}{-7} = \frac{100}{7}$$

We substitute $y = \frac{100}{7}$ into the second of the original equations and solve for x:

$$x + y = 50$$

$$x + \frac{100}{7} = 50$$

$$x = 50 - \frac{100}{7}$$

$$x = \frac{350}{7} - \frac{100}{7}$$

$$x = \frac{250}{7} \qquad \text{Thus, the two numbers are } \boxed{\frac{250}{7} \text{ and } \frac{100}{7}}.$$

Check: $\dfrac{\frac{250}{7}}{\frac{100}{7}} = \frac{250}{7} \cdot \frac{7}{100} = \frac{250}{100} \overset{\checkmark}{=} \frac{5}{2}$ $\qquad \frac{250}{7} + \frac{100}{7} = \frac{350}{7} \overset{\checkmark}{=} 50$

Therefore, the ratio of the two numbers is $\frac{5}{2}$.

Therefore, the sum of the two numbers is 50. ∎

EXAMPLE 3

A total of \$6,000 was split into two investments. Part was invested in a bank account paying 7% interest per year, and the remainder was invested in stocks paying 11% interest per year. If the total yearly interest from the two investments is \$492, how much was invested in the bank account?

Solution

Let $x = $ amount invested at 7% and $y = $ amount invested at 11%. The first relationship is that the total amount invested was \$6,000. This translates to

$$x + y = 6{,}000$$

This same problem was solved using a one-variable approach in Example 4 of Section 4.6.

The second relationship is that the total *interest* from the two investments is \$492. Keeping in mind that interest is equal to principal times rate, we multiply each principal (the x and the y) by its rate. This translates to

$$0.07x + 0.11y = 492$$

We can now proceed to solve this system of equations. We first clear the decimals in the second equation (by multiplying both sides of the second equation by 100). Looking ahead, we see that we will get $11y$ in the second equation, so in the same step we multiply the first equation by -11.

$$
\begin{cases}
x + y = 6{,}000 & \xrightarrow[\text{by } -11]{\textit{Multiply}} & -11x - 11y = -66{,}000 \\
0.07x + 0.11y = 492 & \xrightarrow[\text{by } 100]{\textit{Multiply}} & 7x + 11y = 49{,}200 \\
& & \overline{ -4x = -16{,}800} \\
& & \boxed{x = \$4{,}200}
\end{cases}
$$

Add the two equations.

Substituting $x = 4{,}200$ into the first of our original equations, we can solve for y:

$$x + y = 6{,}000$$

$$4{,}200 + y = 6{,}000$$

$$\boxed{y = \$1{,}800}$$

Check:

$$4,200 + 1,800 \overset{\checkmark}{=} 6,000$$ *The amount invested is $6,000.*

$$0.07(4,200) + 0.11(1,800) = 294 + 198 \overset{\checkmark}{=} 492$$ *The yearly interest is $492.* ∎

As we mentioned in Chapters 3 and 4, in order to solve "value"-type problems we must distinguish between two different relationships. For example,

Mixture problems:	We must distinguish between the number of items and the "value" of each type of item.
Investment problems:	We must distinguish between amount invested and interest earned.
Solution problems:	We must distinguish the amount of the mixture from the amount of pure substance.

Being able to distinguish between the two relationships allows us to write the two equations we need.

Sometimes a problem is worded in such a way that the one-variable approach is difficult, while the two-variable approach is quite straightforward.

EXAMPLE 4

Jameel buys two belts and four ties for a total of $59.70. Kwan buys three belts and five ties for a total of $80.60. Assuming that all the belts are the same price and all the ties are the same price, find the individual costs of a belt and of a tie.

Solution

We translate the given information about each person's purchase into an equation. Let b = the cost of a single belt and t = the cost of a single tie. Jameel's purchase gets translated as

$$2b + 4t = 59.70$$

Kwan's purchase gets translated as

$$3b + 5t = 80.60$$

We now solve the following system of equations:

$$\begin{cases} 2b + 4t = 59.70 \\ 3b + 5t = 80.60 \end{cases}$$

$\xrightarrow{\text{Multiply by 3}}$ $6b + 12t = 179.10$ *Add the two equations.*

$\xrightarrow{\text{Multiply by } -2}$ $\underline{-6b - 10t = -161.20}$

$$2t = 17.90$$
$$t = 8.95$$

We now substitute $t = 8.95$ into the first equation to find b:

$$2b + 4t = 59.70 \quad \text{Substitute } t = 8.95.$$
$$2b + 4(8.95) = 59.70$$
$$2b + 35.80 = 59.70$$
$$2b = 23.90$$

$$b = 11.95 \qquad \boxed{\text{Therefore, a belt costs \$11.95 and a tie costs \$8.95}}.$$

It is left to the student to check this answer in the words of the problem. ∎

⬤www EXERCISES 6.4

Solve each of the following verbal problems algebraically. You may use either a one- or two-variable approach.

1. The sum of two numbers is 130. If their difference is 28, find the two numbers.

2. The difference between two numbers is 3. If the sum of twice the larger and the smaller is 48, find the numbers.

3. Sam has 80 coins consisting of nickels and quarters. If the total value of the coins is $13.60, how many of each type of coin are there?

4. Susan has 92 packages in her truck. Some of the packages weigh 32 lb each, and the rest weigh 12 lb each. If the total weight of all the packages is 1,604 lb, how many of the lighter packages are there on the truck?

5. Two cars start at the same place and time, and travel in opposite directions. One car is traveling 15 kph faster than the other. After 5 hours, the two cars are 275 km apart. Find the speed of each car.

6. The ratio of two positive numbers is 3 to 4. If one of the numbers is 5 more than the other, what are the two numbers?

7. Carmen invests a total of $1,700 in two stocks. One stock pays a yearly dividend of 7%, while the other pays 6%. If Carmen received $110 in combined dividends from the two stocks, how much did she invest in each?

8. A storekeeper is preparing a mixture of peanuts and raisins. If peanuts cost $1.60 per pound and raisins cost $1.85 per pound, how many pounds of each should be used to prepare 50 pounds of a mixture selling at $1.70 per pound?

9. Tim and Marge go into a record shop. Tim buys four cassettes and six CDs for a total of $107.66, while Marge buys five cassettes and three CDs for a total of $76.30. What are the prices of an individual cassette and an individual CD?

10. A discount building supplies store sells both first-quality and second-quality floor tiles. Robin buys three cases of first-quality tiles and one case of second-quality tiles for a total of $66, while Gene buys one case of first-quality tiles and three cases of second-quality tiles for a total of $54. What is the cost per case for each type of tile?

11. The length of a rectangle is twice its width. If the perimeter of the rectangle is 28 in., what are its dimensions?

12. How many 34¢ and 60¢ stamps did Joe buy if he bought 50 stamps and paid $20.64 for them?

13. Pat and Carlos both belong to the same book club. Pat orders two regular selections and three specially discounted ones for a total of $56.90. Carlos orders three regular selections and four specially discounted ones for a total of $80.85. What are the prices of a regular and a specially discounted selection?

14. On Monday John walks for 1 hour, jogs for 2 hours, and covers a total of 28 km. On Tuesday he walks for 2 hours, jogs for 1 hour, and covers a total of 23 km. What are his rate walking and his rate jogging?

15. The ratio of two positive numbers is 6 to 5. If the difference between the two numbers is 8, what are the numbers?

16. Seven thousand tickets worth $137,125 were sold for a concert. General admission tickets cost $22 each, and "standing-room-only" tickets cost $14.50 each. How many of each type were sold?

17. Two airplanes leave an airport at the same time, flying in opposite directions. One plane is flying at twice the speed of the other. If after 4 hours they are 1,800 miles apart, find the speed of each plane.

18. Repeat Exercise 17 if the planes are flying in the *same* direction.

19. Two retailers are ordering from the same source. One retailer orders eight stereo receivers and four turntables at a total cost of $2,060. A second retailer orders five of the same receivers and six of the same turntables at a total cost of $1,690. What are the costs of an individual receiver and an individual turntable?

20. A 25% iodine solution is to be mixed with a 75% iodine solution to produce 5 gallons of a 70% iodine solution. How many gallons of each solution are needed?

21. John goes into a donut shop and buys ten plain donuts and five cream-filled donuts for $3.70. Jane goes into the same shop and buys five plain donuts and ten cream-filled donuts for $4.10. What is the cost of a plain donut?

22. In a recent school election, 571 votes were cast for class president. If the winner received 89 more votes than the loser, how many votes did each receive?

23. A bookstore buys 35 books for $271. Some of the books cost $7 each and the remainder cost $9 each. How many of each type were bought?

24. An artist's supply store sold a total of 20 canvases for $172. If some of the canvases cost $7.50 each and the remainder were $10.25 each, how many of each type were sold?

25. Two cars start traveling directly toward each other at the same time from positions 480 km apart. They meet after 4 hours. If one car travels 40 kph faster than the other car, find the speed of each car.

26. Repeat Exercise 25 if the slower car leaves 2 hours ahead of the faster car and they meet 4 hours after the faster car leaves.

27. A small single-engine plane travels 150 miles per hour with a tailwind, and 90 miles per hour with a headwind. Find the speed of the wind, and the speed of the plane in still air.

28. A 60% acid solution is to be mixed with an 80% acid solution to produce 20 liters of a 65% acid solution. How many liters of each solution are needed?

29. A coffee wholesaler wishes to produce 60 pounds of a coffee blend selling at $3.10 per pound. How many pounds of coffee blends selling at $3.35 per pound and $2.75 per pound should be mixed to produce such a mixture?

30. Mrs. Thomas has $15,000 to invest. She will invest part of it in a corporate bond that pays 8% interest per year and the rest in a stock that pays 11% interest per year. How should she divide up the $15,000 so that her total yearly interest will be 10% of her investments?

31. A company is trying to determine which computer system to install. System A consists of a central minicomputer costing $100,000 plus desktop terminals costing $800 each. System B is a network of desktop personal computers that costs $16,000 to install plus $1,200 for each desktop personal computer. How many desktop setups would the company need in order to make the costs of the two systems equal?

32. Lenore can purchase a car for $15,000, which will require her to spend an average of $80 per month in repairs and maintenance. Or, she can lease a car for $350 per month, which includes all repairs and maintenance. After how many months will the leased car and the purchased car cost the same?

33. A mathematics department has budgeted $10,000 to purchase computers and printers. On this fixed budget they can purchase 10 computers and 10 printers, or they can purchase 12 computers and 2 printers. Find the individual costs of a computer and printer.

34. A small company has budgeted $6,000 per month to lease vehicles. On this budget, the company can lease 12 cars and 4 trucks each month, or 8 cars and 6 trucks. Find the monthly cost to lease a car and to lease a truck.

35. The perimeter of a rectangle is 46 cm. Twice the width is one more than the length. Find the dimensions of the rectangle.

36. The perimeter of a rectangle is 96 ft. Three times the width is eight less than the length. Find the dimensions of the rectangle.

37. A bag contains 36 marbles, some of which are red and the remainder are blue. Twice the number of red marbles is six less than the number of blue marbles. Find the number of marbles of each color.

38. A certain mutual fund contains 100 stocks. On a certain day all of the stocks changed price. Three times the number of stocks that went up is fourteen more than eight times the number of stocks that went down. Find how many stocks went up and how many went down.

CHAPTER 6   SUMMARY

After having completed this chapter, you should be able to:

1. Interpret graphs and read specific information from a graph (Section 6.1).

   For example: Figure 6.12 shows the relationship between the height h, in feet, of an object above the ground and the number t of seconds after the object is thrown into the air.

   Among the things we can see from this graph are the following:

   (a) The object is initially at a height of 30 feet (because when $t = 0$ we have $h = 30$).

   (b) The object increases in height for the first 4 seconds until it reaches its highest point of 60 feet above the ground after 4 seconds ($h = 60$ when $t = 4$).

   (c) The object falls for the next 5 seconds. It hits the ground 9 seconds after it is released ($h = 0$ when $t = 9$).

   (d) After 8 seconds the object is at a height of 15 feet [the point (8, 15) is on the graph].

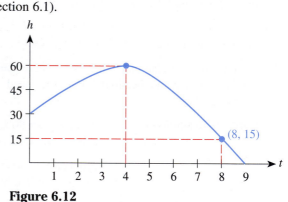

Figure 6.12

2. Solve a system of equations by the graphical method (Section 6.2).

   For example: Solve the following system by the graphical method:

   $$\begin{cases} x - y = 5 \\ 2x + y = 4 \end{cases}$$

   The first equation has intercepts (5, 0) and (0, −5).

   The second equation has intercepts (2, 0) and (0, 4).

   We draw the two lines (see Figure 6.13) and see that they cross at the point (3, −2). We then check that this point satisfies both of the equations.

   Check $3 - (-2) \overset{\checkmark}{=} 5$
   $2(3) + (-2) \overset{\checkmark}{=} 4$

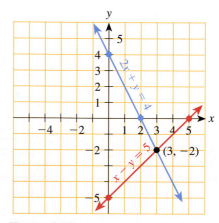

Figure 6.13

3. Solve a system of equations by the elimination or substitution methods (Section 6.3).

   For example: Solve the system

   $$\begin{cases} 3x - 5y = 1 \\ 7x - 2y = 12 \end{cases}$$

   We choose to eliminate y.

   $$\begin{cases} 3x - 5y = 1 \\ 7x - 2y = 12 \end{cases} \quad \begin{array}{l} \xrightarrow{\textit{Multiply by } -2} \\ \xrightarrow{\textit{Multiply by } 5} \end{array} \quad \begin{array}{l} -6x + 10y = -2 \\ \underline{35x - 10y = 60} \quad \textit{Add the resulting equations.} \\ 29x = 58 \\ x = 2 \end{array}$$

   Now we substitute $x = 2$ into one of the original equations (we will use the first one) and solve for y.

   $$3x - 5y = 1$$
   $$3(2) - 5y = 1$$
   $$6 - 5y = 1$$
   $$-5y = -5$$
   $$y = 1 \qquad \text{Thus, our solution is the ordered pair } \boxed{(2,\ 1)}.$$

   Check: We substitute $x = 2$, $y = 1$ into both equations:

   $$\begin{array}{ll} 3x - 5y = 1 & 7x - 2y = 12 \\ 3(2) - 5(1) \overset{?}{=} 1 & 7(2) - 2(1) \overset{?}{=} 12 \\ 6 - 5 \overset{\checkmark}{=} 1 & 14 - 2 \overset{\checkmark}{=} 12 \end{array}$$

4. Solve verbal problems by writing and solving a system of equations (Section 6.4).

   For example: Joe bought eight 60-minute cassettes and five 90-minute cassettes for \$23. Jane bought three 60-minute cassettes and ten 90-minute cassettes for \$26.50. What is the price of a 60-minute cassette?

   Let x = the price of a 60-minute cassette.
   Let y = the price of a 90-minute cassette.

   The information in the problem translates to

   $$\begin{cases} 8x + 5y = 23 \\ 3x + 10y = 26.50 \end{cases}$$

   Since the problem asks for the price of a 60-minute cassette, it is x we are trying to solve for. Therefore, we eliminate y from the system.

   $$\begin{cases} 8x + 5y = 23 \\ 3x + 10y = 26.50 \end{cases} \quad \begin{array}{l} \xrightarrow{\textit{Multiply by } -2} \\ \xrightarrow{\textit{As is}} \end{array} \quad \begin{array}{l} -16x - 10y = -46 \\ \underline{3x + 10y = 26.50} \\ -13x = -19.50 \\ x = 1.50 \end{array} \quad \begin{array}{l} \textit{Add the} \\ \textit{resulting} \\ \textit{equations.} \end{array}$$

   $$\boxed{\text{The price of a 60-minute cassette is \$1.50.}}$$

CHAPTER 6   REVIEW EXERCISES

In Exercises 1–6, solve the system of equations by the graphical method.

1. $\begin{cases} x + y = 4 \\ x - y = 0 \end{cases}$

2. $\begin{cases} 2x + y = 4 \\ 3x - 2y = 6 \end{cases}$

3. $\begin{cases} x - 2y = 8 \\ \quad\quad y = x - 5 \end{cases}$

4. $\begin{cases} y = x + 2 \\ x = y + 2 \end{cases}$

5. $\begin{cases} \quad\quad y = x \\ 3x - 2y = 6 \end{cases}$

6. $\begin{cases} \quad\quad x = 2y \\ 4y - 3x = 12 \end{cases}$

In Exercises 7–20, solve the system of equations algebraically.

7. $\begin{cases} x + y = 4 \\ x - y = 6 \end{cases}$

8. $\begin{cases} 5x - y = 14 \\ 3x - 2y = 0 \end{cases}$

9. $\begin{cases} y = 2x - 3 \\ x = 3y - 2 \end{cases}$

10. $\begin{cases} 3x = y + 4 \\ \quad x = 4y + 2 \end{cases}$

11. $\begin{cases} 2x - y = 10 \\ x + 3y = -16 \end{cases}$

12. $\begin{cases} \dfrac{1}{2}x + 2y = 1 \\ x + \dfrac{1}{3}y = \dfrac{17}{3} \end{cases}$

13. $\begin{cases} x - 5y = 1 \\ 3x - 2y = 3 \end{cases}$

14. $\begin{cases} 2x + 3y = 2 \\ x - 2y = -6 \end{cases}$

15. $\begin{cases} 4x - 3y = 10 \\ 9x + 2y = 5 \end{cases}$

16. $\begin{cases} 2x - 4y = 3 \\ \quad\quad 4x = 8y + 6 \end{cases}$

17. $\begin{cases} \dfrac{x}{2} + y = 5 \\ \quad 2y = 8 - x \end{cases}$

18. $\begin{cases} \dfrac{x + y}{2} = 6 \\ x - y = -4 \end{cases}$

19. $\begin{cases} x + y - 8 = 2x - 4 \\ \quad 2(y - x) = 8 \end{cases}$

20. $\begin{cases} \dfrac{x + y}{3} = x - 2 \\ x + \dfrac{y}{3} = y + 3 \end{cases}$

Write a system of equations, and solve each of the following:

21. Pure water is to be mixed with a 30% solution of alcohol to produce 30 gallons of a 25% solution. How much water should be added?

22. A total of 4,300 tickets valued at $18,800 were sold for a school concert. Adults' tickets cost $4.50 each, and children's tickets cost $4 each. How many of each type were sold?

23. Traci and Gene walk at the same rate and jog at the same rate. Traci walks for 1 hour and jogs for 1 hour, and covers a distance of 17 km. Gene walks for 2 hours and jogs for half an hour, and covers a distance of 16 km. Find the rate at which they both walk and the rate at which they both jog.

24. If *x* oz of a 60% alcohol solution are mixed together with *y* oz of a 40% alcohol solution, the resulting mixture contains 32 oz of alcohol. If *x* oz of a 50% alcohol solution are mixed together with *y* oz of a 30% alcohol solution, the resulting mixture contains 25 oz of alcohol. Find *x* and *y*.

25. The following graph represents the relationship between two quantities w and R:

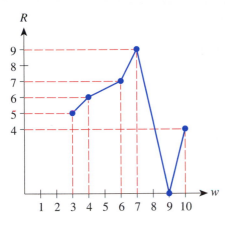

Use this graph to answer the following questions.

(a) What is the value of R when $w = 6$?

(b) As w increases from 7 to 9, are the values of R increasing or decreasing?

(c) What is the smallest value of R? For what value of w does this smallest value of R occur?

(d) What is the largest value of R? For what value of w does this largest value of R occur?

(e) Which is greater: the value of R when $w = 5$ or the value of R when $w = 10$?

26. The following graph shows the change in temperature in degrees Fahrenheit during a 5-hour period on a certain day, where F represents the temperature and h represents the hour.

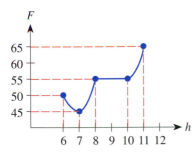

Use this graph to answer the following questions.

(a) What was the maximum temperature during this 5-hour period? At what time did it occur?

(b) What was the minimum temperature during this 5-hour period? At what time did it occur?

(c) Is there any time period during which the temperature remains constant? If so, what is this constant temperature?

(d) Is the temperature increasing or decreasing during the first hour? By how much does the temperature change during the first hour?

CHAPTER 6   PRACTICE TEST

1. The following graph illustrates the price fluctuations of a stock during the 4-hour period from 10:00 A.M. to 2:00 P.M. on a certain day. The horizontal axis represents the hour of the day, and the vertical axis represents the price (in dollars) of a single share of the stock.

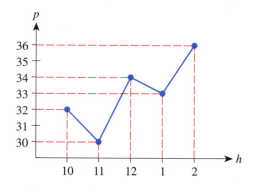

Use the graph to answer the following questions.

(a) What was the price of the stock at 10:00 A.M.?

(b) During what hours was the price of the stock falling?

(c) What was the lowest price of the stock during this 4-hour period?

(d) Did the price of the stock ever hit $32.50 per share? During which hour?

(e) During the fourth hour was the price of the stock rising or falling?

In Problems 2–6, solve the given system of equations algebraically.

2. $\begin{cases} 4x - 3y = 11 \\ 3x + y = 5 \end{cases}$

3. $\begin{cases} 3x - 4 = y - 1 \\ 9 + 3y = x \end{cases}$

4. $\begin{cases} 2x - 7y = 4 \\ 5x - 4y = 10 \end{cases}$

5. $\begin{cases} \dfrac{3x}{2} - y = 6 \\ x - \dfrac{2y}{3} = 5 \end{cases}$

6. $\begin{cases} 6x + 7 = 9y - 5 \\ 4x - 11 = 6y - 19 \end{cases}$

7. Solve the following system of equations by the graphical method. $\begin{cases} 2x - y = -8 \\ x + 2y = 6 \end{cases}$

Solve each of the following problems algebraically.

8. Tim ordered five orchestra tickets and three balcony tickets to a certain Broadway show for a total of $227. Nira ordered six orchestra tickets and two balcony tickets to the same show for a total of $238. What were the prices per orchestra and balcony ticket?

9. A total of $5,550 is split into two investments. Part of the money is invested at 4% and the remainder is invested at 5%. If the total annual interest from the two investments is $259, how much is invested at each rate?

10. James fences in a rectangular garden for which three times its width is one less than twice its length. If the perimeter of the rectangle is 46 feet, find the dimensions of the rectangle.

CHAPTERS 4–6   CUMULATIVE REVIEW

In Exercises 1–4, reduce the given fractions to lowest terms.

1. $\dfrac{-24}{42}$

2. $\dfrac{9x^2}{15x^6}$

3. $\dfrac{36s^8t^9}{20s^9t^8}$

4. $\dfrac{3a - 8a - a}{6a^2 + 2a^2 + 4a^2}$

In Exercises 5–20, perform the indicated operations and simplify as completely as possible.

5. $\dfrac{6x}{25} \cdot \dfrac{10}{x}$

6. $\dfrac{6x}{25} \div \dfrac{10}{x}$

7. $\dfrac{6x}{25} + \dfrac{10}{x}$

8. $\dfrac{7}{2x} - \dfrac{3}{2x}$

9. $\dfrac{3t - 5}{6t^2} + \dfrac{9t + 5}{6t^2}$

10. $\dfrac{6u - 7}{10u^3} - \dfrac{4u - 7}{10u^3}$

11. $\dfrac{12x^3y^2}{35z^2} \div \dfrac{20xy}{14z}$

12. $\dfrac{8uv^3}{9w^6} \cdot \dfrac{27w^2}{36v^6}$

13. $\dfrac{5}{3x} - \dfrac{7}{2x}$

14. $\dfrac{9}{3y} + \dfrac{11}{6z}$

15. $\dfrac{5}{6x^2y} - \dfrac{9}{10xy^3}$

16. $\dfrac{5}{6x^2y} \cdot \dfrac{9}{10xy^3}$

17. $\left(8 \cdot \dfrac{4}{x}\right) \div \dfrac{16}{x^2}$

18. $\left(\dfrac{5}{a} - \dfrac{1}{a}\right) \cdot \dfrac{a^2}{12}$

19. $2 + \dfrac{3}{x} - \dfrac{1}{x^2}$

20. $\dfrac{5}{st} - \dfrac{1}{6s^2} + \dfrac{3}{8st^2}$

In Exercises 21–28, solve the given equation.

21. $\dfrac{x}{3} - \dfrac{x}{4} = \dfrac{x - 4}{6}$

22. $\dfrac{t + 3}{8} + \dfrac{t - 2}{6} = \dfrac{t - 7}{12}$

23. $\dfrac{a}{5} - \dfrac{a}{6} = \dfrac{a}{30}$

24. $z - \dfrac{z}{4} = \dfrac{3z}{8}$

25. $\dfrac{7 - 2y}{4} - \dfrac{5 - 4y}{6} = \dfrac{8y + 5}{9}$

26. $\dfrac{x}{5} - \dfrac{x}{3} = \dfrac{1}{2}$

27. $0.8x - 0.07(x - 5) = 58.75$

28. $\dfrac{2}{3}(x + 1) - \dfrac{1}{2}(x - 7) = 7$

In Exercises 29–36, sketch the graph of the given equation in a rectangular coordinate system. Label the intercepts.

29. $y = 2x - 6$

30. $3x - 5y = 15$

31. $3y - 6x = 12$

32. $4x + 3y = 10$

33. $x - 2 = 0$

34. $y + 3 = 0$

35. $y = 5x$

36. $3y = x$

In Exercises 37–40, find the slope of the line passing through the given pair of points.

37. $(2, -1)$ and $(-3, 4)$

38. $(2, 0)$ and $(4, 5)$

39. $(2, 4)$ and $(-1, 4)$

40. $(3, 1)$ and $(3, -2)$

In Exercises 41–44, write an equation of the line with the given slope that passes through the given point.

41. $m = 4$; $(2, 3)$

42. $m = \dfrac{1}{2}$; $(-4, -1)$

43. $m = -\dfrac{3}{4}$; $(0, 3)$

44. m is undefined; $(0, 3)$

45. Write an equation of the line passing through the points $(-3, 5)$ and $(2, -2)$.

46. Write an equation of the line passing through the points $(0, 4)$ and $(4, 0)$.

In Exercises 47–48, solve the given system of equations graphically.

47. $\begin{cases} 2x - y = 7 \\ x + 2y = 6 \end{cases}$

48. $\begin{cases} 2x + 3y = 3 \\ 5x - 4y = 19 \end{cases}$

In Exercises 49–56, solve each system of equations algebraically.

49. $\begin{cases} 2x - y = 7 \\ x + 2y = 6 \end{cases}$

50. $\begin{cases} 2x + 3y = 3 \\ 5x - 4y = 19 \end{cases}$

51. $\begin{cases} 4x - 3y = 0 \\ 2x - y = \dfrac{1}{3} \end{cases}$

52. $\begin{cases} 5x - 7y = 13 \\ 6x - 4y = 20 \end{cases}$

53. $\begin{cases} y = 5x - 4 \\ x = 3y + 12 \end{cases}$

54. $\begin{cases} x - 8y = -4 \\ 4y - x = 1 \end{cases}$

55. $\begin{cases} x + \dfrac{y}{2} = 5 \\ 2x + y = 10 \end{cases}$

56. $\begin{cases} \dfrac{x}{3} - y = 2 \\ x - 3y = 5 \end{cases}$

Solve each of the following problems algebraically. Be sure to label what the variables represent. Round to the nearest hundredth where necessary.

57. A total of $2,850 was collected on the sale of 360 tickets to an art show. If some of the tickets cost $6.25 each and the rest cost $8.75 each, how many of each type were sold?

58. If there are 454 grams in 1 pound, how many grams are there in 15 ounces? [*Note:* There are 16 ounces in 1 pound.]

59. During an election the ratio of the number of people voting for party A to those voting for party B was 8 to 5. If party B received 15,700 votes, how many votes did party A receive?

60. Irma is going to make three investments paying 8%, 9%, and 10%. She will invest twice as much at 9% as at 8%, and $1,000 more at 10% than at 9%. If the yearly income from the three investments is $3,090, how much is invested altogether?

61. How long would it take someone driving at 80 kph to overtake someone with a 15-minute head start driving at 65 kph?

62. John goes into a clothing store and buys six shirts and two ties for $88.68. Bob buys four of the same priced shirts and three of the same priced ties for $68.27. Find the prices of a single shirt and a single tie.

63. Using local census data, the following graph shows how the population per square mile changed in a certain area from the year 1900 to the year 1990. The horizontal axis represents the year, and the vertical axis represents the population in number of people per square mile.

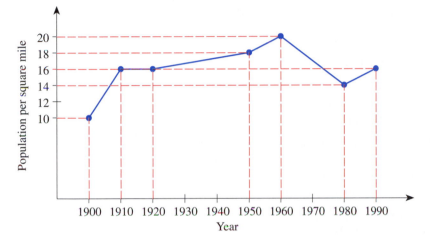

(a) What was the population per square mile in 1950?

(b) Is it accurate to say that the population per square mile increased during every 10-year period illustrated in the graph? Explain.

(c) During which 10-year period did the population per square mile increase the most? What was the increase during this 10-year period?

(d) During which 10-year period did the population per square mile remain constant?

CHAPTERS 4–6   CUMULATIVE PRACTICE TEST

In Exercises 1–6, perform the indicated operations and simplify as completely as possible. Final answers should be reduced to lowest terms.

1. $\dfrac{12s^2t^3}{5d^2} \cdot \dfrac{15d^5}{9st^4}$

2. $\dfrac{16x^3y^5}{9z^4} \div (36xyz)$

3. $\dfrac{11a}{9x} - \dfrac{a}{9x} + \dfrac{5a}{9x}$

4. $\dfrac{3x - 7}{2y} - \dfrac{9x - 7}{2y}$

5. $\dfrac{5}{6ab^2} + \dfrac{4}{9b}$

6. $\left(9 \cdot \dfrac{3}{x}\right) \div \left(\dfrac{30}{x^2}\right)$

7. Solve for a. $\dfrac{a}{6} - \dfrac{a}{9} = 18$

8. Solve for t. $\dfrac{2t - 3}{4} - \dfrac{t - 2}{8} = t + 2$

In Problems 9–11, sketch the graph of the equation in a rectangular coordinate system. Label the intercepts.

9. $x - 3y = 0$

10. $x - 3y = 6$

11. $x - 3 = 6$

12. Find the slope of the line passing through the points $(-2, 2)$ and $(3, 6)$.

13. Write an equation of the line passing through the points $(2, -4)$ and $(-1, 3)$.

In Problems 14–15, solve the system of equations algebraically.

14. $\begin{cases} 3x - 2y = 7 \\ 4x + y = -9 \end{cases}$

15. $\begin{cases} \dfrac{x}{2} - y = 5 \\ -x + 2y = 8 \end{cases}$

Solve each of the following problems algebraically. Be sure to clearly label what your variable represents.

16. The width of a rectangle is 5 more than one-third its length. If the perimeter is 34 m, what are its dimensions?

17. Jamie bought 28 stamps for a total of $10.96. Some were 34¢ stamps and the rest were 50¢ stamps. How many of each were bought?

18. A computer consultant charges $40 per hour for her time and $24 per hour for her assistant's time. On a certain project the assistant worked for 4 hours, after which time he was joined by the consultant and they completed the job together. If the total bill for the job was $480, how many hours did they work together?

19. Terry and Tom leave together from the same location walking in opposite directions. Terry leaves at 11:00 A.M. walking at 6 kph. Twenty minutes later Tom leaves walking at 8 kph. At what time will they be 9 km apart?

20. Sylvia buys four blouses at regular price and three blouses on special sale for a total of $55.30. Barbara buys two blouses at regular price and five on special sale for a total of $50.40. What are the prices of an individual blouse regularly and on special sale?

21. It is a commonly held belief that breakfast is the most important meal of the day. Nutritionists have found that the energy level you maintain after eating breakfast, as measured by the level of your blood sugar, depends on the type of breakfast you eat. If you eat a high-carbohydrate (HC) breakfast, your blood sugar level increases rapidly, reaches a peak, and then decreases rapidly. If you eat a high-protein (HP) breakfast, your blood sugar level increases more slowly, reaches a lower peak, and then decreases more slowly. Which of the following graphs is consistent with this information?

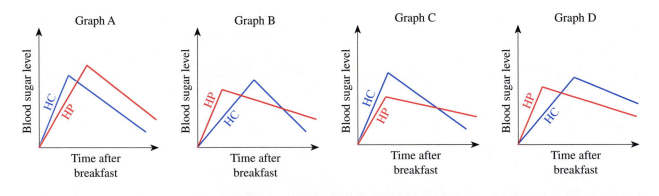

CHAPTER 7

Exponents and Polynomials

In this chapter we will be reviewing much of the material we covered in Chapter 2. We will begin by reviewing our work with exponents, and use our knowledge as a basis on which to extend our understanding of exponents beyond the natural numbers.

As we move on to the material on simplifying algebraic expressions, we will introduce some new terminology as well as extend the scope of examples that we can handle.

7.1 Exponent Rules

Ever since we introduced exponent notation in Section 2.1, we have been continuously using exponents and the first rule for exponents.

Keep in mind that our definition of a^n is for n a positive integer. If an exponent tells us how many times the base appears as a factor in the product, then the exponent must be a counting number (positive integer).

For example, if we have to compute

$$(x^3)(x^4)$$

we immediately get an answer of x^7 often without realizing that we are using the first rule for exponents. After all, the first rule for exponents is just a formal statement of our ability to count factors.

Let's begin our discussion here by reviewing and restating the first rule for exponents.

EXAMPLE 1   Simplify. $a^6 \cdot a^2$

Solution
$$a^6 \cdot a^2 = aaaaaa \cdot aa$$
$$= a^{6+2}$$
$$= \boxed{a^8}$$

     ∎

Exponent Rule 1	$$a^m \cdot a^n = a^{m+n}$$
	In words, this rule says that when we multiply powers of the same base, we keep the base and add the exponents.

Rule 1 is often called the ***product rule for exponents.***
It is proven as follows:

$$a^m \cdot a^n = (aaa \cdots a) \cdot (aaa \cdots a) = (aaa \cdots a) = a^{m+n}$$
$$\underbrace{\qquad}_{m\ factors} \quad \underbrace{\qquad}_{n\ factors} \quad \underbrace{\qquad}_{m\ +\ n\ factors}$$

Of course, as we have seen before, this rule extends to more than two powers of the same base:

$$a^m \cdot a^n \cdot a^p \cdots a^q = a^{m+n+p+\cdots+q}$$

Let's look at several more examples that will enable us to develop several other exponent rules.

EXAMPLE 2   Simplify. $(a^2)^6$

Solution
$$(a^2)^6 = a^2 \cdot a^2 \cdot a^2 \cdot a^2 \cdot a^2 \cdot a^2 \qquad \text{\textit{Use exponent rule 1.}}$$
$$= a^{2+2+2+2+2+2} \qquad\qquad\qquad \text{\textit{Write the repeated addition as}}$$
$$\text{\textit{multiplication.}}$$

$$= a^{6 \cdot 2}$$

$$= \boxed{a^{12}}$$

◼

We can generalize this to the second rule for exponents.

Exponent Rule 2	$(a^m)^n = a^{n \cdot m}$
	In words, this rule says that when we raise a power to a power we keep the base and multiply the exponents.

Rule 2 is often called the ***power of a power rule.***

Sometimes rules 1 and 2 are confused (When do I add the exponents and when do I multiply them?). Keep these differences in mind:

Exponent rule 1 says that when we multiply *two* powers of the same base we keep the base and *add* the exponents.

Exponent rule 2 says that when we raise a power to a power we keep the base and *multiply* the exponents.

Even more basic, however, is the fact that if you are in doubt as to which rule to use, write down the expression and simply count factors. After all, that is where the rules come from in the first place.

EXAMPLE 3   Simplify. $\dfrac{a^6}{a^2}$

Solution $\dfrac{a^6}{a^2} = \dfrac{aaaaaa}{aa}$ *Reduce.*

$$= \frac{\cancel{aa}aaaa}{\cancel{aa}}$$

$$= a^{6-2} = \boxed{a^4}$$

◼

Basically all we have done is cancel as many factors as we have in the denominator with the same number of factors in the numerator.

We can generalize this to the third rule for exponents.

Exponent Rule 3	$\dfrac{a^m}{a^n} = a^{m-n}, \qquad m > n \ (a \neq 0)$
	In words, this rule says that when we divide powers of the same base we keep the base and subtract the bottom exponent from the top exponent.

Rule 3 is often called the ***quotient rule for exponents.***

In order to use rule 2 we must have $m > n$; otherwise, when we subtract the exponents, we will get a zero or negative exponent, which we have not yet defined. This restriction is a nuisance, and we will see how to remove it in the next section.

However, this does not prevent us from reducing an expression like $\dfrac{a^2}{a^6}$ as we usually do:

$$\frac{a^2}{a^6} = \frac{aa}{aaaaaa} = \frac{\cancel{aa}}{\cancel{aa}aaaa} = \boxed{\frac{1}{a^4}}$$

Of course, as always, we restrict all variables from making any denominator equal to 0. Thus, in rule 3 we must have $a \neq 0$.

EXAMPLE 4 | Remove the parentheses. $(ab)^5$

Solution

$$(ab)^5 = (ab)(ab)(ab)(ab)(ab) \qquad \textit{We reorder and regroup using the commutative and}$$
$$= (aaaaa)(bbbbb) \qquad \textit{associative properties.}$$
$$= \boxed{a^5 b^5} \qquad \blacksquare$$

We can generalize this to the fourth rule for exponents.

Exponent Rule 4	$(ab)^n = a^n b^n$
	In words, this says that if a *product* is raised to a power, then each *factor* gets raised to that power.

EXAMPLE 5 | Remove the parentheses. $\left(\dfrac{a}{b}\right)^5$

Solution

$$\left(\frac{a}{b}\right)^5 = \left(\frac{a}{b}\right) \cdot \left(\frac{a}{b}\right) \cdot \left(\frac{a}{b}\right) \cdot \left(\frac{a}{b}\right) \cdot \left(\frac{a}{b}\right) \qquad \textit{Multiply the fractions.}$$
$$= \boxed{\frac{a^5}{b^5}} \qquad \blacksquare$$

We can generalize this to the fifth rule for exponents.

Exponent Rule 5	$\left(\dfrac{a}{b}\right)^n = \dfrac{a^n}{b^n} \qquad b \neq 0$
	In words, this says that if a *quotient* is raised to a power, then the numerator and the denominator each get raised to that power.

For ease of reference we summarize all five exponent rules in the box.

Exponent Rules	For m and n positive integers:	
	Rule 1	$a^m \cdot a^n = a^{m+n}$ *Product rule*
	Rule 2	$(a^m)^n = a^{n \cdot m}$ *Power of a power rule*
	Rule 3	$\left(\dfrac{a^m}{a^n}\right) = a^{m-n} \qquad m > n \quad (a \neq 0) \qquad$ *Quotient rule*
	Rule 4	$(ab)^n = a^n b^n$ *Power of a product rule*
	Rule 5	$\left(\dfrac{a}{b}\right)^n = \dfrac{a^n}{b^n} \qquad b \neq 0 \qquad$ *Power of a quotient rule*

In the next few examples we will see how these rules work together to make it easier for us to simplify expressions involving exponents. Keep in mind that we will be showing *all* the steps in the solutions. As you get the hang of it, you can easily leave out or combine some of the steps.

EXAMPLE 6

Simplify as completely as possible.

(a) $(x^3y^5)^4$ (b) $\left(\dfrac{s^3}{t^7}\right)^6$

Solution

(a) We begin by recognizing that basically we have a *product* raised to a power, and therefore our first step in the solution is to apply exponent rule 4.

$(x^3y^5)^4$ *We have a product raised to a power. Rule 4 says that the outside exponent gets applied to each factor inside. We are viewing x^3 and y^5 as two separate factors.*

$= (x^3)^4(y^5)^4$ *Now for each power of a power, rule 2 says that we keep the base and multiply the exponents.*

$= x^{4\cdot 3}y^{4\cdot 5}$

$= \boxed{x^{12}y^{20}}$

(b) This time we begin with rule 5 because basically we have a *quotient* raised to a power.

$\left(\dfrac{s^3}{t^7}\right)^6$ *We have a quotient raised to a power. Rule 5 says the outside exponent 6 applies to numerator and denominator.*

$= \dfrac{(s^3)^6}{(t^7)^6}$ *Now use rule 2 in both numerator and denominator.*

$= \dfrac{s^{6\cdot 3}}{t^{6\cdot 7}}$

$= \boxed{\dfrac{s^{18}}{t^{42}}}$ ■

EXAMPLE 7

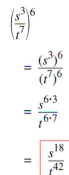

Simplify as completely as possible. $\dfrac{(x^2)^4 x^6}{(xy)^5}$

Solution

We begin by removing the parentheses in the numerator and the denominator.

$\dfrac{(x^2)^4 x^6}{(xy)^5}$ *In the numerator we have a power of a power. Rule 2 says keep the base and multiply the exponents. In the denominator we have a product raised to a power. Rule 4 says the outside exponent of 5 gets applied to each factor.*

$= \dfrac{x^{4\cdot 2}x^6}{x^5 y^5}$

$= \dfrac{x^8 x^6}{x^5 y^5}$ *In the numerator we are multiplying powers of the same base. Rule 1 says we keep the base and add the exponents.*

$$= \frac{x^{8+6}}{x^5 y^5}$$

$$= \frac{x^{14}}{x^5 y^5} = \frac{x^{14}}{x^5} \cdot \frac{1}{y^5} \qquad \textit{Use rule 3 on the x's.}$$

$$= x^{14-5} \cdot \frac{1}{y^5} = x^9 \cdot \frac{1}{y^5}$$

$$= \boxed{\frac{x^9}{y^5}}$$

■

Clearly a number of the steps here are usually mental steps. You should write down as many steps as *you* think are necessary.

Up to this point we have been very detailed in our explanations of the use of the exponent rules. From this point on we will usually refer to the rules by number only. Be sure you understand which rule is being used and why.

Problems involving exponents often lend themselves to more than one method of solution, as the next example illustrates.

EXAMPLE 8

Simplify as completely as possible. $\left(\dfrac{2x^3 y^5}{xy^2} \right)^4$

Solution

We offer two possible approaches.

Solution 1 offers the "inside-out" approach.

$$\left(\frac{2x^3 y^5}{xy^2} \right)^4 \qquad \textit{We work inside the parentheses first. Since this is a quotient, we apply rule 3.}$$

$$= (2x^{3-1} y^{5-2})^4$$

$$= (2x^2 y^3)^4 \qquad \textit{Now we use rule 4 on the product.}$$

$$= 2^4 (x^2)^4 (y^3)^4 \qquad \textit{We evaluate } 2^4 \textit{ and use rule 2.}$$

$$= \boxed{16x^8 y^{12}}$$

Solution 2 offers the "outside-in" approach.

$$\left(\frac{2x^3 y^5}{xy^2} \right)^4 \qquad \textit{We view this expression as a quotient raised to a power, and so we use rule 5 to apply the outside exponent to both numerator and denominator.}$$

$$= \frac{(2x^3 y^5)^4}{(xy^2)^4} \qquad \textit{Now we use rule 4 in the numerator and denominator.}$$

$$= \frac{2^4 (x^3)^4 (y^5)^4}{x^4 (y^2)^4} \qquad \textit{Now we use rule 2 on each power of a power.}$$

$$= \frac{16x^{12} y^{20}}{x^4 y^8} \qquad \textit{Now we use rule 3 on the x's and the y's.}$$

We could reduce this fraction as we discussed in Chapter 2. However, in this chapter we are emphasizing the exponent aspect of these expressions.

$$= 16 \left(\frac{x^{12}}{x^4} \right) \left(\frac{y^{20}}{y^8} \right)$$

$$= 16x^{12-4}y^{20-8}$$

$$= \boxed{16x^8y^{12}} \quad \blacksquare$$

EXAMPLE 9 Simplify as completely as possible. $\dfrac{(-3x^2y^3)^4}{6(xy)^3}$

Solution We begin by using rule 4 in both the numerator and the denominator.

$$\frac{(-3x^2y^3)^4}{6(xy)^3} = \frac{(-3)^4(x^2)^4(y^3)^4}{6x^3y^3}$$

Be careful: The 6 in the denominator is outside the parentheses and therefore does not get raised to the third power. Also, we cannot reduce the −3 with the 6. Why not? We evaluate $(-3)^4$ and use rule 2.

$$= \frac{81x^8y^{12}}{6x^3y^3}$$

Now we use rule 3.

$$= \frac{81}{6}x^{8-3}y^{12-3}$$

$$= \boxed{\frac{27}{2}x^5y^9} \quad \text{which is the same as} \quad \boxed{\frac{27x^5y^9}{2}} \quad \blacksquare$$

Knowing what the rules are does not make you good at using them. The only way to be sure you can apply the rules correctly is by doing the exercises.

EXERCISES 7.1

Simplify each of the following as completely as possible.

1. x^3x^2

2. y^4y^3

3. $(x^3)^2$

4. $(y^4)^3$

5. x^3xx^5

6. a^3a^7a

7. 10^410^5

8. 7^37^8

9. 2^33^4

10. 3^25^3

11. $\dfrac{y^3y^5}{y^2y^4}$

12. $\dfrac{z^2z^7}{z^4z^6}$

13. $\dfrac{9u^9v^8}{3u^3v^4}$

14. $\dfrac{16r^{16}v^4}{8r^8v^8}$

15. $\dfrac{(a^3)^5}{(a^4)^2}$

16. $\dfrac{(x^4)^3}{(x^3)^2}$

17. $(-x^2)^4$

18. $-(x^2)^4$

19. $(x^2y)^2$

20. $(2xy^2)^3$

21. $(x^2y^3)^5$

22. $(a^2b^4)^3$

23. $(2r^3s^5)^4$

24. $2(r^3s^5)^4$

25. $(-x^3y)^3$

26. $(-x^3y)^4$

27. $\left(\dfrac{x^3}{y^2}\right)^4$

28. $\left(\dfrac{u^5}{v^2}\right)^3$

29. $(2x^3)^4(3x^2)^2$

30. $(3a^2)^3(5a^3)^2$

31. $\dfrac{(x^4y^2)^3}{x^5(y^3)^2}$

32. $\dfrac{(a^3b^4)^2}{(a^2)^3b^2}$

33. $\dfrac{(3x^5y^4)^2}{9(x^3y)^3}$

34. $\dfrac{16(ab^2)^5}{(2ab^4)^2}$

35. $\left(\dfrac{x^2y}{4u}\right)^4$

36. $\left(\dfrac{y^3z^5}{3c}\right)^4$

37. $\left(\dfrac{2x^3y^4}{xy^6}\right)^5$

38. $\left(\dfrac{3a^5b^3}{a^3b^6}\right)^4$

39. $\left(\dfrac{-3a^2b^3}{2c}\right)^3$

40. $\left(\dfrac{4u^5v^2}{-5w^2}\right)^3$

41. $\dfrac{-3^2}{(-3)^2}$

42. $\dfrac{(-5)^2}{-5^2}$

43. $\dfrac{-x^2}{(-x)^2}$

44. $\dfrac{(-a)^4}{-a^4}$

45. $\dfrac{-u^3}{(-u)^3}$

46. $\dfrac{(-z)^5}{-z^5}$

47. $\dfrac{(-x^2)^4}{-(x^3)^2}$

48. $\dfrac{-(a^4)^2}{(-a^2)^3}$

49. $\dfrac{-2^4 + 3^2}{(-4 + 3)^2}$

50. $\dfrac{-3^2 + 4^2}{-2^2 - 3}$

51. $\dfrac{-4^2 - (-2)^5}{1 - 3^2}$

52. $\dfrac{(1 - 5)^3 - 2^4}{-2^3 - 2}$

53. $\dfrac{2(-6)^2 - 3(-5)^2}{(5 - 6)^3}$

54. $\dfrac{(3 - 5 - 2 - 6)^2}{(-1 - 2 - 3 - 4)^2}$

In Exercises 55–62, evaluate the given expression for $w = 2$, $x = -3$, $y = 4$, and $z = -5$.

55. $\dfrac{-w^2}{(-w)^2}$

56. $\dfrac{(-z^2)}{-z^2}$

57. $\dfrac{-w^4 + x^2}{(-y + x)^2}$

58. $\dfrac{-x^2 + y^2}{-w^2 - x}$

59. $(w - x + y - z)^2$

60. $\left(\dfrac{w - x}{y - z}\right)^2$

61. $\dfrac{x(-y)}{x - y}$

62. $\dfrac{y(-z)}{w - y}$

? QUESTIONS FOR THOUGHT

63. State in words the difference between the first and second rules for exponents.

64. Explain what x^8 means. In light of this answer, explain why $x^5 + x^3$ is not the same as x^8.

65. If exponent rule 3 is applied to the following expressions, what will you get?

    (a) $\dfrac{x^8}{x^8}$

    (b) $\dfrac{x^4}{x^7}$

66. State what is **wrong** with each of the following:

    (a) $(x^2)^3 \stackrel{?}{=} x^5$

    (b) $x^4 \cdot x^3 \stackrel{?}{=} x^{12}$

    (c) $\dfrac{x^6}{x^2} \stackrel{?}{=} x^3$

    (d) $x^2 + x^3 \stackrel{?}{=} x^5$

◇ **MINI-REVIEW**

In Exercises 67–71, perform the indicated operation and simplify.

67. $\dfrac{2}{x} \cdot \dfrac{8}{y}$

68. $\dfrac{2}{x} \div \dfrac{8}{y}$

69. $\dfrac{2}{x} + \dfrac{8}{y}$

70. $\dfrac{3}{2a} - \dfrac{4}{a^2}$

71. $\dfrac{x}{3} - \dfrac{x}{4}$

72. *Solve for x.* $\dfrac{x}{3} - \dfrac{x}{4} = 3$

73. A car rental agency charges $15 per day plus $0.20 per mile. If a 3-day rental cost $72, how many miles were driven?

74. A collection of 40 coins consists of dimes and quarters. If the total value of the coins is $6.70, how many of each are there?

7.2 Zero and Negative Exponents

Thus far everything we have said about exponents has depended on the fact that our exponents are positive integers. We know, for example, that 5^4 means "take the number 5 as a factor 4 times."

All well and good, but what should 5^0 mean? It certainly cannot mean take the number 5 as a factor zero times—that would give us a blank page. It should be clear that whatever meaning we decide to give to 5^0, the exponent will no longer just be counting the factors for us.

In the last section we discussed the five basic exponent rules. These rules were a direct consequence of our ability to count factors. There (hopefully) was no mystery as to where those rules came from.

Since at this point we have no real motivation as to how we want to define a zero or a negative exponent, we will do an about-face. Instead of our understanding of exponents telling us what the rules are, we will let the rules tell us how to make "sense" out of zero and negative exponents. In other words, whatever meaning we give to these new exponents we certainly do not want to have to learn a new set of exponent rules for them. Consequently, we will let the rules tell us how to define these new exponents in a reasonable way.

Let's be more specific. Consider the expression

$$\dfrac{5^3}{5^3}$$

We know from arithmetic that $\dfrac{5^3}{5^3} = 1$. On the other hand, if we want to be able to apply exponent rule 3, we must have

$$\frac{5^3}{5^3} = 5^{3-3} = 5^0$$

5^0 cannot mean "write 5 as a factor 0 times."

Therefore, in order to be consistent, we must have $5^0 = 1$.

Similarly,

$$\frac{7^6}{7^6} = 1 \quad \text{and we want} \quad \frac{7^6}{7^6} = 7^{6-6} = 7^0$$

Again, to be consistent, we must have $7^0 = 1$.

Thus, it should not be surprising that we make the following definition.

DEFINITION	$a^0 = 1$ for all $a \neq 0$. *Note:* 0^0 is undefined.

In words, this definition says that any nonzero quantity raised to the zero power is always equal to 1.

It is important to realize that a definition is neither right nor wrong—it just is. The proper question to ask about a definition is: "Is it useful?" The answer in this case is yes for many reasons, one of which is the following: Suppose someone decides to define $a^0 = 14$ because 14 happens to be his or her favorite number. This is perfectly legal *but* causes the following difficulty.

Suppose we consider an expression involving exponents such as $a^0 a^4$. Using exponent rule 1, we get

$$a^0 a^4 = a^{0+4} = a^4$$

while using the "other definition," we get

$$a^0 a^4 \stackrel{?}{=} 14 a^4$$

which is not the same answer. The rule and this other definition are not consistent. Something has to give. Either we would have to give up all the exponent rules, or we have to give up this other definition. Clearly, we do not want to give up the rules.

However, with our definition we do not face this problem. Using our definition, we get

$$a^0 a^4 = 1 \cdot a^4 = a^4$$

which is the same answer obtained using rule 1. It can be shown that our definition is consistent with all five of our exponent rules.

The same type of reasoning will lead us to a definition of negative exponents. We know that

$$\frac{5^3}{5^7} = \frac{5 \cdot 5 \cdot 5}{5 \cdot 5 \cdot 5 \cdot 5 \cdot 5 \cdot 5 \cdot 5} = \frac{\cancel{5} \cdot \cancel{5} \cdot \cancel{5}}{\cancel{5} \cdot \cancel{5} \cdot \cancel{5} \cdot 5 \cdot 5 \cdot 5 \cdot 5} = \frac{1}{5^4}$$

On the other hand, using exponent rule 3 on this same expression, we get

$$\frac{5^3}{5^7} = 5^{3-7} = 5^{-4}$$

It is meaningless to say "5^{-4} means write 5 as a factor -4 times."

Thus, for the rules and our definition to be consistent, we want

$$5^{-4} = \frac{1}{5^4}$$

Similarly, we get

$$\frac{8^2}{8^5} = \frac{8 \cdot 8}{8 \cdot 8 \cdot 8 \cdot 8 \cdot 8} = \frac{\cancel{8} \cdot \cancel{8}}{\cancel{8} \cdot \cancel{8} \cdot 8 \cdot 8 \cdot 8} = \frac{1}{8^3}$$

while using exponent rule 3 gives us

$$\frac{8^2}{8^5} = 8^{2-5} = 8^{-3}$$

Based on this analysis we make the following definition.

DEFINITION	For any integer n, $\quad a^{-n} = \dfrac{1}{a^n}, \quad a \neq 0.$

In words, the definition says that a^{-n} is the *reciprocal* of a^n, where a is any nonzero quantity. Remember that the reciprocal of x is $\dfrac{1}{x}$.

In all the examples that follow, we will assume that our variables are not equal to 0 so that we can apply the definitions.

EXAMPLE 1 Write each of the following expressions without zero or negative exponents. (Assume $x \neq 0$.)

(a) 8^0 (b) $4x^0$ (c) $(4x)^0$ (d) x^{-3}

(e) $4x^{-3}$ (f) 3^{-4} (g) -3^4 (h) -2^{-4} (i) $(-2)^{-4}$

Solution

(a) $8^0 = \boxed{1}$ *By the definition of a zero exponent*

(b) $4x^0 = 4 \cdot 1 = \boxed{4}$ *The zero exponent applies only to the x.*

(c) $(4x)^0 = \boxed{1}$ *The zero exponent applies to the entire 4x.*

(d) $x^{-3} = \boxed{\dfrac{1}{x^3}}$ *By the definition of a negative exponent*

(e) $4x^{-3} = 4 \cdot \dfrac{1}{x^3} = \boxed{\dfrac{4}{x^3}}$ *The −3 exponent applies only to the x.*

(f) $3^{-4} = \dfrac{1}{3^4} = \boxed{\dfrac{1}{81}}$ *By the definition of a negative exponent*

(g) $-3^4 = -(3 \cdot 3 \cdot 3 \cdot 3) = \boxed{-81}$ *Note how this differs from part (**f**).*

(h) $-2^{-4} = -\dfrac{1}{2^4} = \boxed{-\dfrac{1}{16}}$ *The −4 exponent applies only to the 2.*

(i) $(-2)^{-4} = \dfrac{1}{(-2)^4} = \boxed{\dfrac{1}{16}}$ *By the definition of a negative exponent*

Note the difference between parts (**h**) and (**i**). ∎

Several comments are in order. First, whenever we are simplifying an expression involving negative exponents, we expect the *final* answer to be expressed with positive exponents only. Second, as we stated earlier, all the exponent rules are totally consistent with our definitions of zero and negative exponents. In particular, exponent rule 3 can be used regardless of whether the exponent in the numerator is larger than the exponent in the denominator.

We are free to use the exponent rules whenever it is appropriate to do so. In reality, *the hardest part of much of algebra is learning what is appropriate for a particular situation.*

Let's summarize all the exponent rules again, this time removing the restriction in rule 3 and including our definitions of zero and negative exponents.

Exponent Rules	For m and n any integers:	
Definition	$a^0 = 1$ for all $a \neq 0$	
Definition	$a^{-n} = \dfrac{1}{a^n}$ for all $a \neq 0$	
Rule 1	$a^m \cdot a^n = a^{m+n}$	*Product rule*
Rule 2	$(a^m)^n = a^{n \cdot m}$	*Power of a power rule*
Rule 3	$\left(\dfrac{a^m}{a^n}\right) = a^{m-n}, \quad a \neq 0$	*Quotient rule*
Rule 4	$(ab)^n = a^n b^n$	*Power of a product rule*
Rule 5	$\left(\dfrac{a}{b}\right)^n = \dfrac{a^n}{b^n}, \quad b \neq 0$	*Power of a quotient rule*

EXAMPLE 2   Simplify. $\dfrac{1}{a^{-3}}$

Solution $\dfrac{1}{a^{-3}}$ *Using the definition of a^{-n} on a^{-3}, we get*

$= \dfrac{1}{\left(\dfrac{1}{a^3}\right)}$ *Using the rule for dividing fractions, we invert the divisor and multiply.*

$= 1 \cdot \dfrac{a^3}{1} = \boxed{a^3}$

We could also have gotten this answer just by thinking about what the example is saying. We have already said that a^{-n} is the reciprocal of a^n. Of course, the reverse is also true because we know that being a reciprocal is a two-way street. Just as $\frac{2}{3}$ is the reciprocal of $\frac{3}{2}$, so too is $\frac{3}{2}$ the reciprocal of $\frac{2}{3}$.

Thus, a^n and a^{-n} are reciprocals *of each other:*

$$a^{-n} = \frac{1}{a^n} \quad \text{and} \quad a^n = \frac{1}{a^{-n}}$$

In other words, we are using the definition of $a^{-n} = \dfrac{1}{a^n}$ for $n = -3$.

$$\frac{1}{a^{-3}} = a^{-(-3)} = a^3$$ ■

EXAMPLE 3   Simplify. $\dfrac{x^3}{x^9}$

Solution $\dfrac{x^3}{x^9} = x^{3-9}$

$$= x^{-6} \qquad \textit{Now use the definition of a negative exponent.}$$

$$= \boxed{\dfrac{1}{x^6}} \qquad\qquad\qquad\qquad\qquad \blacksquare$$

If you are thinking to yourself, "Couldn't Example 3 have been done by simply reducing the fraction, as we did in the past?" you are perfectly correct. We used rule 3 just to illustrate it, not because it necessarily offered any advantages. However, if such an example involves negative exponents, then the use of rule 3 (or the other exponent rules, if they apply) is advisable.

Another important thing to keep in mind is not to confuse a negative exponent with a negative answer. That is, do not confuse the sign of the exponent with the sign of the base. For example:

$$2^{-1} \neq -2$$

$$2^{-1} = \dfrac{1}{2} \qquad \textit{The negative exponent does not make anything negative,}$$
$$\textit{but rather it tells us to take the reciprocal of the base.}$$

```
←——+———+———+———+—●—+———+———→
   −3   −2   −1    0  ↑  1    2
                      2⁻¹
```

EXAMPLE 4

Simplify as completely as possible.

(a) $x^{-3}x^{-9}$ (b) $\dfrac{x^{-3}}{x^{-9}}$ (c) $\dfrac{x^{-9}}{x^{-3}}$ (d) $(x^{-3})^{-9}$ (e) $\dfrac{x^{-3}}{y^{-9}}$

Solution

Many students fall into a "one-over syndrome." That is, as soon as they see a negative exponent, they use the definition

$$a^{-n} = \dfrac{1}{a^n}$$

While this is not wrong, it can often make the expression more messy. It is generally better to apply any appropriate exponent rules *before* using the definition.

(a) $x^{-3}x^{-9}$ *Using rule 1, we get*

$$= x^{-3+(-9)}$$

$$= x^{-12} \qquad \textit{Since we want our final answer with positive exponents only,}$$
$$\textit{\textbf{now} we use the definition of } a^{-n}.$$

$$= \boxed{\dfrac{1}{x^{12}}}$$

(b) $\dfrac{x^{-3}}{x^{-9}}$ *Using rule 3, which says keep the common base and subtract the bottom exponent from the top one, we get*

$$= x^{-3-(-9)} \qquad \textit{Remember we are subtracting a negative number.}$$

$$= x^{-3+9}$$

$$= \boxed{x^6}$$

(c) $\dfrac{x^{-9}}{x^{-3}}$ *Using rule 3, we get*

$$= x^{-9-(-3)}$$

$$= x^{-6} = \boxed{\dfrac{1}{x^6}}$$

(d) $(x^{-3})^{-9}$ *This is a power of a power. Rule 2 says keep the base and*
    $= x^{(-9)(-3)}$ *multiply the exponents.*

    $= \boxed{x^{27}}$

Note that in parts **(b)** and **(d)** of this example we never had to use the definition of a negative exponent at all. Also, look at parts **(b)** and **(c)**. Do you see why the two answers should come out to be reciprocals of each other?

(e) Note that none of the five exponent rules applies to this example. We really have no choice but to begin by using the definition of a negative exponent.

$$\frac{x^{-3}}{y^{-9}} = \frac{\dfrac{1}{x^3}}{\dfrac{1}{y^9}}$$ *This complex fraction means $\dfrac{1}{x^3} \div \dfrac{1}{y^9}$. We use the rule for dividing fractions.*

$$= \frac{1}{x^3} \cdot \frac{y^9}{1}$$

$$= \boxed{\frac{y^9}{x^3}}$$ ■

Let's look at a few more examples similar to those of the last section, but now involving negative exponents as well. Wherever possible we will take the same approach as we did when the examples involved only positive exponents.

EXAMPLE 5   Simplify. $(x^{-3}y^2)^{-4}$

Solution $(x^{-3}y^2)^{-4}$ *We have a product raised to a power, so using rule 4, we get*

    $= (x^{-3})^{-4}(y^2)^{-4}$ *Using rule 2, we get*
    $= x^{12}y^{-8}$ *Using the definition of a^{-n}, we get*

    $= x^{12} \cdot \dfrac{1}{y^8}$

    $= \boxed{\dfrac{x^{12}}{y^8}}$ ■

EXAMPLE 6   Simplify. $\dfrac{x^2y^{-2}}{x^{-4}y^3}$

Solution   We separate the x and y factors and treat this expression as the product of two fractions.

$$\frac{x^2y^{-2}}{x^{-4}y^3} = \left(\frac{x^2}{x^{-4}}\right)\left(\frac{y^{-2}}{y^3}\right)$$ *Use rule 3 in each fraction.*

$$= x^{2-(-4)} \cdot y^{-2-3}$$

$$= x^6y^{-5}$$ *Now we use the definition of a negative exponent.*

$$= x^6 \cdot \frac{1}{y^5}$$

$$= \boxed{\frac{x^6}{y^5}}$$ ■

EXAMPLE 7 Simplify. $\dfrac{(x^2x^{-3})^4}{x^3}$

Solution $\dfrac{(x^2x^{-3})^4}{x^3}$ *We use the product rule (rule 1) inside the parentheses.*

$= \dfrac{(x^{-1})^4}{x^3}$ *We use the power of a power rule (rule 2) in the numerator to get*

$= \dfrac{x^{-4}}{x^3}$ *Now we use the quotient rule (rule 3) to get*

$= x^{-4-3}$

$= x^{-7}$ *Finally, we use the definition of a^{-n}.*

$= \boxed{\dfrac{1}{x^7}}$ ■

EXAMPLE 8 Simplify. $\dfrac{(2x^2y^{-3})^{-1}}{(x^{-3}y)^{-2}}$

Solution $\dfrac{(2x^2y^{-3})^{-1}}{(x^{-3}y)^{-2}}$ *Use rule 4 in numerator and denominator to get*

$= \dfrac{2^{-1}(x^2)^{-1}(y^{-3})^{-1}}{(x^{-3})^{-2}y^{-2}}$ *Using rule 2 in numerator and denominator, we get*

$= \dfrac{2^{-1}x^{-2}y^3}{x^6y^{-2}}$ *Using rule 3, we get*

$= 2^{-1}x^{-2-6}y^{3-(-2)}$

$= 2^{-1}x^{-8}y^5$ *Using the definition of a^{-n}, we get*

$= \dfrac{1}{2} \cdot \dfrac{1}{x^8} \cdot \dfrac{y^5}{1}$

$= \boxed{\dfrac{y^5}{2x^8}}$ ■

EXAMPLE 9 Simplify. $\left(\dfrac{x^{-6}y^3}{x^5y^{-1}}\right)^{-2}$

Solution Working from the inside out, we have

$\left(\dfrac{x^{-6}y^3}{x^5y^{-1}}\right)^{-2}$ *Use rule 3 inside the parentheses to get*

$= (x^{-6-5}y^{3-(-1)})^{-2}$

$= (x^{-11}y^4)^{-2}$ *Using rule 4, we get*

$= (x^{-11})^{-2}(y^4)^{-2}$ *Using rule 2, we get*

$= x^{22}y^{-8}$ *Using the definition of a^{-n}, we get*

$= \boxed{\dfrac{x^{22}}{y^8}}$ ■

EXAMPLE 10 Express using positive exponents only.

     (a) $(x^{-2}y^{-2})^{-1}$ **(b)** $x^{-2} + y^{-2}$ **(c)** $(x^2 + y^2)^{-1}$

Solution | Be careful! It is just as important to know when a particular exponent rule does *not* apply as to know when it does.

(a) $(x^{-2}y^{-2})^{-1}$ *Using rule 4, we get*

$\quad = (x^{-2})^{-1}(y^{-2})^{-1}$

$\quad = \boxed{x^2y^2}$

(b) $x^{-2} + y^{-2} = \boxed{\dfrac{1}{x^2} + \dfrac{1}{y^2}}$ *Obtained by using the definition of a^{-n}*

(c) Rule 4 *does not apply* here because of the addition inside the parentheses. Rule 4 applies only when there is a *product* in the parentheses.

$(x^2 + y^2)^{-1} = \boxed{\dfrac{1}{x^2 + y^2}}$ *Obtained by using the definition of a^{-n}*

Also note that the answers to parts (b) and (c) are not the same.   ■

 EXERCISES 7.2

Exercises 1–6 consist of a number of related parts. Simplify each as completely as possible.

1. (a) $2(-3)$
   (b) x^2x^{-3}
   (c) $(x^2)^{-3}$
   (d) 2^{-3}

2. (a) $-2(-3)$
   (b) $x^{-2}x^{-3}$
   (c) $(x^{-2})^{-3}$
   (d) $(-2)^{-3}$

3. (a) $3(-4)$
   (b) x^3x^{-4}
   (c) $(x^3)^{-4}$
   (d) 3^{-4}

4. (a) $-3(-4)$
   (b) $x^{-3}x^{-4}$
   (c) $(x^{-3})^{-4}$
   (d) -3^{-4}

5. (a) $3 - 4$
   (b) x^3x^{-4}
   (c) $3(-4)$
   (d) $(x^3)^{-4}$
   (e) $3 - (-4)$
   (f) $\dfrac{x^3}{x^{-4}}$
   (g) 3^{-4}

6. (a) $-5 - 3$
   (b) $x^{-5}x^{-3}$
   (c) $-5(-3)$
   (d) $(x^{-5})^{-3}$
   (e) $-5 - (-3)$
   (f) $\dfrac{x^{-5}}{x^{-3}}$
   (g) $(-5)^{-3}$

Simplify each of the following expressions as completely as possible. Final answers should be expressed with positive exponents only. (Assume that all variables represent positive quantities.)

7. 8^0

8. $(-27)^0$

9. $5 \cdot 4^0$

10. $7 \cdot 7^0$

11. xy^0

12. $(xy)^0$

13. 5^{-2}

14. a^{-4}

15. $\dfrac{1}{5^{-2}}$

16. $\dfrac{1}{a^{-4}}$

17. $x^{-4}x^4$

18. x^6x^{-6}

19. $x^{-4}x^{-6}$

20. $(x^{-4})^{-6}$

21. $a^2a^{-4}aa^{-7}$

22. $z^3z^2z^{-4}z^{-7}$

23. $10^{-3}10^8$

24. 10^710^{-5}

25. $10^610^{-5}10^{-4}$

26. $10^{-2}10^{-4}10^9$

27. 10^710^{-7}

28. $x^{-5}x^5$

29. $(xy)^4$

30. $(xy)^{-5}$

31. $2a^{-3}$

32. $(2a)^{-3}$

33. $-3y^{-2}$

34. $(-3y)^{-2}$

35. $-(3y^{-2})$

36. $-(3y)^{-2}$

37. xy^{-1}

38. $(xy)^{-1}$

39. $(a^4b^3)^{-2}$

40. $(a^5b^2)^3$

41. $(a^{-4}b^3)^{-2}$

42. $(a^5b^{-2})^{-3}$

43. $(3x^{-2}y^3z^{-4})^2$

44. $(5x^2y^{-3}z^4)^{-2}$

45. $4(x^{-1}y)^{-3}$

46. $(4x^{-4}y)^3$

47. $\dfrac{x^5}{x^2}$

48. $\dfrac{x^{-5}}{x^{-2}}$

49. $\dfrac{-3a^{-3}}{9a^9}$

50. $\dfrac{-4a^{-4}}{2a^2}$

51. $x^{-2} + y^{-1}$

52. $x^{-1} + y^{-2}$

53. $\dfrac{x^4x^{-10}}{x^{-2}x^{-5}}$

54. $\dfrac{a^{-3}a}{a^6a^{-3}}$

55. $\dfrac{x^4y^{-10}}{x^{-2}y^{-5}}$

56. $\dfrac{a^{-3}b}{a^6b^{-3}}$

57. $\dfrac{10^{-3}10^5}{10^610^{-10}}$

58. $\dfrac{10^210^{-4}}{10^{-6}10^8}$

59. $\dfrac{12(10^{-3})}{4(10^{-7})}$

60. $\dfrac{18(10^{-4})}{3(10^2)}$

61. $\left(\dfrac{a^{-2}}{a^3}\right)^{-3}$

62. $\left(\dfrac{y^{-3}}{y^{-5}}\right)^{-2}$

63. $\dfrac{(x^2y^{-1})^{-1}}{(x^3y^{-2})^2}$

64. $\dfrac{(ay^{-2})^{-3}}{(a^4y^{-3})^2}$

65. $\left(\dfrac{2m^{-2}n^{-3}}{m^{-6}n^{-1}}\right)^{-2}$

66. $\left(\dfrac{3r^{-4}s^{-1}}{r^{-8}s^{-3}}\right)^{-3}$

67. $\left(\dfrac{x^{-1}y^{-2}}{3x^{-2}y^{-3}}\right)^{-1}$

68. $\left(\dfrac{a^{-2}b^{-4}}{5a^{-4}b^{-8}}\right)^{-1}$

69. $\dfrac{(2m^{-2}n^{-3})^{-4}}{(m^{-6}n^{-1})^{-2}}$

70. $\dfrac{(3m^{-4}n^{-1})^{-2}}{(m^{-2}n^{-1})^{-3}}$

71. $\dfrac{(x^5 y)^{-2}(x^{-2}y^3)^2}{(x^{-3}y^{-4})^{-2}}$

72. $\dfrac{(u^2 v^{-3})^{-1}(u^{-1}v^2)^3}{(u^{-3}v)^2}$

In Exercises 73–78, evaluate the given expression for x = 2 and y = –3.

73. x^{-3}

74. y^{-2}

75. $8x^{-1}$

76. $9y^{-1}$

77. $x^{-1} + y^{-1}$

78. $x^{-2} + y^{-2}$

❓ QUESTIONS FOR THOUGHT

79. Discuss the difference between $\dfrac{x^6}{x^4}$ and $\dfrac{x^6}{x^{-4}}$.

80. Discuss the difference between 3^{-1} and -3.

◇ MINI-REVIEW

81. Maria can make five flower arrangements per hour, while Francis can make seven flower arrangements per hour. If Maria starts working at 9:00 A.M. and is joined by Francis at 11:00 A.M., at what time will they have made 70 flower arrangements?

82. A long pole is standing slightly offshore. If one-quarter of the pole is in the sand, 8 ft is in the water, and two-thirds of the pole is in the air, how long is the pole?

7.3   Scientific Notation

There are many occasions (especially in science-related fields) when we may come across either very large or very small numbers. For example, we may read statements such as:

The mass of the earth is approximately 5,980,000,000,000,000,000,000,000 kilograms.

The wavelength of yellow-green light is approximately 0.0000006 meter.

Writing numbers like this in their decimal form can be quite messy and is very prone to errors. Using our knowledge of integer exponents, we can describe an alternative form for writing such numbers that makes them easier to work with. This other form is called *scientific notation.*

Scientific notation is a concise way of expressing very large or very small numbers. Before we describe scientific notation, let's recall some basic facts from arithmetic using the language of exponents.

Multiplying a number by 10 (that is, 10^1) moves the decimal point 1 place to the *right.*

Multiplying a number by 100 (that is, 10^2) moves the decimal point 2 places to the *right.*

Multiplying a number by 1,000 (that is, 10^3) moves the decimal point 3 places to the *right.*

On the other hand,

Dividing a number by 10 (or multiplying by $\frac{1}{10}$, which is the same thing as multiplying by 10^{-1}) moves the decimal point 1 place to the *left*.

Dividing a number by 100 (or multiplying by $\frac{1}{100}$, which is the same as multiplying by 10^{-2}) moves the decimal point 2 places to the *left*.

Dividing a number by 1,000 (or multiplying by $\frac{1}{1,000}$, which is the same thing as multiplying by 10^{-3}) moves the decimal point 3 places to the *left*.

To be more specific, suppose we have a number, say 5.49; let's see what happens to the number as we multiply it by these various powers of 10.

Original number	*Multiplied by*	*Becomes*
5.49	$1 = 10^0$	5.49
5.49	$10 = 10^1$	54.9
5.49	$100 = 10^2$	549.
5.49	$1,000 = 10^3$	5,490.
5.49	$\frac{1}{10} = 10^{-1}$	0.549
5.49	$\frac{1}{100} = 10^{-2}$	0.0549
5.49	$\frac{1}{1,000} = 10^{-3}$	0.00549

Standard notation is the decimal notation we normally use. Numbers such as

$$143.4, \quad 7.956, \quad \text{and} \quad 0.00538$$

are written in standard notation.

DEFINITION	We say that a number is written in ***scientific notation*** if the number is of the form $$a \times 10^n \quad \text{where } 1 \leq a < 10, \quad \text{and } n \text{ is an integer}$$

This is just about the only place in algebra where the symbol "×" is used to indicate multiplication. The reason it is used here is because the "·" can too easily be misread as a decimal point.

Verify for yourself that the two numbers on each line of the following display are equal.

Number in standard notation	*Number in scientific notation*
48,500	4.85×10^4
3,756	3.756×10^3
980	9.8×10^2
72	$7.2 \times 10^1 = 7.2 \times 10$
6.5	$6.5 \times 10^0 = 6.5 \times 1 = 6.5$
0.432	4.32×10^{-1}
0.0999	9.99×10^{-2}
0.005	5×10^{-3}
0.00012	1.2×10^{-4}

Note: A number is in scientific notation if its decimal point is immediately to the right of its first nonzero digit. Thus, the number 6.5 is the same in both columns because any real number between 1 and 10 is already in scientific notation. The 10^0 is usually not written but it is understood.

Here are some numbers not in scientific notation (and why):

0.095×10^4 because 0.095 is not between 1 and 10.   (It should be written 9.5×10^2.)

62×10^{-3} because 62 is not between 1 and 10.   (It should be written 6.2×10^{-2}.)

Scientific notation is also used by many calculators and computers. When a number is either too large or too small to be displayed on a calculator, the number might appear on the display as

$$6.4203 \text{ E } 9 \qquad \text{or} \qquad 2.38 \text{ E } -15$$

(Keep in mind that different calculators may do this differently.) These expressions are actually shorthand forms of scientific notation (the E stands for exponent).

$$6.4203 \text{ E } 9 \quad \text{means} \quad 6.4203 \times 10^9.$$
$$2.38 \text{ E } -15 \quad \text{means} \quad 2.38 \times 10^{-15}.$$

EXAMPLE 1

Convert 2.91×10^5 and 2.91×10^{-5} into standard notation.

Solution

Converting from scientific notation into standard notation is quite straightforward because, as we have already pointed out, multiplying by 10 moves the decimal point one place to the right, while dividing by 10 moves the decimal point one place to the left.

$$2.91 \times 10^5 = 2.91 \times 10 \times 10 \times 10 \times 10 \times 10 = 291,000$$

*Note that the decimal point has moved 5 places to the **right**.*

$$2.91 \times 10^{-5} = 2.91 \times \frac{1}{10} \times \frac{1}{10} \times \frac{1}{10} \times \frac{1}{10} \times \frac{1}{10} = 0.0000291$$

*Note that the decimal point has moved 5 places to the **left**.*

As you can see, we may bypass the middle step by observing that the *number* of places the decimal point is moved is given by the exponent of 10; the *direction* in which it is moved is determined by the sign of the exponent.

In 2.91×10^5, the exponent is +5 and the decimal point moves 5 places to the right.

In 2.91×10^{-5}, the exponent is −5 and the decimal point moves 5 places to the left. ∎

EXAMPLE 2

Convert 2,830 into scientific notation.

Solution

Converting from standard notation to scientific notation requires a little more thought. The decimal point in 2,830 is understood to be to the right of the 0. In order to be in scientific notation the decimal point must be to the immediate right of the first nonzero digit, which in this case means between the 2 and the 8. But moving the decimal point there will, of course, change the value of the number. We compensate for this change by putting in the "correcting" power of 10.

$$2{,}830 \qquad = \qquad 2.83 \qquad \times \qquad 10^3$$

We want to move the decimal point 3 places to the left.   *We put the decimal point where we want it.*   *This power of 10 compensates by putting the decimal point back where it really belongs, 3 places to the right.*

The net effect is that we have *not* changed the value of the number; we have changed only its appearance—which was the whole idea.

Thus, our answer is $\boxed{2.83 \times 10^3}$. ∎

EXAMPLE 3 | Convert 0.072 into scientific notation.

Solution | We want the decimal point to be between the 7 and the 2.

$$0.072 \quad = \quad 7.2 \quad \times \quad 10^{-2}$$

We want to move the decimal point 2 places to the right. | *We put the decimal point where we want it.* | *This power of 10 compensates by putting the decimal point back where it really belongs, 2 places to the left.*

Thus, our answer is $\boxed{7.2 \times 10^{-2}}$. ■

Scientific notation often allows us to carry out complicated arithmetic problems more easily.

EXAMPLE 4 | Compute the following using scientific notation: $\dfrac{4,800}{0.06}$. Express the answer in scientific notation.

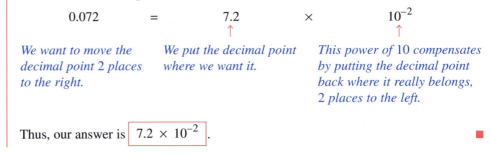

Solution | We first convert each number into scientific notation and then regroup the numbers to make the computation easier.

$$\frac{4,800}{0.06} = \frac{4.8 \times 10^3}{6 \times 10^{-2}}$$

$$= \frac{4.8}{6} \times \frac{10^3}{10^{-2}} \qquad \text{\textit{Divide 6 into 4.8, and use rule 2 on the powers of 10.}}$$

$$= 0.8 \times 10^{3-(-2)}$$

$$= 0.8 \times 10^5 \qquad \text{\textit{This answer is not yet in scientific notation.}}$$

$$= 8 \times 10^{-1} \times 10^5 \qquad \text{\textit{Now use rule 1 on the powers of 10.}}$$

$$= \boxed{8 \times 10^4}$$ ■

EXAMPLE 5 | Compute the following using scientific notation. $\dfrac{(360)(0.004)}{(0.0002)(600,000)}$

Solution | First we convert each number into scientific notation, and then we group the numbers so as to make the computation easiest.

$$\frac{(360)(0.004)}{(0.0002)(600,000)} = \frac{(3.6 \times 10^2)(4 \times 10^{-3})}{(2 \times 10^{-4})(6 \times 10^5)}$$

$$= \frac{(3.6)(4)}{(2)(6)} \times \frac{10^2 \, 10^{-3}}{10^{-4} \, 10^5} \qquad \begin{array}{l} \text{\textit{Reduce the first fraction; use}} \\ \text{\textit{rule 1 on the second.}} \\ 10^2 10^{-3} = 10^{2+(-3)} = 10^{-1} \\ 10^{-4} 10^5 = 10^{-4+5} = 10^1 \end{array}$$

$$= \frac{3.6}{3} \times \frac{10^{-1}}{10^1} \qquad \text{\textit{Use rule 3 on the 10's.}}$$

$$= 1.2 \times 10^{-1-1}$$

$$= \boxed{1.2 \times 10^{-2}} = \boxed{0.012}$$

You might have noticed that this computation would have been simpler if we had written 360 as

$$36 \times 10 \quad \text{rather than} \quad 3.6 \times 10^2$$

Strictly speaking, 36×10 is not scientific notation because 36 is not a number between 1 and 10. Nevertheless, you should feel free to use a "modified" form of scientific notation in your computations. ∎

If we actually needed to compute the value of the expression $\dfrac{(360)(0.004)}{(0.0002)(600{,}000)}$ given in the previous example, we would most likely use a calculator. Keep in mind, however, that using a calculator is not foolproof. If the wrong key is mistakenly pressed, it can radically change the answer. Scientific notation gives us a tool with which we can quickly estimate an answer and recognize whether an answer obtained with a calculator is reasonable. We illustrate this in the next example.

EXAMPLE 6

A manufacturer uses 48 ball bearings in each desk chair. Each ball bearing costs $0.00425. What would be the cost for the ball bearings needed to produce 3,000 chairs?

Solution

Since each chair requires 48 ball bearings, in order to produce 3,000 chairs we need 3,000(48) ball bearings. Since each ball bearing costs $0.00425,

$$\text{Total cost of the ball bearings} = 3{,}000(48)(0.00425)$$

We can easily compute this with a calculator to be $\boxed{\$612}$.

Using scientific notation, we can estimate the correct answer to make sure that the answer of $612 is reasonable. We estimate the cost approximately as

$$(3{,}000)(50)(0.004) = (3 \times 10^3)(5 \times 10)(4 \times 10^{-3})$$
$$= (3)(5)(4) \times 10^{3+1-3} = 60 \times 10 = 600$$

which gives us added confidence in our answer of $612. ∎

EXAMPLE 7

The mass of a hydrogen atom is approximately 1.67×10^{-24} gram. Based on this figure, compute the number of atoms in 1 gram of hydrogen. Express your answer in scientific notation.

Solution

We want to know how many times does 1.67×10^{-24} gram go into 1 gram. Thus, this example is actually a division problem.

$$1 \div (1.67 \times 10^{-24}) = \frac{1}{1.67 \times 10^{-24}}$$

$$= \frac{1}{1.67} \times \frac{1}{10^{-24}}$$

$$= \frac{1}{1.67} \times 10^{24} \qquad \textit{We compute 1 divided by 1.67, and round off to three places.}$$

$$= 0.599 \times 10^{24} \qquad \textit{This answer is not in scientific notation.}$$

$$= \boxed{5.99 \times 10^{23}} \text{ atoms in 1 gram of hydrogen} \qquad ∎$$

Sometimes when faced with a computation, we do not need the exact value but rather we just want to have an idea of the size of the answer. Scientific notation can often be used to aid us in making such an estimate, as illustrated in the next example.

EXAMPLE 8

If we multiply 0.00435 by 598,000, is the result closest to

(a) 100　　　(b) 1,000　　　(c) 2,000　　　(d) 10,000　　　(e) 0.01

Solution

Let's first convert each number into scientific notation.

$$0.00435 = 4.35 \times 10^{-3}$$
$$598,000 = 5.98 \times 10^{5}$$

If we round off 4.35 to 4 and 5.98 to 6, then we can use $(4 \times 10^{-3}) \times (6 \times 10^{5})$ to estimate the answer. We can easily compute this product to be 24×10^{2}, which is 2,400. Therefore of the choices given, 2,000, choice **(c)**, is the closest.

If we use a calculator to do the computation, we find the exact answer to be 2,601.3, so that our estimate of 2,400 gave us a good idea of the size of the answer. ∎

EXERCISES 7.3

In Exercises 1–34, convert each number into scientific notation.

1. 4,530
2. 1,250
3. 0.0453
4. 0.0125
5. 0.00007
6. 0.0004
7. 7,000,000
8. 400,000
9. 85,370
10. 12,340
11. 0.0085370
12. 0.0001234
13. 90
14. 70
15. 9
16. 7
17. 0.9
18. 0.7
19. 0.09
20. 0.07
21. 0.00000003
22. 0.0000002
23. 28
24. 37
25. 47.5
26. 52.4
27. 9,727.3
28. 111.12
29. 56×10^{-2}
30. 872×10^{5}
31. 0.154×10^{4}
32. 234.8×10^{-3}
33. 28.40×10^{6}
34. 0.0025×10^{-4}

In Exercises 35–48, convert each number into standard notation.

35. 2.8×10^{4}
36. 5×10^{3}
37. 2.8×10^{-4}
38. 5×10^{-3}
39. 4.29×10^{7}
40. 1.76×10^{-5}
41. 4.29×10^{-7}
42. 1.76×10^{5}
43. 3.52×10^{-3}
44. 6.81×10^{-2}
45. 3.5286×10^{5}
46. 0.0527×10^{4}
47. 0.026×10^{-3}
48. $78,951 \times 10^{-5}$

In Exercises 49–56, do your computation using scientific notation.

49. (0.004)(250)
50. (600)(0.0015)

51. $\dfrac{0.003}{6,000}$

52. $\dfrac{80}{0.0002}$

53. $\dfrac{(480)(0.008)}{(0.24)(4,000)}$

54. $\dfrac{(0.0075)(6,400)}{(0.032)(250)}$

55. $\dfrac{(0.0036)(0.005)}{(0.01)(0.06)}$

56. $\dfrac{(2,400)(1,500)}{(90,000)(4,000)}$

In Exercises 57–66, estimate the answer without actually carrying out the computation and make the most appropriate choice.

57. If you multiply 1.36×10^8 by 1.18×10^{-4}, the result is closest to

    (a) 100   **(b)** 1,000   **(c)** 10,000   **(d)** 0.001   **(e)** 0.0001

58. If you multiply 1.93×10^5 by 5.12×10^{-3}, the result is closest to

    (a) 10   **(b)** 1,000   **(c)** 10,000   **(d)** 0.01   **(e)** 0.001

59. If you multiply 1.45×10^{-6} by 2.03×10^4, the result is closest to

    (a) 10   **(b)** 100   **(c)** 1   **(d)** 0.1   **(e)** 0.01

60. If you multiply 2.08×10^{-7} by 5.14×10^6, the result is closest to

    (a) 10   **(b)** 100   **(c)** 1   **(d)** 0.1   **(e)** 0.01

61. If you divide 9.28×10^7 by 6.86×10^4, the result is closest to

    (a) 10   **(b)** 100   **(c)** 1,000   **(d)** 0.1   **(e)** 0.01

62. If you divide 7.92×10^{-5} by 2.21×10^{-2}, the result is closest to

    (a) 100   **(b)** 1,000   **(c)** 0.1   **(d)** 0.01   **(e)** 0.001

63. If you multiply 7,250 by 0.004, the result is closest to

    (a) 0.3   **(b)** 3   **(c)** 30   **(d)** 300   **(e)** 3,000

64. If you multiply 0.00032 by 420,000, the result is closest to

    (a) 0.12   **(b)** 1.2   **(c)** 12   **(d)** 120   **(e)** 1,200

65. If you divide 71,340 by 126, the result is closest to

    (a) 0.6   **(b)** 6   **(c)** 60   **(d)** 600   **(e)** 6,000

66. If you divide 2,350 by 48,600, the result is closest to

    (a) 0.05   **(b)** 0.5   **(c)** 5   **(d)** 50   **(e)** 500

In the following exercises, do your computations using scientific notation.

67. The first paragraph of this section states the approximate mass of the earth. Write this number in scientific notation.

68. The first paragraph of this section states the approximate wavelength of yellow-green light. Write this number in scientific notation.

69. If one atom of iron has a mass of 9.3×10^{-23} gram, what is the mass of 80,000 atoms?

70. If the mass of one atom of iron is 9.3×10^{-23} gram, how many atoms are there in 1 gram?

71. Atomic measurements or other very small distances are often measured in units called *angstroms*. One angstrom, which is written 1 Å, is equal to 0.00000001 cm. Write this number in scientific notation.

72. One angstrom is equal to what part of a meter? A kilometer?

73. The thickness of a typical cell wall might be 153 Å. Write this number in scientific notation.

74. The diameter of an atom might typically be 3.47 Å. Write this number in scientific notation.

75. The planet Pluto is 4,250 million miles from Earth. Write this number in scientific notation.

76. Convert the Earth-to-Pluto distance into kilometers. [*Hint:* 1 mile = 1.6 km]

77. If 1 ton is equal to 888.9 kilograms, what is the weight of Earth in tons? (Use the result of Exercise 67.)

78. If 1 in. is equal to 2.54 cm, how many angstroms are there in 1 inch?

79. Light travels at a speed of approximately 186,000 miles per second. How far does light travel in 1 year? (This *distance* is called 1 light-year.)

80. The Mt. Palomar Observatory has a 200-inch mirror telescope that can see a star 5 billion light-years away. Use the result of Exercise 79 to compute this distance in miles.

81. Give the answer to Exercise 80 in kilometers.

82. If light travels 5.87×10^{12} miles in 1 year, how long will it take light to reach us from a star that is 3×10^{22} miles away?

? QUESTIONS FOR THOUGHT

83. Explain how you would multiply 3.74×10^{-5} by 6.38×10^{4} without converting either number into standard notation.

84. What can you say about the sign of the exponent of 10 of a number written in scientific notation if the number is bigger than 1? If the number is smaller than 1? Why?

7.4 Introduction to Polynomials

If a rocket is fired straight up into the air from ground level with an initial velocity of 400 feet per second, then its height *h* (in feet) above the ground *t* seconds after the rocket is fired is given by the equation

$$h = 400t - 16t^2$$

A manufacturer may find that the daily cost *C* (in dollars) of manufacturing *x* tables is given by the equation

$$C = 0.02x^2 - 0.48x + 520$$

Using the language we introduced in Chapter 2, we can say that the height of the rocket is a function of the elapsed time *t*, and the manufacturer's cost is a function of the number of tables made *x*.

Expressions such as $400t - 16t^2$ and $0.02x^2 - 0.48x + 520$ are called *polynomials* and are the topic of discussion in this section.

Terminology

On several occasions we have pointed out the important distinction between *terms* and *factors*. Let's begin by giving a formal definition of a particular type of term.

DEFINITION	A *monomial* is an algebraic expression that is either a constant or a product of constants and one or more variables with whole-number exponents.

The following are all examples of monomials:

$$4x^3y^2 \qquad -5t^6 \qquad \frac{2}{3}a^2 \qquad 6$$

Note that a monomial is simply a special kind of *term*. Adding and subtracting monomials gives us more complex expressions.

DEFINITION	A *polynomial* is the sum of one or more monomials.

For example, all the following are polynomials:

$3x^2 - 5xy$ is often called a *binomial* because it has two terms.

$-5a^3 + 2ab + b^2$ is often called a *trinomial* because it has three terms.

$2y$ is a polynomial with just one term (a monomial). It is usually just called a *term*.

The following are just three examples of *nonpolynomial* expressions:

$\dfrac{3}{x + 1}$ because polynomials do not allow variables in denominators.

x^{-2} because by definition this means $\dfrac{1}{x^2}$ and polynomials do not allow variables in the denominator.

3^x because polynomials do not allow variables as exponents.

Besides classifying polynomials by the number of terms (monomial, binomial, trinomial), we can also classify them by what is known as their *degree*.

DEFINITION	The *degree* of a nonzero *monomial* is the sum of the exponents of the variables. The degree of a nonzero constant is 0. The degree of the number 0 is undefined.

For example:

The degree of the monomial $-2a^4$ is 4.   *(There is only one exponent of a variable.)*

The degree of the monomial $7x^3y^2$ is 5.   *(The sum of the exponents of the variables is $2 + 3$.)*

The degree of the constant monomial 6 is 0 (because we can write $6 = 6x^0$) and the same is true for any nonzero constant.

The degree of the number 0 is undefined.

The reason we must specify "nonzero" in the definition and the reason the degree of the number 0 is undefined is that we can write

$$0 = 0x^8 \quad \text{or} \quad 0 = 0x^{17} \quad \text{or} \quad 0 = 0x^{35}$$

and so we cannot possibly tell what the degree of the number 0 is. Consequently, the degree of 0 is undefined.

DEFINITION	The *degree of a polynomial* is the highest degree of any monomial in it.

For example, the polynomial $5x^4 - 7x^2y + 2x$ has three terms whose individual degrees are

$$\underbrace{5x^4}_{\text{Degree is 4}} \qquad \underbrace{-7x^2y}_{\substack{\text{Degree is 3} \\ 2 + 1 = 3}} \qquad \underbrace{+2x}_{\text{Degree is 1}}$$

Thus, the overall degree of the polynomial is 4, which is the highest degree of any term in it.

The idea of degree provides us with a useful way of categorizing expressions and equations. In fact, we have already used the word *degree* in Chapter 3 when we learned about solving first-*degree* equations.

Most often we are interested in the degree of a polynomial involving only one variable, which is simply the highest exponent of the variable that appears. The expression

$$3x^5 - 4x^3 + 8x^2 + 7$$

is a fifth-degree polynomial, because the highest exponent of the variable x is 5. When a polynomial is written in this way, with the degrees of the terms in descending order, starting with the highest-degree term and ending with the lowest-degree term, the polynomial is said to be in **standard form** for polynomials.

As we have discussed previously, every term has a coefficient. In the polynomial mentioned above, $3x^5 - 4x^3 + 8x^2 + 7$, there are four terms and their coefficients are 3, −4, 8, and 7, respectively. Note that 7, because it is a constant term, is both a term and a coefficient.

When a polynomial is in standard form, the coefficient of the first term is called the **leading coefficient.** The leading coefficient of $3x^5 - 4x^3 + 8x^2 + 7$ is 3.

Sometimes we have occasion to talk about "missing terms." For example, in the polynomial we have been talking about,

$$3x^5 - 4x^3 + 8x^2 + 7$$

the fourth-degree and first-degree terms do not appear.

If we want to think of, say, a fifth-degree polynomial as always containing all the terms from degree 5 down to the constant, then we can say that the fourth-degree and first-degree terms here have a coefficient of 0. In other words, we think of

$$3x^5 - 4x^3 + 8x^2 + 7 \quad \text{as} \quad 3x^5 + 0x^4 - 4x^3 + 8x^2 + 0x + 7$$

When we write the polynomial with the "missing terms" appearing with coefficients of 0, the polynomial is said to be in **complete standard form.**

EXAMPLE 1   Find the degree of the following monomials.   (a) $-4x^3yz^5$ (b) $5^2x^3y^4$

Solution   (a) According to the definition, the degree of a monomial is the sum of the exponents of the variables In $-4x^3yz^5$ the exponents of x, y, and z are 3, 1, and 5, respectively.

Therefore the degree is $3 + 1 + 5 = \boxed{9}$.

(b) Be careful! The definition says that we add the exponents of the *variables*. Since 5 is not a variable its exponent of 2 has no effect on the degree of $5^2x^3y^4$. Therefore

the degree is $3 + 4 = \boxed{7}$. ∎

EXAMPLE 2   Given the polynomial $4x^3 - 5x + 9$:

(a) What is the degree of each term?

(b) What is the degree of the polynomial?

(c) Write the polynomial in complete standard form. What are its coefficients?

Solution

(a) The polynomial $4x^3 - 5x + 9$ consists of three terms.

The degree of the first term, $4x^3$, is 3.

The degree of the second term, $-5x$, is 1.

The degree of the third term, 9, is 0.

(b) According to the definition of the degree of a polynomial, we take the highest degree of any term in it. Thus, the degree of $4x^3 - 5x + 9$ is 3.

(c) We write the polynomial in complete standard form, which means that all the terms from the highest degree down to the constant are accounted for.

$$4x^3 - 5x + 9 = 4x^3 + 0x^2 - 5x + 9$$

Now we see that:

The coefficient of $4x^3$ is 4.

The coefficient of $0x^2$ is 0.

The coefficient of $-5x$ is -5.

9 is both a term and a coefficient.   *Think of 9 as $9 \cdot 1$.* ■

Now let's look at operations with polynomial expressions.

EXAMPLE 3   Add the polynomials. $(5x^2 - 4x + 3) + (3x^2 - 2x - 7)$

Solution   The use of our new terminology does not alter the fact that we have done many examples of this type before (see Section 2.4). Since the parentheses are nonessential, we remove them and reorder and regroup the terms.

$$(5x^2 - 4x + 3) + (3x^2 - 2x - 7) = 5x^2 + 3x^2 - 4x - 2x + 3 - 7$$

Combine like terms.

$$= \boxed{8x^2 - 6x - 4}$$ ■

EXAMPLE 4   Subtract the polynomials. $(3x^2y - 5xy^2) - (x^2 - xy^2 - x^2y)$

Solution   The second set of parentheses has a minus sign in front. We think of it as a -1 multiplying the terms in the set of parentheses.

$$(3x^2y - 5xy^2) - (x^2 - xy^2 - x^2y) = (3x^2y - 5xy^2) - 1(x^2 - xy^2 - x^2y)$$

The first parentheses are nonessential.

$$= 3x^2y - 5xy^2 - x^2 + xy^2 + x^2y$$

Combine like terms.

$$= \boxed{4x^2y - 4xy^2 - x^2}$$ ■

Sometimes examples such as these are stated verbally.

EXAMPLE 5   Subtract $5a^3 - a + 3$ from $a^3 - a^2 - a - 4$.

Solution   As usual, when we deal with subtraction, we have to be careful. The example tells us to subtract the first polynomial, $5a^3 - a + 3$, *from* the second polynomial, $a^3 - a^2 - a - 4$. Remember that subtraction is *not* commutative—the order matters. Algebraically we must write the second polynomial first, as well as put parentheses around the first polynomial.

$$(a^3 - a^2 - a - 4) - (5a^3 - a + 3)$$ *Distribute the understood factor of -1.*

Subtract this from

$$= a^3 - a^2 - a - 4 - 5a^3 + a - 3 \quad \textit{Combine like terms.}$$

$$= \boxed{-4a^3 - a^2 - 7}$$ ∎

It is possible to add and subtract polynomials in a vertical format as well. Since it does not seem to offer any significant advantages, we will illustrate the vertical format in the next example, but we will use primarily the horizontal format we have been using thus far. However, if you happen to prefer the vertical format, feel free to use it.

EXAMPLE 6 (a) Add. $3x^3 - 5x + 2, \quad x^2 - 3x, \quad x^3 - x^2 - 2x + 7$

(b) Subtract $2x^2 - x + 3$ from $x^2 - 2$.

Solution (a) Using the vertical format, we line up the polynomials with like terms directly above each other, then we add up the like terms in each column.

The vertical format will be useful when we discuss dividing polynomials in Section 8.5.

$$
\begin{array}{r}
3x^3 \qquad - 5x + 2 \\
x^2 - 3x \\
\underline{x^3 - x^2 - 2x + 7} \\
4x^3 \qquad - 10x + 9
\end{array}
$$

Thus, our answer is $\boxed{4x^3 - 10x + 9}$.

(b) Subtraction requires a bit more care, since *all* the terms in the bottom polynomial must be subtracted from the top one. Be sure to line up like terms.

$$
\begin{array}{r}
x^2 \qquad - 2 \\
\underline{-(2x^2 - x + 3)}
\end{array}
$$

*The parentheses remind us to subtract **all** the terms in the bottom row.*

We must now change *all* the signs of the polynomial being subtracted and add.

$$
\begin{array}{r}
x^2 \qquad - 2 \\
\underline{-2x^2 + x - 3} \\
-x^2 + x - 5
\end{array}
$$

Thus, the answer is $\boxed{-x^2 + x - 5}$. ∎

In Section 2.2 we discussed evaluating algebraic expressions for certain replacement values of the variables. The process is exactly the same for polynomials.

EXAMPLE 7 Evaluate the polynomial $a^3 + 3a^2 - 5a + 10$ for $a = -3$.

Solution We replace each occurrence of the variable, a, with -3.

$$
\begin{aligned}
a^3 + 3a^2 - 5a + 10 &= (\ )^3 + 3(\ )^2 - 5(\ ) + 10 \\
&= (-3)^3 + 3(-3)^2 - 5(-3) + 10 \qquad \textit{Evaluate pow-} \\
&= -27 + 3(9) + 15 + 10 \qquad\qquad \textit{ers first, then} \\
&= -27 + 27 + 15 + 10 \qquad\qquad \textit{multiply.} \\
&= \boxed{25}
\end{aligned}
$$
∎

EXAMPLE 8 At the beginning of this section we mentioned that if a rocket is fired straight up into the air from ground level with an initial velocity of 400 feet per second, then its height h (in feet) above the ground t seconds after the rocket is fired is given by the equation

$$h = 400t - 16t^2$$

Use this equation to determine the height of the rocket 5 seconds after it is fired.

Solution   This equation tells us that the polynomial $400t - 16t^2$ gives us the height of the rocket t seconds after it is fired. If we want to know the height of the rocket 5 seconds after it is fired, then we want to evaluate the polynomial $400t - 16t^2$ when $t = 5$.

$$h = 400t - 16t^2 \qquad \textit{Substitute } t = 5.$$

$$h = 400(5) - 16(5)^2 = 2{,}000 - 16(25) = 2{,}000 - 400 = \boxed{1{,}600}$$

Therefore, 5 seconds after it is fired, the rocket is 1,600 feet above the ground. ■

EXERCISES 7.4

In Exercises 1–18, answer the following questions:

 (a) How many terms are there?

 (b) What is the degree of each term?

 (c) What is the degree of the polynomial?

1. $3x^5$

2. $-4x^7$

3. $3x + 4$

4. $5y - 7$

5. $x^2 + y^3$

6. $2w^3 - v^5$

7. x^2y^3

8. $2w^3v^5$

9. 8

10. -12

11. $2x^3 - 5x^2 + x$

12. $5a^4 - 3a^3 + 26a$

13. $2x^3 + y^5$

14. $3^6m^2n^5$

15. $2^4x^3y^5$

16. $3m^2 + n^5$

17. $x^5 - x^3y^4 - 2x^2y^3 + y^6$

18. $8r^6 - 2r^4s^3 - t^8$

In Exercises 19–26, answer the following questions:

 (a) What is the degree of each term?

 (b) What is the degree of the polynomial?

 (c) Using *complete standard form,* what is the coefficient of each term?

19. $x^2 - 5x + 6$

20. $y^2 + 3y - 7$

21. $x^2 + 4$

22. $3y^2 - 8$

23. $x^3 - 1$

24. $x^4 - 1$

25. $1 - x^5$

26. $1 + x^6$

In Exercises 27–50, perform the indicated operations and simplify.

27. $(2x^2 - 5) + (3x^2 - 5)$

28. $(a^3 + 7) + (5 - 2a^3)$

29. $(3u^3 - 2u + 7) + (u^3 - u^2 + 7u)$

30. $(5w^4 - w^3 - w) + (2w^3 - 3w^2 - 5w)$

31. $(3u^3 - 2u + 7) - (u^3 - u^2 + 7u)$

32. $(5w^4 - w^3 - w) - (2w^3 - 3w^2 - 5w)$

33. $(4t^3 - t) + (t^2 + t) - (t^3 - t^2)$

34. $(s^4 - 6s^3) - (s^3 - s^2 + s) - (2s^3 - s^4)$

35. $(x^2y + 3xy - x^2y^2) + (x^2y - 5x^2y^2 - xy^2)$

36. $(r^3s^2 - r^2s^3 - 2r^2s^2) + (5r^2s^3 - r^2s^2 - r^3s^2)$

37. $(x^2y + 3xy - x^2y^2) - (x^2y - 5x^2y^2 - xy^2)$

38. $(r^3s^2 - r^2s^3 - 2r^2s^2) - (5r^2s^3 - r^2s^2 - r^3s^2)$

39. $2(y^2 - 4y + 1) + 3(2y^2 - y - 1)$

40. $4(w^3 - w^2 + 7) + 5(w^3 - w - 2)$

41. $5(x^2 - 3x + 2) - 3(2x^2 - 5x - 2)$

42. $8(t^4 - t^3 - 2t) - 2(t^3 - t^4 - 8t)$

43. *Add.* $x^2 + 3x - 7,\ \ 5x - x^2,\ \ 3x^2 - x - 2$

44. *Add.* $x^3 + x^2 - 5x + 9,\ \ 3x^3 - 6x^2 - x - 4,$
    $2x^3 - 5x^2 - 6x - 2$

45. Subtract $x^2 - 7x + 3$ from $2x^2 - 3x + 5$.

46. Subtract $2w^3 + w^2 + 6$ from $w^3 - w^2 + 2w$.

47. Subtract $a^3 - a^2 - b + b^2$ from the sum of
    $a^3 - b^2$ and $a^2b + 2b^2$.

48. Subtract $m^4 - m^3p - 5p$ from the sum of
    $2m^3p - p$ and $4m^4 - 5p + 6$.

49. Subtract the sum of $3x + 6$ and $5x - 8$
    from $2x - 1$.

50. Subtract the sum of $x^2 - 3x + 2$ and $4 - 3x$
    from $9 - x^2$.

In Exercises 51–62, evaluate each polynomial for the given values.

51. $x^2 - x + 3$ for $x = -5$

52. $2a^2 - 5a + 7$ for $a = -2$

53. $y^4 + y^3 + y^2 + y + 1$ for $y = -3$

54. $2s^3 - 4s^2 + 6s - 8$ for $s = 3$

55. $6x^2 - 7x + 3$ for $x = -\dfrac{1}{2}$

56. $-3x^2 + 8x - 1$ for $x = \dfrac{2}{3}$

57. $-\dfrac{1}{2}x^2 - \dfrac{1}{4}x - 2$ for $x = 8$

58. $\dfrac{1}{6}x^3 - \dfrac{2}{3}x + 1$ for $x = -3$

59. $\dfrac{1}{3}x^2 - \dfrac{2}{5}x - 3$ for $x = \dfrac{1}{2}$

60. $-\dfrac{1}{4}x^2 - \dfrac{1}{6}x + 1$ for $x = -\dfrac{1}{3}$

61. $-3x^2y + 5xy^2$ for $x = 2, y = -1$

62. $5w^2v^3 - wv^2 - 8v$ for $w = 2, v = -2$

63. If the perimeter of one rectangle is given by $5x - 12$ and the perimeter of a second (smaller) rectangle is given by $3x + 8$, find the difference between the perimeters of the two rectangles.

64. If we double the length of a square of side x, what is the difference between the area of the bigger square and the smaller square?

65. A sculptor finds that the profit P (in dollars) earned if x pieces are produced is given by the equation $P = 100x - x^2$. Find the profit if the sculptor produces 30 pieces, 50 pieces, and 90 pieces.

66. A concert promoter finds that the profit P (in dollars) is related to the ticket price x (in dollars) according to the equation $P = 20{,}000(-x^2 + 50x - 20)$. Find the profit if the ticket price is \$15, \$25, and \$30.

67. Lamont throws a ball straight down out of an eighth-floor window. The height h (in feet) of the ball above the ground t seconds after it is thrown is given by the equation $h = 120 - 28t - 16t^2$. Find the height of the ball after 1 second, 1.5 seconds, and 2 seconds.

68. A manufacturer finds the daily cost C (in dollars) of manufacturing x tables is given by the equation $C = 0.02x^2 - 0.48x + 5{,}200$. Find the cost to manufacture 100 tables, 150 tables, and 325 tables.

69. Tani throws a ball up into the air so that its height h (in feet) above the ground t seconds after it is thrown is given by the formula $h = 5 + 80t - 16t^2$. Find the height of the ball after 1 second, 2 seconds, and 2.5 seconds.

70. The concentration C of a medication in the bloodstream (in milligrams per milliliter) m minutes after the medication is taken is given by the formula $C = 10 + 20m - 0.1m^2$. Find the concentration of medication in the bloodstream 60, 90, and 120 minutes after the medication is taken.

71. In a certain small town, the number of people N (in thousands) exposed to an influenza virus d days after the first case is diagnosed is given by the equation $N = 2d - 0.01d^2$. Find the number of people exposed 30, 60, 90 days after the first case is diagnosed.

72. A new Web site finds that the number of visits N (in thousands) to the side d days after it opens is given by the formula $N = 6d + 0.02d^2$. Find the number of visits to the site 50, 100 and 150 days after it opens.

 QUESTIONS FOR THOUGHT

73. Explain the difference between a *factor* and a *term*.

74. Is 3 a factor of the expression $6x + 8$? Explain why or why not.

75. Is 2 a factor of the expression $6x + 8$? Explain why or why not.

 **MINI-REVIEW**

In Exercises 76–79, solve for x.

76. $11 - 3x = 19 - 2x$

77. $2(x + 3) - 5(x + 4) = x - 2$

78. $5(2x - 1) - 3(x - 4) = 28 - (x - 11)$

79. $4(x - 1) - 3x = x - 4$

Solve each of the following problems algebraically. Be sure to label what the variable represents.

80. How much 20% iodine solution must be mixed with 80 ml of a 60% iodine solution to produce a 30% solution?

81. Margaret wants to make two investments—one in the amount of $1,800 that pays 8% interest and a second in the amount of $1,400. If the annual interest from the two investments is to be $284, what must the rate of interest for the $1,400 investment be?

7.5 Multiplying Polynomials

In Chapter 2 we learned how to multiply monomials and how to multiply a polynomial by a monomial. Let's begin by reviewing these procedures in Example 1.

EXAMPLE 1   Multiply.   (a) $5x(2x^3 + 3x^2)$ (b) $5x(2x^3)(3x^2)$

Solution   (a) This example calls for the use of the distributive property.

$$5x(2x^3 + 3x^2) = 5x \cdot 2x^3 + 5x \cdot 3x^2$$

$$= \boxed{10x^4 + 15x^3}$$

(b) This is a product of monomials. We use the commutative and associative properties to rearrange and regroup the factors, and then simplify using exponent rule 1.

$$5x(2x^3)(3x^2) = (5 \cdot 2 \cdot 3)(x \cdot x^3 \cdot x^2)$$

$$= \boxed{30x^6}$$

Note that the distributive property does *not* apply here.

Multiplying a polynomial by another polynomial requires the repeated use of the distributive property. However, by analyzing the process, we can develop a mechanical procedure for multiplying any two polynomials. For example, in order to multiply

$$(x + 4)(x^2 + 3x - 7)$$

let's think of this product as $(a + b) \cdot c$. In other words, consider the expression $x^2 + 3x - 7$ as one factor and distribute it over $x + 4$:

$$(x + 4)(x^2 + 3x - 7)$$

is like   *is like*
$(a + b)$ c

In the same way that we can distribute the c in $(a + b)c$ to get $a \cdot c + b \cdot c$, we distribute $(x^2 + 3x - 7)$:

$$(x + 4)(x^2 + 3x - 7) = x(x^2 + 3x - 7) + 4(x^2 + 3x - 7)$$

Now we use the distributive property again on each set of parentheses.

$$= x^3 + 3x^2 - 7x + 4x^2 + 12x - 28$$

Combine like terms.

$$= \boxed{x^3 + 7x^2 + 5x - 28}$$

If we carefully examine our first application of the distributive property, we see that it causes each term in the first polynomial to multiply each term in the second polynomial. That is, each term in $x^2 + 3x - 7$ is multiplied by x *and* is multiplied by 4.

This same procedure works regardless of the number of terms in each polynomial.

Rule for Multiplying Polynomials	In order to multiply two polynomials, multiply each term in the first polynomial by each term in the second polynomial.

EXAMPLE 2

Solution

Multiply and simplify. $(2x - 5)(3x^2 + x - 4)$

We follow the rule for multiplying polynomials and multiply each term in the first polynomial by each term in the second.

$$(2x - 5)(3x^2 + x - 4) = 2x(3x^2 + x - 4) - 5(3x^2 + x - 4)$$

Distribute the 2x and the −5.

$$= 6x^3 + 2x^2 - 8x - 15x^2 - 5x + 20$$

Combine like terms.

$$= \boxed{6x^3 - 13x^2 - 13x + 20}$$ ■

It is also possible to do polynomial multiplication in a vertical format, as follows:

$$
\begin{array}{r}
3x^2 + x - 4 \\
2x - 5 \\
\hline
-15x^2 - 5x + 20 \\
6x^3 + 2x^2 - 8x \\
\hline
6x^3 - 13x^2 - 13x + 20
\end{array}
$$

←—— *Obtained by multiplying top row by −5*
←—— *Obtained by multiplying top row by 2x*
←—— *Obtained by adding like terms, which are lined up in columns*

While the vertical format does sometimes offer advantages, at other times it may not. If the polynomials involve several variables, it may not be clear how to order the polynomials, or the vertical format may involve many columns. Additionally, in the next chapter when we talk about factoring polynomials, we will work exclusively with the horizontal format. Consequently, if you prefer the vertical format, feel free to use it, but we will stick to the horizontal format.

EXAMPLE 3

Multiply and simplify. $(2x^2 + xy - xy^2)(x^2y - xy + y^2)$

Solution

Following our outline, we multiply each term in the first polynomial by each term in the second.

$$(2x^2 + xy - xy^2)(x^2y - xy + y^2)$$

$$= \underbrace{2x^2(x^2y) - 2x^2(xy) + 2x^2(y^2)}_{\text{Multiplying by } 2x^2} + \underbrace{xy(x^2y) - xy(xy) + xy(y^2)}_{\text{Multiplying by } xy}$$

$$\underbrace{-xy^2(x^2y) + xy^2(xy) - xy^2(y^2)}_{\text{Multiplying by } -xy^2}$$

$$= 2x^4y - 2x^3y + 2x^2y^2 + x^3y^2 - x^2y^2 + xy^3 - x^3y^3 + x^2y^3 - xy^4$$

The only like terms are $2x^2y^2$ and $-x^2y^2$.

$$= \boxed{2x^4y - 2x^3y + x^2y^2 + x^3y^2 + xy^3 - x^3y^3 + x^2y^3 - xy^4}$$ ∎

Even though Example 3 was messy because of the "bookkeeping" involved, keep in mind that we are always doing *one* multiplication at a time.

EXAMPLE 4

Multiply and simplify. $(x + 7)(x + 4)$

Solution

While this example is much less complex than the previous one, we proceed in exactly the same way. We multiply each term in the first binomial by each term in the second.

$$(x + 7)(x + 4) = x \cdot x + x \cdot 4 + 7 \cdot x + 7 \cdot 4 \qquad \textit{Combine like terms.}$$

$$= \boxed{x^2 + 11x + 28}$$ ∎

This type of example, in which we multiply two similar binomials, comes up very frequently in algebra—so frequently, in fact, that we need to be able to multiply them rapidly, accurately, and often mentally. In the next chapter, when we discuss factoring, this ability will be particularly important.

In light of this, let's analyze the last example very carefully.

$$(x + 7)(x + 4) = \underbrace{x^2}_{\substack{\textit{Product of first} \\ \textit{terms in each} \\ \textit{set of parentheses}}} + \underbrace{4x + 7x}_{\textit{"Middle" terms}} + \underbrace{28}_{\substack{\textit{Product of second} \\ \textit{or "last" terms in} \\ \textit{each set of parentheses}}}$$

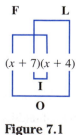

Figure 7.1
FOIL method

The key thing to notice here is that in this type of situation the middle terms are alike and can be combined. Sometimes the middle terms are called the "*outer*" and "*inner*" terms, and the entire multiplication process of *two binomials* is called the ***FOIL method*** (see Figure 7.1).

FOIL   stands for   First   Outer   Inner   Last

Giving the multiplication process a name does not carry it out. It is important to keep in mind that we still have to do the multiplication—each term in the first binomial times each term in the second—whether you call it the FOIL method or not. The name is simply a device to remember the method and help us to be systematic in carrying it out.

It is also very important to remember that the name FOIL applies only to the product of two *binomials*.

Since we will often be multiplying binomials, we would like to shorten the FOIL method by doing part of the multiplication mentally. Let's illustrate what is meant by doing the multiplication mentally.

EXAMPLE 5

Multiply and simplify.

(a) $(x + 5)(x - 3)$ (b) $(3y - 4)(y - 6)$

(c) $(3a - 4)(2a + 7)$ (d) $(x + 2y)(x - 6y)$

Solution

Basically, doing the problem mentally means that we carry the middle terms in our head and write down only the result of combining the like terms.

(a) $(x + 5)(x - 3) = x^2 \ - 3x + 5x \ - 15$ *The shaded portion is the step we do mentally.*

$$= \boxed{x^2 + 2x - 15}$$

(b) $(3y - 4)(y - 6) = 3y^2 \ - 18y - 4y \ + 24$

$$= \boxed{3y^2 - 22y + 24}$$

(c) $(3a - 4)(2a + 7) = 6a^2 \ + 21a - 8a \ - 28$

$$= \boxed{6a^2 + 13a - 28}$$

(d) $(x + 2y)(x - 6y) = x^2 \ - 6xy + 2xy \ - 12y^2$

$$= \boxed{x^2 - 4xy - 12y^2}$$ ■

If we carry the middle terms mentally, multiplying binomials requires writing only the final answer.

EXAMPLE 6

Multiply and simplify. $(x - 3)^3$

Solution

We can choose to begin by multiplying either the first two factors of $(x - 3)$ or the second two factors of $(x - 3)$. We will start with the second two.

$$(x - 3)^3 = (x - 3)(x - 3)(x - 3)$$
$$= (x - 3)(x^2 - 6x + 9)$$ *Obtained by multiplying* $(x - 3)(x - 3)$ *mentally*
$$= x^3 - 6x^2 + 9x - 3x^2 + 18x - 27$$ *Now we combine like terms.*

The result of   The result of
$x(x^2 - 6x + 9)$ $-3(x^2 - 6x + 9)$

$$= \boxed{x^3 - 9x^2 + 27x - 27}$$

Notice that after the first step in the solution, since we were not multiplying binomials any longer, we made no attempt to multiply mentally.   ■

EXAMPLE 7

Multiply and simplify. $(2x + 3)(x - 4) - (x + 5)(x - 2)$

Solution

Following the order of operations, we must multiply both sets of binomials first before subtracting. Watch out for the minus sign between the two products! It applies to the entire result of multiplying $(x + 5)(x - 2)$.

$$(2x + 3)(x - 4) - (x + 5)(x - 2)$$
$$= 2x^2 - 8x + 3x - 12 - (x^2 - 2x + 5x - 10)$$
$$\uparrow$$

This minus sign forces us to put in the parentheses.

$$= 2x^2 - 5x - 12 - (x^2 + 3x - 10)$$

Remove parentheses by distributing the understood factor of –1.

$$= 2x^2 - 5x - 12 - x^2 - 3x + 10$$

Combine like terms.

$$= \boxed{x^2 - 8x - 2} \qquad \blacksquare$$

We began this chapter discussing the various exponent rules, and we pointed out at the time the importance of knowing when it is appropriate to use the rules and when not to. The next example serves to highlight this point.

EXAMPLE 8

Perform the indicated operations and simplify.

(a) $(x^3y)^2$ (b) $(x^3 + y)^2$

Solution

Note that we take entirely different approaches to the two parts because they are entirely different examples.

(a) Since this is *not* a binomial raised to a power, but rather a single term raised to a power, we can apply exponent rule 4 and square each *factor*.

$$(x^3y)^2 = (x^3)^2y^2 \qquad \textit{By exponent rule 4}$$

$$= \boxed{x^6y^2} \qquad \textit{By exponent rule 2}$$

(b) Since this is a binomial being raised to a power, we must use our method for multiplying polynomials (call it FOIL, if you like). Exponent rule 4 *does not* apply.

$$(x^3 + y)^2 = (x^3 + y)(x^3 + y)$$
$$= x^6 + x^3y + x^3y + y^2 \qquad \textit{Combine like terms.}$$
$$= \boxed{x^6 + 2x^3y + y^2} \qquad \blacksquare$$

Many verbal problems involve descriptions that give rise to polynomial expressions. The following example illustrates such a situation.

EXAMPLE 9

A rectangular swimming pool measures 10 ft by 30 ft. If a concrete walk of uniform width x ft is to be built surrounding the pool, write an expression for the total area of the pool and walk in terms of x.

Solution

A diagram would certainly be helpful here. We have drawn a rectangular pool 10 ft by 30 ft and surrounded it with a walk of width x ft all around. From the diagram (see Figure 7.2), we can see that the width of the pool and walk (from bottom to top) is

$$x + 10 + x = 2x + 10$$

We can also see that the length of the pool and walk (from left to right) is

$$x + 30 + x = 2x + 30$$

Figure 7.2
10 ft by 30 ft pool with
surrounding walk

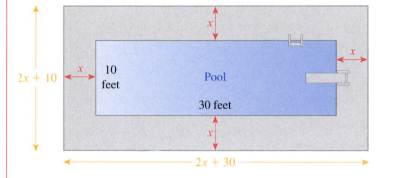

The area, A, of the pool and walk is the total length times the total width. This gives us

$$A = (2x + 30)(2x + 10) = \boxed{4x^2 + 80x + 300 \text{ sq ft}}$$ ∎

www EXERCISES 7.5

*Multiply and simplify each of the following. Whenever possible, do the
multiplication of two binomials mentally.*

1. $3x(5x^3)(4x^2)$

2. $2y^3(5y^2)(3y)$

3. $3x(5x^3 + 4x^2)$

4. $2y^3(5y^2 + 3y)$

5. $4xy(3yz)(-5xz)$

6. $8ab(2ac - 3bc)$

7. $4xy(3yz - 5xz)$

8. $8ab(2ac)(-3bc)$

9. $3x^2(x + 3y) + 4xy(x - 3y)$

10. $5rs(r - 2s) + r^2(3s - 4rs)$

11. $5xy^2(xy - y) - 2y(x^2y^2 - xy^2)$

12. $7r^2s(r^2 - s^2) - 2rs(r^2s - rs^2)$

13. $(x + 2)(x^2 - x + 3)$

14. $(m + 3)(m^2 - 2m + 5)$

15. $(y - 5)(y^2 + 2y - 6)$

16. $(n - 4)(n^2 + 7n + 1)$

17. $(3x - 2)(x^2 + 3x - 5)$

18. $(4a - 3)(a^2 - 7a - 3)$

19. $(5z + 2)(3z^2 + 2z + 8)$

20. $(2c - 1)(6c^2 - 3c - 1)$

21. $(x + y)(x^2 - xy + y^2)$

22. $(m - n)(m^2 + mn + n^2)$

23. $(x^2 + x + 1)(x^2 + x - 1)$

24. $(u^2 + 2u + 1)(u^2 - 2u + 1)$

25. $(x^3 + xy - y^2)(x^3 - 3xy + y^2)$

26. $(v^4 - 4v^2 + 4)(v^4 + 4v^2 + 4)$

27. $(x + 5)(x + 3)$

28. $(y + 4)(y + 6)$

29. $(x - 5)(x - 3)$

30. $(y - 4)(y - 6)$

31. $(x - 5)(x + 3)$

32. $(y + 4)(y - 6)$

33. $(x + 5)(x - 3)$

34. $(y - 4)(y + 6)$

35. $(x + 2y)(x + 3y)$

36. $(y - 5z)(y + 4z)$

37. $(a + 8b)(a - 5b)$

38. $(x - 12y)(x + 3y)$

39. $(3x - 4)(4x - 1)$

40. $(2y - 5)(3y + 4)$

41. $(2r - s)(r + 3s)$

42. $(m + 3p)(2m - 5p)$

43. $(x^2 + 3)(x^2 + 2)$

44. $(t^2 - 5)(t^2 - 4)$

45. $(x + 7)(x + 7)$

46. $(x - 8)(x - 8)$

47. $(x + 7)(x - 7)$

48. $(x - 8)(x + 8)$

49. $(x - 4)^2$

50. $(a + 6)^2$

51. $2(x - 3)(x + 5)$

52. $5(x - 4)(x - 7)$

53. $6(a - 8)(a + 2)$

54. $3(t - 9)(t + 6)$

55. $(x + 2)^3$

56. $(x - 4)^3$

57. $(3x - 5)^2$

58. $(5m - 8)^2$

59. $(2a - 9b)^2$

60. $(4a + 7b)^2$

61. $2x^2(x + 4)(x - 8)$

62. $4z(3z - 5)(2z + 7)$

63. $3x(5x - 6)(3x - 2)$

64. $z^2(2z - 3)(3z - 4)$

65. $(x + 4)(x - 3) + (x - 6)(x - 2)$

66. $(y - 3)(y + 6) + (y + 2)(y - 9)$

67. $(a - 5)(a - 4) - (a - 3)(a - 2)$

68. $(y - 6)(y - 1) - (y - 3)(y - 5)$

69. $(x - 6)^2 - (x + 6)^2$

70. $(2x - 3)^2 - (2x + 3)^2$

71. $(2x - 3)^3$

72. $(3x + 2)^3$

73. $(3a + 4b)^3$

74. $(5r - 2t)^3$

75. $(y - 2)^3 + (y - 2)^2 + y - 2$

76. $2(m + 5)^3 + (m + 5)^2 + m + 5$

77. $(x + 2)^3 - (x + 2)^2 - (x + 2) + 2$

78. $(x - 4)^3 - (x - 4)^2 - (x - 4) - 4$

79. The length of a rectangle is 3 more than twice its width. Using W as the variable, express the perimeter and area of the rectangle in terms of W.

80. The width of a rectangle is 5 less than 3 times its length. Using L as the variable, express the perimeter and area of the rectangle in terms of L.

81. If a square has a side of s in. and the side is increased by 4.5 in., express the change in area in terms of s.

82. If a square has a side of x cm and the side is decreased by 6.1 cm, express the change in area in terms of x.

83. The length of a rectangle is 8 less than 5 times its width. The width is increased by 3 while the length is unchanged. Using a as the variable, express the change in area in terms of a.

84. The length of a rectangle is 4 more than twice its width. The length is increased by 7 and the width is increased by 2. Using y as the variable, express the change in area in terms of y.

85. If a square of side x is cut out of a rectangle whose dimensions are 8 by 10, express the remaining area in terms of x.

86. If an 8 by 10 rectangle is cut out of a square of side x, express the remaining area in terms of x. (We must assume that $x > 10$. Why?)

87. Suppose that the price p of a given item is related to the number of items sold x by the equation $p = 50 - 0.004x$. The revenue R is computed by multiplying the price per item p by the number of items x. Thus, $R = x(50 - 0.004x)$. Find the increase in revenue if x is increased by 500.

88. Use the information given in Exercise 87 to find the decrease in revenue if x is decreased by 250.

② QUESTIONS FOR THOUGHT

89. Given the expressions $(xy)^2$ and $(x + y)^2$ explain:

(a) How are the two expressions similar?

(b) How are they different?

(c) How should each one be multiplied out?

90. For what n is $(x + y)^n$ equal to $x^n + y^n$?

91. Try multiplying out $-4x^2(2x - 7)(3x + 1)$ in three ways:

(a) By first multiplying $-4x^2$ and $(2x - 7)$

(b) By first multiplying $-4x^2$ and $(3x + 1)$

(c) By first multiplying $(2x - 7)(3x + 1)$

Compare your answers. Should they be the same? Why or why not?

◇ MINI-REVIEW

In Exercises 92–95, solve for x.

92. $5 - 2x \geq 17$

93. $8 - 3(x + 1) < 32$

94. $-2 < 3x + 4 \leq 16$

95. $-3 \leq 9 - x < 1$

Solve each of the following problems algebraically. Be sure to label what the variable represents.

96. A large, full tank contains several liquids. If one-fifth of the tank is water, 5 gallons is orange juice, and three-quarters of the tank is wine, how many gallons of wine are in the tank?

97. How many ounces of each of a 35% and 55% alcohol solution must be mixed to produce 40 ounces of a 51% solution?

7.6 Special Products

In the previous section we learned a mechanical procedure for multiplying two polynomials. When this procedure was applied to multiplying out two binomials, we often called it the FOIL method.

Since the structure of products of binomials plays an extremely important role in the discussion to follow, we analyze this structure in detail in the next few examples.

EXAMPLE 1 | Multiply and simplify.

(a) $(x + 5)(x - 3)$ (b) $(x + m)(x + n)$

Solution | Our rule for multiplying polynomials says that each term in the first polynomial multiplies each term in the second.

Recall that to combine $-3x + 5x$, we actually factor out the common x.

(a) $(x + 5)(x - 3) = \underbrace{x \cdot x}_{F} \quad \underbrace{-3 \cdot x}_{O} \quad \underbrace{+5 \cdot x}_{I} \quad \underbrace{-15}_{L}$

$\qquad\qquad\qquad = x^2 - 3x + 5x - 15$

$\qquad\qquad\qquad = x^2 + (-3 + 5)x - 15$

$\qquad\qquad\qquad = \boxed{x^2 + 2x - 15}$

(b) $(x + m)(x + n) = x \cdot x + n \cdot x + m \cdot x + m \cdot n$

We factor out the common factor of x from the middle terms.

$\qquad\qquad\qquad = \boxed{x^2 + (n + m)x + mn}$

Note that the coefficient of x is the *sum* of the second terms in each binomial $(-3 + 5$ or $n + m)$, while the last term is the *product* of the second terms in each binomial $(-3 \cdot 5$ or $m \cdot n)$. ■

Similarly,

$(x + 2)(x + 4) = x^2 + (2 + 4)x + (2)(4) \qquad = x^2 + 6x + 8$

$(x - 5)(x + 3) = x^2 + (-5 + 3)x + (-5)(+3) = x^2 - 2x - 15$

$(x - 7)(x - 3) = x^2 + (-7 - 3)x + (-7)(-3) = x^2 - 10x + 21$

EXAMPLE 2

Multiply and simplify.

(a) $(x - 5)^2$ (b) $(x + p)^2$

Solution

(a) $(x - 5)^2 = (x - 5)(x - 5)$

$\qquad\qquad = x^2 - 5x - 5x + 25$

$\qquad\qquad = \boxed{x^2 - 10x + 25}$

(b) $(x + p)^2 = (x + p)(x + p)$

$\qquad\qquad = x^2 + px + px + p^2 \qquad px + px = 2px$

$\qquad\qquad = \boxed{x^2 + 2px + p^2}$

The expressions $x^2 - 10x + 25$ and $x^2 + 2px + p^2$ are called **perfect square trinomials** (or usually just **perfect squares** for short) because they are the result of squaring a binomial, just as the number 36 is called a perfect square because it is the result of squaring 6. ■

Part **(b)** of Example 2 tells us that

$$(x + p)^2 = x^2 + 2px + p^2$$

If we like, we can think of this as a "formula" for squaring a binomial. In words, it says that the square of a binomial is the square of the first term, plus twice the product of the first and second terms, plus the square of the second term.

That is, we can square binomials such as $(x + 4)^2$ and $(x - 9)^2$ by modeling them after part **(b)**:

$(x + p)^2 = x^2 + 2px + p^2 \qquad\qquad (x + p)^2 = x^2 + 2px + p^2$

$(x + 4)^2 = x^2 + (2 \cdot 4)x + 4^2 \qquad (x - 9)^2 = x^2 + 2(-9)x + (-9)^2$

$\qquad = \boxed{x^2 + 8x + 16} \qquad\qquad\qquad = \boxed{x^2 - 18x + 81}$

On the other hand, we can equally well just multiply out $(x + 4)^2$ or $(x - 9)^2$ by using FOIL the way we usually do. However, regardless of which method we use, we want to be able to write out the final product in one step, *without* writing down the intermediate steps.

The reason this particular product is called "special" is that it comes up frequently enough to make it worth recognizing. In addition, perfect squares will play an important role in Chapter 11 where we discuss solving second-degree equations.

EXAMPLE 3

Multiply and simplify.

(a) $(3x^2y)^2$ (b) $(3x^2 + y)^2$

Solution

It is very important that you recognize the difference between parts **(a)** and **(b)**. Part **(a)** is *not* the square of a binomial. It is a *product* raised to a power and so can be handled by using the exponent rules.

(a) $(3x^2y)^2 = 3^2(x^2)^2y^2$ *Obtained by using exponent rule 4*

$\qquad = \boxed{9x^4y^2}$

Part **(b)**, on the other hand, *is* the square of a *sum,* and so we cannot use exponent rule 4. (The exponent rules pertain only to products and quotients, not to sums and differences.) We must multiply out using FOIL or by following our perfect square form.

(b) $(3x^2 + y)^2 = (3x^2 + y)(3x^2 + y)$

$\qquad = 9x^4 + 2(3x^2)(y) + y^2$ *Using the perfect square form*

$\qquad = \boxed{9x^4 + 6x^2y + y^2}$ ■

Be careful!   Confusing expressions of type **(a)** and **(b)** in Example 3, and therefore improperly applying the exponent rules, is a *very common* error. Always keep in mind that

$$(A + B)^2 \neq A^2 + B^2$$

but rather

$$(A + B)^2 = A^2 + 2AB + B^2$$

EXAMPLE 4

Multiply and simplify.

(a) $(x + 5)(x - 5)$ (b) $(x + a)(x - a)$

Solution

(a) $(x + 5)(x - 5) = x^2 - 5x + 5x - 25$

$\qquad = \boxed{x^2 - 25}$

(b) $(x + a)(x - a) = x^2 - ax + ax - a^2$ *Note that the middle terms add to 0.*

$\qquad = \boxed{x^2 - a^2}$ ■

An expression such as $x^2 - 25$ or $x^2 - a^2$ is called the ***difference of two squares*** (for obvious reasons). This type of expression is given a special name because even though it is the product of two similar binomials (which usually results in a trinomial), we get only two terms due to the middle terms adding to 0.

A note of caution is in order here. It is very easy to confuse expressions such as

$$(x - y)^2 \quad \text{with} \quad x^2 - y^2$$

They are *not* the same:

$$(x - y)^2 = (x - y)(x - y) = x^2 - 2xy + y^2 \qquad \textit{This is a perfect square.}$$

Same signs

which is *not* the same as

$$x^2 - y^2 = (x + y)(x - y) \qquad \textit{This is the difference of two squares.}$$
$$\uparrow \qquad \uparrow$$
$$\textit{Opposite signs}$$

EXAMPLE 5

Multiply and simplify.

(a) $(2x - 3)(5x + 4)$ **(b)** $(ax + b)(cx + d)$

Solution

(a) $(2x - 3)(5x + 4) = (2x)(5x) + 4(2x) - 3(5x) - 3 \cdot 4$
$$= 10x^2 + 8x - 15x - 12$$
$$= \boxed{10x^2 - 7x - 12}$$

(b) $(ax + b)(cx + d) = (ax)(cx) + d(ax) + b(cx) + bd$
$$= acx^2 + adx + bcx + bd \qquad \textit{Factor out x from}$$
$$\qquad\qquad\qquad\qquad\qquad\qquad \textit{the middle terms.}$$
$$= \boxed{acx^2 + (ad + bc)x + bd} \qquad\qquad ■$$

Perhaps part **(b)** of Example 5 should not be called a special product at all, but rather should be called a *general product*. It is not intended that this product be memorized. Nevertheless, we include it because it illustrates how the coefficient of the middle term can be the result of the interaction of the coefficients in the binomials. We will have much more to say about this situation in Section 8.4.

These "special binomial products" are summarized in the box for ease of reference.

General and Special Binomial Products

1. General Products
$$(x + m)(x + n) = x^2 + (n + m)x + mn$$
For example: $(x + 6)(x - 4) = x^2 + 2x - 24$
$$(ax + b)(cx + d) = acx^2 + (ad + bc)x + bd$$
For example: $(3x + 4)(2x + 5) = 6x^2 + 23x + 20$

2. Perfect Square
$$(x + p)^2 = x^2 + 2px + p^2$$
For example: $(x + 7)^2 = x^2 + 14x + 49$

3. Difference of Two Squares
$$(x + r)(x - r) = x^2 - r^2$$
For example: $(x + 7)(x - 7) = x^2 - 49$

While memorizing these forms is certainly *not* necessary, understanding and recognizing them will be very helpful in the work ahead. The best way to develop this understanding and recognition is by working out lots of exercises.

EXERCISES 7.6

Multiply out each of the following. As you work out the problems, identify those exercises that are either a perfect square or the difference of two squares.

1. $(x + 4)(x + 3)$ **2.** $(x + 10)(x + 2)$

3. $(x - 4)(x - 3)$ **4.** $(x - 10)(x - 2)$

5. $(x + 4)(x - 3)$ **6.** $(x + 10)(x - 2)$

7. $x + 4(x - 3)$

8. $x + 10(x - 2)$

9. $(x - 4)(x + 3)$

10. $(x - 10)(x + 2)$

11. $(x + 6)(x + 2)$

12. $(x + 5)(x + 4)$

13. $(x - 6)(x - 2)$

14. $(x - 5)(x - 4)$

15. $(x + 6)(x - 2)$

16. $(x + 5)(x - 4)$

17. $(x - 6)(x + 2)$

18. $(x - 5)(x + 4)$

19. $(x + 12)(x + 1)$

20. $(x + 20)(x + 2)$

21. $(x - 12)(x - 1)$

22. $(x - 20)(x - 2)$

23. $(x + 12)(x - 1)$

24. $(x + 20)(x - 2)$

25. $(x - 12)(x + 1)$

26. $x - 12(x + 1)$

27. $(x - 20)(x + 2)$

28. $x - 20(x + 2)$

29. $(a + 8)(a + 8)$

30. $(t - 6)(t - 6)$

31. $(a - 8)(a - 8)$

32. $(t + 6)(t + 6)$

33. $(a + 8)(a - 8)$

34. $(t + 6)(t - 6)$

35. $(c - 4)^2$

36. $(z + 9)^2$

37. $(c + 4)^2$

38. $(z - 9)^2$

39. $(c + 4)(c - 4)$

40. $(z - 9)(z + 9)$

41. $(3x + 4)(x + 7)$

42. $(2y - 5)(y - 3)$

43. $(3x + 7)(x + 4)$

44. $3x + 7(x + 4)$

45. $(2y - 3)(y - 5)$

46. $2y - 3(y - 5)$

47. $(3x + 4)(x - 7)$

48. $(2y + 5)(y - 3)$

49. $(3x - 4)(x + 7)$

50. $(2y + 3)(y - 5)$

51. $(3x + 4)(5x + 7)$

52. $(2y - 5)(4y - 3)$

53. $(3x + 7)(5x + 4)$

54. $(2y - 3)(4y - 5)$

55. $(3x + 4)(5x - 7)$

56. $(2y - 5)(4y + 3)$

57. $(3x - 4)(5x + 7)$

58. $(2y + 5)(4y - 3)$

59. $(2a + 5)^2$

60. $(3y - 4)^2$

61. $(2a + 5)(2a - 5)$

62. $(3y + 4)(3y - 4)$

63. $(2a - 5)^2$

64. $(x + 3)(x^2 - 4x + 5)$

65. $(x - 4)(x^2 + 6x - 7)$

66. $x - 4(x^2 + 6x - 7)$

67. $(3xy)^2$

68. $(3x + y)^2$

69. $(x^3 + y^2)^2$

70. $(x^3y^2)^2$

71. $(2a + 5y)^2$

72. $(3y - 4z)^2$

73. $(2a + 5y)(2a - 5y)$

74. $(3y - 4z)(3y + 4z)$

? QUESTIONS FOR THOUGHT

75. Multiply out $(x + 6)(x + 4)$ and $(x - 6)(x - 4)$. What is the effect of switching both $+$ signs to $-$ signs?

76. The two examples $(x + 6)(x - 4)$ and $(x - 6)(x + 4)$ also have their signs "switched." What about the middle terms of the resulting trinomials?

77. Look back through this exercise set and identify those *pairs* of exercises that have "switched signs," such as $(x + 6)(x - 4)$ and $(x - 6)(x + 4)$. Are the middle terms always the opposite sign? Will this always be the case? Why?

◇ **MINI-REVIEW**

In Exercises 78–81, perform the indicated operations and simplify.

78. $\dfrac{4x^2y}{15z^3} \cdot \dfrac{12xz}{x^2y}$

79. $\dfrac{20xy}{9z} \div (18xyz)$

80. $\dfrac{a}{3} - \dfrac{a}{2} + \dfrac{a}{5}$

81. *Solve for a.* $\dfrac{a}{3} - \dfrac{a}{2} = \dfrac{a}{5}$

82. A certain sum of money is invested at 10%, and twice that amount is invested at 12%. If the annual interest from the two investments is $408, how much is invested at each rate?

83. Lamont wants to make two investments—one in the amount of $1,400 at 9% interest and a second in the amount of $1,100. If the annual interest from the two investments is $203, what is the rate of interest for the second investment?

CHAPTER 7   SUMMARY

After having completed this chapter, you should be able to:

1. Apply the definition of zero and negative exponents (Section 7.2).

   For example:

   Evaluate.

   (a) $4^0 = \boxed{1}$

   (b) $4^{-3} = \dfrac{1}{4^3} = \boxed{\dfrac{1}{64}}$

2. Apply the various exponent rules to simplify expressions involving integer exponents (Sections 7.1, 7.2).

   For example:

   $$\frac{(4x^{-2}y^3)^2}{2(x^3y^{-5})^{-2}} \qquad \textit{First use exponent rule 4.}$$

   $$= \frac{4^2(x^{-2})^2(y^3)^2}{2(x^3)^{-2}(y^{-5})^{-2}} \qquad \textit{Next use exponent rule 2.}$$

   $$= \frac{16x^{-4}y^6}{2x^{-6}y^{10}} \qquad \textit{Now use exponent rule 3.}$$

   $$= \frac{16}{2}x^{-4-(-6)}y^{6-10}$$

   $$= 8x^2y^{-4} \qquad \textit{Use the definition of } a^{-n}.$$

   $$= \boxed{\dfrac{8x^2}{y^4}}$$

3. Write and use *scientific notation* (Section 7.3).

   For example:

   (a) $28,340 = 2.834 \times 10^4$

   (b) $0.02834 = 2.834 \times 10^{-2}$

   (c) *Compute.*

$$\frac{(0.00008)(2,500)}{0.005} = \frac{(8 \times 10^{-5})(2.5 \times 10^3)}{5 \times 10^{-3}}$$

$$= \frac{8(2.5)}{5} \times \frac{10^{-5} \, 10^3}{10^{-3}}$$

$$= \frac{20}{5} \times \frac{10^{-2}}{10^{-3}}$$

$$= 4 \times 10^{-2-(-3)}$$

$$= 4 \times 10$$

$$= \boxed{40}$$

4. Write a polynomial in *complete standard form,* identify all its coefficients, and find its degree (Section 7.4).

   For example:

   The polynomial $2x^4 - 3x^2 + x^3 - 4$ is written in complete standard form as

$$2x^4 + x^3 - 3x^2 + 0x - 4$$

   The coefficients are 2, 1, –3, 0, and –4, respectively.

   The degree of the polynomial is 4.

5. Add and subtract polynomials (Section 7.4).

   For example:

   (a) $(2x^3 + 3xy - y^2) + (x^3y - xy + 5y^2) = \boxed{2x^3 + x^3y + 2xy + 4y^2}$

   (b) Subtract $x^2 - 4x$ from $3x^2 - 5x$.

$$3x^2 - 5x - (x^2 - 4x) = 3x^2 - 5x - x^2 + 4x$$

$$= \boxed{2x^2 - x}$$

6. Multiply polynomials in general, and binomials mentally (Section 7.5).

   For example:

   (a) $(3x - 4)(2x^2 - 5x - 3)$

   Each term in the first set of parentheses multiplies each term in the second set of parentheses.

$$(3x - 4)(2x^2 - 5x - 3) = 6x^3 - 15x^2 - 9x - 8x^2 + 20x + 12$$

$$= \boxed{6x^3 - 23x^2 + 11x + 12}$$

   (b) *Multiply mentally.* $(2x - 3)(x + 5)$

$$(2x - 3)(x + 5) = 2x \cdot x + 2x \cdot 5 - 3 \cdot x - 3 \cdot 5$$

$$= 2x^2 + 10x - 3x - 15$$

$$= \boxed{2x^2 + 7x - 15}$$

7. Recognize certain "special binomial products" (Section 7.6).

   For example:

   (a) $(x + 5)^2 = (x + 5)(x + 5) = x^2 + 10x + 25$ is a ***perfect square.***

   (b) $(a - 8)(a + 8) = a^2 - 64$ is the ***difference of two squares.***

CHAPTER 7 REVIEW EXERCISES

In Exercises 1–12, simplify the given expression as completely as possible. Express final answers with positive exponents only.

1. 3^{-4}

2. $4^0 + 8 \cdot 4^0 + 4^{-1} + 12 \cdot 4^{-2}$

3. $(3^{-1} + 2^{-2})^2$

4. $(3^{-1} + 2^{-2})^{-1}$

5. $\dfrac{(xy^2)^3}{(x^2y)^4}$

6. $\dfrac{(x^3)^2(y^2)^4}{(x^3y)^2}$

7. $\dfrac{(3x^3y^2)^4}{9(x^2y^4)^3}$

8. $x^{-2}x^{-3}$

9. $(x^{-2})^{-3}$

10. $\dfrac{x^{-2}y^{-5}}{x^{-4}y^{-3}}$

11. $\left(\dfrac{2x^{-2}x^3}{x^{-3}}\right)^{-2}$

12. $\dfrac{(x^2y^{-3})^{-3}}{(x^{-1}y^{-2})^{-4}}$

In Exercises 13–16, write the given number in scientific notation.

13. 58,700,000

14. 0.00587

15. 0.000002

16. 7,000

In Exercises 17–20, write the given number in standard notation.

17. 2.56×10^{-3}

18. 8.79×10^5

19. 5.773×10^8

20. 7.447×10^{-8}

In Exercises 21–24, compute using scientific notation.

21. $(0.008)(250,000)$

22. $(3,600)(0.0005)$

23. $\dfrac{0.001}{0.000025}$

24. $\dfrac{(28,500)(0.004)}{0.0002}$

In Exercises 25–34, answer the following questions:
   (a) How many terms are there?
   (b) What is the degree of each term?
   (c) What is the degree of the polynomial?

25. $x^2 + 3x - 7$

26. $t^3 + t^2 - 3t + 9$

27. $3x^3y - 5y^2 + 6xy$

28. $-5x^5 + 3x^2y^4 - 6x^2 + 2y$

29. $8x - 5$

30. $3 - 4t$

31. 9

32. 0

33. $(3x^5)(2x^3)$

34. $3x^5 + 2x^3$

In Exercises 35–38, write the given polynomial in complete standard form.

35. $2x^3 - 7x^2 + 4$

36. $3t^5 - t^2 - 10$

37. $y^2 + y^5 - 2y - 1$

38. $1 - x^4$

In Exercises 39–94, perform the indicated operations and simplify as completely as possible.

39. $(3x^2 - 5x + 7) + (5x - x^2 - 5)$

40. $(5y^4 - y^2 + 9y) + (2y^2 - y^4 - y)$

41. $(3x^2 - 5x + 7) - (5x - x^2 - 5)$

42. $(5y^4 - y^2 + 9y) - (2y^2 - y^4 - y)$

43. $2(x^2y - xy^2 - 5x^2y^2) + 3(xy^2 + x^2y + x^2y^2)$

44. $4(m^2 - 3m^2n) + 6(m^2n - 2m^2)$

45. $2(x^2y - xy^2) - 5x^2y^2 - 3(xy^2 - x^2y + x^2y^2)$

46. $3(r^2s - rs^2) - r^2s^2 - 4(rs^2 - r^2s^2)$

47. $2a^2(a - 3b) + 4a(a^2 + ab) - 2(a^3 - a^2b)$

48. $8mn(m - mn) - n^2(n - m) - (m^2n - mn^2)$

49. Subtract $x^2 - 4x$ from $x^2 + 4x$.

50. Subtract $3a^2 - b^2$ from $8a^2 - 6b^2$.

51. Subtract $3x - 5$ from the sum of $x^2 + 4x - 3$ and $2x^2 - x - 2$.

52. Subtract the sum of $2a^3 + a + 5$ and $4a - a^2$ from $a^2 - 4$.

53. $(x + 4)(x - 7)$

54. $(a - 5)(a - 4)$

55. $(2x - 3)(4x - 5)$

56. $(5x - 4)(6x - 1)$

57. $(3a - 4b)(2a + 5b)$

58. $(4x - 3y)(7x - 2y)$

59. $3a - 4b(2a + 5b)$

60. $4x - 3y(7x - 2y)$

61. $(x + 2)(x - 3)(x + 1)$

62. $(x - 3)(x - 4)(x - 2)$

63. $(x + 5)(x + 7)$

64. $(a + 6)(a + 3)$

65. $(x - 5)(x - 7)$

66. $(a - 6)(a - 3)$

67. $(x + 5)(x - 7)$

68. $(a - 6)(a + 3)$

69. $(x - 5)(x + 7)$

70. $(a + 6)(a - 3)$

71. $(x - 5)(x - 5)$

72. $(a + 6)(a + 6)$

73. $(x - 5)(x + 5)$

74. $(a + 6)(a - 6)$

75. $(x + 9y)(x - 9y)$

76. $(a - 7b)(a + 7b)$

77. $(2x + 3)(x - 7)$

78. $(3a - 4)(a + 6)$

79. $2x + 3(x - 7)$

80. $3a - 4(a + 6)$

81. $(5x - 2)(3x + 4)$

82. $(4a + 3b)(7a - 2b)$

83. $(x + 6)^2$

84. $(3x - 2y)^2$

85. $(x - 5)^3$

86. $(2x - 1)^3$

87. $3x^2(x - 4)(x + 2)$

88. $2y(3y + 1)(y - 5)$

89. $(x + 8)(x - 8)$

90. $(3x + 2y)(3x - 2y)$

91. $(x + 2)(x^2 - 3x + 4)$

92. $(x - 3y)(x^2 + xy - 4y^2)$

93. $(x^2 + 2x - 1)(x^2 + 2x + 1)$

94. $(y^2 - 3y - 4)(y^2 - 3y + 4)$

95. $(2x - 3)(x + 4) - (x - 2)(x - 1)$

96. $(x - 3)^2 - (x - 2)^2$

97. The length of a rectangle is 5 less than 3 times the width. If we represent the width as w, express the area of the rectangle in terms of w.

98. The side x of a square is increased by 7. Express the increase in area in terms of x.

99. A micron (μm) is 10^{-6} meter. The diameter of a typical cheek cell is 60 μm. How far would 5,000 of these cells stretch if they were lined up in a row?

100. If one atom of hydrogen has a mass of 1.67×10^{-23} g, how many atoms are there in 50 grams of hydrogen?

CHAPTER 7 PRACTICE TEST

1. *Evaluate.* (a) $5^0 + 2^{-2} + 4^{-1}$ (b) $6x^0 - 8x^{-4} + x^{-1}$ for $x = 2$

In Exercises 2–5, simplify as completely as possible. Express final answers with positive exponents only.

2. $\dfrac{(x^4)^2(xy)^3}{x^3 y^5}$

3. $x^{-4}x^{-5}$

4. $(x^{-4})^{-5}$

5. $\dfrac{(2x^{-3}y^4)^4}{4(x^{-2}y^{-1})^3}$

6. Given the polynomial $5x^4 - x^3 + 2x + 7$:

  (a) How many terms are there?

  (b) What is the coefficient of the third-degree term? The second-degree term?

  (c) What is the degree of the polynomial?

In Exercises 7–13, perform the indicated operations and simplify as completely as possible.

7. $-3x^2y(4x^2y)(-2x^3)$

8. $-3x^2y(4x^2y - 2x^3)$

9. $2x(x^2 - y) - 3(x - xy) - (2x^3 - 3x)$

10. $(3x - 2)(4x^2 - 5x + 6)$

11. $3x^2(2x - y) - xy(x + y)$

12. Subtract $x^2 - 4x - 5$ from $2x^2 - 3x - 4$.

13. $(a - 1)^2 - (a + 1)^2$

14. Write in scientific notation.

  (a) 0.00316

  (b) 31,600

15. *Compute using scientific notation.* $\dfrac{(0.24)(5,000)}{0.006}$

16. The side s of a square is decreased by 6. Express the decrease in area in terms of s.

17. One light-year (the distance that light travels in 1 year) is approximately 5.86×10^{12} miles. How many light-years from earth is a star that is 8.32×10^{14} miles from earth? Round your answer to the nearest hundredth.

CHAPTER 8

Factoring

A market research firm finds that the daily revenue R (in dollars) earned by a chemical manufacturing company on the sale of g gallons of mixture KZD is given by $R = g^2 - 80g$. If we want to know how many gallons must be sold each day in order to earn a revenue of \$2,000, we need to be able to solve the equation $2,000 = g^2 - 80g$. The techniques we develop in this chapter will allow us to solve this equation.

This question is answered in Example 8 of Section 8.4.

During our discussion of multiplying polynomials in the previous chapter, we placed particular emphasis on developing the ability to multiply *binomials* quickly and accurately. In this chapter we focus our attention on reversing this process. That is, given a polynomial (particularly of degree 2), we would like to be able to factor it into a product of monomials and/or binomials.

8.1 Common Factors

When we first discussed the distributive property back in Section 2.3, we mentioned and illustrated the fact that it can be used in two ways. We can use it to multiply out, in which case we remove parentheses, or we can use it to factor, in which case we create parentheses.

Even though we know that it makes no difference whether we read an equality statement from left to right or from right to left, when we talk about multiplying out, we write the distributive property as

$$a(b + c) = ab + ac$$

while when we talk about factoring, we write it

$$ab + ac = a(b + c)$$

Factoring an expression changes it from a *sum* into a *product,* and as we have already pointed out, there are numerous situations in which having a product is helpful.

Throughout our discussion, whenever we talk about factoring an expression, we always mean using integers only. In other words, if we list the factors of 5, we would *not* list $\frac{1}{2}$ times 10.

The most basic type of factoring, which we have already discussed briefly in Section 2.3, involves the direct application of the distributive property. For example, if we are interested in factoring the expression $12x + 30$, we look at $12x + 30$ and see that 6 is the largest factor common to both $12x$ and 30. Write (and think)

$$12x + 30 = \boxed{6 \cdot 2x + 6 \cdot 5}$$

$$= \boxed{6(2x + 5)}$$

We have used the distributive property to "factor out" the common factor of 6.

This procedure is usually called *taking out the common factor.* We can, of course, check our answer immediately by multiplying out $6(2x + 5)$ and verifying that we get the original expression, $12x + 30$.

Before we proceed any further, we need to lay down some ground rules. Whenever we are asked to factor an expression, the intention is to factor it as completely as possible. This means that the expression remaining inside parentheses has *no* common factors remaining (other than 1 or –1, of course).

Thus, if we had factored

$$12x + 30 = 2(6x + 15)$$

we have a factorization, but it is *incomplete,* because both $6x$ and 15 still have a common factor of 3. To get the complete factorization, we factor out 6, which is the *greatest common factor.*

EXAMPLE 1

Factor as completely as possible.

(a) $8x^3 + 20x - 28$ (b) $6x^2 - 12x$

Solution

In general, it is probably easiest to begin by first determining the greatest common numerical factor, then the greatest common x factor. If there are other variables, we will look for the greatest common factor of each variable that appears. Then we put all the common factors together to get the overall greatest common factor (GCF for short) of the entire polynomial.

(a) In $8x^3 + 20x - 28$, the GCF of 8, 20, and –28 is 4. Since there is no common x factor (the –28 does not have an x factor in it), the GCF for the entire polynomial is 4.

$$8x^3 + 20x - 28 = \boxed{4 \cdot 2x^3 + 4 \cdot 5x - 4 \cdot 7}$$

Factor out the common factor of 4.

$$= \boxed{4(2x^3 + 5x - 7)}$$

(b) Following the same outline for $6x^2 - 12x$, we see that the GCF for 6 and 12 is 6. Since x^2 is two factors of x, and x is one factor of x, the GCF of x^2 and x is x (they have *one* factor of x in common). Thus, the GCF of $6x^2 - 12x$ is $6x$.

$$6x^2 - 12x = \boxed{6x \cdot x - 6x \cdot 2} \qquad \textit{Factor out the common factor of } 6x.$$

$$= \boxed{6x(x - 2)}$$

As always, we check the factorization by multiplying out. However, remember that this check does not guarantee that we have the *complete* factorization. We should also check the expression inside the parentheses to make sure there are no common factors remaining. ■

Basically, as Example 1 illustrates, factoring out the greatest common factor is a two-step process. First we determine the GCF, and second, we determine what factors remain in each term of the polynomial after the GCF is taken outside the parentheses.

EXAMPLE 2

Factor as completely as possible. $24a^3b - 36a^2c^2 + 48ab^3$

Solution

The GCF of 24, 36, and 48 is 12.

The GCF of a^3, a^2, and a is a.

There is no common factor for b and c because the second term has no b factors and the first and third terms have no c factors.

Thus the GCF of the entire polynomial is $12a$.

$$24a^3b - 36a^2c^2 + 48ab^3 = \boxed{12a \cdot 2a^2b - 12a \cdot 3ac^2 + 12a \cdot 4b^3}$$

$$= \boxed{12a(2a^2b - 3ac^2 + 4b^3)} \qquad ■$$

EXAMPLE 3

Factor as completely as possible. $15x^3y^4 - 5x^2y^3$

Solution

The GCF of 15 and 5 is 5.

The GCF of x^3 and x^2 is x^2.

The GCF of y^4 and y^3 is y^3.

Therefore, the GCF of the entire polynomial is $5x^2y^3$.

$$15x^3y^4 - 5x^2y^3 = \boxed{5x^2y^3 \cdot 3xy - 5x^2y^3 \cdot 1}$$

Do not forget the understood factor of 1.

$$= \boxed{5x^2y^3(3xy - 1)}$$

Note: It is a very common error to neglect putting the 1 into the parentheses, so be careful. (If you check your factorization by multiplying out, then you cannot possibly leave out the 1.) ∎

The next example illustrates that we always have a choice as to the *sign* we choose for the GCF.

EXAMPLE 4

Factor as completely as possible. $-20xy^3 + 35x^2y - 60xy$

Solution

Up to this point we have been speaking about "the" greatest common factor of an expression. However, looking at $-20xy^3 + 35x^2y - 60xy$, we can see that both $5xy$ and $-5xy$ can be the GCF. In other words, we can write

$$-20xy^3 + 35x^2y - 60xy = \boxed{5xy(-4y^2 + 7x - 12)}$$

$$-20xy^3 + 35x^2y - 60xy = \boxed{-5xy(4y^2 - 7x + 12)}$$

We can easily check that both of these factorizations are correct by multiplying out, and that they are complete by recognizing that there is no common factor remaining within the parentheses.

The choice of whether to use $5xy$ or $-5xy$ usually depends on how we are going to use the factored form of the expression. We will have more to say about this as we proceed through the text. ∎

Factoring by Grouping

When we factor out a common factor, it is not necessary that it be a *monomial*.

EXAMPLE 5

Factor as completely as possible. $a(x + 3) - b(x + 3)$

Solution

The entire factor of $(x + 3)$ is a common factor to both terms.

$$a(x + 3) - b(x + 3) = a\ (x + 3)\ - b\ (x + 3)$$

Factor out the common factor of x + 3.

$$= (x + 3)\ (a - b)$$

$$= \boxed{(x + 3)(a - b)}$$ ∎

Sometimes we must group the terms in order to see a common factor.

EXAMPLE 6

Factor as completely as possible. $x^2 + 4x + xy + 4y$

Solution

It is not really apparent how to factor this entire expression, as there is no common factor. However, sometimes when we group the terms we do get common factors.

Let's try grouping the first two terms and the last two terms and see what happens.

$$x^2 + 4x + xy + 4y = (x^2 + 4x) + (xy + 4y)$$

We factor out a common factor of x from the first group and a common factor of y from the second group.

$$= x(x + 4) + y(x + 4)$$

Now there is a common factor of $(x + 4)$.

$$= x\ (x + 4)\ + y\ (x + 4)$$

Factor out the $(x + 4)$.

$$= (x + 4)(x + y)$$ ■

EXAMPLE 7   Factor as completely as possible. $a^2 - ab - 3a + 3b$

Solution   We will again begin by splitting the four terms into two groups.

$$a^2 - ab - 3a + 3b = (a^2 - ab) + (-3a + 3b)$$ *Note how we rewrite $-3a + 3b$.*

If we factor out a common factor of a from the first group, and a common factor of 3 from the second group, we will not readily see that we again have a common factor. In other words, we would get $a(a - b) + 3(-a + b)$.

We do not see any further common factors. Instead, let's factor out -3 from the second group. We get

$$= a(a - b) - 3(a - b)$$
↑
*Watch for **this** sign!*

Now we see a common factor of $(a - b)$.

$$= (a - b)(a - 3)$$ ■

EXERCISES 8.1

In Exercises 1–6, fill in the blank with the factor that produces the required product.

1. _____ $(5x + 10) = 20x^2 + 40x$

2. _____ $(7a + 6) = 35a^3 + 30a^2$

3. _____ $(2x - 3y) = 10x^2 - 15xy$

4. _____ $(7a + 4b) = 21a^2b + 12ab^2$

5. _____ $(4m - 3n + p) = 24m^2np - 18mn^2p + 6mnp^2$

6. _____ $(8x + 6y - 3z) = 72x^2y^2z + 54xy^3z - 27xy^2z^2$

In Exercises 7–58, factor each of the following as completely as possible. If the expression is not factorable, say so. Try factoring by grouping where it might help.

7. $5x + 20$

8. $4x + 28$

9. $6x - 18$

10. $9x - 45$

11. $4y + 27$

12. $6u - 25$

13. $28a - 42$

14. $36y - 48$

15. $12a + 9$

16. $15a + 18$

17. $3a + 6b - 8c$

18. $9m - 12n + 8p$

19. $x^2 + 3x$

20. $y^2 + 6y$

21. $2t^2 + 8t$

22. $4t^2 - 12t$

23. $26y^2 - 39y^3$

24. $50w^4 + 75w^2$

25. $4x^5 + 2x^2 - 8x$

26. $3x^4 - 9x^2 + 6x$

27. $a^2 + a$

28. $t^2 - t$

29. $x^2 - 5x + xy$

30. $a^2 - 3ab - 5a$

31. $3c^6 - 6c^3$

32. $5y^5 - 10y^2$

33. $x^2y - xy^2$

34. $a^3b + ab^3$

35. $24x^2 + 15x$

36. $18u^3 - 63u^4$

37. $38x^3y^2 - 75z^4$

38. $27mn^3 - 62p^5$

39. $12c^3d^5 + 4c^2d^3$

40. $5x^3y - 15x^4y^2$

41. $x^2y^3 - y^2z^4 + x^3z^2$

42. $6a^2b + 10a^3c^2 - 9b^2c^3$

43. $2x^2yz^3 + 8xyz^2 - 10x^2y^2z^2$

44. $6m^3n^2p^4 - 15m^2n^3p^2 + 12mn^2p^3$

45. $6u^3v^2 + 18u^3v^3 - 12u^3v^5$

46. $9w^2z^3 - 3wz + 6wz^4$

47. $x(x - 5) + 4(x - 5)$

48. $a(a + 7) + 3(a + 7)$

49. $y(y + 6) - 3(y + 6)$

50. $z(z - 3) - 5(z - 3)$

51. $x^2 + 8x + xy + 8y$

52. $a^2 - 6a + ab - 6b$

53. $m^2 + mn + 9m + 9n$

54. $w^2 - rw + 10w - 10r$

55. $x^2 - xy - 4x + 4y$

56. $y^2 + wy - 7y - 7w$

57. $3x^2y + 6xy - 5x - 10$

58. $8a^2 - 4ab - 6a + 3b$

? QUESTIONS FOR THOUGHT

59. Which of the following is in completely factored form?

    (a) $x^2 + 5x + 6$, $x(x + 5) + 6$, $(x + 2)(x + 3)$

    (b) $x^3y^2 + x^2y^3$, $xy(x^2y + xy^2)$, $x^2y^2(x + y)$

60. Look at the following and discuss what is happening at each step.

$$x^2 - x - 20$$
$$= x^2 - 5x + 4x - 20$$
$$= x(x - 5) + 4(x - 5)$$
$$= (x + 4)(x - 5)$$

61. What is **wrong** with the following?

$$x^2 - 3x - 5x - 15$$
$$\overset{?}{=} x(x - 3) - 5(x - 3)$$
$$\overset{?}{=} (x - 5)(x - 3)$$

◇ MINI-REVIEW

In Exercises 62–65, simplify the expression and express your final answer with positive exponents only.

62. $(x^3)^4$

63. $(x^{-3})^{-4}$

64. $\dfrac{(3x^2y)^4}{6(xy^3)^2}$

65. $\dfrac{(x^{-2}y^3)^{-1}}{(x^3)^{-2}}$

66. How much pure acid must be added to 40 liters of a 20% acid solution to raise it to a 25% acid solution?

67. How much pure water must be added to 36 liters of an 18% acid solution to dilute it to a 12% acid solution?

8.2 Factoring Trinomials

Factoring a polynomial by taking out a common factor, as we did in the last section, is a fairly straightforward mechanical process.

Now let's turn our attention to factoring trinomials such as

$$x^2 + 5x + 6$$

The first thing we notice is that there is no common factor. If there were a common factor, we would certainly factor that out first.

Were it not for our experience in multiplying out binomials in Sections 7.5 and 7.6, we might simply say that $x^2 + 5x + 6$ cannot be factored. However, we have seen many examples of two binomials multiplying out to give answers of the form $x^2 + 5x + 6$. It is therefore reasonable to ask: Can we construct two binomials so that their product

$$(? \qquad ?)(? \qquad ?) = x^2 + 5x + 6?$$

With a little bit of trial and error, we might very quickly arrive at

$$(x + 3)(x + 2) = x^2 + 5x + 6$$

as our answer. *(Check it!)* However, we want to analyze this example very carefully so that we can develop a systematic approach to factoring trinomials.

One of our "special products" in Section 7.6 was the multiplication of two simple binomials.

$$(x + m)(x + n) = x^2 + nx + mx + mn \qquad \textit{Recall that nx and mx are called}$$
$$= x^2 + (n + m)x + mn \qquad \textit{the middle terms.}$$

Note that the coefficient of x (the first-degree term) is the *sum* of m and n, while the last term is the *product* of m and n.

Let's analyze the factorization of $x^2 + 5x + 6$. To factor $x^2 + 5x + 6$ into the product of two binomials, we know that in order to get x^2 as the first term, the binomials must look like

$$(x \qquad)(x \qquad)$$

Now let's focus on the signs that will go into each set of parentheses without regard to the number that will go into each.

The + sign in front of the 6 tells us that the two signs in the parentheses must be the *same,* either both + signs or both − signs. Do you see why? Since the +6 is the *product* of m and n, m and n must have the same signs (for if their signs were opposite, their product would be negative).

*Remember that when we add two numbers with the **same** sign, we **add** their absolute values and keep the common sign.*

Now the + sign in front of the 5 tells us how the two middle terms add up. Since we already know that the signs are the same, and the +5 tells us that m and n must add up to +5, then both numbers *must* be positive. (If they were both negative, they would add up to a negative number.)

To summarize, looking at $x^2 + 5x + 6$, we see

$$x^2 \qquad + \qquad 5x \qquad + \qquad 6$$

*First, this + sign tells us that the signs in the two parentheses are the **same**.*

Second, this + sign tells us that the signs in the two parentheses are both positive.

Thus far our analysis has told us that

$$x^2 + 5x + 6 = (x \qquad)(x \qquad)$$
$$= (x + \quad)(x + \quad)$$

Now we are ready to find the numbers to be inserted into the parentheses. We are looking for two numbers that *multiply* to 6. The possible pairs of factors are 6 and 1, or 3 and 2. Does one of these pairs *also* add up to 5? Yes. 3 and 2.

Thus, the final result for the factorization is

$$x^2 + 5x + 6 = \boxed{(x + 3)(x + 2)}$$

Of course, $(x + 2)(x + 3)$ is equally correct. The *order* of the factors is irrelevant.

Let's try some examples.

EXAMPLE 1

Factor as completely as possible. $x^2 - 5x - 6$

Solution

As always, we first look for any common factors, but there are none. Therefore, we try to build two binomials that multiply out to $x^2 - 5x - 6$. We know that we must have

$$x^2 - 5x - 6 = (x \qquad)(x \qquad)$$

What do the signs tell us? The minus sign in front of the 6 tells us that the signs in the parentheses must be *opposite*; one must be + and one must be −, because −6 is the product of the two numbers. Thus, we already know that

$$x^2 - 5x - 6 = (x + \quad)(x - \quad)$$

Notice that where we put the + and − signs does not matter, because in $(x \quad)(x \quad)$ the two parentheses are identical.

Since the signs of the two numbers we put into the parentheses are different, when we add the middle terms, we are actually getting a *difference*. The minus sign in front of the 5 tells us that the result of this difference must be negative.

*Remember that when we add two numbers with **opposite** signs, we **subtract** their absolute values and take the sign of the number with the larger absolute value.*

To summarize, looking at $x^2 - 5x - 6$, we see

$$x^2 \qquad - \qquad 5x \qquad - \qquad 6$$

*First, this − sign tells us that the signs in the parentheses are **different**.*

Second, this − sign tells us that the middle terms give us a difference that is negative.

Therefore, we know that the middle term with the larger absolute value must get the minus sign. (Remember that when we add numbers with opposite signs, we keep the sign of the number with the larger absolute value.) As in the last example, since we want the product of the two numbers to be 6 (by putting in the + and − signs we have already taken care of the fact that we want the 6 to be negative), the possible pairs of factors are 6 and 1, or 3 and 2.

Some students say to themselves that they want 5 to be the coefficient of x, and so immediately choose 3 and 2 as the factors. This is incorrect, because it neglects the fact that the 5 must result from a *difference,* and so 3 and 2 do not work. However, 6 and 1 do work. Where we put the 6 and where we put the 1 *do* matter since the two parentheses are not identical (one has a + sign in it, the other a − sign in it). We already know that we want the term with the larger absolute value to be negative. Therefore, we have

$$x^2 - 5x - 6 = (x \quad)(x \quad)$$
$$= (x + \quad)(x - \quad)$$
$$= \boxed{(x + 1)(x - 6)}$$

Again, the order of the factors does not matter as long as you have the same two factors, $x + 1$ and $x - 6$. Be sure to check your factorization by multiplying out the binomials. ∎

If we take the time to analyze the signs, we know whether we are looking for a sum or a difference from the middle terms. This usually makes it easier to find the correct factors (if there are any).

It is possible to look at this last example and simply say we are looking for two numbers whose product is −6 and whose sum is −5, and therefore the two numbers are −6 and +1. This works well for examples where the leading coefficient (that is, the coefficient of x^2) is 1, but not otherwise.

We have taken the time to analyze the signs in order to lay the groundwork for the more complicated factoring work ahead.

EXAMPLE 2

Factor each of the following as completely as possible.

(a) $x^2 - 7x + 6$ (b) $x^2 + 5x - 6$ (c) $x^2 - x - 6$

Solution

Again, we always begin by looking for any common factors. There are none in any of the three examples.

(a) In $x^2 - 7x + 6$, the +6 tells us that the signs are the same; the −7 tells us that they must both be negative.

$$x^2 - 7x + 6 = (x \quad)(x \quad)$$
$$= (x - \quad)(x - \quad)$$
$$= \boxed{(x - 6)(x - 1)}$$

*We need factors of 6 that **add** to 7 (because the signs are the same).* 6 and 1 work.

(b) In $x^2 + 5x - 6$, the −6 tells us that the signs are opposite; the +5 tells us that we want the middle term with larger absolute value to be positive.

$$x^2 + 5x - 6 = (x \quad)(x \quad)$$
$$= (x + \quad)(x - \quad)$$
$$= \boxed{(x + 6)(x - 1)}$$

*We need factors of 6 whose **difference** is 5 (because the signs are opposite).* 6 and 1 work, with the 6 getting the + sign.

(c) In $x^2 - x - 6$, the −6 again tells us that the signs will be opposite, and the −1 coefficient of x tells us that we want the middle term with the larger absolute value to be negative.

$$x^2 - x - 6 = (x \quad)(x \quad)$$
$$= (x + \quad)(x - \quad)$$
$$= \boxed{(x + 2)(x - 3)}$$

*We need factors of 6 whose **difference** is 1 (because the signs are opposite).* 3 and 2 work, with the 3 getting the − sign. ∎

Remember always to check your factorization by multiplying out the factors.

EXAMPLE 3   Factor as completely as possible. $x^2 + 4x + 5$

Solution   First of all, there are no common factors. Next, we try to build two binomials.

$$x^2 + 4x + 5 = (x \quad)(x \quad)$$
$$= (x + \quad)(x + \quad) \qquad \textit{Do you see why there must be two } + \textit{ signs?}$$

The only possible pair of factors for 5 is 5 and 1; therefore, the *only* possible factorization is

$$(x + 5)(x + 1)$$

But this does not work. Check it! Consequently, we say that $x^2 + 4x + 5$ is

$$\boxed{\text{not factorable}}$$

In other words, we cannot find two numbers whose product is 5 and whose sum is 4. ∎

EXAMPLE 4   Factor as completely as possible. $2x + x^2 - 15$

Solution   It is generally easier to factor a trinomial when it is in standard form. You will recall that standard form basically means that the polynomial is written from the highest power of the variable to the lowest. Therefore, we first reorder the terms.

$$2x + x^2 - 15 = x^2 + 2x - 15 \qquad \textit{There are no common factors.}$$
$$= (x \quad)(x \quad) \qquad \textit{The product is } -15 \textit{ so the signs are } + \textit{ and } -.$$
$$= (x + \quad)(x - \quad) \qquad \textit{We need numbers whose product is 15 and whose \textbf{difference} is 2.}$$
$$= \boxed{(x + 5)(x - 3)} \qquad\qquad\qquad\qquad\quad ∎$$

EXAMPLE 5   Factor as completely as possible. $6a^2 - 18a + 12$

Solution   If you immediately attempt to build the two binomials, this example quickly becomes much more complicated than is necessary. In addition, you will probably not get a complete factorization. Our first step should always be to look for any common factors. There is a common factor of 6.

$$6a^2 - 18a + 12 = 6(a^2 - 3a + 2) \qquad \textit{Now we try to factor further.}$$
$$= 6(a \quad)(a \quad)$$
$$= 6(a - \quad)(a - \quad) \qquad \textit{Do you see why both signs must be negative? We need two numbers whose product is 2 and whose \textbf{sum} is 3. They are 2 and 1.}$$
$$= \boxed{6(a - 2)(a - 1)} \qquad \textit{Do not forget the common factor of 6.} \quad ∎$$

EXAMPLE 6   Factor as completely as possible. $x^2 + 6x$

Solution   Resist the temptation to immediately write down $(x \quad)(x \quad)$. We do not need to construct two binomials in this case.

   Remember: Always look for common factors *first.*

$$x^2 + 6x = \boxed{x(x + 6)} \qquad\qquad\qquad\qquad\qquad\qquad ∎$$

EXAMPLE 7   Factor as completely as possible. $x^2 + 8xy + 16y^2$

Solution   We can apply the same basic approach here. However, instead of having an x term and a numerical term in each binomial, we will have an x term and a y term.
Since there is no common factor, we proceed to try to build our two binomials.

$$x^2 + 8xy + 16y^2 = (x\qquad)(x\qquad)\qquad \textit{The second term in each binomial}$$
$$= (x +\quad)(x +\quad)\qquad \textit{will be a y term.}$$

We need two numbers whose product is 16 and whose sum is 8. They are 4 and 4.

$$= \boxed{(x + 4y)(x + 4y)}\quad\text{or}\quad\boxed{(x + 4y)^2}$$

Notice that we did not need to recognize that $x^2 + 8xy + 16y^2$ is a perfect square in order to factor it.   ■

EXAMPLE 8   Factor as completely as possible. $x^2 - 16$

Solution   This is a rather special case in that $x^2 - 16$ is not a trinomial—the middle term is missing. Actually, we can think of the middle term as having a coefficient of 0.

$$x^2 - 16 = x^2 + 0x - 16$$

Thus, we are looking for two numbers whose product is -16 and whose sum is 0.

$$x^2 - 16 = (x\qquad)(x\qquad)\qquad \textit{The signs are opposite because}$$
$$= (x +\quad)(x -\quad)\qquad \textit{the product must be } -16.$$
$$= \boxed{(x + 4)(x - 4)}\qquad \textit{+4 and } -4 \textit{ work.}$$

You may recall that we called this type of expression the *difference of two squares*. It is usually easy to recognize because of its appearance.   ■

EXAMPLE 9   Factor as completely as possible. $4x^2 - 24x + 28$

Solution   As usual, we begin by looking for any common factors. We find that there is a common factor of 4.

$$4x^2 - 24x + 28 = 4(x^2 - 6x + 7)\qquad \textit{We can try to factor the}$$
$$\textit{trinomial further.}$$
$$= 4(x -\quad)(x -\quad)\qquad \textit{We cannot find two factors}$$
$$\textit{of 7 that \textbf{add} to 6.}$$
$$= \boxed{4(x^2 - 6x + 7)}\qquad \textit{The trinomial cannot be}$$
$$\textit{factored any further.}$$ ■

One final comment: It is possible to factor trinomials by listing *all* the possible pairs of binomial factors and then checking to see whether any of them work. Clearly, if this method is chosen, being able to multiply out the binomials mentally is extremely helpful.
In the next section we will look at more complex factoring examples.

www EXERCISES 8.2

Factor each of the following expressions as completely as possible. If an expression is not factorable, say so.

1. $x^2 + 3x$

2. $x^2 + 4x$

3. $x^2 + 3x + 2$

4. $x^2 + 4x + 3$

5. $x^2 - 3x + 2$

6. $x^2 - 4x + 3$

7. $x^2 + 3x - 2$

8. $x^2 - 4x - 3$

9. $x^2 + x - 2$

10. $x^2 - 2x - 3$

11. $x^2 - x - 2$

12. $x^2 + 2x - 3$

13. $a^2 + 8a + 12$

14. $a^2 + 7a + 12$

15. $a^2 - a - 12$

16. $a^2 + 4a - 12$

17. $a^2 - a + 12$

18. $a^2 - 4a - 12$

19. $a^2 - 12a$

20. $a^2 + 12a$

21. $a - 12 + a^2$

22. $12 - 8a + a^2$

23. $y^2 + 11y + 28$

24. $u^2 - 11u + 18$

25. $x^2 - 5x - 36$

26. $t^2 + 5t - 50$

27. $a^2 + 6a - 40$

28. $w^2 - 7w - 60$

29. $z^2 - 17z + 30$

30. $p^2 + 18p + 32$

31. $x^2 - 9x$

32. $w^2 - 4$

33. $x^2 - 9$

34. $w^2 - 4w$

35. $x^2 - 9x - 10$

36. $w^2 - 4w - 32$

37. $x^2 - 3xy + 2y^2$

38. $x^2 + 6xy - 7y^2$

39. $a^2 + 10a + 24$

40. $a^2 - 10a - 24$

41. $y^2 + 12y + 36$

42. $t^2 - 12t + 36$

43. $y^2 - 36$

44. $t^2 + 36$

45. $x^2 - 7x - 18$

46. $m^2 + 6m - 18$

47. $r^2 - 3rs - 10s^2$

48. $r^2 + 9rs - 10s^2$

49. $c^2 - 6c + 5$

50. $c^2 - 13c + 12$

51. $4x^2 + 8x + 4$

52. $6x^2 - 30x + 36$

53. $x^2 - 30 + x$

54. $x^2 - 30 + 7x$

55. $2x^2 - 50$

56. $3x^2 - 27$

57. $x^2 - x - 20$

58. $x^2 - 8x - 20$

59. $x^2 - x + 20$

60. $x^2 - 8x + 20$

61. $y^2 - 11y + 28$

62. $y^2 - 13y - 48$

63. $2y^2 + 2y - 84$

64. $3y^2 - 6y - 72$

65. $49 - d^2$

66. $t^2 - 1$

67. $49 + d^2$

68. $t^2 + 1$

69. $10x^2 - 40xy - 120y^2$

70. $4x^2 + 40xy + 100y^2$

71. $a^2 + 13 + 14a$

72. $8t - 20 + t^2$

73. $6s^2 - 6s - 72$

74. $8y^2 - 24y - 80$

75. $4x^2 - 64$

76. $4x^2 + 64$

Ⓠ QUESTIONS FOR THOUGHT

77. Find *all* integers k so that $x^2 + kx + 10$ can be factored.

78. Find *all* integers b so that $x^2 + bx - 10$ can be factored.

79. Can you find all integers c so that $x^2 + 5x + c$ can be factored? Why or why not?

In Exercises 80–83, perform the indicated operations and simplify.

80. $(5x^2y)^2$

81. $(5x^2 + y)^2$

82. $(3x - 5)(x + 3)$

83. $(2x - 7)(3x + 4)$

In Exercises 84–85, express the given number in scientific notation.

84. 0.00431

85. 28,700

In Exercises 86–87, use scientific notation to compute each of the following.

86. $\dfrac{540}{0.006}$

87. $\dfrac{(2,400)(0.003)}{(0.02)(0.004)}$

88. A sum of $4,000 is to be split into two investments—one paying 6% interest and the other paying 7% interest. If the annual interest from the two investments is $253.60, how much is invested at each rate?

8.3 More Factoring

In order to expand and adapt our factoring skills to more complex situations, let's begin by reviewing a multiplication example.

EXAMPLE 1 Multiply and simplify. $(3x - 4)(2x + 5)$

Solution
$$(3x - 4)(2x + 5) = 6x^2 + 15x - 8x - 20$$
$$= \boxed{6x^2 + 7x - 20}$$

Notice that because the coefficients of x in the original binomial factors were not 1, the coefficient 7, of x, in our final answer is no longer just the sum of the numbers -4 and $+5$. Rather, the 7 results from the interaction of the factors of 6 (3 and 2) with the factors of 20 (4 and 5). ■

The interaction seen in Example 1 is what makes examples where the leading coefficient (that is, the coefficient of x^2) is not 1 more difficult to factor. The following examples illustrate how we can handle these more complicated situations.

EXAMPLE 2 Factor as completely as possible. $2x^2 - 7x + 3$

Solution As usual, the first step is to look for any common factors. In this case there are none.

Because the coefficient of x^2 is 2, and 2 has only one possible pair of factors, 2 and 1, we can begin to analyze this example in the same way that we analyzed examples in the last section. We first insert the possible factors of $2x^2$.

$$2x^2 - 7x + 3 = (2x \quad)(x \quad) \qquad \textit{Next we analyze the signs.}$$
$$= (2x - \quad)(x - \quad)$$

Even though there is only one possible pair of factors for 3, 3 and 1, where we put them does make a difference. This is because of how they will interact with the factors of 2. If we like, we can simply list all the possible binomials and check to see whether any one of them works. The possible pairs of binomials are as follows:

$$(2x - 3)(x - 1) = 2x^2 - 5x + 3$$
$$(2x - 1)(x - 3) = 2x^2 - 7x + 3$$

Thus, the answer is

$$2x^2 - 7x + 3 = \boxed{(2x - 1)(x - 3)}$$

■

EXAMPLE 3

Factor as completely as possible. $3a^2 + 20a + 12$

Solution

There is no common factor.

Analyzing the possible factors of 3 and the signs, we see that if we are to find two binomial factors we must have

$$3a^2 + 20a + 12 = (3a \quad)(a \quad)$$
$$= (3a + \quad)(a + \quad)$$

We are looking for the factors of 3 to interact with the factors of 12 and give us a *sum* of 20. The possible factors of 12 are $12 \cdot 1$, $6 \cdot 2$, and $4 \cdot 3$. Listing the possible binomial pairs, we obtain:

$$(3a + 12)(a + 1) = 3a^2 + 15a + 12$$
$$(3a + 1)(a + 12) = 3a^2 + 37a + 12$$
$$(3a + 6)(a + 2) = 3a^2 + 12a + 12$$
$$(3a + 2)(a + 6) = 3a^2 + 20a + 12$$

Note that we stop when we find the correct factorization. Thus, the correct factorization is

$$3a^2 + 20a + 12 = \boxed{(3a + 2)(a + 6)}$$

■

EXAMPLE 4

Factor as completely as possible. $10x^2 + 7x - 12$

Solution

Again, there are no common factors.

Since *both* 10 and 12 contain several pairs of factors, this example has numerous possibilities to consider right at the beginning. However, our analysis of the signs does tell us that the signs are opposite, so we know we are looking for a *difference* to produce the middle term of $7x$.

We first list the possible binomial pairs using $5x$ and $2x$ to give $10x^2$:

$(5x + 1)(2x - 12) = 10x^2 - 58x - 12$	$(5x - 1)(2x + 12) = 10x^2 + 58x - 12^{\ddagger}$
$(5x + 12)(2x - 1) = 10x^2 + 19x - 12$	$(5x - 12)(2x + 1) = 10x^2 - 19x - 12$
$(5x + 2)(2x - 6) = 10x^2 - 26x - 12$	$(5x - 2)(2x + 6) = 10x^2 + 26x - 12^{\ddagger}$
$(5x + 6)(2x - 2) = 10x^2 + 2x - 12$	$(5x - 6)(2x + 2) = 10x^2 - 2x - 12^{\ddagger}$
$(5x + 3)(2x - 4) = 10x^2 - 14x - 12$	$(5x - 3)(2x + 4) = 10x^2 + 14x - 12^{\ddagger}$
$(5x + 4)(2x - 3) = 10x^2 - 7x - 12$	$(5x - 4)(2x + 3) = 10x^2 + 7x - 12$

Thus, the complete factorization is

$$10x^2 + 7x - 12 = \boxed{(5x - 4)(2x + 3)}$$

Keep in mind that if none of the binomial pairs we have listed is the correct factorization, we must still consider all the possibilities of the form $(10x \quad)(x \quad)$.

In order to make the factorization process as short as possible, several comments are in order here.

First, while the positions of the $+$ and $-$ signs do make a difference, we recognize that a difference that yields $+7$ is also useful. We note that switching the $+$ and $-$ signs in the binomials simply switches the sign of the middle term. If we can get a middle term of $-7x$ from $(5x + 4)(2x - 3)$, then we can get a middle term of $+7x$ by simply switching the signs to $(5x - 4)(2x + 3)$. Therefore, we do not have to write each binomial pair *twice* with the signs interchanged.

Second, notice that even though we wanted to end up with a $+7x$ from our difference, it was not the larger number (4) that got the $+$ sign, but rather the larger middle term (the 3 times $5x$) that got the $+$ sign.

Third, if you are saying to yourself, "Isn't this just a trial-and-error process?", you are absolutely right. We are simply trying to make the procedure as systematic as we can.

Fourth, as you practice more and get better at factoring, you will be able to make an educated guess as to which pairs of factors are most likely to work, and start your list with them. Thus, in this example pairing up 12 with any of the other factors gives a fairly large product as compared to the 7 we are looking for, and so makes it less likely to work. We could have started our list of possible binomial pairs by using $3 \cdot 4$ rather than $12 \cdot 1$ to make a product of 12.

Fifth, the fact that there was no common factor to begin with eliminates some of the lines from the table. Do you see why $(5x - 2)(2x + 6)$ could not possibly work *without* computing the middle term, which is $26x$?

> *Answer:* $2x + 6$ has a common factor of 2, but the original trinomial had no common factor! If $2x + 6$ were one of the factors of $10x^2 + 7x - 12$, then 2 would also have to be a common factor of $10x^2 + 7x - 12$. Thus, since $10x^2 + 7x - 12$ has no common factors, $2x + 6$ cannot possibly be a factor.

In fact, if you look at our list of possible binomial factors, all the lines marked with a "‡" could have been eliminated from consideration for exactly the same reason. Using these ideas, we can substantially narrow down the trial-and-error process. In this example our list could have had two possible pairs instead of twelve! ■

EXAMPLE 5

Factor as completely as possible. $12x^2 + 24x - 36$

Solution

After the last example, you might sigh looking at all the possibilities here. In fact, if you start looking for two binomials immediately, there are 30 possible binomial pairs!

Actually, this example is quite simple if we remember to begin by looking for any common factors. We first take out the common factor of 12.

$$12x^2 + 24x - 36 = 12(x^2 + 2x - 3)$$
$$= 12(x \quad)(x \quad)$$
$$= 12(x + \quad)(x - \quad) \qquad \textit{We need two numbers whose}$$
$$\qquad\qquad\qquad\qquad\qquad\quad \textit{product is 3 and whose}$$
$$= \boxed{12(x + 3)(x - 1)} \qquad \textit{difference is 2.} \qquad ■$$

Remember	Always look for common factors first.

As we saw in Example 4, this "trial-and-error" process for factoring trinomials of the form $Ax^2 + Bx + C$ can be quite laborious, especially when the numbers A and C have numerous factors. Consequently, we now describe a somewhat more systematic method for factoring trinomials. This process is called the *AC*-test and relies heavily on the *factoring by grouping* process discussed in Section 8.1. A proof of why this method works is beyond the scope of this book.

Factoring Trinomials by Grouping—The AC-Test

Another way to factor a trinomial of the form $Ax^2 + Bx + C$ is to use the following outline.

The *AC*-Test for Factoring Trinomials of the Form $Ax^2 + Bx + C$	1. Find the product AC. 2. Find two factors of AC whose sum is B. 3. Rewrite the middle term, Bx, as a sum of terms whose coefficients are the factors found in step 2. 4. Factor by grouping.

Let's illustrate this process with several examples.

EXAMPLE 6

Factor $3x^2 - 11x - 20$ as completely as possible.

Solution

We follow the outline given above:

1. Find the product AC: Since $A = 3$ and $C = -20$, $AC = 3(-20) = -60$.

2. Find two factors of AC whose sum is B. That is, find two factors of -60 whose sum is -11.

   By trial and error we find that the factors are -15 and 4, since $-15(4) = -60$ and $-15 + 4 = -11$.

3. Rewrite the middle term, $-11x$, using the factors we found in step 2.

   Hence, we write $-11x$ as $-15x + 4x$, and we rewrite $3x^2 - 11x - 20$ as $3x^2 - 15x + 4x - 20$.

4. Factor by grouping.

$$3x^2 - 11x - 20 = 3x^2 - 15x + 4x - 20 \qquad \textit{Factor by grouping.}$$
$$= 3x(x - 5) + 4(x - 5) \qquad \textit{There is a common}$$
$$\textit{factor of } x - 5.$$
$$= \boxed{(x - 5)(3x + 4)}$$

Note that we could just as well have written $-11x$ as $4x - 15x$. You should verify for yourself that this leads to the same factorization.

Although we still have to look for factors (of a number larger than A or C), many students find this process a bit more efficient than the standard trial-and-error process. ■

EXAMPLE 7

Factor $6t^2 - 19t + 15$ as completely as possible.

Solution

We first find the product, $6(15) = 90$.

The two factors of 90 that add to -19 are -10 and -9. We rewrite $-19t$ as $-10t - 9t$ and factor by grouping.

$$6t^2 - 19t + 15 = 6t^2 - 10t - 9t + 15$$
$$= 2t(3t - 5) - 3(3t - 5) \qquad \textit{Note that we factor out } -3$$
$$\textit{from the last two terms.}$$
$$= \boxed{(3t - 5)(2t - 3)} \qquad\qquad\qquad ■$$

Not only does the *AC*-test help us factor, but it also tells us when the trinomial is not factorable.

> If there is no pair of factors for AC whose sum is the coefficient of the middle term, then the trinomial does not factor into two binomials.

EXAMPLE 8   Factor $5y^2 + 7y + 9$ as completely as possible.

Solution   The product AC is $5(9) = 45$. The possible pairs of factors of 45 are $45 \cdot 1$, $15 \cdot 3$, and $5 \cdot 9$—none of which has a sum of 7. Therefore, the polynomial is

$$\boxed{\text{not factorable}}$$ ∎

Whether we choose to use the AC-test or not, it is worth repeating that whenever we are trying to factor an expression it is *always* to our advantage to look for a common factor first.

EXAMPLE 9   Factor $20w^2 + 20w - 75$ as completely as possible.

Solution   We begin by looking for the greatest common factor.

$$20w^2 + 20w - 75 \qquad \textit{The greatest common factor is 5.}$$
$$= 5(4w^2 + 4w - 15)$$

We can now use the AC-test to try to factor the trinomial in the parentheses. $AC = 4(-15) = -60$. Since $10(-6) = -60$ and $10 + (-6) = 4$, we rewrite $4w$ as $10w - 6w$.

$$
\begin{aligned}
20w^2 + 20w - 75 &= 5(4w^2 + 4w - 15) \\
&= 5(4w^2 + 10w - 6w - 15) \qquad \textit{Factor by grouping.} \\
&= 5[2w(2w + 5) - 3(2w + 5)] \qquad \textit{Now we factor out} \\
&\qquad\qquad\qquad\qquad\qquad\qquad\quad \textit{the } 2w + 5. \\
&= \boxed{5(2w + 5)(2w - 3)}
\end{aligned}
$$

Note that by taking out the common factor of 5 we were working with a product of $AC = -60$ instead of the original $AC = 20(-75) = -1{,}500$. ∎

We conclude this section with several more factoring examples to illustrate some additional points.

EXAMPLE 10   Factor as completely as possible. $7x - x^2 - 10$

Solution   Since there is no common factor, we can proceed directly to try to build two binomial factors.

As we mentioned in the last section, it is best to rewrite the polynomial in standard form as

$$-x^2 + 7x - 10$$

Since we are more familiar factoring trinomials whose leading coefficient is $+1$ rather than -1, we can first factor out a -1 so that $-x^2$ becomes $+x^2$. That is,

$$-x^2 + 7x - 10 = -(x^2 - 7x + 10) \qquad \textit{Factoring out } -1 \textit{ has the same effect as}$$
$$\textit{multiplying by } -1; \textit{ it changes the sign}$$
$$\textit{of each term inside the parentheses.}$$

Now we can proceed as before. The entire solution appears as follows:

$$
\begin{aligned}
7x - x^2 - 10 &= -x^2 + 7x - 10 \\
&= -(x^2 - 7x + 10)
\end{aligned}
$$

$$= -(x \quad)(x \quad)$$
$$= -(x - \quad)(x - \quad)$$
$$= \boxed{-(x - 5)(x - 2)}$$ *Do not forget the minus sign.* ∎

Having done all these examples, let's pause and outline our approach to factoring.

Outline for Factoring Polynomials	1. Factor out common factors, if any. 2. Make sure that the polynomial is in standard form, preferably with the leading coefficient positive. Factor out −1, if necessary. 3. Factor the remaining polynomial into two binomials, if possible. 4. Check your factorization by multiplying out the factors.

EXAMPLE 11

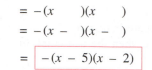

Factor as completely as possible. $5a^3 - 45a^2 - 50a$

Solution

Following our outline, we first factor out the common factor of $5a$.

$$5a^3 - 45a^2 - 50a = 5a(a^2 - 9a - 10)$$
$$= 5a(a \quad)(a \quad)$$
$$= 5a(a + \quad)(a - \quad)$$
$$= \boxed{5a(a + 1)(a - 10)}$$ ∎

EXAMPLE 12

Factor as completely as possible. $3x^3y - 27xy^3$

Solution

The greatest common factor is $3xy$.

$$3x^3y - 27xy^3 = 3xy(x^2 - 9y^2)$$
$$= \boxed{3xy(x + 3y)(x - 3y)}$$ *Within the parentheses we recognize the difference of two squares.* ∎

The ability to factor trinomials (particularly when the leading coefficient is 1 or a prime number) is a very useful skill. We will begin to see how factoring can be used in the next section.

However, what is true for most algebraic skills pertains even more so to factoring. Acquiring the ability to factor accurately and rapidly requires lots of practice. Just because you understand *how* an expression was factored does not necessarily mean that you could factor it yourself. There is absolutely no substitute for doing many exercises.

One final comment: At this point it is easy to get "factorization of the brain" and think that *everything* has to be factored. This is not so. Factoring is just a tool that we will soon learn to apply. However, if you are simplifying an expression such as $3x + 2x + 10$ and get $5x + 10$, this is a completely acceptable answer. You do not have to factor it to $5(x + 2)$, unless you have a reason to do so.

EXERCISES 8.3

Factor each of the following as completely as possible. If the polynomial is not factorable, say so.

1. $x^2 + 5x$

2. $3x^2 - 6x$

3. $x^2 + 5x + 4$

4. $3x^2 - 6x + 3$

5. $x^2 + 5x - 4$

6. $t^2 - 4t + 5$

7. $3x^2 + 8x + 4$

8. $3x^2 - 7x + 4$

9. $2x^2 + 11x + 12$

$2(x^2 + 5x + 6)$
$2(x + 3)(x + 2)$

10. $3x^2 + 13x + 12$

11. $2x^2 + 10x + 12$

12. $3x^2 + 12x + 12$

13. $5x^2 - 27x + 10$

14. $7x^2 - 50x + 7$

15. $5x^2 - 15x + 10$

16. $7x^2 - 56x + 49$

17. $2y^2 - y - 6$

18. $3y^2 - y - 10$

19. $5a^2 + 9a - 18$

20. $5a^2 + 2a - 16$

21. $2t^2 + 7t + 6$

22. $3t^2 + 5t - 12$

23. $2t^2 + 6t + 6$

24. $3t^2 + 6t - 12$

25. $3w^2 - 6w - 30$

26. $5z^2 - 15z - 60$

27. $3x^2 - 4x + 2$

28. $5x^2 - 10x + 4$

29. $3x^2 - 14xy + 15y^2$

30. $3x^2 + 13xy + 12y^2$

31. $6a^2 + 17a + 10$

32. $6a^2 + 19a + 10$

33. $6a^2 + 17a - 10$

34. $6a^2 - 7a - 10$

35. $6a^2 - 18a - 24$

36. $6a^2 - 24a - 30$

37. $x^2 - 36y^2$

38. $x^2 - 16y^2$

39. $4x^2 - 36y^2$

40. $4x^2 - 16y^2$

41. $x^3 + 5x^2 - 24x$

42. $x^3 + 2x^2 - 24x$

43. $x^2 + 5x^2 - 24x$

44. $x^3 + 2x^3 - 24x$

45. $4x^4 - 24x^3 + 32x^2$

46. $6x^5 + 18x^4 - 60x^3$

47. $6x^2y - 8xy^2 + 12xy$

48. $10a^2b^3 + 15ab^2 - 20a^3b^2$

49. $3x^2 - 7x - 48$

50. $3x^2 - 70x - 48$

51. $8x^2 - 32x$

52. $8x^2 - 32$

53. $2x - x^2 + 15$

54. $48 + 8a - a^2$

55. $84xy - 16x^2y - 4x^3y$

56. $27x^2y^2 + 6x^3y^2 - x^4y^2$

57. $-x^2 + 25$

58. $-w^2 - 16$

(?) **QUESTION FOR THOUGHT**

59. As we mentioned in this section, if a polynomial *does not* have a common factor, then we can eliminate any binomial factors that *do* have a common factor. For example, the trinomial $6x^2 - 5x - 4$ has no common factors. Why does this imply that $(3x - 2)(2x + 2)$ should not even be considered as a possible factorization? Of the 12 possible binomial pairs we might list, how many can be eliminated from consideration in this way?

◇ **MINI-REVIEW**

Solve each of the following problems algebraically. Be sure to label what the variable represents.

60. Marlene leaves her house at 7:00 A.M. jogging at 10 kph. One hour later, Sandy leaves the same location walking in the same direction at 6 kph. At what time will the distance between them be 12 km?

61. A street vendor sells franks for $1.25 each and knishes for $0.80 each. If on a certain day she sells 20 more knishes than franks and takes in a total of $169.75, how many of each were sold?

8.4 Solving Polynomial Equations by Factoring

Now that we have learned how to factor polynomial expressions we can apply these factoring techniques to solve certain kinds of polynomial equations of degree higher than 1.

In Chapter 3 we discussed solving first-degree equations in one variable in great detail. Recall that a first-degree equation is one that can be written in the form $ax + b = 0$. We saw that for such an equation there are three possibilities: Either the equation has no solutions (it is a contradiction), exactly one solution (it is conditional), or infinitely many solutions (it is an identity). Solving polynomial equations of higher degree gives rise to other possibilities.

Note that $ax + b$ is a polynomial of the first degree.

An equation involving polynomial expressions of degree 2 is called a **quadratic equation.** Consider the quadratic equation $x^2 + x = 30$. We can easily verify that this equation has at least two solutions since both $x = 5$ and $x = -6$ satisfy this equation.

$$x^2 + x = 30 \qquad \textit{Substitute } x = 5. \qquad\qquad x^2 + x = 30 \qquad \textit{Substitute } x = -6.$$
$$5^2 + 5 \overset{?}{=} 30 \qquad\qquad\qquad\qquad\qquad (-6)^2 + (-6) \overset{?}{=} 30$$
$$25 + 5 \overset{\checkmark}{=} 30 \qquad\qquad\qquad\qquad\qquad\qquad 36 - 6 \overset{\checkmark}{=} 30$$

Some natural questions we might ask include: Are there any other solutions to this equation? How do we actually *find* the solutions to such a quadratic equation? In this section, we will discuss one particular method for solving quadratic equations and see how this method generalizes to higher-degree polynomial equations.

For the most part, we will want to start with a quadratic equation in a particular form.

DEFINITION	A quadratic equation is said to be in ***standard form*** if it is written in the form $ax^2 + bx + c = 0$, with $a > 0$.

Note that standard form for a second-degree *equation* is very similar to standard form for a second-degree *polynomial.* However, do not confuse a polynomial, which is an expression, with an equation.

Also, do not take this definition too literally. It does not mean to say that all the terms are on the left-hand side of the equation and 0 is on the right-hand side. It does mean to say that for a quadratic equation to be in standard form, all the terms must be on one side of the equation with the coefficient of x^2 positive, and 0 on the other side.

EXAMPLE 1 Put each of the following quadratic equations into standard form.

(a) $x^2 + 3x - 5 = x - 3$ (b) $3x - 5 - x^2 = 0$

(c) $(x + 3)^2 = 2x(x - 5)$

Solution (a) In order to get this equation into standard form, we can subtract x and add 3 to both sides of the equation:

$$
\begin{array}{rl}
x^2 + 3x - 5 = & x - 3 \\
\underline{-x + 3} & \underline{-x + 3} \\
\boxed{x^2 + 2x - 2 = 0}
\end{array}
$$

(b) This equation already has all the terms on one side and 0 on the other side, but the coefficient of x^2 is not positive. In order to get the x^2 term to have a positive coefficient, we can multiply both sides of the equation by -1. (While we are at it, we might as well rearrange the terms in the order that we usually prefer—from highest power to lowest.)

$$3x - 5 - x^2 = 0 \qquad \textit{Rearrange the terms to get}$$
$$-x^2 + 3x - 5 = 0 \qquad \textit{Now multiply both sides by } -1.$$
$$-1(-x^2 + 3x - 5) = -1 \cdot 0$$

$$\boxed{x^2 - 3x + 5 = 0} \qquad \textit{This equation is now in standard form.}$$

(c) We begin by multiplying out each side of the equation.

$$(x + 3)^2 = 2x(x - 5)$$
$$(x + 3)(x + 3) = 2x(x - 5)$$
$$x^2 + 6x + 9 = 2x^2 - 10x$$

Keeping in mind that we want the coefficient of x^2 to be positive, we get all the terms on the right-hand side and 0 on the left-hand side. (If we did it the other way around, we would end up with a negative coefficient for the x^2 term, and we would then have the additional step of multiplying both sides of the equation by -1.)

$$x^2 + 6x + 9 = 2x^2 - 10x$$
$$\underline{-x^2 - 6x - 9 \qquad -x^2 - 6x - 9}$$

$$\boxed{0 = x^2 - 16x - 9} \qquad \blacksquare$$

The first method of solution for quadratic equations that we will discuss is based on the following basic fact about real numbers.

Zero-Product Rule	If $a \cdot b = 0$, then either $a = 0$, or $b = 0$, or both a and b are equal to 0.

*In words, this definition says that if a **product** of two factors is equal to zero, then at least one of the factors must be zero.*

Thus, if we want to solve the equation

$$(x - 3)(x + 4) = 0$$

the zero-product rule tells us that if the *product* of two factors is equal to 0, then at least one of the factors must be equal to 0. Thus, we can conclude that

$$x - 3 = 0 \quad \text{or} \quad x + 4 = 0 \qquad \textit{We can now solve these two simple}$$
$$\textit{first-degree equations.}$$

$$\boxed{x = 3} \quad \text{or} \quad \boxed{x = -4}$$

Both of these values are solutions of the original equation, $(x - 3)(x + 4) = 0$.

It is worthwhile noting that this same idea extends to more than two factors. For example, if we want to solve the equation

$$4z(z + 5)(z - 7) = 0$$

we use the fact that in order for the product to be equal to 0, at least one of the factors must be equal to 0. There are four factors: 4, z, $z + 5$, and $z - 7$. We need to find the values of z that make the individual factors equal to 0.

Since the first factor is 4, and 4 can never be equal to 0, we simply ignore it. If you like, you can think of dividing both sides of the equation by 4: 0 divided by 4 is still 0. (We can always ignore *nonzero constant* factors when we are interested in values of the variable that make a product equal to 0.)

This now leaves us with three possibilities:

$$z = 0 \quad \text{or} \quad z + 5 = 0 \quad \text{or} \quad z - 7 = 0$$

$$\boxed{z = 0} \quad \text{or} \quad \boxed{z = -5} \quad \text{or} \quad \boxed{z = 7}$$

As before, we can see that each of these values satisfies the original equation.

Suppose we want to solve the equation $x^2 + x = 30$, which we mentioned at the beginning of this section. We cannot apply the zero-product rule to this equation because we do not have the two necessary ingredients to use it. We do not have a product and we do not have 0.

The difficulty of not having 0 on one side of the equation can be overcome by putting the quadratic equation into standard form. The other difficulty of not having a product can be overcome *if* we can *factor* the nonzero side of our equation. This is exactly what the *factoring* process does: It allows us to express a sum in the form of a product. Thus, our solution is as follows:

$$x^2 + x = 30 \qquad \textit{We begin by putting the equation in standard form.}$$
$$x^2 + x - 30 = 0 \qquad \textit{Now we try to factor the left-hand side.}$$
$$(x + 6)(x - 5) = 0 \qquad \textit{Using the zero-product rule we get}$$
$$x + 6 = 0 \quad \text{or} \quad x - 5 = 0$$
$$\boxed{x = -6} \quad \text{or} \quad \boxed{x = 5}$$

which are the solutions we saw earlier.

The method of solution that we have just illustrated is called the ***factoring method*** for obvious reasons.

There were two basic steps in this solution. The first was to get 0 on one side of the equation, which we can *always* do. The second was to factor the resulting second-degree polynomial that we get on one side of the equation, which we know from experience we cannot always do. Thus, the factoring method has the serious weakness that it does not always work.

In Chapter 11 we will discuss other methods that do not depend on being able to factor, but for the remainder of this section we will look at several more equations that can be solved by the factoring method.

EXAMPLE 2   Solve for y. $3y^2 = 7y$

Solution   In order to solve by the factoring method (which is the only method we have so far), we begin by putting the equation into standard form.

$$3y^2 = 7y \qquad \textit{Subtract 7y from both sides.}$$
$$3y^2 - 7y = 0 \qquad \textit{Factor; remember to look for any common factors first.}$$
$$y(3y - 7) = 0 \qquad \textit{Now we can use the zero-product rule.}$$
$$y = 0 \quad \text{or} \quad 3y - 7 = 0$$
$$\boxed{y = 0} \qquad\qquad 3y = 7$$
$$y = \boxed{\dfrac{7}{3}}$$

Some students attempt to solve this example by dividing both sides of the equation by y. This will lead to an incomplete solution (try it!). The reason is that $y = 0$ may be one of the solutions (as it is in this example), in which case you are dividing by 0, which is not allowed.

Check   We substitute the values $y = 0$ and $y = \frac{7}{3}$ into the original equation.

$$y = 0: \qquad 3y^2 = 7y \qquad\qquad y = \frac{7}{3}: \qquad 3y^2 = 7y$$

$$3(0)^2 \overset{?}{=} 7(0) \qquad\qquad 3\left(\frac{7}{3}\right)^2 \overset{?}{=} 7\left(\frac{7}{3}\right)$$

$$3(0) \overset{?}{=} 0 \qquad\qquad 3\left(\frac{49}{9}\right) \overset{?}{=} 7\left(\frac{7}{3}\right)$$

$$0 \overset{\checkmark}{=} 0 \qquad\qquad \frac{3}{1} \cdot \frac{49}{\underset{3}{9}} \overset{?}{=} \frac{7}{1} \cdot \frac{7}{3}$$

$$\frac{49}{3} \overset{\checkmark}{=} \frac{49}{3} \qquad\blacksquare$$

Let's pause to summarize the factoring method.

Solving Quadratic Equations by the Factoring Method	1. Get the equation into standard form. This may require several steps.
	2. Factor the nonzero side of the equation.
	3. Use the zero-product rule to set each of the factors equal to 0.
	4. Check each of the answers in the original equation.

EXAMPLE 3

Solution

Solve for a. $(2a + 1)(a - 1) = (3a - 2)(2a - 4)$

Following our outline, we begin by multiplying out both sides of the equation, and then get it into standard form.

$$(2a + 1)(a - 1) = (3a - 2)(2a - 4)$$
$$2a^2 - a - 1 = 6a^2 - 16a + 8$$
$$\underline{-2a^2 + a + 1 \quad -2a^2 + \quad a + 1}$$
$$0 = 4a^2 - 15a + 9$$
$$0 = (4a - 3)(a - 3)$$
$$4a - 3 = 0 \quad\text{or}\quad a - 3 = 0$$
$$a = \boxed{\frac{3}{4}} \qquad \boxed{a = 3}$$

We can most easily get standard form by collecting terms on the right-hand side.

Factoring this requires a bit of trial and error.

Now we solve each of these equations.

Check   We substitute $a = 3$ into the original equation.

$$(2a + 1)(a - 1) = (3a - 2)(2a - 4)$$
$$(2(3) + 1)(3 - 1) \overset{?}{=} (3(3) - 2)(2(3) - 4)$$
$$(6 + 1)(2) \overset{?}{=} (9 - 2)(6 - 4)$$
$$(7)(2) \overset{\checkmark}{=} (7)(2)$$

The fact that $a = 3$ checks gives us a great deal of confidence that our other solution is correct as well. Nevertheless, in order to be absolutely sure we must also substitute $a = \frac{3}{4}$ into the original equation. This check is left to the student. $\blacksquare$

EXAMPLE 4

Solution

Solve for y. $(y - 6)(y + 1) = 8$

Be careful not to misinterpret the zero-product rule. We *cannot* set each of the factors $y - 6$ and $y + 1$ equal to 8. If the product of two numbers is equal to 8, we cannot conclude that one of the numbers is 8. The zero-product rule requires that the product of the factors be equal to *zero*.

Therefore, we must again begin by getting the equation into standard form.

$$(y - 6)(y + 1) = 8$$

$$y^2 - 5y - 6 = 8 \qquad \textit{Subtract 8 from both sides.}$$

$$y^2 - 5y - 14 = 0 \qquad \textit{Factor.}$$

$$(y - 7)(y + 2) = 0 \qquad \textit{Now we use the zero-product rule.}$$

$$y - 7 = 0 \quad \text{or} \quad y + 2 = 0$$

$$\boxed{y = 7} \quad \text{or} \quad \boxed{y = -2}$$

Check $y = 7$: $(y - 6)(y + 1) = 8$ $y = -2$: $(y - 6)(y + 1) = 8$

$\qquad\qquad\qquad (7 - 6)(7 + 1) \overset{?}{=} 8 \qquad\qquad\qquad (-2 - 6)(-2 + 1) \overset{?}{=} 8$

$\qquad\qquad\qquad\qquad (1)(8) \overset{\checkmark}{=} 8 \qquad\qquad\qquad\qquad\qquad (-8)(-1) \overset{\checkmark}{=} 8$ ■

As the next example shows, fractional equations can give rise to quadratic equations.

EXAMPLE 5   Solve for x. $x + \dfrac{10}{x} = 7$

Solution   We begin as we would with any fractional equation—by clearing the denominators.

Recall from Section 3.3 that since x cannot be equal to zero (why not?), we may multiply both sides of the equation by x and obtain an equivalent equation.

$$x + \frac{10}{x} = 7 \qquad \textit{Multiply both sides of the equation by the LCD,}$$
$$\qquad\qquad\qquad\qquad \textit{which is } x.$$

$$x\left(x + \frac{10}{x}\right) = x \cdot 7$$

$$x \cdot x + \frac{\cancel{x}}{1} \cdot \frac{10}{\cancel{x}} = 7x$$

$$x^2 + 10 = 7x \qquad \textit{Since this is a quadratic equation, we get it into}$$
$$\qquad\qquad\qquad\qquad \textit{standard form.}$$

$$x^2 - 7x + 10 = 0 \qquad \textit{Factor.}$$

$$(x - 5)(x - 2) = 0$$

$$x - 5 = 0 \quad \text{or} \quad x - 2 = 0$$

$$\boxed{x = 5} \quad \text{or} \quad \boxed{x = 2}$$

Check   We check both solutions in the *original* equation.

$\qquad x = 5: \quad x + \dfrac{10}{x} = 7 \qquad\qquad\qquad x = 2: \quad x + \dfrac{10}{x} = 7$

$\qquad\qquad\qquad 5 + \dfrac{10}{5} \overset{?}{=} 7 \qquad\qquad\qquad\qquad 2 + \dfrac{10}{2} \overset{?}{=} 7$

$\qquad\qquad\qquad\quad 5 + 2 \overset{\checkmark}{=} 7 \qquad\qquad\qquad\qquad\quad 2 + 5 \overset{\checkmark}{=} 7$ ■

In Example 5 we made an important remark. We said that because the equation we obtained as a result of simplifying was a quadratic equation, "we get it into standard form." It is important to keep in mind that if the equation we obtain from our simplifying process is a *first-degree equation,* then we solve it by simply isolating the variable. There would be no need to put the equation into standard form. Standard form is the form we prefer for *quadratic equations* only.

Some polynomial equations for degree higher than 2 can also be solved by the factoring method.

EXAMPLE 6

Solve for t. $t^3 + 35t = 12t^2$

Solution

As we mentioned earlier, the zero-product rule applies to a product regardless of the number of factors. Consequently, we may be able to apply the factoring method to a polynomial equation of degree higher than 2 such as the one given in this example.

$$t^3 + 35t = 12t^2 \qquad \textit{First we get the equation into standard form.}$$
$$t^3 - 12t^2 + 35t = 0 \qquad \textit{Factor as completely as possible.}$$
$$t(t^2 - 12t + 35) = 0$$
$$t(t - 5)(t - 7) = 0 \qquad \textit{Now we use the zero-product rule.}$$
$$t = 0 \quad \text{or} \quad t - 5 = 0 \quad \text{or} \quad t - 7 = 0$$

The check is left to the student.

$$\boxed{t = 0} \quad \text{or} \quad \boxed{t = 5} \quad \text{or} \quad \boxed{t = 7} \qquad \blacksquare$$

Now that we can solve some quadratic equations, we can handle certain applications that give rise to quadratic equations.

EXAMPLE 7

The length of a rectangle is 3 less than twice the width, and its area is equal to 44 sq cm. Find the dimensions of the rectangle.

Solution

We draw a diagram of the rectangle and label the sides according to the information in the problem (see Figure 8.1). Let $x = $ width.

The area of a rectangle is equal to its length times its width. Therefore, our equation is

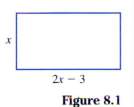

x

$2x - 3$

Figure 8.1

Rectangle for Example 7

$$x(2x - 3) = 44 \qquad \textit{Multiply out and put the equation in standard form.}$$
$$2x^2 - 3x - 44 = 0 \qquad \textit{The factoring method works.}$$
$$(2x - 11)(x + 4) = 0$$
$$2x - 11 = 0 \quad \text{or} \quad x + 4 = 0$$
$$x = \frac{11}{2} \quad \text{or} \quad x = -4$$

Even though $x = -4$ is a solution to the equation, it does not give us a solution to the problem. It makes no sense for the width of a rectangle to have a negative dimension, so we reject the negative answer. Thus, the answer to the problem is

$$\boxed{\text{width} = \frac{11}{2} \text{ cm} = 5\frac{1}{2} \text{ cm}}$$

and the length $= 2\left(\frac{11}{2}\right) - 3 = 11 - 3 = 8$, so

$$\boxed{\text{length} = 8 \text{ cm}}$$

Check   We check to see that the area is equal to 44 square centimeters:

$$\frac{11}{2} \cdot 8 = 11 \cdot 4 = 44 \qquad \blacksquare$$

Let's now return to the situation described at the beginning of this chapter.

EXAMPLE 8

A market research firm finds that the daily revenue R (in dollars) earned by a chemical manufacturing company on the sale of g gallons of mixture KZD is given by the equation $R = g^2 - 80g$. How many gallons must be sold if the revenue from the KZD is to be $2,000 per day?

Solution

If we want to find the number of gallons that must be sold to make the daily revenue $2,000, then we want to find the value of g that makes $R = 2,000$ in the given equation.

$$R = g^2 - 80g \qquad \text{\textit{Substitute } R = 2,000.}$$
$$2,000 = g^2 - 80g \qquad \text{\textit{Since this is a quadratic equation, we}}$$
$$\text{\textit{get it into standard form.}}$$
$$0 = g^2 - 80g - 2,000 \qquad \text{\textit{Factor.}}$$
$$0 = (g - 100)(g + 20)$$
$$g - 100 = 0 \qquad \text{or} \qquad g + 20 = 0$$
$$g = 100 \quad \text{or} \qquad g = -20$$

Both the values $g = 100$ and $g = -20$ satisfy the equation $2,000 = g^2 - 80g$. However, both values do not make sense as answers to the question.

We have found two solutions to the equation, but since it makes no sense for the company to manufacture *negative* twenty gallons of KZD, we reject the answer $g = -20$. Therefore, the answer to the question is that the company must manufacture 100 gallons of KZD daily to produce a corresponding revenue of $2,000. The check is left to the student. ■

Keep in mind that all of the quadratic equations we have solved in this section have been solved using the factoring method. If after getting a quadratic equation into standard form, we cannot factor the nonzero side, then the factoring method fails. In such cases there are other methods available; we will discuss them in Chapter 11.

EXERCISES 8.4

Solve each of the following equations. If the equation is quadratic, use the factoring method. If the factoring method does not apply, say so.

1. $(x - 2)(x + 3) = 0$

2. $(a + 4)(a - 3) = 0$

3. $(x - 2)(x + 3) = 6$

4. $(a + 4)(a - 3) = 8$

5. $x - 2(x + 3) = 6$

6. $a + 4(a - 3) = 8$

7. $y(y - 4) = 0$

8. $w(w - 6) = 0$

9. $y(y - 4) = 12$

10. $w(w - 6) = 27$

11. $x^2 - x - 6 = 0$

12. $t^2 + 2t - 8 = 0$

13. $x^2 - 3x = 10$

14. $t^2 + 10 = 7t$

15. $-m^2 + 2m + 8 = 0$

16. $-w^2 + 9w - 20 = 0$

17. $-m^2 = 8 - 9m$

18. $-w^2 = -w - 20$

19. $p^2 + 3p = p(p + 4)$

20. $2n^2 - 6n = n(2n + 4) + 5$

21. $y^2 = 4y$

22. $a^2 = 5a$

23. $5w^2 = 8w$

24. $4t^2 = 9t$

25. $2a^2 = 11a - 12$

26. $14b = 3b^2 - 5$

27. $2a(a + 3) = 0$

28. $3z(z - 5) = 0$

29. $2a(a + 3) = 20$

30. $3z(z - 5) = -12$

31. $2x^2 + 5x - 4 = x^2 + 3x - 7$

32. $3x^2 - 2x - 6 = 2x^2 - 6x - 3$

33. $a^2 - 4a - 2 = 2a^2 - 9a - 16$

34. $4a^2 - 5a + 3 = 3a^2 + a + 19$

35. $5x^2 = 45$

36. $3x^2 = 48$

37. $(x + 3)^2 = 3x^2 - 10$

38. $(x - 4)^2 = 2x^2 - 11x - 2$

39. $4y = 4y^2 + 1$

40. $9y^2 + 4 = 12y$

41. $(x + 2)^2 = 25$

42. $(z - 3)^2 = 16$

43. $(x - 4)(x + 1) = (x - 3)(x - 2)$

44. $(y + 2)(y + 5) = (y - 1)(y + 6)$

45. $(2x - 4)(x + 1) = (x - 3)(x - 2)$

46. $(y + 2)(y + 5) = (2y - 1)(y + 6)$

47. The width of a rectangle is 2 less than 3 times its length. If the area of the rectangle is 33 sq ft, find its dimensions.

48. The length of a rectangle is 16 more than 5 times its width. If the area of the rectangle is 45 cm^2, find its dimensions.

49. A square painting is surrounded by a 1-inch-wide frame. If the area of the frame alone is 28 sq in., find the dimensions of the painting.

50. A square painting is surrounded by a 3-inch-wide frame. If the total area of the painting plus frame is 225 sq in., find the dimensions of the painting.

51. A rectangular garden measures 5 ft by 7 ft and is surrounded by a uniform path x ft wide. If the total area of the garden plus path is 63 sq ft, find the width of the path.

52. The length of a rectangular garden is 5 more than its width, and is surrounded by a uniform path 2 ft wide. If the total area of the garden plus path is 300 sq ft, find the width of the garden.

53. A carpet store determines that the daily revenue R (in dollars) earned by the store on the sale of y square yards of carpet is given by the equation $R = y^2 - 65y$. How many square yards must be sold if the revenue from the sale of the carpet is to be $1,200 per day?

54. A computer software firm finds that the weekly revenue R (in dollars) earned by the firm on the sale of d compact discs is given by the equation $R = 2d^2 - 190d$. How many CDs must be sold if the revenue from CDs is to be $1,000 per week?

55. If a rocket is shot straight up into the air from ground level with an initial velocity of 240 ft per second, its height h (in feet) above the ground t seconds after it is launched is given by the equation $h = 240t - 16t^2$. How long does it take for the rocket to reach a height of 800 ft? Explain the significance of your answer(s).

56. If a ball is thrown downward from the top of a building 800 ft tall with an initial velocity of 40 ft per second, its height h (in feet) above the ground t seconds after it is thrown is given by the equation $h = 800 - 40t - 16t^2$. How long does it take for the ball to reach a height of 200 ft?

57. In a basketball league of n teams in which each team plays every other team twice, the total number of games played is $n^2 - n$. How many teams are there in a league that plays a total of 90 games?

58. Repeat Exercise 57 if a total number of 210 games are played.

⑦ QUESTIONS FOR THOUGHT

59. Explain what is **wrong** (and why) with the following "solutions."

    (a) $(x - 3)(x - 4) = 7$
    $$x - 3 = 7 \quad \text{or} \quad x - 4 = 7$$
    $$x = 10 \quad \text{or} \quad x = 11$$

    (b) $3x(x - 2) = 0$
    $$x = 3 \quad \text{or} \quad x = 0 \quad \text{or} \quad x - 2 = 0$$
    $$x = 3 \quad \text{or} \quad x = 0 \quad \text{or} \quad x = 2$$

60. Consider the equation $x^2 + 4 = 0$. Is it possible for this equation to have any solutions in the real number system? Why or why not?

8.5 Dividing Polynomials

In Chapter 7 we discussed various procedures with polynomials. We talked about how to add, subtract, and multiply polynomials. We postponed considering division until now so that we could see how division and factoring often go hand in hand.

When we divide a polynomial by a *monomial,* the process is quite straightforward and is basically just an application of the distributive property. That is,

$$\frac{a + b}{c} = \frac{1}{c}(a + b) = \frac{1}{c} \cdot a + \frac{1}{c} \cdot b = \frac{a}{c} + \frac{b}{c}$$

Alternatively, we can say that this is just the reverse of combining fractions with the same denominators.

In any case, the result is the same. Each term in the numerator is divided by the denominator.

EXAMPLE 1   Divide. $\dfrac{4x^5 + 8x^3 - 6x}{2x}$

Solution   We offer two methods of solution.

Method 1.   Just as we can combine several fractions that have the same denominator into a single fraction, we can also break up a single fraction with several terms in the numerator into separate fractions.

$$\frac{4x^5 + 8x^3 - 6x}{2x} = \frac{4x^5}{2x} + \frac{8x^3}{2x} - \frac{6x}{2x} \qquad \textcolor{blue}{\textit{Now we reduce each fraction.}}$$

$$= \frac{\overset{2x^4}{\cancel{4x^5}}}{\cancel{2x}} + \frac{\overset{4x^2}{\cancel{8x^3}}}{\cancel{2x}} - \frac{\overset{3}{\cancel{6x}}}{\cancel{2x}}$$

$$= \boxed{2x^4 + 4x^2 - 3}$$

If you use this method, **be careful.** Be sure to divide *each* term in the numerator by the denominator.

Method 2.   Division can sometimes be accomplished by reducing the original fraction. However, we know that according to the Fundamental Principle of Fractions, we are allowed to cancel common factors only. Thus, we have another application of factoring. We begin by factoring the numerator.

$$\frac{4x^5 + 8x^3 - 6x}{2x} = \frac{2x(2x^4 + 4x^2 - 3)}{2x}$$

$$= \frac{\cancel{2x}(2x^4 + 4x^2 - 3)}{\cancel{2x}}$$

$$= \boxed{2x^4 + 4x^2 - 3} \qquad \blacksquare$$

Let's try another example.

EXAMPLE 2   Divide. $\dfrac{15x^2y^3 - 5xy^2 + 10x^3y}{5x^2y^2}$

Solution   *Method* 1.   We break up the fraction.

$$\frac{15x^2y^3 - 5xy^2 + 10x^3y}{5x^2y^2} = \frac{15x^2y^3}{5x^2y^2} - \frac{5xy^2}{5x^2y^2} + \frac{10x^3y}{5x^2y^2}$$

$$= \frac{\overset{3}{\cancel{15}}\cancel{x^2}y^3}{\cancel{5x^2}\cancel{y^2}} - \frac{\cancel{5}\cancel{x}y^2}{\cancel{5x^2}\cancel{y^2}} + \frac{\overset{2x}{\cancel{10}}\cancel{x^3}\cancel{y}}{\cancel{5x^2}\cancel{y^2}}$$

$$= \boxed{3y - \frac{1}{x} + \frac{2x}{y}}$$

Method 2.   We first factor the numerator. The GCF is $5xy$.

$$\frac{15x^2y^3 - 5xy^2 + 10x^3y}{5x^2y^2} = \frac{5xy(3xy^2 - y + 2x^2)}{5x^2y^2}$$

$$= \frac{\cancel{5xy}(3xy^2 - y + 2x^2)}{\cancel{5}\cancel{x^2}\cancel{y^2}}$$

$$= \boxed{\frac{3xy^2 - y + 2x^2}{xy}}$$

Although the answers obtained by the two methods look different, they are in fact equivalent. In this example, if we combine the fractions in the answer obtained by Method 1 into a single fraction, we will get the answer obtained by Method 2. Try it! Both forms of the answer are acceptable, although for most purposes the form of the answer obtained by Method 2 is preferred. ■

While Method 2 can *sometimes* be used when dividing a polynomial by a *polynomial,* we will postpone considering such an approach until our second meeting with fractions in Chapter 9. Instead, we will outline a mechanical procedure for dividing a polynomial by a polynomial. This procedure basically follows the same lines as long division for numbers does.

For example, if we are to divide $\frac{389}{12}$ as a long division problem, we would write

$$
\begin{array}{r}
32 \\
12\overline{)389} \\
36 \\
\hline
29 \\
24 \\
\hline
5
\end{array}
$$
which is really just shorthand for

$$
\begin{array}{r}
30 + 2 \\
10 + 2\overline{)300 + 80 + 9} \\
300 + 60 \\
\hline
20 + 9 \\
20 + 4 \\
\hline
5
\end{array}
$$

← *We divide 10 into 300.*

← *We divide 10 into 20.*

Note that we focus our attention on dividing by 10, but then we must multiply the number in the quotient by 10 *and* by 2.

We will follow basically the same outline for polynomials. If we wish to divide

$$\frac{x^3 + 5x^2 + 3x - 6}{x + 2}$$

we write

$$\frac{x^3 + 5x^2 + 3x - 6}{x + 2} = x + 2\overline{)x^3 + 5x^2 + 3x - 6}$$

$$x + 2\overline{)x^3 + 5x^2 + 3x - 6}$$

We begin by dividing x^3 by x. That is, we ask: What times x equals x^3?

Answer: x^2 (*because* $x^2 \cdot x = x^3$)

Now we multiply $x + 2$ by x^2.

$$
\begin{array}{r}
x^2 \\
x + 2\overline{)x^3 + 5x^2 + 3x - 6}
\end{array}
$$

$$
\begin{array}{r}
x^2 \\
x + 2\overline{)x^3 + 5x^2 + 3x - 6} \\
-(x^3 + 2x^2) \\
\hline
3x^2 + 3x
\end{array}
$$

Subtract and bring down the next term, $3x$.

Up to here is one complete step in the division process. We repeat this process by dividing $3x^2$ by x. That is, we ask: What times x equals $3x^2$? **Answer:** $3x$

$$
\begin{array}{r}
x^2 + 3x \\
x + 2 \overline{)x^3 + 5x^2 + 3x - 6} \\
-(x^3 + 2x^2) \\
\hline
3x^2 + 3x \\
-(3x^2 + 6x) \\
\hline
-3x - 6
\end{array}
$$

*We multiplied 3x times x + 2. We **subtract** and bring down the next term.*

Next we ask: What times x equals $-3x$? **Answer:** -3

$$
\begin{array}{r}
x^2 + 3x - 3 \\
x + 2 \overline{)x^3 + 5x^2 + 3x - 6} \\
-(x^3 + 2x^2) \\
\hline
3x^2 + 3x \\
-(3x^2 + 6x) \\
\hline
-3x - 6 \\
-(-3x - 6) \\
\hline
0
\end{array}
$$

$\leftarrow$ *Remainder*

Subtract from the line above.

We can, of course, check this answer by multiplying $(x + 2)(x^2 + 3x - 3)$:

$$(x + 2)(x^2 + 3x - 3) = x^3 + 3x^2 - 3x + 2x^2 + 6x - 6 = x^3 + 5x^2 + 3x - 6$$

Before we begin the long division process, we require that the **dividend** (the polynomial we divide into) be in **complete standard form.** Complete standard form (see Section 7.4) basically means that all the terms from highest degree down to lowest are accounted for. We illustrate this in the next example.

EXAMPLE 3

Divide. $\dfrac{4x^3 - 7x - 3}{2x + 3}$

Solution

We will write out the example in long division format with the dividend (numerator) in complete standard form. Note that we include the "missing" x^2 term with its coefficient 0.

$$
\begin{array}{r}
2x^2 - 3x + 1 \\
2x + 3 \overline{)4x^3 + 0x^2 - 7x - 3} \\
-(4x^3 + 6x^2) \\
\hline
-6x^2 - 7x \\
-(-6x^2 - 9x) \\
\hline
2x - 3 \\
-(2x + 3) \\
\hline
-6
\end{array}
$$

*Subtract means change the signs of the line you are subtracting, and **add.***

$\leftarrow$ *Remainder*

Note that the division process ends when the degree of the divisor (in this case, 1) is greater than the degree of the new dividend. Let's check this example. As with arithmetic division we must check that Dividend = (Divisor)(Quotient) + Remainder.

$$
\begin{aligned}
4x^3 - 7x - 3 &\overset{?}{=} (2x + 3)(2x^2 - 3x + 1) - 6 \\
&\overset{?}{=} 4x^3 - 6x^2 + 2x + 6x^2 - 9x + 3 - 6 \\
&\overset{\checkmark}{=} 4x^3 - 7x - 3
\end{aligned}
$$

EXAMPLE 4

Divide. $\dfrac{2x^4 + x^2 + 3}{x - 1}$

Solution

Again, we write out the example in long division format with the numerator in complete standard form.

$$
\begin{array}{r}
2x^3 + 2x^2 + 3x + 3 \\
x - 1\overline{\smash{\big)}\,2x^4 + 0x^3 + x^2 + 0x + 3} \\
\underline{-(2x^4 - 2x^3)} \quad\quad\quad\quad\quad\quad \text{Subtract.}\\
2x^3 + x^2 \\
\underline{-(2x^3 - 2x^2)} \\
3x^2 + 0x \\
\underline{-(3x^2 - 3x)} \\
3x + 3 \\
\underline{-(3x - 3)} \\
6 \quad \leftarrow Remainder
\end{array}
$$

We can leave the answer in this form, or we can write

$$\frac{2x^4 + x^2 + 3}{x - 1} = 2x^3 + 2x^2 + 3x + 3 + \frac{6}{x - 1}$$

We leave the check as an exercise for the student. ∎

EXERCISES 8.5

Divide each of the following. Use the long division process wherever it is appropriate.

1. $\dfrac{3x + 12}{6}$

2. $\dfrac{5a + 20}{15}$

3. $\dfrac{t^2 - 6t}{6t}$

4. $\dfrac{u^2 - 8u}{4u}$

5. $\dfrac{3x^2y - 9xy^2}{3xy}$

6. $\dfrac{6m^3n^2 - 12mn^3}{2mn}$

7. $\dfrac{3x^2y - 9xy^2}{6x^2y^2}$

8. $\dfrac{6m^3n^2 - 12mn^3}{8m^2n^3}$

9. $\dfrac{10a^2b^3c - 15ab^2c^2 - 20a^3b^2c^3}{5ab^2c}$

10. $\dfrac{12x^3y^2z^4 - 24x^2y^3z^2 + 18x^3y^4z^2}{12xyz}$

11. $\dfrac{x^2 - 3x + 2}{x + 2}$

12. $\dfrac{x^2 - 4x - 3}{x - 2}$

13. $\dfrac{t^2 - 3t - 10}{t - 5}$

14. $\dfrac{t^2 + 2t - 8}{t + 4}$

15. $\dfrac{w^2 + 4w - 21}{w + 3}$

16. $\dfrac{z^2 - 6z - 18}{z - 2}$

17. $\dfrac{2x^2 - 3x + 7}{x - 1}$

18. $\dfrac{3x^2 - 4x + 7}{x - 3}$

19. $\dfrac{y^3 + y^2 + y - 14}{y - 2}$

20. $\dfrac{y^3 + 3y^2 - 4y - 12}{y + 2}$

21. $\dfrac{a^2 + 2a^3 - 3a + 2}{a + 1}$

22. $\dfrac{3a^2 + a^3 + 5 - 6a}{a - 1}$

23. $\dfrac{x^3 - x^2 + 36}{x + 3}$

24. $\dfrac{x^3 - 5x + 4}{x - 1}$

25. $\dfrac{x^4 - 16}{x - 2}$

26. $\dfrac{x^4 - 1}{x - 1}$

27. $\dfrac{3x^3 + 14x^2 + 2x - 4}{3x + 2}$

28. $\dfrac{2x^3 - 3x^2 - x + 12}{2x + 3}$

29. $\dfrac{4t^3 - 33t + 24}{2t - 5}$

30. $\dfrac{20t^3 + 33t^2 - 4}{5t + 2}$

? QUESTION FOR THOUGHT

31. Explain how you would generalize the long division process to handle

$$\frac{5x^4 - 8x^3 + 7x^2 - 2x + 1}{x^2 + x - 1}$$

◇ MINI-REVIEW

In Exercises 32–39, evaluate the given expression for $x = -3$ and $y = 5$.

32. x^2

33. $-x^2$

34. xy^2

35. $(xy)^2$

36. y^{-1}

37. x^{-2}

38. xy^{-1}

39. $(xy)^{-1}$

Solve the following problem algebraically. Be sure to label what the variable represents.

40. Maria made three investments paying 5%, 6%, and 8% interest. She invested $5,000 more at 8% than at 5%. If the total amount invested is $10,000 and the annual income from all three investments is $720, how much is invested at each rate?

CHAPTER 8   SUMMARY

After having completed this chapter, you should be able to:

1. Factor polynomials (Sections 8.1, 8.2, 8.3).

   For example:

   (a) $8x^2y - 12xy^3 = \boxed{4xy(2x - 3y^2)}$

   (b) $x^2 - 3x - 10 = (x \quad)(x \quad)$

   $\qquad\qquad\qquad = (x + \quad)(x - \quad)$

   $\qquad\qquad\qquad = \boxed{(x + 2)(x - 5)}$

   (c) $x^2 - 9 = \boxed{(x + 3)(x - 3)}$

(d) $4x^2y + 24xy + 36y = 4y(x^2 + 6x + 9)$
$$= 4y(x \quad)(x \quad)$$
$$= 4y(x + \quad)(x + \quad)$$
$$= \boxed{4y(x + 3)(x + 3)}$$

(e) $2ax^2 + 10ax - 3x - 15$ *Factor by grouping.*
$$= 2ax(x + 5) - 3(x + 5)$$
$$= (x + 5)(2ax - 3)$$

2. Solve a quadratic equation by the factoring method (Section 8.4).

For example: Solve for t.

$$t^2 + 10t = 3t - 10 \qquad \textit{Put the equation in standard form.}$$
$$t^2 + 7t + 10 = 0 \qquad \textit{Factor.}$$
$$(t + 5)(t + 2) = 0 \qquad \textit{Use the zero-product rule.}$$
$$t + 5 = 0 \quad \text{or} \quad t + 2 = 0$$
$$\boxed{t = -5} \quad \text{or} \quad \boxed{t = -2}$$

3. Divide polynomials, using long division where necessary (Section 8.5).

For example:

$$\frac{12x^2 - 8x}{4x} = \frac{12x^2}{4x} - \frac{8x}{4x} \qquad \text{or} \qquad \frac{12x^2 - 8x}{4x} = \frac{4x(3x - 2)}{4x}$$
$$= \frac{12x^2}{4x} - \frac{8x}{4x} \qquad\qquad\qquad\qquad = \boxed{3x - 2}$$
$$= \boxed{3x - 2}$$

CHAPTER 8   REVIEW EXERCISES

In Exercises 1–30, factor the given expressions as completely as possible.

1. $x^2 + 7x + 12$
2. $x^2 - 7x + 12$
3. $x^2 + 7x$
4. $x^2 + 12$
5. $x^2 - 13x + 12$
6. $x^2 - x - 12$
7. $x^2 - 6xy - 27y^2$
8. $r^2 - 8rt + 12t^2$
9. $x^2 - 64$
10. $16x^2 - 64$
11. $2x^2 + 9x + 10$
12. $2x^2 + 8x - 10$
13. $3x^2 - 6x - 24$
14. $3x^2 - 14x - 24$
15. $6a^2 + 36a + 48$
16. $6a^2 + 41a + 48$
17. $5x^3y - 80xy^3$
18. $6m^2n - 8mr^3 + 8n^2r$
19. $x^2 + 9x$
20. $x^2 + 9$
21. $25t^2 - 1$
22. $25t^2 - 100$
23. $30 - x^2 + x$
24. $8 + x - 3x^2$
25. $12x - 3x^2 - 9$
26. $20 + 16x - 4x^2$

27. $x(x - 7) + 3(x - 7)$

28. $a^2 + 5a + ab + 5b$

29. $m^2 + 2mn + 9m + 18n$

30. $x^2 - 5x - xy + 5y$

In Exercises 31–40, solve the given equation.

31. $(x - 5)(x + 4) = 0$

32. $x^2 - 12 = 4x$

33. $(x - 5)(x + 4) = 36$

34. $3x^2 - 7x + 8 = 2x^2 + 4x - 2$

35. $3x^2 + 8x = 35$

36. $(x - 5)^2 = 49$

37. $(2x - 3)(x + 2) = (x + 6)(x - 1)$

38. $(x - 3)(x + 2) = (x + 6)(x - 1)$

39. $5x^2 = 80$

40. $6x^2 - 18x = 24$

41. The length of a rectangle is 3 less than twice its width. If the area of the rectangle is 14 cm^2, find the dimension of the rectangle.

42. A square garden is surrounded by a path 2 ft wide. If the area of the path is 64 sq ft, find the dimensions of the garden.

43. A square garden is surrounded by a path 2 ft wide. If the area of the garden and path together is 144 sq ft, find the dimensions of the garden.

44. A ball is thrown straight up with an initial velocity of 25 ft per second from the top of a building 94 ft tall. Its height h (in feet) above the ground t seconds after the ball is thrown is given by $h = 94 + 25t - 16t^2$. When will the ball be 80 ft above the ground?

45. In a basketball league of n teams in which each team plays every other team twice, the total number of games played is $n^2 - n$. How many teams are there in a league that plays a total of 56 games?

In Exercises 46–53, divide. Use long division where necessary.

46. $\dfrac{x^2 y - xy^2}{xy}$

47. $\dfrac{6r^2 t - 4rt^3 + 10r^2 t^4}{2rt^2}$

48. $\dfrac{x^2 - 4x - 5}{x - 1}$

49. $\dfrac{y^3 - y^2 + y - 1}{y - 2}$

50. $\dfrac{2x^3 - 4x - 4}{x - 3}$

51. $\dfrac{6x^3 - 13x^2 + 11x - 10}{3x - 5}$

52. $\dfrac{x^3 + 8}{x + 2}$

53. $\dfrac{16x^4 - 64}{x - 2}$

CHAPTER 8 PRACTICE TEST

In Exercises 1–12, factor as completely as possible. If the expression is not factorable, say so.

1. $6x^3 + 12x^2 - 15x$

2. $4x^2 y - 8xy^2 - 2xy$

3. $x^2 + 9x + 8$

4. $x^2 - 9xy - 10y^2$

5. $4x^2 - 20x$

6. $5x^3 - 45x$

7. $6x^2 + 24x + 18$

8. $2x^2 - 7x - 15$

9. $x^2 + 4x + 4$

10. $6x^2 + 5x - 6$

11. $x^2y^2 - 9$

12. $2w^2 + 6w + aw + 3a$

13. *Divide.* $\dfrac{12r^3t^2 - 18r^2t^2 + 20r^3t^4}{4r^3t^3}$

14. *Divide.* $\dfrac{2x^3 - 5x + 6}{x - 2}$

In Exercises 15–18, *solve the given equation.*

15. $x^2 - 8x = 20$

16. $(x - 6)^2 = x - 4$

17. $(x - 8)(x - 6) = 3$

18. $5x^2 - 10x + 7 = 2x^2 + x - 3$

19. A square garden 9 ft on a side is surrounded by a concrete path of uniform width x feet. If the area of the path is 40 sq ft, find the width of the path.

20. A ball is dropped from a height of 300 ft. Its height h (in feet) above the ground t seconds after the ball is dropped is given by $h = 300 - 16t^2$. When will the ball be 200 ft above the ground?

CHAPTER 9

More Rational Expressions

An audio equipment manufacturing company determines that when n CD players are produced, the average cost per player is given by the equation

$$C = \frac{85n + 10{,}000}{n}$$

This question is answered in Example 7 of Section 9.4.

Suppose that the company wants the average cost per player to be \$89. How many CD players should be manufactured? In order to answer this question we must solve the equation

$$89 = \frac{85n + 10{,}000}{n}$$

In Chapter 9 we continue our discussion of rational expressions and equations begun in Chapter 4.

9.1   Reducing Rational Expressions

In Chapter 4 we went through a rather detailed discussion of rational expressions and equations. Our presentation there focused on fractions whose denominators were monomials, which simplified our work a great deal.

Now that we know how to factor polynomial expressions, we are going to repeat much of what we did in Chapter 4, but this time we will be working with fractions whose numerators and/or denominators are polynomials of more than one term. For example, we will be working with fractions such as

$$\frac{3x^2}{x^2 + 3} \quad \text{and} \quad \frac{x + 2}{x^2 - x - 6}$$

As you will see very shortly, the factoring techniques we learned in Chapter 8 will play an extremely important role. As we stressed in Section 4.1, the basic idea underlying almost all our work with fractions is the Fundamental Principle of Fractions.

Fundamental Principle of Fractions	$\dfrac{a}{b} = \dfrac{a \cdot k}{b \cdot k} \qquad$ where $b,\ k \neq 0$

In words again, the Fundamental Principle says two things:

- Reading from left to right, it says that the value of a fraction is not changed when both the numerator and denominator are *multiplied* by the same nonzero *factor.* We call this process **building the fraction.**

- Reading from right to left, it says that the value of a fraction is not changed when both the numerator and the denominator are divided by the same nonzero factor. We call this process **reducing the fraction,** and we frequently say that we are canceling the common factor.

   The crucial idea is that we can build or reduce with common factors, but *not* with common terms.

EXAMPLE 1

Reduce to lowest terms. $\dfrac{3x^2 + 12x}{12x}$

Solution   This type of example is the scene of one of the most common errors in algebra—that of improperly reducing a fraction. The following common "solutions" or variations thereof are *incorrect!*

   Incorrect solution 1: $\dfrac{3x^2 + 12x}{12x} = \dfrac{3x^2 + \cancel{12x}}{\cancel{12x}} = 3x^2 \quad \text{or} \quad 3x^2 + 1$

Incorrect solution 2: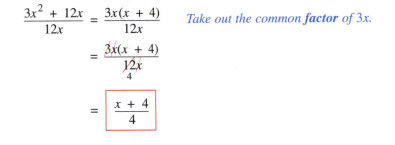

The Fundamental Principle says that we can reduce a fraction using *common factors only.* These "solutions" are wrong because they both attempt to reduce a *term* with a *factor.* In this fraction, the $12x$ in the denominator is a factor, but the $12x$ in the numerator is a term.

Now that we have seen why these two "solutions" are incorrect, what is the correct approach? Since the Fundamental Principle requires us to reduce a fraction by using common factors, we should try to factor the numerator.

$$\frac{3x^2 + 12x}{12x} = \frac{3x(x + 4)}{12x} \qquad \textit{Take out the common \textbf{factor} of } 3x.$$

$$= \frac{\cancel{3}x(x + 4)}{\underset{4}{\cancel{12}x}}$$

$$= \boxed{\dfrac{x + 4}{4}}$$

Note that this is the final answer. This answer cannot be reduced any further because the 4 in the *numerator* is not a factor. ■

It is important to point out that the various procedures we learn are not isolated ideas but often merely variations on a theme. For instance, we have just completed an example on reducing fractions. We could also have looked at Example 1 and thought of it as an example on dividing polynomials. So we could have attempted a solution as was described in Section 8.5.

$$\frac{3x^2 + 12x}{12x} = \frac{3x^2}{12x} + \frac{12x}{12x} \qquad \textit{We divided each term in the numerator by } 12x.$$

$$= \frac{\overset{x}{\cancel{3}x^2}}{\underset{4}{\cancel{12}x}} + \frac{12x}{12x} \qquad \textit{Reduce each fraction individually.}$$

$$= \boxed{\dfrac{x}{4} + 1}$$

Is this answer of $\dfrac{x}{4} + 1$ equivalent to the first answer of $\dfrac{x + 4}{4}$? If we take our second answer and combine the two terms by finding a common denominator, we get

$$\frac{x}{4} + 1 = \frac{x}{4} + \frac{4}{4} = \frac{x + 4}{4}$$

and we see that our two answers are in fact equivalent. Recognizing that they are actually two forms of the same expression makes it even less likely that you will try to reduce $\dfrac{x + 4}{4}$.

Remember	Fractions can be reduced using common *factors* only, not common terms.

EXAMPLE 2   Reduce to lowest terms. $\dfrac{2y^2 + 4y}{y^2 - 4}$

Solution   As you can see, it almost seems as if these examples are designed to lead you astray (look how neatly the y^2 is lined up over the y^2, and the $4y$ over the 4). *You must be alert and resist the inclination to reduce this fraction in its present form.*

We must have both the numerator and the denominator in factored form *before* we can reduce. You may recall our mentioning in Chapter 8 that there are a number of situations in which it is advantageous to have an expression in factored form. Working with fractions is clearly one such situation.

$$\frac{2y^2 + 4y}{y^2 - 4}$$

In the numerator we factor out the common factor of 2y.
The denominator factors as the difference of two squares.

$$= \frac{2y(y + 2)}{(y + 2)(y - 2)}$$

We reduce the fractions using the entire common factor of y + 2 in numerator and denominator.

$$= \boxed{\frac{2y}{y - 2}}$$

■

As we have seen previously, the factoring process can sometimes involve more than one step.

EXAMPLE 3

Reduce to lowest terms. $\dfrac{2x^2 - 12x + 18}{x^3 - x^2 - 6x}$

Solution

Whenever we try to factor, we always look for common factors first.

$$\frac{2x^2 - 12x + 18}{x^3 - x^2 - 6x} = \frac{2(x^2 - 6x + 9)}{x(x^2 - x - 6)}$$

Now we try to factor further.

$$= \frac{2(x - 3)(x - 3)}{x(x - 3)(x + 2)} = \frac{2(x - 3)(x - 3)}{x(x - 3)(x + 2)}$$

$$= \boxed{\frac{2(x - 3)}{x(x + 2)}} \quad \text{or} \quad \boxed{\frac{2x - 6}{x^2 + 2x}}$$

Either answer is acceptable.

■

EXAMPLE 4

Reduce to lowest terms. $\dfrac{x^2 + 6x + 8}{x^2 - 3x - 4}$

Solution

We cannot reduce this fraction in its present form, so we try to factor it first.

$$\frac{x^2 + 6x + 8}{x^2 - 3x - 4} = \frac{(x + 4)(x + 2)}{(x - 4)(x + 1)}$$

This *cannot* be reduced since there are no common factors. Therefore, the original fraction is already in lowest terms. Our answer is either

$$\boxed{\frac{x^2 + 6x + 8}{x^2 - 3x - 4}} \quad \text{or} \quad \boxed{\frac{(x + 4)(x + 2)}{(x - 4)(x + 1)}}$$

■

EXAMPLE 5

Reduce to lowest terms. $\dfrac{x^3 + 2x^2 + 3x^2}{x^2 + 3x - 5 + x}$

Solution

Before plunging ahead we should take a moment to recognize that both the numerator and the denominator can be simplified by combining like terms.

$$\frac{x^3 + 2x^2 + 3x^2}{x^2 + 3x - 5 + x} = \frac{x^3 + 5x^2}{x^2 + 4x - 5}$$

Now we factor the numerator and denominator.

$$= \frac{x^2(x + 5)}{(x + 5)(x - 1)}$$

$$= \frac{x^2(x + 5)}{(x + 5)(x - 1)}$$

$$= \boxed{\frac{x^2}{x - 1}}$$ ∎

EXAMPLE 6 Reduce to lowest terms. $\dfrac{x^3 y^2}{x^3 y - x y^3}$

Solution We see that the numerator consists of factors but that the denominator does not. We begin by taking out the common factor of xy from the denominator.

$$\frac{x^3 y^2}{x^3 y - x y^3} = \frac{x^3 y^2}{xy(x^2 - y^2)} \qquad \textit{We can factor the denominator further.}$$

$$= \frac{x^3 y^2}{xy(x - y)(x + y)}$$

$$= \frac{x^3 y^2}{xy(x - y)(x + y)}$$

$$= \boxed{\frac{x^2 y}{(x - y)(x + y)}} \quad \text{or} \quad \boxed{\frac{x^2 y}{x^2 - y^2}}$$

If you are looking at this solution and then asking yourself what was the point of factoring $x^2 - y^2$, give yourself a pat on the back. In fact, if we recognize in the second step that there is no possibility of canceling one of the *binomial* factors of $x^2 - y^2$ with a *monomial* factor in the numerator, then we can omit this step entirely. However—*be careful.* It is better to factor an expression and not use the factorization, than not to factor and consequently miss an opportunity to reduce it. ∎

One last comment: Keep in mind that in Chapter 8, just factoring an expression was an entire problem in and of itself. We have now reached the point where factoring is just one part of a larger problem.

EXERCISES 9.1

Reduce each of the following as completely as possible.

1. $\dfrac{8x^3 y^{10}}{10x^6 y^5}$

2. $\dfrac{12a^8 b^3}{9a^4 b^9}$

3. $\dfrac{5x - 7x}{x^2 - 7x^2}$

4. $\dfrac{3m^3 - 8m^3}{6m^2 - 4m^2}$

5. $\dfrac{6x^2(x + 4)^5}{9x^3(x + 4)}$

6. $\dfrac{15y^4(y - 2)^2}{10y(y - 2)^6}$

7. $\dfrac{12a^2 b + 6c^3}{8a^2 b + 4c^3}$

8. $\dfrac{12a^2 b(6c^3)}{8a^2 b(4c^3)}$

9. $\dfrac{3x - 6}{5x - 10}$

10. $\dfrac{4x + 8}{7x + 14}$

11. $\dfrac{3x - 6}{6x - 12}$

12. $\dfrac{4x + 8}{12x - 24}$

13. $\dfrac{3x - 6}{6x + 12}$

14. $\dfrac{4x + 8}{12x + 24}$

15. $\dfrac{5y}{10y + 20}$

16. $\dfrac{-3a}{6a - 18}$

17. $\dfrac{6x + 18}{x^2 - 9}$

18. $\dfrac{8x - 16}{x^2 - 4}$

19. $\dfrac{6x^2 + 18}{x^2 - 9}$

20. $\dfrac{8x^2 - 16}{x^2 - 4}$

21. $\dfrac{t^2 + 3t}{t^2 + 3t - 10}$

22. $\dfrac{m^2 - 2m}{m^2 - 2m - 8}$

23. $\dfrac{2x^2 + x + x}{4x^3 + 4x^2}$

24. $\dfrac{7x - x}{3x^2 + x^2}$

25. $\dfrac{y^2 - 5y - 6}{y^2 - 12y + 36}$

26. $\dfrac{y^2 - 8y + 16}{y^2 - 10y + 24}$

27. $\dfrac{s^2 - 2s - 15}{s^2 - 6s + 5}$

28. $\dfrac{z^2 + 10z + 24}{z^2 + 5z + 4}$

29. $\dfrac{x^2(x + 3)(x - 4)}{x^4 - 4x^3}$

30. $\dfrac{2y(y + 2)(y - 5)}{4y^3 + 8y^2}$

31. $\dfrac{3a^2 + a - 2}{a^2 - a - 2}$

32. $\dfrac{2c^2 - 5c + 3}{c^2 - 2c - 3}$

33. $\dfrac{4x^2 + 7x - 2}{x^2 + 4x + 4}$

34. $\dfrac{6x^2 - 7x - 5}{4x^2 - 1}$

35. $\dfrac{x^2 - x + 3x - 8}{x^2 + 4x}$

36. $\dfrac{a^2 - 2a - 10 - a}{a^2 - 2a + a^2 - 8a}$

37. $\dfrac{6x^2 - 12x - 18}{3x^2 - 9x - 30}$

38. $\dfrac{4x^2 + 12x - 16}{2x^2 + 8x}$

39. $\dfrac{6x^2 - 5x^2 - 4}{x^2 - 6x + 8}$

40. $\dfrac{6x^2 - 23x - 4}{x^2 - 6x + 8}$

41. $\dfrac{x^2 - 7x + 10}{x^2 - 7x + 12}$

42. $\dfrac{x^2 - 9x + 18}{x^2 - 9x + 14}$

43. $\dfrac{y^3 - y^2 - 2y}{6y^2 - 24}$

44. $\dfrac{z^3 + 2z^2 - 3z}{12z^2 + 36z}$

45. $\dfrac{c^2 - 9c}{c^3 - 9c^2}$

46. $\dfrac{x^4 - 9x^2}{4x^2 - 8x - 12}$

② QUESTION FOR THOUGHT

47. A student was asked to reduce two fractions and proceeded as follows:

(a) $\dfrac{x^2 + 5}{x} \overset{?}{=} \dfrac{\overset{x}{x^2} + 5}{\cancel{x}} \overset{?}{=} x + 5$

(b) $\dfrac{x^2 + y^2}{x + y} \overset{?}{=} \dfrac{\overset{x}{\cancel{x^2}} + \overset{y}{\cancel{y^2}}}{\cancel{x + y}} \overset{?}{=} x + y$

Discuss what the student did in each case and whether or not it was correct.

◇ **MINI-REVIEW**

In Exercises 48–51, perform the indicated operations and simplify.

48. $\dfrac{x^2}{6y} \cdot \dfrac{8y^2}{x}$

49. $\dfrac{9x^2}{25y^6} \div \dfrac{xy}{15}$

50. $5 \cdot \dfrac{x}{10}$

51. $\dfrac{5}{x} \div 10$

52. Solve for x. $\dfrac{x}{3} = \dfrac{x + 6}{2}$

53. Solve for a. $\dfrac{2a}{5} = \dfrac{a - 2}{3}$

9.2 Multiplying and Dividing Rational Expressions

In the last section we saw how to put our factoring techniques to use in reducing rational expressions. Here we will use these techniques to take another look at multiplying and dividing algebraic fractions.

We will continue to operate under the same ground rules we established earlier. That is, whenever we work with fractions, we expect our final answers to be reduced to lowest terms.

As always when discussing fractions, variables are allowed only those values for which the denominator is not equal to 0.

EXAMPLE 1 Multiply and simplify. $\dfrac{x^2}{6} \cdot \dfrac{10}{x^2 - 5x}$

Solution As we saw in Chapter 4, we much prefer to reduce *before* we multiply. This preference is even stronger now that the expressions that appear are more complex than before.

$\dfrac{x^2}{6} \cdot \dfrac{10}{x^2 - 5x}$ *We begin by factoring the second denominator.*

$= \dfrac{x^2}{6} \cdot \dfrac{10}{x(x - 5)}$ *Reduce.*

$= \dfrac{\overset{x}{\cancel{x^2}}}{\underset{3}{\cancel{6}}} \cdot \dfrac{\overset{5}{\cancel{10}}}{x(x - 5)}$

$= \boxed{\dfrac{5x}{3(x - 5)}}$ or $\boxed{\dfrac{5x}{3x - 15}}$

Note that even though we could have reduced the 6 with the 10 immediately, we prefer to have everything in factored form before we begin to reduce. ■

EXAMPLE 2 Multiply and simplify. $\dfrac{4x}{x^2 - 4} \cdot \dfrac{x^2 - 5x + 6}{4x^2 - 12x}$

Solution | We begin by factoring wherever possible so that we can reduce before we multiply out the polynomials.

$$\frac{4x}{x^2 - 4} \cdot \frac{x^2 - 5x + 6}{4x^2 - 12x} = \frac{4x}{(x + 2)(x - 2)} \cdot \frac{(x - 3)(x - 2)}{4x(x - 3)}$$

$$= \frac{4x}{(x + 2)(x - 2)} \cdot \frac{(x - 3)(x - 2)}{4x(x - 3)}$$

$$= \boxed{\frac{1}{x + 2}}$$ *Do not forget the understood factor of 1 in the numerator.* ■

EXAMPLE 3 | Multiply and simplify. $5x \cdot \dfrac{x^2}{x^2 - 5x}$

Solution | Since this example involves fractions, we write down the understood denominator of 1.

$$5x \cdot \frac{x^2}{x^2 - 5x} = \frac{5x}{1} \cdot \frac{x^2}{x^2 - 5x}$$ *Factor the denominator.*

$$= \frac{5x}{1} \cdot \frac{x^2}{x(x - 5)}$$ *Reduce.*

$$= \frac{5x}{1} \cdot \frac{x^2}{x(x - 5)}$$

$$= \boxed{\frac{5x^2}{x - 5}}$$

Note that we chose to reduce using the factor of x in the denominator and the factor of x in the first numerator. We could just as well have reduced using the x^2 in the second numerator. ■

Now let's turn our attention to division.

EXAMPLE 4 | Divide and simplify. $\dfrac{x^2 - y^2}{x^2 - 2xy + y^2} \div \dfrac{12x + 12y}{6x - 6y}$

Solution | We follow the rule for dividing fractions, which says "invert the divisor and multiply."

$$\frac{x^2 - y^2}{x^2 - 2xy + y^2} \div \frac{12x + 12y}{6x - 6y} = \frac{x^2 - y^2}{x^2 - 2xy + y^2} \cdot \frac{6x - 6y}{12x + 12y}$$ *Next factor.*

$$= \frac{(x + y)(x - y)}{(x - y)(x - y)} \cdot \frac{6(x - y)}{12(x + y)}$$ *Now reduce.*

$$= \frac{(x + y)(x - y)}{(x - y)(x - y)} \cdot \frac{6(x - y)}{\underset{2}{12}(x + y)}$$

$$= \boxed{\frac{1}{2}}$$ ■

EXAMPLE 5 | Divide and simplify. $\dfrac{2x^2 + x - 15}{x^2 + 6x + 9} \div (8x - 20)$

Solution | We think of $8x - 20$ as $\dfrac{8x - 20}{1}$, and follow the rule for division.

$$\frac{2x^2 + x - 15}{x^2 + 6x + 9} \div \frac{8x - 20}{1} = \frac{2x^2 + x - 15}{x^2 + 6x + 9} \cdot \frac{1}{8x - 20} \qquad \textit{Next factor.}$$

$$= \frac{(2x - 5)(x + 3)}{(x + 3)(x + 3)} \cdot \frac{1}{4(2x - 5)} \qquad \textit{Now reduce.}$$

$$= \frac{(2x - 5)(x + 3)}{(x + 3)(x + 3)} \cdot \frac{1}{4(2x - 5)}$$

$$= \boxed{\frac{1}{4(x + 3)}} \quad \text{or} \quad \boxed{\frac{1}{4x + 12}} \qquad \blacksquare$$

As we have pointed out on numerous occasions, it is very important to read a problem carefully before plunging into it. See what the problem involves so that you can apply the *appropriate* procedures. Keep this in mind as you study the next two examples and their solutions.

EXAMPLE 6 Perform the indicated operations and simplify. $\left(\dfrac{8}{x} \cdot \dfrac{x}{2} \right) \div \dfrac{x + 4}{16}$

Solution We work inside the parentheses first—the operation within the parentheses is *multiplication*.

$$\left(\frac{8}{x} \cdot \frac{x}{2} \right) \div \frac{x + 4}{16} = \left(\frac{\overset{4}{\cancel{8}}}{\cancel{x}} \cdot \frac{\cancel{x}}{\cancel{2}} \right) \div \frac{x + 4}{16}$$

$$= 4 \div \frac{x + 4}{16}$$

$$= \frac{4}{1} \cdot \frac{16}{x + 4} \qquad \textit{This cannot be reduced, so we multiply.}$$

$$= \boxed{\frac{64}{x + 4}} \qquad \blacksquare$$

EXAMPLE 7 Perform the indicated operations and simplify. $\left(\dfrac{8}{x} - \dfrac{x}{2} \right) \div \dfrac{x + 4}{16}$

Solution Again we begin to work inside the parentheses—this time the operation within the parentheses is *subtraction*, which requires a common denominator. The common denominator within the parentheses is $2x$.

$$\left(\frac{8}{x} - \frac{x}{2} \right) \div \frac{x + 4}{16} = \left(\frac{8(2)}{2x} - \frac{x(x)}{2x} \right) \div \frac{x + 4}{16} \qquad \textit{Subtract the fractions.}$$

$$= \frac{16 - x^2}{2x} \div \frac{x + 4}{16} \qquad \textit{Next use the rule for division.}$$

$$= \frac{16 - x^2}{2x} \cdot \frac{16}{x + 4} \qquad \textit{Now factor.}$$

$$= \frac{(4 + x)(4 - x)}{2x} \cdot \frac{16}{x + 4} \qquad \textit{Next reduce.}$$

$$= \frac{(4 + x)(4 - x)}{2x} \cdot \frac{\overset{8}{\cancel{16}}}{x + 4} \qquad \textit{Note that } x + 4 \textit{ and } 4 + x \\ \textit{are the same.}$$

$$= \boxed{\frac{8(4 - x)}{x}} \quad \text{or} \quad \boxed{\frac{32 - 8x}{x}} \qquad \blacksquare$$

Even though Examples 6 and 7 look very similar, they are actually quite different. Again we emphasize that you must read problems carefully to understand what the problem is asking and what procedures are involved.

If you had difficulty with carrying out the subtraction in the first step of Example 7, it is strongly suggested that you review Section 4.3 before going on to the next section.

EXERCISES 9.2

In each of the following exercises, perform the indicated operations and simplify as completely as possible.

1. $\dfrac{8x^2y^3}{9yz^3} \cdot \dfrac{12a^2z}{2ax^2}$

2. $\dfrac{10r^3s}{6mr} \cdot \dfrac{9m^4}{4mrs}$

3. $\dfrac{3st^2}{5p} \div \dfrac{15t^2}{s^2}$

4. $\dfrac{4s^3t}{7a} \div \dfrac{14s^2}{t}$

5. $\dfrac{12x^2y^3}{5z} \cdot (30xyz)$

6. $\dfrac{18x^3y^2}{7z^4} \div (14x^2y^2z^2)$

7. $(28a^2b^3z^4) \div \dfrac{4a}{7b}$

8. $(28a^2b^3z^4) \cdot \dfrac{4a}{7b}$

9. $\dfrac{x^2 + 4x}{x^2} \cdot \dfrac{x}{x^2 + 6x + 5}$

10. $\dfrac{y^2 - 4y}{y^4} \cdot \dfrac{y^3}{y^2 - 3y - 4}$

11. $\dfrac{x^2 + 4x}{x^2 + 4x + 4} \cdot \dfrac{x^2 - 4}{x^2 - 4x + 4}$

12. $\dfrac{a^2 - 4a}{a^2 - 4a - 12} \cdot \dfrac{a^2 + 5a + 6}{a^2 - a - 12}$

13. $\dfrac{r^2 - 4r - 5}{2r - 10} \div \dfrac{r^2 - 3r + 2}{4r^2}$

14. $\dfrac{m^2 + 3m - 10}{5m^2} \div \dfrac{m^2 + 5m}{m^2 + 10m + 25}$

15. $\dfrac{m^2}{m^2 + 3m} \div \dfrac{3m^2}{m^2 + 6m}$

16. $\dfrac{3x^2}{x^2 + 3x} \div \dfrac{x^2 + 6x}{6x^2}$

17. $\dfrac{x^2 + 3x + 2}{x^2 + 2x} \cdot \dfrac{x}{x^2 + 2}$

18. $\dfrac{x^2 - 3x + 2}{x^2 - 2x} \cdot \dfrac{2x}{x^2 - 2x}$

19. $\dfrac{y^2 - 3y - 4}{y^2 - 2y - 8} \cdot \dfrac{y^2 + 4y + 4}{y^2 - 8y + 16}$

20. $\dfrac{y^2 - 3y - 4}{y^2 - 2y - 8} \div \dfrac{y^2 + 4y + 4}{y^2 - 8y + 16}$

21. $\dfrac{x^2 + 2x}{x^2 - x - 2} \cdot \dfrac{x - 2}{x}$

22. $\dfrac{x^2 - 4x - 5}{2x^2 - 10x} \div \dfrac{x + 1}{8x}$

23. $\dfrac{2x^2 + x - 15}{x^2 - 9} \cdot \dfrac{6x^2 + 7x + 1}{2x^2 - 3x - 5}$

24. $\dfrac{4z^2 + 12z + 5}{2z^2 + z - 1} \div \dfrac{4z^2 - 25}{2z^2 - 3z + 1}$

25. $\dfrac{4a}{a + 4} \cdot \dfrac{a + 5}{5a} \div \dfrac{a^2 + 6a + 5}{a^2 + 5a + 4}$

26. $\dfrac{3y}{y - 3} \div \dfrac{y^2 - 3y + 2}{y^2 - 4y + 3} \cdot \dfrac{y - 2}{2y}$

27. $\dfrac{x}{2} \div \dfrac{2}{x} \cdot \dfrac{x^2 - 16}{x^2 - 4x}$

28. $\dfrac{a}{8} \div \left(\dfrac{8}{a^2} \div \dfrac{a^2 - 64}{a^2 - 8a} \right)$

29. $\dfrac{t^2 + 2t}{t^2 + 2t + 1} \div \dfrac{2t^2 + 7t + 6}{t^2 + t}$

30. $\dfrac{u^2 + 6u + 9}{u^2 + 9} \cdot \dfrac{4u^2 + 36}{2u + 6}$

31. $\dfrac{2x^2 + 6x + 4x}{x^2 - 25} \cdot \dfrac{(x + 5)^2}{4x - 2x}$

32. $\dfrac{z^2 + 5z + z}{(z + 6)^2} \div \dfrac{z^2 + z}{z^2 + 7z + 6}$

33. $\dfrac{x^3y - xy^3}{8x^2y + 4xy^2} \div \dfrac{(x - y)^2}{2x^2 + 3xy + y^2}$

34. $\dfrac{r^3t - rt^2}{rt^2 - r^2t} \cdot \dfrac{t^2 - 2rt + r^2}{r^2 - t}$

35. $\left(\dfrac{x}{3} \cdot \dfrac{x}{4}\right) \cdot \dfrac{12}{x^2 - 3x}$

36. $\left(\dfrac{x}{3} - \dfrac{x}{4}\right) \cdot \dfrac{12}{x^2 - 3x}$

37. $\left(\dfrac{c}{2} + \dfrac{c}{5}\right) \div \dfrac{c^2 + 7c}{10}$

38. $\left(\dfrac{c}{2} \cdot \dfrac{c}{5}\right) \div \dfrac{c^2 + 7c}{10}$

39. $\left(\dfrac{w}{4} - \dfrac{9}{w}\right) \cdot \left(\dfrac{w + 10}{2w} - \dfrac{5}{w}\right)$

40. $\left(\dfrac{5t + 8}{4t} - \dfrac{t + 2}{t}\right) \div \left(\dfrac{t}{8} - \dfrac{2}{t}\right)$

 **QUESTIONS FOR THOUGHT**

41. Make up a verbal problem involving rate, time, and distance that would give rise to the following equation (be sure to indicate what the variable would represent):

$$20t + 30(t + 2) = 310$$

42. Make up a verbal problem involving investments that would give rise to the following equation (be sure to indicate what the variable would represent):

$$0.08x + 0.12(x + 500) = 260$$

◆ **MINI-REVIEW**

In Exercises 43–46, perform the indicated operations and simplify.

43. $\dfrac{5}{6x} + \dfrac{3}{8x^2}$

44. $\dfrac{1}{10x^2} - \dfrac{7}{6xy}$

45. $9 - \dfrac{6}{x}$

46. $x + \dfrac{2}{3x} - \dfrac{1}{2y}$

47. The Drama Department wants to put on a production. They plan to sell 250 orchestra seats and 150 balcony seats. They also plan to charge $4 more for orchestra seats than for balcony seats and they want to collect a total of $3,400. How much should they charge for an orchestra ticket?

48. Ricardo can run at the rate of 8 mph and walk at the rate of 4 mph. If he wants to run and walk a total of 13 miles in 2 hours, how far should he run?

9.3   Adding and Subtracting Rational Expressions

In this section we extend the procedures we developed in Section 4.3. You are strongly urged to review that section before continuing.

You will recall that addition and subtraction of fractions are quite straightforward when the fractions have the same denominator. We have the general rule:

$$\dfrac{a}{d} + \dfrac{b}{d} - \dfrac{c}{d} = \dfrac{a + b - c}{d}$$

In other words, if the denominators are the same, just combine the numerators and keep the common denominator.

It was mentioned only briefly then that even when the denominators are the same there may still be some things to watch out for. We elaborate on this idea a bit more now.

EXAMPLE 1

Combine and simplify. $\dfrac{2x - 3}{3x - 6} + \dfrac{x + 3}{3x - 6}$

Solution

Since the denominators are the same, we can proceed directly using the rule for addition of fractions.

$$\frac{2x - 3}{3x - 6} + \frac{x + 3}{3x - 6} = \frac{2x - 3 + x + 3}{3x - 6}$$ *Combine like terms.*

$$= \frac{3x}{3x - 6}$$ *We want to reduce this fraction if we can. Therefore, we must first factor the denominator.*

$$= \frac{3x}{3(x - 2)}$$

$$= \frac{\cancel{3}x}{\cancel{3}(x - 2)}$$

$$= \boxed{\frac{x}{x - 2}}$$ ■

EXAMPLE 2

Combine and simplify. $\dfrac{5x^2 - 12}{2x^2 + 8x} - \dfrac{2x^2 + 7}{2x^2 + 8x} + \dfrac{3 - 2x^2}{2x^2 + 8x}$

Solution

Since the denominators are the same, there is a tendency to "copy" across the numerators of the example. This is incorrect because the minus sign in front of the second fraction must be applied to the *entire* numerator of the second fraction.

$$\frac{5x^2 - 12}{2x^2 + 8x} - \frac{2x^2 + 7}{2x^2 + 8x} + \frac{3 - 2x^2}{2x^2 + 8x} = \frac{5x^2 - 12 - (2x^2 + 7) + 3 - 2x^2}{2x^2 + 8x}$$

The parentheses around the $2x^2 + 7$ are inserted to make sure we do the subtraction correctly.

$$\downarrow$$

$$= \frac{5x^2 - 12 - 2x^2 - 7 + 3 - 2x^2}{2x^2 + 8x}$$

The arrow indicates where a sign error often occurs if you do not insert the parentheses in the previous step. Now we combine like terms in the numerator.

$$= \frac{x^2 - 16}{2x^2 + 8x}$$

Next we factor in the hope that we can reduce.

$$= \frac{(x + 4)(x - 4)}{2x(x + 4)}$$

$$= \frac{(x + 4)(x - 4)}{2x(x + 4)}$$

$$= \boxed{\frac{x - 4}{2x}}$$

Be careful: Watch out for the sign at the spot indicated by the arrow in this example. Whenever the numerator of a fraction contains more than one term and the fraction is preceded by a minus sign, the understood coefficient of -1 must be distributed to *each* term in the numerator. ■

EXAMPLE 3

Combine and simplify. $\dfrac{2x^2}{10x} + \dfrac{7}{10x}$

Solution

Your first instinct might be to reduce the first fraction immediately:

$$\frac{2x^2}{10x} = \frac{2\overset{x}{\cancel{x^2}}}{\underset{5}{\cancel{10x}}} = \frac{x}{5}$$

This is not wrong, it is just inappropriate. Since the example is telling us to combine the fractions, having both denominators the same is of greater importance than having each fraction reduced to lowest terms. If we reduce first, we lose the common denominator. Always take a moment to think about what the problem is asking before deciding on the steps to follow. The solution to this example is

$$\frac{2x^2}{10x} + \frac{7}{10x} = \boxed{\frac{2x^2 + 7}{10x}}$$ ∎

When we are asked to combine fractions whose denominators are not the same, we first *build* the fractions so that they have a common denominator, and then we proceed as before.

Let's review the mechanical procedure we developed in Chapter 4 for finding the least common denominator (LCD).

| **To Find the LCD** | **Step 1.** Factor each denominator as completely as possible. |
| | **Step 2.** The LCD consists of the product of each *distinct* factor taken the maximum number of times it appears in any one denominator. |

Now that we are working with polynomial denominators, we have to carry out this process very carefully.

EXAMPLE 4

Combine and simplify. $\dfrac{7}{3x} - \dfrac{2}{x + 3}$

Solution

Since the denominators are not the same, our first step is to build the fractions to have the same denominators.

As usual, since we prefer to use the LCD, we follow the outline given in the box. The first step in our outline is unnecessary in this example since the denominators are already in factored form.

In order to carry out the second step, we need to identify the distinct factors. Consequently, it is crucial to be able to distinguish terms from factors.

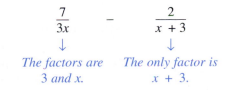

$$\frac{7}{3x} \qquad - \qquad \frac{2}{x + 3}$$
$$\downarrow \qquad\qquad\qquad \downarrow$$
The factors are   The only factor is
3 and x.   x + 3.

Therefore, the distinct factors are 3, x, and $x + 3$.

In $3x$ $\begin{cases} 3 \text{ appears as a } factor \text{ once.} \\ x \text{ appears as a } factor \text{ once.} \\ x + 3 \text{ appears as a } factor \text{ zero times.} \end{cases}$

In $x + 3$ $\begin{cases} 3 \text{ appears as a } \textit{factor} \text{ zero times.} \\ x \text{ appears as a } \textit{factor} \text{ zero times.} \\ x + 3 \text{ appears as a } \textit{factor} \text{ once.} \end{cases}$

Again we must emphasize that x and 3 are not *factors* of $x + 3$.

Following our outline, we take each of 3, x, and $x + 3$ as a factor *once* for the LCD.

The LCD is $3x(x + 3)$.

Next we use the Fundamental Principle to build each of the original fractions into an equivalent one with the LCD as the denominator. We do this by filling in the missing factors in both the numerator and the denominator.

We put the LCD in each denominator. Whatever factor we have inserted in the denominator must also be inserted in the numerator.

$$\frac{7}{3x} - \frac{2}{x + 3}$$

$$= \frac{7 \, (x + 3)}{3x \, (x + 3)} - \frac{2 \, (3x)}{3x \, (x + 3)}$$

Here the missing factor was $x + 3$. *Here the missing factor was $3x$.*

$$= \frac{7(x + 3) - 2(3x)}{3x(x + 3)}$$ *In the numerator, multiply out and combine like terms.*

$$= \frac{7x + 21 - 6x}{3x(x + 3)}$$

$$= \boxed{\frac{x + 21}{3x(x + 3)}}$$ ∎

EXAMPLE 5   Combine and simplify. $\dfrac{2}{7x^2} + \dfrac{3}{x^2 - 7x}$

Solution   As before, since the denominators are not the same, we want to find the LCD. However, since the denominators are not both in factored form, our first step is to factor the second denominator.

$$\frac{2}{7x^2} + \frac{3}{x^2 - 7x} = \frac{2}{7x^2} + \frac{3}{x(x - 7)}$$

Now we are ready to determine the LCD.

$$\frac{2}{7x^2} \qquad + \qquad \frac{3}{x(x - 7)}$$
$$\downarrow \qquad\qquad\qquad \downarrow$$

The factors are   *The factors are*
7 and x.   *x and $x - 7$.*

Therefore, the distinct factors are 7, x, and $x - 7$.

In $7x^2$ $\begin{cases} 7 \text{ appears as a } \textit{factor} \text{ once.} \\ x \text{ appears as a } \textit{factor} \text{ twice.} \\ x - 7 \text{ appears as a } \textit{factor} \text{ zero times.} \end{cases}$

In $x(x - 7)$ $\begin{cases} 7 \text{ appears as a } \textit{factor} \text{ zero times.} \\ x \text{ appears as a } \textit{factor} \text{ once.} \\ x - 7 \text{ appears as a } \textit{factor} \text{ once.} \end{cases}$

Following our outline, each distinct factor is taken the *maximum* number of times it appears. Therefore, we take 7 once, x twice, and $x - 7$ once.

The LCD is $7x^2(x - 7)$.

Go over this process of finding the LCD to make sure you follow the procedure. Next we build the fractions.

$$\frac{2}{7x^2} + \frac{3}{x^2 - 7x} = \frac{2}{7x^2} + \frac{3}{x(x - 7)}$$

Whatever factors we have inserted in the denominator must also be inserted in the numerator.

$$= \frac{2\ (x - 7)}{7x^2(x - 7)} + \frac{3\ (7x)}{7x^2(x - 7)}$$

Here the missing factor was x − 7.   Here the missing factor was 7x.

$$= \frac{2(x - 7) + 3(7x)}{7x^2(x - 7)}$$ *In the numerator, multiply out and combine like terms.*

$$= \frac{2x - 14 + 21x}{7x^2(x - 7)}$$

$$= \boxed{\frac{23x - 14}{7x^2(x - 7)}}$$ ■

EXAMPLE 6   Combine and simplify. $\dfrac{4}{y - 4} - \dfrac{3}{y + 3}$

Solution   The denominators cannot be factored any further. We see that the LCD is $(y - 4)(y + 3)$.

$$\frac{4}{y - 4} - \frac{3}{y + 3} = \frac{4\ (y + 3)}{(y - 4)\ (y + 3)} - \frac{3\ (y - 4)}{(y - 4)\ (y + 3)}$$ *Now combine.*

$$= \frac{4(y + 3) - 3(y - 4)}{(y - 4)(y + 3)}$$ *Do not forget to distribute the coefficient of −3.*

$$= \frac{4y + 12 - 3y + 12}{(y - 4)(y + 3)}$$

$$= \boxed{\frac{y + 24}{(y - 4)(y + 3)}}$$ ■

EXAMPLE 7   Combine and simplify. $c + \dfrac{4}{c - 5}$

Solution   Since we are working with fractions, we make a point of writing the first c as $\dfrac{c}{1}$. The LCD is $c - 5$.

$$c + \frac{4}{c - 5} = \frac{c}{1} + \frac{4}{c - 5}$$

$$= \frac{c\ (c - 5)}{1 \cdot\ (c - 5)} + \frac{4}{c - 5}$$ *Now combine fractions.*

$$= \frac{c(c - 5) + 4}{c - 5}$$

$$= \frac{c^2 - 5c + 4}{c - 5}$$ *The numerator does factor, however.*

$$= \frac{(c-4)(c-1)}{c-5}$$

The fraction does not reduce. Therefore we may leave the final answer in either form.

$$= \boxed{\frac{c^2 - 5c + 4}{c - 5}} = \boxed{\frac{(c-4)(c-1)}{c-5}}$$ ∎

EXAMPLE 8

Combine and simplify. $\dfrac{7}{x^2 + x - 12} - \dfrac{4}{x^2 + 4x} + \dfrac{3}{x^2 - 3x}$

Solution

To find the LCD we begin by factoring each denominator.

$$\frac{7}{x^2 + x - 12} - \frac{4}{x^2 + 4x} + \frac{3}{x^2 - 3x}$$

$$= \frac{7}{(x+4)(x-3)} - \frac{4}{x(x+4)} + \frac{3}{x(x-3)}$$

Following our outline, we find that the LCD is $x(x+4)(x-3)$.

$$= \frac{7 \cdot \boxed{x}}{\boxed{x} \cdot (x+4)(x-3)} - \frac{4 \cdot \boxed{(x-3)}}{x(x+4) \cdot \boxed{(x-3)}} + \frac{3 \cdot \boxed{(x+4)}}{x \cdot \boxed{(x+4)} \cdot (x-3)}$$

Here the missing factor was x.   *Here the missing factor was x − 3.*   *Here the missing factor was x + 4.*

$$= \frac{7x - 4(x-3) + 3(x+4)}{x(x+4)(x-3)}$$

$$= \frac{7x - 4x + 12 + 3x + 12}{x(x+4)(x-3)}$$

Watch the sign here! Next combine like terms.

$$= \frac{6x + 24}{x(x+4)(x-3)}$$

Then factor the numerator.

$$= \frac{6(x+4)}{x(x+4)(x-3)}$$

Next reduce.

$$= \frac{6(x+4)}{x(x+4)(x-3)}$$

$$= \boxed{\frac{6}{x(x-3)}} \quad \text{or} \quad \boxed{\frac{6}{x^2 - 3x}}$$

It is worthwhile to go back and look at the LCD to see that it contains within it all of the original denominators.

As this example shows, it clearly does not pay to multiply out the LCD until we are sure that we cannot reduce. In fact, we may leave our answer in factored form. ∎

EXAMPLE 9

Perform the indicated operations and simplify. $\dfrac{x^2 - 3x}{x^2 - 6x + 9} \cdot \dfrac{3x - 9}{9x}$

Solution

We do begin by factoring, but there is no LCD necessary in this example because it is a multiplication problem, *not* addition or subtraction. Remember to look at the example before you start working on it.

$$\frac{x^2 - 3x}{x^2 - 6x + 9} \cdot \frac{3x - 9}{9x} = \frac{x(x-3)}{(x-3)(x-3)} \cdot \frac{3(x-3)}{9x}$$

$$= \frac{x(x-3)}{(x-3)(x-3)} \cdot \frac{3(x-3)}{9x} = \boxed{\frac{1}{3}} \quad \blacksquare$$

Complex Fractions

Recall that a complex fraction is a fraction that contains fractions within it. We can apply the methods we have learned in this chapter to simplify complex fractions.

EXAMPLE 10 Simplify as completely as possible. $\dfrac{\dfrac{x+3}{3}}{1 - \dfrac{9}{x^2}}$

Solution Keep in mind that a complex fraction is just an alternative way of writing the division of fractional expressions. The complex fraction given here means

$$\left(\frac{x+3}{3}\right) \div \left(1 - \frac{9}{x^2}\right)$$

We begin by performing the subtraction $1 - \dfrac{9}{x^2}$. The LCD is x^2.

$$\frac{\dfrac{x+3}{3}}{1 - \dfrac{9}{x^2}} = \frac{\dfrac{x+3}{3}}{\dfrac{x^2}{x^2} - \dfrac{9}{x^2}}$$

$$= \frac{\dfrac{x+3}{3}}{\dfrac{x^2-9}{x^2}} \qquad \textit{Keep in mind that this means}$$
$$\qquad\qquad \frac{x+3}{3} \div \frac{x^2-9}{x^2}. \textit{ We follow the rule for division.}$$

$$= \frac{x+3}{3} \cdot \frac{x^2}{x^2-9} \qquad \textit{We factor and reduce.}$$

$$= \frac{x+3}{3} \cdot \frac{x^2}{(x+3)(x-3)}$$

$$= \boxed{\frac{x^2}{3(x-3)}} \qquad\qquad \blacksquare$$

EXAMPLE 11 Simplify as completely as possible. $\dfrac{\dfrac{4}{y} + \dfrac{4}{x}}{\dfrac{1}{y^2} - \dfrac{1}{x^2}}$

Solution The complex fraction given in this example means

$$\left(\frac{4}{y} + \frac{4}{x}\right) \div \left(\frac{1}{y^2} - \frac{1}{x^2}\right)$$

We offer two possible solutions.

Solution 1. We basically do what the example tells us to do. Following the order of operations, we add the two fractions in the numerator and subtract the two fractions in the denominator. We then obtain a straightforward division problem to which we apply the rule for dividing fractions. In order to add the fractions in the numerator, we use an LCD of xy; in order to subtract the fractions in the denominator, we use an LCD of x^2y^2.

$$\frac{\dfrac{4}{y} + \dfrac{4}{x}}{\dfrac{1}{y^2} - \dfrac{1}{x^2}} = \frac{\dfrac{4(x)}{xy} + \dfrac{4(y)}{xy}}{\dfrac{x^2}{x^2y^2} - \dfrac{y^2}{x^2y^2}}$$

The LCD for the numerator is xy; for the denominator it is x^2y^2.

$$= \frac{\dfrac{4x + 4y}{xy}}{\dfrac{x^2 - y^2}{x^2y^2}}$$

Remember:

This means $\dfrac{4x + 4y}{xy} \div \dfrac{x^2 - y^2}{x^2y^2}.$

We use the rule for dividing fractions: "Invert and multiply."

$$= \frac{4x + 4y}{xy} \cdot \frac{x^2y^2}{x^2 - y^2}$$

Factor and reduce.

$$= \frac{4(x + y)}{xy} \cdot \frac{x^2y^2}{(x + y)(x - y)}$$

$$= \frac{4(\cancel{x + y})}{\cancel{xy}} \cdot \frac{\overset{x \ y}{\cancel{x^2y^2}}}{(\cancel{x + y})(x - y)}$$

$$= \boxed{\frac{4xy}{x - y}}$$

Solution 2. We can do this example a bit differently by applying the Fundamental Principle of Fractions to the entire complex fraction.

We will call the denominators of fractions *within* the complex fraction **minor denominators.** If we multiply the numerator and denominator of the complex fraction by the LCD of all the minor denominators, we will obtain a simple fraction. We can then proceed to reduce it (if possible).

$$\frac{\dfrac{4}{y} + \dfrac{4}{x}}{\dfrac{1}{y^2} - \dfrac{1}{x^2}}$$

*There are four minor denominators here: y, x, y^2, x^2. The LCD of **all** the minor denominators is x^2y^2.*

$$= \frac{\dfrac{x^2y^2}{1}\left(\dfrac{4}{y} + \dfrac{4}{x}\right)}{\dfrac{x^2y^2}{1}\left(\dfrac{1}{y^2} - \dfrac{1}{x^2}\right)}$$

We multiply the numerator and denominator of the complex fraction by x^2y^2.

Use the distributive property.

$$= \frac{\dfrac{x^2y^2}{1} \cdot \dfrac{4}{y} + \dfrac{x^2y^2}{1} \cdot \dfrac{4}{x}}{\dfrac{x^2y^2}{1} \cdot \dfrac{1}{y^2} - \dfrac{x^2y^2}{1} \cdot \dfrac{1}{x^2}}$$

Reduce.

$$= \frac{\dfrac{x^2\overset{y}{\cancel{y^2}}}{1} \cdot \dfrac{4}{\cancel{y}} + \dfrac{\overset{x}{\cancel{x^2}}y^2}{1} \cdot \dfrac{4}{\cancel{x}}}{\dfrac{x^2\cancel{y^2}}{1} \cdot \dfrac{1}{\cancel{y^2}} - \dfrac{\cancel{x^2}y^2}{1} \cdot \dfrac{1}{\cancel{x^2}}}$$

$$= \frac{4x^2y + 4xy^2}{x^2 - y^2}$$

Factor and reduce.

$$= \frac{4xy(x + y)}{(x + y)(x - y)}$$

$$= \boxed{\frac{4xy}{x - y}}$$

The basic difference in the two approaches is that the first method works on the numerator and denominator of the complex fraction separately, whereas the second method works on the entire complex fraction at once.

Be careful using the method of Solution 2, which requires us to use the Fundamental Principle to multiply the numerator and denominator of the complex fraction by the *same* quantity (the LCD of all minor denominators).

You should choose the method with which you feel most comfortable. ■

Complex fractions come up in some real-life applications.

EXAMPLE 12

City A and city B are 150 miles apart. A car travels from city A to city B at 60 mph and then returns to city A at 40 mph. What is the average rate of the car for the entire trip?

Solution

Many students simply average the two speeds and think that the correct answer is 50 mph. As we shall discover in a moment, this answer is not correct!

Keeping the distance-rate-time relationship in mind, we compute the average rate as follows:

$$\text{Average rate for entire trip } = \frac{\text{Total distance}}{\text{Total time}} = \frac{\text{Distance going } + \text{ Distance returning}}{\text{Time going } + \text{ Time returning}}$$

We know the total distance is 150 + 150.

We compute the time for each part of the trip as $\dfrac{Distance}{Rate}$.

We have

$$Time\ going\ = \frac{150}{60} = \frac{5}{2} \quad \text{and} \quad Time\ returning\ = \frac{150}{40} = \frac{15}{4}$$

Therefore, we have

$$Average\ rate\ for\ entire\ trip\ = \frac{Total\ distance}{Total\ time} = \frac{150 + 150}{\dfrac{5}{2} + \dfrac{15}{4}} \qquad \textit{We simplify the complex fraction.}$$

$$= \frac{300}{\dfrac{10}{4} + \dfrac{15}{4}} = \frac{300}{\dfrac{25}{4}} = \frac{300}{1} \cdot \frac{4}{25} = 48$$

Thus, the average rate of speed for the entire trip is $\boxed{48 \text{ mph}}$. ■

EXERCISES 9.3

In each of the following exercises, perform the indicated operations and simplify your answer as completely as possible.

1. $\dfrac{5x - 2}{4x + 8} + \dfrac{3x + 2}{4x + 8}$

2. $\dfrac{10x + 3}{5x - 10} + \dfrac{5x - 3}{5x - 10}$

3. $\dfrac{5x - 2}{4x + 8} - \dfrac{3x + 2}{4x + 8}$

4. $\dfrac{10x + 3}{5x - 10} - \dfrac{5x - 3}{5x - 10}$

5. $\dfrac{x + 7}{x + 2} + \dfrac{x + 3}{x + 2} + \dfrac{x - 2}{x + 2}$

6. $\dfrac{a - 6}{a + 3} + \dfrac{a + 9}{a + 3} + \dfrac{a + 6}{a + 3}$

7. $\dfrac{x + 7}{x + 2} - \dfrac{x + 3}{x + 2} + \dfrac{x - 2}{x + 2}$

8. $\dfrac{a - 6}{a + 3} + \dfrac{a + 9}{a + 3} - \dfrac{a + 6}{a + 3}$

9. $\dfrac{y^2}{4y} + \dfrac{2y}{4y}$

10. $\dfrac{2z}{10z^2} + \dfrac{5}{10z^2}$

11. $\dfrac{5}{2x} + \dfrac{4}{x+2}$

12. $\dfrac{8}{3x} + \dfrac{6}{x+3}$

13. $\dfrac{5}{2x} \cdot \dfrac{4}{x+2}$

14. $\dfrac{8}{3x} \cdot \dfrac{6}{x+3}$

15. $\dfrac{2}{x+2} + \dfrac{3}{x+3}$

16. $\dfrac{4}{x-4} + \dfrac{2}{x-2}$

17. $\dfrac{7}{a+7} - \dfrac{5}{a+5}$

18. $\dfrac{9}{a-3} - \dfrac{4}{a+4}$

19. $\dfrac{4}{3x^2} - \dfrac{2}{x^2+3x}$

20. $\dfrac{5}{6r^2} + \dfrac{1}{r^2-6r}$

21. $\dfrac{3}{4a^2} + \dfrac{1}{6a-18}$

22. $\dfrac{5}{6p} - \dfrac{4}{9p-18}$

23. $\dfrac{4}{x^2+4x} - \dfrac{2}{x^2-4x}$

24. $\dfrac{3}{x^2-6x} - \dfrac{2}{x^2+6x}$

25. $2 - \dfrac{x}{x-1}$

26. $3 - \dfrac{x}{x-2}$

27. $2 \cdot \dfrac{x}{x-1}$

28. $3 \cdot \dfrac{x}{x-2}$

29. $\dfrac{3}{x^2-16} - \dfrac{3}{2x^2+8x}$

30. $\dfrac{2}{x^2-9} - \dfrac{2}{3x^2+9x}$

31. $\dfrac{12}{x^2+x-2} - \dfrac{4}{x^2-x}$

32. $\dfrac{4}{x^2-2x-3} - \dfrac{2}{x^2-4x+3}$

33. $\dfrac{x}{x^2+6x+9} + \dfrac{1}{x^2+4x+3}$

34. $\dfrac{y}{y^2-3y-10} + \dfrac{3}{y^2+4y+4}$

35. $\dfrac{3a+6}{3a+2} \cdot \dfrac{a+1}{a+2}$

36. $\dfrac{3a+6}{3a+2} + \dfrac{a+1}{a+2}$

37. $5 - \dfrac{1}{3x-6} + \dfrac{3}{x^2-2x}$

38. $7 + \dfrac{3}{4x+12} - \dfrac{2}{x^2+3x}$

39. $\dfrac{5}{x^2+x-6} - \dfrac{3}{x^2+3x} + \dfrac{2}{x^2-2x}$

40. $\dfrac{3}{4x^2-36} + \dfrac{7}{6x^2-18x}$

41. $\left(\dfrac{x}{2} - \dfrac{2}{x}\right) \div \dfrac{x^2-2x}{x^2}$

42. $\left(\dfrac{2}{4y-3} - \dfrac{1}{3y}\right) \cdot \dfrac{20y-15}{4y^2-9}$

43. $\dfrac{\dfrac{1}{x}-1}{\dfrac{2}{x}}$

44. $\dfrac{\dfrac{5}{a^2}}{2+\dfrac{3}{a}}$

45. $\dfrac{1-\dfrac{1}{a}}{\dfrac{1}{a}-\dfrac{1}{a^2}}$

46. $\dfrac{2-\dfrac{3}{c}}{2+\dfrac{3}{c}}$

47. $\dfrac{\dfrac{a}{2}-\dfrac{b}{4}}{\dfrac{4}{b^2}-\dfrac{1}{a^2}}$

48. $\dfrac{\dfrac{1}{3}-\dfrac{1}{x}}{\dfrac{2}{9x}-\dfrac{2}{3x^2}}$

49. $\dfrac{\dfrac{1}{x} + \dfrac{1}{x+2}}{\dfrac{x+1}{x+2}}$

50. $\dfrac{\dfrac{2}{a} - \dfrac{1}{a+3}}{\dfrac{a+6}{a}}$

51. $\dfrac{\dfrac{1}{t+1} - \dfrac{1}{t+2}}{1 - \dfrac{1}{t+2}}$

52. $\dfrac{\dfrac{1}{w-1} - \dfrac{1}{w+3}}{1 - \dfrac{1}{w+3}}$

53. Evaluate the formula $\dfrac{X-a}{\dfrac{s}{n}}$ for $X = 78$, $a = 70$, $s = 6.2$, $n = 20$.

54. The *harmonic mean* of three numbers a, b, and c is defined to be

$$\frac{3}{\dfrac{1}{a} + \dfrac{1}{b} + \dfrac{1}{c}}$$

Use this formula to find the exact value of the harmonic mean and the value rounded to two decimal places for $a = 4$, $b = 5$, and $c = 8$.

 **QUESTIONS FOR THOUGHT**

55. Is the solution to Example 12 dependent on the distance between city A and city B? [*Hint:* Rework the solution using x for the distance between the two cities.]

56. Multiply out $\left(x + \dfrac{2}{x}\right)\left(x - \dfrac{3}{x}\right)$ in the following two ways.

   (a) First, multiply out the two binomials using FOIL.

   (b) Second, combine the fractions with the parentheses, and then multiply the resulting fractions.

   Which method did you find easier? If the instructions were to express your answer as a single fraction, which method would you choose in general?

MINI-REVIEW

In Exercises 57–60, solve the given equation.

57. $\dfrac{x-1}{2} = \dfrac{x}{3}$

58. $\dfrac{x}{2} - 1 = \dfrac{x}{3}$

59. $\dfrac{a}{5} - \dfrac{a}{6} = \dfrac{1}{2}$

60. $0.25m + 0.1m = 56$

9.4   Fractional Equations

In this section we will apply the techniques we developed in the previous section to solving fractional equations.

In Section 4.4, we saw that we can use the LCD to clear the denominators from an equation and give us an equivalent equation that is much easier to solve. Let's review that process in our first example.

EXAMPLE 1 Solve for t. $\dfrac{3t - 5}{4} - \dfrac{t + 1}{6} = \dfrac{t - 2}{3}$

Solution Since we want to solve for t, we want to produce (if possible) an equivalent equation without fractional expressions. Therefore, we multiply both sides of the equation by the LCD to "clear" the denominators.

$$\frac{3t - 5}{4} - \frac{t + 1}{6} = \frac{t - 2}{3}$$

The LCD is 12. We multiply both sides of the equation by 12.

$$12\left(\frac{3t - 5}{4} - \frac{t + 1}{6}\right) = 12\left(\frac{t - 2}{3}\right)$$

$$\frac{12}{1} \cdot \frac{(3t - 5)}{4} - \frac{12}{1} \cdot \frac{(t + 1)}{6} = \frac{12}{1} \cdot \frac{(t - 2)}{3}$$

*We distribute the 12 to each term and reduce. If you want to shorten the amount of writing by doing these steps mentally, **watch out** for a sign error where the arrow points.*

$$\frac{\overset{3}{\cancel{12}}}{1} \cdot \frac{(3t - 5)}{\cancel{4}} - \frac{\overset{2}{\cancel{12}}}{1} \cdot \frac{(t + 1)}{\cancel{6}} = \frac{\overset{4}{\cancel{12}}}{1} \cdot \frac{(t - 2)}{\cancel{3}}$$

$$3(3t - 5) - 2(t + 1) = 4(t - 2)$$

Remove parentheses.

$$9t - 15 - 2t - 2 = 4t - 8$$

Combine like terms.

$$7t - 17 = 4t - 8$$

Isolate t.

$$3t = 9$$

$$\boxed{t = 3}$$

Check $t = 3$: $\dfrac{3(3) - 5}{4} - \dfrac{(3) + 1}{6} \overset{?}{=} \dfrac{(3) - 2}{3}$

$$\frac{9 - 5}{4} - \frac{4}{6} \overset{?}{=} \frac{1}{3}$$

$$\frac{4}{4} - \frac{2}{3} \overset{?}{=} \frac{1}{3}$$

$$1 - \frac{2}{3} \overset{\checkmark}{=} \frac{1}{3}$$

We take basically the same approach when there are variables in the denominator.

EXAMPLE 2 Solve for x. $\dfrac{5}{6} + \dfrac{1}{x - 4} = \dfrac{8}{3x - 12}$

Solution In order to find the LCD, we first factor each denominator.

$$\frac{5}{6} + \frac{1}{x - 4} = \frac{8}{3x - 12}$$

$$\frac{5}{6} + \frac{1}{x - 4} = \frac{8}{3(x - 4)}$$

We multiply both sides of the equation by the LCD, which is $6(x - 4)$, to clear denominators.

$$6(x - 4)\left(\frac{5}{6} + \frac{1}{x - 4}\right) = 6(x - 4) \cdot \frac{8}{3(x - 4)}$$

Each term gets multiplied by $6(x - 4)$.

$$\frac{6(x - 4)}{1} \cdot \frac{5}{6} + \frac{6(x - 4)}{1} \cdot \frac{1}{x - 4} = \frac{6(x - 4)}{1} \cdot \frac{8}{3(x - 4)}$$

$$\frac{6(x - 4)}{1} \cdot \frac{5}{6} + \frac{6(x - 4)}{1} \cdot \frac{1}{x - 4} = \frac{\overset{2}{6}(x - 4)}{1} \cdot \frac{8}{3(x - 4)}$$

$$5(x - 4) + 6 = 16$$
$$5x - 20 + 6 = 16$$
$$5x - 14 = 16$$
$$5x = 30$$
$$\boxed{x = 6}$$

Check $x = 6$: $\dfrac{5}{6} + \dfrac{1}{(6) - 4} \overset{?}{=} \dfrac{8}{3(6) - 12}$

$$\frac{5}{6} + \frac{1}{2} \overset{?}{=} \frac{8}{18 - 12}$$

$$\frac{5}{6} + \frac{3}{6} \overset{\checkmark}{=} \frac{8}{6}$$

■

EXAMPLE 3   Combine and simplify. $\dfrac{2}{x} - \dfrac{3}{x + 1} + \dfrac{5}{4x}$

Solution   We must repeat the warning we gave in Section 4.4 about not confusing equations with expressions. This example is *not* an equation, and therefore we cannot clear the denominators. Notice that even the instructions are telling us that this is not an equation (it does not say solve). We will use the LCD in this example, but to *build* the fractions, not clear the denominators. We proceed as we did in the last section.

$$\frac{2}{x} - \frac{3}{x + 1} + \frac{5}{4x} \qquad \textit{The LCD is } 4x(x + 1).$$

$$= \frac{2 \cdot 4(x + 1)}{4 \; x \; (x + 1)} - \frac{3 \cdot 4x}{4x \; (x + 1)} + \frac{5 \cdot (x + 1)}{4x \; (x + 1)}$$

$$= \frac{8(x + 1) - 12x + 5(x + 1)}{4x(x + 1)}$$

$$= \frac{8x + 8 - 12x + 5x + 5}{4x(x + 1)} = \boxed{\frac{x + 13}{4x(x + 1)}}$$

■

EXAMPLE 4   Solve for x. $\dfrac{6}{x - 3} + 5 = \dfrac{x + 3}{x - 3}$

Solution   We proceed as before to clear the denominators by multiplying both sides of the equation by the LCD, which is $x - 3$.

$$\frac{6}{x - 3} + 5 = \frac{x + 3}{x - 3}$$

$$(x - 3)\left(\frac{6}{x - 3} + 5\right) = (x - 3)\left(\frac{x + 3}{x - 3}\right)$$

$$\frac{(x - 3)}{1} \cdot \frac{6}{x - 3} + \frac{(x - 3)}{1} \cdot 5 = \frac{(x - 3)}{1} \cdot \frac{(x + 3)}{(x - 3)}$$

$$\frac{(x - 3)}{1} \cdot \frac{6}{x - 3} + \frac{(x - 3)}{1} \cdot \frac{5}{1} = \frac{(x - 3)}{1} \cdot \frac{(x + 3)}{(x - 3)}$$

$$6 + 5(x - 3) = x + 3$$

$$6 + 5x - 15 = x + 3$$

$$5x - 9 = x + 3$$

$$4x = 12$$

$$x = 3 \qquad \textit{Are you wondering why}$$
$$\textit{there is no box?}$$

Check $x = 3$: $\dfrac{6}{3 - 3} + 5 \stackrel{?}{=} \dfrac{3 + 3}{3 - 3}$

$$\frac{6}{0} + 5 \stackrel{?}{=} \frac{6}{0}$$

$\frac{6}{0}$ is undefined. We are never allowed to divide by 0. Therefore, $x = 3$ is *not* a solution.

What happened? Our logic tells us that *if* there is a solution to the original equation, then it must be $x = 3$. Since we see that $x = 3$ is not a solution, that must mean

> the original equation has *no solutions*   ■

How is it possible for us to solve an equation properly (we really did not make any mistakes in Example 4) and yet get an answer that does not work?

The difficulty lies in the first step of the solution, where we multiplied both sides of the equation by $x - 3$. Back in Section 3.2 we saw that if we want to be sure that we obtain an *equivalent* equation, then we cannot multiply both sides of the equation by 0. If x is equal to 3, then $x - 3 = 0$. Therefore, when we multiplied by $x - 3$, we were actually multiplying by 0, and we got an equation that was no longer equivalent to the original equation.

Thus, when we multiply both sides of an equation by a variable quantity that might be equal to 0, we must be sure to check our answers in the *original* equation. ***This check is not optional***—it is a necessary step in the solution. We are not checking for errors, but rather to see if we have obtained a valid solution.

An alternative way of saying this is that since $x - 3$ is a denominator of the original equation, and since we are never allowed to divide by 0, $x = 3$ was disqualified from consideration as a solution *at the outset*.

EXAMPLE 5   Solve for y. $\dfrac{2}{y - 1} - \dfrac{4}{3y} = \dfrac{1}{y^2 - y}$

Solution   We will use the LCD to clear denominators. In order to find the LCD we make sure each denominator is in factored form.

$$\frac{2}{y - 1} - \frac{4}{3y} = \frac{1}{y^2 - y}$$

$$\frac{2}{y - 1} - \frac{4}{3y} = \frac{1}{y(y - 1)} \qquad \begin{array}{l} \textit{LCD} = 3y(y - 1). \\ \textit{We multiply each term by} \\ \textit{the LCD.} \end{array}$$

$$\frac{3y(y - 1)}{1} \cdot \frac{2}{y - 1} - \frac{3y(y - 1)}{1} \cdot \frac{4}{3y} = \frac{3y(y - 1)}{1} \cdot \frac{1}{y(y - 1)}$$

$$\frac{3y(y - 1)}{1} \cdot \frac{2}{y - 1} - \frac{3y(y - 1)}{1} \cdot \frac{4}{3y} = \frac{3y(y - 1)}{1} \cdot \frac{1}{y(y - 1)}$$

$$6y - 4(y - 1) = 3$$
$$6y - 4y + 4 = 3$$
$$2y + 4 = 3$$
$$2y = -1$$
$$\boxed{y = -\frac{1}{2}}$$

Check $y = -\frac{1}{2}$: $\dfrac{2}{-\dfrac{1}{2} - 1} - \dfrac{4}{3\left(-\dfrac{1}{2}\right)} \overset{?}{=} \dfrac{1}{\left(-\dfrac{1}{2}\right)^2 - \left(-\dfrac{1}{2}\right)}$

$$\dfrac{2}{-\dfrac{3}{2}} - \dfrac{4}{-\dfrac{3}{2}} \overset{?}{=} \dfrac{1}{\dfrac{1}{4} + \dfrac{1}{2}}$$

$$\dfrac{2}{1} \cdot \dfrac{-2}{3} - \dfrac{4}{1} \cdot \dfrac{-2}{3} \overset{?}{=} \dfrac{1}{\dfrac{3}{4}}$$

$$\dfrac{-4}{3} + \dfrac{8}{3} \overset{?}{=} 1 \cdot \dfrac{4}{3}$$

$$\dfrac{4}{3} \overset{\checkmark}{=} \dfrac{4}{3}$$

■

A final comment about the word "check" is in order. A check can serve two purposes. We can check an equation in order to see whether we have made any algebraic or arithmetic errors. As far as this purpose is concerned, you may think the word "check" was misapplied in Example 5, because the check was more difficult than the original problem. If the check had turned out wrong, you might have believed it more likely that your error occurred in the check rather than in the process of solving the equation. However, since the fact remains that the only way to be sure we have a correct answer is to check it, we must weigh the confidence we have in our check against the confidence we have in our solution.

A second purpose a check can serve is to ensure that we have obtained a valid answer to our original problem. Thus, in Example 5 we must *at least* verify that $y = -\frac{1}{2}$ is a valid solution in that it does not make any of the denominators in the original equation equal to 0. Of course, just because we verify that $y = -\frac{1}{2}$ is a valid answer does *not* mean that it is a correct answer as well. The only way to verify that an answer actually works is by substituting it into the original equation.

EXAMPLE 6   Find the value of t so that the line passing through the points $(0, -2)$ and $(3, t)$ is parallel to the line passing through the points $(-1, 2)$ and $(1, 7)$.

Solution   Even though we will solve this problem algebraically, it is usually a good idea to draw a diagram whenever we can to aid us in understanding the problem. Figure 9.1 illustrates the given information.

Figure 9.1

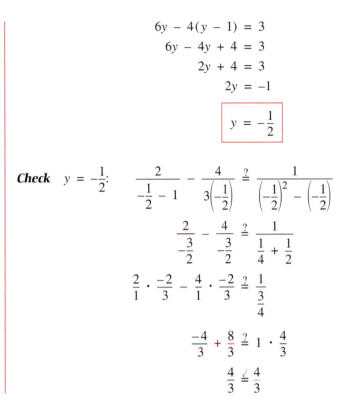

Recall that the slope of a line passing through two points is the "change in y over the change in x."

We have drawn line L through the points $(-1, 2)$ and $(1, 7)$, and line M parallel to line L through the point $(0, -2)$ by approximating the location of the point $(3, t)$ visually. Now we will find the value of t algebraically.

Based on our work in Chapter 5, we know that if the two lines are parallel then their slopes must be equal, so we set up an equation that says exactly that.

The slope of line M = The slope of line L

$$\frac{t - (-2)}{3 - 0} = \frac{7 - 2}{1 - (-1)}$$

$$\frac{t + 2}{3} = \frac{5}{2} \qquad \textit{Multiply both sides of the equation by 6.}$$

$$\frac{6}{1} \cdot \frac{t + 2}{3} = \frac{6}{1} \cdot \frac{5}{2}$$

$$2(t + 2) = 15$$

$$2t + 4 = 15$$

$$2t = 11$$

$$t = 5.5$$

We note that this answer agrees very well with the location of the point $(3, t)$ in Figure 9.1. ∎

Let's now return to the problem posed at the beginning of this chapter.

EXAMPLE 7

An audio equipment manufacturing company determines that when n CD players are produced, the average cost per player is given by the equation

$$C = \frac{85n + 10{,}000}{n}$$

Suppose that the company wants the average cost per player to be \$89. How many CD players should be manufactured?

Solution

In order to answer the question, we must substitute $C = 89$ in the given equation and then solve the resulting equation for n.

$$C = \frac{85n + 10{,}000}{n} \qquad \textit{Substitute C = 89.}$$

$$89 = \frac{85n + 10{,}000}{n} \qquad \textit{Multiply both sides of the equation by n.}$$

$$89n = \frac{85n + 10{,}000}{n} \cdot \frac{n}{1}$$

$$89n = 85n + 10{,}000$$

$$4n = 10{,}000$$

$$n = 2{,}500$$

See Exercise 65 for some related questions.

Thus, based on the company's mathematical model, in order to achieve an average price of \$89, the company should manufacture 2,500 CD players. ∎

 **EXERCISES 9.4**

In each of the following, if the exercise is an equation, solve it and check. Otherwise, perform the indicated operations and simplify.

1. $\dfrac{x}{2} + \dfrac{x}{3} = 10$

2. $\dfrac{t}{2} - \dfrac{t}{5} = 6$

3. $\dfrac{y}{6} - \dfrac{y}{4} = 2$

4. $\dfrac{z}{9} - \dfrac{z}{6} = 2$

5. $\dfrac{a+1}{3} + \dfrac{a+3}{4} = 4$

6. $\dfrac{w-2}{5} + \dfrac{w-1}{3} = 3$

7. $\dfrac{u-2}{2} - \dfrac{u+4}{8} = 1$

8. $\dfrac{m+3}{4} - \dfrac{m-4}{8} = 2$

9. $\dfrac{x}{2} + \dfrac{x}{3} + \dfrac{x}{4}$

10. $\dfrac{x}{2} + \dfrac{x}{3} = \dfrac{x}{4}$

11. $\dfrac{x}{2} \cdot \dfrac{x}{3} \cdot \dfrac{x}{4}$

12. $\dfrac{x}{2} - \dfrac{x}{3} - \dfrac{4}{x}$

13. $\dfrac{x+1}{2} - \dfrac{3}{x} = \dfrac{x}{2}$

14. $\dfrac{x+1}{2} - \dfrac{2}{x} - \dfrac{x}{2}$

15. $\dfrac{x+1}{2} \cdot \dfrac{3}{x} = \dfrac{2}{x}$

16. $\dfrac{x+1}{2} \cdot \dfrac{3}{x} \div \dfrac{2}{x}$

17. $\dfrac{x}{5} + \dfrac{x-1}{4} + \dfrac{x-3}{2} = 3$

18. $\dfrac{x}{6} + \dfrac{x-3}{3} + \dfrac{x-2}{4} = 3$

19. $\dfrac{a+5}{2} = \dfrac{a+2}{5}$

20. $\dfrac{a-3}{7} = \dfrac{a-7}{3}$

21. $\dfrac{x+2}{3} + 1 = \dfrac{x+3}{2} - 1$

22. $\dfrac{y-5}{5} - 3 = \dfrac{y+7}{6} - 4$

23. $\dfrac{2r+1}{3} - \dfrac{r+1}{5} = \dfrac{r+8}{6}$

24. $\dfrac{5a-8}{4} - \dfrac{a+4}{8} = \dfrac{a}{2}$

25. $\dfrac{3}{x} - \dfrac{2}{3} = \dfrac{2}{x}$

26. $\dfrac{5}{y} - \dfrac{3}{5} = \dfrac{4}{y}$

27. $\dfrac{4}{t-2} + \dfrac{3}{t} = \dfrac{1}{2t}$

28. $\dfrac{2}{a+3} + \dfrac{1}{a} = \dfrac{4}{3a}$

29. $\dfrac{5}{2x} - \dfrac{8}{x+2}$

30. $\dfrac{5}{2x} \cdot \dfrac{8}{x+2}$

31. $\dfrac{4}{x-1} - \dfrac{5}{8} = \dfrac{3}{2x-2}$

32. $\dfrac{10}{y+3} - \dfrac{3}{5} = \dfrac{10y+1}{3y+9}$

33. $\dfrac{8}{x-2} + 3 = \dfrac{x+6}{x-2}$

34. $4 + \dfrac{5}{a+3} = \dfrac{a+8}{a+3}$

35. $\dfrac{4}{x+2} - \dfrac{3}{x+6}$

36. $\dfrac{4}{x+2} = \dfrac{3}{x+6}$

37. $\dfrac{3}{x+4} + \dfrac{2}{5} = \dfrac{x-1}{x+4}$

38. $\dfrac{3}{x+4} + \dfrac{2}{5} = \dfrac{x}{x+4}$

39. $\dfrac{7}{x+3} - \dfrac{1}{4} = \dfrac{x+10}{x+3}$

40. $\dfrac{7}{x+3} - \dfrac{1}{4} = \dfrac{x}{x+3}$

41. $\dfrac{m+3}{6} - \dfrac{m+4}{8} = m$

42. $\dfrac{5}{4t} - \dfrac{2}{t} - \dfrac{1}{2} = -\dfrac{1}{8}$

43. $\dfrac{5}{x^2-x} - \dfrac{1}{2x-2} = \dfrac{1}{x}$

44. $\dfrac{11}{x^2-9} - \dfrac{7}{2x+6} = \dfrac{2}{x+3}$

45. $\dfrac{x}{x-5} + \dfrac{3}{2} = \dfrac{5}{x-5}$

46. $\dfrac{c-5}{10} - \dfrac{c-2}{4} = c$

47. $\dfrac{a}{2} + \dfrac{2}{a-2} = \dfrac{a-4}{2}$

48. $\dfrac{y}{3} - \dfrac{3}{y} = \dfrac{y-3}{3}$

49. $\dfrac{2}{x^2 - x} + \dfrac{8}{x^2 - 1}$

50. $\dfrac{x}{x + 2} \div \dfrac{x^2}{x^2 - 4}$

51. $\dfrac{1}{x + 3} \div \dfrac{x + 1}{x^2 - 9}$

52. $\dfrac{1}{x + 3} = \dfrac{x + 1}{x^2 - 9}$

53. $\dfrac{5}{x^2 - 2x - 3} = \dfrac{4}{x^2 - 3x - 4}$

54. $\dfrac{6}{r^2 - 9} = \dfrac{3}{r^2 - 3r}$

55. $\dfrac{9}{x^2 - 3x + 2} - \dfrac{2}{x - 1} = \dfrac{1}{x - 2}$

56. $\dfrac{5}{y^2 - 3y - 10} + \dfrac{3}{y - 5} = \dfrac{-1}{4y + 8}$

57. The length of a rectangle is 3 more than twice the width. If the ratio of width to length is 2 to 5, find the dimensions of the rectangle.

58. The width of a rectangle is 10 less than three times its length. If the ratio of width to length is 7 to 4, find the dimensions of the rectangle.

59. Find the value of c so that the line passing through the points $(1, c)$ and $(-5, 3)$ is parallel to the line passing through the points $(4, 3)$ and $(7, -2)$.

60. Find the value of t so that the line passing through the points $(t, 4)$ and $(3, t)$ is parallel to the line passing through the points $(0, -1)$ and $(4, -3)$.

61. Use the mathematical model given in Example 7 to find the number of CD players to be manufactured so that the average price is $90.

62. Use the mathematical model given in Example 7 to find the number of CD players to be manufactured so that the average price is $86.

63. An ecological study group suggests that under certain conditions the following equation relates the number n of deer that can reasonably be supported on x acres of foraging land:

$$n = \dfrac{100x}{0.6x + 5}$$

According to this model, how many acres would it take to support 160 deer?

64. An insect colony is growing according to the equation

$$n = \dfrac{800(2t + 3)}{0.2t + 5}$$

where n is the number of insects in the colony t hours after the initial formation of the colony. How many hours does it take until the colony has 2,300 insects?

② QUESTIONS FOR THOUGHT

65. In Example 7 we used the mathematical model $C = \dfrac{85n + 10,000}{n}$ to give the average cost per CD player when n CD players are produced. Find the average cost if 5,000, 8,000, 12,000, and 15,000 CD players are manufactured. While we usually expect the average cost per player to decrease as more players are manufactured, what do you notice happening to the average cost as more and more CD players are produced?

66. Determine whether the values $x = 3, 8, -5$, and 10 satisfy the equation

$$\dfrac{x + 3}{2} - \dfrac{2x + 5}{4} = \dfrac{1}{4}$$

Try some other values of x. Why do you think there are so many solutions to this equation? What type of equation is this? How would you prove this?

67. Make up a verbal problem involving a rectangle that would give rise to the following equation. Be sure to indicate what the variable represents.

$$2x + 2(3x - 5) = 62$$

◇ **MINI-REVIEW**

Solve each of the problems algebraically. Be sure to label what the variable represents.

68. In a certain voting district the ratio of Republicans to Democrats is 4 to 7. If there are 2,520 Democrats, how many Republicans are there?

69. At a certain college the ratio of those who graduate in 4 years to those who take longer to graduate is 13 to 3. If in the last 5 years 2,600 students graduated in 4 years, how many graduates were there altogether during the last 5 years?

9.5 Literal Equations

Thus far when we have been asked to solve an equation that has a unique solution, our answer has been a number. Solving such an equation means finding its *numerical* solution—that is, the number that satisfies the equation. (Keep in mind that we have restricted our attention to first- and second-degree equations in *one* variable.) For example, to solve the equation

$$3x - 5 = x + 7$$

we apply the procedures we have learned and obtain the solution $x = 6$.

However, if an equation has more than one variable, then solving the equation takes on an entirely different meaning. An equation that contains more than one variable (letter) is often called a **literal equation.** The reason is that, as we shall soon see, when we solve such an equation for one of its variables, we will get a literal solution rather than a numerical one.

If a literal equation is of the form where one of the variables is totally isolated on one side of the equation, then we say that the equation is solved **explicitly** for that variable. For example, the equation

$$z = 3x - 4$$

is solved explicitly for z, while the *same* equation written in the form

$$z - 3x + 4 = 0$$

is not solved explicitly for either variable.

While we cannot be asked to solve a literal equation in order to get a numerical solution, we can be asked to solve it explicitly for a particular variable.

EXAMPLE 1   Solve the equation $3x + 5y = 15$

(a) Explicitly for x (b) Explicitly for y

Solution   (a) To solve the equation explicitly for x means we want to isolate x on one side of the equation.

$$3x + 5y = 15$$
$$\underline{\quad -5y \qquad\quad -5y\quad}$$
$$3x \qquad = 15 - 5y$$

We collect all x terms alone on one side of the equation by subtracting 5y.

Now we solve for x. We can divide the entire right side by 3 or we can divide each term on the right side by 3.

$$\frac{3x}{3} = \frac{15 - 5y}{3} = \frac{15}{3} - \frac{5y}{3}$$

$$\boxed{x = \frac{15 - 5y}{3}} \quad \text{or} \quad \boxed{x = 5 - \frac{5}{3}y}$$

Both forms of the answer are acceptable.

(b) To solve explicitly for y means we want to isolate y on one side of the equation.

$$
\begin{array}{rcl}
3x + 5y & = & 15 \\
-3x & & - 3x \\
\hline
5y & = & 15 - 3x \\
\dfrac{5y}{5} & = & \dfrac{15 - 3x}{5}
\end{array}
$$

$$\boxed{y = \frac{15 - 3x}{5}} \quad \text{or} \quad \boxed{y = 3 - \frac{3}{5}x}$$

Keep in mind that the solutions to both parts **(a)** and **(b)** are equivalent to the original equation $3x + 5y = 15$. ■

EXAMPLE 2

Solution

Solve the equation $r + 4s = s - 4r + 2$ for r.

When asked to solve for r, it is understood that we want to solve explicitly for r. Just as we did with numerical equations, we isolate the variable we are solving for. In this case we focus our attention on isolating r.

$$
\begin{array}{rcll}
r + 4s & = & s - 4r + 2 & \\
- 4s & = & -4s & \\
\hline
r & = & -3s - 4r + 2 & \textit{Do not stop here!} \\
+4r & & + 4r & \textit{r must appear on one side only.} \\
\hline
5r & = & -3s + 2 & \\
\dfrac{5r}{5} & = & \dfrac{-3s + 2}{5} &
\end{array}
$$

$$\boxed{r = \frac{-3s + 2}{5}}$$

If we had decided to isolate r on the right-hand side of the equation, we could have proceeded as follows:

$$
\begin{array}{rcll}
r + 4s & = & s - 4r + 2 & \textit{Subtract s and 2 from both sides.} \\
- s - 2 & = & -s - 2 & \\
\hline
r + 3s - 2 & = & -4r & \\
-r & & - r & \\
\hline
3s - 2 & = & -5r & \\
\dfrac{3s - 2}{-5} & = & \dfrac{-5r}{-5} &
\end{array}
$$

$$\boxed{-\frac{3s - 2}{5} = r}$$

Recall that in a fraction we prefer to have the minus sign in front of the fraction rather than in the denominator.

Which side you decide to isolate the variable on is not important. What is important is that you recognize that the two answers are equivalent.

Our first answer is

$$\frac{-3s + 2}{5} = \frac{-(3s - 2)}{5} = -\frac{3s - 2}{5}$$

which is our second answer. Keep this in mind when you check your answers with those in the answer key. ■

Sometimes literal equations that have some real-life interpretation are called *formulas.* For instance, the following formula for *simple* interest

$$A = P(1 + rt)$$

allows us to compute the amount A of money in an account if the original amount, P (P stands for principal), is invested at a rate of $r\%$ per year (r is written as a decimal) for t years.

For example, if $1,000 is invested at 8% for 3 years, then

$$P = 1,000 \qquad r = 0.08 \qquad t = 3$$

and the amount of money, A, in the account after 3 years is

$$A = 1,000[1 + 0.08(3)]$$
$$A = 1,000(1.24)$$
$$A = \$1,240$$

This formula is fine if we are given P, r, t, and want to compute A. But what if we are given A, P, t and we want to compute r? In that case we would much prefer to have a formula that is solved explicitly for r rather than A.

EXAMPLE 3

Solve $A = P(1 + rt)$ for r.

Solution

To isolate r, we can begin by first multiplying out the parentheses.

$$A = P(1 + rt)$$
$$A = P + Prt \qquad \textit{Isolate the r term.}$$
$$A - P = Prt \qquad \textit{Divide both sides by what is multiplying r.}$$
$$\frac{A - P}{Pt} = \frac{Prt}{Pt}$$

$$\boxed{\frac{A - P}{Pt} = r}$$ ■

Naturally, this same procedure can be applied to some literal inequalities as well.

EXAMPLE 4

Solve for a. $2a - 5b < 6a + 8c$

Solution

$$
\begin{array}{r}
2a - 5b < 6a + 8c \\
\underline{-6a \qquad\qquad -6a} \\
-4a - 5b < 8c \\
\underline{+ 5b \qquad\quad + 5b} \\
-4a < 8c + 5b \\
\frac{-4a}{-4} > \frac{8c + 5b}{-4} \\
\uparrow
\end{array}
$$

*Remember that when we divide both sides of an inequality by a negative number, we must **reverse** the inequality symbol.*

$$a > -\frac{8c + 5b}{4}$$ ■

Sometimes additional steps are necessary in order to isolate a particular variable.

EXAMPLE 5 | Solve for x. $ax + b = cx + d$

Solution | Let's compare solving this equation for x with solving $6x + 5 = 4x + 7$ for x. Notice that we will follow virtually the same procedure in both cases. The steps in the two solutions correspond to each other. First we collect all the x terms on one side of the equation and all the other terms on the opposite side of the equation.

$$
\begin{array}{rcl}
6x + 5 &=& 4x + 7 \\
-4x & & -4x \\
\hline
6x - 4x + 5 &=& 7 \\
-5 & & -5 \\
\hline
6x - 4x &=& 7 - 5
\end{array}
\qquad
\begin{array}{rcl}
ax + b &=& cx + d \\
-cx & & -cx \\
\hline
ax - cx + b &=& d \\
& & -b \quad -b \\
\hline
ax - cx &=& d - b
\end{array}
$$

The next step we usually do mentally.   *Now what? We want to isolate x.*
Let's show it, without doing the   *Let's try factoring out x.*
arithmetic.

$$(6 - 4)x = 7 - 5$$ $$x(a - c) = d - b$$

$$\frac{(6 - 4)x}{6 - 4} = \frac{7 - 5}{6 - 4}$$ $$\frac{x(a - c)}{a - c} = \frac{d - b}{a - c}$$

$$\boxed{x = 1}$$ $$\boxed{x = \frac{d - b}{a - c}}$$

Since we are not allowed to divide by 0, this answer assumes that $a \neq c$. ■

EXERCISES 9.5

Solve each of the following equations or inequalities explicitly for the indicated variable.

1. $4x - 3y = 12$ for x

2. $4x - 3y = 12$ for y

3. $3x + 6y = 18$ for y

4. $3x + 6y = 18$ for x

5. $a - 3b = 2a - b + 5$ for a

6. $a - 3b = 2a - b + 5$ for b

7. $3(m + 2p) = 4(p - m)$ for m

8. $3(m + 2p) > 4(p - m)$ for p

9. $5x - 7z + 12 = 3y - 7x + 2z$ for z

10. $5x - 7z + 12 = 3y - 7x + 2z$ for y

11. $2r + 3(r - 5) = r - 5s + 1$ for r

12. $3h - 6(h + g) = g + 3h - 4$ for g

13. $a(b + c) = d$ for a

14. $u(v + w) - 3 = 5$ for u

15. $\frac{x}{2} + \frac{y}{3} = \frac{x}{3} + \frac{y}{4} - \frac{1}{6}$ for x

16. $\frac{x + y}{3} = \frac{x}{3} + \frac{y}{4} - \frac{x}{6} + 2$ for y

17. $4(x - y) - \frac{x - y}{2} = 3 - (y - x)$ for x

18. $4(x - y) - \frac{x - y}{2} = 3 - (y - x)$ for y

19. $ax + b = 2x + d$ for x

20. $ma + r = 4a - s$ for a

21. $2pn - \frac{y}{3} = 3n + ay + p$ for y

22. $2pn - \frac{y}{3} = 3n + ay + p$ for n

23. $y = \frac{u + 1}{u - 1}$ for u

24. $x = \frac{2t - 3}{3t - 2}$ for t

25. $(x + a)(y + b) = c$ for x

26. $(2x + a)(3x - b) = c$ for a

Each of the following is a formula either from mathematics or the physical or social sciences. Solve each of the formulas for the indicated variable.

27. $y = mx + b$ for x

28. $y = mx + b$ for m

29. $A = \frac{1}{2}h(b_1 + b_2)$ for h

30. $A = \frac{1}{2}h(b_1 + b_2)$ for b_2

31. $A = P(1 + rt)$ for P

32. $A = P(1 + rt)$ for t

33. $C = \frac{5}{9}(F - 32)$ for F

34. $F = \frac{9}{5}C + 32$ for C

35. $S = S_0 + v_0 t + \frac{1}{2}gt^2$ for v_0

36. $S = S_0 + v_0 t + \frac{1}{2}gt^2$ for g

37. $\frac{x - \mu}{s} < 2$ for x (Assume $s > 0$)

38. $\frac{x - \mu}{s} < 2$ for μ (Assume $s > 0$)

39. $\frac{1}{f} = \frac{1}{f_1} + \frac{1}{f_2}$ for f

40. $\frac{1}{f} = \frac{1}{f_1} + \frac{1}{f_2}$ for f_1

(?) QUESTION FOR THOUGHT

41. Make up a verbal problem involving a ratio that would give rise to the following equation (be sure to indicate what the variable would represent):

$$\frac{x}{600} = \frac{9}{10}$$

◇ MINI-REVIEW

In Exercises 42–45, simplify the given expression as completely as possible. Final answers should be expressed with positive exponents only.

42. $\dfrac{4x^4}{36x^{36}}$

43. $x^3 x^4 (x^3)^4$

44. $\dfrac{(x^{-2})^{-5}}{x^{-2} x^{-5}}$

45. *Evaluate.* $2^{-3} + 3^{-2}$

9.6   Applications

We close this chapter with yet another look at verbal problems. We will look at a variety of examples; some relate to problems we have already studied but extend the ideas a bit further.

You may find it helpful to review the material in Sections 3.3 and/or 4.6 before continuing here.

EXAMPLE 1   A rectangle has a perimeter of 52 cm. The ratio of its length to its width is 9 to 4. Find its dimensions.

Solution

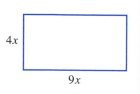

$4x$

$9x$

Figure 9.2
Rectangle for Example 1

The fact that the ratio of the length to the width is 9 to 4 means that we can label the rectangle as shown in Figure 9.2. Notice that

$$\frac{9x}{4x} = \frac{9\cancel{x}}{4\cancel{x}} = \frac{9}{4} \qquad \textit{In effect, we are letting 9x = length and 4x = width.}$$

Since the perimeter is given as 52 cm, our equation is

$$2L + 2W = 52$$

$$2(9x) + 2(4x) = 52$$

$$18x + 8x = 52$$

$$26x = 52$$

$$x = 2$$

Be careful! $x = 2$ is not the answer to the problem. Looking at the figure, we see that

$$\boxed{\text{Length} = 18 \text{ cm}} \quad \text{and} \quad \boxed{\text{Width} = 8 \text{ cm}}$$

Check $\dfrac{18 \text{ cm}}{8 \text{ cm}} = \dfrac{18}{8} = \dfrac{9}{4}$ Thus the ratio is correct.

$$2(18) + 2(8) = 36 + 16 = 52 \text{ cm} \quad \text{So the perimeter is also correct.} \quad \blacksquare$$

EXAMPLE 2

A metal bar weighing 160 kg is divided into two parts, so that the ratio of the weight of the lighter part to that of the heavier part is 3 to 5. Find the weight of each part.

Solution

This example can be done in exactly the same way as the previous one was. (Try it.) However, we offer an alternative solution.

If we let

$$x = \text{Number of kilograms in the lighter part of the bar,}$$

then

$$160 - x = \text{Number of kilograms in the heavier part of the bar}$$

because Heavier part = Total − Lighter part.

The information given in the problem is that

$$\frac{\text{Weight of lighter part}}{\text{Weight of heavier part}} = \frac{3}{5}$$

Therefore, our equation is

$$\frac{x}{160 - x} = \frac{3}{5}$$

The LCD is 5(160 − x). We multiply both sides of the equation by the LCD to clear the denominators.

$$\frac{5(160 - \cancel{x})}{1} \cdot \frac{x}{160 - \cancel{x}} = \frac{3}{5} \cdot \frac{5(160 - x)}{1}$$

$$5x = 3(160 - x)$$

$$5x = 480 - 3x$$

$$8x = 480$$

$$x = 60$$

Therefore, the lighter part weighs

$$\boxed{60 \text{ kg}}$$

and the heavier part weighs

$$\boxed{100 \text{ kg}}$$

Check $\dfrac{60}{100} = \dfrac{3}{5}$ Thus the ratio checks, and clearly the weights add up to 160 kg.

There was no reason we had to let x represent the weight of the lighter part. It is important to realize that we could have let x represent the number of kilograms in the heavier part. In that case we would have gotten a different answer for x (x would have come out to be 100 kg instead of 60 kg), *but* our final answer to the verbal problem would have been the same. ∎

EXAMPLE 3

Susie Slugger has compiled a batting average of .300 by getting 90 hits in 300 at bats ($\frac{90}{300} = \frac{3}{10} = .300$). How many consecutive hits must Susie get in order to raise her batting average to .400? (The answer may surprise you.)

Solution

A hitter's batting average is computed as

$$\frac{\text{Number of hits}}{\text{Number of at bats}}$$

Let x = number of consecutive hits Susie needs to raise her average to .400. Then Susie will have $300 + x$ at bats, so her average will be

$$\frac{\text{Number of hits}}{\text{Number of at bats}} = \frac{90 + x}{300 + x}$$

and since we want her average to be $.400 = \frac{400}{1,000} = \frac{2}{5}$, our equation is

$$\frac{90 + x}{300 + x} = \frac{2}{5} \qquad \textcolor{blue}{\textit{The LCD is } 5(300 + x).}$$

$$\frac{5(300 + x)}{1} \cdot \frac{90 + x}{300 + x} = \frac{5(300 + x)}{1} \cdot \frac{2}{5}$$

$$5(90 + x) = 2(300 + x)$$

$$450 + 5x = 600 + 2x$$

$$3x = 150$$

$$\boxed{x = 50}$$

Susie needs 50 hits in a row to raise her average to .400.

Check   If Susie should perform the incredibly remarkable feat of getting 50 hits in a row, she would have $90 + 50 = 140$ hits, in $300 + 50 = 350$ at bats, which would give her a batting average of $\frac{140}{350} = \frac{14}{35} = \frac{2}{5} = .400$. ∎

EXAMPLE 4

One clerk can perform a job in 5 hours, while a second clerk can do the same job in 7 hours. How long would it take them to do the job if they work together?

Solution

We must make two assumptions here. The first is that the clerks work in a totally cooperative fashion. The second is that they are working at a constant rate. (These may not be totally realistic assumptions, but we make them nonetheless.)

Our basic approach is to analyze what *part* of the job each person does.

Let x = number of hours it takes for them to do the job together. Since it takes the first clerk 5 hours to do the job, he or she does

$\dfrac{1}{5}$ of the job in 1 hour

$\dfrac{2}{5}$ of the job in 2 hours

$\frac{3}{5}$ of the job in 3 hours

$\frac{x}{5}$ of the job in x hours

Similarly, since it takes the second clerk 7 hours to do the same job, he or she does

$\frac{1}{7}$ of the job in 1 hour

$\frac{2}{7}$ of the job in 2 hours

$\frac{3}{7}$ of the job in 3 hours

$\frac{x}{7}$ of the job in x hours

If we add the parts of the job done by each clerk, we must get 1. That is, together they have 1 complete job.

$$\left(\begin{array}{c}\text{Part of job done}\\\text{by first clerk}\end{array}\right) + \left(\begin{array}{c}\text{Part of job done}\\\text{by second clerk}\end{array}\right) = 1 \text{ complete job}$$

Thus, our equation is

$$\frac{x}{5} + \frac{x}{7} = 1 \qquad \textit{The LCD is 35.}$$

$$\overset{7}{\frac{35}{1}} \cdot \frac{x}{5} + \overset{5}{\frac{35}{1}} \cdot \frac{x}{7} = 35 \cdot 1$$

$$7x + 5x = 35$$

$$12x = 35$$

$$x = \frac{35}{12} = 2\frac{11}{12} \text{ hours} = \boxed{2 \text{ hours and } 55 \text{ minutes}}$$

Check The check is left to the student. ■

EXAMPLE 5

Robert and Jan were both competing in long-distance races. Robert's race was 150 km long, while Jan's race was 180 km long. They both completed their races in the same amount of time. If Jan's pace was 2 kph faster than Robert's, how fast was Robert's pace?

Solution In order to write and solve an equation to answer the question, let's analyze this problem carefully.

THINKING OUT LOUD

What do we need to find?	Robert's pace—that is, his speed
Where do we start?	In order to write an equation, we must first identify two quantities that are equal. Reading the problem carefully, we see that the time it took Robert to run his race was equal to the time it took Jan to run his race.

(continued)

THINKING OUT LOUD

(continued)

What is our equation going to say?	Based on our observation, one possibility is that the equation will say Robert's time = Jan's time
What's next?	Since we are looking for Robert's rate of speed, it seems appropriate to involve the rates in the time equation.
What information are we given?	We are given the distance of each race, and that Jan's rate was 2 kph faster than Robert's rate.

We know that distance is equal to rate times time. Based on this analysis it seems reasonable to proceed as follows.

We will use the relationship

$$d = r \cdot t$$

Let

$$r = \text{rate for Robert}$$

Then

$$r + 2 = \text{rate for Jan}$$

Since their *times* are the same, we have

$$t_{\text{Robert}} = t_{\text{Jan}}$$

We will use the formula $d = rt$ in the form $\dfrac{d}{r} = t$.

$$\frac{d_{\text{Robert}}}{r_{\text{Robert}}} = \frac{d_{\text{Jan}}}{r_{\text{Jan}}} \qquad \textit{This equation says that Robert's time = Jan's time.}$$

$$\frac{150}{r} = \frac{180}{r + 2} \qquad \textit{The LCD is } r(r + 2). \textit{ We multiply both sides of the equation by } r(r + 2).$$

$$\frac{r(r + 2)}{1} \cdot \frac{150}{r} = \frac{180}{r + 2} \cdot \frac{r(r + 2)}{1}$$

$$150(r + 2) = 180r$$

$$150r + 300 = 180r$$

$$300 = 30r$$

$$10 = r$$

Thus, Robert's rate was $\boxed{10 \text{ kph}}$.

Check If Robert's rate was 10 kph, he completed his 150-kilometer race in $\frac{150}{10} = 15$ hours. Jan's rate was 2 kph faster, so Jan's rate was 12 kph. Therefore, Jan completed the 180-kilometer race in $\frac{180}{12} = 15$ hours as well. ■

Sometimes a verbal problem consists of nothing more than reading a problem and substituting values.

EXAMPLE 6

A basic formula from physics states that the height s of a free-falling object (neglecting air resistance) is given by the formula

$$s = s_0 + v_0 t + \frac{1}{2}gt^2$$

where

$$s_0 = \text{Initial height}$$
$$v_0 = \text{Initial velocity}$$
$$t = \text{Time in seconds}$$
$$g = \text{Acceleration due to gravity} = -32\frac{\text{ft}}{\text{sec}^2}$$

What is the height of a ball dropped from an initial height of 2,000 ft after 10 seconds?

Solution

If we strip away all the extra information in the problem, what the problem is asking us is

Find s when $s_0 = 2,000$, $t = 10$, and $g = -32$.

But what about v_0? We are told that the ball is being dropped, which implies that the initial velocity is zero. In other words, $v_0 = 0$.

We substitute all these values into the given formula and we get

$$s = s_0 + v_0 t + \frac{1}{2}gt^2$$

$$s = 2,000 + 0(10) + \frac{1}{2}(-32)(10)^2$$

$$s = 2,000 - 1,600$$

$$\boxed{s = 400 \text{ ft}}$$

The ball is 400 ft above the ground after 10 seconds. ■

In the beginning of Chapter 4, we introduced the lens equation

$$\frac{1}{f} = \frac{1}{d_s} + \frac{1}{d_i}$$

which relates the focal length of the lens f, the distance d_s from the subject to the lens, and the distance d_i from the image to the lens. Now that we know how to solve fractional equations we can solve the following problem.

EXAMPLE 7

A lens used in close-up photography has a focal length of 36 millimeters (mm). If the image distance must be 90 mm, what must the subject distance be?

Solution

We are told that the image distance is 90 mm and the focal length is 36 mm, which means that $d_i = 90$ and $f = 36$.

$$\frac{1}{f} = \frac{1}{d_s} + \frac{1}{d_i} \qquad \textit{Substitute } f = 36 \textit{ and } d_i = 90.$$

$$\frac{1}{36} = \frac{1}{d_s} + \frac{1}{90} \qquad \textit{We clear the denominators; the LCD is } 180d_s.$$

$$\overset{5}{\cancel{180d_s}} \cdot \frac{1}{36} = \overset{}{\cancel{180d_s}} \cdot \frac{1}{d_s} + \overset{2}{\cancel{180d_s}} \cdot \frac{1}{90}$$

$$5d_s = 180 + 2d_s$$

$$3d_s = 180$$

$$\boxed{d_s = 60 \text{ mm}} \qquad \textit{The subject distance is 60 mm.} \qquad ■$$

Solve each of the following problems algebraically.

1. The ratio of the width of a rectangle to its length is 3 to 7, and its perimeter is 100 meters. What are its dimensions?

2. The length of a rectangle is 3 more than its width. The ratio of the length of the rectangle to the *perimeter* is 1 to 3. Find the dimensions of the rectangle.

3. Of 200 people surveyed, the proportion of those who preferred brand X to those who did not was 13 to 12. How many people preferred brand X?

4. In a recent survey 300 more people preferred brand X than the famous national brand. If the ratio of those who preferred the national brand to those who preferred brand X was 5 to 7, how many people were surveyed altogether?

5. Suppose Joe Slugger's batting average is .300 as the result of 120 hits in 400 at bats. How many hits must he get in his next 50 at bats to raise his average to .400? Look at your answer. What does it imply?

6. Joe Slugger has 90 hits in 300 at bats, resulting in a batting average of .300. How many hits must he get in his next 100 at bats in order to raise his average to .400? (You might guess an answer first; the actual answer may surprise you.)

7. Bill can mow a lawn in 3 hours, while Sandy can do the same job in 2 hours. How long will it take them to mow the lawn together?

8. Judy can paint her house in 8 days, while Jane can do the same job in 10 days. How long will it take them to paint the house working together?

9. An electrician works twice as fast as his apprentice. Together they can complete a rewiring job in 6 hours. How long would it take each of them working alone?

10. A physics professor can perform an experiment 3 times as fast as her graduate assistant. Together they can perform the experiment in 3 hours. How long would it take each of them working alone?

11. A plane travels 100 kph faster than a train. The plane covers 500 km in the same time that the train covers 300 km. Find the speed of each.

12. A swimmer can swim freestyle 2 meters per second faster than she can swim the breaststroke. If she covers 96 meters swimming freestyle in the same time she covers 64 meters doing the breaststroke, how fast does she swim each stroke?

13. A man can row at the rate of 4 mph in still water. He can row 8 miles upstream (against the current) in the same time that he can row 24 miles downstream (with the current). What is the speed of the current?

14. A family drives to its vacation home at the rate of 90 kph, and returns home at the rate of 80 kph. If the trip returning takes 2 hours longer than the trip going, how far is the trip to the vacation home?

15. Ronnie walks over to a friend's house at the rate of 6 kph and jogs home at the rate of 14 kph. If the total time, walking and jogging, is 3 hours, how far is it to the friend's house?

16. A plane's air speed (speed in still air) is 500 kph. The plane covers 1,120 km with a tailwind in the same time it covers 800 km with a headwind (against the wind). What is the speed of the wind?

17. One number is 3 times another, and the sum of their reciprocals is $\frac{5}{3}$. Find the numbers.

18. One number is twice another, and the sum of their reciprocals is 2. Find the numbers.

19. If the same number is added to the numerator and the denominator of $\frac{3}{5}$, the resulting fraction has the value $\frac{5}{6}$. Find the number.

20. What number must be added to both the numerator and the denominator of $\frac{7}{9}$ to obtain a fraction whose value is $\frac{2}{3}$?

21. The denominator of a fraction is 2 more than its numerator, and the reciprocal of the fraction is equal to itself. Find the fraction.

22. One number is 5 times another and the difference of their reciprocals is $\frac{2}{5}$. Find the numbers.

23. The surface area S of a closed right circular cylinder is given by the formula $S = 2\pi rh + 2\pi r^2$, where r is the radius and h is the height. Find S if $r = 6$ cm and $h = 15$ cm. (You may leave your answer in terms of π.)

24. Using the same formula as in Exercise 23, find h if $S = 351.68$ sq cm, $r = 4$ cm, and $\pi \approx 3.14$. ("$\approx$" means approximately equal.)

25. In electronics, the total resistance R_T of a circuit (measured in ohms) consisting of two wires with resistance R_1 and R_2 connected in parallel is given by the formula

$$\frac{1}{R_T} = \frac{1}{R_1} + \frac{1}{R_2}$$

If $R_1 = 20$ ohms and $R_2 = 30$ ohms, find the total resistance.

26. Use the formula given in Exercise 25 to find R_1 if $R_2 = 12$ ohms and the total resistance is 4.8 ohms.

27. In photography the following formula is frequently used:

$$\frac{1}{f} = \frac{1}{d_s} + \frac{1}{d_i}$$

This is called the *lens equation* and it relates the focal length of the lens, f, the distance d_s from the subject to the lens, and the distance d_i from the image to the lens. Find the focal length of a lens if the object distance is 6 cm and the image distance is 3 cm.

28. Use the formula given in Exercise 27 to find the image distance if the object distance is 4 in. and the focal length is 2.4 in.

29. Nora can paint a room alone in 5 hours and Taisha can paint the same room alone in 8 hours. Nora paints for 2 hours alone and then quits, at which time Taisha finishes painting the room alone. How long does it take to paint the room together?

30. Jim can build a storage shed in 4 days and Steve can build the same shed in 6 hours. Jim starts working on the shed alone but quits before finishing, at which time Steve finishes building the shed alone. If it takes a total of 5 hours to build the shed, how long does Jim work before quitting?

31. Tina and Latifa are given the job of inspecting 30 computers. Tina can inspect the computers alone in 5 hours and Latifa can inspect the computers alone in 7 hours. Tina starts the job alone at 9:00 A.M. and works for one hour, at which time she is joined by Latifa, and they finish the inspection together. At what time do they finish?

32. Mr. Chung and Ms. Furman work for an accounting firm. They are given the job of doing a company audit. Mr. Chung can do the audit alone in 125 hours and Ms. Furman can do the audit alone in 100 hours. Mr. Chung starts the audit and works alone for 50 hours, at which time he is joined by Ms. Furman, and they finish the audit together. How long does the audit take altogether?

33. Sean and Laura are painting an apartment together. Sean paints twice as fast as Laura. Sean paints for 8 hours and Laura paints for 5 hours in order to paint the entire apartment. How long would it take Sean to paint the apartment alone?

34. Malik and Cherise run a part-time cleaning service. It would take Malik three hours longer to clean an apartment alone than it would take Cherise to clean it alone. If they work together and clean the apartment in 2 hours, how long would it take each of them alone?

 QUESTION FOR THOUGHT

35. Explain how Example 4, about the two clerks working on a job together, can be thought of as a rate–time problem.

CHAPTER 9 SUMMARY

After having completed this chapter, you should be able to:

1. Reduce a fraction by first factoring numerator and denominator where necessary (Section 9.1).

   For example:

   $$\frac{2x^2 - 6x}{x^2 + x - 12} = \frac{2x(x - 3)}{(x + 4)(x - 3)}$$

   $$= \frac{2x(x - 3)}{(x + 4)(x - 3)}$$

   $$= \boxed{\frac{2x}{x + 4}}$$

2. Multiply and divide fractions whose numerators and/or denominators are polynomials (Section 9.2).

   For example:

   $$\frac{3x^2}{9 - x^2} \div \frac{6x - 2x^2}{x^2 + 4x + 3} = \frac{3x^2}{(3 - x)(3 + x)} \cdot \frac{(x + 3)(x + 1)}{2x(3 - x)}$$

   We have factored where possible, changed division to multiplication, and inverted the divisor.

   $$= \frac{3\overset{x}{x^2}}{(3 - x)(3 + x)} \cdot \frac{(x + 3)(x + 1)}{2x(3 - x)}$$

   $$= \boxed{\frac{3x(x + 1)}{2(3 - x)^2}}$$

3. Find the LCD for polynomial denominators (Section 9.3).

   For example: If the expressions $3x^2$, $x^2 + 3x$, and $x^2 + 2x - 3$ appear as denominators, what is their LCD?

   We first put each polynomial in factored form.

   $$3x^2 = 3x^2$$
   $$x^2 + 3x = x(x + 3)$$
   $$x^2 + 2x - 3 = (x + 3)(x - 1)$$

   The distinct factors are 3, x, $x + 3$, and $x - 1$.

   According to our outline, the LCD is $\boxed{3x^2(x + 3)(x - 1)}$.

4. Combine fractions with polynomial denominators (Section 9.3).

For example:

$$\frac{3}{x^2 + 2x} + \frac{4}{x^2 - 4} = \frac{3}{x(x + 2)} + \frac{4}{(x + 2)(x - 2)} \qquad LCD = x(x + 2)(x - 2).$$

$$= \frac{3 \cdot (x - 2)}{x \cdot (x - 2) \cdot (x + 2)} + \frac{4 \cdot (x)}{x \cdot (x + 2)(x - 2)}$$

Here the missing factor was x − 2.   *Here the missing factor was x.*

$$= \frac{3x - 6 + 4x}{x(x + 2)(x - 2)}$$

$$= \boxed{\frac{7x - 6}{x(x + 2)(x - 2)}}$$

5. Solve and check a fractional equation involving polynomial denominators (Section 9.4).

For example: Solve for a and check. $\dfrac{5}{6} - \dfrac{3}{a + 2} = \dfrac{8}{15}$ $LCD = 30(a + 2)$

$$\frac{30(a + 2)}{1} \cdot \frac{5}{6} - \frac{30(a + 2)}{1} \cdot \frac{3}{a + 2} = \frac{30(a + 2)}{1} \cdot \frac{8}{15}$$

$$\frac{\overset{5}{\cancel{30}}(a + 2)}{1} \cdot \frac{5}{6} - \frac{30(\cancel{a + 2})}{1} \cdot \frac{3}{\cancel{a + 2}} = \frac{\overset{2}{\cancel{30}}(a + 2)}{1} \cdot \frac{8}{\cancel{15}}$$

$$25(a + 2) - 90 = 16(a + 2)$$
$$25a + 50 - 90 = 16a + 32$$
$$25a - 40 = 16a + 32$$
$$9a = 72$$

$$\boxed{a = 8}$$

Check: $a = 8$:

$$\frac{5}{6} - \frac{3}{(8) + 2} \overset{?}{=} \frac{8}{15}$$

$$\frac{5}{6} - \frac{3}{10} \overset{?}{=} \frac{8}{15}$$

$$\frac{25}{30} - \frac{9}{30} \overset{?}{=} \frac{16}{30}$$

$$\frac{16}{30} \overset{\checkmark}{=} \frac{16}{30}$$

6. Solve a literal equation explicitly for a specified variable (Section 9.5).

For example: Solve for y. $3x - 4y = r - 2(y + x + 5)$

$$3x - 4y = r - 2(y + x + 5)$$
$$3x - 4y = r - 2y - 2x - 10 \qquad \textit{Get the y terms on one side.}$$

$$\underline{ + 4y = + 4y}$$
$$3x = r + 2y - 2x - 10 \qquad \textit{We isolate the y term.}$$

$$\underline{+2x - r + 10 = -r + 2x + 10}$$
$$5x - r + 10 = 2y$$

$$\boxed{\frac{5x - r + 10}{2} = y}$$

7. Solve verbal problems that give rise to fractional equations (Section 9.6).

CHAPTER 9   REVIEW EXERCISES

In Exercises 1–8, reduce the fraction to lowest terms. If the fraction cannot be reduced, then say so.

1. $\dfrac{x^2}{x^2 + 2}$

2. $\dfrac{2x^2 + 12x}{6x^2}$

3. $\dfrac{x^2 + 3x - 4}{x^2 - 16}$

4. $\dfrac{2x^2 + 7x - 15}{4x^2 - 9}$

5. $\dfrac{a^2 + 8a + 16}{a^2 + 6a + 8}$

6. $\dfrac{x^2y - xy^2}{x^2 + xy - 2y^2}$

7. $\dfrac{3z^2 - 12}{3z^2 + 9z + 6}$

8. $\dfrac{c^2 + 16}{3c^3 - 12c^2}$

In Exercises 9–18, perform the indicated operations and simplify as completely as possible.

9. $\dfrac{x}{x + 2} + \dfrac{x}{2}$

10. $\dfrac{a^2}{a^2 + 4a} \cdot \dfrac{a^2 - 16}{4a^2}$

11. $\dfrac{3}{2x + 4} + \dfrac{6}{x^2 + 2x}$

12. $\dfrac{3}{2x + 4} \div \dfrac{6}{x^2 + 2x}$

13. $\dfrac{x^2 - 5x - 6}{2x - 12} \div \dfrac{x^2 + 2x + 1}{8x^2}$

14. $\dfrac{1}{y} + \dfrac{2}{y + 2} - \dfrac{3}{y + 6}$

15. $\dfrac{5}{z^2 + z - 6} - \dfrac{3}{z^2 + 3z}$

16. $\dfrac{x^2 + xy - 2y^2}{x^2 + 4xy + 4y^2} \div (x^2 - y^2)$

17. $2 + \dfrac{3}{x + 2} - \dfrac{1}{x}$

18. $\left(\dfrac{x}{3} - \dfrac{3}{x}\right) \div \dfrac{2x^2 - 6x}{x^2}$

In Exercises 19–26, solve the given equations.

19. $\dfrac{5}{x} - \dfrac{2}{3x} = \dfrac{13}{6}$

20. $\dfrac{3}{a + 2} - \dfrac{5}{2a + 4} = \dfrac{1}{2}$

21. $\dfrac{y + 2}{y} + \dfrac{4}{y + 2} = \dfrac{y}{y + 2}$

22. $\dfrac{2}{x^2 - 4} - \dfrac{3}{x + 2} = \dfrac{4}{x - 2}$

23. $\dfrac{x + 2}{x - 3} + \dfrac{4}{3} = \dfrac{5}{x - 3}$

24. $\dfrac{7}{6} - \dfrac{3}{a + 2} = \dfrac{11}{12}$

25. $3x - 4y + 7 = 8x - 7y + 3$ for x

26. $\dfrac{1}{x} - \dfrac{2}{a} = \dfrac{3}{c}$ for a

Solve each of the following verbal problems algebraically.

27. A bag contains 80 more red marbles than black ones. If the ratio of red marbles to black ones is 7 to 5, how many marbles are there in the bag altogether?

28. Leslie can row at the rate of 4 mph. If a trip 8 miles upstream takes twice as long as the trip downstream, what is the rate of the current?

29. Together John and Susan can complete a job of posting 1,500 notices in 4 hours. If Susan alone could do the job in 6 hours, how long would it take John to do the job alone?

30. What number must be added to the numerator and the denominator of $\frac{8}{13}$ so that the resulting fraction is equal to $\frac{4}{5}$?

31. The length of a rectangle is 5 less than twice its width. If the ratio of length to width is 5 to 3, find the dimensions of the rectangle.

32. Find the value of c so that the line passing through the points $(c, 4)$ and $(2, c)$ has slope $-\frac{3}{4}$.

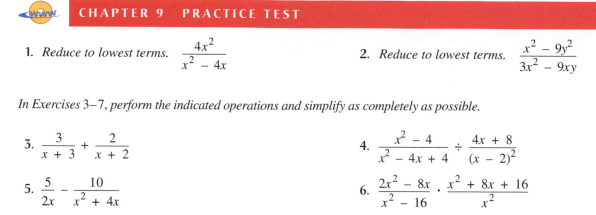

CHAPTER 9 PRACTICE TEST

1. *Reduce to lowest terms.* $\dfrac{4x^2}{x^2 - 4x}$

2. *Reduce to lowest terms.* $\dfrac{x^2 - 9y^2}{3x^2 - 9xy}$

In Exercises 3–7, perform the indicated operations and simplify as completely as possible.

3. $\dfrac{3}{x + 3} + \dfrac{2}{x + 2}$

4. $\dfrac{x^2 - 4}{x^2 - 4x + 4} \div \dfrac{4x + 8}{(x - 2)^2}$

5. $\dfrac{5}{2x} - \dfrac{10}{x^2 + 4x}$

6. $\dfrac{2x^2 - 8x}{x^2 - 16} \cdot \dfrac{x^2 + 8x + 16}{x^2}$

7. $\dfrac{2x - 5}{x^2 - 3x} - \dfrac{3x + 7}{x^2 - 3x} + \dfrac{6x - 3}{x^2 - 3x}$

8. *Solve for x.* $\dfrac{3}{x - 2} - \dfrac{4}{x} = \dfrac{1}{2x}$

9. *Solve for t.* $at + b = \dfrac{3t}{2} + 7$

10. *Solve for c.* $\dfrac{3c}{c - 2} + 4 = \dfrac{c + 4}{c - 2}$

Solve each of the following problems algebraically.

11. A person travels from town A to town B and back in a total of 14 hours. If the average speed going was 45 kph and the average speed returning was 60 kph, how far is it from town A to town B?

12. The formula for the area, A, of a trapezoid is

$$A = \frac{1}{2}h(b_1 + b_2)$$

where h = height, and b_1, b_2 are the bases. If $A = 105$ sq cm, $h = 3$ cm, and $b_1 = 10$ cm, find b_2.

13. Haleema can assemble a puzzle alone in 8 hours and Andrea can assemble the same puzzle alone in 12 hours. Haleema works alone for a time and then quits, at which time Andrea finishes the puzzle alone. If it takes a total of 9 hours to complete the puzzle, how long does each person work on the puzzle?

14. Suppose that the formula

$$N = \frac{20p}{100 - p}$$

gives the number of quality control checks it takes to ensure that p percent of defective items are found in a certain factory. What percentage of defective items are found if 180 quality control checks are made?

CHAPTERS 7–9 CUMULATIVE REVIEW

In Exercises 1–10, perform the indicated operations and simplify as completely as possible. Fractions should be reduced to lowest terms.

1. $(x + 8)(x - 5)$

2. $(x + 8)(x^2 + x - 5)$

3. $(a + b + c)(a + b - c)$

4. $2z(z - 3)(z + 6)$

5. $(5a - 3c)(4a + 3c)$

6. $2(t^2 - 3t) - t(t - 5)$

7. $(x + 3)(x - 12) + (x + 6)^2$

8. $(x - 8)(x - 2) - (x - 4)^2$

9. $(a - 3)(2a + 3)(2a - 3)$

10. $(x + y)(x - 1)(x - y)(x + 1)$

11. Find the sum of $x^2 - xy + 3y^2$, $5x^2 - 8y^2$, and $y^2 - 6x^2$.

12. Subtract $3a^3 - 4a + 7$ from $a^3 - a - 2$.

13. (a) What is the degree of the polynomial $5x^4 - 3x^2 + 6x - 1$?

    (b) What is the coefficient of the second-degree term?

14. Write the polynomial $4 - x + 3x^3$ in complete standard form.

In Exercises 15–20, divide the given polynomials. Use long division where necessary.

15. $\dfrac{x^2 + 8x}{2x}$

16. $\dfrac{8r^2s - 12rs^3 - 4r^3s^2}{6r^2s^2}$

17. $\dfrac{y^2 - 3y + 4}{y - 2}$

18. $\dfrac{2a^3 - 5a^2 - 5a + 6}{a - 3}$

19. $\dfrac{18x^3 - 5x - 28}{3x - 4}$

20. $\dfrac{t^4 - t^2 - 6}{t + 2}$

In Exercises 21–28, simplify the expression as completely as possible. Final answers should be expressed with positive exponents only.

21. $5^0 + 2^{-3} + 2^{-4}$

22. $3^{-1} + 6 \cdot 3^{-2}$

23. $\dfrac{(x^2)^3}{x^2 x^3}$

24. $\dfrac{a^{-5}}{a^{-4}}$

25. $\dfrac{(2x^3)^4}{4(x^5)^3}$

26. $\dfrac{(x^{-2}y^3)^{-4}}{(xy^{-2})^{-5}}$

27. $\dfrac{(3a^{-3}t^2)^{-3}}{(a^{-1}t^{-2})^2}$

28. $\left(\dfrac{5x^{-2}y^3}{x^{-1}y^{-2}}\right)^{-3}$

29. *Write in scientific notation.*   0.000439

30. *Write in scientific notation.*   578,000

31. *Simplify.*

    $\dfrac{(4 \times 10^{-3})(5 \times 10^4)}{2 \times 10^{-3}}$

32. *Use scientific notation to compute.*

    $\dfrac{(0.0006)(4,000)}{(0.024)(50,000)}$

In Exercises 33–52, factor the polynomial as completely as possible.

33. $x^2 + 6x + 5$

34. $x^2 + 6x$

35. $x^2 - 5x + 6$

36. $x^2 - 5x - 6$

37. $6x^3y - 12xy^2 - 9x^2y$

38. $10m^3n^5 - 5m^2n^3$

39. $u^2 - 49$

40. $4a^2 - 24a + 36$

41. $2r^2 + r - 15$

42. $t^4 - 36t^2$

43. $5t^2 + 10t + 15$

44. $x^3y - xy^3$

45. $6x^2 - 17xy + 12y^2$

46. $24 + 10x - x^2$

47. $x^2 + 16x$

48. $x^2 + 16$

49. $x^2 + ax + xy + ay$

50. $a^2 - 3a + az - 3z$

51. $x^2 - 4x - ax + 4a$

52. $x^8 - y^8$

In Exercises 53–54, simplify the fraction.

53. $\dfrac{x^2 - 4x}{x^2 - 16}$

54. $\dfrac{t^2 - 5t + 6}{t^2 - 6t + 9}$

In Exercises 55–63, perform the indicated operations and simplify as completely as possible.

55. $\dfrac{3}{4x} + \dfrac{5}{x + 4}$

56. $\dfrac{5}{6xy^3} - \dfrac{7}{4x^2}$

57. $\dfrac{x^2 - 5x}{10x} \cdot \dfrac{x^2}{x^2 - 25}$

58. $\dfrac{6rt}{r^2 - 2rt + t^2} \div \dfrac{t^2}{r^2 - t^2}$

59. $\dfrac{6}{x^2 + 2x} - \dfrac{4}{x^2 - 2x}$

60. $\dfrac{6}{u + 3} - \dfrac{4}{3u} + \dfrac{1}{2u + 6}$

61. $\dfrac{\dfrac{a}{2} - \dfrac{8}{a}}{\dfrac{a^2 - 8a + 16}{4}}$

62. $8 \cdot \dfrac{t}{t + 2} - \dfrac{5}{3t^2 + 6t} \cdot (6t^2 + 18t)$

63. $\dfrac{2z + 9}{4z + 12} - \dfrac{5z + 8}{4z + 12} + \dfrac{3z + 1}{4z + 12}$

In Exercises 64–75, solve the given equation. If it contains more than one variable, solve for the indicated variable.

64. $\dfrac{11}{x} - \dfrac{2}{3} = \dfrac{25}{3x}$

65. $\dfrac{4}{3a + 6} - \dfrac{3}{2} = \dfrac{5}{6a + 12}$

66. $5x - 7t = 9x - 4t + 12$ for t

67. $\dfrac{3y}{5} - a = 4y + 3a - 6$ for y

68. $\dfrac{8}{z - 2} + 5 = \dfrac{4z}{z - 2}$

69. $\dfrac{x + 8}{8} - \dfrac{x + 6}{6} = \dfrac{x + 3}{3} - \dfrac{x + 4}{4}$

70. $\dfrac{3}{4}(x - 6) + \dfrac{2}{5}(x - 7) = 1 - (x + 4)$

71. $\dfrac{3c + 1}{3c - 2} = \dfrac{3c}{3c - 1}$

72. $(x + 5)(x - 3) = 20$

73. $3x^2 - 5x - 7 = x^2 - 6x + 8$

74. $(x + 3)^2 = 12x + 1$

75. $(x + 4)(x - 6) = (x - 2)(x + 5)$

Solve each of the following problems algebraically. Be sure to label what the variable represents.

76. If Bob can plow a field in 6 days and Martha can plow the same field in 4 days, how long will it take them to plow the field working together?

77. If Roger can overhaul an engine in 8 hours working alone or in 5 hours working with Pat, how long will it take Pat to overhaul the engine working alone?

78. The numerator of a fraction is 4 less than the denominator. If the numerator is increased by 3 and the denominator is increased by 1, the value of the fraction is $\frac{1}{2}$. Find the original fraction.

CHAPTERS 7–9   CUMULATIVE PRACTICE TEST

In Problems 1–6, perform the indicated operations. Simplify your answers and express final answers using positive exponents only.

1. $(x - 2y)(x^2 - 3xy - y^2)$

2. $\dfrac{4x^{-8}y^6}{6x^{-4}y^{-2}}$

3. $2a(3a - 5) + (a - 6)(a - 4)$

4. $(x - 5)^2 - (x + 5)^2$

5. $\dfrac{(2x^3)^{-4}}{(x^{-3})^3}$

6. $\dfrac{12s^2t^5 - 8s^3t^2}{6s^3t^3}$

7. Use long division to find the quotient and remainder.
$$\dfrac{4x^3 - 3x^2 + 5x - 20}{x - 2}$$

8. Use long division to find the quotient and remainder.
$$\dfrac{x^4 - x^2 - 12}{x + 2}$$

9. Write in scientific notation.

(a) 0.000916   (b) 916,000

10. Compute using scientific notation.
$$\dfrac{(0.008)(25,000)}{(6,000)(0.00015)}$$

11. Subtract the sum of $x^3 - 3x$ and $x^2 - 5x$ from $x^3 - x^2 - x + 1$.

In Problems 12–19, factor the given polynomial as completely as possible.

12. $x^2 - 10x - 24$

13. $6a^2b^5 - 3ab^3$

14. $2t^2 + 5t - 12$

15. $6x^2 - 36x + 72$

16. $3x^3y - 12xy^3$

17. $a(a + 5) - 7(a + 5)$

18. $x^2 - 3x - xy + 3y$

19. $2u^4 - 32$

In Problems 20–21, reduce to lowest terms.

20. $\dfrac{3x^2 - 12x}{x^2 - x - 12}$

21. $\dfrac{t^2 - t - 6}{t^2 + t - 6}$

In Problems 22–26, perform the indicated operations and simplify as completely as possible.

22. $\dfrac{5}{x + 5} - \dfrac{4}{x + 4}$

23. $\dfrac{w^2 - 3w - 10}{4w^2 + 8w} \cdot \dfrac{w^2}{w^2 - 10w + 25}$

24. $\dfrac{6}{a^2 + 3a} - \dfrac{3}{a^2 - 3a}$

25. $\dfrac{u^2 - 9u}{u^2} \div (u^2 - 81)$

26. $\dfrac{2 + \dfrac{1}{x}}{4 - \dfrac{1}{x^2}}$

In Problems 27–30, solve the given equation.

27. $\dfrac{9}{4t - 12} - \dfrac{2}{3} = \dfrac{11}{12t - 36}$

28. *Solve for u.* $\dfrac{2}{5}u - 4x = au - x + 7$

29. $\dfrac{10}{x + 4} + \dfrac{3}{5} = \dfrac{6 - x}{x + 4}$

30. $(x + 8)(x - 3) = 2x - 14$

31. Roni drives 140 km in the same time that Lamar drives 160 km. If Roni's speed is 5 kph slower than Lamar's speed, find Roni's driving time.

32. The 228 students in a psychology class are divided into two parts so that the ratio of the larger group to the smaller group is 12 to 7. How many students are in each part?

CHAPTER 10

Radical Expressions and Equations

10.1   Definitions and Basic Notation

In Section 1.6 we introduced the idea of an *irrational* number. Recall that an irrational number is a number that is associated with a point on the real number line, but that cannot be expressed as the quotient of two integers. We also mentioned that the decimal representation of an irrational number is a nonterminating, nonrepeating decimal. As a result of this fact, if we want to discuss the exact value of an irrational number, we can never write down the answer precisely as a decimal.

For example, we indicated in Section 1.6 (without proof) that a number whose square is equal to 2 (called a square root of 2) is an irrational number. Thus, its decimal representation is an infinite nonrepeating decimal. This does not mean that every square root is an irrational number. If we ask for a square root of 100, we can give an exact answer of 10, because $10^2 = 100$. Certainly 10 is not an irrational number!

As we progress through this section, we will elaborate much more on the similarities and differences between square roots that are irrational numbers and square roots that are rational numbers.

Nevertheless, there are times when we want to talk about the "exact value" of an irrational number, such as a square root of 2. In order to do that we must have a symbolic way of referring to it. Let us begin with a definition.

DEFINITION	If $a^2 = b$, then a is called a ***square root of b.***

For example:

$$3^2 = 9 \quad \text{so 3 is a square root of 9.}$$
$$(-3)^2 = 9 \quad \text{so } -3 \text{ is also a square root of 9.}$$

Thus, we see that 9 has two square roots—one positive, the other negative.

Similarly, 25 has two square roots, 5 and −5. In general, every positive number has two square roots. However, a negative number does *not* have any square roots in the real number system. For example, the number −16 has no square roots because it is impossible to find a real number that when squared gives *negative* 16. When you square any real number, whether it is positive, negative, or 0, the result cannot be negative.

Remember	Square roots of negative numbers are not defined in the real number system.

Numbers such as 9 and 25, whose square roots are integers, are called ***perfect squares.*** If, however, b is an integer that is not a perfect square, such as 7, then its square roots are irrational numbers.

Since a number such as a square root of 7 is an irrational number, we need a symbol with which we can denote its exact value. However, whatever symbol we choose, we must make sure that it distinguishes between the positive and negative square root. We certainly cannot allow the same symbol to stand for two different numbers. Therefore, we give the following definition.

DEFINITION *The symbol $\sqrt{\ }$ is called a* **radical sign.**	For any nonnegative real number b, $\sqrt{b}$ represents the *nonnegative* quantity that when squared gives b. ($\sqrt{b}$ is called the *principal square root of b.*) $\qquad\qquad \sqrt{b} \quad$ is read $\quad$ *the square root of b.* This is called the ***radical notation*** for a square root.

For instance: $\sqrt{25} = 5$ *because* $5 \cdot 5 = 25$.

Other examples:

Keep in mind that we are working within the framework of the real number system. In the real number system $\sqrt{-4}$ does not exist.

$\sqrt{64} = 8$ *because* $8 \cdot 8 = 64$

$-\sqrt{4} = -2$ The minus sign is attached after you compute $\sqrt{4}$.

$\sqrt{-4}$ does not exist.   There is no real number that when squared is equal to -4.

$\sqrt{0} = 0$ *because* $0 \cdot 0 = 0$

Note that $\sqrt{25} \neq -5$, even though $(-5) \cdot (-5) = 25$. Part of the definition of the expression $\sqrt{25}$ is that it be nonnegative.

Remember	The symbol $\sqrt{b}$ stands for the *nonnegative* square root of b.

$\sqrt{25}$ is read as "*the* square root of 25."

$-\sqrt{25}$ is read as "the *negative* square root of 25."

Also note that 0, which is neither positive nor negative, has only one square root, which is 0.

Keep in mind that the square root is defined in terms of its inverse operation—squaring—just as subtraction and division are defined in terms of their inverse operations—addition and multiplication. Although you learned addition and multiplication tables, you never learned a subtraction or a division table. When you subtracted 3 from 8, you thought to yourself, "What do I *add* to 3 to get 8?" When you divided 28 by 4, you thought, "What do I *multiply* by 4 to get 28?"

To find square roots you have to think in terms of squares. To find $\sqrt{144}$ you think, "What number squared is equal to 144?" We know the answer is 12 because $12 \cdot 12 = 144$. This is often a trial-and-error process.

As we continue with our discussion of square roots, keep the following in mind: Consider the two numbers $\sqrt{4}$ and $\sqrt{7}$. Even though they are numerically different, *conceptually* $\sqrt{4}$ and $\sqrt{7}$ are the same: $\sqrt{4}$ asks for a positive number whose square is 4, while $\sqrt{7}$ asks for a positive number whose square is 7. The only difference is that for $\sqrt{4}$ we get a nice neat numerical answer, while for $\sqrt{7}$ we do not.

On occasion, however, we may need to do computations involving square roots that are irrational numbers. In such situations you may see statements such as "$\sqrt{7} = 2.65$" or "$\sqrt{7} = 2.646$." These are only approximate values (remember that $\sqrt{7}$ is an irrational number) and should really be written as $\sqrt{7} \approx 2.65$ or $\sqrt{7} \approx 2.646$, where the symbol "$\approx$" means approximately equal.

When we write $\sqrt{7} = 2.65$, we mean that 2.65 is an approximate value for $\sqrt{7}$ that is correct to one decimal place. Alternatively, we may say that 2.65 is $\sqrt{7}$ rounded off to the nearest hundredth. In other words, in using 2.65 as an approximate value for $\sqrt{7}$, the digits 2 and 6 are accurate; the 5 is obtained by rounding off.

Similarly, if we write $\sqrt{7} = 2.646$, we mean that 2.646 is an approximate value for $\sqrt{7}$ that is correct to two decimal places. That is, it is rounded off to the nearest thousandth, so that the digits 2, 6, and 4 are accurate; the 6 is obtained by rounding off. (The last "rounded" digit might, in fact, be accurate, but usually we do not know.)

If we use a calculator to obtain a value for $\sqrt{7}$, we would typically see the value 2.6457513, which is correct to seven decimal places.

Let's now return to our discussion of radical notation.

EXAMPLE 1   Evaluate each of the following.

(a) $\sqrt{49}$ (b) $-\sqrt{49}$ (c) $\sqrt{-49}$

Solution   (a) $\sqrt{49} = \boxed{7}$ because $7 \cdot 7 = 49$.

(b) $-\sqrt{49} = \boxed{-7}$ First we evaluate $\sqrt{49}$, and then we place the minus sign in front.

(c) $\sqrt{-49}$ does not exist in the real number system. There is *no* real number that when squared is equal to -49. ∎

EXAMPLE 2

Evaluate. $(\sqrt{5})^2$

Solution Even though we do not know the exact value of $\sqrt{5}$, we do know what it means: $\sqrt{5}$ means the nonnegative number whose square is equal to 5.

Therefore, we have $(\sqrt{5})^2 = \boxed{5}$ by the definition of the square root. ∎

We will use the following basic fact over and over again.

> For all nonnegative numbers a,
> $$\sqrt{a} \cdot \sqrt{a} = a \quad \text{or} \quad (\sqrt{a})^2 = a$$

This is simply the definition of the square root written symbolically.

EXAMPLE 3

Evaluate. (a) $(\sqrt{7})^4$ (b) $(\sqrt{6})^5$

Solution (a) Again you might ask how we can raise $\sqrt{7}$ to the 4th power when we do not know the exact value of $\sqrt{7}$. The point is that even though we do not know its exact value, we do know exactly the kind of number it is. We know that $\sqrt{7}$ is a number that when squared is equal to 7. That is, $\sqrt{7} \cdot \sqrt{7} = 7$. Therefore,

$$(\sqrt{7})^4 = \underbrace{\sqrt{7} \cdot \sqrt{7}} \cdot \underbrace{\sqrt{7} \cdot \sqrt{7}}$$
$$= \quad 7 \quad \cdot \quad 7$$
$$= \boxed{49}$$

(b) We proceed in a similar fashion.

$$(\sqrt{6})^5 = \underbrace{\sqrt{6} \cdot \sqrt{6}} \cdot \underbrace{\sqrt{6} \cdot \sqrt{6}} \cdot \sqrt{6}$$
$$= \quad 6 \quad \cdot \quad 6 \quad \cdot \sqrt{6}$$
$$= \boxed{36\sqrt{6}}$$ ∎

EXAMPLE 4

Evaluate. (a) $\sqrt{36 + 64}$ (b) $\sqrt{36}\,\sqrt{64}$

Solution (a) $\sqrt{36 + 64} = \sqrt{100} = \boxed{10}$

Note that the answer is *not* obtained by computing the square root of each number under the radical sign separately! $\sqrt{36 + 64} \neq 6 + 8 = 14$!

(b) As with other algebraic expressions, when we write two radicals next to each other, it automatically means multiplication.

$$\sqrt{36}\,\sqrt{64} = 6 \cdot 8 = \boxed{48}$$ ∎

EXAMPLE 5 Estimate $\sqrt{41}$ between two consecutive integers without a calculator, then use a calculator to find $\sqrt{41}$ rounded to two decimal places to confirm the estimate.

Solution We recognize that since 41 is not a perfect square, $\sqrt{41}$ will be an irrational number. We know

$$6^2 = 36 \qquad \text{and} \qquad 7^2 = 49 \qquad \textit{Since 41 is between 36 and 49, we know}$$
$$6 < \quad \sqrt{41} \quad < 7$$

Therefore, we know that $\sqrt{41}$ is between 6 and 7.

Using a calculator, we find $\sqrt{41}$ with the following keystrokes:

the calculator will display $\boxed{6.403124}$, which gives 6.40 rounded to two decimal places. This confirms our estimate that $\sqrt{41}$ lies between 6 and 7. ■

 **EXERCISES 10.1**

Evaluate or simplify each of the following. Assume that all variables represent nonnegative quantities.

1. $\sqrt{4}$

2. $\sqrt{9}$

3. $-\sqrt{4}$

4. $-\sqrt{9}$

5. $\sqrt{-4}$

6. $\sqrt{-9}$

7. $\sqrt{25}$

8. $\sqrt{49}$

9. $-\sqrt{100}$

10. $-\sqrt{36}$

11. $\sqrt{64}$

12. $\sqrt{81}$

13. $\sqrt{121}$

14. $\sqrt{144}$

15. $\sqrt{169}$

16. $\sqrt{196}$

17. $\sqrt{225}$

18. $\sqrt{256}$

19. $\sqrt{289}$

20. $\sqrt{324}$

21. $\sqrt{361}$

22. $\sqrt{400}$

23. $\sqrt{3}\,\sqrt{3}$

24. $\sqrt{10}\,\sqrt{10}$

25. $\sqrt{29}\,\sqrt{29}$

26. $\sqrt{67}\,\sqrt{67}$

27. $(\sqrt{11})^2$

28. $(\sqrt{5})^2$

29. $(\sqrt{33})^2$

30. $(\sqrt{41})^2$

31. $\sqrt{x}\,\sqrt{x}$

32. $(\sqrt{x})^2$

33. $(\sqrt{7})^4$

34. $(\sqrt{3})^6$

35. $(\sqrt{7})^5$

36. $(\sqrt{3})^7$

37. $\sqrt{25-9}$

38. $\sqrt{100}-\sqrt{36}$

39. $\sqrt{25}-\sqrt{9}$

40. $\sqrt{100-36}$

41. $(\sqrt{25}-\sqrt{9})^2$

42. $(\sqrt{100-36})^2$

43. $(\sqrt{25-9})^2$

44. $(\sqrt{100}-\sqrt{36})^2$

In Exercises 45–54, use a calculator to find the following square roots. Round off your answers to the nearest hundredth.

45. $\sqrt{425}$

46. $\sqrt{983}$

47. $\sqrt{637}$

48. $\sqrt{730}$

49. $\sqrt{73.6}$

50. $\sqrt{58.9}$

51. $\sqrt{2.09}$

52. $\sqrt{7.85}$

53. $\sqrt{0.037}$

54. $\sqrt{0.00049}$

In Exercises 55–58, use a calculator to give decimal approximations rounded to one, two, and three places.

55. $\sqrt{17}$

56. $\sqrt{23}$

57. $\sqrt{110}$

58. $\sqrt{260}$

In Exercises 59–66, estimate the given square root between two consecutive integers without using a calculator, then use a calculator to find the square root rounded to two decimal places to confirm your estimate.

59. $\sqrt{73}$

60. $\sqrt{115}$

61. $\sqrt{217}$

62. $\sqrt{59}$

63. $\sqrt{673}$

64. $\sqrt{318}$

65. $\sqrt{285}$

66. $\sqrt{872}$

QUESTIONS FOR THOUGHT

67. Determining whether a number is a perfect square is not as hard as it might at first seem. For instance, suppose we wanted to know whether the number 648 is a perfect square. Do you see why the square root of 648 must be between 20 and 30? Can you think of an easy way to check if 648 can possibly be a perfect square? [*Hint:* Think about the possible final digit a perfect square can have.] Is 648 a perfect square?

68. Use the result of Exercise 67 to check if 841 can possibly be a perfect square. Is it?

69. Discuss the correctness of the following steps:

$$2 \overset{?}{=} \sqrt{1 + 1}$$
$$\overset{?}{=} \sqrt{1} + \sqrt{1}$$
$$\overset{?}{=} 1 + 1$$
$$\overset{?}{=} 2$$

70. Describe the following statement in words:

$$\sqrt{a}\,\sqrt{a} = a$$

MINI-REVIEW

In Exercises 71–74, perform the indicated operations and simplify.

71. $\dfrac{x^2 - 6x}{x^2} \cdot \dfrac{2x}{x^2 - 12x + 36}$

72. $\dfrac{y^2 + 2y}{y^2 - 4} \div \dfrac{y^2}{y^2 - 6y + 8}$

73. $\dfrac{3}{2x} + \dfrac{5}{x+2}$

74. *Solve for x.* $\dfrac{6}{x-6} = \dfrac{2}{x-2}$

Solve the following problems algebraically. Be sure to label what the variable represents.

75. A hardware store sells 40 lightbulbs. Some were regular bulbs selling for 75¢ each, and the rest were long-life bulbs selling for 89¢ each. If $31.68 was collected for the bulbs, how many of each type were sold?

76. Xavier made three investments at 6.5%, 7.6%, and 9.2%. The amount invested at 7.6% is $1,000 less than the amount invested at 9.2%, and the amount invested at 6.5% is twice the amount invested at 7.6%. If the annual income from the three investments is $837, how much is invested altogether?

10.2 Properties of Radicals and Simplest Radical Form

Just as we did with all the different types of algebraic expressions we have studied up to now, such as polynomials and rational expressions, we will discuss how to add, subtract, multiply, and divide radical expressions.

It should come as no surprise that we will want our radical expressions to be in "simplified" form. Consequently, we need to describe what it means for a radical expression to be in its simplest form.

Properties of Radicals

Let's look at a few numerical examples to help us recognize some of the properties of radicals that we will use in our simplifying process.

$$\sqrt{4 \cdot 100} = \sqrt{400} = 20 \quad \text{and} \quad \sqrt{4}\,\sqrt{100} = 2 \cdot 10 = 20$$

Therefore, we see that $\sqrt{4 \cdot 100} = \sqrt{4}\,\sqrt{100}$.

$$\sqrt{9 \cdot 16} = \sqrt{144} = 12 \quad \text{and} \quad \sqrt{9}\,\sqrt{16} = 3 \cdot 4 = 12$$

Therefore, we see that $\sqrt{9 \cdot 16} = \sqrt{9}\,\sqrt{16}$.

Similarly,

$$\frac{\sqrt{100}}{\sqrt{4}} = \frac{10}{2} = 5 \quad \text{and} \quad \sqrt{\frac{100}{4}} = \sqrt{25} = 5$$

Therefore, we see that $\dfrac{\sqrt{100}}{\sqrt{4}} = \sqrt{\dfrac{100}{4}}$

$$\frac{\sqrt{9}}{\sqrt{16}} = \frac{3}{4} \quad \text{and} \quad \sqrt{\frac{9}{16}} = \frac{3}{4} \quad \text{because} \quad \frac{3}{4} \cdot \frac{3}{4} = \frac{9}{16}$$

Therefore, we see that $\dfrac{\sqrt{9}}{\sqrt{16}} = \sqrt{\dfrac{9}{16}}$.

These examples suggest the following properties of radicals:

Properties of Radicals	For all nonnegative numbers a and b,
Property 1	$\sqrt{ab} = \sqrt{a}\,\sqrt{b}$
Property 2	$\sqrt{\dfrac{a}{b}} = \dfrac{\sqrt{a}}{\sqrt{b}}, \quad b \neq 0$

In words, these properties say that the square root of a *product* (*quotient*) is the *product* (*quotient*) of the square roots. But these properties say nothing about the square root of a *sum* or *difference*.

Keep in mind that we have not proved these properties. We have merely observed some numerical evidence that suggests they are true. Let's prove property 1.

We want to show that $\sqrt{a}\sqrt{b} = \sqrt{ab}$. We know $\sqrt{ab}$ means the nonnegative quantity that when squared is equal to ab. Let's see what happens when we square $\sqrt{a}\sqrt{b}$.

$$(\sqrt{a}\sqrt{b})^2 = (\sqrt{a}\sqrt{b})(\sqrt{a}\sqrt{b}) \qquad \textit{Reorder and regroup the factors.}$$
$$= \sqrt{a}\sqrt{a}\sqrt{b}\sqrt{b}$$
$$= \quad a \quad \cdot \quad b$$
$$= ab$$

Thus, we see that when we square $\sqrt{a}\sqrt{b}$ we get ab. This is precisely the definition of ab, and we conclude that $\sqrt{ab} = \sqrt{a}\sqrt{b}$. Property 2 is proven in an analogous way.

EXAMPLE 1   Evaluate. $\sqrt{9 \cdot 16} + \sqrt{9 + 16}$

Solution   *Be careful!* The misuse of the properties of radicals is a very common error.

$$\sqrt{9 \cdot 16} = \sqrt{9}\sqrt{16} \qquad \textit{By property 1}$$

$$\textbf{\textit{but}}$$

$$\sqrt{9 + 16} \neq \sqrt{9} + \sqrt{16} \quad \textbf{\textit{because}} \quad \sqrt{25} \neq 3 + 4$$

Returning to our example, we have

$$\sqrt{9 \cdot 16} + \sqrt{9 + 16} = \sqrt{9}\sqrt{16} + \sqrt{25}$$
$$= (3)(4) + 5$$
$$= 12 + 5$$
$$= \boxed{17} \qquad\qquad \blacksquare$$

Remember | The square root of a *sum* is not equal to the sum of the square roots.

$$\sqrt{a + b} \neq \sqrt{a} + \sqrt{b} \qquad (a, \, b > 0)$$

Given an example such as $\sqrt{3}\sqrt{3}$, students often use property 1 mentally to think

$$\sqrt{3}\sqrt{3} = \sqrt{3 \cdot 3} = \sqrt{9} = 3$$

While this is, of course, perfectly correct, it is severely missing the point.

$\sqrt{3}\sqrt{3} = 3$ because that is what $\sqrt{3}$ *means*—$\sqrt{3}$ is the positive number that when multiplied by itself gives 3.

To emphasize this point, think about how you would compute

$$\sqrt{4{,}583}\,\sqrt{4{,}583}$$

Hopefully, you do not want to do the following:

$$\sqrt{4{,}583}\,\sqrt{4{,}583} = \sqrt{4{,}583 \cdot 4{,}583} = \sqrt{21{,}003{,}889} = 4{.}583$$

Understanding what the radical sign *means* allows us to do computations such as $\sqrt{4{,}583}\,\sqrt{4{,}583} = 4{,}583$. It is just a specific example of the basic fact about square roots, that for all nonnegative numbers a,

$$\boxed{\sqrt{a}\sqrt{a} = a}$$

We need to clarify one last detail before we move on to define what we mean by the simplest radical form.

There is a natural tendency to accept the statement

$$\sqrt{a^2} = a \quad \text{because} \quad a \cdot a = a^2$$

However, this is not quite true, as the following example illustrates.

$$\sqrt{4^2} = \sqrt{16} = 4$$

$$\textit{but} \quad \sqrt{(-4)^2} = \sqrt{16} = 4 \neq -4$$

So we see that $\sqrt{a^2}$ is not equal to a when a is negative.

In order to avoid this complication, we will agree from here on to *assume that all variables appearing under square root signs are nonnegative,* unless otherwise specified.

Up to this point we have used the phrase *algebraic expression* without having formally defined it. Now that we have defined radicals, we can give the following definition.

| **DEFINITION** | An *algebraic expression* is an expression obtained by adding, subtracting, multiplying, dividing, and taking radicals of constants and/or variables. |

Some examples of algebraic expressions that we have had, or soon will have, are the following:

$$-17 \qquad 3x^2 - 5x + 3 \qquad \frac{4x - 3}{x^2 - 5} \qquad \sqrt{2x - 7} \qquad \frac{3\sqrt{x} - 4}{\sqrt{x} + 9}$$

Simplest Radical Form

As with the other types of expressions we have worked with, we want algebraic expressions involving radicals to be in simplest form. Consequently, we must define what we mean by simplest radical form.

| **DEFINITION** | An expression is said to be in *simplest radical form* if it satisfies the following three conditions:

 1. The expression under the radical sign does not contain any perfect square factors. In other words, we want the expression under the radical sign to be as "small" as possible. As we shall soon see, expressions such as $\sqrt{12}$ and $\sqrt{a^7}$ violate this condition.

 2. There are no fractions under the radical sign. For example, $\sqrt{\dfrac{3}{5}}$ violates this condition.

 3. There are no radicals in denominators. For example, $\dfrac{2}{\sqrt{3}}$ violates this condition. |

Radical expressions that satisfy these three conditions are generally easier to work with.

Let's see how the two properties of radicals help us transform radical expressions into simplest radical form.

EXAMPLE 2   Express in simplest radical form.

(a) $\sqrt{12}$ (b) $\sqrt{72}$

Solution   (a) If the expression appearing under the radical sign (which is called the *radicand*) has a factor that is a perfect square, we can use property 1 of radicals to simplify it.

$$\sqrt{12} = \sqrt{4 \cdot 3} = \sqrt{4}\sqrt{3} = \boxed{2\sqrt{3}}$$

Note that it would not have helped us to write $2 = 6 \cdot 2$, because neither 6 nor 2 is a perfect square.

(b) We will work out the solution via two slightly different paths.

First path: *Second path:*

$\sqrt{72} = \sqrt{9 \cdot 8}$ $\leftarrow$ *Use radical property 1.* $\rightarrow$ $\sqrt{72} = \sqrt{36 \cdot 2}$

$\qquad = \sqrt{9}\,\sqrt{8}$ $\qquad\qquad\qquad\qquad\qquad = \sqrt{36}\,\sqrt{2}$

$\qquad = 3\sqrt{8}$ $\qquad\qquad\qquad\qquad\qquad\qquad = \boxed{6\sqrt{2}}$

$\qquad = 3\sqrt{4 \cdot 2}$ $\leftarrow$ *Use radical property 1 again.*

$\qquad = 3\sqrt{4}\,\sqrt{2}$

$\qquad = 3(2)\sqrt{2}$

$\qquad = \boxed{6\sqrt{2}}$

Clearly, it is much more efficient to try to find the *largest* perfect square you can at the beginning. However, even if you do not, keep applying radical property 1 until the quantity under the radical sign has no perfect square factors remaining. ■

As we pointed out previously, in order to avoid complications, we will assume that all variables appearing under radical signs are nonnegative.

EXAMPLE 3

Simplify.

(a) $\sqrt{x^2}$ **(b)** $\sqrt{x^4}$ **(c)** $\sqrt{x^6}$ **(d)** $\sqrt{x^{16}}$

Solution

(a) $\sqrt{x^2} = \boxed{x}$ *Because $x \cdot x = x^2$*

(b) $\sqrt{x^4} = \boxed{x^2}$ *Because $x^2 \cdot x^2 = x^4$*

(c) $\sqrt{x^6} = \boxed{x^3}$ *Because $x^3 \cdot x^3 = x^6$*

(d) $\sqrt{x^{16}} = \boxed{x^8}$ *Because $x^8 \cdot x^8 = x^{16}$*

Watch out! We are looking for $\sqrt{x^{16}}$, not $\sqrt{16}$. ■

In Example 3, we notice that in each case finding the square root of a power involves looking for two equal numbers that *add* up to the power. We can always do this when the power is *even*.

$$\boxed{\text{Even powers are always perfect squares.}}$$

Note that when we square an expression in exponential form we end up multiplying the exponent by 2 according to exponent rule 2.

$$(x^5)^2 = x^5 \cdot x^5 = x^{2 \cdot 5} = x^{10}$$
$$(x^n)^2 = x^n \cdot x^n = x^{2n}$$

Thus, finding the square root of a power requires dividing the exponent by 2, as we just saw in Example 3. This is fine for even powers, but what about odd powers?

EXAMPLE 4

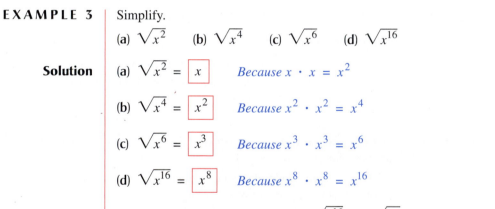

Simplify as completely as possible.

(a) $\sqrt{x^7}$ **(b)** $\sqrt{18x^6}$ **(c)** $\sqrt{20x^7 y^{10}}$

Solution

(a) Since 7 is not even, x^7 is *not* a perfect square. In order to satisfy condition 1 for the simplest radical form, we factor out the biggest perfect square (meaning even power) in x^7, which is x^6. If we factor out only x^2 or x^4, which are also perfect squares, we will not have simplified *completely,* since the expression under the radical sign will still have a perfect square factor.

$$\sqrt{x^7} = \sqrt{x^6 \cdot x} \qquad \textit{This step can be thought of as a mental step.}$$
$$= \sqrt{x^6}\sqrt{x} \qquad \textit{We use radical property 1.}$$
$$= \boxed{x^3\sqrt{x}}$$

(b) Let's visualize the radical in two factors. The first part consists of the perfect square factors, and the second part consists of whatever factors remain.

$$\sqrt{18x^6} = \sqrt{9x^6 \cdot 2}$$
$$= \underbrace{\sqrt{9x^6}}_{\substack{Perfect\ square \\ factors}} \cdot \underbrace{\sqrt{2}}_{\substack{Remaining \\ factors}}$$

Now use radical property 1.

$$= \sqrt{9}\sqrt{x^6}\sqrt{2}$$
$$= \boxed{3x^3\sqrt{2}}$$

(c) Let's show how we factor the radical in steps.

$$\sqrt{20x^7y^{10}} = \sqrt{}\ \sqrt{}$$

We want two radicals: the first for the perfect square factors, the second for what is left. First we factor 20.

$$\sqrt{4}\ \sqrt{5} \qquad \textit{Next we factor } x^7.$$
$$\sqrt{4x^6}\ \sqrt{5x} \qquad \textit{The } y^{10} \textit{ is a perfect square, so it goes in the first radical.}$$

$$= \sqrt{4x^6y^{10}}\ \sqrt{5x} \quad \leftarrow \quad \textit{This radical contains the remaining factors.}$$
$$\uparrow$$
This radical contains perfect square factors.

$$= \sqrt{4}\sqrt{x^6}\sqrt{y^{10}}\sqrt{5x}$$
$$= \boxed{2x^3y^5\sqrt{5x}}$$

EXAMPLE 5

Simplify as completely as possible.

(a) $\sqrt{x^2y^2}$ (b) $\sqrt{x^2 + y^2}$

Solution

(a) Using radical property 1, we get:

$$\sqrt{x^2y^2} = \sqrt{x^2}\sqrt{y^2} = \boxed{xy}$$

(b) $\sqrt{x^2 + y^2}$ *cannot be simplified!*

$$\sqrt{x^2 + y^2} \neq x + y \quad \text{because} \quad (x + y)^2 = (x + y)(x + y) = x^2 + 2xy + y^2$$

Watch out! This is a very common error.

EXAMPLE 6

Express in simplest radical form.

(a) $\sqrt{\dfrac{3}{4}}$ (b) $\sqrt{\dfrac{3}{5}}$ (c) $\dfrac{10}{\sqrt{6}}$

Solution

(a) This is not in simplified form because the fraction inside the radical violates condition 2 of the simplest radical form. Using radical property 2, we get:

$$\sqrt{\frac{3}{4}} = \frac{\sqrt{3}}{\sqrt{4}} = \boxed{\frac{\sqrt{3}}{2}} \qquad \textit{This is now in simplest radical form.}$$

(b) This also violates condition 2. Again we use radical property 2.

$$\sqrt{\frac{3}{5}} = \frac{\sqrt{3}}{\sqrt{5}}$$

However, the radical in the denominator still violates condition 3 of simplest radical form. In order to get rid of the radical sign in the denominator, we apply the Fundamental Principle of Fractions in a clever way.

$$\sqrt{\frac{3}{5}} = \frac{\sqrt{3}}{\sqrt{5}} = \frac{\sqrt{3}}{\sqrt{5}} \cdot \frac{\sqrt{5}}{\sqrt{5}} \qquad \textit{We want to multiply the denominator by } \sqrt{5} \textit{ to get rid of the radical } (\sqrt{5}\sqrt{5} = 5), \textit{ but then we must also multiply the numerator by } \sqrt{5}.$$

$$= \boxed{\frac{\sqrt{15}}{5}} \qquad \sqrt{3}\,\sqrt{5} = \sqrt{3 \cdot 5} = \sqrt{15}$$

This process is called **rationalizing the denominator**, and $\dfrac{\sqrt{5}}{\sqrt{5}}$ is called the **rationalizing factor.**

(c) Condition 3 for simplest radical form is violated because of the radical in the denominator. We rationalize the denominator using the rationalizing factor of $\dfrac{\sqrt{6}}{\sqrt{6}}$.

$$\frac{10}{\sqrt{6}} = \frac{10}{\sqrt{6}} \cdot \frac{\sqrt{6}}{\sqrt{6}}$$

$$= \frac{10\sqrt{6}}{6} \qquad \textit{Because } \sqrt{6} \cdot \sqrt{6} = 6; \textit{ now reduce } \dfrac{10}{6}.$$

$$= \frac{\overset{5}{\cancel{10}}\sqrt{6}}{\underset{3}{\cancel{6}}} \qquad \textit{We can reduce the 10 with the 6 because they are \textbf{factors} of the numerator and denominator, respectively.}$$

$$= \boxed{\frac{5\sqrt{6}}{3}} \qquad\qquad\qquad\qquad\qquad\qquad ■$$

EXAMPLE 7

Simplify as completely as possible. $\dfrac{21}{2\sqrt{3}}$

Solution

Condition 3 of simplest radical form is again violated.

What is the rationalizing factor? You may use $\dfrac{2\sqrt{3}}{2\sqrt{3}}$ (try it!), but that is really not necessary since the 2 in the denominator is not bothering us. It is the $\sqrt{3}$ we want to rationalize. The simplest rationalizing factor is $\dfrac{\sqrt{3}}{\sqrt{3}}$.

$$\frac{21}{2\sqrt{3}} = \frac{21}{2\sqrt{3}} \cdot \frac{\sqrt{3}}{\sqrt{3}}$$

$$= \frac{21\sqrt{3}}{2 \cdot 3} \qquad \textit{Because } \sqrt{3} \cdot \sqrt{3} = 3; \textit{ now reduce the 3 and the 21.}$$

$$= \boxed{\frac{7\sqrt{3}}{2}} \qquad\qquad\qquad\qquad\qquad\qquad ■$$

In the next section we will begin our discussion of how we perform the various arithmetic operations with radical expressions.

As we proceed through this chapter we will see a variety of real-life situations that can be described using radical expressions.

EXAMPLE 8

After 2 years, P dollars will grow to A dollars if it is invested at a simple annual interest rate r satisfying the formula

$$r = \sqrt{\frac{A}{P}} - 1$$

Use this formula to compute the interest rate required in order for $5,000 to grow to $6,000 in 2 years. Round the answer as a percent to two decimal places.

Solution

According to the given formula and the information in the example, we want to compute the value r when $A = 6,000$ and $P = 5,000$.

$$r = \sqrt{\frac{A}{P}} - 1 \qquad \textit{Substitute } A = 6,000 \textit{ and } P = 5,000.$$

$$r = \sqrt{\frac{6,000}{5,000}} - 1 \qquad \textit{Using a calculator, we compute this to be}$$

$$\boxed{r = 0.0954451 = 9.54\%}$$

Thus, in order for $5,000 to grow to $6,000 in 2 years, it must be invested at an annual interest rate of 9.54%. ∎

EXAMPLE 9

Police often use mathematics to determine the speed of a vehicle in an accident by measuring the length of the skid marks left by the car. Suppose that for a particular model car the police know that the speed s of the vehicle (in mph) is given by the formula

$$s = 8.3\sqrt{L}$$

where L is the length of the skid marks (in feet). A driver of such a car is involved in an accident in which the police find skid marks measuring 95 ft. The driver claims that he was driving at the legal speed limit of 60 mph. Is his claim plausible?

Solution

Using the given formula we can substitute $L = 95$ and compute the corresponding speed s.

$$s = 8.3\sqrt{L} \qquad \textit{Substitute } L = 95.$$

$$s = 8.3\sqrt{95} \qquad \textit{Using a calculator we compute this to be}$$

$$s = 80.89839 \qquad \textit{Which we round to approximately 81 mph}$$

Thus, according to the formula, 95-foot skid marks would mean that the car was traveling at approximately 81 mph. Assuming the validity of the given formula, we would have to conclude that the driver's claim of driving at 60 mph does not seem plausible. ∎

EXERCISES 10.2

Simplify each of the following expressions as completely as possible. Be sure your answers are in simplest radical form. Assume that all variables appearing under radical signs are nonnegative.

1. $\sqrt{64}$

2. $\sqrt{144}$

3. $\sqrt{18}$

4. $\sqrt{20}$

5. $\sqrt{32}$

6. $\sqrt{48}$

7. $\sqrt{50}$

8. $\sqrt{54}$

9. $\sqrt{400}$

10. $\sqrt{900}$

11. $\sqrt{x^6}$

12. $\sqrt{x^{10}}$

13. $\sqrt{x^7}$

14. $\sqrt{x^{11}}$

15. $\sqrt{16x^{16}}$

16. $\sqrt{36x^{36}}$

17. $\sqrt{9x^9}$

18. $\sqrt{25x^{25}}$

19. $\sqrt{40x^8}$

20. $\sqrt{24a^{12}}$

21. $\sqrt{25x^7}$

22. $\sqrt{16a^{11}}$

23. $\sqrt{12x^5}$

24. $\sqrt{20y^3}$

25. $\sqrt{a^2b^4}$

26. $\sqrt{a^2 + b^4}$

27. $\sqrt{x^6 + y^8}$

28. $\sqrt{x^6y^8}$

29. $\sqrt{49a^8b^{12}}$

30. $\sqrt{36m^6n^{18}}$

31. $\sqrt{28x^9y^6}$

32. $\sqrt{4x^{10}y^7}$

33. $\sqrt{50m^7n^{11}}$

34. $\sqrt{60r^{13}t^5}$

35. $\sqrt{81 + x^6 + y^{10}}$

36. $\sqrt{100 + a^8 + b^4}$

37. $\sqrt{48x^6y^8z^9}$

38. $\sqrt{54x^7y^{14}z^{14}}$

39. $\sqrt{\dfrac{4}{9}}$

40. $\sqrt{\dfrac{16}{49}}$

41. $\sqrt{\dfrac{7}{25}}$

42. $\sqrt{\dfrac{13}{64}}$

43. $\sqrt{\dfrac{5}{6}}$

44. $\sqrt{\dfrac{3}{10}}$

45. $\sqrt{\dfrac{1}{2}}$

46. $\sqrt{\dfrac{1}{3}}$

47. $\dfrac{1}{\sqrt{2}}$

48. $\dfrac{1}{\sqrt{3}}$

49. $\dfrac{18}{\sqrt{10}}$

50. $\dfrac{12}{\sqrt{14}}$

51. $\dfrac{3}{\sqrt{x}}$

52. $\dfrac{4}{\sqrt{y}}$

53. $\dfrac{15}{2\sqrt{7}}$

54. $\dfrac{11}{3\sqrt{5}}$

55. $\dfrac{12}{5\sqrt{6}}$

56. $\dfrac{24}{7\sqrt{3}}$

57. $\dfrac{8x}{\sqrt{2x}}$

58. $\dfrac{12y}{\sqrt{3y}}$

59. $\dfrac{5}{2\sqrt{x}}$

60. $\dfrac{1}{3\sqrt{y}}$

61. $\dfrac{x^2}{\sqrt{xy}}$

62. $\dfrac{a^3}{\sqrt{2a}}$

63. $\dfrac{\sqrt{8}}{\sqrt{6}}$

64. $\dfrac{\sqrt{12}}{\sqrt{18}}$

65. Use the formula given in Example 8 to find the rate r (rounded to two decimal places) that will make $8,000 grow to $10,000 in 2 years.

66. Use the formula given in Example 8 to find the rate r (rounded to two decimal places) that will make $4,000 grow to $5,000 in 2 years.

67. Use the formula given in Example 9 to find the speed s (rounded to the nearest mph) of the vehicle if it leaves a skid mark 50 ft long.

68. Use the formula given in Example 9 to find the speed s (rounded to the nearest mph) of the vehicle if it leaves a skid mark 100 ft long. Is your answer double the speed of Exercise 67? Should it be?

69. The time t (in seconds) it takes a free-falling object to fall h feet is given by the formula

$$t = \sqrt{\frac{h}{16}}$$

Use this formula to determine how long it would take an object to reach the ground if it is dropped from the top of the Empire State Building, which is 1,250 ft tall. Round your answer to the nearest tenth.

70. Use the formula given in Exercise 69 to determine how long it would take an object to reach the ground if it is dropped from the top of the Sears Tower, which is 1,450 ft tall. Round your answer to the nearest tenth.

71. At an altitude of m miles above the earth, the approximate distance d to the horizon in miles is given by the formula

$$d = \sqrt{8,000m}$$

Use this formula to find the distance to the horizon from a plane at an altitude of 6 miles. Round your answer to the nearest mile.

72. Use the formula in Exercise 71 to find the distance to the horizon from the top of Mt. McKinley, which is approximately 3.8 miles high.

73. In an electrical circuit the amount of current is measured in *amperes* (I), the amount of power in *watts* (W), and the resistance in *ohms* (R). The following formula relates these three quantities:

$$I = \sqrt{\frac{W}{R}}$$

Use this formula to compute the current in a circuit with 200 watts and a resistance of 70 ohms.

74. Use the formula of Exercise 73 to compute the current in a circuit with 145 watts and a resistance of 32 ohms.

? QUESTIONS FOR THOUGHT

75. Complete the following table.

x	y	$\sqrt{xy}$	$\sqrt{x}\sqrt{y}$	$\sqrt{x+y}$	$\sqrt{x}+\sqrt{y}$
0	1				
4	9				
9	16				
4	5				
9	7				

Based on your results, what can you say about the relationship between $\sqrt{xy}$ and $\sqrt{x}\sqrt{y}$, and the relationship between $\sqrt{x+y}$ and $\sqrt{x}+\sqrt{y}$?

76. A student suggested the following procedure for rationalizing the denominator:

   To rationalize the denominator in the fraction $\dfrac{2}{\sqrt{5}}$ "square the numerator and denominator" to get

   $$\frac{2}{\sqrt{5}} \stackrel{?}{=} \frac{2^2}{(\sqrt{5})^2} \stackrel{?}{=} \frac{4}{5}$$

   What is wrong with this "procedure"?

77. Historically, the conditions for simplest radical form were motivated by arithmetic considerations. Use a calculator to find $\sqrt{5}$ rounded to three decimal places, then compute $\dfrac{1}{\sqrt{5}}$ by hand.

   Now rationalize the denominator in $\dfrac{1}{\sqrt{5}}$ and then do the computation again. Which computation was easier? Why?

78. Repeat Exercise 77 for $\dfrac{\sqrt{3}}{\sqrt{7}}$.

◇ **MINI-REVIEW**

In Exercises 79–82, solve the given equation for the indicated variable.

79. $4x + 3y = 8$ for x

80. $5x + 2y = 9x - 3y$ for y

81. $4(a + 3b) - 2(a - 5b) = 6 - a + b$ for a

82. $\dfrac{u}{2} - \dfrac{v}{3} = u + v$ for u

Solve the following problems algebraically. Be sure to label what the variable represents.

83. A train leaves town A at 11:00 A.M. traveling at 110 kph, while a second train leaves town B at 1:00 P.M. traveling at 140 kph. If the two towns are 595 km apart and the two trains are traveling toward each other, at what time will they pass each other?

84. The U.S. House of Representatives has 435 members, all of whom voted on a certain bill. The ratio of those who voted in favor of the bill to those who voted against was 8 to 7. How many were in favor of the bill?

10.3 Adding and Subtracting Radical Expressions

It is very important to keep in mind that the procedures and principles we are going to use for combining radical expressions are the same as those we have used for other types of expressions—only the objects we are working with are different.

For example, we know that $5x + 3x = 8x$. We called this addition process *combining like terms*, and it is derived from the distributive property:

$$5x + 3x = (5 + 3)x = 8x$$
$$\text{Similarly,} \quad 5\sqrt{2} + 3\sqrt{2} = (5 + 3)\sqrt{2} = 8\sqrt{2}$$

On the other hand, neither $5x + 3y$ nor $5\sqrt{2} + 3\sqrt{7}$ can be combined since in both cases we are dealing with unlike terms.

In order not to misread $\sqrt{2}x$ as $\sqrt{2x}$, we usually write $\sqrt{2}x$ as $x\sqrt{2}$.

What about an expression such as $5x + \sqrt{2}x$? According to our definition of like terms, these are like terms since their variable parts are identical. However, to combine them we would have to write

$$5x + \sqrt{2}x = (5 + \sqrt{2})x$$

which is not of much help. Consequently, in such a case we will allow ourselves to leave the answer in the form $5x + \sqrt{2}x$.

EXAMPLE 1

Simplify as completely as possible.

(a) $5\sqrt{3} - 7\sqrt{3}$ (b) $4\sqrt{7} + 2\sqrt{7} - \sqrt{7}$

(c) $3\sqrt{2} - 4\sqrt{3} + 6\sqrt{2} + 2\sqrt{3} + \sqrt{5}$

Solution

(a) $5\sqrt{3} - 7\sqrt{3}$ *Think: "This is just like $5x - 7x$."*

$$= (5 - 7)\sqrt{3}$$

$$= -2\sqrt{3}$$

(b) $4\sqrt{7} + 2\sqrt{7} - \sqrt{7} = (4 + 2 - 1)\sqrt{7} = 5\sqrt{7}$

(c) $3\sqrt{2} - 4\sqrt{3} + 6\sqrt{2} + 2\sqrt{3} + \sqrt{5}$ *We look for and combine like terms by adding their coefficients.*

$$= (3 + 6)\sqrt{2} + (-4 + 2)\sqrt{3} + \sqrt{5}$$

$$= 9\sqrt{2} - 2\sqrt{3} + \sqrt{5}$$ ∎

EXAMPLE 2

Simplify.

(a) $\sqrt{3}\,\sqrt{3}$ (b) $\sqrt{3} + \sqrt{3}$

Solution

Read the example carefully. Do not confuse multiplication with addition.

(a) $\sqrt{3}\,\sqrt{3} = 3$ *By the definition of square root*

(b) $\sqrt{3} + \sqrt{3} = 2\sqrt{3}$ *This is just like $x + x = 2x$.* ∎

Often, like terms involving radicals are not immediately apparent.

EXAMPLE 3

Simplify.

(a) $\sqrt{75} - \sqrt{12}$ (b) $\sqrt{40} + \sqrt{60}$ (c) $\sqrt{16} + \sqrt{18}$

Solution

At first glance we do not see any like terms that can be combined. However, these expressions are *not* in simplest radical form. Let's see what happens if we first put each term into simplest radical form.

(a) Using radical property 1, we get:

$$\sqrt{75} - \sqrt{12} = \sqrt{25}\,\sqrt{3} - \sqrt{4}\,\sqrt{3}$$

$$= 5\sqrt{3} - 2\sqrt{3}$$ *Now combine like terms.*

$$= 3\sqrt{3}$$

Actually, we had like terms all along; they just were not obvious.

(b) Using radical property 1, we get:

$$\sqrt{40} + \sqrt{60} = \sqrt{4}\,\sqrt{10} + \sqrt{4}\,\sqrt{15}$$

$$= \boxed{2\sqrt{10} + 2\sqrt{15}}$$

There are no like terms, and so we cannot simplify any further.

(c) We evaluate $\sqrt{16}$ and simplify $\sqrt{18}$.

$$\sqrt{16} + \sqrt{18} = 4 + \sqrt{9}\,\sqrt{2}$$

$$= \boxed{4 + 3\sqrt{2}}$$

Note that $4 + 3\sqrt{2} \neq 7\sqrt{2}$; $4 + 3\sqrt{2}$ is just like $4 + 3x$. Since these are not like terms, we cannot combine them. ∎

The next few examples illustrate several more variations on the same theme.

EXAMPLE 4

Simplify as completely as possible. $3\sqrt{28y^3} + y\sqrt{63y}$

Solution

We begin by using radical property 1 to get each radical into simplest radical form.

$$3\sqrt{28y^3} + y\sqrt{63y} = 3\sqrt{4y^2}\,\sqrt{7y} + y\sqrt{9}\,\sqrt{7y}$$

$$= 3(2y)\sqrt{7y} + y(3)\sqrt{7y}$$

$$= 6y\sqrt{7y} + 3y\sqrt{7y} \qquad \textit{Combine like terms.}$$

$$= \boxed{9y\sqrt{7y}} \qquad\qquad\qquad ∎$$

EXAMPLE 5

Multiply and simplify. $3(\sqrt{2} - \sqrt{7}) - 5(2\sqrt{7} - \sqrt{2})$

Solution

We work out this example as if it were "$3(x - y) - 5(2y - x)$." We begin by multiplying out using the distributive property.

$$3(\sqrt{2} - \sqrt{7}) - 5(2\sqrt{7} - \sqrt{2}) = 3\sqrt{2} - 3\sqrt{7} - 5(2)\sqrt{7} - 5(-\sqrt{2})$$

$$= 3\sqrt{2} - 3\sqrt{7} - 10\sqrt{7} + 5\sqrt{2} \quad \textit{Combine like}$$
$$\textit{terms.}$$

$$= \boxed{8\sqrt{2} - 13\sqrt{7}} \qquad\qquad ∎$$

EXAMPLE 6

Combine. $\sqrt{12} + \dfrac{1}{\sqrt{3}}$

Solution

We first put each term into simplest radical form, which means something different for each term. In the first term we factor the $\sqrt{12}$, and in the second term we rationalize the denominator.

$$\sqrt{12} + \frac{1}{\sqrt{3}} = \sqrt{4}\,\sqrt{3} + \frac{1}{\sqrt{3}} \cdot \frac{\sqrt{3}}{\sqrt{3}}$$

$$= 2\sqrt{3} + \frac{\sqrt{3}}{3} \qquad\qquad \textit{Now we combine; the LCD is 3.}$$

$$= \frac{3 \cdot 2\sqrt{3}}{3} + \frac{\sqrt{3}}{3}$$

$$= \frac{6\sqrt{3} + \sqrt{3}}{3} \qquad\qquad \textit{Combine like terms.}$$

$$= \boxed{\frac{7\sqrt{3}}{3}} \qquad\qquad\qquad ∎$$

EXAMPLE 7   Combine. $\sqrt{\dfrac{2}{3}} + \sqrt{\dfrac{3}{2}}$

Solution   There are a number of reasonable ways to proceed, all of which will lead us to the correct answer. We offer the one we think is the most "natural." We begin by using radical property 2.

$$\sqrt{\frac{2}{3}} + \sqrt{\frac{3}{2}} = \frac{\sqrt{2}}{\sqrt{3}} + \frac{\sqrt{3}}{\sqrt{2}} \qquad \textit{Next we rationalize each denominator.}$$

$$= \frac{\sqrt{2}}{\sqrt{3}} \cdot \frac{\sqrt{3}}{\sqrt{3}} + \frac{\sqrt{3}}{\sqrt{2}} \cdot \frac{\sqrt{2}}{\sqrt{2}}$$

$$= \frac{\sqrt{6}}{3} + \frac{\sqrt{6}}{2} \qquad \textit{Add the fractions; the LCD is 6.}$$

$$= \frac{2\sqrt{6}}{2 \cdot 3} + \frac{3\sqrt{6}}{3 \cdot 2}$$

$$= \frac{2\sqrt{6}}{6} + \frac{3\sqrt{6}}{6} = \boxed{\frac{5\sqrt{6}}{6}} \qquad \blacksquare$$

Keep in mind that our basic approach to these problems has been to first put each term into simplest radical form before adding and/or subtracting.

EXERCISES 10.3

In each of the following, perform the indicated operations and simplify as completely as possible. Assume all variables appearing under radical signs are nonnegative.

1. $x + 2x + 3x$
2. $3a + a + 5a$
3. $\sqrt{5} + 2\sqrt{5} + 3\sqrt{5}$
4. $3\sqrt{7} + \sqrt{7} + 5\sqrt{7}$
5. $4x - x$
6. $7a - a$
7. $4\sqrt{6} - \sqrt{6}$
8. $7\sqrt{11} - \sqrt{11}$
9. $3x + 5y$
10. $6a + b$
11. $3\sqrt{5} + 5\sqrt{3}$
12. $6\sqrt{2} + \sqrt{5}$
13. $x \cdot x$
14. $a \cdot a$
15. $\sqrt{5} \cdot \sqrt{5}$
16. $\sqrt{13} \cdot \sqrt{13}$
17. $x + x$
18. $a + a$
19. $\sqrt{5} + \sqrt{5}$
20. $\sqrt{13} + \sqrt{13}$
21. $\sqrt{5} + 3\sqrt{7} - 4\sqrt{5} - 5\sqrt{7}$
22. $\sqrt{11} - 8\sqrt{2} - \sqrt{2} - 3\sqrt{11}$
23. $3\sqrt{3} + 4\sqrt{3} - \sqrt{2}$
24. $5\sqrt{5} + 2\sqrt{5} - 4\sqrt{3}$
25. $2(x - y) + 3(y - x)$
26. $4(a - b) + 5(b - a)$
27. $2(\sqrt{5} - \sqrt{3}) + 3(\sqrt{3} - \sqrt{5})$
28. $4(\sqrt{11} - \sqrt{7}) + 5(\sqrt{7} - \sqrt{11})$
29. $5(3\sqrt{6} + 2\sqrt{7}) - 4(\sqrt{7} - 2\sqrt{6})$
30. $3(2\sqrt{5} + 4\sqrt{10}) - 5(\sqrt{5} - 3\sqrt{10})$
31. $6(\sqrt{m} - \sqrt{n}) - (3\sqrt{m} + 6\sqrt{n})$
32. $2(\sqrt{x} - \sqrt{y}) - (4\sqrt{x} - 2\sqrt{y})$
33. $\sqrt{8} + \sqrt{18}$
34. $\sqrt{12} + \sqrt{27}$
35. $\sqrt{25} + \sqrt{24}$
36. $\sqrt{49} + \sqrt{50}$
37. $\sqrt{20} - \sqrt{5}$
38. $\sqrt{28} - \sqrt{7}$

39. $4\sqrt{12} - \sqrt{75}$

40. $6\sqrt{8} - \sqrt{98}$

41. $\sqrt{20} + \sqrt{40} + \sqrt{60}$

42. $\sqrt{18} + \sqrt{27} + \sqrt{45}$

43. $3\sqrt{72} - 5\sqrt{32}$

44. $4\sqrt{96} - 5\sqrt{24}$

45. $5\sqrt{36} + 4\sqrt{30}$

46. $8\sqrt{25} - 3\sqrt{21}$

47. $\sqrt{25x} + \sqrt{36x}$

48. $\sqrt{16m} - \sqrt{64m}$

49. $\sqrt{12w} + \sqrt{27w}$

50. $\sqrt{18z} + \sqrt{8z}$

51. $\sqrt{45x} - \sqrt{20x}$

52. $\sqrt{54x} - \sqrt{24x}$

53. $\sqrt{20y^3} - \sqrt{45y^3}$

54. $\sqrt{50t^7} - \sqrt{32t^7}$

55. $x\sqrt{28xy^3} + y\sqrt{63x^3y}$

56. $b^2\sqrt{40a^5b^2} + a\sqrt{90a^3b^6}$

57. $\dfrac{\sqrt{32x^3y^2}}{2xy} - \sqrt{8x}$

58. $\dfrac{\sqrt{48x^5y^4}}{4x^2y^2} - \sqrt{12x}$

59. $\sqrt{2} + \sqrt{\dfrac{1}{2}}$

60. $\sqrt{5} + \sqrt{\dfrac{1}{5}}$

61. $\sqrt{27} + \dfrac{4}{\sqrt{3}}$

62. $\sqrt{24} + \dfrac{5}{\sqrt{6}}$

63. $\dfrac{\sqrt{8}}{7} + \dfrac{7}{\sqrt{2}}$

64. $\dfrac{\sqrt{12}}{5} + \dfrac{5}{\sqrt{3}}$

65. $\sqrt{\dfrac{2}{7}} + \sqrt{\dfrac{7}{2}}$

66. $3\sqrt{\dfrac{x}{3}} + x\sqrt{\dfrac{3}{x}}$

Ⓠ QUESTIONS FOR THOUGHT

67. Use a calculator to find $\sqrt{80}$. Then simplify $\sqrt{80}$ and again use the calculator to compute the value of the simplified form. Are the results the same? Should they be?

68. Repeat Exercise 67 for $\sqrt{150}$.

 MINI-REVIEW

In Exercises 69–72, sketch the graph of each equation in a rectangular coordinate system. Label the intercepts.

69. $4x - 3y = 12$

70. $5y - 2x = 10$

71. $x = 2$

72. $y + 2 = 0$

Solve the following problems algebraically. Be sure to label what the variable represents.

73. A law firm charges $120 per hour for an associate's time and $40 per hour for a law clerk's time. On a certain case a client was billed $1,200 for a total of 14 hours. How many hours of each type were billed?

74. Lynette can run at the rate of 10 kph and bike at the rate of 24 kph. If she runs for one-half hour longer than she bikes and she covers a total of 56 km, how long was she biking?

10.4 Multiplying and Dividing Radical Expressions

Just as we did in the last section in the case of adding and subtracting radical expressions, we will multiply and divide them by applying the same basic procedures that we use for polynomials. We will constantly be using the properties of radicals, which we repeat in the box for easy reference.

	For all nonnegative numbers a and b,
Property 1	$\sqrt{ab} = \sqrt{a}\,\sqrt{b}$
Property 2	$\sqrt{\dfrac{a}{b}} = \dfrac{\sqrt{a}}{\sqrt{b}}$

EXAMPLE 1 Multiply. $\sqrt{2}\,\sqrt{3}\,\sqrt{5}$

Solution In the same way that we use radical property 1 to "break up" a radical (over multiplication!), so too we can use it to write the product of two or more radicals as a single radical.

$$\sqrt{2}\,\sqrt{3}\,\sqrt{5} = \sqrt{2 \cdot 3 \cdot 5} = \boxed{\sqrt{30}} \qquad \blacksquare$$

EXAMPLE 2 Multiply. $\sqrt{12} \cdot \sqrt{18}$

Solution We offer two approaches.

First approach:

$$
\begin{aligned}
\sqrt{12} \cdot \sqrt{18} &= \sqrt{12 \cdot 18} && \textit{By radical property 1}\\
&= \sqrt{216} && \textit{We look for the largest perfect square factor}\\
&= \sqrt{36}\,\sqrt{6} && \textit{of 216, which is 36.}\\
&= \boxed{6\sqrt{6}}
\end{aligned}
$$

Second approach: We first put each radical into simplest radical form.

$$
\begin{aligned}
\sqrt{12} \cdot \sqrt{18} &= \sqrt{4}\,\sqrt{3} \cdot \sqrt{9}\,\sqrt{2}\\
&= 2\sqrt{3} \cdot 3\sqrt{2} && \textit{Rearrange the factors.}\\
&= 2 \cdot 3 \,\sqrt{3}\,\sqrt{2}\\
&= \boxed{6\sqrt{6}} && \blacksquare
\end{aligned}
$$

Note that the second approach used in Example 2 kept the numbers much smaller. The arithmetic was easier when we simplified the radical first. Since this is usually the case, we will generally use the following outline for working with radicals.

Outline for Working with Radicals	1. Put each radical into simplest radical form.
	2. Perform any indicated operations, if possible.
	3. Make sure the final answer is also in simplest radical form.

EXAMPLE 3

Multiply.

(a) $x(2x + y)$ (b) $\sqrt{3}(2\sqrt{3} + \sqrt{5})$

Solution

At first glance, part (b) may look a bit difficult. However, it is built along exactly the same lines as part (a), which should be quite familiar by now.

(a) $x(2x + y) = \boxed{x \cdot 2x + x \cdot y} = \boxed{2xx + xy} = \boxed{2x^2 + xy}$

(b) $\sqrt{3}(2\sqrt{3} + \sqrt{5}) = \sqrt{3} \cdot 2\sqrt{3} + \sqrt{3}\sqrt{5} = 2\underbrace{\sqrt{3}\sqrt{3}}_{3} + \sqrt{3 \cdot 5}$

$$= 2 \cdot 3 + \sqrt{15} = \boxed{6 + \sqrt{15}}$$

While the first two steps in part (a) are surely mental steps by now, in part (b) they may not be. In fact, because the objects we are working with in part (b) are still somewhat new, it might be a good idea to write the "mental steps" for as long as you feel it is necessary. ∎

EXAMPLE 4

Multiply and simplify. $2\sqrt{x}(\sqrt{x} - 3) - 4(3 - 5\sqrt{x})$

Solution

We proceed as we would if there were no radicals—by using the distributive property to remove the parentheses.

$2\sqrt{x}(\sqrt{x} - 3) - 4(3 - 5\sqrt{x}) = 2\underbrace{\sqrt{x}\sqrt{x}}{} - 6\sqrt{x} - 12 + 20\sqrt{x}$

$$= 2x - 6\sqrt{x} - 12 + 20\sqrt{x} \qquad \textit{Combine like terms.}$$

$$= \boxed{2x + 14\sqrt{x} - 12}$$ ∎

EXAMPLE 5

Multiply and simplify. $(\sqrt{x} - 3)(\sqrt{x} + 5)$

Solution

We handle this example just as we would $(a - 3)(a + 5)$. Each term in the first set of parentheses multiplies each term in the second set. In the case of two binomials, this process was called the FOIL method.

$(\sqrt{x} - 3)(\sqrt{x} + 5) = \underbrace{\sqrt{x}\sqrt{x}}{} + 5\sqrt{x} - 3\sqrt{x} - 15$

$$= x + 5\sqrt{x} - 3\sqrt{x} - 15 \qquad \textit{Combine like terms.}$$

$$= \boxed{x + 2\sqrt{x} - 15}$$ ∎

EXAMPLE 6

Multiply and simplify. $(\sqrt{7} - \sqrt{3})^2$

Solution

Watch out! Avoid the temptation to square each term separately.

$$\boxed{\textit{Remember:}\quad (a + b)^2 \neq a^2 + b^2}$$

$(\sqrt{7} - \sqrt{3})^2 = (\sqrt{7} - \sqrt{3})(\sqrt{7} - \sqrt{3})$

$$= \sqrt{7}\sqrt{7} - \sqrt{7}\sqrt{3} - \sqrt{7}\sqrt{3} + \sqrt{3}\sqrt{3}$$

$$= 7 - \sqrt{21} - \sqrt{21} + 3 \qquad \textit{Combine like terms. } -\sqrt{21} - \sqrt{21}$$
$$\textit{is like } -x - x = -2x.$$

$$= \boxed{10 - 2\sqrt{21}}$$ ∎

EXAMPLE 7   Multiply and simplify. $(\sqrt{a} - 3)^2 - (\sqrt{a - 3})^2$

Solution   Note the difference between the two expressions being squared. The first is made up of *two* terms; the second is not.

$$(\sqrt{a} - 3)^2 - (\sqrt{a - 3})^2 = (\sqrt{a} - 3)(\sqrt{a} - 3) \qquad -\sqrt{a - 3}\sqrt{a - 3}$$

$$= \sqrt{a}\sqrt{a} - 3\sqrt{a} - 3\sqrt{a} + 9 - \qquad (a - 3)$$

Note that the parentheses around $a - 3$ are essential.
Combine like terms and remove the parentheses.

$$= a - 6\sqrt{a} + 9 - a + 3$$

$$= \boxed{-6\sqrt{a} + 12} \qquad \blacksquare$$

EXAMPLE 8   Simplify.

(a) $\dfrac{\sqrt{72}}{\sqrt{6}}$ (b) $\dfrac{\sqrt{6b^7}}{\sqrt{30ab}}$

Solution   (a) We offer two solutions to illustrate a point.

Solution 1.   We first simplify $\sqrt{72}$.

$$\frac{\sqrt{72}}{\sqrt{6}} = \frac{\sqrt{36}\sqrt{2}}{\sqrt{6}}$$

$$= \frac{6\sqrt{2}}{\sqrt{6}} \qquad \textit{Rationalize.}$$

$$= \frac{6\sqrt{2}}{\sqrt{6}} \cdot \frac{\sqrt{6}}{\sqrt{6}}$$

$$= \frac{6\sqrt{12}}{6} \qquad \textit{Reduce.}$$

$$= \sqrt{12}$$

$$= \sqrt{4}\sqrt{3}$$

$$= \boxed{2\sqrt{3}}$$

Solution 2.   We first make one radical using property 2.

$$\frac{\sqrt{72}}{\sqrt{6}} = \sqrt{\frac{72}{6}}$$

$$= \sqrt{12}$$

$$= \sqrt{4}\sqrt{3} = \boxed{2\sqrt{3}}$$

Clearly, the second method is more efficient. If you have the quotient of two radical expressions and you see common factors that can be reduced, it is usually a better strategy to use property 2 first to make a single radical and reduce the fraction within the radical sign. Then proceed to simplify the remaining expression.

(b) Since we see that there are common factors (the 6 and the 30 will reduce, etc.), we begin by using property 2.

$$\frac{\sqrt{6b^7}}{\sqrt{30ab}} = \sqrt{\frac{6b^7}{30ab}} \qquad \textit{Reduce.}$$

$$= \sqrt{\frac{b^6}{5a}} \qquad \textit{Use property 2 again.}$$

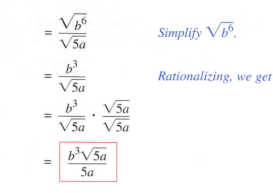

$$= \frac{\sqrt{b^6}}{\sqrt{5a}} \qquad \textit{Simplify } \sqrt{b^6}.$$

$$= \frac{b^3}{\sqrt{5a}} \qquad \textit{Rationalizing, we get}$$

$$= \frac{b^3}{\sqrt{5a}} \cdot \frac{\sqrt{5a}}{\sqrt{5a}}$$

$$= \boxed{\frac{b^3\sqrt{5a}}{5a}}$$

EXAMPLE 9   Multiply and simplify. $(\sqrt{13} - 3)(\sqrt{13} + 3)$

Solution   Multiply out.

$$(\sqrt{13} - 3)(\sqrt{13} + 3) = \sqrt{13}\,\sqrt{13} + 3\sqrt{13} - 3\sqrt{13} - 9 \qquad \textit{The middle terms}$$
$$= 13 - 9 \qquad\qquad\qquad\qquad \textit{sum to } 0.$$

$$= \boxed{4} \qquad \textit{Note that this answer does not involve radicals.}$$

We will soon make good use of this fact.

In the last section we saw that in order to rationalize the denominator of an expression such as $\dfrac{5}{\sqrt{13}}$, we multiply the fraction by the rationalizing factor of $\dfrac{\sqrt{13}}{\sqrt{13}}$.

$$\frac{5}{\sqrt{13}} = \frac{5}{\sqrt{13}} \cdot \frac{\sqrt{13}}{\sqrt{13}} = \frac{5\sqrt{13}}{13}$$

In this way we have removed the radical from the denominator, and the expression now satisfies *all* the conditions of simplest radical form.

But what if we had to simplify $\dfrac{5}{\sqrt{13} - 3}$? If we again multiply the numerator and denominator by $\sqrt{13}$ (which is mathematically legal), we do not get the job done. We still end up with a radical in the denominator. *Try it!*

If we look back to Example 9, we can see what does work. The rationalizing factor is $\dfrac{\sqrt{13} + 3}{\sqrt{13} + 3}$.

$$\frac{5}{\sqrt{13} - 3} = \frac{5}{\sqrt{13} - 3} \cdot \frac{\sqrt{13} + 3}{\sqrt{13} + 3} \qquad \textit{In Example 9 we saw that}$$
$$\hphantom{\frac{5}{\sqrt{13} - 3} = } \qquad\qquad\qquad\qquad (\sqrt{13} - 3)(\sqrt{13} + 3) = 4.$$

$$= \boxed{\frac{5(\sqrt{13} + 3)}{4}}$$

The expressions $\sqrt{13} - 3$ and $\sqrt{13} + 3$ are called **conjugates** of each other. To get a conjugate of an expression made up of *two* terms, just change the sign of the second term. Thus, a conjugate of $1 + \sqrt{5}$ is $1 - \sqrt{5}$.

EXAMPLE 10   Rationalize the denominator. $\dfrac{20}{\sqrt{7} + \sqrt{2}}$

Solution   The rationalizing factor for this expression is $\dfrac{\sqrt{7} - \sqrt{2}}{\sqrt{7} - \sqrt{2}}$.

$$\frac{20}{\sqrt{7} + \sqrt{2}} = \frac{20}{\sqrt{7} + \sqrt{2}} \cdot \frac{\sqrt{7} - \sqrt{2}}{\sqrt{7} - \sqrt{2}}$$

In the denominators, the conjugates multiply as the difference of two squares. We get the square of the first term minus the square of the last term. The middle terms drop out.

$$= \frac{20(\sqrt{7} - \sqrt{2})}{7 - 2}$$

$$= \frac{20(\sqrt{7} - \sqrt{2})}{5}$$

Note that we do not multiply out the top; we keep the factored form to see if we can reduce the fraction. We reduce the 20 and the 5.

$$= \frac{\overset{4}{\cancel{20}}(\sqrt{7} - \sqrt{2})}{\cancel{5}}$$

$$= \boxed{4(\sqrt{7} - \sqrt{2})} \quad \text{or} \quad \boxed{4\sqrt{7} - 4\sqrt{2}} \qquad \blacksquare$$

EXAMPLE 11

Simplify as completely as possible. $\dfrac{8}{3 - \sqrt{5}} - \dfrac{10}{\sqrt{5}}$

Solution

We begin by rationalizing each denominator. Keep in mind that each fraction has its own rationalizing factor.

$$\frac{8}{3 - \sqrt{5}} - \frac{10}{\sqrt{5}} = \frac{8}{3 - \sqrt{5}} \cdot \frac{3 + \sqrt{5}}{3 + \sqrt{5}} - \frac{10}{\sqrt{5}} \cdot \frac{\sqrt{5}}{\sqrt{5}}$$

$$= \frac{8(3 + \sqrt{5})}{9 - 5} - \frac{10\sqrt{5}}{5}$$

$$= \frac{8(3 + \sqrt{5})}{4} - \frac{10\sqrt{5}}{5} \qquad \textit{Reduce each fraction.}$$

$$= \frac{\overset{2}{\cancel{8}}(3 + \sqrt{5})}{\cancel{4}} - \frac{\overset{2}{\cancel{10}}\sqrt{5}}{\cancel{5}}$$

$$= 2(3 + \sqrt{5}) - 2\sqrt{5}$$

$$= 6 + 2\sqrt{5} - 2\sqrt{5}$$

$$= \boxed{6} \qquad \blacksquare$$

EXAMPLE 12

Simplify as completely as possible. $\dfrac{12 + \sqrt{18}}{6}$

Solution

We begin by simplifying the radical.

$$\frac{12 + \sqrt{18}}{6} = \frac{12 + \sqrt{9}\sqrt{2}}{6}$$

$$= \frac{12 + 3\sqrt{2}}{6} \qquad \textit{Factor out the common factor of 3 in the numerator.}$$

$$= \frac{3(4 + \sqrt{2})}{6} \qquad \textit{Now reduce the 3 and the 6.}$$

$$= \frac{\cancel{3}(4 + \sqrt{2})}{\underset{2}{\cancel{6}}}$$

$$= \boxed{\frac{4 + \sqrt{2}}{2}} \qquad \textit{Note that we \textbf{cannot} reduce the 4 with the 2.} \quad \blacksquare$$

Being able to work with radical expressions allows us to broaden the scope of the types of questions we can answer.

EXAMPLE 13

Verify that $3 + \sqrt{2}$ is a solution to the equation $x^2 - 6x + 7 = 0$.

Solution

We must replace each x in the equation by the value $3 + \sqrt{2}$ and show that it satisfies the equation.

$$x^2 - 6x + 7 = 0$$

$$(3 + \sqrt{2})^2 - 6(3 + \sqrt{2}) + 7 \overset{?}{=} 0$$

$$(3 + \sqrt{2})(3 + \sqrt{2}) - 6(3 + \sqrt{2}) + 7 \overset{?}{=} 0$$

$$9 + 3\sqrt{2} + 3\sqrt{2} + 2 - 18 - 6\sqrt{2} + 7 \overset{?}{=} 0 \qquad \textit{Combine like terms.}$$

$$0 \overset{\checkmark}{=} 0 \qquad\qquad\qquad \blacksquare$$

EXERCISES 10.4

Perform the indicated operations. Simplify all answers as completely as possible.
Assume that all variables appearing under radical signs are nonnegative.

1. $\sqrt{3}\,\sqrt{11}$

2. $\sqrt{5}\,\sqrt{7}$

3. $\sqrt{3}\,\sqrt{5}\,\sqrt{13}$

4. $\sqrt{2}\,\sqrt{7}\,\sqrt{11}$

5. $\sqrt{6} + \sqrt{24}$

6. $\sqrt{5} + \sqrt{45}$

7. $\sqrt{6}\,\sqrt{24}$

8. $\sqrt{5}\,\sqrt{45}$

9. $\sqrt{3}\,\sqrt{5}\,\sqrt{6}$

10. $\sqrt{2}\,\sqrt{6}\,\sqrt{10}$

11. $\sqrt{3}(\sqrt{5} + \sqrt{6})$

12. $\sqrt{2}(\sqrt{6} + \sqrt{10})$

13. $\sqrt{18}\,\sqrt{32}$

14. $\sqrt{24}\,\sqrt{28}$

15. $\sqrt{3}(2\sqrt{3} - 3\sqrt{2})$

16. $\sqrt{5}(3\sqrt{5} - 4\sqrt{3})$

17. $5\sqrt{x}(\sqrt{x} - 2\sqrt{5})$

18. $4\sqrt{a}(2\sqrt{a} + 3\sqrt{7})$

19. $3\sqrt{2}(\sqrt{2} - 4) + \sqrt{2}(5 - \sqrt{2})$

20. $5\sqrt{3}(\sqrt{3} - 2) + \sqrt{3}(7 - \sqrt{3})$

21. $4\sqrt{x}(\sqrt{x} - \sqrt{2}) - \sqrt{x}(3\sqrt{x} - 2\sqrt{2})$

22. $2\sqrt{y}(\sqrt{y} - \sqrt{3}) - \sqrt{y}(3\sqrt{y} - 4\sqrt{3})$

23. $(\sqrt{11} + 3)(\sqrt{11} - 6)$

24. $(\sqrt{10} - 5)(\sqrt{10} + 2)$

25. $(\sqrt{x} + \sqrt{3})^2$

26. $(\sqrt{a} - \sqrt{5})^2$

27. $(\sqrt{x} + \sqrt{3})(\sqrt{x} - \sqrt{3})$

28. $(\sqrt{a} - \sqrt{5})(\sqrt{a} + \sqrt{5})$

29. $(3\sqrt{2} - 2\sqrt{5})^2$

30. $(4 + 5\sqrt{3})^2$

31. $(3\sqrt{x} - \sqrt{7})(3\sqrt{x} + \sqrt{7})$

32. $(2\sqrt{y} + 3\sqrt{6})(2\sqrt{y} - 3\sqrt{6})$

33. $(2\sqrt{x} - 3)(3\sqrt{x} + 4)$

34. $(4\sqrt{a} + 1)(3\sqrt{a} - 1)$

35. $(\sqrt{28} - \sqrt{24})(\sqrt{7} - \sqrt{6})$

36. $(\sqrt{32} - \sqrt{20})(\sqrt{2} + \sqrt{5})$

37. $(\sqrt{t + 9})^2 + (\sqrt{t} + 9)^2$

38. $(\sqrt{z} - 1)^2 + (\sqrt{z - 1})^2$

39. $(\sqrt{x} + 2)^2 - (\sqrt{x + 2})^2$

40. $(\sqrt{m - 4})^2 - (\sqrt{m} - 4)^2$

41. $\dfrac{10}{\sqrt{11}}$

42. $\dfrac{12}{\sqrt{5}}$

43. $\dfrac{\sqrt{54}}{\sqrt{3}}$

44. $\dfrac{\sqrt{200}}{\sqrt{5}}$

45. $\dfrac{\sqrt{xy^3}}{\sqrt{x^3 y}}$

46. $\dfrac{\sqrt{3a^7}}{\sqrt{27a^5}}$

47. $\dfrac{\sqrt{a^2 b^5}}{\sqrt{ab^8}}$

48. $\dfrac{\sqrt{12a^2 b^3}}{\sqrt{3a^3 b}}$

49. $\dfrac{10}{4 - \sqrt{11}}$

50. $\dfrac{12}{3 + \sqrt{5}}$

51. $\dfrac{6}{\sqrt{x} + \sqrt{3}}$

52. $\dfrac{8}{\sqrt{a} - \sqrt{2}}$

53. $\dfrac{\sqrt{3}}{2 + \sqrt{3}}$

54. $\dfrac{\sqrt{5}}{\sqrt{6} - \sqrt{5}}$

55. $\dfrac{\sqrt{5} + \sqrt{3}}{\sqrt{5} - \sqrt{3}}$

56. $\dfrac{\sqrt{11} - \sqrt{7}}{\sqrt{11} + \sqrt{7}}$

In Exercises 57–62, rationalize the denominators and simplify.

57. $\dfrac{8}{\sqrt{5} - \sqrt{3}} - \dfrac{12}{\sqrt{3}}$

58. $\dfrac{16}{\sqrt{6} + \sqrt{2}} + \dfrac{8}{\sqrt{2}}$

59. $\dfrac{6}{3 - \sqrt{7}} - \dfrac{21}{\sqrt{7}}$

60. $\dfrac{12}{\sqrt{15} + \sqrt{3}} - \dfrac{30}{\sqrt{15}}$

61. $(3 + \sqrt{5})^2 + \dfrac{8}{3 - \sqrt{5}}$

62. $(4 - \sqrt{7})^2 - \dfrac{18}{4 + \sqrt{7}}$

In Exercises 63–66, reduce to lowest terms.

63. $\dfrac{4 + \sqrt{8}}{6}$

64. $\dfrac{6 + \sqrt{12}}{8}$

65. $\dfrac{12 - \sqrt{20}}{10}$

66. $\dfrac{10 - \sqrt{24}}{12}$

67. Verify that $2 + \sqrt{10}$ is a solution to the equation $x^2 - 4x - 6 = 0$.

68. Determine whether $3 - \sqrt{13}$ is a solution to the equation $x^2 - 6x = 3$.

② QUESTION FOR THOUGHT

69. Discuss what (if anything) is **wrong** with the following:

(a) $\dfrac{3\sqrt{10}}{6} \stackrel{?}{=} \dfrac{3\sqrt{10}}{\underset{2}{\cancel{6}}} \stackrel{?}{=} \dfrac{\sqrt{10}}{2} \stackrel{?}{=} \dfrac{\sqrt{5 \cdot 2}}{2} \stackrel{?}{=} \dfrac{\sqrt{5 \cdot 2}}{\cancel{2}} \stackrel{?}{=} \sqrt{5}$

(b) $\dfrac{2 + \sqrt{5}}{4} \stackrel{?}{=} \dfrac{2 + \sqrt{5}}{\underset{2}{\cancel{4}}} \stackrel{?}{=} \dfrac{1 + \sqrt{5}}{2}$

◇ MINI-REVIEW

In Exercises 70–77, factor the given expression as completely as possible.

70. $x^2 - 6x$

71. $x^2 - 6x - 7$

72. $2x^2 - 50$

73. $x^2 + 10x - 16$

74. $a^2 + 6a - 40$

75. $w^3 - 8w^2 + 16w$

76. $2y^2 + 7y + 3$

77. $6u^2 - 5u - 14$

Solve the following problems algebraically. Be sure to label what the variable represents.

78. Marta leaves her home at 9:00 A.M. driving at 90 kph, while Sarah leaves the same location at 10:00 A.M. following along the same route at 80 kph. At what time will they be 120 km apart?

79. An orchestral society put on a concert. The members sold 200 tickets in advance and 75 tickets at the door. They charged $1.50 more for tickets at the door than for advance-sale tickets. If they collected a total of $1,075, how much did they charge for tickets at the door?

10.5   Radical Equations

The time t (in seconds) it takes a free-falling object to fall h feet is given by the formula

$$t = \frac{\sqrt{h}}{4}$$

If a ball is dropped from the top of a building and it takes 5.2 seconds to hit the ground, we could determine the height of the building by substituting $t = 5.2$ into this formula and solving for h. In other words, we would solve the equation

$$5.2 = \frac{\sqrt{h}}{4} \quad \text{This type of equation is called a \textbf{radical equation.}}$$

Radical equations are the subject of this section. However, in order to develop a method for solving radical equations, we must first take a moment to discuss another property of equality.

In Sections 3.1 and 3.2 we discussed the four basic properties of equality. Briefly summarized, these say that we obtain an equivalent equation if we add or subtract the same quantity to both sides of an equation, or if we multiply or divide both sides of an equation by the same *nonzero* quantity. These properties do not aid us in trying to solve an equation such as

$$\sqrt{x} = 3$$

However, if two quantities are equal, then their squares are also equal. Algebraically, we have

$$\boxed{\text{If } a = b, \text{ then } a^2 = b^2.}$$

Keeping in mind that for $a \geq 0$, we have the basic property of radicals

$$\boxed{(\sqrt{a})^2 = a}$$

and recognizing that in an equation such as $\sqrt{x} = 3$ our basic goal is still the same—to get x alone on one side of the equation—we can use the basic property of radicals to *square both sides of the equation.* Therefore, our solution is as follows:

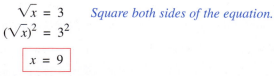

$$\sqrt{x} = 3 \qquad \textit{Square both sides of the equation.}$$
$$(\sqrt{x})^2 = 3^2$$

$$\boxed{x = 9}$$

We can easily check that $x = 9$ satisfies the original equation.

Let's look at several examples illustrating this procedure before we formulate an outline for solving radical equations.

EXAMPLE 1   Solve for x. $\sqrt{x + 4} = 6$

Solution

$\sqrt{x + 4} = 6$ *Square both sides of the equation.*

$(\sqrt{x + 4})^2 = 6^2$

$x + 4 = 36$ *Now we can solve for x.*

$\underline{-4 \quad -4}$

$\boxed{x = 32}$

Check $x = 32$: $\sqrt{x + 4} = 6$

$\sqrt{32 + 4} \stackrel{?}{=} 6$

$\sqrt{36} \stackrel{\checkmark}{=} 6$ ■

EXAMPLE 2   Solve for x. $\sqrt{x} + 4 = 6$

Solution

Be careful! We do want to square both sides of the equation in order to obtain an equation without radicals. However, squaring both sides of an equation is not the same as squaring each term. For example,

$$3 + 5 = 8 \quad \textbf{\textit{but}} \quad 3^2 + 5^2 \neq 8^2 \quad \text{because} \quad 9 + 25 \neq 64$$

We may *not* square each term separately.

If we do in fact square both sides of the original equation properly, we get the following:

$\sqrt{x} + 4 = 6$ *Square both sides.*

$(\sqrt{x} + 4)^2 = 6^2$

$(\sqrt{x} + 4)(\sqrt{x} + 4) = 6^2$ *Multiply out the left-hand side as the*

$x + 8\sqrt{x} + 16 = 36$ *product of two binomials.*

Note that we still have a radical in the equation, so that while these steps were mathematically correct, they did not get us any closer to a solution to the equation.

In order for the squaring process to work most efficiently, we would like to *isolate* the radical first, *before* we square both sides of the equation. We should proceed as follows:

$\sqrt{x} + 4 = 6$ *Subtract 4 from both sides.*

$\underline{-4 \quad -4}$

$\sqrt{x} = 2$ *Now we square both sides.*

$(\sqrt{x})^2 = 2^2$

$\boxed{x = 4}$

Check $x = 4$: $\sqrt{x} + 4 = 6$

$\sqrt{4} + 4 \stackrel{?}{=} 6$

$2 + 4 \stackrel{\checkmark}{=} 6$ ■

Look over the last two examples very carefully to see the differences between them. Note how and when the squaring process is applied.

EXAMPLE 3   Solve for a. $6 - \sqrt{3a} = 15$

Solution

$6 - \sqrt{3a} = 15$ *We first isolate the radical by subtracting 6 from both sides.*

$\underline{-6 -6}$

$-\sqrt{3a} = 9$ *Square both sides.*

$$(-\sqrt{3a})^2 = 9^2 \qquad \textit{Be careful. When you square a negative, you get a positive.}$$

$$3a = 81$$

$$a = 27 \qquad \textit{Are you wondering why there is no box around the answer?}$$

Check $a = 27$: $\qquad 6 - \sqrt{3a} = 15$

$$6 - \sqrt{3(27)} \overset{?}{=} 15$$

$$6 - \sqrt{81} \overset{?}{=} 15$$

$$6 - 9 \overset{?}{=} 15$$

$$-3 \neq 15 \qquad \textit{So } a = 27 \textit{ does \textbf{not} satisfy the \textbf{original}}$$
$$\textit{equation.}$$

What happened? Our logic tells us that *if* there is a solution to the original equation, then it must be $a = 27$. Since we see that $a = 27$ is not a solution, that must mean that

$$\boxed{\text{the original equation has no solutions.}} \qquad\qquad \blacksquare$$

Did we make an algebraic error in solving Example 3? The answer is no! In Section 9.4, when we were solving fractional equations, we came across a similar situation. That is, we saw that some procedures we perform in order to solve an equation do not yield an equivalent equation.

We can see this quite clearly with a simple example. The equation $x = -5$ has only one solution, but if we square both sides, we get $x^2 = (-5)^2$ or $x^2 = 25$, which has two solutions: $x = 5$ and $x = -5$. Thus, squaring both sides of an equation does not necessarily yield an equivalent equation. Therefore, when we use the squaring property to solve an equation, we *must* check the possible solution back in the original equation.

Let's summarize what we have learned from the last few examples.

Outline for Solving Radical Equations	1. Isolate the radical term on one side of the equation.
	2. Square both sides of the equation.
	3. Solve for the variable.
	4. Always check your answer in the original equation.

EXAMPLE 4

Solve for u. $\quad 4 = \sqrt{1 - 2u} + 1$

Solution

Following our outline, we begin by isolating the radical:

$$4 = \sqrt{1 - 2u} + 1$$

$$\underline{-1 \qquad\qquad\qquad -1}$$

$$3 = \sqrt{1 - 2u} \qquad \textit{Square both sides.}$$

$$3^2 = (\sqrt{1 - 2u})^2$$

$$9 = 1 - 2u$$

$$8 = -2u \qquad\qquad \textit{Divide both sides of the equation by } -2.$$

$$\frac{8}{-2} = \frac{-2u}{-2}$$

$$-4 = u \qquad\qquad \textit{We must check in the original equation.}$$

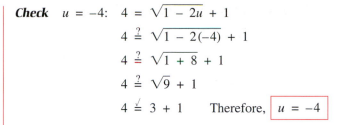

Check $u = -4$: $4 = \sqrt{1 - 2u} + 1$

$$4 \stackrel{?}{=} \sqrt{1 - 2(-4)} + 1$$

$$4 \stackrel{?}{=} \sqrt{1 + 8} + 1$$

$$4 \stackrel{?}{=} \sqrt{9} + 1$$

$$4 \stackrel{\checkmark}{=} 3 + 1 \qquad \text{Therefore, } \boxed{u = -4}$$

Notice that there is no objection to u being a negative number, as long as the expression under the radical sign is not negative. ∎

EXAMPLE 5

Solve for a. $15\sqrt{a} + 2 = 10\sqrt{a} + 6$

Solution

Since this equation has radical terms on both sides of the equation, we must first decide on which side to isolate the radical term. As we saw with first-degree equations, it does not make any difference which side we decide on.

$$15\sqrt{a} + 2 = 10\sqrt{a} + 6 \qquad \textit{We isolate the radical term on the left-hand side.}$$

$$\underline{-10\sqrt{a} \qquad\quad -10\sqrt{a}}$$

$$5\sqrt{a} + 2 = 6 \qquad \textit{Now subtract 2 from both sides.}$$

$$\underline{\qquad -2 \ -2}$$

$$5\sqrt{a} = 4 \qquad \textit{Next divide both sides by 5.}$$

$$\sqrt{a} = \frac{4}{5} \qquad \textit{Square both sides.}$$

$$(\sqrt{a})^2 = \left(\frac{4}{5}\right)^2$$

$$a = \frac{16}{25} \qquad \textit{Again, we must check.}$$

Check $a = \frac{16}{25}$: $15\sqrt{a} + 2 = 10\sqrt{a} + 6$

$$15\sqrt{\frac{16}{25}} + 2 \stackrel{?}{=} 10\sqrt{\frac{16}{25}} + 6$$

$$15\left(\frac{4}{5}\right) + 2 \stackrel{?}{=} 10\left(\frac{4}{5}\right) + 6$$

$$\frac{\overset{3}{\cancel{15}}}{1} \cdot \frac{4}{\cancel{5}} + 2 \stackrel{?}{=} \frac{\overset{2}{\cancel{10}}}{1} \cdot \frac{4}{\cancel{5}} + 6$$

$$3 \cdot 4 + 2 \stackrel{?}{=} 2 \cdot 4 + 6$$

$$12 + 2 \stackrel{\checkmark}{=} 8 + 6 \qquad \text{Therefore, } \boxed{a = \frac{16}{25}} \qquad ∎$$

EXAMPLE 6

Solve for x. $2\sqrt{3x + 1} + 3 = 11$

Solution

$$2\sqrt{3x + 1} + 3 = 11 \qquad \textit{We first isolate the radical.}$$

$$\underline{\qquad\quad -3 \ -3}$$

$$2\sqrt{3x + 1} = 8 \qquad \textit{Divide both sides by 2.}$$

$$\frac{2\sqrt{3x + 1}}{2} = \frac{8}{2} = 4 \qquad \textit{Now square both sides.}$$

$$(\sqrt{3x + 1})^2 = 4^2$$

$$3x + 1 = 16 \qquad \textit{Now we isolate } x.$$

$$3x = 15$$

$$x = 5$$

Check $x = 5$:

$$2\sqrt{3x + 1} + 3 = 11$$

$$2\sqrt{3(5) + 1} + 3 \stackrel{?}{=} 11$$

$$2\sqrt{15 + 1} + 3 \stackrel{?}{=} 11$$

$$2\sqrt{16} + 3 \stackrel{?}{=} 11$$

$$2(4) + 3 \stackrel{?}{=} 11$$

$$8 + 3 \stackrel{\checkmark}{=} 11 \qquad \text{Therefore,} \quad \boxed{x = 5} \qquad \blacksquare$$

As we saw previously, some real-life situations involve radical formulas and equations.

EXAMPLE 7

The time t (in seconds) it takes a free-falling object to fall h feet is given by the formula

$$t = \frac{\sqrt{h}}{4}$$

If a ball is dropped from the top of a building and it takes 5.2 seconds for the ball to hit the ground, estimate the height of the building to the nearest foot.

Solution

In terms of the given formula, we are given that $t = 5.2$ and we want to find h.

$$t = \frac{\sqrt{h}}{4} \qquad \textit{Substitute } t = 5.2.$$

$$5.2 = \frac{\sqrt{h}}{4} \qquad \textit{Multiply both sides by 4.}$$

$$20.8 = \sqrt{h} \qquad \textit{Square both sides.}$$

$$(20.8)^2 = (\sqrt{h})^2$$

$$432.64 = h$$

Thus, rounding to the nearest foot, the height of the building is 433 ft. $\qquad \blacksquare$

EXERCISES 10.5

Solve the given equation.

1. $\sqrt{x} = 5$
2. $\sqrt{y} = 10$
3. $9 = \sqrt{a}$
4. $6 = \sqrt{b}$
5. $\sqrt{u} = 1$
6. $\sqrt{u} = -6$
7. $\sqrt{u} = -8$
8. $\sqrt{u} = 8$
9. $\sqrt{x + 3} = 10$
10. $\sqrt{x - 4} = 7$
11. $\sqrt{x} + 3 = 10$
12. $\sqrt{x} - 4 = 7$
13. $\sqrt{2t - 1} = 5$
14. $\sqrt{3t + 4} = 10$
15. $\sqrt{2t} - 1 = 5$
16. $\sqrt{3t} + 4 = 10$
17. $\sqrt{5 - 4x} = 7$
18. $\sqrt{6 - 5x} = 6$
19. $8 = \sqrt{8 - 14u}$
20. $4 = \sqrt{4 - 3u}$
21. $\sqrt{1 + x} + 8 = 4$
22. $\sqrt{1 + x} - 8 = 4$
23. $\sqrt{3 + x} + 8 = 3$
24. $\sqrt{3 + x} - 8 = 3$

25. $4 - \sqrt{x} = 0$

26. $5 - \sqrt{x} = 0$

27. $4 - \sqrt{x} = 3$

28. $5 - \sqrt{x} = 2$

29. $10 - 2\sqrt{y} = 4$

30. $12 - 3\sqrt{y} = 6$

31. $10 - \sqrt{2y} = 2$

32. $12 - \sqrt{3y} = 3$

33. $6 - \sqrt{a} = 11$

34. $8 = 3 - \sqrt{u}$

35. $7 - \sqrt{a} = 9$

36. $7 = 9 + \sqrt{a}$

37. $6\sqrt{c} - 3 = 3\sqrt{c}$

38. $8\sqrt{m} - 10 = 6\sqrt{m}$

39. $\sqrt{2x + 1} - 1 = 4$

40. $\sqrt{4x + 5} - 2 = 5$

41. $6 + \sqrt{5x + 1} = 12$

42. $4 + \sqrt{7x - 3} = 9$

43. $5\sqrt{w} + 3 = 4\sqrt{w} + 11$

44. $6\sqrt{t} + 5 = 4\sqrt{t} + 9$

45. $2\sqrt{w} + 5 = 7\sqrt{w} - 10$

46. $3\sqrt{t} + 10 = 9\sqrt{t} - 2$

47. $2\sqrt{2x - 1} + 7 = 10$

48. $3\sqrt{4x - 1} + 6 = 8$

49. $4\sqrt{7 - 8x} + 3 = 15$

50. $5\sqrt{4 - 18x} - 8 = 22$

51. $\sqrt{2x - 1} = \sqrt{x + 5}$

52. $\sqrt{3x + 4} = \sqrt{5x - 8}$

53. $\sqrt{5x - 1} = \sqrt{3x + 9}$

54. $\sqrt{6x + 1} = \sqrt{4x + 11}$

55. In an electrical circuit the amount of current is measured in *amperes* (I), the amount of power in *watts* (W), and the resistance in *ohms* (R). The formula relating these three quantities is

$$I = \sqrt{\frac{W}{R}}$$

Find the wattage in a system with 12 amperes and 9 ohms.

56. Use the formula in Exercise 55 to find the amperage in a system with 1,200 watts and 5 ohms.

57. Use the formula given in Example 7 to find the height of the building if it takes 3.4 seconds for the ball to hit the ground.

58. Use the formula given in Example 7 to find the height of the building if it takes 6.8 seconds for the ball to hit the ground. Is your answer twice the height of the building in Exercise 57? Should it be?

59. In an airplane flying above the earth, the approximate distance d (in miles) to the horizon is given by the equation

$$d = \sqrt{8,000a}$$

where a is the altitude of the plane in miles. Determine the altitude of a plane (to the nearest tenth of a mile) if its distance to the horizon is 150 miles.

60. Use the formula in Exercise 59 to determine the altitude of a plane (to the nearest tenth of a mile) if its distance to the horizon is 225 miles.

61. After 2 years, P dollars will grow to A dollars if it is invested at an annual interest rate r satisfying the formula

$$r = \sqrt{\frac{A}{P}} - 1$$

Use this formula to determine how much must be invested at 6% in order to grow to $5,000 in 2 years.

62. Use the formula in Exercise 61 to determine how much must be invested at 8% to grow to $7,500 in 2 years.

⑦ QUESTIONS FOR THOUGHT

63. If we start with the equation $x = 5$ and we square both sides, we get $x^2 = 25$. Are these two equations equivalent? Explain why or why not. In general, does squaring both sides of an equation yield an equivalent equation? Explain.

64. Does the equation $\sqrt{x} + 3 = 2$ have any solutions? Explain.

◇ MINI-REVIEW

In Exercises 65–68, divide (use long division where necessary).

65. $\dfrac{8x^3 + 12x - 4x^2}{2x}$

66. $\dfrac{9ab^2 - 15ab + 6a^2b}{12ab}$

67. $\dfrac{3x^2 - 8x + 4}{x - 2}$

68. $\dfrac{8x^3 - 6x + 3}{2x - 1}$

CHAPTER 10 SUMMARY

After having completed this chapter, you should be able to:

1. Recognize and understand basic radical notation (Section 10.1).

   For example:

   (a) $\sqrt{36}$ means the nonnegative number that when squared is equal to 36.

   Therefore, $\sqrt{36} = 6$ because $6 \cdot 6 = 36$.

   (b) $\sqrt{x^8} = x^4$ because $x^4 \cdot x^4 = x^8$.

2. Apply the basic properties of radicals to obtain an expression in *simplest radical form* (Section 10.2).

   For example:

   (a) $\sqrt{18x^9} = \sqrt{9x^8}\sqrt{2x} = 3x^4\sqrt{2x}$

   (b) $\dfrac{5}{\sqrt{x}} = \dfrac{5}{\sqrt{x}} \cdot \dfrac{\sqrt{x}}{\sqrt{x}} = \boxed{\dfrac{5\sqrt{x}}{x}}$

3. Add and subtract radical expressions (Section 10.3).

   For example:

   (a) $3\sqrt{7} - 8\sqrt{7} = \boxed{-5\sqrt{7}}$

   (b) $\sqrt{40} + \sqrt{50} + \sqrt{90} = \sqrt{4}\sqrt{10} + \sqrt{25}\sqrt{2} + \sqrt{9}\sqrt{10}$

   $$= 2\sqrt{10} + 5\sqrt{2} + 3\sqrt{10}$$

   $$= \boxed{5\sqrt{10} + 5\sqrt{2}}$$

4. Multiply radical expressions (Section 10.4).

   For example:

   (a) $2\sqrt{3}(\sqrt{3} - 4\sqrt{5}) = 2\sqrt{3}\sqrt{3} - 8\sqrt{3}\sqrt{5}$

   $$= \boxed{6 - 8\sqrt{15}}$$

   (b) $(2\sqrt{x} - \sqrt{3})(3\sqrt{x} - 4\sqrt{3}) = 6\sqrt{x}\sqrt{x} - 8\sqrt{3x} - 3\sqrt{3x} + 4\sqrt{3}\sqrt{3}$

   $$= 6x - 11\sqrt{3x} + 4 \cdot 3$$

   $$= \boxed{6x - 11\sqrt{3x} + 12}$$

5. Divide radical expressions, which frequently involves rationalizing denominators (Section 10.4).

For example:

$$\frac{12}{\sqrt{5} - \sqrt{2}} = \frac{12}{\sqrt{5} - \sqrt{2}} \cdot \frac{\sqrt{5} + \sqrt{2}}{\sqrt{5} + \sqrt{2}}$$

$$= \frac{12(\sqrt{5} + \sqrt{2})}{5 - 2}$$

$$= \frac{12(\sqrt{5} + \sqrt{2})}{3} \qquad \textit{Reduce.}$$

$$= \boxed{4(\sqrt{5} + \sqrt{2})}$$

6. Solve some radical equations (Section 10.5).

For example: Solve for a. $\sqrt{2a} - 3 = 7$

$$\sqrt{2a} - 3 = 7 \qquad \textit{Isolate the radical.}$$

$$\underline{\phantom{\sqrt{2a}} +3 \quad +3}$$

$$\sqrt{2a} = 10 \qquad \textit{Square both sides.}$$

$$(\sqrt{2a})^2 = 10^2$$

$$2a = 100$$

$$\boxed{a = 50}$$

Check $\sqrt{2(50)} - 3 \overset{?}{=} 7$

$$\sqrt{100} - 3 \overset{?}{=} 7$$

$$10 - 3 \overset{\checkmark}{=} 7$$

CHAPTER 10 REVIEW EXERCISES

In Exercises 1–12, simplify the given expression as completely as possible. Assume that all variables appearing under radical signs are nonnegative.

1. $\sqrt{49}$

2. $-\sqrt{100}$

3. $\sqrt{-16}$

4. $\sqrt{90}$

5. $\sqrt{96}$

6. $\sqrt{16x^{16}}$

7. $\sqrt{9x^9}$

8. $\sqrt{20x^7y^{10}}$

9. $\sqrt{\dfrac{4}{9}}$

10. $\sqrt{\dfrac{3}{4}}$

11. $\sqrt{\dfrac{3}{5}}$

12. $\sqrt{\dfrac{4}{5}}$

In Exercises 13–37, perform the indicated operations. Be sure to express your answer in simplest radical form. Assume that all variables appearing under radical signs are nonnegative.

13. $8\sqrt{7} - 5\sqrt{7} - \sqrt{7}$

14. $3\sqrt{5} - 4\sqrt{3} + 3\sqrt{3} - 7\sqrt{5}$

15. $\sqrt{45} - \sqrt{20}$

16. $8\sqrt{32} - 5\sqrt{18} - \sqrt{8}$

17. $\sqrt{75x} + \sqrt{12x}$

18. $x\sqrt{54x^3} - \sqrt{24x^5}$

19. $\dfrac{\sqrt{12x^3y^2}}{xy} + \sqrt{27x}$

20. $\sqrt{\dfrac{5}{7}} + \sqrt{\dfrac{7}{5}}$

21. $\sqrt{5}(3\sqrt{5} + \sqrt{2})$

22. $(4\sqrt{x} - \sqrt{3})(\sqrt{x} - 2\sqrt{3})$

23. $(3\sqrt{7} - 2\sqrt{3})(2\sqrt{7} + 5\sqrt{3})$

24. $(\sqrt{36} - \sqrt{16})^2$

25. $(\sqrt{x} - 3)^2$

26. $(\sqrt{a} - \sqrt{10})(\sqrt{a} + \sqrt{10})$

27. $\dfrac{7}{\sqrt{3}}$

28. $\dfrac{10}{\sqrt{6}}$

29. $\dfrac{x^2}{\sqrt{x}}$

30. $\dfrac{7}{3\sqrt{6}}$

31. $\dfrac{18}{\sqrt{12}}$

32. $\dfrac{\sqrt{8m^5}}{\sqrt{18m}}$

33. $\dfrac{14}{3 - \sqrt{2}}$

34. $\dfrac{14}{3\sqrt{2}}$

35. $\dfrac{2 + \sqrt{5}}{6 + \sqrt{5}}$

36. $\dfrac{12}{\sqrt{10} - \sqrt{6}} - \dfrac{18}{\sqrt{6}}$

37. $(\sqrt{x + 7})^2 - (\sqrt{x} + \sqrt{7})^2$

38. Show that $2 + \sqrt{3}$ is a solution to the equation $x^2 - 4x + 1 = 0$.

In Exercises 39–44, solve the given equation.

39. $\sqrt{x + 5} = 7$

40. $\sqrt{x} + 5 = 7$

41. $10 - \sqrt{4u} = 2$

42. $2\sqrt{a} - 4 = 4\sqrt{a} - 12$

43. $8 - \sqrt{y - 5} = 11$

44. $4 - \sqrt{3z + 6} = 2$

45. The time T (in seconds) it takes a pendulum to complete one cycle (that is, to swing through a complete arc and return to its original position) is given by the formula

$$T = 2\pi \sqrt{\dfrac{L}{980}}$$

where L is the length of the pendulum in centimeters. Use this formula to find the length of a pendulum that takes 1.5 seconds to complete one cycle. Use 3.14 for π and round your answer to the nearest tenth.

CHAPTER 10 PRACTICE TEST

Perform the indicated operations. Make sure your final answer is in simplest radical form. Assume that all variables appearing under radical signs are nonnegative.

1. $\sqrt{25x^{16}y^6}$

2. $2\sqrt{27} - 3\sqrt{12} + \sqrt{300}$

3. $\sqrt{50x^3} - x\sqrt{32x}$

4. $\dfrac{\sqrt{48x^5y^{12}}}{\sqrt{8xy^2}}$

5. $\sqrt{20x^8y^9} + 3x^4y^4\sqrt{5y}$

6. $\dfrac{3x^2}{\sqrt{6x}}$

7. $(2\sqrt{x} - \sqrt{5})(\sqrt{x} + 3\sqrt{5})$

8. $(\sqrt{100} - \sqrt{36})^2$

9. $(\sqrt{x} - 4)^2 - (\sqrt{x - 4})^2$

10. $\dfrac{10}{\sqrt{7} - \sqrt{3}}$

11. Determine whether $1 - \sqrt{2}$ is a solution to the equation $x^2 - 2x = 1$.

12. *Solve for a.* $12 - \sqrt{3a} = 6$

13. A company finds that its cost C (in dollars) per unit item is given by the formula

$$C = 1.8\sqrt{1,000 - n}$$

where n is the number of items produced. If the cost per unit item is $18, find the number of items produced.

CHAPTER 11

Quadratic Equations

In Section 8.4 we began to discuss second-degree (quadratic) equations. In this chapter we begin with a review of the factoring method and then expand our discussion to include quadratic equations that cannot be solved by the factoring method. In the course of our discussion we will return to the following situation first introduced in Chapter 8.

This problem is solved in Example 7 of Section 11.3.

A market research firm finds that the daily revenue R (in dollars) earned by a chemical manufacturing company on the sale of g gallons of mixture KZD is given by the equation $R = g^2 - 80g$. If we want to know how many gallons must be sold each day in order to earn a daily revenue of $3,000, then we need to substitute the given value for R and solve the resulting quadratic equation.

The techniques we develop in this chapter will allow us to deal with this situation.

11.1 The Factoring and Square Root Methods

In Section 8.4 we discussed the factoring method for solving quadratic equations, which we review in Example 1.

EXAMPLE 1 Solve for x. $(x - 5)(x + 3) = 20$

Solution Up to this point the only method we have to solve a quadratic equation is the factoring method. Recall that the factoring method requires us to get the equation into standard form, factor the nonzero side of the equation, and then finally apply the zero-product rule.

$$(x - 5)(x + 3) = 20 \qquad \textit{Multiply out the left-hand side.}$$
$$x^2 - 2x - 15 = 20 \qquad \textit{Subtract 20 from both sides to get the equation into standard form.}$$
$$x^2 - 2x - 35 = 0 \qquad \textit{Factor the left-hand side.}$$
$$(x - 7)(x + 5) = 0 \qquad \textit{Now we use the zero-product rule.}$$
$$x - 7 = 0 \quad \text{or} \quad x + 5 = 0$$
$$\boxed{x = 7} \quad \text{or} \quad \boxed{x = -5}$$

Check $x = 7$: $\quad (x - 5)(x + 3) = 20 \qquad x = -5$: $\quad (x - 5)(x + 3) = 20$
$$(7 - 5)(7 + 3) \overset{?}{=} 20 \qquad\qquad (-5 - 5)(-5 + 3) \overset{?}{=} 20$$
$$(2)(10) \overset{\checkmark}{=} 20 \qquad\qquad\qquad (-10)(-2) \overset{\checkmark}{=} 20 \qquad ∎$$

As we have seen previously, fractional equations can give rise to quadratic equations.

EXAMPLE 2 Solve for x. $x + \dfrac{4x}{x - 1} = \dfrac{4}{x - 1}$

Solution We begin as we would with any fractional equation—by clearing the fractions.

$$x + \frac{4x}{x - 1} = \frac{4}{x - 1} \qquad \textit{Multiply both sides of the equation by the LCD, which is } x - 1.$$
$$x(x - 1) + \frac{x - 1}{1} \cdot \frac{4x}{x - 1} = \frac{x - 1}{1} \cdot \frac{4}{x - 1}$$
$$x(x - 1) + 4x = 4$$
$$x^2 - x + 4x = 4 \qquad \textit{Combine like terms.}$$
$$x^2 + 3x = 4$$

Since this last equation is quadratic, we get it into standard form.

$$x^2 + 3x - 4 = 0 \qquad \textit{Factor.}$$
$$(x + 4)(x - 1) = 0$$
$$x + 4 = 0 \quad \text{or} \quad x - 1 = 0$$
$$x = -4 \quad \text{or} \qquad x = 1 \qquad \textit{Notice that we have not put a box around}$$
$$\textit{our answers.}$$

Remember that for a fractional equation, we *must* check to make sure that we do not have any extraneous solutions (ones that do not satisfy the *original* equation).

Check: We check in the *original* equation.

$$x = -4: \qquad x + \frac{4x}{x - 1} = \frac{4}{x - 1} \qquad\qquad x = 1: \qquad x + \frac{4x}{x - 1} = \frac{4}{x - 1}$$

$$-4 + \frac{4(-4)}{-4 - 1} \overset{?}{=} \frac{4}{-4 - 1} \qquad\qquad 1 + \frac{4(1)}{1 - 1} \overset{?}{=} \frac{4}{1 - 1}$$

$$-4 + \frac{-16}{-5} \overset{?}{=} \frac{4}{-5} \qquad\qquad 1 + \frac{4}{0} \overset{?}{=} \frac{4}{0}$$

$$\frac{-20}{5} + \frac{16}{5} \overset{\checkmark}{=} -\frac{4}{5} \qquad\qquad x = 1 \textit{ does not work.}$$
$$\textit{We cannot divide by } 0.$$

Therefore, we have only one solution: $\boxed{x = -4}$.   ■

Again it is worth repeating that we must be alert to recognize what type of equation we are solving. In Example 2, we got the equation into standard form because it is a quadratic equation. This step would be unnecessary if it was a first-degree equation.

As we have already noted, the factoring method is limited. If we get a quadratic equation into standard form and we cannot factor the nonzero side, then the factoring method will not work. The remainder of this chapter is devoted to developing other methods for solving quadratic equations.

The Square Root Method

As we have already noted, the factoring method has the drawback of being a "special" method, in that it works only when we can factor.

The next method we will discuss may also seem somewhat limited at first glance, but as we shall see in this section and the next, we will be able to generalize it to a method that works for *all* quadratic equations.

If asked to solve the equation $x^2 = 9$, we could proceed as follows (using the factoring method):

$$x^2 = 9 \qquad \textit{Get the equation into standard form by subtracting 9}$$
$$\textit{from both sides.}$$
$$x^2 - 9 = 0 \qquad \textit{Factor.}$$
$$(x - 3)(x + 3) = 0$$
$$x - 3 = 0 \quad \text{or} \quad x + 3 = 0$$
$$\boxed{x = 3} \quad \text{or} \quad \boxed{x = -3}$$

If we think a moment about the equation $x^2 = 9$, it should certainly come as no surprise that the solutions are $+3$ and -3. After all, the equation $x^2 = 9$ is asking for a number whose square is equal to 9. In other words, the equation requires that x be a square root of 9. As we saw in Chapter 10, every positive number has two square roots, one positive and the other negative. Thus, solving this equation involves finding the square roots of 9, which are $\sqrt{9}$ and $-\sqrt{9}$, or 3 and -3.

More generally, we can state the following theorem.

| **THEOREM** | If $u^2 = d$, then $u = \sqrt{d}$ or $u = -\sqrt{d}$, for $d \geq 0$. |

A shorter way of writing the two solutions $u = \sqrt{d}$ and $u = -\sqrt{d}$ is to write

$$u = \pm\sqrt{d} \qquad \text{(read "plus or minus } \sqrt{d}\text{")}$$

We use the symbol "$\pm$" instead of writing the equation twice, once with a plus sign and once with a minus sign. The reason we must insist on $d \geq 0$ is that otherwise we will get the square root of a negative number, which is not defined in the real number system.

When we invoke this theorem, we will say that we are "taking square roots" of the equation. This procedure of taking square roots is called the *square root method.*

It is important to remember that the symbol $\sqrt{d}$ stands for the *positive* square root of d. However, when we are looking for a solution to the equation $u^2 = d$, we have no prior knowledge as to whether u is positive or negative, and so we must include both square roots in our solution set.

While the square root method is often shorter than the factoring method, if that were its only advantage, we would not bother with "another" method. In fact, the square root method works in some situations where we cannot factor.

EXAMPLE 3 Solve for a. $a^2 = 11$

Solution

$a^2 = 11$ *This equation is in exactly the form that our theorem requires. Take square roots.*

$\boxed{a = \pm\sqrt{11}}$

We can see that these solutions satisfy the equation. (Throughout the remainder of this chapter, whenever the solutions to an equation are not checked, the check is left to the student.)

Note that had we tried to solve this equation by the factoring method, we would have subtracted 11 from both sides, and gotten $a^2 - 11 = 0$. However, this does not factor (with integers), and so the factoring method fails. ■

*If a quadratic equation has no x term, we will refer to it as a **pure quadratic.***

Recall that the standard form for a quadratic equation is $ax^2 + bx + c = 0$. If we analyze the square root method, we see that it works because there is no x term. Thus, we will normally use it when $b = 0$, since we can then isolate the x^2 term and take square roots.

EXAMPLE 4 Solve for x.
(a) $x^2 + 2 = 5 - 3x^2$
(b) $3x^2 - 2 = 8$

Solution (a) Since we see no x term we will use the square root method.

$$x^2 + 2 = 5 - 3x^2 \qquad \textit{Isolate the } x^2 \textit{ term.}$$
$$4x^2 = 3$$
$$x^2 = \frac{3}{4} \qquad \textit{Take square roots.}$$

$$x = \pm\sqrt{\frac{3}{4}} = \pm\frac{\sqrt{3}}{\sqrt{4}} = \boxed{\pm\frac{\sqrt{3}}{2}}$$

Notice that we always put our final answer in simplest radical form.

(b) Again we see no x term, so we use the square root method.

$$3x^2 - 2 = 8$$
$$3x^2 = 10$$
$$x^2 = \frac{10}{3} \qquad \textit{Take square roots.}$$
$$x = \pm\sqrt{\frac{10}{3}} \qquad \textit{Simplify the radical.}$$
$$x = \pm\frac{\sqrt{10}}{\sqrt{3}} = \pm\frac{\sqrt{10}}{\sqrt{3}} \cdot \frac{\sqrt{3}}{\sqrt{3}} = \pm\frac{\sqrt{30}}{3} \qquad \blacksquare$$

EXAMPLE 5

Solve for z.

(a) $(z - 3)(z - 2) = 13 - 5z$

(b) $2z^2 + 7 = 1$

Solution

(a) We begin by getting the equation into standard form.

$$(z - 3)(z - 2) = 13 - 5z$$
$$z^2 - 5z + 6 = 13 - 5z$$
$$\underline{+5z - 13 \qquad -13 + 5z}$$
$$z^2 - 7 = 0 \qquad \textit{Since this equation is a pure}$$
$$z^2 = 7 \qquad \textit{quadratic, take square roots.}$$

$$\boxed{z = \pm\sqrt{7}}$$

Of course, if you notice that the z term is going to drop out, you would not bother subtracting 13 from both sides. Rather, you would subtract 6 from both sides in order to isolate the second-degree term more quickly. Thus, we see that if we are going to use the square root method, it is not important that the equation be in standard form.

(b) $2z^2 + 7 = 1$
$$2z^2 = -6$$
$$z^2 = -3 \qquad \textit{Take square roots.}$$
$$z = \pm\sqrt{-3} \qquad \text{There are} \boxed{\textbf{\textit{no real solutions}}}.$$

Since $\sqrt{-3}$ does not exist in the real number system, our equation has *no real solutions*. $\qquad \blacksquare$

We mentioned above that the square root method is used primarily in those cases when the standard form of the equation has no x term. There is one very important exception to this rule. If we look carefully at the content of the theorem quoted earlier in this section, we can paraphrase it as follows:

If $(something)^2 = d$, then $something = \pm\sqrt{d}$

The theorem gives us the solutions to an equation that is in the form

Perfect square = Nonnegative number

In other words, it is not necessary that the *something* be a single letter. In fact, the *something* can itself be an expression. In particular, we are interested in the case when it is a binomial.

EXAMPLE 6

Solve for y. $(y - 5)^2 = 9$

Solution

The given equation is in the form $(\quad)^2 = 9$ and thus according to our theorem, $(\quad) = \pm\sqrt{9}$. In other words, the *something* in this problem is the binomial $y - 5$.

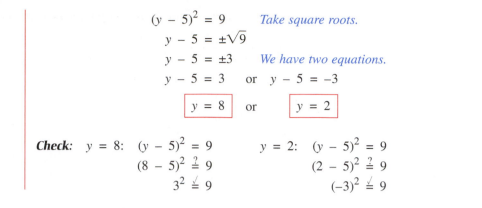

$$(y - 5)^2 = 9 \qquad \textit{Take square roots.}$$
$$y - 5 = \pm\sqrt{9}$$
$$y - 5 = \pm 3 \qquad \textit{We have two equations.}$$
$$y - 5 = 3 \quad \text{or} \quad y - 5 = -3$$
$$\boxed{y = 8} \quad \text{or} \quad \boxed{y = 2}$$

Check: $y = 8$: $(y - 5)^2 = 9$ $\qquad$ $y = 2$: $(y - 5)^2 = 9$
$$(8 - 5)^2 \overset{?}{=} 9 \qquad\qquad\qquad (2 - 5)^2 \overset{?}{=} 9$$
$$3^2 \overset{\checkmark}{=} 9 \qquad\qquad\qquad\quad (-3)^2 \overset{\checkmark}{=} 9 \qquad\qquad ■$$

You might be asking yourself, "Could we have solved the equation in Example 6 by the factoring method as well?" Let's try it and see.

$$(y - 5)^2 = 9 \qquad \textit{Multiply out.}$$
$$(y - 5)(y - 5) = 9$$
$$y^2 - 10y + 25 = 9$$
$$y^2 - 10y + 16 = 0$$
$$(y - 8)(y - 2) = 0$$
$$y - 8 = 0 \quad \text{or} \quad y - 2 = 0$$
$$\boxed{y = 8} \quad \text{or} \quad \boxed{y = 2}$$

So we see that the factoring method works as well, although the square root method was a bit simpler.

It is worth repeating that if the only advantage of the square root method were that it made some solutions a bit simpler, we would not bother ourselves with learning yet another method. However, as the next example further illustrates, the square root method offers us an approach that works for cases in which the factoring method fails totally.

EXAMPLE 7 $\quad$ Solve for x. $(x + 3)^2 = 7$

Solution $\quad$ Suppose we try to solve this equation by the factoring method.

$$(x + 3)^2 = 7$$
$$(x + 3)(x + 3) = 7$$
$$x^2 + 6x + 9 = 7$$
$$x^2 + 6x + 2 = 0 \qquad \textit{Try to factor.}$$

We are stuck! However, the original equation is in the form $(\textit{binomial})^2 = \text{number}$. Therefore, let's try the square root method.

$$(x + 3)^2 = 7 \qquad\qquad\qquad\qquad \textit{Take square roots.}$$
$$x + 3 = \pm\sqrt{7} \qquad\qquad\qquad\qquad \textit{We get two equations.}$$
$$x + 3 = \sqrt{7} \qquad \text{or} \quad x + 3 = -\sqrt{7} \qquad \textit{Subtract 3 from both sides of}$$
$$\textit{each equation.}$$
$$x = -3 + \sqrt{7} \quad \text{or} \qquad x = -3 - \sqrt{7} \qquad\qquad\qquad ■$$

Note that unlike Example 6, separating the equation $x + 3 = \pm\sqrt{7}$ into two equations in Example 7 did not yield any simplification of our answers. In such a case we might as well write

$$x + 3 = \pm\sqrt{7} \qquad \textit{Subtract 3 from both sides.}$$
$$\underline{-3 \quad -3}$$
$$x = -3 \pm \sqrt{7} \qquad \textit{We could write } x = \pm\sqrt{7} - 3, \textit{ but } -3 \pm \sqrt{7} \textit{ is the preferred form.}$$

These are the exact answers, and whenever we have a radical that is an irrational number, we may leave the answer in this form. However, if we are asked to approximate irrational answers involving radicals, we use a calculator, as illustrated in the next example.

EXAMPLE 8

Solve for x and approximate the solutions to two decimal places. $\quad 17x^2 - 8 = x^2 + 7$

Solution

Recognizing that this equation is a pure quadratic (no x terms), we use the square root method.

$$17x^2 - 8 = x^2 + 7$$
$$16x^2 = 15$$
$$x^2 = \frac{15}{16} \qquad \textit{Take square roots.}$$
$$x = \pm\sqrt{\frac{15}{16}} = \pm\frac{\sqrt{15}}{\sqrt{16}} = \boxed{\frac{\pm\sqrt{15}}{4}}$$

In order to approximate these solutions to two decimal places, we can use the following sequence of keystrokes on many calculators. (Remember that different calculators require different keystroke sequences.)

$$\boxed{1}\ \boxed{5}\ \boxed{\sqrt{\ }}\ \boxed{\div}\ \boxed{4}\ \boxed{=}$$

The display will read $\boxed{.9682458}$. Therefore, rounding to the nearest hundredth, we have

$$\boxed{x = \pm 0.97}$$

It is left to the reader to check both the exact and approximate answers in the original equation. ■

If the square root method yields a radical that is not an irrational number, then we are expected to give each answer individually, as in the following example.

EXAMPLE 9

Solve for x. $\quad \left(x - \dfrac{2}{3}\right)^2 = \dfrac{1}{4}$

Solution

Using the square root method, we get

$$\left(x - \frac{2}{3}\right)^2 = \frac{1}{4} \qquad \textit{Take square roots.}$$
$$x - \frac{2}{3} = \pm\sqrt{\frac{1}{4}} = \pm\frac{\sqrt{1}}{\sqrt{4}}$$
$$x - \frac{2}{3} = \pm\frac{1}{2} \qquad \textit{Add } \frac{2}{3} \textit{ to both sides.}$$
$$\underline{+\frac{2}{3} \qquad +\frac{2}{3}}$$
$$x = \frac{2}{3} \pm \frac{1}{2} \qquad \textit{We do not leave the answer in this form.}$$

$$x = \frac{2}{3} + \frac{1}{2} \quad \text{or} \quad x = \frac{2}{3} - \frac{1}{2}$$

$$\boxed{x = \frac{7}{6}} \quad \text{or} \quad \boxed{x = \frac{1}{6}} \qquad \blacksquare$$

To summarize then, the square root method allows us to find the solutions (if any) of quadratic equations that are in the form

$$(x + p)^2 = d$$

where p and d are numbers.

In the next section we will carry the square root method one step further. We will learn how to adapt it to situations in which the "x term is not missing"—that is, to situations in which we do not have a perfect square.

EXERCISES 11.1

Solve each of the following equations. If the equation is quadratic, use the factoring or square root method. If the equation has no real solutions, say so.

1. $x^2 + 4x = 32$

2. $2x^2 - 4x - 7 = x^2 + 3x + 1$

3. $(x + 3)(x + 5) = 80$

4. $(x + 1)(x - 4) = (x - 1)(x + 4)$

5. $x^2 = 25$

6. $a^2 = 81$

7. $b^2 - 16 = 0$

8. $y^2 - 36 = 0$

9. $9b^2 - 16 = 0$

10. $4y^2 - 36 = 0$

11. $b^2 + 16 = 0$

12. $y^2 + 36 = 0$

13. $25x^2 = 4$

14. $49x^2 = 100$

15. $36x^2 - 15 = 0$

16. $64x^2 - 30 = 0$

17. $3b^2 = 11$

18. $11w^2 - 7 = 0$

19. $3b^2 = 12$

20. $11w^2 - 11 = 0$

21. $9a^2 = 20$

22. $4c^2 = 27$

23. $3y^2 = 32$

24. $5z^2 = 48$

25. $7y^2 - 4 = 5y^2 + 6$

26. $8x^2 - 7 = 3x^2 + 3$

27. $5a^2 - 3a + 4 = 2a^2 - 3a + 13$

28. $2a^2 - 4a + 1 = 1 - 4a$

29. $3a^2 - 18 = 5a^2 - 10$

30. $1 - 4w^2 = w^2 + 11$

31. $(y - 3)(y + 4) = y$

32. $(2x + 3)(x - 1) = x + 3$

33. $(2v - 1)(v + 2) = 3(v + 4)$

34. $(3x - 2)(x + 4) = 5(2x + 8)$

35. $(x + 2)^2 = 4(x + 7)$

36. $(x - 4)^2 = 4(9 - 2x)$

37. $(t - 2)^2 = 9$

38. $(t + 2)^2 = 25$

39. $(a + 5)^2 = 7$

40. $(y - 4)^2 = 13$

41. $(x - 6)^2 = 12$

42. $(x + 1)^2 = 18$

43. $(x + 5)^2 = 10$

44. $(x + 5)^2 = 10x$

45. $(x - 6)^2 = 3x$

46. $(x - 6)^2 = 20$

47. $(2x - 3)(x + 4) = x(x + 9)$

48. $(2x - 3)(x + 4) = x(2x + 9)$

49. $\left(m - \frac{2}{3}\right)^2 = \frac{4}{9}$

50. $\left(d + \frac{1}{4}\right)^2 = \frac{9}{16}$

51. $\left(x + \dfrac{2}{5}\right)^2 = \dfrac{3}{25}$

52. $\left(x - \dfrac{3}{7}\right)^2 = \dfrac{5}{49}$

53. $\left(y - \dfrac{1}{2}\right)^2 = \dfrac{2}{3}$

54. $\left(x + \dfrac{1}{3}\right)^2 = \dfrac{3}{5}$

55. $x + \dfrac{1}{x} = 2$

56. $x + \dfrac{3}{x} = \dfrac{7}{2}$

57. $\dfrac{x - 1}{x + 1} = \dfrac{x}{x + 3}$

58. $\dfrac{x - 1}{x} = \dfrac{3x}{x - 2}$

59. $\dfrac{2x}{x + 2} + 1 = x$

60. $\dfrac{2}{x - 1} + x = 4$

61. $a - \dfrac{5a}{a + 1} = \dfrac{5}{a + 1}$

62. $a + \dfrac{3a}{a - 3} = \dfrac{9}{a - 3}$

63. $\dfrac{3}{x - 2} + \dfrac{7}{x + 2} = \dfrac{x + 1}{x - 2}$

64. $\dfrac{2}{x - 1} + \dfrac{3x}{x + 2} = \dfrac{2(5x + 9)}{x^2 + x - 2}$

In Exercises 65–74, solve the given equation. For quadratic equations, choose either the factoring method or the square root method, whichever you think is the easier to use.

65. $2x^2 + 7x - 5 = 3x^2 + 9x - 4$

66. $x^2 + 4x + 9 = 3x^2 + 4x + 1$

67. $(y - 2)(y + 3) = y + 10$

68. $(3y - 1)(2y + 3) = 7(y^2 + y - 2)$

69. $(y - 2)(y + 3) = (2y - 7)(y + 4)$

70. $(y + 3)(y - 5) = (2y + 5)(y - 3)$

71. $\dfrac{2x}{x + 1} = \dfrac{x}{x - 1}$

72. $\dfrac{a + 1}{a + 3} = \dfrac{a}{a + 1}$

73. $4(x + 1) = \dfrac{9}{x + 1}$

74. $\dfrac{1}{x} + \dfrac{6}{x + 2} = 2$

In Exercises 75–84, solve the equations using the square root method. Round off your answers to the nearest hundredth.

75. $x^2 = 7$

76. $t^2 = 21$

77. $3k^2 = 20$

78. $5u^2 = 18$

79. $4x^2 = 19.7$

80. $6w^2 = 1.6$

81. $(2a - 0.3)^2 = 7.5$

82. $(3x - 0.5)^2 = 10.4$

83. $(4.6y + 2.2)^2 = 10.6$

84. $(3.1n + 4.9)^2 = 38.2$

? **QUESTIONS FOR THOUGHT**

85. Justify each step in the following method of solving the equation $x^2 + 4x - 7 = 0$.

$$x^2 + 4x - 7 = 0$$
$$x^2 + 4x = 7$$
$$x^2 + 4x + 4 = 7 + 4$$
$$x^2 + 4x + 4 = 11$$
$$(x + 2)^2 = 11$$
$$x + 2 = \pm\sqrt{11}$$
$$\boxed{x = -2 \pm \sqrt{11}}$$

86. Verify that $2 + \sqrt{3}$ is a solution to the equation $x^2 - 4x + 1 = 0$. Is $2 - \sqrt{3}$ also a solution?

◇ **MINI-REVIEW**

In Exercises 87–90, solve the system of equations algebraically.

87. $\begin{cases} 2x + y = 6 \\ 3x - 2y = 23 \end{cases}$

88. $\begin{cases} 3x + 4y = 12 \\ 5x + 3y = 20 \end{cases}$

89. $\begin{cases} 6x - 4y = 9 \\ 3x - 2y = 2 \end{cases}$

90. $\begin{cases} x - \dfrac{y}{2} = 0 \\ 2x + y = -4 \end{cases}$

91. Lorraine can prepare a gourmet meal in 4 hours, while Renaldo can prepare the same meal in 3 hours. How long will it take them to prepare the meal if they work together?

92. A survey of 600 people showed that the ratio of those who favored candidate A to candidate B was 7 to 5. Of 24,000 actual votes cast, candidate A received 13,187. Was this more or less than the number of votes expected for candidate A based on the survey?

11.2   The Method of Completing the Square

Thus far we have learned two methods for solving quadratic equations: the factoring method, which is limited to those equations that can be factored; and the square root method, which is limited to equations of the special form $(x + p)^2 = d$.

In this section we will see how to convert *any* quadratic equation into one that is of this special form and can therefore be solved by the square root method. Let's begin by analyzing an example. (All checks in the next few sections are left to the student.)

EXAMPLE 1   Solve for x. $(x - 3)^2 = 17$

Solution   In the last section we saw that an equation in this form can be solved quite easily by the square root method.

$$(x - 3)^2 = 17 \quad \textit{Take square roots.}$$
$$x - 3 = \pm\sqrt{17}$$
$$\boxed{x = 3 \pm \sqrt{17}}$$

But what if this equation had been given in any of the following alternate *equivalent* forms?

$$(x - 3)(x - 3) = 17 \quad \textit{Written out}$$
$$x^2 - 6x + 9 = 17 \quad \textit{Multiplied out}$$
$$x^2 - 6x - 8 = 0 \quad \textit{In standard form}$$

It is possible that if we were looking for it, we might recognize the first two of these alternate forms as perfect squares. However, it is unreasonable for us to be expected to recognize the last of these as being equivalent to a perfect square. Keep these three forms of the equation $(x - 3)^2 = 17$ in mind, as we will refer to them again. ■

How do we proceed if we are given an equation such as $x^2 - 6x - 8 = 0$, which we cannot factor and which is not a perfect square? Before we proceed, let us analyze the structure of a perfect square such as $(x + 7)^2$ or $(x + p)^2$.

$$\begin{aligned} (x + 7)^2 &= (x + 7)(x + 7) \\ &= x^2 + 7x + 7x + 49 \\ &= x^2 + 14x + 49 \end{aligned} \qquad \begin{aligned} (x + p)^2 &= (x + p)(x + p) \\ &= x^2 + px + px + p^2 \\ &= x^2 + 2px + p^2 \end{aligned}$$

Do not get hung up on the letters. In most examples, p will have some numerical value. The key thing to keep in mind is that the coefficient of the x (the first-degree term) is $2p$, twice the number in the binomial; the numerical term is p^2, the square of the number in the binomial.

We know that we can solve a quadratic equation of the form $(x + p)^2 = d$ by the square root method. So let's use $x^2 + 2px + p^2$ as a model for our equations and see if we can transform them into perfect squares.

Model of a Perfect Square	$(x + p)^2 = x^2 + 2px + p^2$

An example should make this idea clear. Suppose we want to solve the equation $x^2 - 6x - 8 = 0$. We cannot factor it, so the factoring method fails, and it is not a perfect square, so the square root method fails. Let's see if looking at our model of a perfect square can help us.

Our first step is to add 8 to both sides of the equation. Remember that having an equation in standard form is important for the factoring method, where we need 0 on one side of the equation. Since we *cannot* factor here, 0 is no longer important.

$$\begin{aligned} x^2 - 6x - 8 &= 0 \\ x^2 - 6x &= 8 \qquad \textit{Notice that we left a space where the } -8 \textit{ used to be.} \end{aligned}$$

Now we compare $x^2 - 6x$ to our model perfect square $x^2 + 2px + p^2$. If our model and example are to match *exactly*, then they must match up term by term.

- The x^2 in our model matches the x^2 in our example exactly.

- If the x terms are to match as well, then their coefficients must be the same.

  The coefficient of x in our model is $2p$ (it will always be the same in the model), while the coefficient of x in this example is -6.

  Thus, in order for the x coefficients to be the same, we must have $2p = -6$, which implies that $p = -3$.

- Finally, our model contains the quantity p^2.

  We have just determined that $p = -3$; therefore, $p^2 = (-3)^2 = 9$.

  In order to have a perfect square as in the model, we want our example to have $x^2 - 6x + 9$ on the left-hand side of the equation. In order to get this, we add 9 to *both* sides of the equation.

Thus far, our solution looks as follows:

$$\begin{aligned} x^2 - 6x - 8 &= 0 \\ x^2 - 6x &= 8 \qquad \textit{By looking at the model we determine that } 2p = -6; \\ & \textit{therefore, } p = -3, \textit{ and so the "missing" term is } p^2 = 9. \\ \underline{+9 \qquad +9}& \qquad \textit{Therefore, we add 9 to both sides of the equation.} \\ x^2 - 6x + 9 &= 17 \qquad \textit{The left-hand side of this equation is now a perfect square.} \\ (x - 3)^2 &= 17 \qquad \textit{Notice that the left-hand side of the equation conforms} \\ & \textit{to the model. Since } p = -3, \textit{ we have} \\ & (x + p)^2 = (x + (-3))^2 = (x - 3)^2. \end{aligned}$$

p goes in here according to our model.

If you look back now at Example 1, you will see that we have taken one of the "unrecognizable" forms of that equation and reconstructed the "nice" perfect square form of that equation. We have not finished yet because we have not solved the equation. From here on we proceed exactly as we did in Example 1, by taking square roots and solving for x.

The process we have just carried out is called the ***method of completing the square.*** It consists of determining the number that is needed to make a perfect square on one side of the equation.

Our analysis of the perfect square form $(x + p)^2 = x^2 + 2px + p^2$ and our use of $2p$, p, and p^2 were just devices to help us determine what the "missing" term was. Since $2p$ is always the coefficient of x in our model and p^2 is always the constant term in our model, we will always be computing p (which is one-half of $2p$) and then squaring that to get p^2. In other words, the term needed to complete the square is always *the square of one-half the coefficient of x (the first-degree term)*.

Let's use the method of completing the square in the next example.

EXAMPLE 2

Solution

Solve by completing the square. $y^2 + 8y - 5 = 0$

We first add 5 to both sides of the equation to get $y^2 + 8y = 5$. Now we compare the left-hand side of this equation to our model. Do not get hung up on the letters. Instead of the variable x in our model, this equation happens to have the variable y.

- The y^2 in this example matches the x^2 in our model.

- The coefficient of x in our model is $2p$ and that must match the coefficient of the first-degree term in this example, which is 8.

  Thus, we have $2p = 8$ and so $p = 4$.

- Therefore, the missing p^2 term is $p^2 = 4^2 = 16$, and so we add 16 to both sides of the equation. Thus far our solution looks as follows:

Equivalently, we could compute
$p^2 = (\frac{1}{2} \cdot 8)^2 = 4^2 = 16$

$$y^2 + 8y - 5 = 0 \qquad \textit{Add 5 to both sides.}$$
$$y^2 + 8y \quad = 5 \qquad \textit{Since } 2p = 8; p = 4; p^2 = 16$$

Add 16 to both sides of the equation to complete the square.

$$y^2 + 8y + 16 = 5 + 16 \qquad \textit{The left-hand side is now a perfect square.}$$
It matches our model with $p = 4$.
$(y + p)^2 = (y + 4)^2$

$$(y + 4)^2 = 21 \qquad \textit{Now we can take square roots.}$$
$$y + 4 = \pm\sqrt{21} \qquad \textit{Subtract 4 from both sides.}$$

$$\boxed{y = -4 \pm \sqrt{21}}$$

Remember that the whole purpose of completing the square is to be able to solve the equation by the square root method. Do not forget to solve the equation after you have completed the square.

EXAMPLE 3

Solution

Solve by completing the square. $2x^2 + 94 = 24x$

In order to complete the square we want to start with the equation having the constant (numerical) term isolated and, of course, with the leading coefficient being positive. Therefore, our first step is to rewrite the equation.

$$2x^2 + 94 = 24x$$
$$2x^2 - 24x = -94$$

Comparing this to our model we see that we have $2x^2$ instead of x^2. Therefore, we divide both sides of the equation by 2.

$$x^2 - 12x = -47$$

We compute $\left[\frac{1}{2}(-12)\right]^2 = (-6)^2 = 36$. To

$$x^2 - 12x + 36 = -47 + 36$$

complete the square, we add 36 to both sides.

$$(x - 6)^2 = -11$$

Now take square roots.

$$x - 6 = \pm\sqrt{-11}$$

No real solutions *since the square root of a negative number is not a real number.*

Notice that the method of completing the square also tells us when an equation has no real solutions. ∎

Let's outline the method of completing the square.

Method of Completing the Square	1. Isolate the constant (numerical term), making sure that the second-degree term has a positive coefficient. 2. If the coefficient of the second-degree term is not 1, divide both sides of the equation by that coefficient. 3. Take one-half the coefficient of the first-degree term and square it. Add that quantity to both sides of the equation. 4. Now having a perfect square on one side of the equation, we can solve the equation by the square root method.

Keep in mind that the computation we have been doing with $2p$, p, and p^2 has simply been a device to help us understand and carry out step 3 in the outline.

EXAMPLE 4   Solve by completing the square. $x^2 - 5x + 4 = 0$

Solution   Following our outline, we proceed as follows:

$$x^2 - 5x + 4 = 0$$

We isolate the constant.

$$x^2 - 5x = -4$$

We compute $\left[\frac{1}{2}(-5)\right]^2 = \left(\frac{-5}{2}\right)^2 = \frac{25}{4}$.

To complete the square, we must add $\frac{25}{4}$ to both sides of the equation.

$$x^2 - 5x + \frac{25}{4} = -4 + \frac{25}{4}$$

Because we built it, we know that the left-hand side is a perfect square. We know it is

$$(x + p)^2 = \left[x + \left(-\frac{5}{2}\right)\right]^2 = \left(x - \frac{5}{2}\right)^2.$$

Thus, the left-hand side becomes

$$\left(x - \frac{5}{2}\right)^2 = -4 + \frac{25}{4}$$

Combine -4 and $\frac{25}{4}$; the LCD is 4.

$$\left(x - \frac{5}{2}\right)^2 = -\frac{16}{4} + \frac{25}{4}$$

Combine fractions.

$$\left(x - \frac{5}{2}\right)^2 = \frac{9}{4}$$

Take square roots.

$$x - \frac{5}{2} = \pm\sqrt{\frac{9}{4}} = \pm\frac{3}{2}$$

Add $\frac{5}{2}$ to both sides.

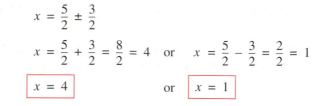

$$x = \frac{5}{2} + \frac{3}{2} = \frac{8}{2} = 4 \quad \text{or} \quad x = \frac{5}{2} - \frac{3}{2} = \frac{2}{2} = 1$$

$$\boxed{x = 4} \qquad \text{or} \qquad \boxed{x = 1}$$

The fact that we get rational answers tells us that we could have used the factoring method in the original equation.

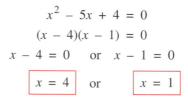

$$x = 4 \quad \text{or} \quad x = 1$$

Comparing the two methods of solution, it is quite obvious that the factoring method is easier and quicker for this equation. This example simply gave us an opportunity to practice the method of completing the square. ■

At this point you can probably see that while the method of completing the square has the advantage of *always working,* it also has the potential for creating some messy arithmetic.

The method of completing the square has numerous applications in mathematics. In the next section we will see how we can use the method of completing the square to derive a formula that also works for all quadratic equations, but which is generally much easier to use.

EXERCISES 11.2

In Exercises 1–34, solve the given equation by the method of completing the square.

1. $x^2 + 8x + 6 = 0$

2. $x^2 + 10x + 20 = 0$

3. $x^2 - 4x - 3 = 0$

4. $x^2 - 2x - 5 = 0$

5. $x^2 - 10x = 15$

6. $x^2 - 6x = 19$

7. $a^2 - 8a - 20 = 0$

8. $a^2 - 4a - 5 = 0$

9. $-x^2 - 12x = 6$

10. $-x^2 + 10x = 5$

11. $2z^2 - 12z + 4 = 0$

12. $3z^2 + 6z - 18 = 0$

13. $10 = 5y^2 + 20y$

14. $24 = 4y^2 - 8y$

15. $u^2 + 5u - 2 = 0$

16. $u^2 - 3u - 1 = 0$

17. $z^2 + 5 = 7z$

18. $10 - 9z = z^2$

19. $w^2 - 3w = 2w^2 - 7w + 2$

20. $2m^2 + 6 = 3m^2 + 6m - 2$

21. $x^2 + 4x + 5 = 2x - 3$

22. $x^2 + 8x + 10 = 2x - 2$

23. $(x - 3)(x + 2) = 9x - 1$

24. $(x - 5)(x - 4) = 7(x + 2)$

25. $(a - 2)(a + 1) = 6$

26. $(a + 2)(a - 1) = 5$

27. $(x - 4)(x + 3) = 1 - x$

28. $(y + 5)(y - 2) = 2 + 3y$

29. $2x^2 + 3 = 6x$

30. $3x^2 + 1 = 6x$

31. $6t^2 - 5t = t^2 + 3t - 1$

32. $y^2 - 2 = 7y - y^2$

33. $4z^2 + 20z + 19 = 0$

34. $2w^2 + 5w = 13$

In Exercises 35–42, solve the given equation by either the factoring method or the square root method (completing the square where necessary). Choose whichever method you think is more appropriate.

35. $(x + 3)^2 = 6$

36. $(x - 2)^2 = 10x$

37. $(x + 3)^2 = 6x$

38. $(x - 2)^2 = 10$

39. $x^2 + 8x - 9 = 0$

40. $x^2 + 8x - 7 = 0$

41. $3x^2 + 4 = 8x$

42. $5x^2 + 6x = 8$

 QUESTIONS FOR THOUGHT

43. State the relationship between the coefficient of the middle term and the constant of a perfect square.

44. Solve *and* check the equation.

$$\frac{x}{x - 1} = \frac{2}{x - 2}$$

 **MINI-REVIEW**

In Exercises 45–52, simplify as completely as possible. (Assume $x \geq 0$.)

45. $\sqrt{48} - \sqrt{75}$

46. $\sqrt{20x^3} + x\sqrt{45x}$

47. $\sqrt{\dfrac{2}{7}}$

48. $\dfrac{10}{\sqrt{6}}$

49. $\dfrac{3}{3 - \sqrt{2}}$

50. $\dfrac{12}{\sqrt{5} - \sqrt{3}}$

51. $\dfrac{6 + 4\sqrt{3}}{2}$

52. $\dfrac{10 - \sqrt{20}}{6}$

Solve the following problem algebraically. Be sure to indicate what the variable represents.

53. Yvonne can assemble a 1,000-piece jigsaw puzzle in 8 hours, while Bill can assemble the same puzzle in 10 hours. Bill starts working on the puzzle alone and quits after 3 hours. How long will it take Yvonne to finish the puzzle on her own?

11.3 The Quadratic Formula

Completing the square is a useful algebraic technique that will be needed again in several places in intermediate algebra, precalculus, and calculus. It is the most powerful of the methods for solving quadratic equations that we have learned so far, because unlike the others it can be applied to *all* quadratic equations. However, as we mentioned in the last section, it is also potentially the messiest and most tedious to use.

Let us begin by solving a quadratic equation by the method of completing the square.

EXAMPLE 1 Solve by completing the square. $2x^2 + 7x + 4 = 0$

Solution We will follow the outline we presented in the last section, with one slight addition. We will number the basic steps in the process. These numbers appear at the right-hand side of the page. We will use these numbers to make it easier to refer back to the steps in the solution.

$2x^2 + 7x + 4 = 0$ *Subtract 4 from both sides.*   **(1)**

$2x^2 + 7x = -4$ *Divide both sides of the equation by 2.*   **(2)**

$\dfrac{2x^2}{2} + \dfrac{7x}{2} = \dfrac{-4}{2}$

$x^2 + \dfrac{7}{2}x = -2$ *We want to take one-half the coefficient of x and square it.*

$$\left[\dfrac{1}{2}\left(\dfrac{7}{2}\right)\right]^2 = \left(\dfrac{7}{4}\right)^2 = \dfrac{49}{16}$$

Note that this means $p = \dfrac{7}{4}$.

We add $\dfrac{49}{16}$ to both sides.   **(3)**

$x^2 + \dfrac{7}{2}x + \dfrac{49}{16} = -2 + \dfrac{49}{16}$ *On the left-hand side we know that we have $(x + p)^2$ with $p = \dfrac{7}{4}$. On the right-hand side we combine -2 and $\dfrac{49}{16}$; the LCD is 16.*

$\left(x + \dfrac{7}{4}\right)^2 = \dfrac{-32}{16} + \dfrac{49}{16}$ **(4)**

$\left(x + \dfrac{7}{4}\right)^2 = \dfrac{17}{16}$ *Take square roots.*   **(5)**

$x + \dfrac{7}{4} = \pm\sqrt{\dfrac{17}{16}} = \pm\dfrac{\sqrt{17}}{\sqrt{16}} = \pm\dfrac{\sqrt{17}}{4}$ *Subtract $\dfrac{7}{4}$ from both sides.*   **(6)**

$x = -\dfrac{7}{4} \pm \dfrac{\sqrt{17}}{4}$ *Since the denominators are the same, we can easily combine.*

$x = \dfrac{-7 \pm \sqrt{17}}{4}$ **(7)**

■

We certainly do not relish the thought of making this our usual method for solving quadratic equations. Instead, we can *use* algebra to carry out the process of completing the square *once* and obtain a formula that we can apply to all quadratic equations.

We begin with the general quadratic equation in standard form. That is, we start with the equation $ax^2 + bx + c = 0$, where $a > 0$, and carry out the process of completing the square for this general equation. We will number the steps just as we did in Example 1, so that you can see that the process is the same. The only difference is that we will be working with letters instead of numbers.

$ax^2 + bx + c = 0$ *Subtract c from both sides.*   **(1)**

$ax^2 + bx = -c$ *Divide both sides of the equation by a.*

$\dfrac{ax^2}{a} + \dfrac{bx}{a} = \dfrac{-c}{a}$ **(2)**

$x^2 + \dfrac{b}{a}x = -\dfrac{c}{a}$ *We want to take one-half the coefficient of x and square it.*

$$\left[\frac{1}{2}\left(\frac{b}{a}\right)\right]^2 = \left(\frac{b}{2a}\right)^2 = \frac{b^2}{4a^2}$$

Note this means that $p = \frac{b}{2a}$.

We add $\frac{b^2}{4a^2}$ *to both sides.* **(3)**

$$x^2 + \frac{b}{a}x + \frac{b^2}{4a^2} = -\frac{c}{a} + \frac{b^2}{4a^2}$$

On the left-hand side we know that we have $(x + p)^2$ *with* $p = \frac{b}{2a}$.

On the right-hand side we combine $-\frac{c}{a}$ *and* $\frac{b^2}{4a^2}$; *the LCD is* $4a^2$.

$$\left(x + \frac{b}{2a}\right)^2 = -\frac{4ac}{4a^2} + \frac{b^2}{4a^2}$$ **(4)**

$$\left(x + \frac{b}{2a}\right)^2 = \frac{-4ac + b^2}{4a^2}$$ *Take square roots.* **(5)**

$$x + \frac{b}{2a} = \pm\sqrt{\frac{b^2 - 4ac}{4a^2}} = \pm\frac{\sqrt{b^2 - 4ac}}{\sqrt{4a^2}} = \pm\frac{\sqrt{b^2 - 4ac}}{2a}$$

Subtract $\frac{b}{2a}$ *from both sides.* **(6)**

$$x = -\frac{b}{2a} \pm \frac{\sqrt{b^2 - 4ac}}{2a}$$ *Since the denominators are the same, we can combine.*

$$\boxed{x = \frac{-b \pm \sqrt{b^2 - 4ac}}{2a}}$$ **(7)**

We have assumed that a is positive. In fact, a similar derivation works when a is negative.

Thus, we have derived what is called the ***quadratic formula.***

In words, the quadratic formula tells us that if we have a quadratic equation in standard form, then all we have to do is substitute the values of a, b, and c into the formula to get the solutions (if any).

The Quadratic Formula	The solutions to the quadratic equation $ax^2 + bx + c = 0$ are given by the formula $$x = \frac{-b \pm \sqrt{b^2 - 4ac}}{2a} \qquad (a \neq 0)$$

EXAMPLE 2   Solve by using the quadratic formula. $x^2 - 3x - 5 = 0$

Solution   This equation is already in standard form. In order to use the formula we must identify a, b, and c. Remember that a is the coefficient of the second-degree term, b is the coefficient of the first-degree term, and c is the constant. Therefore, we have

$$a = 1, \quad b = -3, \quad \text{and} \quad c = -5$$

Substituting these values into the formula, we get

$$x = \frac{-b \pm \sqrt{b^2 - 4ac}}{2a}$$ *Our first step is just substituting the values.*

$$x = \frac{-(-3) \pm \sqrt{(-3)^2 - 4(1)(-5)}}{2(1)}$$

$$x = \frac{3 \pm \sqrt{9 + 20}}{2}$$

$$\boxed{x = \frac{3 \pm \sqrt{29}}{2}}$$

   ■

Notice that the quadratic formula is much easier to use than the method of completing the square. This is because the formula contains within it all the steps that we would have to do if we were completing the square.

Here are some things to watch out for when using the quadratic formula:

1. If b is a negative number, then the $-b$ that appears in the formula will be positive.

2. If a is positive (as it will be if the equation is in standard form) and c is negative, then you will end up *adding* the numbers under the radical sign because $-4ac$ is positive.

3. Do not forget that the quantity $2a$ is the denominator of the *entire* expression $-b \pm \sqrt{b^2 - 4ac}$.

EXAMPLE 3

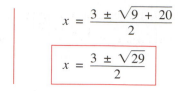

Solve by using the quadratic formula and approximate the solutions to two decimal places. $3u^2 - 4u = 5$

Solution

We begin by putting the equation into standard form.

$$3u^2 - 4u = 5$$

$$3u^2 - 4u - 5 = 0 \qquad \textit{We substitute into the formula } a = 3, b = -4, c = -5.$$
$$\textit{Thus, from the formula we get the solutions}$$

$$u = \frac{-(-4) \pm \sqrt{(-4)^2 - 4(3)(-5)}}{2(3)}$$

$$u = \frac{4 \pm \sqrt{16 + 60}}{6}$$

$$u = \frac{4 \pm \sqrt{76}}{6} \qquad \textit{Simplify the radical.}$$

$$u = \frac{4 \pm \sqrt{4}\sqrt{19}}{6} = \frac{4 \pm 2\sqrt{19}}{6} \qquad \textit{Now factor out a 2 in the numerator.}$$

$$u = \frac{2(2 \pm \sqrt{19})}{6} \qquad \textit{Reduce.}$$

$$u = \frac{\cancel{2}(2 \pm \sqrt{19})}{\underset{3}{\cancel{6}}}$$

$$\boxed{u = \frac{2 \pm \sqrt{19}}{3}} \qquad \text{These are the exact solutions.}$$

In order to approximate the solutions to two decimal places, we can evaluate the exact answers with the following keystroke sequence:

The display will read $\boxed{2.119633}$. Repeating this keystroke sequence with a $\boxed{-}$ in place of $\boxed{+}$ gives $\boxed{-.7862996}$. Therefore, the solutions are

$$\boxed{u = 2.12, -0.79}$$

rounded off to two decimal places.

In fact, if we are using a calculator to approximate the solutions, we can just as easily use the unsimplified form $u = \dfrac{4 + \sqrt{76}}{6}$, which a calculator also computes to be $\boxed{2.119633}$. ∎

EXAMPLE 4

Solve by using the quadratic formula. $3 - 2t = t^2$

Solution

We put the equation into standard form and identify a, b, and c.

$$3 - 2t = t^2$$
$$0 = t^2 + 2t - 3 \qquad a = 1, b = 2, \text{ and } c = -3$$
$$t = \frac{-2 \pm \sqrt{2^2 - 4(1)(-3)}}{2(1)}$$
$$t = \frac{-2 \pm \sqrt{4 + 12}}{2} = \frac{-2 \pm \sqrt{16}}{2} = \frac{-2 \pm 4}{2}$$
$$t = \frac{-2 + 4}{2} = \frac{2}{2} = 1 \quad \text{or} \quad t = \frac{-2 - 4}{2} = \frac{-6}{2} = -3$$
$$\boxed{t = 1} \qquad\qquad \text{or} \qquad \boxed{t = -3}$$

As with completing the square, when we get rational solutions, it means that we could have solved the equation by the factoring method. *Try it!* ∎

EXAMPLE 5

Solve for t. $2t^2 - 5t + 7 = t(2t - 3)$

Solution

Be careful! Do not automatically assume that just because there is a second-degree term in the initial equation that this must be a quadratic equation.

We begin by multiplying out the right-hand side.

$$2t^2 - 5t + 7 = t(2t - 3)$$

$$\begin{array}{r} 2t^2 - 5t + 7 = 2t^2 - 3t \\ \underline{-2t^2 + 3t -2t^2 + 3t} \\ -2t + 7 = 0 \\ -2t = -7 \end{array}$$ *Put the equation in standard form.*

This is not a quadratic equation at all! This is a first-degree equation. We simply isolate t.

$$\boxed{t = \frac{7}{2}}$$

∎

As Example 5 clearly shows, the method of solution we choose for an equation depends on the *type* of equation with which we are dealing. Look carefully at the equation before deciding on a method of solution.

EXAMPLE 6

Solve by using the quadratic formula. $x^2 + 3x + 4 = 0$

Solution

Since the equation is already in standard form, we begin by identifying a, b, and c.

$$a = 1, \quad b = 3, \quad \text{and} \quad c = 4$$

$$x = \frac{-b \pm \sqrt{b^2 - 4ac}}{2a}$$

$$x = \frac{-3 \pm \sqrt{3^2 - 4(1)(4)}}{2(1)}$$

$$x = \frac{-3 \pm \sqrt{9 - 16}}{2}$$

$$x = \frac{-3 \pm \sqrt{-7}}{2} \qquad \boxed{\textit{No real solutions}}$$

As soon as we see that the answer involves the square root of a negative number, we can stop and say that the equation has *no real solutions*. ■

Let's take another look at an example we have examined before.

EXAMPLE 7

A market research firm finds that the daily revenue R (in dollars) earned by a chemical manufacturing company on the sale of g gallons of mixture KZD is given by the equation $R = g^2 - 80g$. How many gallons must be manufactured in order to earn a daily revenue of $3,000? Round the answer to the nearest tenth of a gallon.

Solution

In order to find the number of gallons that need to be manufactured to produce a revenue of $3,000 means we must substitute $R = 3,000$ into the given equation and then solve for g.

$$R = g^2 - 80g \qquad \textit{Substitute } R = 3,000.$$

$$3,000 = g^2 - 80g \qquad \textit{Get the equation in standard form.}$$

$$0 = g^2 - 80g - 3,000 \qquad \textit{We will use the quadratic formula.}$$
$$\textit{We have } a = 1, b = -80, \textit{ and } c = -3,000.$$

$$g = \frac{-(-80) \pm \sqrt{(-80)^2 - 4(1)(-3,000)}}{2(1)}$$

$$g = \frac{80 \pm \sqrt{6,400 + 12,000}}{2}$$

$$g = \frac{80 \pm \sqrt{18,400}}{2} \qquad \textit{Using a calculator we obtain}$$

$$g = 107.82330 \quad \text{or} \quad g = -27.82330$$

Obviously, the negative answer makes no sense for the number of gallons. Therefore, in order to produce a revenue of $3,000 the company needs to manufacture $\boxed{107.8 \text{ gallons}}$ of KZD. ■

In the next section we will discuss how to decide which method to choose for solving a particular quadratic equation.

EXERCISES 11.3

In Exercises 1–8, identify a, b, and c as used in the quadratic formula.

1. $x^2 + 3x - 5 = 0$

2. $x^2 + 5x - 2 = 0$

3. $t^2 - 7t = 6$

4. $t^2 + 7 = 6t$

5. $2u^2 = 8u$

6. $5u = 10u^2$

7. $3x^2 - 11 = 0$

8. $4z^2 = 7$

In Exercises 9–40, solve the equation by using the quadratic formula where appropriate.

9. $x^2 + 3x - 5 = 0$

10. $x^2 + 5x - 2 = 0$

11. $y^2 + 4y - 6 = 0$

12. $y^2 + 2y - 5 = 0$

13. $u^2 - 2u + 3 = 0$

14. $u^2 - 3u + 3 = 0$

15. $t^2 - 7t = 6$

16. $t^2 + 6 = 6t$

17. $2x^2 - 3x - 1 = 0$

18. $3x^2 + 5x + 2 = 0$

19. $5x^2 - x = 2$

20. $7x^2 - 3 = x$

21. $t^2 - 3t + 4 = 2t^2 + 4t - 3$

22. $2t^2 + 4t + 1 = 3t^2 + 7t + 4$

23. $(5w + 2)(w - 1) = 3w + 1$

24. $(3w - 1)(2w - 3) = w$

25. $(x - 1)^2 = x(x - 5)$

26. $(x + 2)^2 = x(x + 4)$

27. $3x^2 - 5x + 7 = 2x(x - 5) + 9x + 5$

28. $t^2 + 4t = t(2t - 2) + 13$

29. $2u^2 = 6u + 3$

30. $8r - 2 = 5r^2$

31. $x^2(x - 1) = (x - 1)^3$

32. $y^2(y + 2) = (y + 2)^3$

33. $4x = 9x^2$

34. $5u^2 = 3u$

35. $4 = 9x^2$

36. $5u^2 = 3$

37. $\dfrac{w}{2} = \dfrac{3}{w + 2}$

38. $\dfrac{t}{6} = \dfrac{3}{t + 4}$

39. $\dfrac{y}{y + 1} = \dfrac{y + 2}{3y}$

40. $\dfrac{z}{2} + \dfrac{3}{z} = z$

In Exercises 41–48, solve the equation and round off your answers to the nearest hundredth.

41. $x^2 - 1 = x$

42. $x^2 - 3 = 4x$

43. $t^2 - 7t = 10$

44. $u^2 - 5u + 5 = 0$

45. $3w^2 + 7w = 21$

46. $2z^2 = 13 - 31z$

47. $1.7x^2 - 3.2x = 6.1$

48. $3.4x^2 + 7.3x + 2.05 = 0$

In Exercises 49–52, round your answer to the nearest tenth.

49. A manufacturing firm determines that the revenue R (in dollars) earned on the manufacture of s square feet of plastic is given by

$$R = s^2 - 250s + 600$$

Determine the number of square feet that must be manufactured to produce a revenue of $50,000.

50. The daily profit P (in dollars) earned by a company on the sale of x gallons of machine lubricant is given by

$$P = -x^2 + 160x - 3,400$$

Determine the number of gallons of lubricant that must be sold to produce a daily profit of $2,000.

51. A ball is thrown upward from the top of a building so that its height h (in feet) above the ground t seconds after it is thrown is given by

$$h = 120 + 40t - 16t^2$$

How many seconds does it take for the ball to be 140 ft above the ground?

52. A ball is thrown downward from the top of a cliff 1,280 ft high so that the distance d the ball has fallen (in feet) t seconds after it is thrown is given by

$$d = 75t + 16t^2$$

How many seconds does it take for the ball to hit the ground?

QUESTIONS FOR THOUGHT

53. Solve Example 1 of this section, which was solved by the method of completing the square, by the quadratic formula. Which method of solution do you think is easier?

54. Compare and contrast the various methods we have discussed for solving quadratic equations.

55. Discuss what is **wrong** with each of the following "solutions."

(a) $x^2 - 3x - 1 = 0$

$$x = \frac{-3 \pm \sqrt{9 - 4(1)}}{2}$$

$$x = \frac{-3 \pm \sqrt{5}}{2}$$

(b) $x^2 - 5x - 3 = 0$

$$x = \frac{5 \pm \sqrt{25 - 12}}{2}$$

$$x = \frac{5 \pm \sqrt{13}}{2}$$

(c) $x^2 - 5x + 3 = 0$

$$x = 5 \pm \frac{\sqrt{25 - 12}}{2}$$

$$x = 5 \pm \frac{\sqrt{13}}{2}$$

(d) $x^2 - 6x - 3 = 0$

$$x = \frac{6 \pm \sqrt{36 + 12}}{2} = \frac{6 \pm \sqrt{48}}{2}$$

$$x = \frac{6 \pm \sqrt{16\sqrt{3}}}{2} = \frac{6 \pm 4\sqrt{3}}{2}$$

$$x = 6 \pm 2\sqrt{3}$$

MINI-REVIEW

In Exercises 56–59, perform the indicated operations and simplify.

56. $4(\sqrt{3} - 2\sqrt{5}) - 3(\sqrt{5} + 4\sqrt{3})$

57. $3\sqrt{2}(\sqrt{2} + \sqrt{7}) + \sqrt{7}(5\sqrt{2} - \sqrt{7})$

58. $(\sqrt{x} - 3)^2$

59. $(2\sqrt{x} - \sqrt{5})(3\sqrt{x} - 4\sqrt{5})$

Solve each of the following problems algebraically. Be sure to label what the variable(s) represent(s).

60. Lynne bought 2 pounds of pistachio nuts and 3 pounds of cashews for a total of $23, while Gail bought 3 pounds of pistachio nuts and 2 pounds of cashews for a total of $22. What is the price per pound for each?

61. Sam paid $42 for the five pastrami sandwiches and six tuna sandwiches he ordered for his office staff. Bernie paid $47.25 for the four pastrami sandwiches and nine tuna sandwiches he ordered for his office staff from the same delicatessen. What is the price for each type of sandwich?

11.4   Choosing a Method

At this point we have three methods for solving a quadratic equation:

1. The factoring method

2. The square root method

   Within this method we include the process of completing the square, which is often necessary before we can take square roots.

3. The quadratic formula

   We know that both completing the square and the quadratic formula work for *all* quadratic equations. On the other hand, we have seen that the factoring method, when it works, is often the easiest method. How then should we decide on which method to choose for a particular equation?

   Based on what we have seen up to now, it is fairly safe to say that unless an equation is given in the form of a perfect square, it is generally easier to solve a quadratic equation by the factoring method or by using the quadratic formula than it is to solve by completing the square. Consequently, our first step in the process of solving a quadratic equation will usually be to get the equation into standard form, since standard form is required for both the factoring method and the quadratic formula.

   Let's look at several examples and analyze why we choose a particular method in each case. Then based on this analysis we can try to offer some general guidelines as to how to choose the "best method."

EXAMPLE 1   Solve for x.

(a) $3x^2 = 5x$ (b) $x^2 = 3x + 28$

Solution   (a) We begin by putting the equation into standard form.

$$3x^2 = 5x$$

$$3x^2 - 5x = 0 \qquad \textit{We \textbf{should} notice a common factor of x; the factoring}$$

$$x(3x - 5) = 0 \qquad \textit{method is the most convenient.}$$

$$x = 0 \quad \text{or} \quad 3x - 5 = 0$$

$$\boxed{x = 0} \quad \text{or} \quad \boxed{x = \frac{5}{3}}$$

Clearly then, in such a situation where there is no constant term ($c = 0$) but there is a first-degree term ($b \neq 0$), there will be a common variable factor, and so the factoring method will work.

(b) Again we begin by putting the equation in standard form.

$$x^2 = 3x + 28$$

$$x^2 - 3x - 28 = 0$$

$$(x - 7)(x + 4) = 0$$

Since the leading coefficient is 1, we should feel fairly confident that if the left-hand side factors we will find the correct factorization quickly.

$$x - 7 = 0 \quad \text{or} \quad x + 4 = 0$$

$$\boxed{x = 7} \quad \text{or} \quad \boxed{x = -4}$$

This example is further evidence of the fact that the factoring method, when it works, is usually the easiest method to use, particularly if there is a common factor of the variable as in part **(a)**. ■

EXAMPLE 2

Solve for x.

(a) $x(x - 2) = 6x - 2$

(b) $10x^2 + 39x - 100 = 20 - 5x - 14x^2$

Solution

(a) We begin by multiplying out the left-hand side and putting the equation into standard form.

$$x(x - 2) = 6x - 2$$

$$x^2 - 2x = 6x - 2$$

$$x^2 - 8x + 2 = 0$$

At a glance we should be able to see that this cannot be factored. Therefore, we use the quadratic formula with $a = 1$, $b = -8$, and $c = 2$.

$$x = \frac{-b \pm \sqrt{b^2 - 4ac}}{2a}$$

$$x = \frac{-(-8) \pm \sqrt{(-8)^2 - 4(1)(2)}}{2(1)}$$

$$x = \frac{8 \pm \sqrt{64 - 8}}{2} = \frac{8 \pm \sqrt{56}}{2} \qquad \textit{Simplify the radical.}$$

$$x = \frac{8 \pm \sqrt{4}\sqrt{14}}{2} = \frac{8 \pm 2\sqrt{14}}{2} \qquad \textit{Factor and reduce.}$$

$$x = \frac{2(4 \pm \sqrt{14})}{2} = \frac{\cancel{2}(4 \pm \sqrt{14})}{\cancel{2}}$$

$$\boxed{x = 4 \pm \sqrt{14}}$$

(b) We begin by getting the equation into standard form.

$$10x^2 + 39x - 100 = 20 - 5x - 14x^2$$

$$24x^2 + 44x - 120 = 0 \qquad \textit{Factor out the common factor of 4.}$$

$$4(6x^2 + 11x - 30) = 0 \qquad \textit{Divide both sides of the equation by 4.}$$

$$\frac{\cancel{4}(6x^2 + 11x - 30)}{\cancel{4}} = \frac{0}{4}$$

$$6x^2 + 11x - 30 = 0 \qquad \textit{This looks like it may be difficult to factor, or may not factor at all. Let's use the formula. } a = 6, b = 11, \textit{ and } c = -30$$

$$x = \frac{-11 \pm \sqrt{11^2 - 4(6)(-30)}}{2(6)}$$

$$x = \frac{-11 \pm \sqrt{121 + 720}}{12} = \frac{-11 \pm \sqrt{841}}{12} \qquad \sqrt{841} = 29^*$$

*We have previously mentioned how we can quickly determine whether a reasonably small number is a perfect square. Because 841 ends in 1, the only possible candidates for its square roots are 21 and 29.

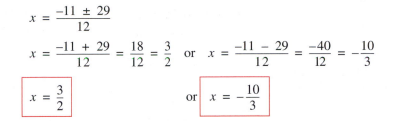

$$x = \frac{-11 \pm 29}{12}$$

$$x = \frac{-11 + 29}{12} = \frac{18}{12} = \frac{3}{2} \quad \text{or} \quad x = \frac{-11 - 29}{12} = \frac{-40}{12} = -\frac{10}{3}$$

$$\boxed{x = \frac{3}{2}} \qquad \text{or} \qquad \boxed{x = -\frac{10}{3}}$$

This example illustrates several things. First, even though using the quadratic formula is a straightforward mechanical procedure, the arithmetic can get messy, and there are numerous opportunities to make careless errors.

Second, factoring out the common factor of 4 was useful even though we used the quadratic formula and not the factoring method. (Imagine what the arithmetic would have been like if we had used the formula with the values $a = 24$, $b = 44$, and $c = -120$.)

Third, the fact that we obtained rational solutions tells us that we could have used the factoring method. (*Try it!*) Nevertheless, it is not clear whether the factoring method (which requires a fair amount of trial and error) or the quadratic formula (which is a bit messy) is the "better" method. ∎

Basically, if you are good at factoring, it probably pays to invest a minute or two in trying to use the factoring method. If that fails, then generally you would use the quadratic formula.

EXAMPLE 3 Solve for x. $(3x - 5)^2 = 17$

Solution Since this equation is given in perfect square form, we might as well take advantage of that fact and use the square root method.

$$(3x - 5)^2 = 17 \qquad \textit{Take square roots.}$$

$$3x - 5 = \pm\sqrt{17}$$

$$3x = 5 \pm \sqrt{17}$$

$$\boxed{x = \frac{5 \pm \sqrt{17}}{3}}$$

Had we wanted to, we could have multiplied out $(3x - 5)^2$, put the equation into standard form, and then used the formula. (*Try it!*) However, if an equation is given in the form of a perfect square, it seems foolish not to take advantage of the square root method. ∎

Unless you are given specific instructions as to which method of solution to use, you are free to choose whichever method *you* find the most efficient.

EXAMPLE 4 Solve for t. $3t^2 + 2 = 5t^2 - 4$

Solution We can begin by getting this equation into standard form.

$$3t^2 + 2 = 5t^2 - 4 \qquad \textit{We recognize that this equation is a pure quadratic.}$$

$$0 = 2t^2 - 6 \qquad \textit{Therefore, the square root method will work.}$$

$$6 = 2t^2 \qquad \textit{Divide both sides by 2.}$$

$$3 = t^2 \qquad \textit{Take square roots.}$$

$$\boxed{\pm\sqrt{3} = t}$$

As soon as we see that there is no first-degree term ($b = 0$), we know that we can use the square root method. ∎

Based on our discussion thus far, we offer the following outline for solving quadratic equations.

Outline for Solving Quadratic Equations	1. Simplify both sides of the equation as completely as possible and put the equation into standard form.
	2. Factor out any common factors, and divide both sides of the equation by any common *numerical* factor to eliminate it.
	3. If there is a common variable factor, use the factoring method. In other words, the factoring method works when $c = 0$.
	4. If there is no first-degree term ($b = 0$), then the square root method will work.
	5. If it looks like the nonzero side of the equation can be factored fairly easily, then try the factoring method.
	6. If it looks too complicated to factor or does not factor, then use the quadratic formula.
	7. Check the equation either by substituting into the original equation or by solving the equation using a different method.

EXERCISES 11.4

*In each of the following exercises, use the method **you** think is the most appropriate to solve the given equation. Check your answers by using a different method.*

1. $x^2 + 6x + 5 = 0$

2. $y^2 + 4y + 3 = 0$

3. $x^2 + 6x - 5 = 0$

4. $y^2 + 4y - 3 = 0$

5. $2r^2 + 1 = 3r$

6. $2r^2 + 1 = 4r$

7. $w^2 = 4w + 5$

8. $w^2 = 4w$

9. $(x + 1)(x + 2) = (x + 3)(x + 4)$

10. $(z + 3)(z - 2) = (z - 3)(z + 2)$

11. $(a + 1)^2 - (a + 3)(a - 2) = 2a + 6$

12. $y(y + 6) = (y + 4)(y + 3)$

13. $4x^2 = 16x - 28$

14. $6x^2 = 24x - 30$

15. $(x - 1)^2 = 5$

16. $(m - 3)^2 = 7m$

17. $(x - 1)^2 = 5x$

18. $(m - 3)^2 = 7$

19. $(u + 2)^2 = 4(u + 5)$

20. $(u + 2)^2 = u(u + 5)$

21. $y^2 - 4y + 10 = 5(y + 2)$

22. $3y^2 - 7y - 12 = 4(2y^2 - 3)$

23. $(t + 4)(t - 8) = 13$

24. $(c - 3)(c + 6) = 22$

25. $(n + 2)(n + 1) = 3$

26. $(n + 2)(n + 1) = 3n$

27. $z^2 - 3z = 3z - 9$

28. $4z + 5 = z^2 + 12z + 21$

29. $16z + 12 = 3z^2$

30. $17z - 12 = 5z^2$

31. $x + \dfrac{1}{x} = 2$

32. $w + \dfrac{2}{w} = 1$

33. $x^2 + 1 = \dfrac{5}{2}x$

34. $x^2 + \dfrac{2}{3}x = 4$

35. $\dfrac{x}{x + 1} = \dfrac{4}{x + 4}$

36. $\dfrac{x + 2}{x} = \dfrac{5}{x - 4}$

37. $\dfrac{x}{x+1} = \dfrac{x+2}{x+3}$

38. $\dfrac{x+2}{x+4} = \dfrac{x+3}{x+6}$

39. $\dfrac{3x}{x+1} + \dfrac{2}{x-1} = 4$

40. $\dfrac{x+4}{x+1} - \dfrac{4}{x+2} = 2$

? QUESTIONS FOR THOUGHT

41. Solve the equation $2x^2 + 3x = 20$ by all three methods. Which was the easiest? The most difficult? Why?

42. Solve the equation $3x^2 - 5x - 1 = 0$. Check your answer by substituting one of your solutions back into the original equation. Also check your answer by solving the equation using a different method.

    Which method of checking was easier? Why? Which check do you have more confidence in? Why?

◇ MINI-REVIEW

43. Find the slope of the line passing through each of the following sets of points:

    (a) $(-1, 3)$ and $(3, -4)$

    (b) $(-4, 7)$ and $(2, 7)$

    (c) $(-5, 2)$ and $(-5, -1)$

44. Sketch the graph of the line with slope $\frac{2}{3}$ that passes through the point $(1, 2)$.

45. Write an equation of the line that passes through the points $(-2, -5)$ and $(1, -1)$.

11.5 Applications

Now that we have the ability to solve quadratic equations, we can solve an even wider variety of problems. Keep in mind that when you are solving a verbal problem you do not know whether the resulting equation will be linear or quadratic. Be sure to look carefully at the equation before deciding on a method of solution.

EXAMPLE 1   The sum of a number and its reciprocal is $\frac{29}{10}$. Find the number.

Solution   We translate the words in the problem as follows:

Let x = the number; then $\dfrac{1}{x}$ = its reciprocal.

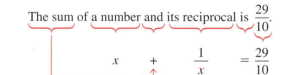

Thus, our equation is

$$x + \frac{1}{x} = \frac{29}{10}$$

Multiply both sides of the equation by the LCD, which is 10x, to clear the fractions.

$$\frac{10x}{1}\left(x + \frac{1}{x}\right) = \frac{10x}{1}\left(\frac{29}{10}\right)$$

$$10x \cdot x + \frac{10x}{1} \cdot \frac{1}{x} = \frac{10x}{1} \cdot \frac{29}{10}$$

$$10x^2 + \frac{10x}{1} \cdot \frac{1}{x} = \frac{10x}{1} \cdot \frac{29}{10}$$

$$10x^2 + 10 = 29x \qquad \textit{We recognize this as a quadratic equation.}$$
$$\textit{Put the equation in standard form.}$$

$$10x^2 - 29x + 10 = 0 \qquad \textit{The factoring method works.}$$

$$(5x - 2)(2x - 5) = 0$$

$$5x - 2 = 0 \quad \text{or} \quad 2x - 5 = 0$$

$$\boxed{x = \frac{2}{5}} \quad \text{or} \quad \boxed{x = \frac{5}{2}}$$

We leave it to the student to check that both solutions satisfy the statement of the problem. Notice that the two solutions are reciprocals of each other. ■

EXAMPLE 2

Samantha and Jenna are both driving to a tennis tournament 200 miles away. They both leave at the same time, but Samantha arrives 1 hour ahead of Jenna because she was driving 10 miles per hour faster than Jenna. Find the rate at which each drove to the tournament.

Solution

We should recognize that our starting point in such a problem is the relationship $d = rt$. How do we use it here? The problem asks us to find the rates at which they drive.

Let r = Jenna's rate.

Then $r + 10$ = Samantha's rate.   *She was driving 10 mph faster than Jenna.*

The problem also tells us that Samantha arrived 1 hour ahead of Jenna, which means that (since they left at the same time) Samantha's driving time was 1 hour *less*. Thus, we have the following "time relationship":

$$t_{\text{Samantha}} = t_{\text{Jenna}} - 1$$

We use the fact that $d = rt$ in the equivalent form of $t = \dfrac{d}{r}$ by substituting into the time relationship:

$$\frac{d_{\text{Samantha}}}{r_{\text{Samantha}}} = \frac{d_{\text{Jenna}}}{r_{\text{Jenna}}} - 1$$

However, we know that they both traveled the same distance, which is given to be 200 miles, and we have represented the rates above. Let's substitute these quantities into the last equation.

$$\frac{200}{r + 10} = \frac{200}{r} - 1 \qquad \textit{We multiply both sides of the equation by the LCD of } r(r + 10) \textit{ to clear fractions.}$$

$$\frac{r(r + 10)}{1} \cdot \frac{200}{r + 10} = \frac{r(r + 10)}{1} \cdot \frac{200}{r} - r(r + 10)$$

$$200r = 200(r + 10) - r(r + 10)$$

$$200r = 200r + 2,000 - r^2 - 10r \qquad \textit{Put the equation in standard form.}$$

$$r^2 + 10r - 2,000 = 0$$

$$(r + 50)(r - 40) = 0$$

$$r + 50 = 0 \quad \text{or} \quad r - 40 = 0$$

$$r = -50 \quad \text{or} \qquad r = 40$$

Since it makes no sense for a rate to be negative, the answer to the problem is that $r = 40$.

Thus, we have ⟨Jenna's rate is 40 mph⟩ and ⟨Samantha's rate is 50 mph⟩. The check is left to the student. ■

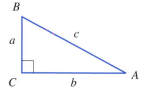

A **_right triangle_** is a triangle with a right (90°) angle. The sides that form the right angle are called the **_legs._** The side opposite the right angle (which is the longest side of the triangle) is called the **_hypotenuse._**

There is a famous theorem of Pythagoras that says the sum of the squares of the legs of a right triangle is equal to the square of the hypotenuse. Symbolically, we draw our right triangle and label the lengths of the legs a and b and the length of the hypotenuse c, as indicated in the accompanying figure.

PYTHAGOREAN THEOREM	In a right triangle with legs a and b and hypotenuse c, $$a^2 + b^2 = c^2$$

EXAMPLE 3 Find the value of x in the following triangle.

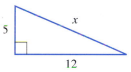

Solution We apply the Pythagorean theorem.

$$5^2 + 12^2 = x^2$$
$$25 + 144 = x^2$$
$$169 = x^2 \quad \textit{Take square roots.}$$
$$\pm 13 = x \quad \textit{Since the length of a side of a triangle cannot be negative}$$

$$\boxed{x = 13}$$ ■

EXAMPLE 4 The diagonal of a square is 10 centimeters long. Find the length of a side of the square.

Solution A square has the properties that all its sides have the same length and that all its angles are right angles. If we let s represent the length of a side of the square, then we have the situation shown in Figure 11.1.

We see that the diagonal forms two right triangles. We use the Pythagorean theorem to obtain the equation

$$s^2 + s^2 = 10^2$$
$$2s^2 = 100$$
$$s^2 = 50 \quad \textit{Take square roots.}$$
$$s = \pm\sqrt{50} = \pm\sqrt{25}\sqrt{2} = \pm 5\sqrt{2}$$

Since it makes no sense for a length to be negative, we reject the negative solution.

The answer to the question is that the length of a side of the square is

$$5\sqrt{2} \text{ cm} \approx 7 \text{ cm}$$ ■

Figure 11.1
Diagram for Example 4

EXAMPLE 5 The Weave-A-Website Company determines that the profit P (in dollars) made for each Web site designed is given by the equation

$$P = 0.95x^2 + 5.25x + 110 \quad \text{for} \quad 1 \le x \le 5$$

where x represents the number of Web sites (in hundreds).

(a) Find the profit per Web site (to the nearest dollar) if 150 Web sites are designed.

(b) How many Web sites (to the nearest ten) must be designed if the profit per Web site is to be $150?

Solution

Remember that x is the number of Web sites in hundreds, so that x = 2.5 means 250 Web sites.

(a) The given information tells us that in order to compute the profit per Web site when 200 Web sites are designed, we substitute $x = 2$ into the equation for P. Similarly, if 250 Web sites are designed, we substitute $x = 2.5$ into the equation.

Therefore, if 150 Web sites are designed we compute the profit per Web site by substituting $x = 1.5$ into the formula for P.

$$P = 0.95x^2 + 5.25x + 110 \qquad \text{\textit{Substitute } } x = 1.5.$$
$$= 0.95(1.5)^2 + 5.25(1.5) + 110 \qquad \text{\textit{Using a calculator, we get}}$$
$$P = 120.0125$$

Thus to the nearest dollar, the profit per Web site is $120.

(b) We are asked to find the number of Web sites that will produce a profit of $150 per Web site. In other words, we are given the value of P and we are required to solve the equation for x.

$$P = 0.95x^2 + 5.25x + 110 \qquad \text{\textit{Substitute } } P = 150.$$
$$150 = 0.95x^2 + 5.25x + 110 \qquad \text{\textit{We recognize this as a quadratic equation.}}$$

We put the equation in standard form and identify a, b, and c.

$$0 = 0.95x^2 + 5.25x - 40 \qquad \text{\textit{We substitute } } a = 0.95, b = 5.25,$$
and c = −40 into the quadratic formula.

$$x = \frac{-5.25 \pm \sqrt{(5.25)^2 - 4(0.95)(-40)}}{2(0.95)} \qquad \text{\textit{The formula gives two values for x.}}$$

$$x = 4.29 \quad \text{or} \quad x = -9.82 \qquad \text{\textit{Rounded to the nearest hundredth}}$$
Obviously, we reject x = −9.82 since it does not make sense.

Substitute x = 4.29 into the equation for P to check this answer.

$x = 4.29$ means that if 429 Web sites are designed, the profit for each Web site will be $150. Thus the answer to the question is 430 Web sites (to the nearest ten).

Note the difference between parts **(a)** and **(b).** In part **(a),** we are substituting a value for x into the equation and directly computing the value of P (fairly straightforward), whereas in part **(b)** we are substituting a value for P and we must then solve the resulting equation for x (substantially more work). ■

EXERCISES 11.5

Solve each of the following exercises algebraically.

1. The sum of a number and its reciprocal is $\frac{13}{6}$. Find the number.

2. The sum of a number and 3 times its reciprocal is $\frac{79}{10}$. Find the number.

3. The sum of two numbers is 20, and their product is 96. Find the two numbers.

4. One number is 5 more than 3 times a second number. If their product is -2, what are the numbers?

5. The length of a rectangle is 3 more than twice its width, and its area is 90 sq m. Find its dimensions.

6. The width of a rectangle is one-third its length. If the area of the rectangle is 20 sq in., what are the dimensions of the rectangle?

7. One leg of a right triangle is 7 cm long and the hypotenuse is 15 cm long. What is the length of the other leg?

8. A rectangle has dimensions 10 mm by 20 mm. What is the length of the diagonal of the rectangle?

9. Find the length of the diagonal of a square whose side is 8 in.

10. Find the length of the sides of a square whose diagonal is 8 in.

11. Harold leans a 30-ft ladder against a building (see the accompanying figure). The base of the ladder is 8 ft from the building. How high up the building does the ladder reach?

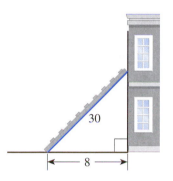

12. Sam leans a 50-ft ladder against a building so that the top of the ladder is 45 ft above the ground. How far is the base of the ladder from the building?

13. The lengths of the sides of a right triangle are three consecutive integers. Find them.

14. The lengths of the sides of a right triangle are three consecutive even integers. Find them.

15. The denominator of a fraction is 1 more than the numerator. If the numerator is increased by 3, the resulting fraction is 1 more than the original fraction. Find the original fraction.

16. The numerator of a fraction is 1 less than the denominator. If $\frac{7}{12}$ is added to the fraction, the result is the reciprocal of the original fraction. Find the original fraction.

17. A concert hall contains 768 seats. If the number of rows of seats is 8 less than the number of seats in each row, how many seats are there in each row?

18. A box contains 800 marbles in small bags, all containing the same number of marbles. There are half as many bags as there are marbles in each bag. How many bags are there?

19. A motorist completes a trip covering 150 km in 2 hours. She covers the first 120 km at a certain rate of speed and then decreases her speed by 20 kph for the remaining 30 km. Find her speed during the first 120 km.

20. Arnold travels from town A to town B, which are 300 km apart. His rate going is twice as fast as his rate returning. If his total traveling time was $7\frac{1}{2}$ hours, what was his rate of speed going from A to B?

21. Suppose the profit (P) made on the sale of theater tickets is related to the price (x) of the tickets according to the following equation:

$$P = 1,000(-x^2 + 15x - 35)$$

How much profit is earned if the price per ticket is $5? $4?

22. Suppose a ball is thrown up into the air in such a way that its height (H) above the ground t seconds after it is thrown is given by the equation

$$H = -16t^2 + 80t + 10$$

How high is the ball after 3 seconds? In how many seconds will it be 5 feet above the ground?

In Exercises 23–36, solve each problem algebraically. Round off your answers to the nearest tenth where necessary.

23. The three points P, Q, and R form a right triangle with the right angle at Q. If the distance from P to Q is 8.6 m and the distance from Q to R is 4.9 m, find the distance from P to R.

24. The length of a rectangle is 1 less than twice the width. If the area of the rectangle is 10 sq ft, find the dimensions of the rectangle.

25. Neva and Yomar are both driving a distance of 180 miles. Neva covers the distance in one half hour less than Yomar does because she drives 12 mph faster than Yomar. Find Neva's rate of speed.

26. Alma can do a certain job 1 hour faster than Kim can. In order to complete the job, Alma works for 4 hours and Kim works for 5 hours. How long does it take Alma to do the job alone?

27. If a rock is dropped from a height of 1,000 feet, its height h after t seconds is given by the formula $h = 1,000 - 16t^2$.

    (a) Find the height of the object after 3.5 seconds.

    (b) How long will it take for the rock to reach a height of 700 feet?

    (c) How long will it take for the rock to hit the ground? [*Hint:* When is $h = 0$?]

28. If a rocket is launched upward from an initial height of 80 m with an initial velocity of 120 meters per second, then its height h after t seconds is given by

$$h = 80 + 120t - 9.8t^2$$

    (a) Find the height of the ball after 2.4 seconds.

    (b) Approximately how long will it take the rocket to reach a height of 400 meters?

    (c) Approximately how long will it take the rocket to hit the ground?

29. The profit P (in dollars) on each television manufactured by Videocom, Inc., is related to the number of TV sets produced each day according to the equation

$$P = 0.82x^2 + 4.25x + 95$$

where x is the number of TV sets produced each day (in hundreds) and $2 \le x \le 8$.

    (a) Find the profit per set if 350 sets are produced each day.

    (b) How many sets (to the nearest ten) should be produced each day to make the profit per set $130?

30. The cost C (in dollars) on each cellular telephone manufactured by Clearvoice, Inc., is related to the number of phones produced each week according to the equation

$$C = 0.06x^2 - 3.2x + 75.6$$

where x is the number of phones produced each week (in thousands) and $10 \le x \le 25$.

(a) Find the cost per phone if 15,000 phones are produced each week.

(b) How many phones should be produced each week to make the cost per phone $36?

31. A baseball "diamond" is actually a square 90 ft on a side. Find the distance from home plate to second base (to the nearest foot).

32. The size of a television is given by the diagonal size of its rectangular screen. What is the size of a television screen that is 20 in. by 25 in. (to the nearest inch)?

33. A rectangular garden measuring 10 ft by 15 ft is to be surrounded on three sides by a uniform tile path, as illustrated in the accompanying figure. If 100 sq ft of tile is to be used, how wide should the path be?

34. A rectangular garden measuring 10 ft by 15 ft is to be surrounded on all four sides by a uniform tile path. If there are 100 sq ft of tile available, how wide should the path be?

35. A father and son can paint a house together in 6 days. Painting alone, it takes the son 9 days longer than it takes the father. How long would it take each person painting alone?

36. Janet and Irena can mow a lawn together in 2 hours. Mowing alone, it takes Janet one hour longer than it takes Irena. How long would it take each person to mow the lawn mowing alone?

 QUESTIONS FOR THOUGHT

37. The Greeks believed that the rectangle that had the most pleasing form was one that had the ratio of its width to length the same as the ratio of its length to its width plus length. Such a rectangle is called a *golden rectangle*. Algebraically, this means that

$$\frac{w}{l} = \frac{l}{w + l}$$

(a) If the width of a rectangle is equal to 1 inch, what must the length be in order for it to be a golden rectangle? (The answer is called the *golden ratio* or *golden mean*.)

(b) Repeat part (a) for the general width and length by solving the equation $\frac{w}{l} = \frac{l}{w + l}$ for l in terms of w. That is, treat l as the variable in the equation, and treat w as if it were a constant.

38. An alternative way to define the golden ratio is as follows. If the width of a rectangle is 1, what must the length be (call it x) so that if we remove a square of side 1 from one end of the rectangle, the remaining rectangle (which has dimensions of width $= x - 1$ and length $= 1$) has the same ratio of length to width as the original rectangle? (See the accompanying figure.) Express this relationship algebraically, and solve for x. Did you get the same answer as for part (a) of Exercise 37?

39. Can you solve Exercise 14 for three consecutive *odd* integers? Why or why not?

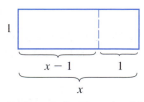

Diagram for Exercise 38

11.6 Graphing Quadratic Equations

In Chapter 3 we discussed solving first-degree equations in one variable. In Chapter 5 we recognized that to graph a first-degree equation in two variables we need a two-dimensional coordinate system. In this process, we repeatedly used our ability to solve first-degree equations in one variable to find points that were on the graph and, in particular, to find the x- and y-intercepts of the graph.

Now that we have developed methods for solving the general quadratic equation in one variable of the form $ax^2 + bx + c = 0$, we apply these techniques to graph the general quadratic equation in two variables of the form $y = ax^2 + bx + c$. We will develop a method for obtaining the graph of such a general quadratic equation through a series of simple examples.

We begin by examining the two quadratic equations $y = x^2$ and $y = -x^2$. To get an idea of what these graphs look like, we construct a table of values, plot the corresponding points, and connect them with a smooth curve.

x	$y = x^2$	y	x	$y = -x^2$	y
-3	$y = (-3)^2$	9	-3	$y = -(-3)^2$	-9
-2	$y = (-2)^2$	4	-2	$y = -(-2)^2$	-4
-1	$y = (-1)^2$	1	-1	$y = -(-1)^2$	-1
$-\frac{1}{2}$	$y = \left(-\frac{1}{2}\right)^2$	$\frac{1}{4}$	$-\frac{1}{2}$	$y = -\left(-\frac{1}{2}\right)^2$	$-\frac{1}{4}$
0	$y = (0)^2$	0	0	$y = -(0)^2$	0
$\frac{1}{2}$	$y = \left(\frac{1}{2}\right)^2$	$\frac{1}{4}$	$\frac{1}{2}$	$y = -\left(\frac{1}{2}\right)^2$	$-\frac{1}{4}$
1	$y = (1)^2$	1	1	$y = -(1)^2$	-1
2	$y = (2)^2$	4	2	$y = -(2)^2$	-4
3	$y = (3)^2$	9	3	$y = -(3)^2$	-9

The graphs appear in Figures 11.2 and 11.3. These graphs are called **parabolas,** and every graph of a quadratic equation will have a shape similar to these. The graphs

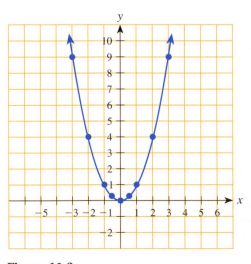

Figure 11.2

The graph of $y = x^2$

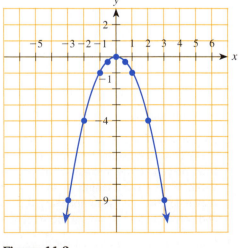

Figure 11.3

The graph of $y = -x^2$

of $y = x^2$ and $y = -x^2$ are typical of the graphs of $y = ax^2$. If a is positive, the graph of $y = ax^2$ will be similar to the graph of $y = x^2$. If a is negative, the graph of $y = ax^2$ will be similar to the graph of $y = -x^2$. (The effect of the coefficient a on the shape of the graph is discussed in Exercises 47–50.)

The lowest or highest point on the graph of a parabola is called the **vertex** of the parabola. One of the most important aspects of a parabola is that the graph is symmetric with respect to the vertical line through the vertex. This vertical line is called the **axis of symmetry.** In the graphs in Figures 11.2 and 11.3, the y-axis is the axis of symmetry. This simply means that if a mirror were placed on the y-axis, the left-hand portion of the graph would be a reflection of the right-hand portion.

EXAMPLE 1

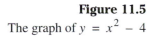

Sketch the graphs and label the intercepts of each of the following.

(a) $y = x^2 + 3$ (b) $y = x^2 - 4$

Solution

Rather than compute entirely new tables of values for these equations, let's try to analyze the equations to figure out what the graphs will look like.

(a) If we compare the equation $y = x^2 + 3$ to the equation $y = x^2$, we see that for each x value, the y value on the graph of $y = x^2 + 3$ is 3 more than the corresponding value on the graph of $y = x^2$. This means that for each x value the point on the graph of $y = x^2 + 3$ is 3 units higher than the corresponding point on the graph of $y = x^2$. This suggests that the graph of $y = x^2 + 3$ can be obtained by shifting the graph of $y = x^2$ up 3 units. The graph appears in Figure 11.4. The graph of $y = x^2$ appears as the red curve.

Based on Figure 11.4 we can see that the graph of $y = x^2 + 3$ has no x-intercepts but does have a y-intercept of $y = 3$. It is left to the student to verify those intercepts algebraically: Substitute $y = 0$ to find the x-intercepts and substitute $x = 0$ to find the y-intercept.

Figure 11.4

The graph of $y = x^2 + 3$

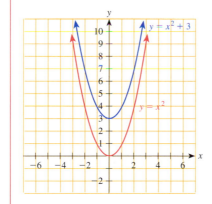

(b) A similar analysis leads us to the conclusion that for each x value the point on the graph of $y = x^2 - 4$ is 4 units lower than the corresponding point on the graph of $y = x^2$. The graph appears in Figure 11.5.

Figure 11.5

The graph of $y = x^2 - 4$

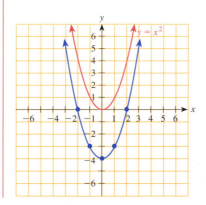

Recall that to find the x-intercepts we set $y = 0$ and solve for x. To find the y-intercepts we set $x = 0$ and solve for y.

To find the *x*-intercepts, we set $y = 0$ and solve for *x*:

$y = x^2 - 4$ *Set $y = 0$.*

$0 = x^2 - 4$

$4 = x^2$ *Take square roots.*

$x = \pm\sqrt{4} = \pm2$ *Therefore, the graph crosses the x-axis at -2 and 2.*

From the graph we can see that the *y*-intercept is -4.   ■

Based on the discussion so far, we can see that the graph of $y = ax^2 + c$ will be very similar to the graphs of Example 1, but it will be shifted up or down depending on whether *c* is positive or negative.

EXAMPLE 2   Sketch the graph of $y = 2x^2 - 8x$. Label the intercepts.

Solution   We have already stated that the graph of this equation will be a parabola. Let's begin by finding the *x*-intercepts.

$y = 2x^2 - 8x$ *We set $y = 0$ to find the x-intercepts.*

$0 = 2x^2 - 8x$ *We can solve this quadratic equation by factoring.*

$0 = 2x(x - 4)$

$x = 0$ or $x = 4$

Thus, the *x*-intercepts are 0 and 4, which are plotted in Figure 11.6.

Finding the vertex would be very helpful in obtaining the graph. Based on our previous discussion, we know that the vertex will be on the axis of symmetry. Since the axis of symmetry divides the parabola into two equal parts, it must fall midway between the two *x*-intercepts. To get the *x* value midway between the intercepts of 0 and 4, we would average those two values: $\frac{0+4}{2} = 2$.

Figure 11.6

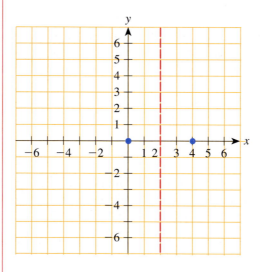

Therefore, the vertex will have an *x*-coordinate of 2. To get the *y*-coordinate of the vertex, we substitute $x = 2$ into the original equation:

$y = 2x^2 - 8x$ *Substitute $x = 2$.*

$y = 2(2)^2 - 8(2) = 8 - 16 = -8$

Therefore, the vertex of this parabola is $(2, -8)$. With the vertex, we can now complete the graph of the parabola, which appears in Figure 11.7.

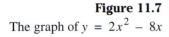

Figure 11.7

The graph of $y = 2x^2 - 8x$

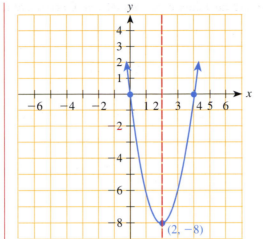

$(2, -8)$

Based on this example, we recognize that whenever we can find the x-intercepts of a parabola, we can find the x-coordinate of the vertex. Let's now consider the parabola with equation $y = ax^2 + bx$. We can find the x-intercepts by factoring:

$$y = ax^2 + bx \qquad \textit{Set } y = 0 \textit{ and solve by factoring.}$$
$$0 = x(ax + b)$$
$$x = 0 \quad \text{or} \quad ax + b = 0$$
$$x = 0 \quad \text{or} \qquad x = -\frac{b}{a}$$

Now the x-coordinate of the vertex of this parabola will be midway between these two x-intercepts, and so again we find the average of the x-intercepts. The x-coordinate of the vertex is

$$\frac{0 + \dfrac{-b}{a}}{2} = \frac{-b}{a} \cdot \frac{1}{2} = \frac{-b}{2a}$$

We have thus determined the x-coordinate of the vertex of any parabola with equation $y = ax^2 + bx$. Our last step is to use the results of Example 1 to recognize that to get the graph of $y = ax^2 + bx + c$ we will simply shift the graph of $y = ax^2 + bx$ up or down (depending on whether c is positive or negative), which has *no effect* on the axis of symmetry or on the x-coordinate of the vertex. We can therefore state the following.

> The x-coordinate of the vertex of the parabola $y = ax^2 + bx + c$ is given by
>
> $$x = \frac{-b}{2a}$$

EXAMPLE 3

Sketch the graph of $y = -x^2 - 2x + 4$. Find the x-intercepts.

Solution

We begin by finding the vertex. For this parabola $a = -1$ and $b = -2$; therefore, the x-coordinate of the vertex is

$$x = \frac{-b}{2a} = \frac{-(-2)}{2(-1)} = -1$$

Keep in mind that $x = -1$ is also the equation of the axis of symmetry. (Recall from Section 5.2 that the equation of a vertical line is of the form $x = k$ where k is a constant.)

To find the y-coordinate of the vertex, we substitute the x-coordinate into the original equation:

$$y = -x^2 - 2x + 4 \qquad \text{\textit{Substitute }} x = -1.$$
$$y = -(-1)^2 - 2(-1) + 4 = -1 + 2 + 4 = 5 \qquad \text{\textit{Therefore the vertex is }} (-1, 5).$$

Next we can easily find the y-intercept by substituting $x = 0$ into the original equation:

$$y = -x^2 - 2x + 4 \qquad \text{\textit{Substitute }} x = 0.$$
$$y = -(0)^2 - 2(0) + 4 = 4 \qquad \text{\textit{Therefore the y-intercept is 4.}}$$

These two points (the vertex and the y-intercept) give us a reasonable sense of what the graph looks like. See Figure 11.8. It seems appropriate to find the x-intercepts as well so that we can locate them accurately.

$$y = -x^2 - 2x + 4 \qquad \text{\textit{To find the x-intercept(s), we set }} y = 0.$$
$$0 = -x^2 - 2x + 4 \qquad \text{\textit{We can multiply both sides of the}}$$
$$\text{\textit{equation by }} -1.$$
$$0 = x^2 + 2x - 4 \qquad \text{\textit{The left-hand side does not factor,}}$$
$$x = \frac{-(2) \pm \sqrt{(2)^2 - 4(1)(-4)}}{2(1)} \qquad \text{\textit{so we use the quadratic formula}}$$
$$\text{\textit{with }} a = 1, b = 2, \text{\textit{ and }} c = -4.$$
$$x = \frac{-2 \pm \sqrt{4 + 16}}{2}$$
$$x = \frac{-2 \pm \sqrt{20}}{2} = \frac{-2 \pm 2\sqrt{5}}{2} = \frac{2(-1 \pm \sqrt{5})}{2} = -1 \pm \sqrt{5}$$

We can approximate these x-intercepts with a calculator as $x = -1 + \sqrt{5} \approx 1.24$ and $x = -1 - \sqrt{5} \approx -3.24$, which agree with the graph in Figure 11.8.

Figure 11.8

The graph of
$y = -x^2 - 2x + 4$

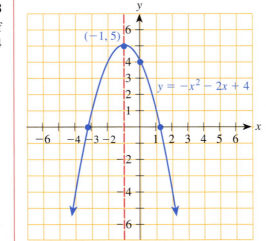

The four points we have found are usually enough to get a reasonably accurate graph. However, you can find additional points on the graph if you want to increase the accuracy. ■

We conclude this section by offering an outline for graphing parabolas.

Outline for Graphing Parabolas	1. Find the *vertex* and the axis of symmetry.
	2. Find the *y-intercept*.
	3. Find the *x-intercepts* (if any).
	4. Find *additional points* if necessary.
	5. *Sketch* the graph.

It is very important to keep in mind that this outline pertains to graphing *quadratic equations*. When asked to sketch a graph, you must take a moment to determine whether the equation is linear or quadratic.

Let's conclude by returning one last time to an example we have looked at several times before.

EXAMPLE 4

A market research firm finds that the daily revenue R (in dollars) earned by a chemical manufacturing company on the sale of g gallons of mixture KZD is given by the equation $R = g^2 - 80g$.

(a) Sketch the graph of this equation and use the graph to describe the relationship between the number of gallons produced and the revenue earned.

(b) Describe graphically what it means to find the number of gallons of KZD that must be produced to earn a daily revenue of $6,400, and use the graph to approximate this value.

(c) Find the value approximated in part (b) correct to the nearest tenth of a gallon.

Solution

(a) In order to graph the equation $R = g^2 - 80g$, it helps a great deal to recognize this equation as being like the equation $y = x^2 - 80x$. As we have just discussed, the graph of such an equation is a parabola. Consequently, we will sketch the graph by labeling the horizontal axis g and the vertical axis R.

We first find the vertex. From the equation $R = g^2 - 80g$ we have $a = 1$ and $b = -80$. Therefore the g-coordinate of the vertex will be

$$g = \frac{-b}{2a} = \frac{-(-80)}{2(1)} = \frac{80}{2} = 40$$

We now substitute $g = 40$ into the equation for R to find the R-coordinate of the vertex.

$R = g^2 - 80g$ *Substitute $g = 40$.*

$R = (40)^2 - 80(40) = 1{,}600 - 3{,}200 = -1{,}600$ *Therefore the vertex is $(40, -1{,}600)$.*

Next we find the g-intercepts by setting $R = 0$.

$R = g^2 - 80g$ *Substitute $R = 0$.*

$0 = g^2 - 80g$ *Factor.*

$0 = g(g - 80)$

$g = 0$ or $g - 80 = 0$

$g = 0$ or $g = 80$ *Therefore the g-intercepts are 0 and 80.*

The graph appears in Figure 11.9. Note that we have not drawn any part of the graph to the left of the R-axis since it makes no sense for g (the number of gallons) to be negative. However, we have drawn the graph below the g-axis (where R is negative) since a negative revenue does make sense. We interpret a negative revenue as a *loss*.

Figure 11.9

The graph of $R = g^2 - 80g$

We recognize that the revenue is dependent on the number of gallons produced. Thus our choice of labeling the vertical axis as R agrees with the convention of having the dependent variable represented on the vertical axis.

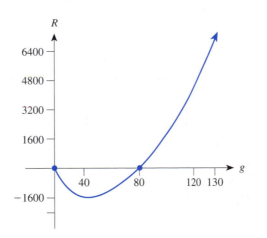

From the graph we can see that for g between 0 and 80 the graph is below the g-axis. This means that for g between 0 and 80 the corresponding R-values are negative. In other words, if the number of gallons produced is between 0 and 80, the company will experience a loss. We can even say that the most the company will *lose* is \$1,600 (which corresponds to the lowest point on the graph) and this will happen if the company manufactures 40 gallons.

We also see that for $g > 80$ the graph is above the g-axis, which means that R is positive. Thus we can also say that if the company produces more than 80 gallons the revenue will be positive and will continue to increase as more gallons are manufactured.

(b) Asking how many gallons must be manufactured to produce a revenue of \$6,400 is the same as asking what g-value corresponds to an R-value of 6,400. Looking at the graph in Figure 11.10, we can see that a revenue of \$6,400 results from the manufacture of between 125 and 130 gallons. Consequently we could estimate that 128 gallons will result in revenue of \$6,400.

Figure 11.10

Estimating on the graph of $R = g^2 - 80g$

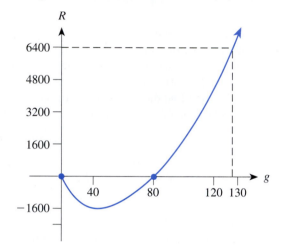

(c) In order to find the "exact" value, we need to substitute $R = 6,400$ in the equation $R = g^2 - 80g$ and solve for g.

$$R = g^2 - 80g \qquad \textit{Substitute } R = 6,400.$$
$$6,400 = g^2 - 80g \qquad \textit{Get the equation in standard form.}$$
$$0 = g^2 - 80g - 6,400 \qquad \textit{We will use the quadratic formula.}$$
$$\textit{We have } a = 1, b = -80, \textit{ and } c = -6,400.$$

$$g = \frac{-(-80) \pm \sqrt{(-80)^2 - 4(1)(-6,400)}}{2(1)}$$

$$g = \frac{80 \pm \sqrt{6,400 + 25,600}}{2}$$

$$g = \frac{80 \pm \sqrt{32,000}}{2} \qquad \textit{Using a calculator, we obtain}$$

$$g = 129.44272 \quad \text{or} \quad g = -49.44272$$

Obviously, the negative answer makes no sense for the number of gallons. Therefore, in order to produce a revenue of \$6,400, the company needs to manufacture 129.4 gallons of KZD. This agrees quite well with our estimate in part **(b)**. ∎

EXERCISES 11.6

In Exercises 1–40, sketch the graph of the given equation. Find the intercepts; approximate to the nearest tenth where necessary.

1. $y = 2x^2$

2. $y = -3x^2$

3. $y = x^2 + 2$

4. $y = x^2 - 1$

5. $y = x^2 - 9$

6. $y = x^2 + 1$

7. $y = -x^2 + 1$

8. $y = -x^2 - 3$

9. $y = x^2 - 2x$

10. $y = 2x - x^2$

11. $y = 6x - 2x^2$

12. $y = 3x^2 + 6x$

13. $y = x^2 - 2$

14. $y = x - 2$

15. $y = 2x - 10$

16. $y = 2x^2 - 10$

17. $y = x^2 - 4x + 3$

18. $y = x^2 - 6x + 5$

19. $y = x^2 + 2x - 8$

20. $y = x^2 - 4x - 12$

21. $y = -x^2 + 2x + 3$

22. $y = -x^2 - 2x + 3$

23. $y = x^2 + 4x + 4$

24. $y = x^2 - 6x + 9$

25. $y = (x - 1)^2$

26. $y = -(x + 3)^2$

27. $y = x^2 + 3x + 2$

28. $y = x^2 - 5x - 6$

29. $y = -x^2 + x + 6$

30. $y = -x^2 + 3x + 10$

31. $y = x^2 - 4x + 2$

32. $y = -x^2 - 6x + 3$

33. $y = -x^2 - 8x - 5$

34. $y = x^2 + 2x - 6$

35. $y = x^2 - 4x + 5$

36. $y = x^2 + 2x + 3$

37. $y = 3x + 6$

38. $y = -4x + 5$

39. $y = 4x^2 - 8x - 5$

40. $y = 2x^2 - 3x - 2$

41. If an object is thrown upward into the air with an initial velocity of 40 ft/sec, its height h (in feet) above the ground after t seconds is given by the formula

$$h = 40t - 16t^2 \qquad \text{where } t \geq 0$$

(a) Sketch the graph of this equation. Let t be the horizontal axis and h be the vertical axis.

(b) Use the graph to determine the highest point the object reaches.

(c) After how many seconds does the object reach this highest point?

42. Suppose that a company's profit P, in thousands of dollars, on the sale of x items is given by the equation

$$P = x^2 - 6x - 40 \qquad \text{where } x \geq 0$$

(a) Sketch the graph of this equation. Let x be the horizontal axis and P be the vertical axis.

(b) Use the graph to determine the minimum profit the company earns.

(c) How many items does the company sell if it experiences this minimum profit?

In Exercises 43–46, use the sketch of the given parabola to identify

(a) *the vertex*

(b) *the axis of symmetry*

(c) *the x values for which the y values are increasing*

(d) *the x values for which the y values are decreasing*

43. 44. 45. 46.

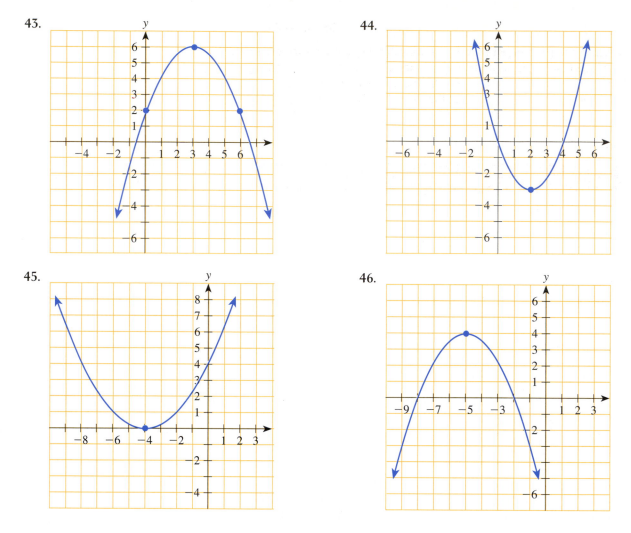

? QUESTIONS FOR THOUGHT

47. Sketch the graphs of $y = x^2$, $y = 2x^2$, and $y = 3x^2$ on the same coordinate system. How would you describe the effect the coefficients 2 and 3 have on the graph of $y = x^2$?

48. Sketch the graphs of $y = x^2$, $y = \frac{1}{2}x^2$, and $y = \frac{1}{3}x^2$ on the same coordinate system. How would you describe the effect the coefficients $\frac{1}{2}$ and $\frac{1}{3}$ have on the graph of $y = x^2$?

49. Sketch the graphs of $y = -x^2$ and $y = -3x^2$ on the same coordinate system. How would you describe the effect the coefficient –3 has on the graph of $y = x^2$?

50. Sketch the graphs of $y = -x^2$ and $y = -\frac{1}{3}x^2$ on the same coordinate system. How would you describe the effect the coefficient $-\frac{1}{3}$ has on the graph of $y = x^2$?

◇ MINI-REVIEW

In Exercises 51–56, multiply and simplify.

51. $(x - 5)^2$

52. $(x + 8)^2$

53. $(x + 5)(x - 5)$

54. $(x + 8)(x - 8)$

55. $(x + p)^2$ **56.** $(x - h)^2$

57. A handyman charges $19 per hour for his time and $11 per hour for his assistant's time. On a certain job the assistant begins a job alone at 10:00 A.M. and is joined at 11:00 A.M. by the handyman, at which time they work together to complete the job. If the total bill was $146, at what time was the job finished?

CHAPTER 11   SUMMARY

After having completed this chapter, you should be able to:

1. Solve quadratic equations by the factoring and square root methods (Section 11.1).

   For example:

   (a) Solve for x: $x^2 - 5x = 14$ *Get the quadratic equation into standard form.*

   $$x^2 - 5x - 14 = 0 \qquad \textit{Factor.}$$
   $$(x - 7)(x + 2) = 14$$
   $$x - 7 = 0 \quad \text{or} \quad x + 2 = 0$$
   $$\boxed{x = 7} \quad \text{or} \quad \boxed{x = -2}$$

   (b) Solve for z: $4z^2 = 11$ *Isolate the z^2 term.*

   $$z^2 = \frac{11}{4} \qquad \textit{Take square roots.}$$
   $$z^2 = \pm\sqrt{\frac{11}{4}} = \pm\frac{\sqrt{11}}{\sqrt{4}}$$
   $$\boxed{z = \pm\frac{\sqrt{11}}{2}}$$

3. Solve a quadratic equation by the square root method where it is necessary to first complete the square (Section 11.2).

   For example:

   $x^2 - 8x + 4 = 0$ *Subtract 4 from both sides.*

   $x^2 - 8x \qquad = -4$ *Take one-half the coefficient of the first-degree term and square it. That is, take one-half of -8, which is -4, and square it; the result is 16. Therefore, we add 16 to both sides of the equation.*

   $x^2 - 8x + 16 = -4 + 16$ *The left-hand side is now a perfect square.*

   $(x - 4)^2 = 12$ *Take square roots.*

   $x - 4 = \pm\sqrt{12} = \pm\sqrt{4}\sqrt{3} = \pm 2\sqrt{3}$ *Add 4 to both sides.*

   $$\boxed{x = 4 \pm 2\sqrt{3}}$$

4. Solve a quadratic equation by using the quadratic formula (Section 11.3), which says:

   The solutions to the equation $ax^2 + bx + c = 0$ are given by the formula

   $$x = \frac{-b \pm \sqrt{b^2 - 4ac}}{2a} \qquad (a \neq 0)$$

For example:

$$3x^2 + 5 = 9x \quad \textit{Get the equation in standard form.}$$
$$3x^2 - 9x + 5 = 0 \quad \textit{a = 3, b = -9, and c = 5}$$
$$x = \frac{-(-9) \pm \sqrt{(-9)^2 - 4(3)(5)}}{2(3)}$$
$$x = \frac{9 \pm \sqrt{81 - 60}}{6}$$
$$\boxed{x = \frac{9 \pm \sqrt{21}}{6}}$$

5. Solve verbal problems that give rise to quadratic equations (Section 11.5).

   For example: The product of a number and 2 more than itself is 4. Find the number.

$$\text{Let } x = \text{the number.}$$
$$\text{Then } x + 2 = \text{"2 more than itself."}$$

The problem tells us that the product of the two numbers is 4. Therefore, our equation is

$$x(x + 2) = 4 \quad \textit{We put the equation in standard form.}$$

$$x^2 + 2x - 4 = 0 \quad \textit{The left-hand side does not factor; use the formula.}$$

$$x = \frac{-2 \pm \sqrt{2^2 - 4(1)(-4)}}{2(1)}$$
$$x = \frac{-2 \pm \sqrt{4 + 16}}{2} = \frac{-2 \pm \sqrt{20}}{2} = \frac{-2 \pm \sqrt{4}\sqrt{5}}{2} = \frac{-2 \pm 2\sqrt{5}}{2}$$
$$x = \frac{2(-1 \pm \sqrt{5})}{2} = \frac{2(-1 \pm \sqrt{5})}{2}$$
$$x = -1 \pm \sqrt{5}$$

We apparently have two solutions: $x = -1 + \sqrt{5}$ and $x = -1 - \sqrt{5}$. Let's check one of these, say the first one.

Check: If the first number is $-1 + \sqrt{5}$, then "2 more than itself" is $-1 + \sqrt{5} + 2 = 1 + \sqrt{5}$. The product of the two numbers is supposed to be 4:

$$(-1 + \sqrt{5})(1 + \sqrt{5}) \overset{?}{=} 4 \quad \textit{Multiply out.}$$
$$-1 - \sqrt{5} + \sqrt{5} + 5 \overset{?}{=} 4 \quad \textit{Combine like terms.}$$
$$4 \overset{\checkmark}{=} 4$$

We leave it to the student to verify that $-1 - \sqrt{5}$ also checks.

6. Be able to recognize the equation of a parabola and sketch its graph (Section 11.6).

   The graph of an equation of the form $y = ax^2 + bx + c$ is a parabola.

   The x-coordinate of the vertex of a parabola is given by the formula $x = \frac{-b}{2a}$.

   For example:

   The equation $y = x^2 - 4x + 3$ is of this form, and so its graph is a parabola.

   Therefore, the x-coordinate of its vertex is $x = \frac{-(-4)}{2(1)} = 2$.

   We find the y-coordinate of the vertex by substituting $x = 2$ into the equation $y = x^2 - 4x + 3$, obtaining $y = (2)^2 - 4(2) + 3 = -1$. Therefore, the vertex is $(2, -1)$.

We find the y-intercept by setting $x = 0$ and obtaining $y = 3$.

We find the x-intercepts by setting $y = 0$ and solving for x:

$$0 = x^2 - 4x + 3$$
$$0 = (x - 1)(x - 3)$$
$$x = 1 \quad \text{or} \quad x = 3 \qquad \text{Therefore, the } x\text{-intercepts are 1 and 3.}$$

Putting all this information together, we obtain the graph of the parabola that appears in Figure 11.11.

Figure 11.11

The graph of $y = x^2 - 4x + 3$

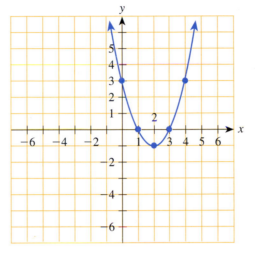

CHAPTER 11   REVIEW EXERCISES

Solve each of the following equations.

1. $x^2 - 7x - 6 = 0$

2. $x^2 - 10x - 24 = 0$

3. $x^2 + 5 = 4x$

4. $x^2 + 4x = 5$

5. $(u - 6)^2 = 13$

6. $(z + 3)^2 = -9$

7. $2y^2 + 7y = 15$

8. $2y^2 - 1 = 3y$

9. $18x^2 - 24x + 6 = 0$

10. $20x^2 - 8x - 16 = 0$

11. $(x - 6)^2 = (x + 3)(x - 5)$

12. $2x^2 - 13x + 5 = x(x - 3)$

13. $u^2 + 1 = \dfrac{13u}{6}$

14. $\dfrac{z^2}{9} - \dfrac{z}{3} = 2$

15. $3x(x - 2) = (x - 3)^2$

16. $2w = 4 + \dfrac{3}{w}$

17. $\dfrac{x + 3}{x + 6} = \dfrac{x + 2}{x + 4}$

18. $\dfrac{x}{x - 1} = \dfrac{9}{x} - \dfrac{5}{2}$

Solve the next two equations by completing the square.

19. $x^2 - 7x + 3 = 0$

20. $3x^2 - 12x = 6$

In Exercises 21–26, sketch the graph of the given equation. Label the intercepts. Round off to the nearest tenth where necessary.

21. $y = x^2 - 6x$

22. $y = x^2 - 6$

23. $y = -x^2 - 2x + 8$

24. $y = 4x - 2$

25. $y = 2x^2 - 4x + 3$

26. $y = 2x^2 - x - 3$

CHAPTER 11 PRACTICE TEST

Solve each of the following equations. Choose any method you like.

1. $(x + 5)(x - 2) = 18$

2. $2x^2 - 3x - 3 = 0$

3. $x^2 - x + 14 = 2x(x - 3)$

4. $(x + 5)^2 = 10$

5. $(x + 5)^2 = 10x$

6. $4x^2 - 8x + 2 = x^2 - 21x + 12$

7. $5x^2 + 15 = 30x$

8. $\dfrac{x}{2} - \dfrac{8}{x} = x - 4$

9. Solve the following equation by completing the square, and check your answer by using the quadratic formula: $3x^2 - 12x = 7$

10. The length of a rectangle is 7 more than twice its width. If the area of the rectangle is 30 sq cm, find its dimensions.

11. Sketch the graph of $y = 6x - x^2$. Label the intercepts.

CHAPTERS 10–11   CUMULATIVE REVIEW

In Exercises 1–8, simplify the expression as completely as possible. Fractions should be reduced to lowest terms.

1. $\sqrt{36x^{16}y^{12}}$

2. $\sqrt{\dfrac{45x^9y^5}{5x^2y}}$

3. $\dfrac{7}{\sqrt{6}}$

4. $\dfrac{9}{\sqrt{6} - 2}$

5. $\dfrac{20}{3 - \sqrt{5}}$

6. $\dfrac{15}{\sqrt{5}}$

7. $\sqrt{120}$

8. $\sqrt{18x^3y^8}$

In Exercises 9–17, perform the indicated operations and simplify as completely as possible.

9. $\sqrt{45t} - \sqrt{20t}$

10. $3\sqrt{2}(\sqrt{3} - \sqrt{6}) - 5(\sqrt{12} - \sqrt{6})$

11. $\sqrt{\dfrac{3}{7}} + \sqrt{21}$

12. $(2\sqrt{x} - 3)(\sqrt{x} + 5)$

13. $\sqrt{27x^6y^5} - xy\sqrt{12x^4y^3}$

14. $\dfrac{6}{\sqrt{5}} + \dfrac{3}{\sqrt{20}}$

15. $(3\sqrt{2} - 4\sqrt{5})(4\sqrt{2} - 2\sqrt{5})$

16. $(\sqrt{x} + 3)^2 + (\sqrt{x + 3})^2$

17. $\dfrac{15}{\sqrt{7} - \sqrt{2}} - \dfrac{10}{\sqrt{2}}$

In Exercises 18–33, solve the given equation.

18. $x^2 - 3x = 0$

19. $x^2 - 3x = 10$

20. $1 - \dfrac{1}{x} = \dfrac{6}{x^2}$

21. $\dfrac{x}{x - 2} = \dfrac{x + 1}{x - 3}$

22. $x^2 + 3x = 5$

23. $3x^2 - 6x - 2 = x^2 - x - 4$

24. $(x + 3)(x - 4) = 0$

25. $(x + 3)(x - 4) = 8$

26. $(2x + 3)(x + 4) = 9$

27. $(2x - 3)(x - 4) = (x - 2)(2x - 1)$

28. $(a + 5)^2 = 9$

29. $(s - 2)^2 = 10$

30. $\sqrt{x - 3} = 4$

31. $\sqrt{x} - 3 = 4$

32. $9 - \sqrt{2a} = 12$

33. $3\sqrt{t + 4} - 1 = 7$

34. Solve by completing the square: $x^2 - 6x + 7 = 0$

35. Solve by completing the square: $2x^2 - 10x + 6 = 0$

In Exercises 36–39, sketch the graph of the given equation. Label the intercepts.
Round to the nearest tenth where necessary.

36. $y = x^2 - 4x - 5$

37. $y = (2x - 3)^2$

38. $y = 9 - x^2$

39. $y = x^2 - x - 1$

40. The length of a rectangle is 3 more than twice the width. If the area of the rectangle is 20 sq in., find the dimensions of the rectangle.

41. Find the perimeter of the right triangle in the accompanying figure.

42. Find the length of a diagonal of a square of side 10 cm.

43. Kim is driving to a destination 280 miles away and follows the same route there and back. On the way back, traffic congestion caused her to slow down so that her average speed on the return trip was 30 mph less than her average speed on the way there. If the total traveling time there and back was 11 hours, find her average speed on the way there.

44. The sum of a number and its reciprocal is $\frac{5}{2}$. Find the number(s).

CHAPTERS 10–11   CUMULATIVE PRACTICE TEST

In Problems 1–10, perform the indicated operations and simplify as completely as
possible. Assume that all variables appearing under radical signs are nonnegative.

1. $\sqrt{72}$

2. $\sqrt{48x^3y^4}$

3. $3\sqrt{20} - 4\sqrt{45}$

4. $\sqrt{27x^5y^6} - 2x^2y\sqrt{3xy^4}$

5. $\dfrac{8}{\sqrt{5}}$

6. $\sqrt{36y^{36}}$

7. $\dfrac{24}{4 - \sqrt{7}}$

8. $(2\sqrt{3} - 5)^2$

9. $(5\sqrt{t} + \sqrt{3})(2\sqrt{t} - 4\sqrt{3})$

10. $\dfrac{9}{3 - \sqrt{6}} - \dfrac{18}{\sqrt{6}}$

In Problems 11–17, solve the given equation.

11. $(x + 3)(x - 5) = 9$

12. $(x - 2)^2 = 7$

13. $\dfrac{x}{2x - 5} = \dfrac{x + 4}{2x}$

14. $\sqrt{4x - 3} - 2 = 3$

15. $2x^2 + 1 = 5x$

16. $\dfrac{t^2}{2} - \dfrac{2}{3} = 1$

17. $(2x + 5)^2 = 4x(x - 3)$

18. *Solve by completing the square.* $x^2 - 10x = 3$

19. Find the area of a rectangle with diagonal 10 in. and width 5 in.

20. Two trucks are following the same route to make a delivery 400 miles away. One truck averages 10 mph faster than the other one. If the slower truck took 2 hours longer to make the trip than the faster truck, find the average speed of each truck.

In Problems 21–22, sketch the graph of the given equation. Label the intercepts. Round to the nearest tenth where necessary.

21. $y = x^2 + 4x - 5$

22. $y = 7 - x^2$

G

Geometry

In this chapter we are attempting to give neither a rigorous introduction to geometry nor even a summary of a traditional high school geometry course. Rather, we will try to give a brief overview that highlights basic definitions, terminology, theorems, and formulas. We pay particular attention to those results that are often encountered in algebra, pre-calculus, and calculus courses.

G.1 Angles

We begin by assuming that certain undefined terms are intuitively understood. For example, we all understand what is meant by a point and by a straight line. Points and lines are often named with letters; lines are also frequently named by the points through which they pass. Thus, in Figure G.1(a), the line passing through the points A and B can be referred to as line L and is usually denoted by $\overleftrightarrow{AB}$. The arrows indicate that the line goes on forever in both directions. On the other hand, the straight *line segment* from A to B is denoted by $\overline{AB}$. A and B are called the *endpoints* of the line segment. See Figure G.1(b).

Figure G.1

The length of the line segment $\overline{AB}$ is denoted as $|\overline{AB}|$. Thus in Figure G.1(b), $|\overline{AB}| = 5$.

(a) The line L or $\overleftrightarrow{AB}$ (b) The line segment $\overline{AB}$

We may also have occasion to talk about a *half-line* or *ray*. Ray $\overrightarrow{AB}$ appears in Figure G.2. Ray $\overrightarrow{AB}$ has only one endpoint, the point A.

Figure G.2

The ray $\overrightarrow{AB}$

An *angle* is formed by two rays with a common endpoint. The common endpoint is called the *vertex.* Figure G.3(a) shows angle A, which is often written $\angle A$. Note that $\angle A$ can also be referred to as $\angle BAC$, or $\angle CAB$, or $\angle 1$. When we name an angle using the three-letter designation, the middle letter must be the vertex.

Figure G.3

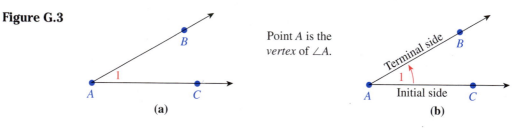

Point A is the vertex of $\angle A$.

(a)   (b)

It is helpful to think of the rays that form an angle in the following way: We keep one ray of the angle fixed (this is called the *initial side*) and allow the second ray (this is called the *terminal side*) to rotate to form the angle. The little arrow in Figure G.3(b) indicates that in $\angle A$, $\overrightarrow{AC}$ is the initial side and $\overrightarrow{AB}$ is the terminal side.

Angles are usually measured using units called *degrees.* If we keep in mind that when we measure an angle we are trying to measure what part of a complete rotation we have, then we define one degree (written $1°$) to be $\frac{1}{360}$ of an entire rotation. Figure G.4(a) shows an angle formed by one complete rotation; Figure G.4(b) shows several angles and their degree measurements.

Figure G.4

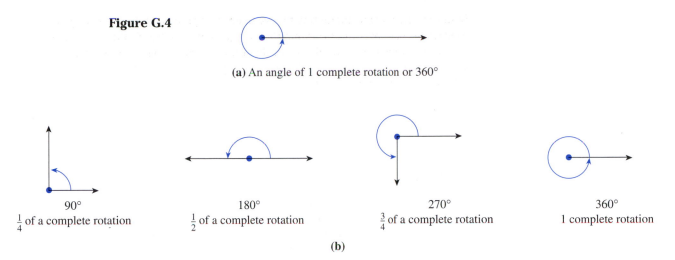

(a) An angle of 1 complete rotation or 360°

90°
$\frac{1}{4}$ of a complete rotation

180°
$\frac{1}{2}$ of a complete rotation

270°
$\frac{3}{4}$ of a complete rotation

360°
1 complete rotation

(b)

An *acute* angle is an angle whose measure is less than 90°.

A *right* angle is an angle whose measure is equal to 90°.

An *obtuse* angle is an angle whose measure is greater than 90° but less than 180°.

A *straight* angle is an angle whose measure is equal to 180°.

An example of each kind of angle is shown in Figure G.5(a). Note that a right angle is indicated by a little square at the vertex.

Lines that intersect at right angles are called *perpendicular lines.*

Figure G.5a

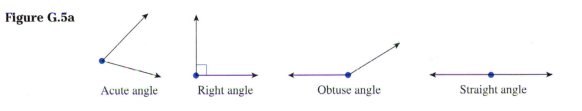

Acute angle   Right angle   Obtuse angle   Straight angle

Two angles are called *complementary* if together they form a right angle (90°).

Two angles are called *supplementary* if together they form a straight angle (180°).

In Figure G.5b, angles ∠CBD and ∠DBE are complementary since together they form right angle ∠CBE, whereas angles ∠ABD and ∠DBE are supplementary since together they form straight angle ∠ABE.

Figure G.5b

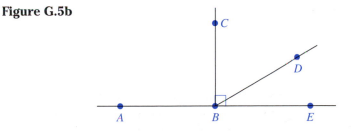

Numerically, a 30° angle and a 60° angle are complementary since they add up to 90°, whereas a 45° angle and a 135° angle are supplementary since they add up to 180°. (In order to keep the definitions of complementary and supplementary straight, you might find it helpful to remember that "c" comes before "s" and 90 comes before 180.)

EXAMPLE 1

In Figure G.6, find x.

Figure G.6

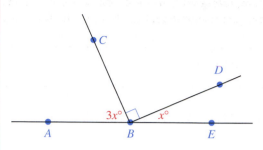

Solution

Since $\angle ABE$ is a straight angle, and a straight angle is 180°, we have the following:

$\angle ABC + \angle CBD + \angle DBE = 180$ *$\angle ABC = 3x$, $\angle DBE = x$, and since $\angle CBD$ is a right angle, $\angle CBD = 90°$ and we have*

$$3x + 90 + x = 180 \quad \textit{Solve for } x$$

$$4x + 90 = 180$$

$$4x = 90 \quad \textit{Divide each side by 4 to get}$$

$$x = \frac{90}{4}$$

$$x = \frac{45}{2} = \boxed{22.5}$$

EXAMPLE 2

An angle is 18° less than twice its complement. Find the angle.

Solution

Let

$$x = \text{the angle}$$

then

$$90 - x = \text{the complement of the angle} \quad \textit{Since their sum is } 90°$$

The given information is translated as

$$x = 2(90 - x) - 18$$
$$x = 180 - 2x - 18$$
$$x = 162 - 2x$$
$$3x = 162$$
$$x = 54°$$

Check If the angle is 54°, then its complement is 90° − 54° = 36°. Also 54° is 18° less than twice 36°.

Vertical Angles

When two lines intersect, four angles are formed (see Figure G.7). The angles on opposite sides of the point of intersection are called *vertical angles*.

Figure G.7

$\angle 1$ and $\angle 3$ are vertical angles, as are $\angle 2$ and $\angle 4$.

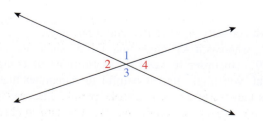

From Figure G.6 we can see that $\angle 1 + \angle 2 = 180°$ and $\angle 2 + \angle 3 = 180°$. Therefore, it follows that $\angle 1 + \angle 2 = \angle 2 + \angle 3$ and hence $\angle 1$ must be equal to $\angle 3$. We have just demonstrated the truth of the following theorem.

THEOREM G.1	Vertical angles are equal.

EXAMPLE 3

In Figure G.8, lines L and R intersect as indicated. Find the measures of the remaining angles.

Figure G.8

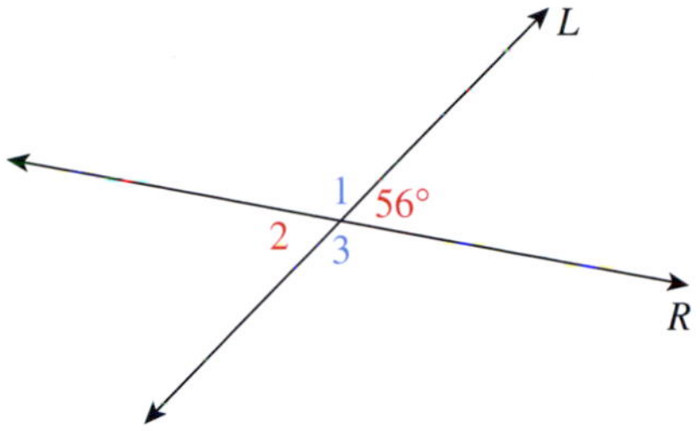

Solution

Since $\angle 2$ and the angle of 56° are vertical angles, and vertical angles are equal, we have $\boxed{\angle 2 = 56°}$.

We can see that $\angle 1$ and the 56° angle form a straight angle. Therefore, $\angle 1$ is supplementary to a 56° angle and so $\angle 1 = 124°$. Since $\angle 1$ and $\angle 3$ are vertical angles, we have $\boxed{\angle 1 = \angle 3 = 124°}$. ■

EXAMPLE 4

In Figure G. 9, find the measure of each angle.

Figure G.9

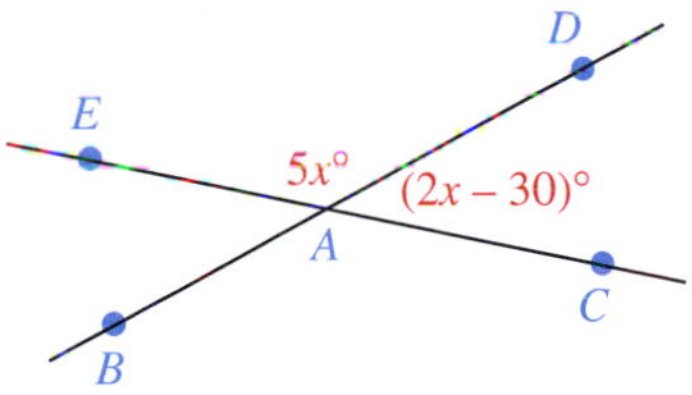

Solution

Since $\angle EAC$ is a straight angle, and a straight angle is 180°, we have the following:

$$\angle EAD + \angle DAC = 180 \qquad \textit{Since } \angle EAD = 5x, \textit{ and } \angle DAC = 2x - 30, \textit{ we have}$$
$$5x + 2x - 30 = 180 \qquad \textit{Solve for x.}$$
$$7x - 30 = 180$$
$$7x = 210$$
$$x = 30$$

Since $x = 30$, we have

$$\angle EAD = 5x = 5(30) = 150°$$
$$\angle DAC = 2x - 30 = 2(30) - 30 = 30°$$
$$\angle BAC = 150°, \quad \text{since } \angle BAC \text{ and } \angle EAD \text{ are vertical angles.}$$
$$\angle EAB = 30°, \quad \text{since } \angle EAB \text{ and } \angle DAC \text{ are vertical angles.} \qquad ■$$

Parallel Lines

In everyday usage we may describe parallel lines as "lines that never meet." However, this is not a practical definition since it is hard to check whether two given lines ever meet. Instead, given two lines, we call any line (or line segment) that crosses these two lines a **transversal.** We can then say that two lines are parallel if they are going in the same direction with respect to any transversal, which means that they make the same angles with the transversal.

In Figure G.10 we see lines L_1 and L_2 being crossed by the transversal M. We can see that requiring the lines to go in the same direction requires that $\angle 1 = \angle 5$ or $\angle 3 = \angle 5$. The arrows on the lines indicate that line L_1 is parallel to L_2. Angles 1 and 5 are called **corresponding angles.** Angles 3 and 5 are called **alternate interior angles.**

Figure G.10

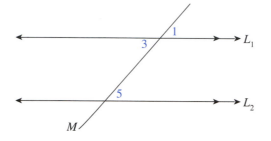

In general, when two lines are crossed by a transversal, four pairs of corresponding angles are formed and two pairs of alternate interior angles are formed. See Figure G.11.

Figure G.11

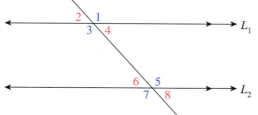

Referring to Figure G.11, the pairs of corresponding angles and alternate interior angles are as follows:

Pairs of corresponding angles	Pairs of alternate interior angles
$\angle 1$ and $\angle 5$	$\angle 3$ and $\angle 5$
$\angle 2$ and $\angle 6$	$\angle 4$ and $\angle 6$
$\angle 3$ and $\angle 7$	
$\angle 4$ and $\angle 8$	

We now state the following theorem.

THEOREM G.2 | In the situation described by Figure G.11, lines L_1 and L_2 are parallel if and only if *any* pair of corresponding angles or alternate interior angles is equal.

EXAMPLE 5

Remember that the arrows on the lines tell us that the lines are parallel.

Fill in all the missing angles in the following diagram.

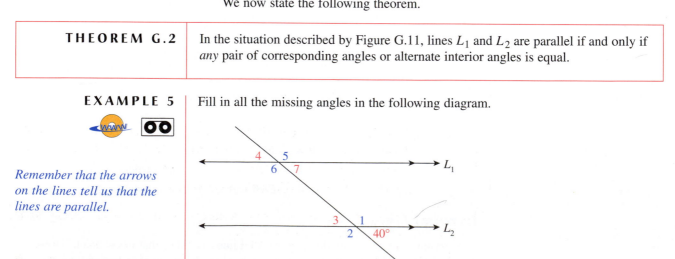

Solution | Since ∠3 is a vertical angle to the angle of 40°, $\boxed{\angle 3 = 40°}$.

Since ∠1 is supplementary to 40° (together they form a straight angle), we have $\boxed{\angle 1 = 140°}$.

Similarly, since ∠2 is vertical to ∠1, $\boxed{\angle 2 = 140°}$.

Since the two lines L_1 and L_2 are parallel, the corresponding angles are equal. Therefore, we have ∠1 = ∠5, and so $\boxed{\angle 5 = 140°}$.

Finally, since ∠6 corresponds to ∠2, and ∠4 corresponds to ∠3 and is vertical to ∠7. $\boxed{\angle 6 = 140°, \angle 4 = \angle 7 = 40°}$. ■

EXAMPLE 6 | In Figure G.12, find the measure of y.

Figure G.12

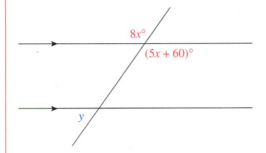

Solution | The angles labeled $8x$ and $5x + 60$ are vertical angles and therefore are equal, so we have

$$8x = 5x + 60$$
$$3x = 60$$
$$x = 20$$

Since $x = 20$, the angle designated $8x$ is $8(20) = 160°$. Note the angle designated $5x + 60$ is also $160°$ since it is vertical to the angle designated $8x$.

Let's call z the angle corresponding to the angle designated y (see Figure G.13).

Figure G.13

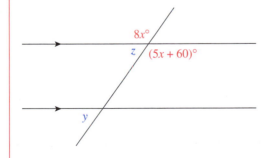

The angle designated z is supplementary to the angle designated $8x$ [and to the angle designated $(5x + 60)$], and since the angle designated $8x$ has a measure of $160°$, we have

$$z + 160 = 180 \quad \Rightarrow \quad z = 20°$$

But the angle designated z and the angle designated y are a pair of corresponding angles, and thus are equal by Theorem G.2.

Therefore $\boxed{y = 20°}$. ■

EXERCISES G.1

1. Given the following figure, which pairs of angles are complementary?

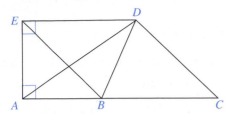

2. Given the same figure as in Exercise 1, which pairs of angles are supplementary?

3. Find the complement of each of the following angles:
   (a) 30°   (b) 60°   (c) 45°   (d) 18°   (e) 89°

4. Find the supplement of each of the following angles:
   (a) 15°   (b) 80°   (c) 24°   (d) 110°   (e) 90°

5. If an angle is 28° less than three times its supplement, how large is the angle?

6. If an angle is 28° less than three times its complement, how large is the angle?

7. How large is an angle if it is 12° more than twice its complement?

8. How large is an angle if it is 12° more than twice its supplement?

In Exercises 9–12, find x.

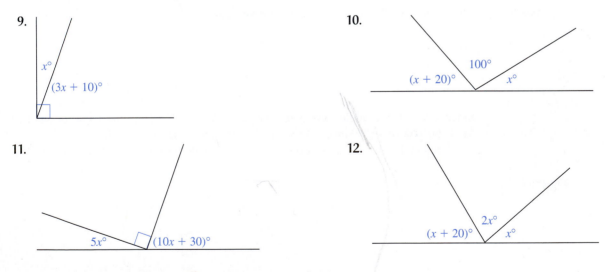

9.

10.

11.

12.

13. In the following figure ∠4 = 35°. Find the remaining angles.

14. Find x.

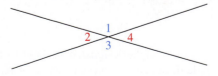

15. In the following figure $\angle 4 = 25°$. Find the measures of angles 1, 2, and 3.

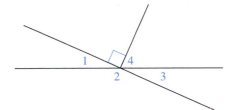

16. Find x.

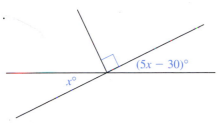

17. In the following figure $\angle 4 = 125°$. Find all the remaining angles.

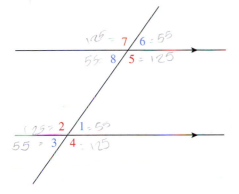

18. In the following figure $\angle 8 = 32°$. Find all the remaining angles.

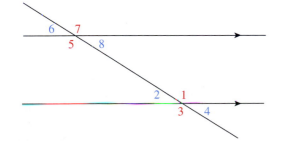

19. Find x.

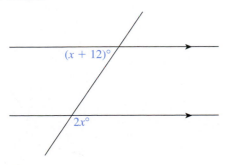

20. Find x.

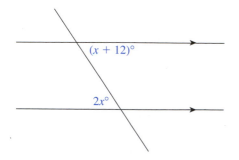

21. In the following figure, L_1 is parallel to L_2, L_3 is parallel to L_4, and $\angle 2 = 40°$. Find all the remaining angles.

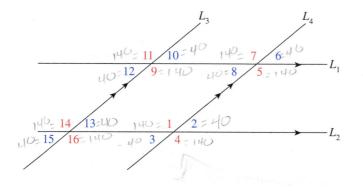

22. In the following figure, L_1 is parallel to L_2 and L_3 is parallel to L_4. Find x.

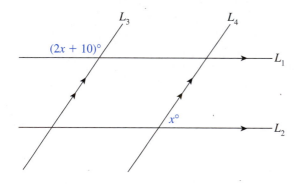

23. In the following figure, L_1 is parallel to L_2 and L_3 is parallel to L_4. Find the measures of angles 1, 2, 3, and 4.

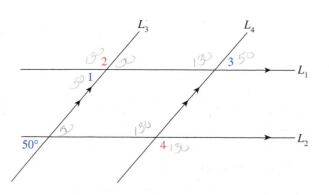

24. Find the measures of angles 1, 2, and 3.

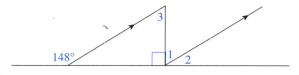

G.2 Triangles

When a portion of the plane is totally enclosed by straight line segments, the enclosing figure is called a *polygon.* Polygons are usually named according to the number of sides that they have. A polygon of three sides is called a *triangle;* of four sides is called a *quadrilateral;* of five sides is called a *pentagon;* etc. In this section we will concentrate on triangles.

Triangles are often categorized by the number of sides of equal length.

An *equilateral* triangle has three equal sides.

An *isosceles* triangle has two equal sides.

A *scalene* triangle has all three sides unequal.

One perhaps somewhat surprising fact is that every triangle, regardless of its size or shape, has the property that the sum of its three angles is 180°.

We can prove this fact by considering any triangle *ABC* (often denoted as △*ABC*) with line *L* drawn through *C* and parallel to $\overline{AB}$, as in Figure G.14.

Figure G.14

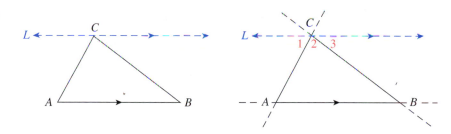

We can see that ∠1 = ∠A and ∠3 = ∠B because they are alternate interior angles. ∠2 is just another name for ∠C of the triangle.

Thus, we see that ∠A + ∠B + ∠C = ∠1 + ∠2 + ∠3 = 180°, and we can state the following theorem.

THEOREM G.3	The sum of the angles of a triangle is 180°.

Triangles have the property that within a specific triangle, the longer the side, the larger the opposite angle and vice versa. It then follows that if any of the sides of the triangle are of equal length, the angles opposite those sides will also be equal. Thus a triangle is equilateral (all three sides are equal), if and only if all three angles are equal; the triangle is isosceles (two sides are equal) if and only if two angles are equal.

The angles opposite the equal sides of an isosceles triangle are called the *base angles* of the isosceles triangle.

The base angles of an isosceles triangle are equal.

A word about notation: In figures, when we want to indicate that two line segments are the same size, we put a small slash through each of the segments that are equal. If there are other groups of equal segments, we use a double slash, or a triple slash as shown in Figure G.15. For angles that are the same size, we use arcs, double arcs, and triple arcs as shown in the same figure.

Figure G.15

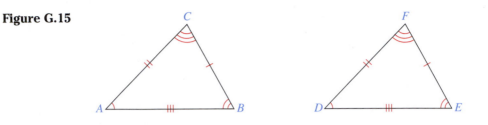

EXAMPLE 1	Find the number of degrees in each angle of an equilateral triangle.
Solution	From the comments we have just made, the fact that the triangle is equilateral means that all the sides are equal and so all the angles must be equal. Therefore, we have labeled all three angles with the same letter, x. Since the angles must all be equal, and the sum of the angles is 180°, we can write

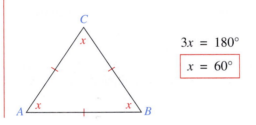

$$3x = 180°$$
$$\boxed{x = 60°}$$

Note that we indicated that the sides of the triangle are of equal length by putting the same number of slashes through them.

EXAMPLE 2	Find x.

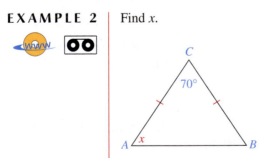

Solution	Since the triangle is isosceles (because the diagram tells us that $|\overline{AC}| = |\overline{BC}|$), $\angle A$ must be equal to $\angle B$, since they are base angles. Thus we label both $\angle A$ and $\angle B$ as x in the figure below, and write the following equation:

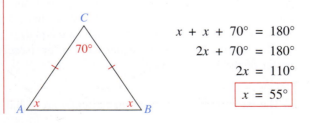

$$x + x + 70° = 180°$$
$$2x + 70° = 180°$$
$$2x = 110°$$
$$\boxed{x = 55°}$$

Suppose two angles of one triange are equal to two angles of another triangle, such as the example given in the figure below: $\angle 1 = \angle 4$ and $\angle 2 = \angle 5$. Since the sum of the three angles in any triangle is 180°, we can clearly see that the remaining two angles, $\angle 3$ and $\angle 6$, are equal to each other.

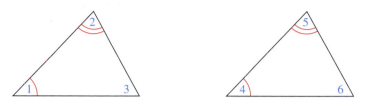

Thus we have the following:

> If two angles in one triangle are equal to two angles in another triangle, then the third angles in each triangle are equal to each other.

If one of the angles of a triangle is a right angle, then the triangle is called a **right triangle.** The sides that form the right angle are called the **legs,** and the side opposite the right angle is called the **hypotenuse.** Figure G.16 shows right triangle $\triangle ABC$. The legs are $\overline{AC}$ and $\overline{BC}$, and the hypotenuse is $\overline{AB}$.

Figure G.16

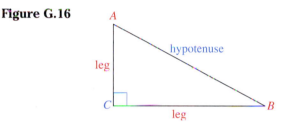

Pythagorean Theorem

One of the most famous theorems in all of mathematics is named for the Greek mathematician Pythagoras.

THEOREM G.4 **THE PYTHAGOREAN** **THEOREM**	In right $\triangle ABC$, $a^2 + b^2 = c^2$. 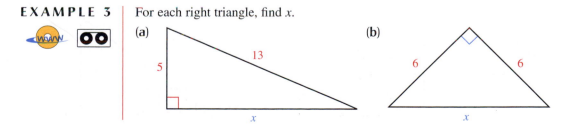

In words, the Pythagorean theorem says that "the sum of the squares of the legs of a right triangle is equal to the square of the hypotenuse."

EXAMPLE 3

For each right triangle, find x.

(a)

5

13

x

(b)

6   6

x

Solution (a) By the Pythagorean theorem we have

$$x^2 + 5^2 = 13^2$$
$$x^2 + 25 = 169$$
$$x^2 = 144 \qquad \textit{Take square roots.}$$
$$x = \pm\sqrt{144} = \pm 12$$

Since x represents the length of one leg of the triangle, x must be a positive number. Thus, $\boxed{x = 12}$.

(b) By the Pythagorean theorem we have

$$x^2 = 6^2 + 6^2$$
$$x^2 = 72 \qquad \textit{Take square roots.}$$
$$x = \pm\sqrt{72} \qquad \textit{Simplify the radical to get}$$
$$x = \pm\sqrt{36 \cdot 2} = \pm\sqrt{36}\sqrt{2} = \pm 6\sqrt{2}$$

Since x represents a length and must be a positive $\boxed{x = 6\sqrt{2}}$. ∎

We also point out that the converse of the Pythagorean theorem is true. That is, if we have a triangle in which $a^2 + b^2 = c^2$, then the triangle must be a right triangle.

For example, if we have the triangle shown in Figure G.17, we can see that $3^2 + 4^2 = 5^2$ because $9 + 16 = 25$. Therefore, $\triangle ABC$ must be a right triangle with $\angle C$ being the right angle.

Figure G.17

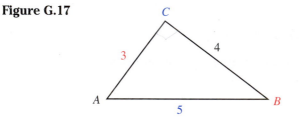

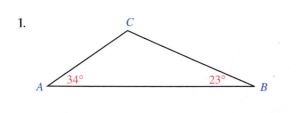 **EXERCISES G.2**

In Exercises 1–4, find all the missing angles.

1.

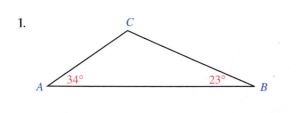

2.

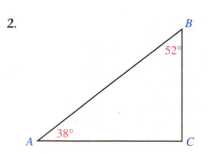

3.

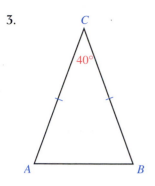

4.

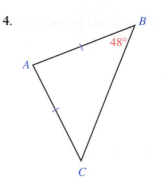

In Exercises 5–12, find x.

5.

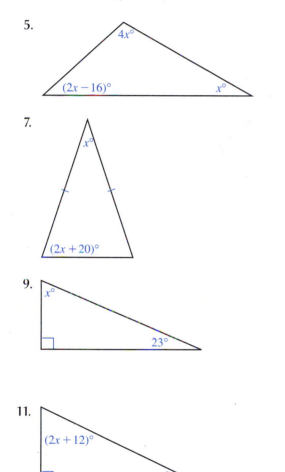

6.

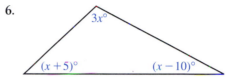

7.

8.

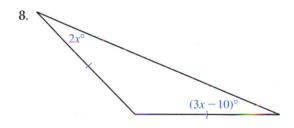

9.

10.

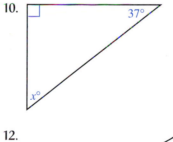

11.

12.

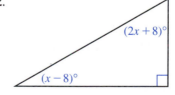

In Exercises 13–14, find the missing side of each triangle.

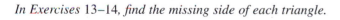

13. *B*

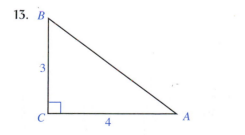

14. *B*

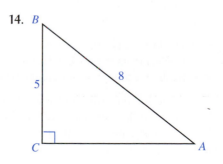

In Exercises 15–18, find the missing side (or sides) of each triangle(s).

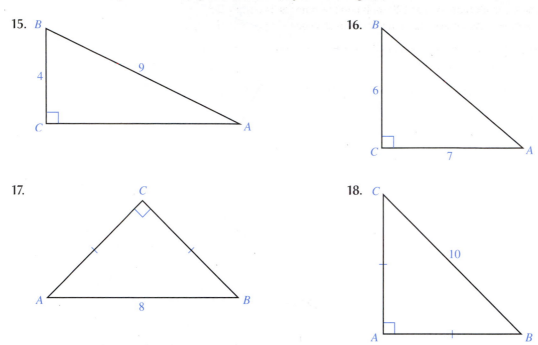

15.

16.

17.

18.

In Exercises 19–20, find the length of the indicated segment.

19. Find $|\overline{AC}|$.

20. Find $|\overline{CB}|$.

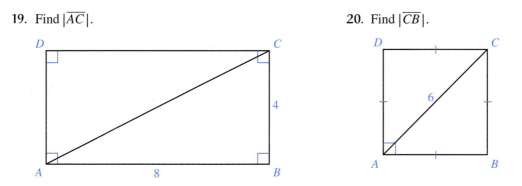

21. A 30-ft ladder is leaning against a building. If the foot of the ladder is 10 ft away from the base of the building, how far up the building does the ladder reach?

22. A 40-ft ladder is leaning against a wall. If the ladder reaches 20 ft up the wall, how far away from the base of the wall is the foot of the ladder?

23. Two airplanes leave an airport at the same time and at a 90° angle from each other. After an hour of flying at the same altitude, one plane is 160 miles from the airport, and the other is 180 miles from the airport. To the nearest tenth of a mile, how far are the planes from each other?

24. Two boats leave a dock at the same time and at a 90° angle from each other. After 3 hours one boat is 30 miles from the dock, while the other is 50 miles from the dock. To the nearest tenth of a mile, how far are the boats from each other?

25. Two boats leave a dock at the same time and at a 90° angle from each other. One boat travels at 20 (nautical) miles per hour, while the other travels at 32 (nautical) miles per hour. To the nearest tenth of a mile, how far are the boats from each other after 3 hours?

26. Two airplanes leave an airport at the same time and at a 90° angle from each other. Both planes fly at the same altitude, one at 120 mph and the other at 140 mph. To the nearest tenth of a mile, how far are the planes from each other after 3 hours?

27. Will a 15-inch ruler fit in an $8\frac{1}{2}$- by 11-inch envelope? Explain your answer. (Assume the width of the ruler is negligible.)

28. Will a 13-inch ruler fit in an $8\frac{1}{2}$- by 11-inch envelope? Explain your answer. (Assume the width of the ruler is negligible.)

29. Find x.

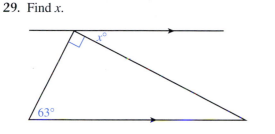

30. Find x.

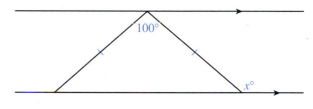

31. Find x.

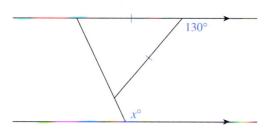

32. Find x.

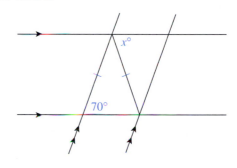

33. Find x.

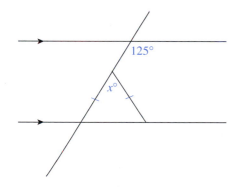

34. Find x.

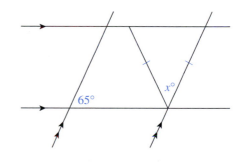

35. Find x.

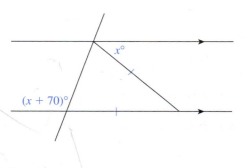

36. Find x.

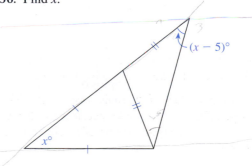

G.3   Congruence

When two geometric figures are *identical* (like two identical pieces of a puzzle that can be placed exactly on top of each other), the figures are called **congruent**. Congruent figures are exact duplicates of each other.

There are many everyday situations in which we are interested in making congruent (identical) copies of objects. For example, tracing over an existing figure creates a congruent copy; using a dress pattern to cut out pieces of fabric creates congruent pieces of material; mass producing automobile or computer parts depends on the idea that component parts must be congruent, in order for all the parts to fit together.

Consider the two figures shown in Figure G.18

Figure G.18

Recall that a polygon is a figure made up of straight line segments.

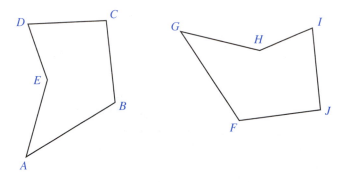

In order to demonstrate that these two figures are congruent, we could cut out one of the figures and superimpose it onto the other figure to show that they are identical. Is there any other way to establish the congruence of two figures? A moment's thought should convince us that two polygons are congruent if and only if the sides and angles of one are individually congruent to the sides and angles of the other. Referring back to Figure G.18, instead of cutting out one figure and superimposing it on the other to establish congruence, we could simply measure the sides and angles of the two figures and compare them. In other words, if we establish all of the following:

$$\overline{AB} \cong \overline{GF} \qquad \angle A \cong \angle G$$
$$\overline{BC} \cong \overline{FJ} \qquad \angle B \cong \angle F$$
$$\overline{CD} \cong \overline{JI} \qquad \angle C \cong \angle J$$
$$\overline{DE} \cong \overline{IH} \qquad \angle D \cong \angle I$$
$$\overline{EA} \cong \overline{HG} \qquad \angle E \cong \angle H$$

The symbol $\cong$ is read "is congruent to."

then the two figures must be congruent. After all, this is exactly what congruence means—that all the parts of the two figures match up exactly.

The sides and angles of one figure that match up with the sides and angles of a congruent figure are called the **corresponding parts** of the congruent figures.

> Two figures are congruent if and only if their corresponding parts are congruent.

If the two polygons happen to be triangles, the equality of three sides and three angles of one triangle with those of the second triangle would certainly establish the congruence of the two triangles. A natural question that arises is: Is it necessary to determine the equality of all six corresponding parts of one triangle (three sides and three angles) with the six parts of the other triangle? In order to answer this question, consider Figure G.19, which illustrates two sides and one angle of a triangle.

Figure G.19

Specifying two sides and the
included angle of a triangle

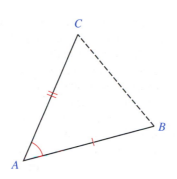

*How many different triangles
can be made given sides $\overline{AB}$,
$\overline{AC}$, and $\angle A$?*

How many triangles can be formed with the two specified sides and the angle (usually called the included angle) between them? The answer is clearly "one!" Our only choice is to connect points *B* and *C* to complete the triangle, as indicated by the dashed line in Figure G.19. In other words, if two triangles have two sides and the included angle of one equal to two sides and the included angle of another, then the triangles must be congruent. This congruence property of triangles is generally referred to as SAS (side-angle-side) and stands for the following statement:

> Two triangles are congruent if two sides and the included angle of one triangle are congruent to two sides and the included angle of the other.

In other words, we are saying: If we know that three particular parts of the two triangles are identical, then all six parts of the two triangles must be identical.

Figure G.20 illustrates two additional situations in which three parts of a triangle are sufficient to completely determine the triangle.

Figure G.20

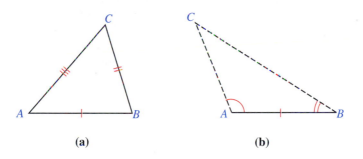

(a)   (b)

(a) Figure G.20(a) illustrates that when all three sides of a triangle are specified, only one triangle is possible. This congruence property of triangles is generally referred to as SSS (side-side-side) and stands for the following statement:

> Two triangles are congruent if three sides of one triangle are congruent to three sides of the other.

(b) Figure G.20(b) illustrates that when two angles and the included side of a triangle are specified, only one triangle is possible. This congruence property of triangles is generally referred to as ASA (angle-side-angle) and stands for the following statement:

> Two triangles are congruent if two angles and the included side of one triangle are congruent to two angles and the included side of the other.

EXAMPLE 1

Suppose that Lamar and Cherise both want to enclose a triangular garden. They both purchase three straight pieces of wooden fencing of lengths 5 ft, 8 ft, and 10 ft. How many possible different configurations can they create for the triangular enclosure?

Solution

What we are actually asking here is: How many triangles can we form if the lengths of the three sides are specified? From our previous discussion, we know from SSS that there is only one possible triangle that can be constructed. Therefore, both Lamar and Cherise must build the same triangular garden, as illustrated in Figure G.21.

Figure G.21

The enclosed triangular gardens of Example 1.

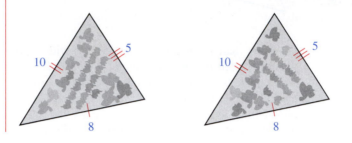

EXAMPLE 2

Establish the congruence of the following triangles and determine which other parts of the two triangles are congruent

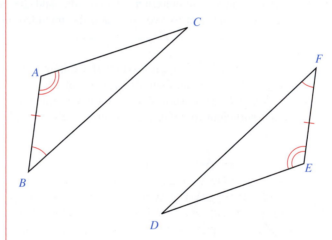

Solution

Looking at the two triangles carefully, we see that two angles and the included side of one triangle are congruent to two angles and the included side of the second triangle. Therefore the two triangles are congruent by ASA.

Because $\angle C$ and $\angle D$ are corresponding parts of congruent triangles, $\angle C \cong \angle D$. $\overline{AC} \cong \overline{ED}$ because these sides are opposite equal angles. Similarly, we have $\overline{BC} \cong \overline{FD}$ because these sides are opposite equal angles.

EXAMPLE 3

In Figure G.22, $\overline{AB} \parallel \overline{DC}$ and $\overline{BC} \parallel \overline{AD}$. Explains why $\triangle ADC \cong \triangle CBA$.

Figure G.22

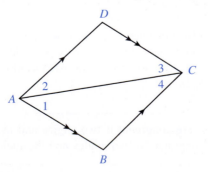

Solution   We can view $\overline{AC}$ as a transversal cutting across the parallel line segments $\overline{AB}$ and $\overline{DC}$. Hence $\angle 1$ and $\angle 3$ are alternate interior angles to this pair of parallel lines, and are therefore congruent: $\angle 1 \cong \angle 3$ (see Figure G.23(a)). On the other hand, we can view $\overline{AC}$ as a transversal cutting across the parallel line segments $\overline{AD}$ and $\overline{BC}$. Hence $\angle 2$ and $\angle 4$ are alternate interior angles to this pair of parallel lines, and therefore congruent: $\angle 2 \cong \angle 4$ (see Figure G.23(b)). Both triangles $\triangle ABC$ and $\triangle CDA$ share the common side $\overline{AC}$. By ASA, $\triangle ADC \cong \triangle CBA$ (see Figure G.23(c)).

Figure G.23

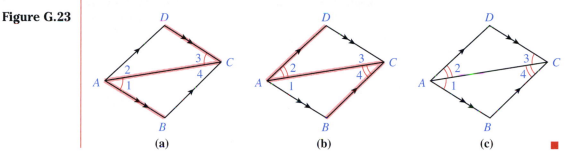

(a)   (b)   (c)

EXAMPLE 4

In Figure G.24, explain why $\overline{ED} \cong \overline{BC}$.

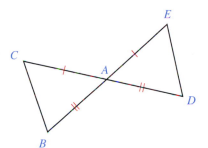

Figure G.24

Solution   We are given that $\overline{AE} \cong \overline{AC}$ and $\overline{AD} \cong \overline{AB}$. Since $\angle CAB$ and $\angle EAD$ are vertical angles, they are congruent. We can now see that two sides and the included angle of $\triangle ABC$ are congruent to two sides and the included angle of $\triangle ADE$. See Figure G.25. By SAS, $\triangle ABC$ and $\triangle ADE$ are congruent. Since side $\overline{ED}$ corresponds to side $\overline{BC}$, and corresponding parts of congruent triangles are congruent, we have $\overline{ED} \cong \overline{BC}$.

Figure G.25

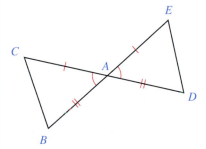

EXAMPLE 5   In Figure G.26, given $\triangle ABC$ is isosceles with $\overline{AC} \cong \overline{BC}$ and $\overline{AB} \perp \overline{CD}$. Explain why $\angle ACD \cong \angle BCD$ and $\overline{AD} \cong \overline{DB}$.

Figure G.26

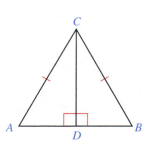

Solution We are given that $\overline{AC} \cong \overline{BC}$. It follows that $\angle A \cong \angle B$, since they are the base angles of an isosceles triangle. Since $\overline{AB} \perp \overline{CD}$, $\angle ADC$ and $\angle BDC$ are both right angles and therefore $\angle ADC \cong \angle BDC$. See Figure G.27(a). Let's now focus on the two triangles $\triangle ADC$ and $\triangle BDC$. Since two angles of $\triangle ADC$ are congruent to two angles of $\triangle BDC$, by our discussion in the last section, the third angles of each triangle must be congruent to each other. Hence we have $\angle ACD \cong \angle BCD$.

Since we have two angles and the included side of $\triangle ADC$ congruent to two angles and the included side of $\triangle BDC$, by ASA we have $\triangle ADC \cong \triangle BDC$. Therefore $AD \cong DB$, since they are corresponding parts of congruent triangles. See Figure G.27(b).

Figure G.27

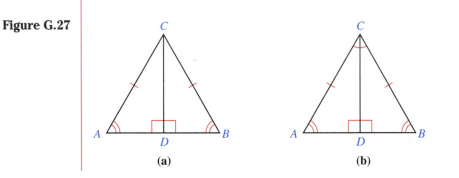

(a)   (b)

EXERCISES G.3

In Exercises 1–6, explain why the two triangles are congruent.

1.

2.

3.

4.

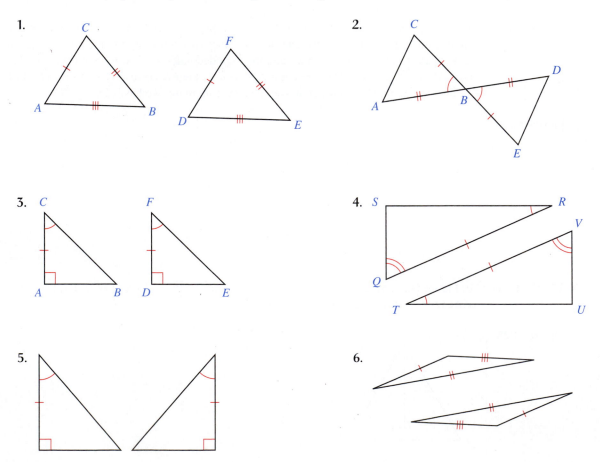

5.

6.

In Exercises 7–14, explain why the two triangles are congruent.

7.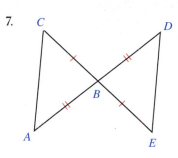

8.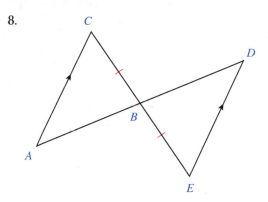

9.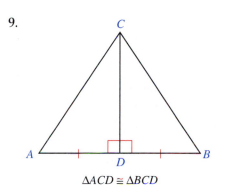

$\triangle ACD \cong \triangle BCD$

10.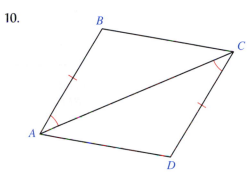

11.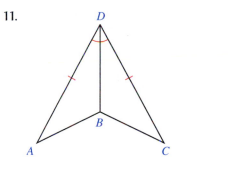

12. $\triangle ADC \cong \triangle BDC$

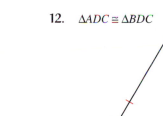

13.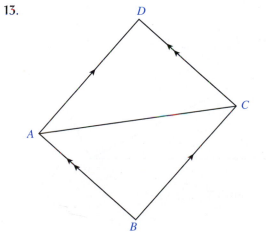

14.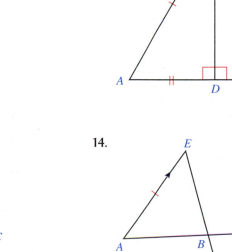

15. Given the figure below with $\overline{CE} \cong \overline{CA}$ and $\angle E \cong \angle A$. Show $\angle D \cong \angle B$.

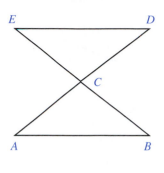

16. Given the figure below with $\angle ACD \cong \angle BCD$ and $\overline{CD} \perp \overline{AB}$. Show $\triangle ACB$ is isosceles.

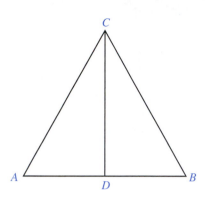

17. Given the figure below with $\overline{AC} \cong \overline{CB}$ and $\angle CAD \cong \angle CBE$. Show $\triangle ADC \cong \triangle BEC$.

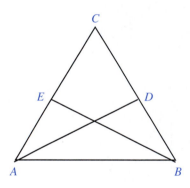

18. Given the figure below with $\overline{AB} \perp \overline{EA}$, $\overline{DC} \perp \overline{BC}$, $\overline{AB} \cong \overline{BC}$, and $\overline{EA} \cong \overline{DC}$. Show that $\angle EBA \cong \angle DBC$.

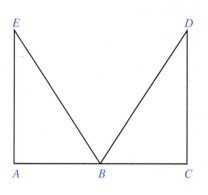

19. Given the figure below with $\overline{CE} \cong \overline{CD}$ and $\angle CAD \cong \angle CBE$. Show $\overline{EB} \cong \overline{AD}$.

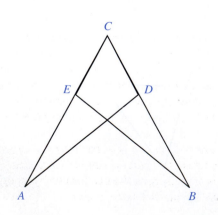

20. Given the figure below with $\angle A \cong \angle C$, $\angle AFB \cong \angle CED$, and $\overline{FB} \cong \overline{DE}$. Show that $\overline{AB} \cong \overline{DC}$.

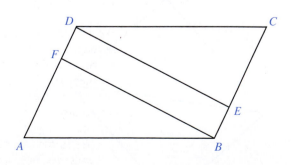

In Exercise 21–24, determine what additional information is necessary in order to show that the triangles are congruent by the given theorems.

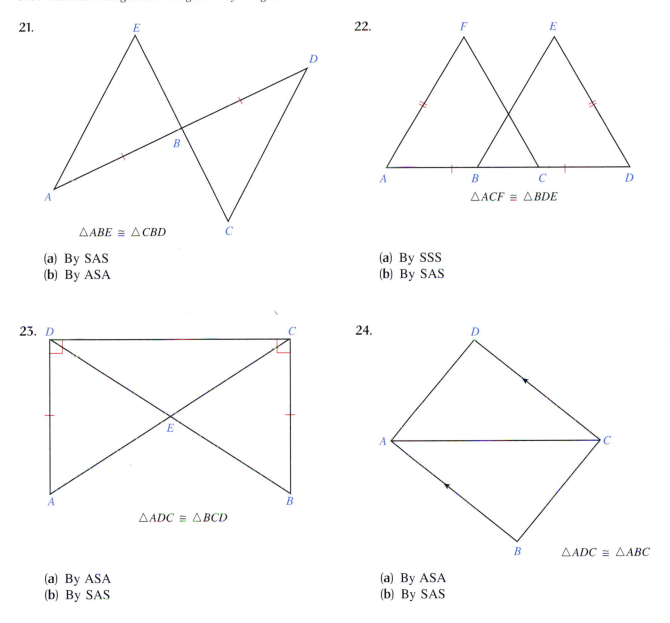

21.

△*ABE* ≅ △*CBD*

(a) By SAS
(b) By ASA

22.

△*ACF* ≅ △*BDE*

(a) By SSS
(b) By SAS

23.

△*ADC* ≅ △*BCD*

(a) By ASA
(b) By SAS

24.

△*ADC* ≅ △*ABC*

(a) By ASA
(b) By SAS

G.4   Similarity

We are familiar with instruments that enlarge or reduce the observed size of objects. A magnifying glass, a telescope, and a movie projector are familiar examples of instruments that enlarge the observed size of an object. Scale drawings, maps, and architectural plans are examples of images that are obtained by shrinking the observed size of an object. What all these instruments and situations have in common is that the enlarged or shrunken image maintains the same shape as the original object, but not necessarily the same size.

When two geometric figures have the same shape we say that they are *similar.* For example, Figure G.28 on page 538 illustrates two pairs of similar geometric figures. In particular, when the geometric figures are polygons (figures formed by straight line segments), having the same shape means having the same angles.

Figure G.28

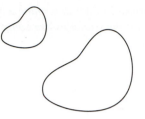

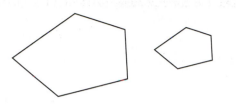

Next let's turn our attention to similar triangles.

Similar Triangles

In light of the previous discussion, we have the following:

DEFINITION	Two triangles are *similar* if the three angles of one are equal to the three angles of the other.

As we noted in section G.2, if two angles of one triangle are equal to two angles of another triangle, then the third angle of each must be equal as well. Thus we have

> Two triangles are similar if two angles of one are equal to two angles of the other.

When two triangles are similar we use the symbol ~. In Figure G.29, we have △*ABC* ~ △*DEF*, meaning △*ABC* is similar to △*DEF*.

Figure G.29

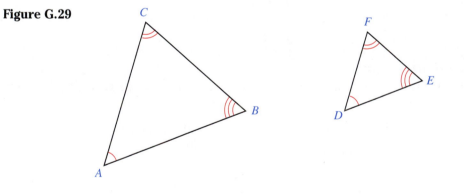

As with congruent triangles, the pairs of congruent angles are called *corresponding angles.* In Figure G.30, ∠*A* ≅ ∠*A'*, ∠*B* ≅ ∠*B'*, and ∠*C* ≅ ∠*C'*. With similar triangles, we refer to the *corresponding sides* as the sides opposite the congruent angles. In Figure G.30, △*ABC* ~ △*A'B'C'*: The side of length 3 corresponds to the side of length 6 (since they are both opposite equal angles). Similarly, the other pairs of corresponding sides are 5 and 10, and 7 and 14.

Figure G.30

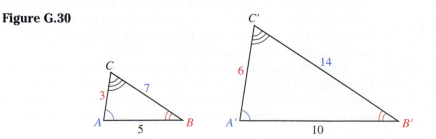

Note that the corresponding sides are in proportion:

$$\frac{3}{6} = \frac{5}{10} = \frac{7}{14}$$

All similar triangles share this property, which is stated formally in the following theorem.

THEOREM G.5	The corresponding sides of similar triangles are in proportion.

EXAMPLE 1 In Figure G.31, find *x*.

Figure G.31

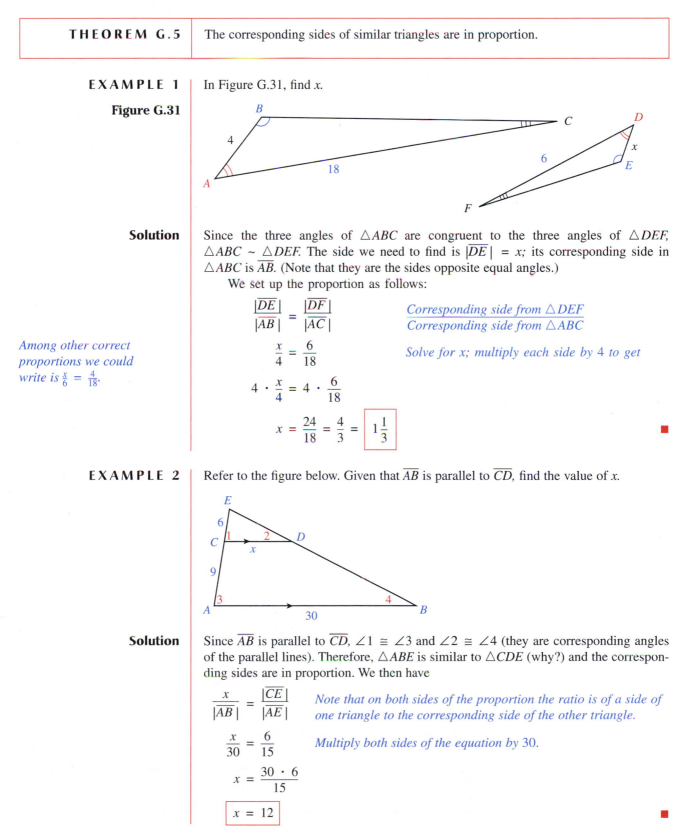

Solution Since the three angles of $\triangle ABC$ are congruent to the three angles of $\triangle DEF$, $\triangle ABC \sim \triangle DEF$. The side we need to find is $|\overline{DE}| = x$; its corresponding side in $\triangle ABC$ is $\overline{AB}$. (Note that they are the sides opposite equal angles.)

We set up the proportion as follows:

Among other correct proportions we could write is $\frac{x}{6} = \frac{4}{18}$.

$$\frac{|\overline{DE}|}{|\overline{AB}|} = \frac{|\overline{DF}|}{|\overline{AC}|} \qquad \begin{array}{l} \textit{Corresponding side from } \triangle DEF \\ \textit{Corresponding side from } \triangle ABC \end{array}$$

$$\frac{x}{4} = \frac{6}{18} \qquad \textit{Solve for x; multiply each side by 4 to get}$$

$$4 \cdot \frac{x}{4} = 4 \cdot \frac{6}{18}$$

$$x = \frac{24}{18} = \frac{4}{3} = \boxed{1\frac{1}{3}} \qquad \blacksquare$$

EXAMPLE 2 Refer to the figure below. Given that $\overline{AB}$ is parallel to $\overline{CD}$, find the value of *x*.

Solution Since $\overline{AB}$ is parallel to $\overline{CD}$, $\angle 1 \cong \angle 3$ and $\angle 2 \cong \angle 4$ (they are corresponding angles of the parallel lines). Therefore, $\triangle ABE$ is similar to $\triangle CDE$ (why?) and the corresponding sides are in proportion. We then have

$$\frac{x}{|\overline{AB}|} = \frac{|\overline{CE}|}{|\overline{AE}|} \qquad \begin{array}{l}\textit{Note that on both sides of the proportion the ratio is of a side of} \\ \textit{one triangle to the corresponding side of the other triangle.}\end{array}$$

$$\frac{x}{30} = \frac{6}{15} \qquad \textit{Multiply both sides of the equation by 30.}$$

$$x = \frac{30 \cdot 6}{15}$$

$$\boxed{x = 12} \qquad \blacksquare$$

EXAMPLE 3

Matt measures the height of a tree in the following way. He stands 50 feet away from the tree and asks his friend Al to walk toward him, starting from the tree. Matt lies down on the (level) ground, and watches Al walk toward him. As soon as the top of Al's head is in the same line of sight as the top of the tree, Matt tells Al to stop. See Figure G.32. Matt then measures the distance from where he was lying to where Al stopped. If this distance is 6 feet, and he knows Al is 5'6″, how tall is the tree (to the nearest foot)?

Figure G.32

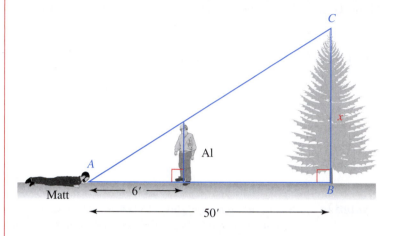

Solution

You can see by the figure that there is a right triangle with the tree as one side, and a right triangle with Al as one side. Both right triangles share angle A, the angle formed by the ground and Matt's line of sight. Each triangle also has a right angle. This means that two of the angles of one triangle are equal to two of the angles of the other, and the two right triangles are therefore similar. Since the triangles are similar, the sides are in proportion. We are trying to find the height of the tree, so we form a proportion using the height of the tree and the sides we are given as follows (see Figure G.33):

Figure G.33

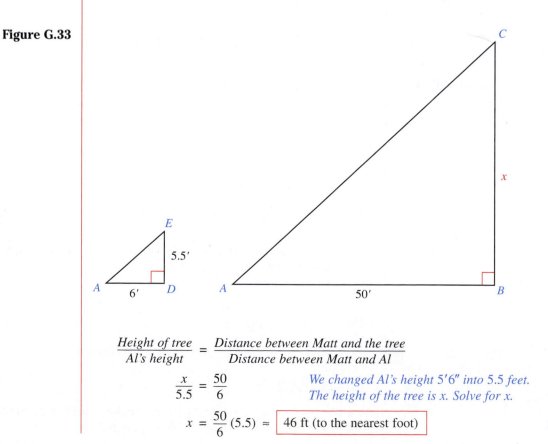

$$\frac{\textit{Height of tree}}{\textit{Al's height}} = \frac{\textit{Distance between Matt and the tree}}{\textit{Distance between Matt and Al}}$$

$$\frac{x}{5.5} = \frac{50}{6} \qquad \textit{We changed Al's height 5'6″ into 5.5 feet.}$$
$$\textit{The height of the tree is x. Solve for x.}$$

$$x = \frac{50}{6}(5.5) \approx \boxed{46 \text{ ft (to the nearest foot)}}$$ ∎

EXAMPLE 4

In Figure G.34, find x.

Figure G.34

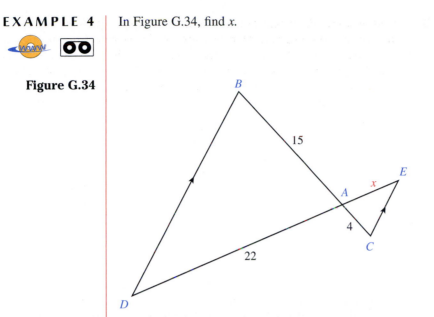

Solution

By the figure, we see $\overline{BD} \parallel \overline{EC}$. Line segment $\overline{BC}$ is a transversal cutting across the parallel lines. Since $\angle B$ and $\angle C$ are alternate interior angles, they are congruent: $\angle B \cong \angle C$. $\angle 1$ is congruent to $\angle 2$ since they are vertical angles. This is enough to establish that $\triangle ABD$ and $\triangle ACE$ are similar. See Figure G.35.

Figure G.35

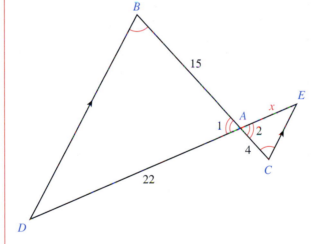

Since $\triangle ABD \sim \triangle ACE$, their corresponding sides are proportional.

$$\frac{|\overline{AE}|}{|\overline{AD}|} = \frac{|\overline{AC}|}{|\overline{AB}|}$$ *Corresponding side from $\triangle ACE$*
Corresponding side from $\triangle ABD$

$$\frac{x}{22} = \frac{4}{15}$$ *Solve for x; multiply each side by 22 to get*

$$x = \frac{4}{15}(22)$$

$$x = \frac{88}{15} = \boxed{5\frac{13}{15}}$$ ■

Scaling Factors

Let's look at the idea of similarity from a slightly different perspective. Consider the three similar triangles in Figure G.36 on page 542.

Figure G.36

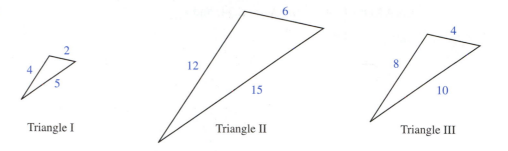

Triangle I   Triangle II   Triangle III

The ratio of the corresponding sides in triangles I and II is $\frac{1}{3}$. We can interpret this ratio to mean that the sides of triangle II are 3 times the lengths of the corresponding sides in triangle I. We say that the number 3 is the *scaling factor* for these two similar triangles. Equivalently, we can say that the sides of triangle I are $\frac{1}{3}$ times the length of the sides in triangle II, in which case the scaling factor is $\frac{1}{3}$.

Looking at triangles I and III, we can see that the scaling factor is 2. This scaling factor is similar to the way the power of a telescope or a pair of binoculars is described. If we are told that a telescope has a "100 power" lens, this means that the observed size of the object through the telescope is 100 times larger than the size of the object as observed with the naked eye. In other words, the image size has been scaled up by a factor of 100. Similarly, if a scale model of an airplane is built on a scale of $\frac{1}{60}$, this means that 1 inch on the model represents 60 inches (or 5 feet) of the actual airplane.

EXAMPLE 5

A blueprint is drawn with a scaling factor of $\frac{1}{30}$ (in feet). If the length and width of a rectangular room on the blueprint are 1.6 ft and 2.3 ft, respectively, find the actual dimensions of the room.

Solution

The fact that the scaling factor is $\frac{1}{30}$ means that the actual dimensions are 30 times the blueprint dimensions. Therefore

$$\text{length} = 30(1.6) = 48 \text{ ft}$$
$$\text{width} = 30(2.3) = 69 \text{ ft}$$

Alternatively, we could have found the length using the proportion
$$\frac{1.6}{length} = \frac{1}{30}$$

Thus, the actual room is $\boxed{48 \text{ ft by } 69 \text{ ft}}$. ■

Two Special Triangles

Two special triangles play important roles in mathematics. They are the **isosceles right triangle** and the **30°–60° right triangle.**

The Isosceles Right Triangle Figure G.37 shows an isosceles right triangle. Note that because the legs are equal, the base angles must be equal, and since the base angles must have a sum of 90°, they must be 45° each. We have labeled each leg s and the hypotenuse x. We solve for x by using the Pythagorean theorem.

Figure G.37
An isosceles right triangle

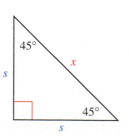

$$x^2 = s^2 + s^2$$
$$x^2 = 2s^2$$

Take square roots to get

$$x = \pm\sqrt{2s^2}$$

Simplify the radical. Since s is positive, $\sqrt{s^2} = s$.

$$x = \pm\sqrt{s^2}\sqrt{2} = \pm s\sqrt{2}$$

Since x is a length, we reject the negative solution, and so $x = s\sqrt{2}$. We have just derived the following.

The Isosceles (45°) Right Triangle

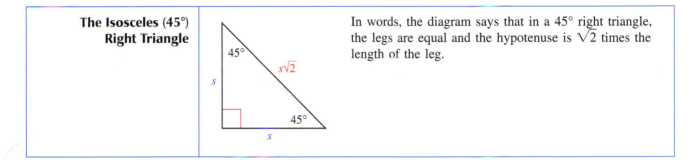

In words, the diagram says that in a 45° right triangle, the legs are equal and the hypotenuse is $\sqrt{2}$ times the length of the leg.

The 30°–60° Right Triangle Figure G.38(a) shows a 30°–60° right triangle. We label the hypotenuse h. If we duplicate the triangle as indicated by the dotted lines in Figure G.38(b), we can see that $\triangle ABD$ is equilateral (because each angle is 60°), so that $|\overline{AD}|$ is also h and $|\overline{AC}|$ must be $\frac{h}{2}$.

Figure G.38

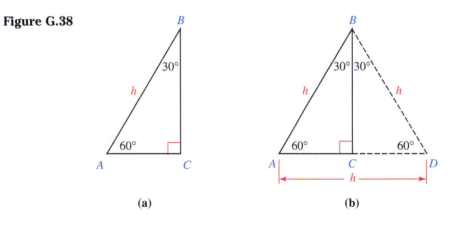

(a)   (b)

In Figure G.39 we redraw Figure G.38(a). We have labeled the hypotenuse h, $|\overline{AC}|$ as $\frac{h}{2}$, and the unknown side $|\overline{BC}|$ as x. We find x by again using the Pythagorean theorem.

Figure G.39

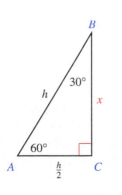

Why is the name "30°–60° right triangle" actually redundant?

$$x^2 + \left(\frac{h}{2}\right)^2 = h^2$$

$$x^2 + \frac{h^2}{4} = h^2 \qquad \textit{Isolate } x^2. \textit{ Subtract } \frac{h^2}{4} \textit{ from each side.}$$

$$x^2 = h^2 - \frac{h^2}{4} \qquad \textit{Combine.}$$

$$x^2 = \frac{3h^2}{4} \qquad \textit{Take square roots.}$$

$$x = \pm\sqrt{\frac{3h^2}{4}} \qquad \textit{Simplify the radical.}$$

$$x = \pm\frac{\sqrt{3h^2}}{\sqrt{4}}$$

$$x = \pm\frac{h\sqrt{3}}{2} = \frac{h}{2}\sqrt{3}$$

As before, we reject the negative solution, so $x = \frac{h}{2}\sqrt{3}$. We have derived the following.

The 30°–60° Right Triangle	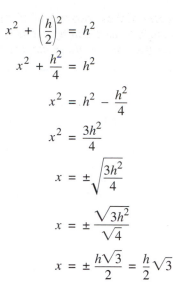	In words, the diagram says that in a 30°–60° right triangle, the side opposite the 30° angle is one-half the hypotenuse, and the side opposite the 60° angle is one-half the hypotenuse times $\sqrt{3}$.

EXAMPLE 6

Find the missing sides and angles in each of the following triangles.

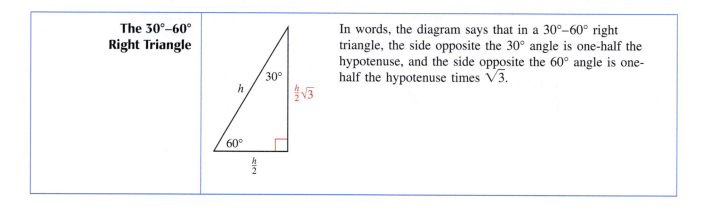

(a)   (b)

In words, describe the relationships among the sides of an isosceles right triangle.

Solution

(a) Since the triangle is an isosceles right triangle, $\angle A = \angle B = 45°$. Since the hypotenuse of a 45° right triangle is $\sqrt{2}$ times the leg, $|\overline{AB}| = 8\sqrt{2}$.

In words, describe the relationships among the sides of a 30°–60° right triangle.

(b) From the diagram, we can see that $\angle B$ must be 60°. Since this is a 30°–60° right triangle, the side opposite the 30° angle is equal to half the hypotenuse. Therefore, $|\overline{BC}| = 10 = \frac{1}{2}|\overline{AB}|$, and so $|\overline{AB}| = 20$. $\overline{AC}$ is the side opposite 60° and is, therefore, one-half the hypotenuse times $\sqrt{3}$, and so $|\overline{AC}| = 10\sqrt{3}$. ■

The simplest examples of these two special triangles are obtained by choosing the leg of the 45° right triangle to be 1 and by choosing the hypotenuse of the 30°–60° right triangle to be 2. We then obtain the prototypes illustrated in the following box.

Prototypes for the Isosceles and 30°–60° Right Triangles

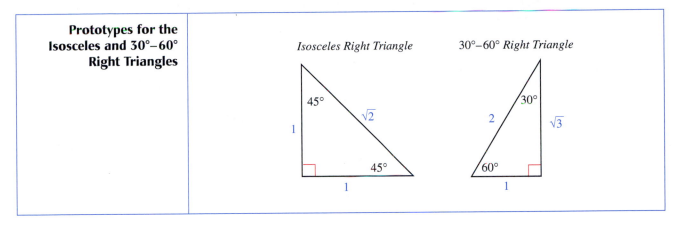

Isosceles Right Triangle

30°–60° Right Triangle

EXERCISES G.4

In Exercises 1–6, explain why the given pairs of triangles are similar. Identify the corresponding angles.

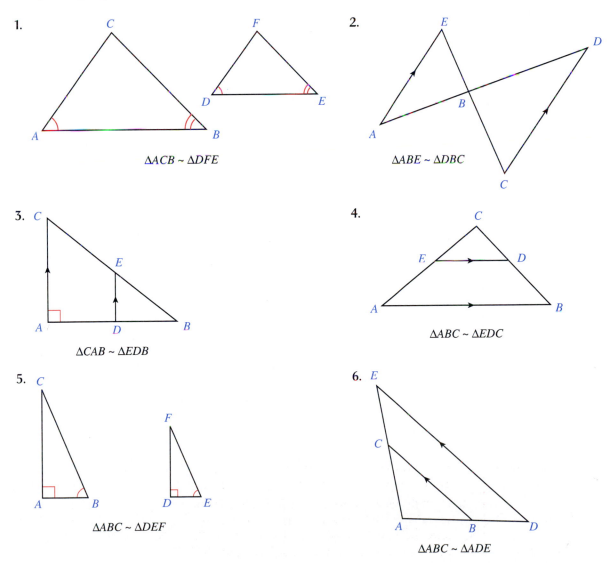

1.

ΔACB ~ ΔDFE

2.

ΔABE ~ ΔDBC

3.

ΔCAB ~ ΔEDB

4.

ΔABC ~ ΔEDC

5.

ΔABC ~ ΔDEF

6.

ΔABC ~ ΔADE

In Exercises 7–10, △ABC is similar to △DEF. Find the missing sides in △DEF.

7.

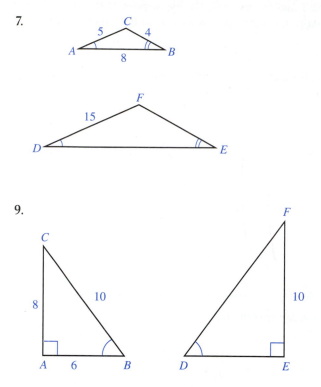

8.

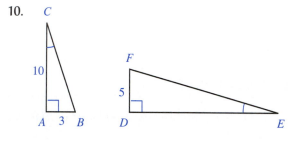

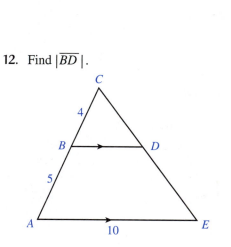

9.

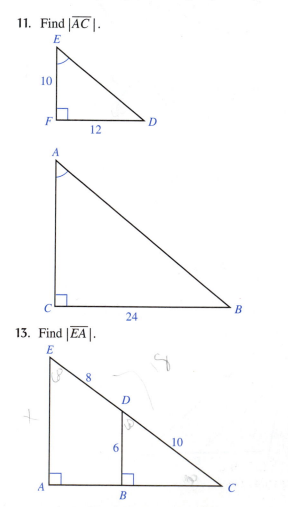

10.

In Exercises 11–14, find the length of the indicated side.

11. Find $|\overline{AC}|$.

12. Find $|\overline{BD}|$.

13. Find $|\overline{EA}|$.

14. Find $|\overline{DE}|$.

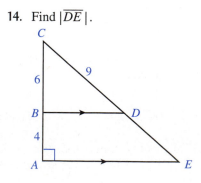

In Exercises 15–18, find the missing sides.

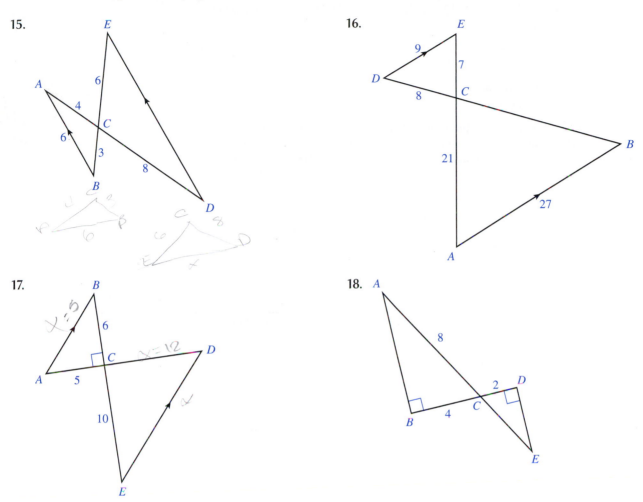

15.

16.

17.

18.

19. Suppose that a man 6 ft tall casts a shadow 4 ft long. Determine the height of a flagpole that casts a shadow 18 ft long. See the accompanying figure.

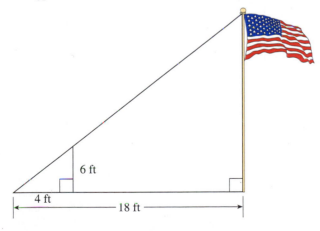

20. Suppose that a bush 3 ft tall casts a shadow 5 ft long. Determine the length of the shadow cast by a tree 20 ft tall.

In Exercises 21–26, round your answer to the nearest tenth where necessary.

21. The corresponding sides of two similar triangles are in the ratio of 4 to 7. If a side of the smaller triangle is 5.8 cm, find the length of the corresponding side of the larger triangle.

22. The corresponding sides of two similar geometric figures are in the ratio of 9 to 4. If a side of the larger figure is 15.3 meters, find the length of the corresponding side of the smaller triangle.

23. A scale drawing uses a scale of $\frac{1}{25}$. If the scale drawing places a door 1.4 inches from a wall, how far from the wall will the actual door be placed?

24. A scale model of an airplane uses a scale of $\frac{1}{40}$. If the model exhibits a wingspan of 8 inches, what is the actual wingspan of the plane?

25. A scale model of a very small piece of machinery uses a scale of $\frac{28}{1}$. If the widest part of the model measures 8.3 cm and the narrowest part of the model measures 3.2 cm, find the actual widest and narrowest dimensions of the actual machine part.

26. Shawna is planning a trip in which she drives from city A to city B to city C and then returns to city A. On a map that uses a scale where 1 inch represents 50 miles, she finds that city A is 1.8 inches from city B, which is 2.2 inches from city C, which is 0.9 inch from city A. Find the actual driving distance for this trip.

In Exercises 27–34, Find the length of the missing sides of the given right triangles.

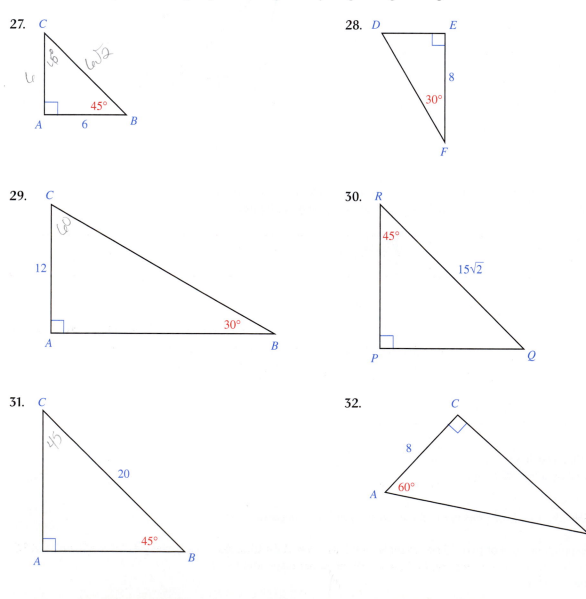

27.

28.

29.

30.

31.

32.

33.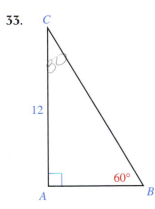

34.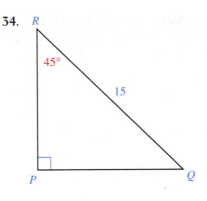

G.5 Quadrilaterals

We have already seen that the sum of the angles of a triangle is 180°. As we can see in Figure G.40, any quadrilateral can be divided into two triangles. In △ABD, ∠1 + ∠2 + ∠3 = 180° and in △BCD, ∠4 + ∠5 + ∠6 = 180°. But all these angles together are the angles of the quadrilateral; therefore we can see that

$$\angle A + \angle B + \angle C + \angle D = 360°$$

Figure G.40

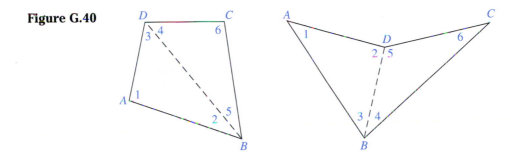

We have just proved the following theorem.

THEOREM G.6	The sum of the angles of a quadrilateral is 360°.

Parallelograms

A quadrilateral in which both pairs of opposite sides are parallel is called a ***parallelogram.***

A ***diagonal*** is a line segment that joins two nonadjacent vertices.

Let's examine the parallelogram $ABCD$ in Figure G. 41:

Figure G.41

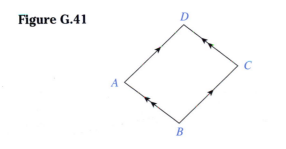

Let's draw a diagonal from A to C to form two triangles: △ADC and △ABC. See Figure G.42(a) on page 550.

Figure G.42

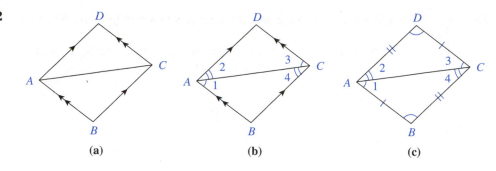

(a)   (b)   (c)

Viewing $\overline{AC}$ as a transversal cutting across the parallel line segments $\overline{AB}$ and $\overline{CD}$, we see that $\angle 1 \cong \angle 3$ since they are alternate interior angles to this pair of parallel lines. On the other hand, viewing $\overline{AC}$ as a transversal cutting across the parallel line segments $\overline{AD}$ and $\overline{BC}$, we see that $\angle 2 \cong \angle 4$ since they are alternate interior angles to this pair of parallel lines (Figure G.42(b)). Both triangles share the common side $\overline{AC}$ (Figure G.42.(c)), and by ASA, the two triangles are congruent. Since corresponding parts of congruent triangles are congruent, $\overline{AB} \cong \overline{CD}$ and $\overline{BC} \cong \overline{AD}$; and $\angle B \cong \angle D$. Since $\angle 1 \cong \angle 3$ and $\angle 2 \cong \angle 4$, we have $\angle 1 + \angle 2 = \angle 3 + \angle 4$. This means that $\angle DAB \cong \angle BCD$. We have the following:

THEOREM G.7	1. The opposite angles of a parallelogram are equal. 2. The opposite sides of a parallelogram are equal.

The content of Theorem G.7 is illustrated in Figure G.43.

Figure G.43

The opposite sides and angles of parellelogram *ABCD* are equal.

Figure G.44

Rhombus *ABCD*

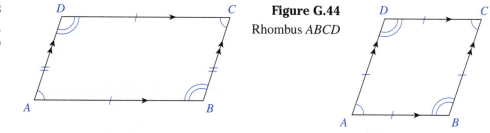

If all four sides of a parallelogram are equal, it is called a ***rhombus*** (see Figure G.44).

If a parallelogram contains a right angle (and therefore it follows that all its angles must be right angles), it is called a ***rectangle*** (see Figure G.45(a)).

Figure G.45

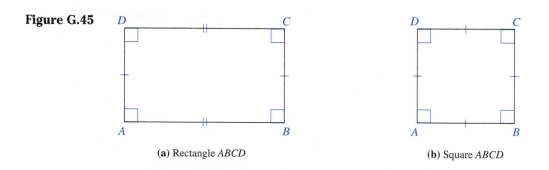

(a) Rectangle *ABCD*   (b) Square *ABCD*

If the adjacent sides of a rectangle are equal (and therefore all four sides are equal), it is called a ***square*** (see Figure G.45(b)).

EXAMPLE 1 Show that the diagonals of a parallelogram bisect each other.

Solution We start by drawing a picture of a parallelogram *ABCD* with diagonals $\overline{AC}$ and $\overline{BD}$ intersecting at *E*. See Figure G.46.

Figure G.46

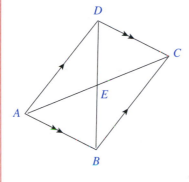

This time we focus on △*AEB* and △*CED*. Viewing $\overline{AC}$ as a transversal cutting across the parallel line segments $\overline{AB}$ and $\overline{DC}$, we see that ∠*BAE* ≅ ∠*DCE* since they are alternate interior angles to this pair of parallel lines (Figure G.47(a)). Viewing $\overline{DB}$ as a transversal cutting across the same parallel line segments, we see that ∠*CDE* ≅ ∠*EBA* since they are alternate interior angles to this pair of parallel lines (Figure G.47(b)).

Figure G.47

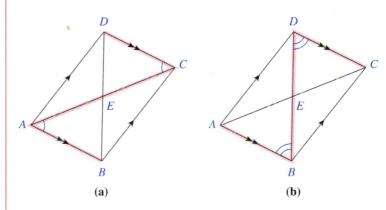

(a)   (b)

Since opposite sides of a parallelogram are congruent (Theorem G.7), by ASA, we have △*AEB* ≅ △*CED*. See Figure G.48(a). $\overline{DE}$ ≅ $\overline{BE}$ since they are corresponding parts of congruent triangles. This shows that diagonal $\overline{AC}$ divides diagonal $\overline{DB}$ into equal parts. Hence diagonal $\overline{AC}$ bisects diagonal $\overline{DB}$. See Figure G.48(b).

Figure G.48

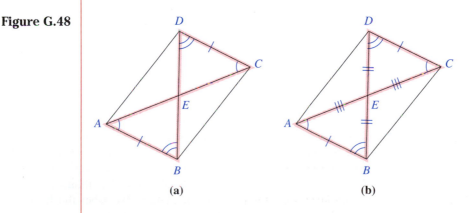

(a)   (b)

$\overline{AE}$ ≅ $\overline{EC}$ since they are also corresponding parts of congruent triangles. This shows that diagonal $\overline{DB}$ divides diagonal $\overline{AC}$ into equal parts. Hence diagonal $\overline{DB}$ bisects diagonal $\overline{AC}$. See Figure G.48(b). ■

EXAMPLE 2

To the nearest tenth of a centimeter, find the length of the side of a square with diagonal 15 cm.

Solution

We draw the figure (see Figure G.49). We note that a diagonal divides the square into two right triangles. the diagonal is 15 cm and is the hypotenuse of each right triangle. We label each side x, and use the Pythagorean theorem to find x, the length of the sides:

Figure G.49

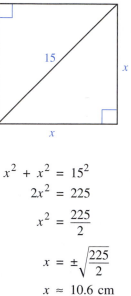

$$x^2 + x^2 = 15^2$$
$$2x^2 = 225$$
$$x^2 = \frac{225}{2} \qquad \textit{Take square roots to get}$$
$$x = \pm\sqrt{\frac{225}{2}} \qquad \textit{Ignore the negative answer. We use a calculator to get}$$
$$x \approx 10.6 \text{ cm}$$

The length of the sides of the square are $\boxed{10.6 \text{ cm}}$, rounded to the nearest tenth. ■

EXAMPLE 3

In a rhombus, the diagonals are perpendicular bisectors of each other. Find the length of the sides of a rhombus with diagonals 12 inches and 16 inches.

Solution

Again we start by drawing a figure (Figure G.50(a)). We draw the diagonals of the rhombus so that they bisect each other and meet at right angles.

Figure G.50

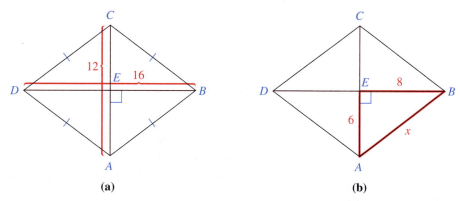

(a)   (b)

We note that now we have four right triangles. Let's examine $\triangle ABE$. Since the diagonals of the rhombus bisect each other, each leg of $\triangle ABE$ is half the length of each diagonal; therefore the legs of $\triangle ABE$ are 6 inches and 8 inches. See Figure G.50(b). (We note that all four triangles are congruent to each other since they each have the same-size legs, and all sides of a rhombus are equal.) The side of the rhombus is the hypotenuse of $\triangle ABE$. We label the hypotenuse x and use the Pythagorean theorem to find it:

$$6^2 + 8^2 = x^2$$
$$36 + 64 = x^2$$

$$100 = x^2 \qquad \textit{Take square roots to get}$$
$$10 = x \qquad \textit{Use the positive root.}$$

The length of the sides of the rhombus is 10 inches. ∎

EXAMPLE 4

Given the following diagram, find the measures of the four angles in parallelogram *ABCD*.

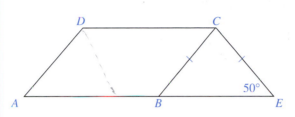

Solution

Since △*BEC* is isosceles, ∠*CBE* = 50° (because the base angles of an isosceles triangle are equal).

Since ∠*ABC* is supplementary to ∠*CBE*, ∠*ABC* = 130° .

Since the opposite angles of a parallelogram are equal, ∠*D* = 130° .

Now, since the sum of the angles of the parallelogram is 360°, the sum of the other two angles is 360° − 130° − 130° = 100°. Therefore, ∠*A* = ∠*BCD* = 50° . ∎

There is one more special type of quadrilateral that comes up frequently. If a quadrilateral has only one pair of parallel sides it is called a ***trapezoid.*** The two parallel sides are called the ***bases*** of the trapezoid. In Figure G.51 the bases are labeled b_1 and b_2.

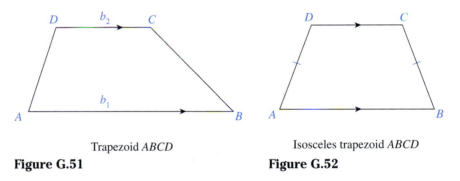

Trapezoid *ABCD*

Figure G.51

Isosceles trapezoid *ABCD*

Figure G.52

If the nonparallel sides of a trapezoid are of equal length, the trapezoid is called an ***isosceles trapezoid.*** Note that the base angles of an isosceles trapezoid are equal (think of extending sides $\overline{AD}$ and $\overline{BC}$ to form an isosceles triangle that would have the same base angles as the given isosceles trapezoid). See Figure G.52.

EXAMPLE 5

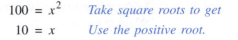

Given the trapezoid *ABCD* in figure G.53, find the length of the missing side.

Figure G.53

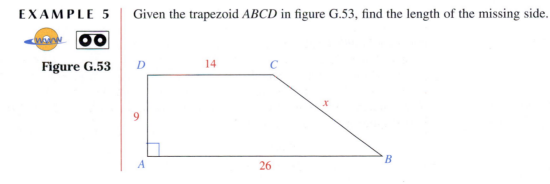

Solution Let's draw a perpendicular line from vertex C to base AB, and label the intersection of this line segment and the bottom base E. See Figure G.54. We have divided trapezoid $ABCD$ into rectangle $AECD$ and right triangle BEC. First we focus on rectangle $AECD$. We can see $|\overline{EC}| = 9$ since it is the same length as the side of the rectangle opposite $\overline{EC}$; $|\overline{AE}| = 14$, since it is the same length as the side of the rectangle opposite $\overline{AE}$. Since $|\overline{AB}| = 26$, that leaves $|\overline{EB}| = 12$. Now we focus on the right triangle EBC. This triangle has legs 9 and 12. We can use the Pythagorean theorem to find the length of the missing side, which in this case, is the hypotenuse of the right triangle.

Figure G.54

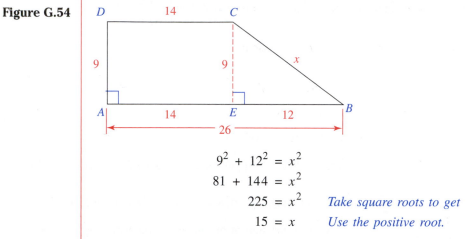

$$9^2 + 12^2 = x^2$$
$$81 + 144 = x^2$$
$$225 = x^2 \qquad \textit{Take square roots to get}$$
$$15 = x \qquad \textit{Use the positive root.}$$

The length of the missing side of the trapezoid is 15. ∎

EXERCISES G.5

In Exercises 1–4, find the missing angles of the quadrilateral.

1.

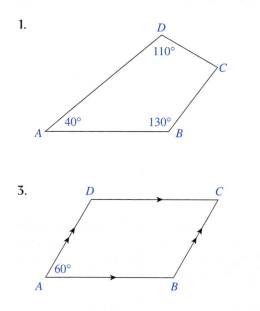

2.

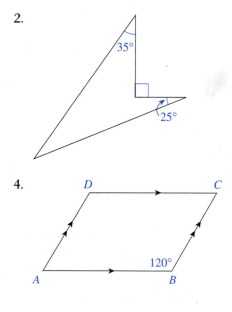

3.

4.

In Exercises 5–8, find x. Each figure is a parallelogram.

5.

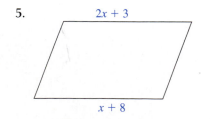

6.

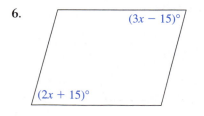

7.

(5x − 18)°   6x°

8.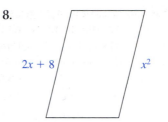

2x + 8 x^2

9. Find the length of the diagonal of a rectangle whose base is 8 cm and whose height is 5 cm.

10. Find the length of the diagonal of a rectangle whose base is 6 cm and whose height is 9 cm.

11. Find the length of the diagonal of a square with side 4 in.

12. Find the length of the diagonal of a square with side 8 in.

13. Find the length of the side of a square with diagonal 5 cm.

14. Find the length of the side of a square with diagonal 8 cm.

The following figure is for Exercises 15–17. It is an isosceles trapezoid ABCD with $\overline{AD} \cong \overline{CB}$.

15. Referring to the figure at the right, show $\triangle ADB \cong \triangle BCA$.

16. Referring to the figure at the right, show $\triangle AEB \sim \triangle CED$.

17. Referring to the figure at the right, if $\angle DCB \cong \angle CDA$, show $\triangle ADC \cong \triangle BCD$.

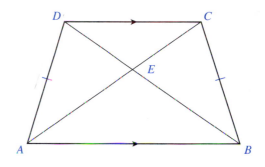

18. The figure below is rectangle ABCD. E is the midpoint of $\overline{DC}$. Show $\angle DAE \cong \angle CBE$.

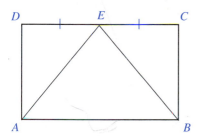

19. The figure below is parallelogram ABCD. Side $\overline{AB}$ is extended and meets the line segment from vertex C at a right angle at E. Side $\overline{DC}$ is extended and meets the line segment from vertex A at a right angle at F. Show $\triangle AFD \cong \triangle CEB$.

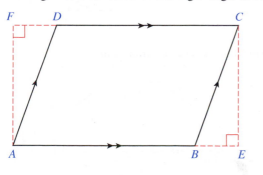

20. The figure to the right is rhombus *ABCD* with diagonals *AC* and *BD* intersecting *E*. Note that the diagonals divide the rhombus into four triangles: △*AEB*, △*BEC*, △*CED*, and △*DEA*. Use the fact that the diagonals of a parallelogram bisect each other to show that all four triangles are congruent to each other. If all four triangles are congruent to each other, what can you conclude about the angles formed by the intersecting diagonals?

21. Find the length of the sides of a rhombus with diagonals 12″ and 18″.

22. If the side of a rhombus is 10″ and a diagonal is 15″, find the length of the other diagonal to the nearest tenth.

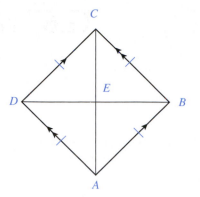

In Exercises 23–30, find x. Round to the nearest tenth where necessary.

23.

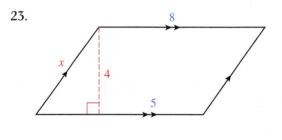

24.

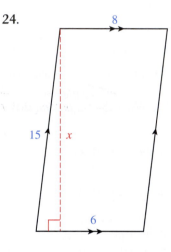

25. *ABCD* is a square.

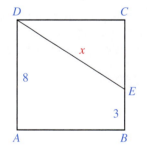

26. *ABCD* is a rectangle.

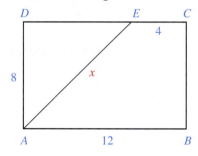

27. *ABCD* is a trapezoid.

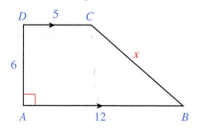

28. *ABCD* is a trapezoid.

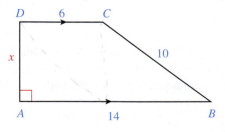

29. *ABCD* is an isosceles trapezoid.

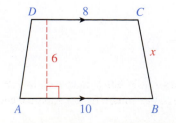

30. *ABCD* is an isosceles trapezoid.

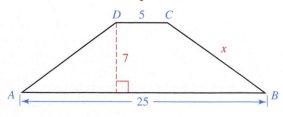

G.6   Perimeter and Area

The perimeter of a polygon is simply the sum of the lengths of all its sides.

While there are a number of "formulas" for the perimeters of various geometric figures, they are just formal statements of this basic fact: To compute the perimeter of a polygon, we simply add the lengths of all its sides.

EXAMPLE 1   Find the perimeter of the following rectangle.

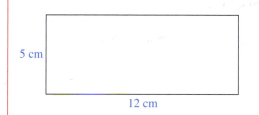

Solution   Since the opposite sides of a rectangle are equal, we have two sides of length 12 cm and two sides of length 5 cm. If we let P = the perimeter, we have

$$P = 2(12) + 2(5) = 24 + 10 = \boxed{34 \text{ cm}}$$ ■

Area

Most people have an intuitive idea of what we mean by the area of a geometric figure. However, making our intuitive ideas mathematically precise is not quite so easy. Fortunately, the situation for parallelograms, triangles, and trapezoids is much simpler than for some other figures.

The basic unit we will use for measuring area is 1 *square unit*—that is, a square with each side one unit long. The unit of length is arbitrary. A square unit can be 1 cm by 1 cm or 1 foot by 1 foot or 1 mile by 1 mile. Figure G.55 shows 1 square unit.

Figure G.55
1 square unit

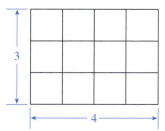

The area of a geometric figure is defined to be the number of square units needed to precisely cover that figure.

If we consider a rectangle whose dimensions are 3 by 4 units (see Figure G.56), we can see that it contains 12 square units.

Figure G.56

A 3 by 4 rectangle has an area of 12 square units.

Be sure you do not confuse length, which is measured in basic units such as feet or meters, with area, which is measured in square units such as square feet or square meters.

Rather than call the sides of the rectangle the length and the width, we will refer to them as *base* and *height* (you will soon see why these names are preferable). The base and height are usually labeled as *b* and *h*, respectively, as in Figure G.57.

Figure G.57

Base and height of
a rectangle

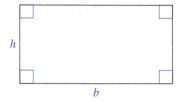

Based on our analysis of the case of the 3 by 4 rectangle, which clearly has an area of 12 square units, we can generalize and say that the area of a rectangle is the product of its base and height.

The area of a rectangle is given by $A = bh$.

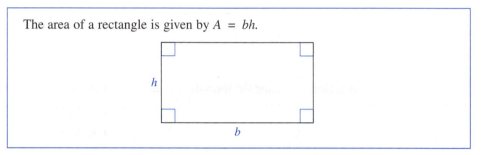

Let us now consider a parallelogram *ABCD*, and let *b* = the length of side $\overline{AB}$. We draw a perpendicular from *D* to side $\overline{AB}$ and also from *C* to the extension of side $\overline{AB}$ at *F*. Such a perpendicular line from a point to a line is often called an *altitude*. Since they are of equal length, we label both of these altitudes as *h*. All of this is shown in Figure G.58.

Figure G.58

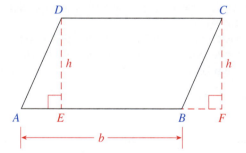

Looking very carefully at Figure G.58, we can see several things. First, we see that *EFCD* is a rectangle. Second, we can see that $\triangle ADE$ is congruent to $\triangle BCF$. We can think of cutting off $\triangle ADE$ from the left side of the parallelogram and pasting it on the right side as $\triangle BCF$. In this way we change the figure from a parallelogram into a rectangle but we *do not* change the area. Third, the length of $\overline{AB}$ is *b*, which is also the length of $\overline{EF}$. Therefore, the area of parallelogram *ABCD* is equal to the area of rectangle *EFCD*, which is *bh*. We have thus established the following:

The area of a parallelogram is given by $A = bh$.

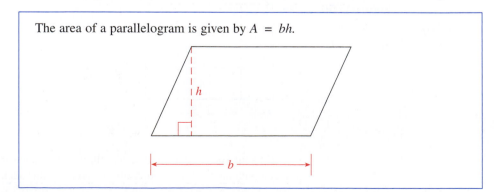

Be sure to recognize that for a rectangle the base and height are the two sides of the rectangle, whereas for a parallelogram the base is one side of the parallelogram but the height is the altitude drawn to that base from the opposite side.

Note that once a side is chosen as the base, *b*, the altitude to that base is the same length no matter whether it falls within the parallelogram or outside it, as can be seen in Figure G.58.

EXAMPLE 2   Find the area of the following parallelogram.

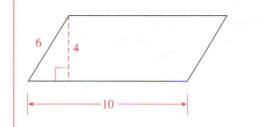

Solution   Using the formula for the area of a parallelogram, we have

$$A = bh$$
$$A = (10)(4)$$

$$\boxed{A = 40 \text{ square units}}$$

Note that we did *not* need to know the fact that the length of the other side of the parallelogram is 6.   ■

We next consider the area of a triangle. All we need to do is look at Figure G.59, in which we take an arbitrary $\triangle ABC$ and duplicate it (as $\triangle BCD$) to produce parallelogram *ABCD*.

Figure G.59

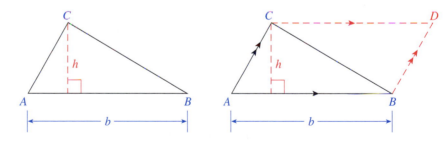

Since $\triangle ABC$ is congruent to $\triangle BCD$, their areas are identical. Hence, the area of the triangle is one-half the area of the parallelogram, and so we have the following formula.

The area of a triangle is given by $A = \dfrac{1}{2}bh$.

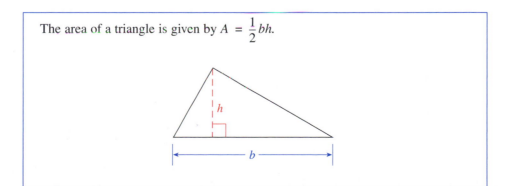

EXAMPLE 3 | Find the area of the following triangle.

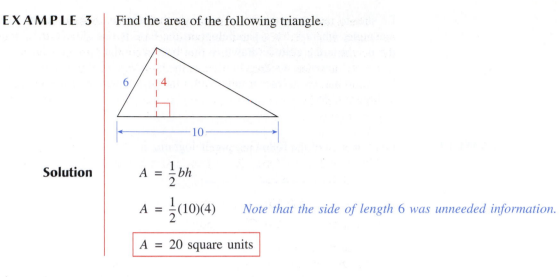

Solution

$$A = \frac{1}{2}bh$$

$$A = \frac{1}{2}(10)(4) \qquad \textit{Note that the side of length 6 was unneeded information.}$$

$$\boxed{A = 20 \text{ square units}}$$ ■

EXAMPLE 4 | Find the area of the following triangle.

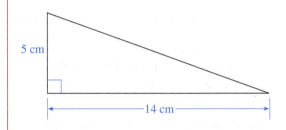

Solution Finding the area of a *right* triangle can be particularly easy if we know the lengths of both legs. In such a case, we can use one of the legs as the base and the other leg as the height.

$$A = \frac{1}{2}bh$$

$$A = \frac{1}{2}(14)(5)$$

$$\boxed{A = 35 \text{ sq cm}}$$ ■

Perimeter and Area of Scaled Figures

In Section G.4 we discussed the idea of similarity. Let's examine the perimeter and area of similar triangles and quadrilaterals.

Consider the following two similar triangles in Figure G.60.

Figure G.60

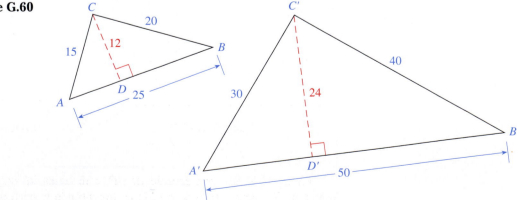

The ratio of the corresponding sides of $\triangle ABC$ to those of $\triangle A'B'C'$ is $\frac{1}{2}$. The perimeter of $\triangle ABC$ is 60 and the perimeter of $\triangle A'B'C'$ is 120, so that the ratio of the perimeters is also $\frac{1}{2}$. Of course this should come as no surprise since each side of $\triangle A'B'C'$ is twice the length of the corresponding side of $\triangle ABC$.

In effect, we have demonstrated that the perimeters of similar figures are in the same ratio as their sides. Alternatively, we can say that the same scaling factor that applies to the sides of the similar triangles also applies to the perimeters.

Let's now look at the area of these two similar triangles. Returning to Figure G.60, we have drawn in the heights $\overline{CD}$ and $\overline{C'D'}$ of the two triangles. We note that $\triangle ABC$ and $\triangle A'B'C'$ are also similar and therefore the two heights are in the same ratio as the sides we have

$$\text{The area of } \triangle ABC = \tfrac{1}{2}(25)(12) = 150$$

$$\text{The area of } \triangle A'B'C' = \tfrac{1}{2}(50)(24) = 600 \qquad \textit{Note that } 600 = 4 \cdot 150.$$

Since the base and height of $\triangle A'B'C'$ are *each* 2 times the base and height of $\triangle ABC$, the area of $\triangle A'B'C'$ is 4 times the area of $\triangle ABC$.

In general, if the scaling factor for two similar triangles is K, then both the base and the height are getting multiplied by K, and so the scaling factor for the *area* is K^2.

> If the scaling factor for two similar triangles is K, then the scaling factor for their *perimeter* is also K and the scaling factor for their *area* is K^2.

EXAMPLE 5 Suppose a side of one triangle is 20 inches and the corresponding side of a second similar triangle is 50 inches. If the perimeter of the first triangle is 100 inches and its area is 600 square inches, find the perimeter and area of the similar triangle.

Solution From the ratio of the similar sides we determine that the scaling factor is $\frac{50}{20} = \frac{5}{2} = 2.5$. Thus, to obtain the perimeter of the second triangle, we multiply by 2.5 to obtain

$$2.5(100) = \boxed{250 \text{ inches}}.$$

To obtain the area of the second triangle, we multiply by $(2.5)^2 = 6.25$ to obtain

$$6.25(600) = \boxed{3{,}750 \text{ square inches}}. \qquad \blacksquare$$

The final figure we consider in this section is the trapezoid. Recall that a trapezoid is a quadrilateral with one pair of opposite sides parallel. See Figure G.61(a). If the nonparallel sides of a trapezoid are of equal length, the trapezoid is called an *isosceles trapezoid.* (See Figure G.61(b).)

Figure G.61

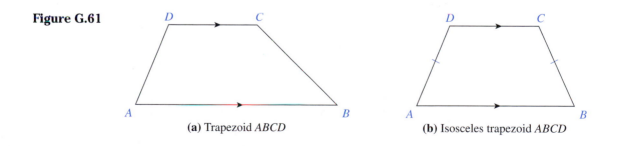

(a) Trapezoid *ABCD*   (b) Isosceles trapezoid *ABCD*

Figure G.62 shows trapezoid $ABCD$ with bases b_1 and b_2, with diagonal $\overline{BD}$ drawn in. We have also used h to label the altitude to each base.

Figure G.62

Trapezoid *ABCD*

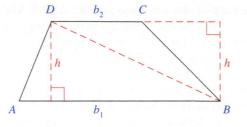

We note that the diagonal divides the trapezoid into two triangles, $\triangle ABD$ and $\triangle BCD$. The area of each triangle is one-half the base times the height. We thus have the following:

$$
\begin{aligned}
\text{Area of trapezoid } ABCD &= \textit{Area } \triangle ABD + \textit{Area of } \triangle BCD \\
&= \tfrac{1}{2}b_1 h \qquad\quad + \tfrac{1}{2}b_2 h \\
&= \tfrac{1}{2}h(b_1 + b_2)
\end{aligned}
$$

Factor out the common factor of $\tfrac{1}{2}h$.

We have thus derived the following:

The area of a trapezoid is given by $A = \tfrac{1}{2}h(b_1 + b_2)$.

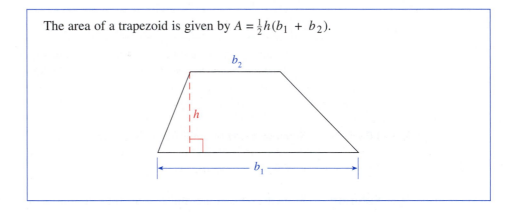

It is interesting to note that this formula can also be written as

$$
A = h\left(\frac{b_1 + b_2}{2}\right)
$$

When written this way, the formula is saying "the area of a trapezoid is the height times the *average* of the two bases."

EXAMPLE 6

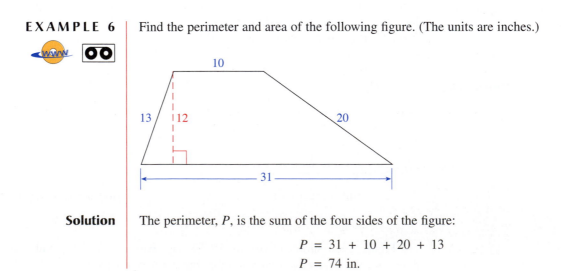

Find the perimeter and area of the following figure. (The units are inches.)

Solution

The perimeter, *P*, is the sum of the four sides of the figure:

$$
\begin{aligned}
P &= 31 + 10 + 20 + 13 \\
P &= 74 \text{ in.}
\end{aligned}
$$

To find the area of the trapezoid, we use the formula

$$A = \tfrac{1}{2}h(b_1 + b_2)$$
$$A = \tfrac{1}{2}(12)(31 + 10)$$
$$A = (6)(41)$$

$$\boxed{A = 246 \text{ sq in.}}$$

■

EXAMPLE 7

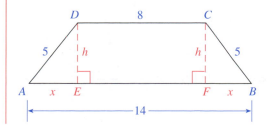

Find the area of a rectangle whose height is of length 12 in. and whose diagonal is of length 20 in.

Solution

We draw the accompanying diagram. In order to find the area of the rectangle, we would like to find the length of the base, which we have labeled b. We note that $\triangle ABD$ is a right triangle, and so we can apply the Pythagorean theorem:

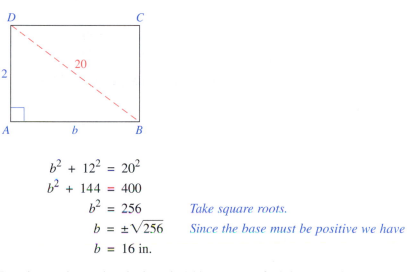

$$b^2 + 12^2 = 20^2$$
$$b^2 + 144 = 400$$
$$b^2 = 256 \qquad \textit{Take square roots.}$$
$$b = \pm\sqrt{256} \qquad \textit{Since the base must be positive we have}$$
$$b = 16 \text{ in.}$$

Now that we know that the base is 16 in., we can find the area of the rectangle:

$$A = bh = (16)(12) = \boxed{192 \text{ sq in.}}$$

■

EXAMPLE 8

Find the area of the isosceles trapezoid. (The units are meters.)

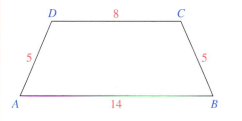

Solution

In order to find the area of this trapezoid, we need to find the length of altitude DE. Let's call this altitude h. See the next figure.

Since this is an isosceles trapezoid, the base angles are equal and so $\triangle ADE$ and $\triangle BCF$ are congruent. Thus, $\overline{AE}$ and $\overline{BF}$ are of equal length. Let's call this length x. Since the length of $\overline{AB}$ is 14 and $|\overline{EF}| = |\overline{DC}| = 8$, we have

$$x + x + 8 = 14$$
$$2x + 8 = 14$$
$$2x = 6$$
$$x = 3$$

Now we can look at $\triangle AED$; since it is a right triangle, we can apply the Pythagorean theorem:

$$3^2 + h^2 = 5^2$$
$$9 + h^2 = 25$$
$$h^2 = 16$$
$$h = \pm 4$$

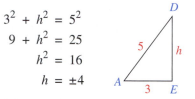

Since h represents a length, we must reject the negative answer, and so we have $h = 4$.

Now that we know the altitude, we can find the area of trapezoid $ABCD$:

$$A = \tfrac{1}{2}h(b_1 + b_2)$$

$$A = \tfrac{1}{2}(4)(14 + 8)$$

$$A = 2(22)$$

$$\boxed{A = 44 \text{ square meters}}$$

■

 **EXERCISES G.6**

1. Find the perimeter and area of a rectangle whose width is 5 ft and whose length is 8 ft.

2. Find the perimeter and area of a rectangle whose width is 7 meters and whose length is 9 meters.

3. Find the perimeter and area of a rectangle whose dimensions are 6 in. by 12 in.

4. Find the perimeter and area of a rectangle with base 9 cm and height 8 cm.

5. Find the perimeter and area of a square with side 3 in.

6. Find the perimeter and area of a square with side 8 ft.

In Exercises 7–18, find the area of the given figure.

7.

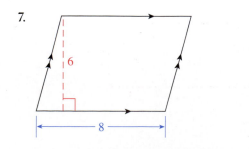

8.

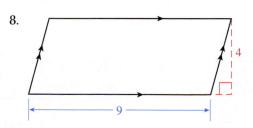

9.

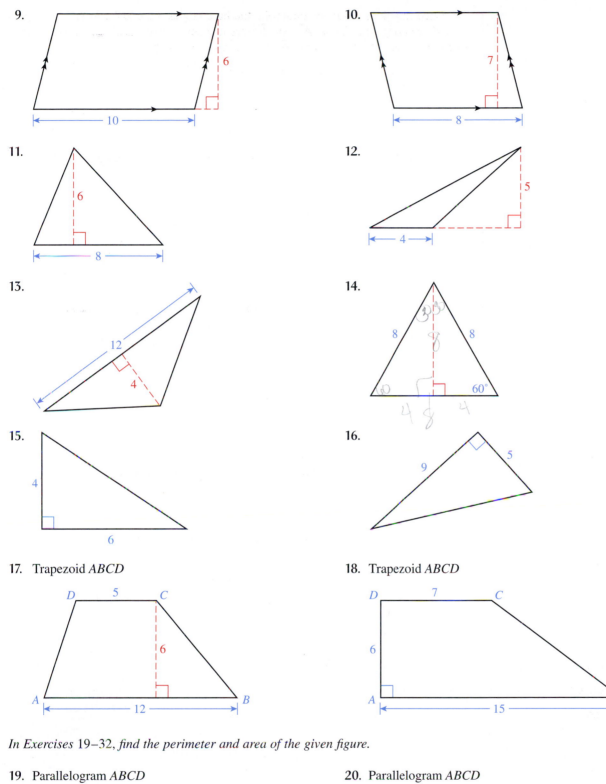

10.

11.

12.

13.

14.

15.

16.

17. Trapezoid *ABCD*

18. Trapezoid *ABCD*

In Exercises 19–32, find the perimeter and area of the given figure.

19. Parallelogram *ABCD*

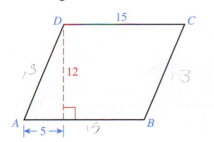

20. Parallelogram *ABCD*

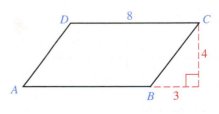

21.

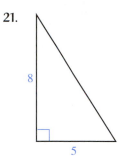

22.

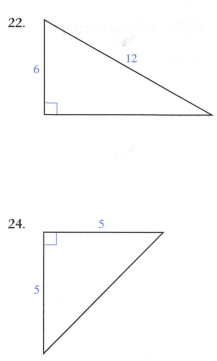

23.

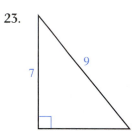

24.

25. A rectangle with base 8 cm and diagonal 12 cm

26. A rectangle with height 5 in. and diagonal 9 in.

27. A square with diagonal 10 mm

28. A square with diagonal 15 meters

29. Trapezoid *ABCD*

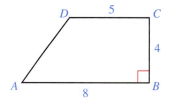

30. Trapezoid *ABCD*

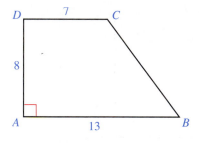

31. Isosceles trapezoid *ABCD*

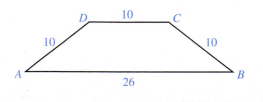

32. Isosceles trapezoid *ABCD*

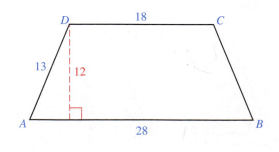

33. Find the area of *ABCDE*. *ABDE* is a rectangle.

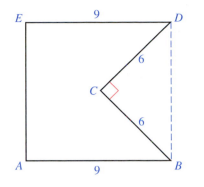

34. Find the area of *ABCDE*.

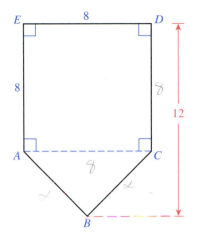

35. Find the area of *AEFG*. *ABCD* is a rectangle.

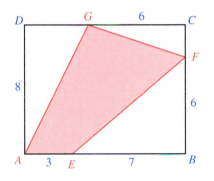

36. Find the area of △*ABE*. *ABCD* is a rectangle.

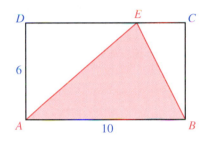

37. $|\overline{AF}| = |\overline{BE}|$. Find the area of the figure. *ABCD* is a rectangle.

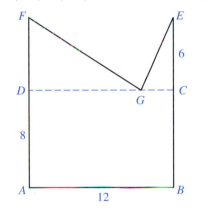

38. Find the area of the figure.

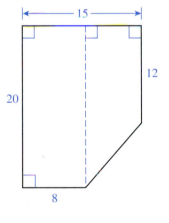

39. Find the area of the shaded region. *ACDF* is a rectangle.

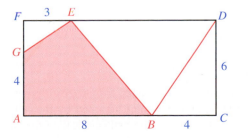

40. Find the area of the shaded region. *ABCD* is a rectangle.

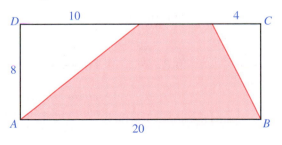

41. A side of one triangle is 28 inches and the corresponding side of a second similar triangle is 42 inches. If the perimeter of the first triangle is 98 inches and its area is 420 square inches, find the perimeter and area of the similar triangle.

42. A side of one rectangle is 25 feet and the corresponding side of a second similar rectangle is 10 feet. If the perimeter of the first rectangle is 150 feet and its area is 1,250 square feet, find the perimeter and area of the similar rectangle.

43. A side of one polygon is 15 inches and the corresponding side of a second similar polygon is 24 feet. If the perimeter of the first polygon is 80 inches and its area is 2,000 square inches, find the perimeter and area of the similar polygon.

44. A side of one polygon is 15 feet and the corresponding side of a second similar polygon is 2 feet. If the perimeter of the first polygon is 80 feet and its area is 2,000 square feet, find the perimeter and area of the similar polygon.

45. A scale drawing uses a scale of $\frac{1}{20}$ for the floor plan of a house. If the area of the first floor on the scale drawing is 378 square inches, what is the actual area of the first floor in square feet?

46. A scale drawing uses a scale of $\frac{1}{30}$ for the floor plan of a house. If the perimeter of the house on the scale drawing is 62 inches, what is the actual perimeter of the house in feet?

❓ QUESTION FOR THOUGHT

47. The following is an outline of a proof of the Pythagorean theorem. Consider a right triangle ABC and the following figure, which is created by arranging four copies of $\triangle ABC$ as indicated.

    Justify and/or explain each of the following statements.

    (a) The entire figure and the figure in the middle are squares.

    (b) The area of the outer square is $(a + b)^2$.

    (c) The area of the inner square is c^2.

    (d) The area of each triangle is $\frac{1}{2}ab$.

    (e) The area of the outer square is equal to the area of the inner square plus the area of the four inner triangles.

    (f) Use these results to prove the Pythagorean theorem.

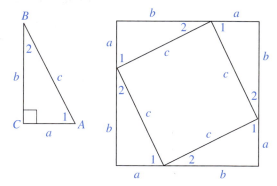

G.7   Circles

A *circle* is a set of points all of which are the same distance from a given point. The given point is called the *center* of the circle and is usually denoted by the letter O. The common distance from all points on the circle to the center is called the *radius* (plural *radii*). Figure G.53 shows circle O with radius r.

Figure G.63

Circle O with radius r

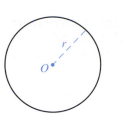

A line segment from one point on a circle to another is called a **chord.**

A chord that passes through the center of the circle is called a **diameter.**

A **secant** is a line (or line segment) that passes through two points on the circle.

A **tangent** is a line (or line segment) outside a circle that intersects the circle in exactly one point.

Figure G.64 shows a circle with diameter $\overline{AB}$, chord $\overline{CD}$, secant S, and tangent T drawn in.

Figure G.64

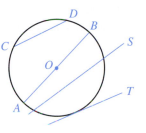

A **central angle** of a circle is an angle whose vertex is at the center of the circle.

An **arc** of a circle is a portion of the circle that lies between two points on the circle. The arc of a circle connecting the points A and B is denoted by $\overset{\frown}{AB}$.

In Figure G.65 we have central angle BOA intercepting $\overset{\frown}{AB}$. The degree measure of an arc is defined to be the degree measure of its central angle. Thus, central angle AOB of 60° will intercept $\overset{\frown}{AB}$ of 60 degrees (since both are one-sixth of the entire circle).

Figure G.65

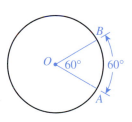

Two circles are called **concentric** if they have the same center.

Figure G.66 shows two concentric circles with central angles of 90°. We can see that both intercepted arcs $\overset{\frown}{AB}$ and $\overset{\frown}{CD}$ are 90° (they are both one-quarter of the entire circle). However, while both arcs have the same number of degrees, they are clearly not the same length.

Figure G.66

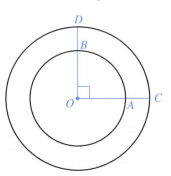

Be careful not to confuse the *length* of an arc (which we have not yet discussed) with its degree measure.

Circumference

The distance around a circle, or its perimeter, is called its **circumference.** Unlike the situation for triangles and quadrilaterals, which have sides to measure, we have no easy way to measure the distance around the circle. It is generally difficult to measure curved lines accurately. However, there *is* a rather simple formula for the circumference of a circle that is a direct consequence of a rather remarkable fact.

The ancient Greeks discovered that for any circle, large or small, the ratio of its circumference, C, to its diameter, d, is constant. They called this constant π (the Greek letter pi).* Symbolically, we may write

$$\frac{C}{d} = \pi$$

If we multiply both sides of this equation by d we obtain the following formula.

> The circumference of a circle is given by
>
> $$C = \pi d$$

Since the diameter of a circle is twice the radius we may also write

$$\boxed{C = 2\pi r}$$

EXAMPLE 1 Find the circumference of a circle whose diameter is 9 in. Give your answer in terms of π and also to the nearest tenth.

Solution Using the formula for the circumference of a circle, we get

$$C = \pi d = \pi \cdot 9$$

$$\boxed{C = 9\pi \text{ in.}}$$

Most scientific calculators have a $\boxed{\pi}$ key. Using this key, we could do this computation as $\boxed{9}$ $\boxed{\times}$ $\boxed{\pi}$ $\boxed{=}$ and the display will read $\boxed{28.274334}$, which we would round off to 28.3.

If we want a numerical answer, we generally use an approximate value for π, $\pi \approx 3.14$. Thus,

$$C = \pi d = (3.14)(9)$$
$$C \approx 28.26 \quad \textit{Round off to the nearest tenth.}$$
$$\boxed{C = 28.3}$$

A *sector* of a circle is that portion of a circle enclosed by a central angle. A sector is the shaded portion illustrated in Figure G.67.

Figure G.67

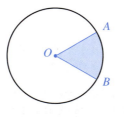

---

*The number π is an irrational number, which means that its decimal representation never stops and never repeats. The great mathematician Archimedes is often credited with obtaining an early accurate estimate for the value of π. He estimated π to be between $3\frac{1}{7}$ and $3\frac{10}{71}$. Today, π has been computed to hundreds of thousands of decimal places. The value of π accurate to five decimal places is 3.14159.

EXAMPLE 2

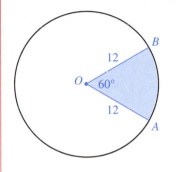

Given the following sector with a central angle of 60° in a circle whose radius is 12 cm, find the following:

(a) The length of $\overparen{AB}$ (b) The perimeter of sector AOB

Solution

We have drawn a diagram containing the given information and shaded in the sector whose perimeter we are trying to find. The perimeter of the sector is equal to the sum of the lengths of the two radii plus the length of the arc $\overparen{AB}$.

(a) Since the central angle of the sector is 60°, which is one-sixth of 360°, the length of arc $\overparen{AB}$ is going to be $\frac{1}{6}$ of the circumference of the entire circle. Therefore, the length of arc $\overparen{AB}$ is

$$\frac{1}{6}(\text{Circumference of entire circle}) = \frac{1}{6}\pi d = \frac{1}{6}(\pi)24 = \boxed{4\pi \text{ cm}}$$

(b) The perimeter of the sector is $12 + 12 + 4\pi = \boxed{24 + 4\pi \text{ cm}}$ ■

The *length* of arc $\overparen{AB}$ is actually a fractional portion of the circumference of the circle. That fraction is determined by the size of the central angle, x. Since the circumference of a circle is $2\pi r$, and a circle is 360°, we have

The length, L, of an arc with central angle $x°$ is

$$L = \frac{x}{360}(2\pi r)$$

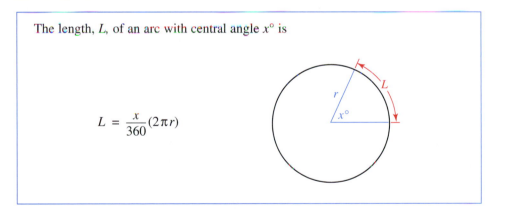

Area

Finding the area of a circle presents an even greater problem than finding its perimeter. How are we to find out how many square units can be fitted into a region with a curved boundary?

Fortunately, we have a formula for the area of a circle. The Greeks found that the ratio of the area of a circle to the square of its radius is also constant. The really amazing fact is that the ratio is the same constant, π. That is,

$$\frac{A}{r^2} = \pi$$

Thus, we have the following formula.

> The area of a circle is given by
> $$A = \pi r^2$$

EXAMPLE 3 Find the area of a circle whose diameter is 32 in. Give your answer in terms of π and to the nearest hundredth.

Solution The formula for the area of a circle requires the length of the radius. Since the diameter of the circle is 32 in., the radius is 16 in. Then

$$A = \pi r^2 = \pi(16)^2 = \boxed{256\pi \text{ sq in.}}$$

To compute the area to the nearest hundredth, we multiply 256π and get

$$\boxed{804.25 \text{ sq in.}}$$ *Rounded to the nearest hundredth*   ■

EXAMPLE 4 Find the area of the following figure, composed of a rectangle and a semicircle. Round off your answer to the nearest square foot.

9 ft

20 ft

Solution The dashed line is not actually part of the figure; it is drawn in just to make the figure clearer. The total area of this figure, A, is the area of the rectangle plus the area of the semicircle. To compute the area of the semicircle, we need to know its radius. Since the opposite sides of a rectangle are equal, the diameter of the semicircle is 9, and its radius is $\frac{9}{2}$. Therefore, we have

$$A = A_{\text{rectangle}} + A_{\text{semicircle}}$$
$$= bh + \frac{1}{2}\pi r^2$$
$$= (20)(9) + \frac{1}{2}\pi\left(\frac{9}{2}\right)^2$$
$$= 180 + \frac{1}{2}\pi\left(\frac{81}{4}\right)$$
$$\boxed{A = 180 + \frac{81\pi}{8} \text{ sq ft}}$$ *Using the approximate value 3.14 for π and rounding to the nearest square foot, we have*

$$\boxed{A \approx 212 \text{ sq ft}}$$ ■

EXAMPLE 5 Find the area of the shaded portion of the following figure. Arc *ACB* is a semicircle.

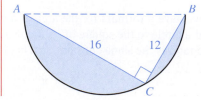

A B

16   12

C

Solution | We can visualize the required area as the area of the semicircle minus the area of the right triangle ABC. In order to get the area of the semicircle, we need to find its radius. $\overline{AB}$ is the diameter of the semicircle and the hypotenuse of right triangle ABC. Therefore, we can find the length $\overline{AB}$ by using the Pythagorean theorem:

$$|\overline{AB}|^2 = 12^2 + 16^2$$
$$|\overline{AB}|^2 = 144 + 256 = 400 \qquad \textit{Take square roots.}$$
$$|\overline{AB}| = 20 \qquad \textit{Since a length must be positive}$$

Now we can compute the required area:

$$\text{Area of shaded portion of figure} = \text{Area of semicircle} - \text{Area of triangle}$$
$$= \frac{1}{2}\pi r^2 - \frac{1}{2}bh \qquad \textit{Since the diameter is 20, the radius is 10.}$$
$$= \frac{1}{2}\pi(10)^2 - \frac{1}{2}(12)(16)$$
$$= \boxed{50\pi - 96 \text{ square units}} \qquad \blacksquare$$

The area of a *sector* is a fractional part of the area of the circle. That fraction is determined by the size of the central angle, x. Since the area of a circle is πr^2, and a circle is 360°, we have

The area of a sector, A_s, with central angle $x°$ is

$$A_s = \frac{x}{360}(\pi r^2)$$

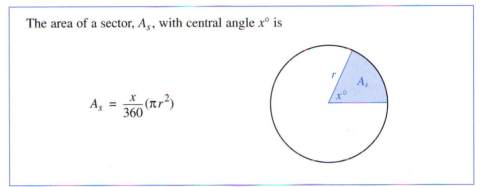

EXAMPLE 6 | To the nearest hundredth of a centimeter, find the area of a sector with central angle 75° and radius 8 cm. (See the figure below.)

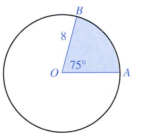

Solution | A sector of 75° is $\frac{75}{360}$ of a circle. Hence the area of sector AOB is $\frac{75}{360}$ of the area of the circle.

$$A_s = \frac{x°}{360°}(\pi r^2) \qquad \textit{Since the radius is 8 and the central angle is 75°, we have}$$

$$A_s = \frac{75}{360}\pi(8^2)$$

$$A_s = \frac{75\pi(64)}{360} \qquad \textit{Which simplifies to}$$

$$A_s = \frac{40\pi}{3} \approx \boxed{41.89 \text{ cm}^2}$$

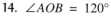 **EXERCISES G.7**

In Exercises 1–4, find the number of degrees in $\overset{\frown}{AB}$.

1.

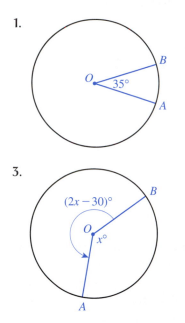

2.

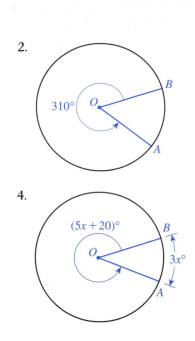

3.

4.

5. Find the circumference of a circle with diameter 6 in.

6. Find the circumference of a circle with diameter 8 cm.

7. Find the circumference of a circle with radius 4 ft.

8. Find the circumference of a circle with radius 10 mm.

9. Find the arc length of a 30° sector of a circle with a radius of 9 in.

10. Find the arc length of a 60° sector of a circle with a radius of 9 in.

In Exercises 11–14, find the perimeter and area of the indicated sector.

11. $\angle AOB = 65°$

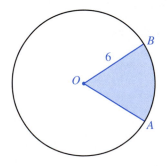

12. $\angle AOB = 80°$

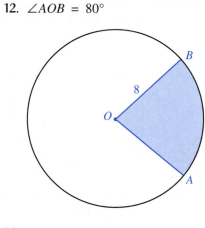

13. $\angle AOB = 10°$

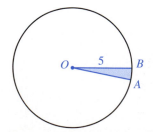

14. $\angle AOB = 120°$

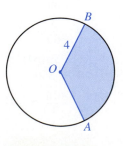

15. Find the area of a circle with radius 3 in.

16. Find the area of a circle with radius 8 cm.

17. Find the area of a circle with diameter 12 ft.

18. Find the area of a circle with diameter 9 meters.

19. Find the perimeter and area of a semicircle with radius 10 in.

20. Find the perimeter and area of a semicircle with diameter 10 in.

21. Find the area of a sector with a central angle of 80° and a radius of 5 in.

22. Find the area of a sector with a central angle of 40° and a radius of 8 cm.

In Exercises 23–24, find the area and perimeter of the given figure. All the arcs are semicircles.

23.

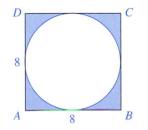

24.

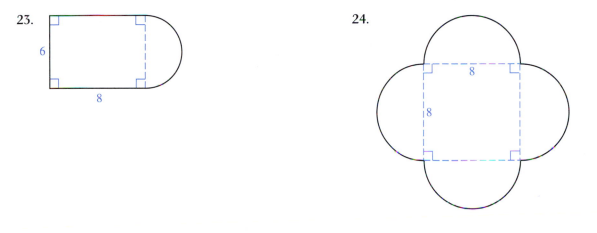

In Exercises 25–26, find the area of the shaded region.

25. *ABCD* is a square of side 8 cm.

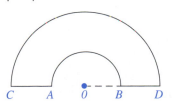

26. *ABCD* is a square of side 6 cm.

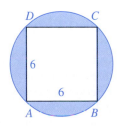

27. Find the area of the following figure. $\overparen{AB}$ and $\overparen{CD}$ are concentric semicircles with center *O*. $|\overline{OB}|$ = 6 and $|\overline{BD}|$ = 8.

28. Find the area of region *ACDB*. $\overparen{AB}$ and $\overparen{CD}$ are arcs of concentric circles with center *O*. $|\overline{OA}|$ = 8 and $|\overline{AC}|$ = 2.

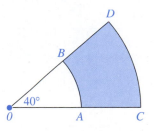

29. Find the perimeter of the following figure. $\overset{\frown}{AB}$ is a semicircle.

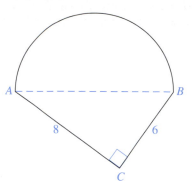

30. Find the area of the following region. $\overset{\frown}{AB}$ and $\overset{\frown}{BC}$ are congruent semicircles. *ACDE* is a rectangle.

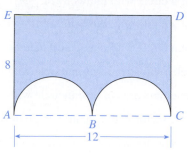

31. Find the perimeter and area of the given figure. *BCDE* is a rectangle. $\overset{\frown}{AB}$ is a semicircle.

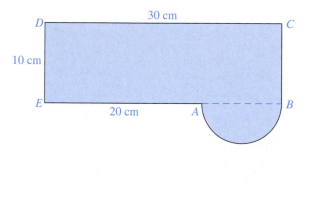

32. Find the area of the shaded region. *ABC* is a right triangle. $\overset{\frown}{DE}$ is a semicircle of radius 2 in.

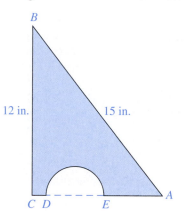

33. Which contains more pizza: one round 12-in. (diameter is 12 in.) pie or two round 8-in. pies? If a 12-in. pie costs $9 and an 8-in. pie costs $4, which is the better buy?

34. Repeat Exercise 33 for one 15-in. pie that costs $12 or two 10-in. pies that cost $8 each.

G.8 Solid Geometry

Our study of solid geometry will deal with the surface area and volume of certain three-dimensional objects. By the **surface area** of an object we mean the area of the exterior of the object. For example, consider the solid box pictured in Figure G.68 with length *L*, width *W*, and height *H*. The values of *L*, *W*, and *H* are called the *dimensions* of the rectangular box. The sides of the solid are called its *faces*, and because these faces are all rectangles, it is called a *rectangular solid*. The line segments forming the sides of the rectangles are called the *edges* of the solid.

Figure G.68
A rectangular solid

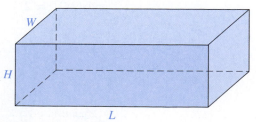

The surface area of a rectangular solid is the sum of the areas of its six faces. Since the top and bottom faces are identical, the front and back faces are identical, and the left and right faces are identical, we can compute the surface area, S, of the rectangular solid by adding up the area of its six faces to obtain the following formula.

The Surface Area of a Rectangular Solid	
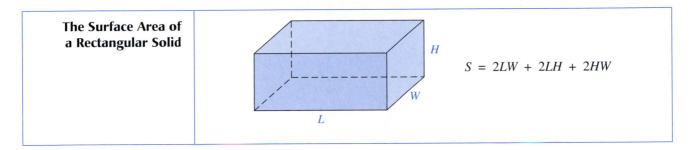	$S = 2LW + 2LH + 2HW$

If the length, width, and height of a rectangular solid are all equal, the solid is called a *cube.*

By the *volume* of a solid we mean the space occupied by the object. The unit we use to measure volume is the cubic unit. By *one cubic unit* (also called a *unit cube*) we mean a cube with each edge of length 1 unit (1 inch, 1 centimeter, 1 mile, etc.). See Figure G.69.

Figure G.69

The basic unit for measuring volume: One cubic unit

Measuring the volume of a solid means calculating how many cubic units it takes to fill up the solid. In the case of a rectangular solid, this is not very difficult. For example, Figure G.70 illustrates how a rectangular solid of dimensions 6 cm by 2 cm by 3 cm can be sliced up into 36 unit cubes, and so its volume is 36 cubic centimeters (often written 36 cm^3).

Figure G.70

Computing the volume of a rectangular solid

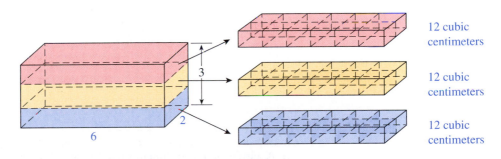

We can generalize the result of Figure G.70 to obtain the following formula.

The Volume of a Rectangular Solid	
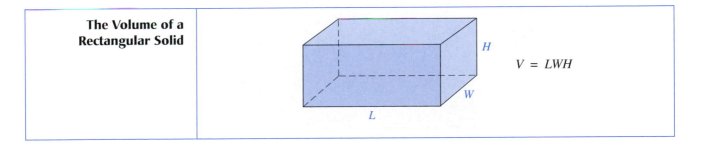	$V = LWH$

EXAMPLE 1

Find the surface area and volume of the following rectangular box.

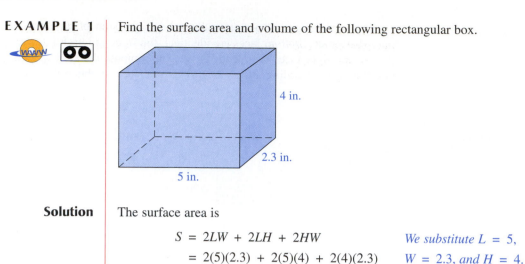

4 in.

2.3 in.

5 in.

Solution

The surface area is

$$S = 2LW + 2LH + 2HW$$

$$= 2(5)(2.3) + 2(5)(4) + 2(4)(2.3)$$

$$= \boxed{81.4 \text{ sq in.}}$$

We substitute $L = 5$, $W = 2.3$, and $H = 4$.

The volume is

$$V = LWH$$

$$= 5(2.3)(4)$$

$$= \boxed{46 \text{ cubic in.}}$$

The Cylinder

Recall that in order to find the area of a circle with radius 3 in., we use the formula $A = \pi r^2$ to obtain $A = \pi(3)^2 = 9\pi \approx 28.3$ sq in. This means that approximately 28.3 unit squares are needed to cover a circle of radius 3. See Figure G.71.

Figure G.71

A circle of radius 3 in. has an area of approximately 28.3 sq in.

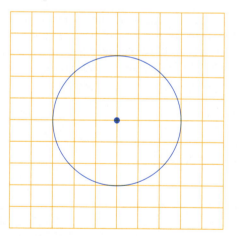

A circle is a two-dimensional figure. The three-dimensional figure formed by moving the circle parallel to itself, as illustrated in Figure G.72, is called a ***right circular cylinder.***

Figure G.72

Forming a right circular cylinder

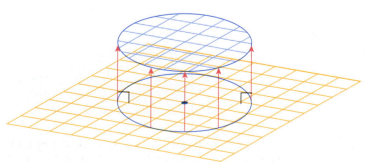

As with a rectangular solid, finding the volume of a cylinder means determining the number of *cubic* units it takes to fill up the cylinder. Let's examine the cylinder formed by moving a circle of radius 3 inches parallel to itself to a height of 1 inch. See Figure G.73.

Figure G.73

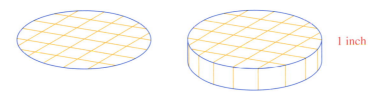

1 inch

As we can see from Figure G.73, each of the 28.3 square inches in the area of the circle will generate a cubic inch. Hence, a circle of area 28.3 sq in. will generate a cylinder made up of 28.3 cubic inches.

Similarly, Figure G.74 illustrates a cylinder generated by a circle of radius r inches moved parallel to itself through a height of 6 inches. Examining the individual layers, we recognize that each of the πr^2 inches in the area of the circle will generate a cubic inch. The volume of each layer is πr^2 times a height of 1 inch. Thus, the volume of this cylinder is $6(\pi r^2) = 6\pi r^2$ cubic inches.

Figure G.74

The volume of a cylinder of radius r inches and height 6 inches

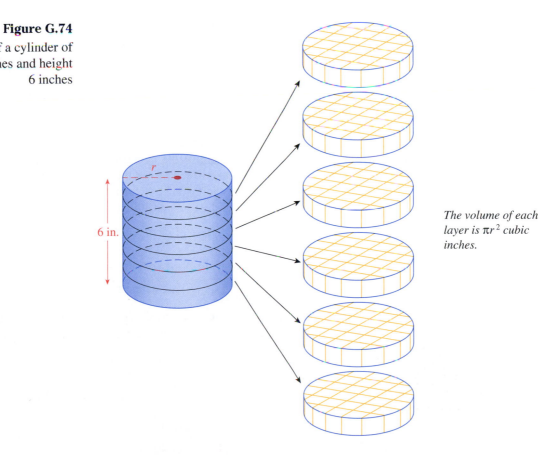

The volume of each layer is πr^2 cubic inches.

6 in.

Hence, we find the volume of a cylinder by multiplying the area of the circle (called the base of the cylinder) by the height of the cylinder. We have the following formula.

The Volume of a Right Circular Cylinder	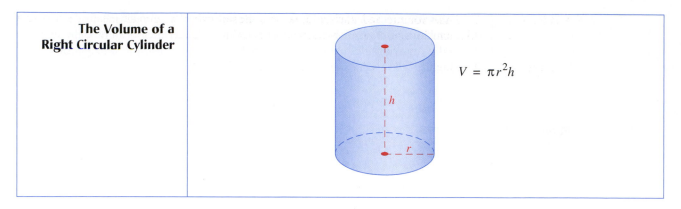 $V = \pi r^2 h$

The surface area of a closed right circular cylinder consists of the area of the two circles, called the **bases,** and the area of the curved surface, called the **lateral surface area.** See Figure G.75.

Figure G.75

The surface area of a closed right circular cylinder

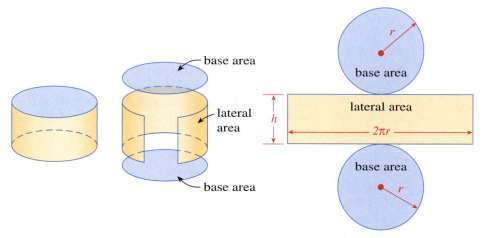

As illustrated in Figure G.75, if we cut the cylinder open and lay it flat, we can find its surface area by adding the area of the two bases (which are circles) and the lateral area (which is a rectangle). The area of each circle is πr^2, where r is the radius of the circle. To find the lateral area, the area of the rectangle, we note that the height of the rectangle is h, the height of the cylinder. Let's look at the width of the rectangle. Note that if we were to reconstruct the cylinder, the width of the rectangle wraps around the bases of the cylinder (which are circles). Therefore, the width of the rectangle is equal to the circumference of the circle, which is $2\pi r$. Hence,

Lateral surface area = (Base of rectangle)(Height of rectangle) = $(2\pi r)(h)$

We have shown that

$$\begin{array}{ccc} \text{Total surface area of} & = & \text{Area of the two bases} + \text{Lateral surface area} \\ \text{a right circular cylinder} & & \end{array}$$

$$S = \qquad 2\pi r^2 \qquad + \qquad 2\pi rh$$

The Total Surface Area of a Closed Right Circular Cylinder	$S = 2\pi r^2 + 2\pi rh$

EXAMPLE 2

Find the volume and surface area of a closed cylinder of radius 8 cm and height 15.5 cm. Give your answers in terms of π and rounded to the nearest tenth.

Solution

To compute the volume, we use the formula $V = \pi r^2 h$:

$V = \pi r^2 h$ *Substitute $r = 8$ and $h = 15.5$.*

$= \pi(8)^2(15.5) = \boxed{992\pi \text{ cm}^3}$ *Using $\pi \approx 3.14$ and rounding to the nearest tenth, we get*

$= \boxed{3{,}114.9 \text{ cm}^3}$

To compute the surface area, we use the formula $S = 2\pi r^2 + 2\pi rh$.

$S = 2\pi r^2 + 2\pi rh$ *Substitute $r = 8$ and $h = 15.5$.*

$= 2\pi(8)^2 + 2\pi(8)(15.5) = \boxed{376\pi \text{ cm}^2}$ *Using $\pi \approx 3.14$ and rounding to the nearest tenth, we get*

$= \boxed{1{,}180.6 \text{ cm}^2}$ ■

The following boxes contain the formulas for the surface area and volume of some commonly occurring geometric solids. For ease of reference, we have included the formulas for a rectangular solid and a right circular cylinder.

Rectangular Solid

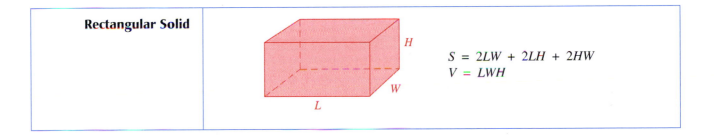

$S = 2LW + 2LH + 2HW$
$V = LWH$

Sphere

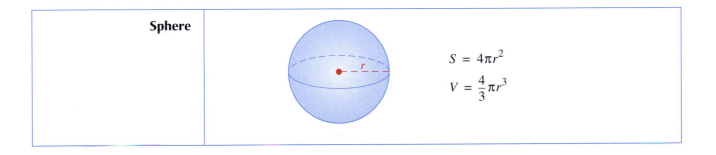

$S = 4\pi r^2$
$V = \dfrac{4}{3}\pi r^3$

Right Circular Cylinder

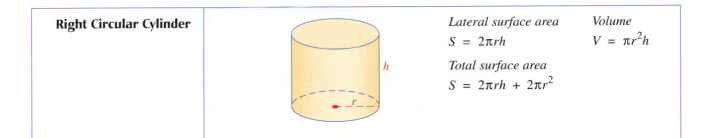

Lateral surface area   *Volume*
$S = 2\pi rh$ $V = \pi r^2 h$

Total surface area
$S = 2\pi rh + 2\pi r^2$

Right Circular Cone

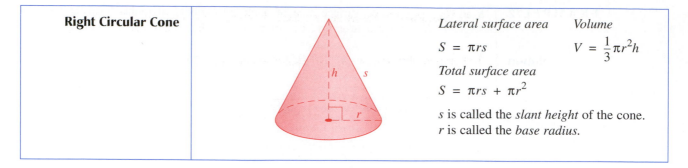

Lateral surface area   *Volume*

$S = \pi rs$ $V = \dfrac{1}{3}\pi r^2 h$

Total surface area

$S = \pi rs + \pi r^2$

s is called the *slant height* of the cone.
r is called the *base radius.*

EXAMPLE 3

To the nearest tenth, find the volume and surface area for the figure below. The figure is a hemisphere (half a sphere) on top of a right circular cylinder.

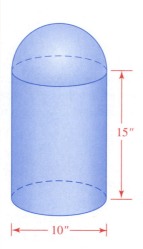

15"

|← —— 10" —— →|

Solution

We see that the diameter of the hemisphere and the cylinder is 10 inches, and the height of the cylinder is 15 inches.

To find the volume of this solid, we find the volume of each of the hemisphere and the cylinder:

$V = V_{\text{hemisphere}} + V_{\text{cylinder}}$ *Since a hemisphere is half a sphere,*

$V = \dfrac{1}{2}V_{\text{sphere}} + V_{\text{cylinder}}$ *Use the formula to get*

$V = \dfrac{1}{2}\left(\dfrac{4}{3}\pi r^3\right) + \pi r^2 h$ *The radius of the sphere and the cylinder is 5 inches (half the diameter); the height of the cylinder is 15 inches; hence, r = 5 and h = 15.*

$V = \dfrac{1}{2}\left(\dfrac{4}{3}\pi(5)^3\right) + \pi(5)^2(15)$

$V = \dfrac{1}{2}\left(\dfrac{4}{3}\pi(125)\right) + \pi(25)(15)$ *Simplify.*

$V = \dfrac{250\pi}{3} + 375\pi = \dfrac{250\pi}{3} + \dfrac{1{,}125\pi}{3} = \dfrac{1{,}375\pi}{3}$ *Using a calculator, we get*

$V = 1{,}439.9$

Hence the volume is $\boxed{1{,}439.9 \text{ cubic in.}}$

To find the surface area, we note that we want only the outside portion of the hemisphere, the lateral area of the cylinder, and the bottom base of the cylinder.

$$S = S_{\text{hemisphere}} + S_{\text{lateral area of cylinder}} + S_{\text{bottom base of cylinder}}$$ *Since a hemisphere is half a sphere,*

$$S = \frac{1}{2}S_{\text{sphere}} + S_{\text{lateral area of cylinder}} + S_{\text{bottom base of cylinder}}$$ *Use the formula to get*

$$S = \frac{1}{2}(4\pi r^2) + 2\pi rh + \pi r^2$$ *The radius of the sphere and the cylinder is 5 inches; the height of the cylinder is 15 inches; hence, r = 5 and h = 15.*

$$S = \frac{1}{2}(4\pi(5)^2) + 2\pi(5)(15) + \pi(5)^2$$

$$S = 50\pi + 150\pi + 25\pi$$ *Simplify.*

$$= 225\pi$$

$$S \approx 706.9$$

The surface area is $\boxed{706.9 \text{ square in.}}$ ∎

EXAMPLE 4

A party hat is to be made in the shape of a right circular cone, as illustrated in Figure G.76. If the material used to make the hat costs $0.175 per square foot, find the material cost of manufacturing one such party hat.

Figure G.76

The party hat of Example 4

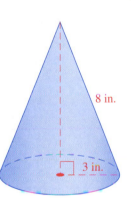

8 in.

3 in.

Solution

The material used in making the party hat goes into creating the lateral surface area of the cone. To compute the cost of one hat, we must compute its lateral surface area in square feet and then multiply this surface area by the cost per square foot for the material.

The lateral surface area is $A = \pi rs = \pi(3)(8) \approx 75.4$ sq in. To convert this area into square feet, we divide by 144, which is the number of square inches in 1 square foot. Thus,

$$75.4 \text{ sq in.} = \frac{75.4}{144} \approx 0.52 \text{ sq ft}$$

Finally, to compute the cost of the hat, we multiply this area by the cost of the material per square foot:

$$\text{Cost of one hat} = 0.52(0.175) \approx 0.09$$

Thus, the cost is about $\boxed{\$0.09 \text{ or } 9 \text{ cents}}$ per hat for the material. ∎

www EXERCISES G.8

In this exercise set, answers may be left in terms of π unless other instructions are given. In Exercises 1–8, use the given figure to find the total surface area and volume of the solid.

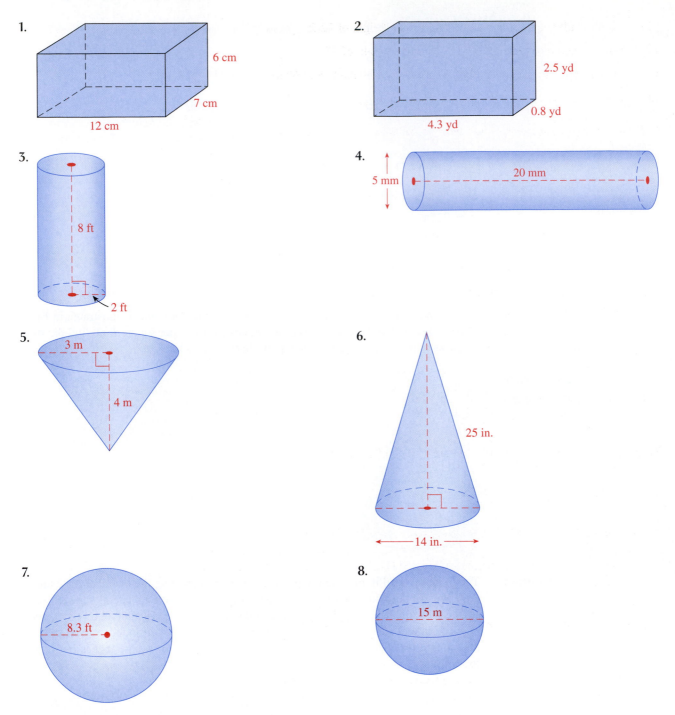

1. 6 cm, 7 cm, 12 cm

2. 2.5 yd, 0.8 yd, 4.3 yd

3. 8 ft, 2 ft

4. 5 mm, 20 mm

5. 3 m, 4 m

6. 25 in., 14 in.

7. 8.3 ft

8. 15 m

In Exercises 9–18, use the formulas given in this section to compute the total surface area and the volume of the figure described. All answers should be rounded to the nearest tenth.

9. A sphere of radius 3 ft

10. A sphere of diameter 3 ft

11. A closed right circular cylinder of height 5 cm and radius 4 cm

12. A right circular cylinder, closed at only one end, of height 4 cm and radius 5 cm

13. A rectangular solid of dimensions 2.4 meters by 15 meters by 1.8 meters

14. A cube of side 1.6 mm

15. A right circular cone of radius 8 ft and a slant height of 12 ft

16. A right circular cone of radius 15.2 cm and a height of 25 cm

17. An open right circular cylinder of diameter 25 mm and height 60 mm

18. A closed right circular cylinder of diameter 60 mm and height 25 mm

In Exercises 19–38, answers should be rounded to the nearest tenth unless otherwise indicated.

19. A closed right circular cylinder has a radius of 3 meters. Find the volume of the cylinder if its lateral surface area is 96π square meters. Leave your answer in terms of π.

20. A closed right circular cylinder has a radius of 3 meters. Find the lateral surface area of the cylinder if its volume is 96π cubic meters. Leave your answer in terms of π.

21. Find the volume of a sphere if its surface area is 100π sq cm. Leave your answer in terms of π.

22. If the volume of a sphere is 100 cm^3, find the surface area of the sphere. Round your answer to the nearest hundredth.

23. Find the lateral surface area of a right circular cone of height 8 in. and volume 96π cubic in. Leave your answer in terms of π.

24. Find the volume of a right circular cone of slant height 13 mm and lateral surface area 204.2 sq mm.

25. How many cubic yards of concrete are needed to pave a driveway that is to be 20 yd long, 3.75 yd wide, and 0.2 yd thick? Answer to the nearest tenth.

26. How much water is needed to fill a cylindrical swimming pool of diameter 4 meters to a height of 1.4 meters?

27. A rectangular box is to be used to store wet materials. The inside dimensions of the box are 3.8 ft long, 16 ft wide, and 4.5 ft high. If the inside of the box is to be lined with a waterproof material that costs $1.26 per sq ft, find the cost of lining the box.

28. A company charges $12.36 per cubic meter for a certain type of solid cylindrical metal rod. What is the cost for such a metal rod that is 18 meters long and 0.08 meter in diameter?

29. Assuming that there is no waste of material, how many solid steel cylinders of diameter 2 in. and height 8 in. can be made from a solid rectangular block of steel that measures 60 in. by 12 in. by 30 in.?

30. How many of the steel cylinders described in Exercise 29 can be made from a solid rectangular block that measures 8 ft by 3 ft by 20 ft?

31. The cover of a barbecue grill is in the shape of a hemisphere (one-half of a sphere) and is made from a material that costs $1.85 per square foot. Find the cost of such a cover if its diameter is 3.5 ft.

32. The surface area of a basketball is approximately 285 sq in. Find the radius of the basketball to the nearest tenth of an inch.

33. A conical funnel of radius 15 cm and height 7.5 cm is to be lined with a paper filter. What area of filter paper is needed?

34. A hot water heater is in the shape of a right circular cylinder with a radius of 1.2 ft and a height of 5.6 ft. How many square feet of insulation is needed to cover the top and sides of the heater?

35. A silo is built in the form of a cylinder surmounted by a hemisphere, as indicated in the figure. Find the number of cubic feet of grain this silo can hold.

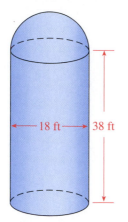

18 ft — 38 ft

36. A cylindrical tube of length 40 in. and diameter 6 in. is surrounded by another tube of diameter 6.8 in., as illustrated in the accompanying figure. The space between the two tubes is used to hold a coolant for the contents of the inner tube. How much coolant can the space between the two tubes hold?

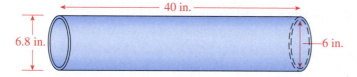

40 in.

6.8 in. 6 in.

37. The accompanying figure illustrates a right circular cylinder with a right circular cone inside it. The radius of both the cylinder and the cone is 6 cm, and the height of both is 15 cm. Find the volume contained in the space outside the cone and inside the cylinder. Leave your answer in terms of π.

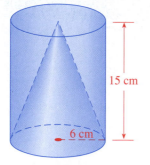

15 cm

6 cm

38. An "ice cream cone" is formed by placing a hemisphere of diameter 5 cm on top of a right circular cone of height 10 cm, as illustrated in the accompanying figure.

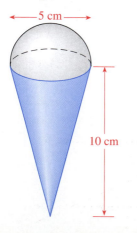

5 cm

10 cm

(a) If this ice cream cone is completely filled with ice cream, how much ice cream will it contain?

(b) If one ounce of ice cream has a volume of approximately 29.6 cubic centimeters, how many ounces of ice cream will this cone contain?

(?) QUESTIONS FOR THOUGHT

39. Consider a cube of side x.

(a) Show that the surface area of a cube of side x is $S = 6x^2$.

(b) If the edge of a cube is doubled in length, what happens to the surface area? To the volume? [*Hint:* Consider the ratio of the original surface area to the new surface area, and similarly for the volumes.]

(c) If the edge of a cube is tripled in length, what happens to the surface area? To the volume? [*Hint:* Consider the ratio of the original surface area to the new surface area, and similarly for the volumes.]

(d) Can you generalize the results of parts **(b)** and **(c)** to describe what happens to the surface area and volume of a cube if the length of its edge is multiplied by k?

40. Describe the similarities in the process of computing area as compared with the process of computing volume.

REVIEW EXERCISES

Throughout the following set of review exercises, round your answer to the nearest tenth where necessary.

1. Find the measures of $\angle 1$, $\angle 2$, $\angle 3$, and $\angle 4$.

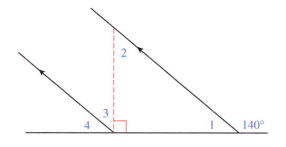

2. Find x and y.

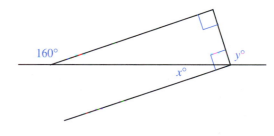

In Exercises 3–8, find x.

3.

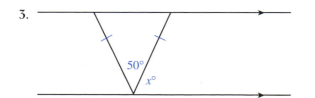

4.

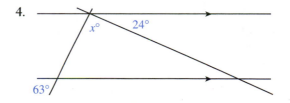

5.

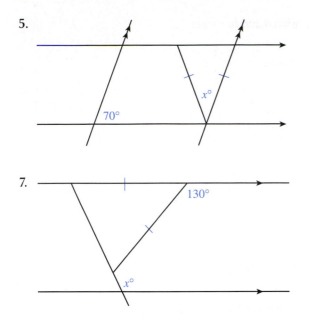

6.

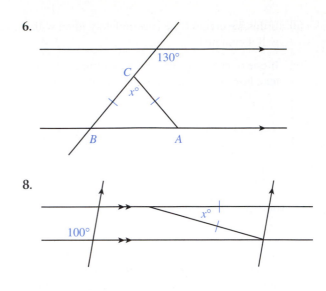

7.

8.

In Exercises 9–12, find the missing sides of the triangle(s).

9.

10.

11.

12.

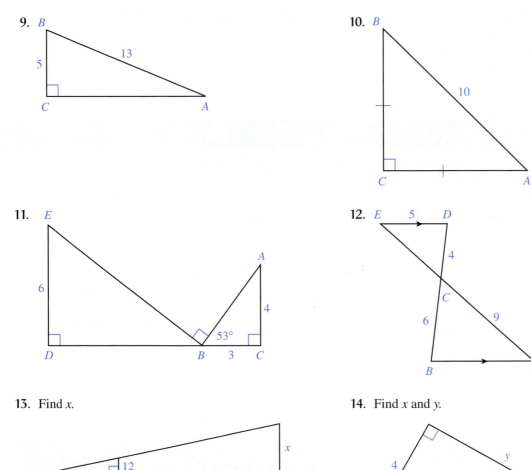

13. Find *x*.

14. Find *x* and *y*.

15. Find a and b.

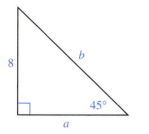

16. Find a and b.

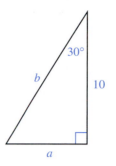

17. Find x.

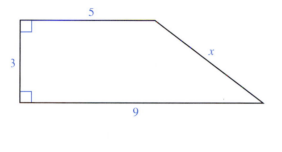

18. Find x.

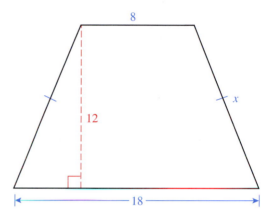

19. Find the perimeter and area of a square whose diagonal is 12 inches.

20. Find the perimeter and area of a rectangle with side 8 cm and diagonal 10 cm.

21. Find the perimeter and area of a right triangle with one leg 5 feet and hypotenuse 9 feet.

22. The corresponding sides of two similar triangles are 3 inches and 2 inches, respectively. If the perimeter and area of the smaller triangle are 20 inches and 24 square inches, respectively, find the perimeter and area of the larger triangle.

In Exercises 23–27, find the perimeter and area of the given figure.

23. Parallelogram *ABCD*

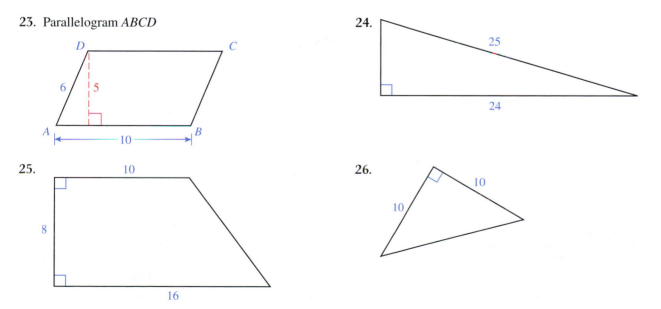

24.

25.

26.

27.

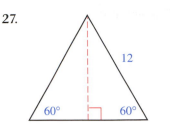

28. Find the area of the following figure.

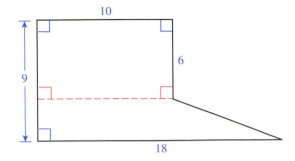

29. Find the area of the shaded region. *ABCD* is a rectangle.

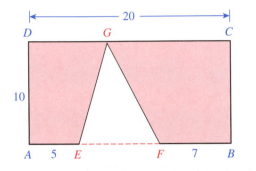

30. Find the area of the following figure. *ABCD* is a rectangle and $|\overline{AF}| = |\overline{BE}|$.

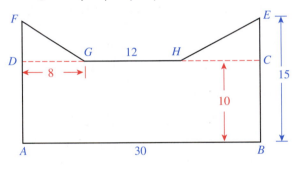

31. Find the area of the following figure. *ABCD* is a rectangle and $|\overline{AF}| = |\overline{BE}|$.

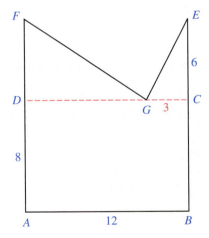

32. Given rectangle *ABCD* as indicated:

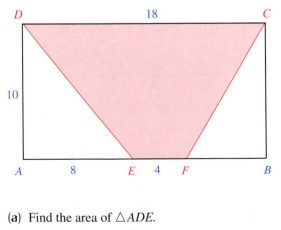

    (a) Find the area of △*ADE*.

    (b) Find the area of △*BCF*.

    (c) What kind of figure is *CDEF*?

    (d) Find the area of *CDEF*.

33. Given rectangle *ACEG* as indicated:

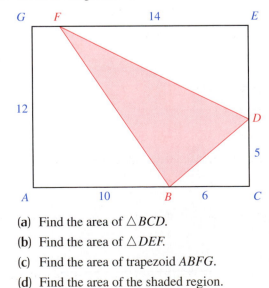

    (a) Find the area of △*BCD*.

    (b) Find the area of △*DEF*.

    (c) Find the area of trapezoid *ABFG*.

    (d) Find the area of the shaded region.

34. Given rectangle *ABCD* as indicated:

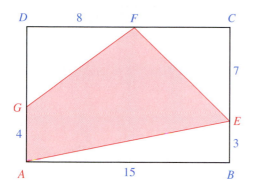

(a) Find the area of △*ABE*.

(b) Find the area of △*ECF.*

(c) Find the area of △*FDG*.

(d) Find the area of *ABCD*.

(e) Find the area of the shaded region *AEFG*.

35. Given rectangle *ACDF* as indicated:

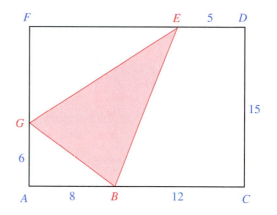

(a) Find the area of △*ABG*.

(b) Find the area of △*EFG*.

(c) Find the area of trapezoid *BCDE*.

(d) Find the area of *ACDF.*

(e) Find the area of the shaded region.

36. Given rectangle *ACEG* as indicated:

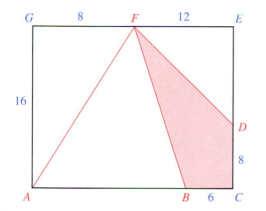

(a) Find the area of △*AGF.*

(b) Find the area of △*DEF.*

(c) Find the area of △*ABF.*

(d) Find the area of the shaded region *BCDF.*

37. Given rectangle *ACDF* as indicated:

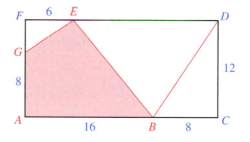

(a) Find the area of △*BCD*.

(b) Find the area of △*BDE*.

(c) Find the area of △*EFG*.

(d) Find the area of the shaded region *ABEG*.

38. Find the circumference and area of a circle with diameter 20 inches.

39. Find the circumference and area of a circle with radius 16 inches.

40. Find the length of $\overset{\frown}{AB}$.

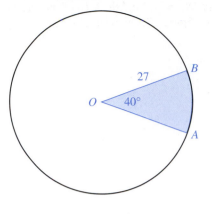

41. Find the area of the shaded sector.

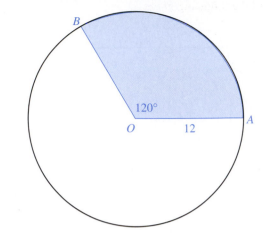

42. Find the perimeter of sector *AOB*.

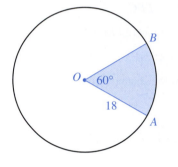

43. Find the area of the shaded region.

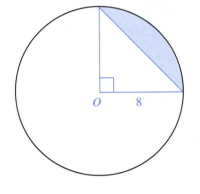

44. *ABCD* is a square. Find the area of the shaded portion of the following figure.

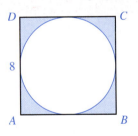

45. Find the perimeter of the following figure. *ABCD* is a rectangle. Arcs $\overset{\frown}{CE}$ and $\overset{\frown}{DF}$ are congruent semicircles.

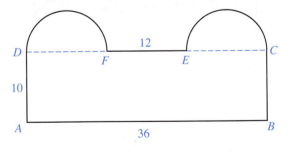

46. Find the area and perimeter of the shaded region. *ABCD* is a square and $\overset{\frown}{AB}$ is a semicircle.

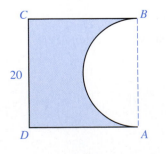

47. *ABCD* is a square of side 12 inches. The two arcs are congruent semicircles. Find the area of the shaded region. The answer may be left in terms of π.

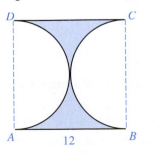

48. *ABEF* is a rectangle and $\overset{\frown}{CD}$ is a semicircle. Find the area of the following figure.

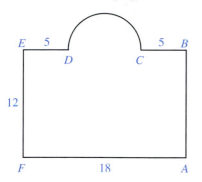

49. $\overset{\frown}{AB}$ and $\overset{\frown}{CD}$ are concentric semicircles. $\overline{OA} = 6$ and $\overline{OC} = 4$. Find the area of the shaded portion of the figure. The answer may be left in terms of π.

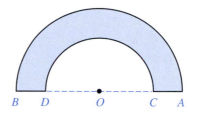

50. Given that *BCDE* is a square, △*ABC* is isosceles, and $\overset{\frown}{DE}$ is a semicircle, find the area of the following figure.

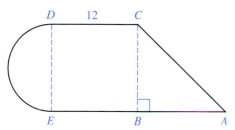

51. Find the surface area and volume of a rectangular solid whose edges are 3, 5, and 8 cm.

52. Find the surface area and volume of a sphere of radius 8 inches.

53. Find the surface area and volume of a closed right circular cylinder of radius 3 feet and height 10 feet.

54. Find the total surface area and volume of a closed right circular cone of base radius 3 meters and height 4 meters.

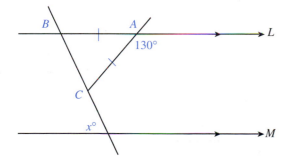

CHAPTER 11 PRACTICE TEST

In Exercises 1–10, round your answer to the nearest tenth where necessary.

1. Find *x*.

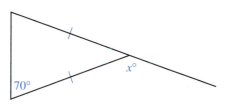

2. Find the value of *x*. *L* ‖ *M* and $|\overline{AB}| = |\overline{AC}|$.

3. Use the accompanying figure to find $\angle 1$, $\angle 2$, $\angle 3$, and the length of line segment $\overline{AC}$ labeled as x.

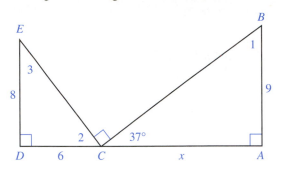

4. Find the perimeter and area of the following triangle.

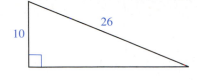

5. Find the area of an equilateral triangle of side 6 inches.

6. Find the perimeter of a 60° sector of a circle with radius 12 cm.

7. Given that *ABDE* is a rectangle, arc $\overparen{AE}$ is a semicircle, and arcs $\overparen{BC}$ and $\overparen{CD}$ are congruent semicircles, find the area of the figure.

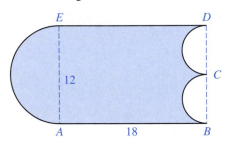

8. Find the area of the shaded portion of the following figure. *ADEF* is a rectangle.

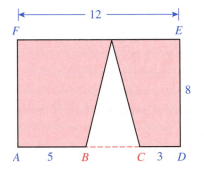

9. Find the area of the shaded portion of the following figure. $\overparen{AB}$ is a semicircle.

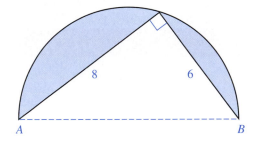

10. Find the cost of constructing the following rectangular box from material that costs $2.50 per square foot.

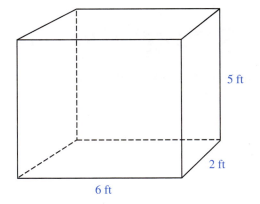

A Review of Arithmetic

Before proceeding with this appendix, it is important that you read through the introductory material in Section 1.1.

A.1 Fractions

In Section 1.1 we mentioned that there are various ways to indicate multiplication. Similarly, there are two popular ways to indicate division. If we wish to indicate "20 divided by 4," we may write

$$20 \div 4 \quad \text{or} \quad \frac{20}{4}$$

The fraction bar is an alternative way of indicating division and is the notation most frequently used in algebra.

In general, a ***fraction*** is an expression of the form $\frac{p}{q}$, which means p divided by q. In the expression $\frac{p}{q}$, p is called the ***numerator*** of the fraction and q is called the ***denominator*** of the fraction.

We can easily locate a fraction on the number line. For example, if we want to locate $\frac{3}{4}$, we would simply divide the interval from 0 to 1 into 4 equal parts and count 3 of the 4 equal parts as shown here:

$$
\begin{array}{cccccc}
& + & + & + & + & + \\
0 & \frac{1}{4} & \frac{2}{4} & \frac{3}{4} & \frac{4}{4} = 1 &
\end{array}
$$

While this book assumes no previous knowledge of algebra, we do assume a basic familiarity with fractions, decimals, and percent. Nevertheless, in the remainder of this appendix we will review the arithmetic of fractions, decimals, and percent for those of you who need to brush up, as these skills will be used consistently throughout your study of algebra.

We know from previous experience with fractions that two fractions may look different but in fact have the same value. This is illustrated in Figure A.1. The figure illustrates that the fractions $\frac{3}{4}$ and $\frac{6}{8}$ both represent the same amount. Having 3 out of 4 equal parts of a certain whole is the same as having 6 out of 8 equal parts of the same whole.

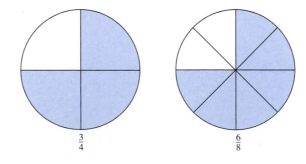

Figure A.1
Equality of two fractions

$$\frac{3}{4} \qquad\qquad \frac{6}{8}$$

Instead of drawing pictures to determine if two fractions are equivalent, we can use the Fundamental Principle of Fractions, stated in the next box.

Fundamental Principle of Fractions	The value of a fraction is unchanged if the numerator and denominator are multiplied or divided by the same nonzero number.

Algebraically we can write this as follows:

Fundamental Principle of Fractions	$\dfrac{a}{b} = \dfrac{a \cdot k}{b \cdot k}$ $b, k \neq 0$

Two fractions that have the same value are called *equivalent fractions.*

When we multiply both the numerator and the denominator by the same number we say that we are **building fractions,** whereas when we divide both numerator and denominator by the same number, we say that we are **reducing fractions.** For example:

$$\frac{3}{4} = \frac{3 \cdot 2}{4 \cdot 2} = \frac{6}{8}$$ *This is an example of building fractions.*

$$\frac{6}{8} = \frac{3 \cdot 2}{4 \cdot 2} = \frac{3}{4}$$ *This is an example of reducing fractions.*

The slashes indicate that the numerator and denominator were both divided by 2.

When we use the Fundamental Principle in this way, we say that we have *canceled* the common factor of 2.

DEFINITION   We say that a fraction is reduced to *lowest terms* if the numerator and denominator have no common factor other than 1.

We will always expect our final answers to be reduced to lowest terms.

EXAMPLE 1   Reduce to lowest terms.

(a) $\dfrac{10}{35}$ (b) $\dfrac{48}{60}$ (c) $\dfrac{14}{42}$ (d) $\dfrac{18}{25}$

Solution   (a) $\dfrac{10}{35} = \dfrac{2 \cdot 5}{5 \cdot 7} = \boxed{\dfrac{2}{7}}$

(b) There are several ways we can proceed. Since the Fundamental Principle of Fractions tells us that we can reduce a fraction by canceling *common factors* only, one way to begin is to factor the numerator and denominator as completely as possible.

$$\frac{48}{60} = \frac{2 \cdot 2 \cdot 2 \cdot 2 \cdot 3}{2 \cdot 2 \cdot 3 \cdot 5}$$

$$= \frac{2 \cdot 2 \cdot 2 \cdot 2 \cdot 3}{2 \cdot 2 \cdot 3 \cdot 5}$$

$$= \frac{2 \cdot 2}{5}$$

$$= \boxed{\frac{4}{5}}$$

Alternatively, we may see that 12 is the greatest common factor of both the numerator and denominator.

$$\frac{48}{60} = \frac{4 \cdot 12}{5 \cdot 12} = \frac{4 \cdot 12}{5 \cdot 12} = \boxed{\frac{4}{5}}$$

This is often written in the following shorthand fashion:

$$\frac{48}{60} = \frac{\overset{4}{\cancel{48}}}{\underset{5}{\cancel{60}}} = \boxed{\frac{4}{5}}$$

Note that this shorthand method "hides" the fact that the common factor is 12. Nevertheless, because it is very efficient it is often used.

(c) $\dfrac{14}{42} = \dfrac{2 \cdot 7}{6 \cdot 7} = \dfrac{2 \cdot \cancel{7}}{\underset{3}{\cancel{6}} \cdot \cancel{7}} = \boxed{\dfrac{1}{3}}$

Do not forget the understood factor of 1 in the numerator.

(d) $\dfrac{18}{25} = \dfrac{2 \cdot 3 \cdot 3}{5 \cdot 5}$ cannot be reduced since there are no common factors. ■

At this point it is important to point out again that the Fundamental Principle pertains to *factors* and *not* common *terms*. Avoid the following type of common error:

$$\frac{20 + 4}{4} \neq \frac{20 + \cancel{4}}{\cancel{4}} \neq 20 \quad \text{because} \quad \frac{20 + 4}{4} = \frac{24}{4} = 6$$

Remember	The Fundamental Principle of Fractions allows us to reduce by using common *factors*, not common terms.

The process of building fractions is the reverse of reducing. We obtain an equivalent fraction by multiplying both the numerator and the denominator by the same number.
For example:

$$\frac{2}{3} = \frac{2 \cdot 5}{3 \cdot 5} = \frac{10}{15}$$

$$\frac{2}{3} = \frac{2 \cdot 11}{3 \cdot 11} = \frac{22}{33}$$

$$\frac{2}{3} = \frac{2 \cdot 2 \cdot 7}{3 \cdot 2 \cdot 7} = \frac{2 \cdot 14}{3 \cdot 14} = \frac{28}{42}$$

Hence, $\dfrac{2}{3}, \dfrac{10}{15}, \dfrac{22}{33}$, and $\dfrac{28}{42}$ all represent the same number.

EXAMPLE 2   Write $\dfrac{5}{12}$ as an equivalent fraction with a denominator of 72.

Solution $\dfrac{5}{12} = \dfrac{?}{72}$ *We notice that $72 = 12 \cdot 6$.*

$\dfrac{5}{12} = \dfrac{?}{12 \cdot 6}$ *Since the denominator was multiplied by 6, the Fundamental Principle requires us to multiply the numerator by 6.*

$\dfrac{5}{12} = \dfrac{5 \cdot 6}{12 \cdot 6} = \boxed{\dfrac{30}{72}}$ ■

EXAMPLE 3   An archer shoots 80 arrows at a target and hits the target 72 times. What fractional part of her total shots hit the target?

Solution   The archer hit the target 72 out of 80 times.

$$\frac{72}{80} = \frac{8 \cdot 9}{8 \cdot 10} = \boxed{\frac{9}{10}}$$

The archer hit the target 9 out of 10 times. ■

EXERCISES A.1

In Exercises 1–10, write two fractions equivalent to the given fraction.

1. $\dfrac{6}{9}$ 2. $\dfrac{12}{15}$

3. $\dfrac{14}{35}$ 4. $\dfrac{21}{18}$

5. $\dfrac{33}{44}$ 6. $\dfrac{39}{65}$

7. 5   8. 9

9. 1   10. 0

In Exercises 11–26, reduce the given fraction to lowest terms.

11. $\dfrac{8}{10}$ 12. $\dfrac{15}{18}$

13. $\dfrac{6}{42}$ 14. $\dfrac{9}{45}$

15. $\dfrac{18}{36}$ 16. $\dfrac{32}{48}$

17. $\dfrac{20}{49}$ 18. $\dfrac{25}{36}$

19. $\dfrac{28}{72}$ 20. $\dfrac{24}{32}$

21. $\dfrac{90}{15}$ 22. $\dfrac{80}{16}$

23. $\dfrac{220}{80}$ 24. $\dfrac{175}{30}$

25. $\dfrac{81}{100}$ 26. $\dfrac{30}{77}$

In Exercises 27–34, fill in the question mark to make the fractions equivalent.

27. $\dfrac{5}{6} = \dfrac{?}{24}$ 28. $\dfrac{3}{8} = \dfrac{?}{40}$

29. $\dfrac{15}{7} = \dfrac{45}{?}$ 30. $\dfrac{28}{9} = \dfrac{56}{?}$

31. $\dfrac{16}{25} = \dfrac{?}{100}$ 32. $\dfrac{5}{12} = \dfrac{?}{96}$

33. $\dfrac{1}{18} = \dfrac{?}{108}$ 34. $\dfrac{1}{14} = \dfrac{?}{42}$

35. A businesswoman spent $1,500 at a computer store. She spent $1,250 on a computer and $250 on software. What fractional part of the total did she spend on the computer? What fractional part of the total did she spend on the software?

36. A fisherman caught 30 fish but threw 16 back because they were smaller than the minimum legal size. What fractional part of the total number of fish caught were thrown back? What fractional part of the total number of fish caught were kept?

37. A marksman hit the bull's-eye with 54 of 60 shots. What fractional part of the total number of shots hit the bull's-eye? What fractional part of the total number of shots missed the bull's-eye?

38. A car salesperson sold 24 cars during the month of February, of which 15 were new cars and the rest were used cars. What fractional part of the total number of cars were new? What fractional part of the total number of cars were used?

39. A student got 24 out of 28 problems correct on a math test. What fractional part of the test was correct? What fractional part of the test was incorrect?

40. A small-town police force has 26 members, 10 of whom are women. What fractional part of the police force is female? What fractional part of the police force is male?

41. A bakery sells 45 cheese danish, 30 blueberry danish, and 25 cherry danish. What fraction of all danish sold were cheese? Blueberry? Cherry?

42. An ice cream vendor sells 65 chocolate pops, 40 vanilla pops, and 20 strawberry pops. What fraction of all the pops sold were chocolate? Vanilla? Strawberry?

43. A record store sold 48 albums on compact discs, 56 albums on cassettes, and 16 albums on LPs. What fraction of the albums were sold on LPs? On compact discs? On cassettes?

44. One year a masonry contractor had 35 bricklaying jobs, 40 waterproofing jobs, and 15 pointing jobs. What fraction of the jobs involved bricklaying? Waterproofing? Pointing?

$A.2$ Multiplying and Dividing Fractions

We know that when we want to compute twice a number we multiply the number by 2. Similarly, if we want to compute $\frac{1}{2}$ of a number we multiply by $\frac{1}{2}$. How then should we compute $\frac{1}{2}$ of $\frac{1}{3}$ (which we would write as $\frac{1}{2} \cdot \frac{1}{3}$)?

We can visualize $\frac{1}{2}$ of $\frac{1}{3}$ as shown in Figure A.2.

Figure A.2

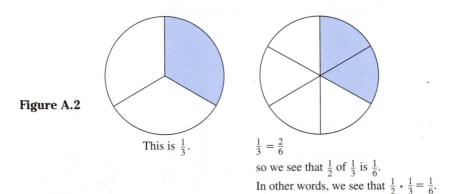

This is $\frac{1}{3}$.

$\frac{1}{3} = \frac{2}{6}$

so we see that $\frac{1}{2}$ of $\frac{1}{3}$ is $\frac{1}{6}$.

In other words, we see that $\frac{1}{2} \cdot \frac{1}{3} = \frac{1}{6}$.

Note that we obtain $\frac{1}{6}$ if we multiply $\frac{1 \cdot 1}{2 \cdot 3}$. We can generalize this to the multiplication of any two fractions, as indicated in the box.

Rule for Multiplying Fractions	$\dfrac{a}{b} \cdot \dfrac{c}{d} = \dfrac{a \cdot c}{b \cdot d} \qquad b, d \neq 0$

In other words, the product of two fractions is the product of the numerators over the product of the denominators.

EXAMPLE 1   Multiply. $\dfrac{3}{7} \cdot \dfrac{9}{10}$

Solution $\dfrac{3}{7} \cdot \dfrac{9}{10} = \dfrac{3 \cdot 9}{7 \cdot 10} = \boxed{\dfrac{27}{70}}$ ■

EXAMPLE 2   Multiply. $\dfrac{5}{8} \cdot \dfrac{16}{25}$

Solution $\dfrac{5}{8} \cdot \dfrac{16}{25} = \dfrac{5 \cdot 16}{8 \cdot 25} = \dfrac{80}{200} = \dfrac{2 \cdot 40}{5 \cdot 40} = \boxed{\dfrac{2}{5}}$

In obtaining this answer, we followed the rule for multiplying fractions as stated. However, the rule for multiplying fractions says in effect that $\dfrac{a}{b} \cdot \dfrac{c}{d}$ is really just one fraction with the factors of the numerator and denominator separated by a multiplication sign. Thus, it is easier to reduce *before* we multiply, and so we could have proceeded as follows:

$$\frac{5}{8} \cdot \frac{16}{25} = \frac{5}{8} \cdot \frac{\overset{2}{\cancel{16}}}{\underset{5}{\cancel{25}}} = \frac{2}{5}$$

In general, it is usually to our advantage to try to reduce *before* we multiply.   ■

EXAMPLE 3   Multiply. $\dfrac{49}{24} \cdot \dfrac{32}{28}$

Solution $\dfrac{49}{24} \cdot \dfrac{32}{28} = \dfrac{\overset{7}{\cancel{49}}}{\underset{3}{\cancel{24}}} \cdot \dfrac{\overset{4}{\cancel{32}}}{\underset{4}{\cancel{28}}} = \boxed{\dfrac{7}{3}}$ ■

EXAMPLE 4   Multiply. $8 \cdot \dfrac{5}{12}$

Solution   It helps to think of 8 as $\dfrac{8}{1}$.

$$8 \cdot \frac{5}{12} = \frac{8}{1} \cdot \frac{5}{12} = \frac{\overset{2}{\cancel{8}}}{1} \cdot \frac{5}{\underset{3}{\cancel{12}}} = \boxed{\frac{10}{3}}$$
   ■

Division

In order to understand the rule for dividing fractions, let's keep in mind that division is defined in terms of multiplication. For example, $20 \div 4 = 5$ because $5 \cdot 4 = 20$.

Rule for Dividing Fractions	$\dfrac{a}{b} \div \dfrac{c}{d} = \dfrac{a}{b} \cdot \dfrac{d}{c} \qquad b, c, d \neq 0$

EXAMPLE 5 Divide. $\dfrac{4}{7} \div \dfrac{3}{5}$

Solution Following the rule for dividing fractions, we get

$$\frac{4}{7} \div \frac{3}{5} = \frac{4}{7} \cdot \frac{5}{3} = \boxed{\frac{20}{21}}$$

Note that in order to check this division we would compute

$$\frac{20}{21} \cdot \frac{3}{5} \overset{?}{=} \frac{4}{7}$$

$$\frac{\overset{4}{\cancel{20}}}{\underset{7}{\cancel{21}}} \cdot \frac{\cancel{3}}{\cancel{5}} \overset{?}{=} \frac{4}{7}$$

$$\frac{4}{7} \overset{\checkmark}{=} \frac{4}{7}$$

In other words, the rule for dividing fractions is formulated in such a way that the multiplication checks. ■

EXAMPLE 6 Divide. $\dfrac{16}{45} \div \dfrac{12}{25}$

Solution $\dfrac{16}{45} \div \dfrac{12}{25} = \dfrac{16}{45} \cdot \dfrac{25}{12} = \dfrac{\overset{4}{\cancel{16}}}{\underset{9}{\cancel{45}}} \cdot \dfrac{\overset{5}{\cancel{25}}}{\underset{3}{\cancel{12}}} = \boxed{\dfrac{20}{27}}$ ■

EXAMPLE 7 Divide. $\dfrac{8}{9} \div 72$

Solution We think of 72 as $\dfrac{72}{1}$.

$$\frac{8}{9} \div 72 = \frac{8}{9} \div \frac{72}{1} \qquad \textit{Follow the rule for division.}$$

$$= \frac{8}{9} \cdot \frac{1}{72} = \frac{8}{9} \cdot \frac{1}{\underset{9}{\cancel{72}}} = \boxed{\frac{1}{81}}$$ ■

There is a very useful idea related to the division of fractions.

DEFINITION	If we let x represent a nonzero number, then the *reciprocal* of a number x is $\dfrac{1}{x}$.

Thus, we have the following:

The reciprocal of 8 is $\frac{1}{8}$.

The reciprocal of $\frac{3}{5}$ is $\dfrac{1}{\frac{3}{5}} = 1 \div \frac{3}{5} = 1 \cdot \frac{5}{3} = \frac{5}{3}$.

In general, we have that the reciprocal of $\frac{a}{b}$ is $\frac{b}{a}$.

In light of this definition, we can state the rule for dividing fractions: To divide by a fraction, multiply by its reciprocal.

EXAMPLE 8   A political pollster sent out 600 questionnaires, $\frac{5}{8}$ of which were returned. How many questionnaires were returned?

Solution   Computing a fraction of a number is similar to computing a multiple of a number. For instance, if a pound of apples costs 60¢, then we get the cost of 4 pounds of apples by multiplying 60¢ by 4 to get $2.40. Similarly, if we want the cost of $\frac{3}{4}$ of a pound of apples, we multiply 60¢ by $\frac{3}{4}$ to get 45¢.

In this example we want to compute $\frac{5}{8}$ of 600, and so we multiply 600 by $\frac{5}{8}$.

$$\frac{5}{8} \cdot 600 = \frac{5}{8} \cdot \frac{\overset{75}{\cancel{600}}}{1} = 375$$

Therefore, $\boxed{375 \text{ questionnaires were returned}}$.   ■

EXAMPLE 9   A gourmet coffee shop has 30 pounds of a certain coffee blend. How many $\frac{3}{4}$-pound bags can be filled with this blend?

Solution   In order to solve the problem we need to find out how many times $\frac{3}{4}$ goes into 30; therefore we divide 30 by $\frac{3}{4}$:

$$30 \div \frac{3}{4} = \frac{30}{1} \cdot \frac{4}{3}$$
$$= 40$$

Therefore, the shop can fill $\boxed{40 \text{ bags}}$.   ■

EXERCISES A.2

Perform the indicated operations and simplify as completely as possible.

1. $\dfrac{2}{3} \cdot \dfrac{5}{7}$

2. $\dfrac{3}{7} \cdot \dfrac{2}{5}$

3. $\dfrac{1}{4} \cdot \dfrac{1}{9}$

4. $\dfrac{1}{3} \cdot \dfrac{1}{8}$

5. $\dfrac{8}{9} \cdot \dfrac{3}{4}$

6. $\dfrac{9}{16} \cdot \dfrac{4}{3}$

7. $\dfrac{5}{6} \cdot \dfrac{4}{15}$

8. $\dfrac{6}{7} \cdot \dfrac{14}{9}$

9. $\dfrac{5}{6} \div \dfrac{4}{15}$

10. $\dfrac{6}{7} \div \dfrac{14}{9}$

11. $\dfrac{15}{18} \cdot \dfrac{24}{25}$

12. $\dfrac{12}{18} \cdot \dfrac{20}{36}$

13. $\dfrac{15}{18} \div \dfrac{24}{25}$

14. $\dfrac{12}{18} \div \dfrac{20}{36}$

15. $\dfrac{3}{5} \cdot \dfrac{5}{3}$

16. $\dfrac{4}{7} \div \dfrac{7}{4}$

17. $\dfrac{3}{5} \div \dfrac{5}{3}$

18. $\dfrac{4}{7} \cdot \dfrac{7}{4}$

19. $18 \cdot \dfrac{3}{2}$

20. $12 \cdot \dfrac{3}{4}$

21. $18 \div \dfrac{3}{2}$

22. $12 \div \dfrac{3}{4}$

23. $\dfrac{3}{2} \div 18$

24. $\dfrac{3}{4} \div 12$

25. $\dfrac{28}{45} \cdot \dfrac{63}{40}$

26. $\dfrac{28}{45} \div \dfrac{63}{40}$

27. $\dfrac{3}{5} \cdot \dfrac{12}{7} \cdot \dfrac{10}{9}$

28. $\dfrac{4}{15} \cdot \dfrac{6}{13} \cdot \dfrac{26}{36}$

29. $\dfrac{12}{25} \cdot \dfrac{5}{2} \div \dfrac{6}{5}$

30. $\dfrac{18}{35} \div \dfrac{9}{14} \cdot \dfrac{5}{4}$

31. What is the reciprocal of $\frac{5}{9}$?

32. What is the reciprocal of $\frac{3}{10}$?

33. $\frac{4}{7}$ is the reciprocal of what number?

34. 8 is the reciprocal of what number?

35. Ruby earns an annual salary of $22,500. If Allison earns $\frac{4}{5}$ of Ruby's salary, how much does Allison earn?

36. Lamar earns $54 per week at a part-time job. If he saves $\frac{2}{9}$ of his salary, how much does he save each week?

37. How many pieces of rope $\frac{5}{6}$ foot long can be cut from a 12-foot-long rope?

38. A farmer packs peaches in baskets that hold $\frac{1}{3}$ bushel each. How many baskets can be filled with 15 bushels of peaches?

39. A small computer manufacturing company has budgeted for 300 hours of labor during a certain month. If $\frac{1}{4}$ of the hours are for assembly and $\frac{3}{5}$ of the hours are for testing, how many hours are left over for packing?

40. A jewelry designer has 120 grams of gold. She uses $\frac{1}{5}$ of the gold to make a ring and $\frac{2}{3}$ of the gold to make a necklace. How much gold is left over?

41. A pharmacy has 600 grams of a certain compound; 400 grams are used to make $\frac{2}{5}$-gram tablets, and the other 200 grams are used to make $\frac{5}{8}$-gram tablets. How many tablets are made altogether?

42. A publishing company prints 12,000 hardback copies and 15,000 paperback copies of a certain book. If $\frac{5}{6}$ of the hardbacks and $\frac{9}{10}$ of the paperbacks are sold, how many copies of the book are left?

A.3 Adding and Subtracting Fractions

We begin by defining addition and subtraction of two fractions with the same denominator. We can state the rule as follows:

Rule for Adding and Subtracting Fractions	$\dfrac{a}{c} + \dfrac{b}{c} = \dfrac{a + b}{c}$ $\qquad$ $\dfrac{a}{c} - \dfrac{b}{c} = \dfrac{a - b}{c}$ $\qquad$ $c \neq 0$

In other words, this rule says that we can add or subtract fractions with common denominators by adding or subtracting the numerators and putting the result over the common denominator.

EXAMPLE 1   Add. $\dfrac{5}{14} + \dfrac{7}{14}$

Solution
$$\frac{5}{14} + \frac{7}{14} = \frac{5 + 7}{14}$$

$$= \frac{12}{14} = \frac{\overset{6}{\cancel{12}}}{\underset{7}{\cancel{14}}}$$

$$= \boxed{\frac{6}{7}} \qquad \textit{Remember that final answers should be reduced to lowest terms.} \qquad \blacksquare$$

In order to add or subtract two fractions with different denominators, we first use the Fundamental Principle of Fractions to change each fraction into an equivalent fraction with the same denominator. In general, we will find it most convenient to use the smallest possible common denominator—called the **least common denominator** or LCD for short.

We will find the following outline useful for finding the LCD.

Outline for Finding the LCD	**Step 1** Factor each denominator as completely as possible.
	Step 2 The LCD consists of the product of each *distinct* factor the *maximum* number of times it appears in any one denominator

Let's illustrate how to use this outline in Example 2.

EXAMPLE 2   Add. $\dfrac{7}{12} + \dfrac{1}{30}$

Solution   We need to convert both fractions into equivalent fractions with the same denominator. Using our outline, we proceed as follows:

Step 1 $\left.\begin{array}{l} 12 = 2 \cdot 2 \cdot 3 \\ 30 = 2 \cdot 3 \cdot 5 \end{array}\right\}$ *Notice that the **distinct** factors are 2, 3, 5.*

Step 2   We choose each distinct factor the *maximum* number of times it appears in any one denominator.

In $12 = 2 \cdot 2 \cdot 3$ $\begin{cases} 2 \text{ appears as a factor twice.} \\ 3 \text{ appears as a factor once.} \end{cases}$

In $30 = 2 \cdot 3 \cdot 5$ $\begin{cases} 2 \text{ appears as a factor once} \\ 3 \text{ appears as a factor once.} \\ 5 \text{ appears as a factor once.} \end{cases}$

Following our outline, we need 2 as a factor twice, 3 as a factor once, and 5 as a factor once.

Thus, the LCD is $2 \cdot 2 \cdot 3 \cdot 5 = 60$.

Now we use the Fundamental Principle to build fractions into equivalent fractions with an LCD of 60 as a denominator. We examine each denominator and determine which factors are missing from the LCD:

$$\frac{7}{12} = \frac{7}{2 \cdot 2 \cdot 3} = \frac{7 \cdot 5}{2 \cdot 2 \cdot 3 \cdot 5} \qquad \textit{The missing factor was 5.}$$

$$\frac{1}{30} = \frac{1}{2 \cdot 3 \cdot 5} = \frac{1 \cdot 2}{2 \cdot 2 \cdot 3 \cdot 5} \qquad \textit{The missing factor was 2.}$$

Therefore, we have

$$\frac{7}{12} + \frac{1}{30} = \frac{7 \cdot 5}{60} + \frac{1 \cdot 2}{60} = \frac{35}{60} + \frac{2}{60} = \boxed{\frac{37}{60}} \qquad ■$$

EXAMPLE 3   Add. $\dfrac{3}{10} + \dfrac{15}{4}$

Solution   You can probably "see" that the LCD is 20. If not, you can use the outline to obtain $2 \cdot 2 \cdot 5 = 20$ as the LCD. Try it!

$$\frac{3}{10} + \frac{15}{4} = \frac{3(2)}{20} + \frac{15(5)}{20} = \frac{6 + 75}{20} = \boxed{\frac{81}{20}} \qquad ■$$

EXAMPLE 4   Multiply. $\dfrac{3}{10} \cdot \dfrac{15}{4}$

Solution   Be careful! This is a multiplication problem and so no LCD is needed.

$$\frac{3}{10} \cdot \frac{15}{4} = \frac{3}{\underset{2}{\cancel{10}}} \cdot \frac{\overset{3}{\cancel{15}}}{4} = \boxed{\frac{9}{8}} \qquad ■$$

Looking at the answer to Example 4, it seems like an appropriate time to comment on the idea of an "improper" fraction versus a mixed number. Some texts refer to a fraction in which the numerator is greater than or equal to the denominator as an *improper fraction*. For our purposes, however, all we require is that the fraction be reduced to lowest terms.

Thus, in the last example, our answer of $\dfrac{9}{8}$ is acceptable. On some occasions we may prefer to express $\dfrac{9}{8}$ as a ***mixed number*** by dividing 8 into 9, giving $1\dfrac{1}{8}$.

EXAMPLE 5   Convert $3\dfrac{5}{6}$ from a mixed number to a fraction.

Solution $3\dfrac{5}{6}$ means $3 + \dfrac{5}{6} = \dfrac{3}{1} + \dfrac{5}{6} \qquad \textit{The LCD is 6.}$

$$= \frac{3(6)}{6} + \frac{5}{6}$$

$$= \frac{18}{6} + \frac{5}{6}$$

$$= \boxed{\frac{23}{6}} \qquad ■$$

The idea of a common denominator is also useful in deciding which of several fractions is largest or smallest.

EXAMPLE 6 Which of the following fractions is the largest?

$$\frac{3}{5}, \quad \frac{5}{9}, \quad \frac{17}{30}$$

Solution One way to compare the sizes of various fractions is to convert them to equivalent fractions with the same denominator. The LCD for 5, 9, and 30 is 90.

$$\frac{3}{5} = \frac{3(18)}{90} = \frac{54}{90}$$

$$\frac{5}{9} = \frac{5(10)}{90} = \frac{50}{90}$$

$$\frac{17}{30} = \frac{17(3)}{90} = \frac{51}{90}$$

Therefore, we can see that the largest fraction is $\frac{54}{90}$, which is $\boxed{\dfrac{3}{5}}$. ■

EXAMPLE 7 How many $2\frac{1}{2}$-gallon water cans can be filled completely from a tank containing 112 gallons of water?

Solution In order to answer this question we want to know how many times $2\frac{1}{2}$ goes into 112. Therefore, we divide 112 by $2\frac{1}{2}$.

$$112 \div 2\frac{1}{2} \qquad \textit{We first convert } 2\frac{1}{2} \textit{ from a mixed number to a fraction.}$$

$$= 112 \div \frac{5}{2} \qquad \textit{Follow the rule for division.}$$

$$= \frac{112}{1} \cdot \frac{2}{5}$$

$$= \frac{224}{5} \qquad \textit{We divide 224 by 5.}$$

$$= 44\frac{4}{5}$$

The $\frac{4}{5}$ represents four-fifths of a can, and since the example asks how many cans can be *completely* filled, we ignore the fraction. Therefore,

$\boxed{\text{44 of the water cans can be filled completely}}$. ■

EXERCISES A.3

In Exercises 1–26, perform the indicated operations and simplify as completely as possible.

1. $\dfrac{2}{7} + \dfrac{3}{7}$

2. $\dfrac{5}{9} + \dfrac{2}{9}$

3. $\dfrac{5}{8} - \dfrac{3}{8} + \dfrac{1}{8}$

4. $\dfrac{8}{15} - \dfrac{4}{15} + \dfrac{3}{15}$

5. $\dfrac{5}{12} + \dfrac{1}{12}$

6. $\dfrac{11}{20} + \dfrac{3}{20}$

7. $\dfrac{3}{10} + \dfrac{7}{10}$

8. $\dfrac{23}{18} + \dfrac{13}{18}$

9. $\dfrac{1}{2} + \dfrac{1}{4}$

10. $\dfrac{2}{3} - \dfrac{1}{6}$

11. $\dfrac{5}{6} - \dfrac{3}{8}$

12. $\dfrac{3}{10} - \dfrac{1}{4}$

13. $\dfrac{5}{6} \cdot \dfrac{3}{8}$

14. $\dfrac{3}{10} \div \dfrac{1}{4}$

15. $\dfrac{7}{12} + \dfrac{1}{3}$

16. $\dfrac{8}{15} + \dfrac{2}{5}$

17. $\dfrac{1}{2} + \dfrac{1}{3} + \dfrac{1}{4}$

18. $\dfrac{4}{9} + \dfrac{2}{3} + \dfrac{1}{6}$

19. $\dfrac{4}{15} + \dfrac{7}{20}$

20. $\dfrac{5}{24} + \dfrac{7}{60}$

21. $\dfrac{3}{28} + \dfrac{2}{35}$

22. $\dfrac{11}{18} + \dfrac{3}{20}$

23. $5 + \dfrac{3}{5}$

24. $5 \cdot \dfrac{3}{5}$

25. $5 \div \dfrac{3}{5}$

26. $\dfrac{3}{5} \div 5$

In Exercises 27–30, convert the given mixed number to a fraction.

27. $3\dfrac{3}{4}$

28. $4\dfrac{5}{8}$

29. $12\dfrac{4}{5}$

30. $10\dfrac{1}{9}$

In Exercises 31–34, convert the given fraction to a mixed number.

31. $\dfrac{23}{5}$

32. $\dfrac{12}{7}$

33. $\dfrac{43}{6}$

34. $\dfrac{100}{9}$

In Exercises 35–38, find the smallest of the three fractions.

35. $\dfrac{3}{4}, \ \dfrac{4}{5}, \ \dfrac{2}{3}$

36. $\dfrac{3}{10}, \ \dfrac{4}{15}, \ \dfrac{7}{20}$

37. $\dfrac{5}{12}, \ \dfrac{7}{15}, \ \dfrac{9}{20}$

38. $\dfrac{9}{10}, \ \dfrac{7}{8}, \ \dfrac{8}{9}$

39. A cement truck makes two deliveries of $29\dfrac{1}{3}$ and $17\dfrac{1}{3}$ cubic yards of cement to a construction site. If $36\dfrac{1}{2}$ cubic yards of the cement were used, how much is left over?

40. A water tank contains 250 gallons of water. If Renaldo removes $48\dfrac{1}{3}$ gallons and then removes $29\dfrac{1}{5}$ gallons, how much water is left in the tank?

41. A car averages $22\dfrac{1}{2}$ miles per gallon. How many gallons of gasoline are used on a 240-mile trip?

42. Wilma bikes at the rate of $10\dfrac{2}{3}$ km per hour. How long would it take her to travel 20 km?

43. A real estate developer is selling $2\frac{1}{4}$-acre parcels of land. If she sells 28 plots, how much land has she sold?

44. A gardener uses a wheelbarrow that can hold $1\frac{1}{5}$ cubic yards of dirt. How much dirt is transported in 15 trips with a full wheelbarrow?

45. If Raul runs $27\frac{1}{2}$ miles in $4\frac{1}{2}$ hours, what is his rate of speed in miles per hour?

46. Find the area of a rectangle whose dimensions are $2\frac{1}{2}$ meters by $7\frac{1}{3}$ meters.

A.4 Decimals

A *decimal fraction* or *decimal* is another way of writing a fraction. For example, the decimal 0.4 is another way of writing $\frac{4}{10}$. Similarly, 0.709 is another way of writing $\frac{709}{1,000}$, and 0.0037 is another way of writing $\frac{37}{10,000}$.

To convert a decimal to a fraction, we simply write the decimal as a fraction and reduce it, if possible. For instance,

$$0.458 = \frac{458}{1,000} = \frac{229}{500}$$

To convert a fraction to a decimal, we divide the numerator by the denominator. For example, to convert $\frac{3}{8}$ to a decimal, we divide 3 by 8:

$$
\begin{array}{r}
0.375 \\
8\overline{)3.000} \\
\underline{2\,4} \\
60 \\
\underline{56} \\
40 \\
\underline{40} \\
0
\end{array}
$$

Therefore, $\frac{3}{8} = 0.375$.

In performing operations with fractions, we often find the decimal form more convenient to use. (If you use a calculator, you may be able to input fractions, but the answer that appears in the display is usually a decimal.) However, be aware that if you convert a fraction into a decimal in order to perform some arithmetic operation and the decimal is not exact (because you rounded off), then your answer will not be exact, as is generally required by the exercises in this text.

Addition and Subtraction

To add or subtract decimals, we arrange the numbers in a column with the decimal points lined up, then add or subtract as usual. The decimal point in the answer is directly beneath the decimal point in the column of numbers.

EXAMPLE 1   Add. $2.07 + 81.6 + 0.084$

Solution   We write

$$
\begin{array}{r}
2.070 \\
81.600 \\
+0.084 \\
\hline
83.754
\end{array}
$$

← *It is helpful to put extra zeros at the end of these numbers so that they are all the same length.*

Line up the decimal point.

EXAMPLE 2 | Subtract. $21.7 - 0.095$

Solution | We write

$$
\begin{array}{r}
21.700 \\
-\ \ 0.095 \\
\hline
21.605
\end{array}
$$

∎

Multiplication

To multiply two decimals, we multiply the numbers as if they were whole numbers. The *number* of decimal places in the answer is the *total* number of decimal places in the numbers being multiplied.

EXAMPLE 3 | Multiply. 43.6×2.87

Solution |

$$
\begin{array}{r}
4\,3.6 \quad \leftarrow \quad \textit{1 decimal place} \\
2.8\,7 \quad \leftarrow \quad \textit{2 decimal places} \\
\hline
3\,0\,5\,2 \\
3\,4\,8\,8 \\
8\,7\,2 \\
\hline
1\,2\,5.1\,3\,2 \quad \leftarrow \quad \textit{3 decimal places}
\end{array}
$$

∎

Division

EXAMPLE 4 | Divide. $78.3 \div 5$

Solution | Since we are dividing by a whole number, we set this problem up in the usual division format and put a decimal in the quotient directly above the decimal in the dividend.
Now we divide as usual:

$$
\begin{array}{r}
15.66 \\
5\,)\overline{78.30} \\
5 \\
\hline
28 \\
25 \\
\hline
33 \\
30 \\
\hline
30 \\
30 \\
\hline
0
\end{array}
$$

Note that we place a zero at the end of 78.3 so that we can complete the division process.

The answer is $\boxed{15.66}$.

∎

EXAMPLE 5 | Divide. $716.68 \div 16.4$

Solution | Our basic approach is to convert this problem of dividing by a decimal into an equivalent problem of dividing by a whole number. We move the decimal point in 16.4 one place to the right so that it becomes 164 (a whole number). We must then also move the decimal point in 716.68 the same number of places to the right, making it 7,166.8. (In effect what we are doing is viewing $716.68 \div 16.4$ as a fraction, $\dfrac{716.68}{16.4}$, and then multiplying both the numerator and denominator of the fraction by 10 to change the denominator 16.4 into a whole number.) That is,

$$\frac{716.68}{16.4} = \frac{(716.68)(10)}{(16.4)(10)} = \frac{7,166.8}{164}$$

which we write as

$$16.4\overline{)716.6.8}$$

We now put the decimal point in our answer *directly* above the decimal point in 7,166.8 and divide as we usually do with whole numbers:

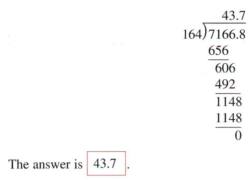

The answer is $\boxed{43.7}$. ■

![www] EXERCISES A.4

In Exercises 1–34, perform the indicated operations.

1. 4.7 + 3.5 + 21.7

2. 2.65 + 8.37 + 11.69

3. 15.87 − 6.35

4. 228.7 − 139.8

5. 21.62 + 4.1 + 57.236

6. 128.05 + 96.341 + 27.2

7. 6.5 + 0.003 + 2.08

8. 12.86 + 2.015 + 1.3

9. 9.27 − 7.85

10. 29.46 − 12.77

11. (3.9)(5.8)

12. (26.4)(15.7)

13. 6 − 0.03

14. 11 − 5.2

15. (2.63)(13.05)

16. (31.6)(19.52)

17. (0.37)(5.16)

18. (0.0025)(150.3)

19. 39.6 ÷ 5

20. 928.42 ÷ 6

21. 630 ÷ 1.5

22. 80 ÷ 2.5

23. 3.28 ÷ 0.16

24. 205.63 ÷ 0.05

25. (20.05)(0.004)

26. (46.8)(3.06)

27. (0.032)(0.05)

28. (56)(0.003)

29. 0.28 ÷ 0.04

30. 0.096 ÷ 0.4

31. (0.01)(0.02)(0.03)

32. (1.2)(2.1)(3.2)

33. 3.9 ÷ 0.015

34. 385.57 ÷ 0.285

35. How much does a manufacturer spend if he buys 1,200 of a certain item at a price of $0.0825 per item?

36. A recycling plant pays $0.425 for each pound of recyclable paper. How much does the plant pay for 210 pounds of recyclable paper?

37. A customer's electric bill for a certain month was $102. If the electric company charges $0.12 per kilowatt-hour, how many kilowatt-hours did the customer use that month?

38. If a telephone company customer was charged $8.32 for 128 message units, what is the company's rate per message unit?

39. An automobile manufacturer claims that a certain model gets 24.2 miles per gallon on the highway. An owner of this model drives 275 miles and uses 12.8 gallons of gas. Is this mileage per gallon as good as claimed?

40. Kanita decides to retile her floor. She knows that she needs to cover 157.5 sq ft. If each tile covers 0.5625 sq ft, how many tiles does she need to buy?

41. A certain chemical process uses 35.4 liters of a very expensive mixture that costs $18.65 per liter. How much does the mixture used for this process cost?

42. A company leases a copying machine and pays $0.012 per copy. If the leasing charge for a certain month was $58.20, how many copies did the company make on the machine?

A.5 Percents

An important idea that is related to decimals is the idea of percent. The word **percent** means "part of a hundred." The symbol for percent is %. Thus, 8 percent is written 8%.

$$1\% \quad \text{means} \quad \frac{1}{100} \quad \text{which is} \quad 0.01$$

$$45\% \quad \text{means} \quad \frac{45}{100} \quad \text{which is} \quad 0.45$$

$$8\% \quad \text{means} \quad \frac{8}{100} \quad \text{which is} \quad 0.08$$

$$137\% \quad \text{means} \quad \frac{137}{100} \quad \text{which is} \quad 1.37$$

$$0.4\% \quad \text{means} \quad \frac{0.4}{100} \quad \text{which is} \quad 0.004$$

Note the pattern of movement of the decimal point as we change from a percent to a decimal or from a decimal to a percent.

Rule for Changing from Percent to Decimal	To convert a number from a percent to a decimal, we drop the percent sign and move the decimal point two places to the *left* (we are dividing by 100).

Rule for Changing from Decimal to Percent	To convert from a decimal to a percent, we move the decimal point two places to the *right* (we are multiplying by 100) and insert the percent sign.

EXAMPLE 1 Convert 28% to a decimal.

Solution In order to use this rule we must be aware of the location of the decimal point in the given percent. In 28% the decimal point is understood to follow the 8.

$$28\% = 28.\% = \boxed{0.28}$$ *Note that we moved the decimal point two places to the left.* ∎

EXAMPLE 2 Convert 0.365 to a percent.

Solution $0.365 = \boxed{36.5\%}$ *Note that we moved the decimal point two places to the right.*   ■

Computing Percentages

Since a percent is a fraction, computing a percentage of a number is the same as taking a fraction of a number, and therefore we use the operation of multiplication. However, we must first convert the percent into its decimal equivalent.

EXAMPLE 3   Compute 42% of 75.

Solution   We convert 42% into a decimal, 42% = 0.42, and then we multiply:

$$0.42(75) = \boxed{31.5}$$ ■

EXAMPLE 4   Compute 107.5% of 1,200.

Solution   We convert 107.5% into a decimal, 107.5% = 1.075, and then we multiply:

$$1.075(1,200) = \boxed{1,290}$$ ■

EXAMPLE 5   Convert $\frac{3}{5}$ into percent.

Solution   We convert $\frac{3}{5}$ into a decimal by dividing 5 into 3, which gives us 0.6. We then convert 0.6 into percent.

$$0.6 = 0.60 = \boxed{60\%}$$ ■

EXAMPLE 6   24 is what percent of 60?

Solution   The problem is asking us to convert the fraction $\frac{24}{60}$ into percent. We first convert $\frac{24}{60}$ into a decimal. We can begin by reducing the fraction $\frac{24}{60} = \frac{2}{5}$. We then divide 5 into 2, giving us 0.4. Finally, we convert 0.4 into percent:

$$0.4 = 0.40 = \boxed{40\%}$$ ■

EXAMPLE 7   80 is 40% of what number?

Solution   This problem is asking us to find a number such that 40% of it is 80. In other words, we are looking for a number that, when *multiplied* by 0.40, is equal to 80. Therefore, to find the number, we *divide* 80 by 0.40.

$$\frac{80}{0.40} = 0.4.\overline{)80.0.}\ ^{200.}$$

Therefore, the answer is $\boxed{200}$.   ■

EXAMPLE 8   A vacuum cleaner is tagged to sell at $97. If the store advertises a 15% discount off the tagged price, what is the discounted price of the vacuum cleaner?

Solution   In order to answer the question, we will first find the amount of the discount by computing 15% of $97. We can then find the discounted price by subtracting this amount from the original price.

Original price − Amount of discount = Discounted price

The discount is 15% of $97 = 0.15(97) = $14.55

The discounted price = $97 − $14.55 = $82.45 ∎

EXAMPLE 9

If the price of a pound of apples goes up from 25¢ per pound to 40¢ per pound, what is the percent increase?

Solution

The *amount* of the increase is 40¢ − 25¢ = 15¢.

In order to compute the *percent increase,* we compare the amount of the increase to the *original* price. That is,

$$\% \text{ increase} = \frac{\text{Amount of increase}}{\text{Original price}}$$

Therefore, we have

$$\% \text{ increase} = \frac{15}{25} = \frac{3}{5} = 0.6 = 60\%$$

Whenever we compute a percent increase or decrease, we always divide the amount of increase or decrease by the *original* amount. ∎

EXERCISES A.5

In Exercises 1–24, convert the percents into decimals and the decimals into percents.

1. 25%

2. 0.38

3. 0.78

4. 43%

5. 0.05

6. 7%

7. 9%

8. 0.03

9. 150%

10. 2.4

11. 28%

12. 0.53

13. 0.67

14. 78%

15. 2

16. 40

17. 137%

18. 5%

19. 0.007

20. 0.23%

21. 62.4%

22. 0.089

23. 8.6%

24. 10.4%

In Exercises 25–30, compute the given percentages.

25. 30% of 70

26. 65% of 250

27. 7.2% of 35

28. 26.5% of 900

29. 0.8% of 5

30. 1% of 314

In Exercises 31–33, convert the given fraction to a percentage.

31. $\frac{1}{4}$

32. $\frac{3}{5}$

33. $\dfrac{5}{8}$

34. 12 is what percent of 30?

35. 8 is what percent of 20?

36. 25 is what percent of 75?

37. 9 is 6% of what number?

38. 75 is 8% of what number?

39. 24.6 is 30% of what number?

40. 75 is 120% of what number?

41. What is the price of a $1.50 pen sold at a 20% discount?

42. What is the price of a $950 stereo system being sold at a 35% discount?

43. The population of a town is 12,500. It is anticipated that the population will increase by 6% in the next 5 years. If so, what will the population be 5 years from now?

44. A car manufacturer plans a price increase of 4.5%. What will the new price of a $15,000 car be?

45. According to the 1980 census, the population of a small city was 140,000. If the 1990 census shows the population to be 133,000, what is the percent decrease in the population?

46. The price of a barrel of oil went from $32 to $27.52. What was the percent decrease in the price?

47. In the 1988 presidential election, 52% of the 8,600 eligible voters in a small town actually voted. In the 1992 presidential election, there were 9,250 eligible voters in the same town, of whom 48% actually voted. In which election were more votes actually cast in this town?

48. Two stores are both selling the same set of dishes. Store A is advertising the set at an original price of $125 with an 18% discount, while store B is advertising the set at an original price of $132 with a 20% discount. At which store are the dishes cheaper? By how much?

Using a Scientific Calculator

The hand-held calculator has become almost as commonplace as pencil and paper, and while the calculator is a wonderful tool to help us do tedious calculations, it is not a substitute for thinking.

For instance, a calculator can take the two numbers 236.8 and 12.6 and easily multiply or divide them. However, the calculator cannot decide which operation to perform or which number to divide into which; we must make those decisions by understanding what question we are trying to answer.

The figure below illustrates a typical scientific calculator.

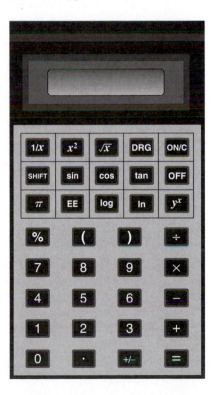

The basic pattern to execute each of the arithmetic operations is:

1. Enter the first number.
2. Hit the operation key.
3. Enter the second number.
4. Hit the $=$ key.
5. Read the answer on the display.

When you hit the $=$ key, you are instructing the calculator to complete the entered calculation.

Keep in mind that calculators may differ in the keystroke sequence necessary to perform a particular calculation. The keys discussed here and their use are merely an illustration. You must read the instruction manual for your calculator to become familiar with the proper use of the keys on your calculator.

EXAMPLE 1

Use a calculator to do the following computations.

(a) 26(345)   (b) 137.57 − 26.084   (c) $\dfrac{146.16}{0.025}$

Solution

Note that when you enter a whole number, the calculator automatically inserts a decimal point at the end of the number *unless* you insert the decimal point yourself by using the $\cdot$ key.

(a) To compute 26(345), the keystroke sequence would be

$$\boxed{26} \quad \boxed{\times} \quad \boxed{345} \quad \boxed{=}$$

and the answer appearing on the display would read $\boxed{8970}$.

(b) To compute 137.57 − 26.084, the keystroke sequence would be

$$\boxed{137.57} \quad \boxed{-} \quad \boxed{26.084} \quad \boxed{=}$$

and the answer appearing on the display would read $\boxed{111.486}$.

Note that the calculator automatically inserts the decimal point in the correct place. This is called a ***floating decimal point.***

(c) To compute $\dfrac{146.16}{0.025}$, the keystroke sequence would be

$$\boxed{146.16} \quad \boxed{\div} \quad \boxed{.025} \quad \boxed{=}$$

and the answer appearing on the display would read $\boxed{5846.4}$. ■

Calculators may vary as to the use of some of their special keys. It is very important that you read the instruction manual that comes with your calculator so that you know how to use these special keys. Here are some special keys that are particularly important.

1. The $\boxed{\text{ON/C}}$ key is the *on/clear* key. If the calculator is off, this key turns it on. If the calculator is on, pressing this key will clear the display of the previously entered number or operation.

   Pressing this key twice clears all previously entered numbers and operations.

2. The $\boxed{+/-}$ key is the *change of sign* key. When this key is pressed, the sign of the number on the display or the result of the calculation just performed is *changed*. If it was positive, it becomes negative and vice versa.

   Be careful! To enter a negative number you must use the $\boxed{+/-}$ key, not the subtraction key.

3. The $\boxed{x^2}$ key is the *squaring* key. Pressing this key squares the number on the display.

4. The $\boxed{\sqrt{x}}$ key is the *square root* key. Pressing this key takes the square root of the number on the display.

5. The $\boxed{1/x}$ key is the *reciprocal* key. Pressing this key takes the reciprocal of the number on the display.

6. The $\boxed{\%}$ key is the *percent* key. Pressing this key automatically converts the number on the display into a percent.

7. The $\boxed{y^x}$ key is the *y to the power x* key. This key is used as follows: To compute 3^5 you would press $\boxed{3}$ $\boxed{y^x}$ $\boxed{5}$ and the display will read $\boxed{243}$.

Before we illustrate some common computations it is important to note that different brands of calculators vary as to the exact keystroke sequence required. In particular, it is important to determine whether or not your calculator uses *algebraic logic,* which means that the calculator carries out computations according to the algebraic order of operations.

We know that we compute $4 + 5 \times 3$ as $4 + 15 = 19$ because according to the order of operations we perform the multiplication before the addition. If we want to compute $4 + 5 \times 3$ on a calculator with the following keystrokes

$$\boxed{4}\ \boxed{+}\ \boxed{5}\ \boxed{\times}\ \boxed{3}\ \boxed{=}$$

and the display reads $\boxed{19}$, then we know that the calculator is using algebraic logic. If the display does not read $\boxed{19}$, then the calculator does not use algebraic logic, and you will need either to use the grouping symbols available on your calculator or to read your calculator's instructional manual to determine how your calculator determines which operation to perform first.

The majority of the most popular scientific calculators use algebraic logic, and so in the illustrative examples that follow we will assume that the calculator is following algebraic logic.

EXAMPLE 2   How much does Jim spend if he buys 5 hot dogs at $1.65 each and 3 sodas at $0.85 each?

Solution   We use the following keystrokes:

$$\boxed{5}\ \boxed{\times}\ \boxed{1.65}\ \boxed{+}\ \boxed{3}\ \boxed{\times}\ \boxed{.85}\ \boxed{=}$$

The display will read $\boxed{10.8}$. Thus, Jim spent $\boxed{\$10.80}$ on the hot dogs and soda. ■

As we commented previously, the calculator makes the computation easier but it does not tell us what to do.

EXAMPLE 3   Yolanda wants to compute her car's gas mileage over a 2-week period. She records the following data: The first week she drives 206.8 miles and uses 11.6 gallons of gasoline; the second week she drives 384.6 miles and uses 22.5 gallons of gasoline. What was her average gas mileage (in miles per gallon) over this 2-week period? Round your answer to the nearest tenth.

Solution   In order to compute the average number of miles per gallon, we divide the total number of miles driven by the total number of gallons used:

$$\text{Miles per gallon} = \frac{\text{Total number of miles driven}}{\text{Total number of gallons used}}$$

We carry out this calculation with the following sequence of keystrokes:

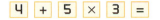

$$\boxed{(}\ \boxed{206.8}\ \boxed{+}\ \boxed{384.6}\ \boxed{)}\ \boxed{\div}\ \boxed{(}\ \boxed{11.6}\ \boxed{+}\ \boxed{22.5}\ \boxed{)}\ \boxed{=}$$

How would the calculator carry out the computation if we did not use the grouping symbols?

and the display will read $\boxed{17.343108}$. Rounding this to the nearest tenth, Yolanda's average gas mileage for the 2-week period was

$$\boxed{17.3}\ \text{mpg}$$

Note that we had to use grouping symbols in this computation to instruct the calculator to add the number of miles and add the number of gallons *before* dividing the two quantities. ■

There is one more important thing to keep in mind when using a calculator. It is quite easy to press the wrong key accidentally and get an incorrect answer. While we cannot always detect errors of this sort, sometimes a quick estimate of the expected answer can expose a gross error.

For instance, in Example 3 we could have estimated the total mileage as 600 miles (200 miles plus 400 miles) and the total gas used as 30 gallons (10 gallons plus 20 gallons). Thus we can estimate the gas mileage as

$$\frac{600 \text{ miles}}{30 \text{ gallons}} = 20 \text{ mpg}$$

Consequently, our answer of 17.3 mpg seems reasonable. Had we obtained an answer of 71.9 mpg (as we would have had we accidentally entered the mileage as 2,068 instead of 206.8), we would immediately recognize that something was wrong.

Always keep this idea in mind as you work out verbal problems.

EXAMPLE 4   How much annual interest is paid on $1,250 in an account earning 6.3% interest per year?

Solution   The annual interest is computed by multiplying the amount of money in the account by the percentage rate (written as a decimal). However, pressing the % key converts the previously entered number into a percent and does the computation after automatically converting the percent into a decimal. Thus the following keystroke sequence will carry out the required computation:

1250 × 6.3 % =

The display reads 78.75 . Thus the annual interest is $78.75 . ■

EXAMPLE 5   Randy drives a car bought in Europe that has an odometer that gives distance in kilometers only. If 1 kilometer is equivalent to 0.6215 mile, and Randy drives 57 kilometers to work every day, how far does Randy drive to work in miles? Round the answer to the nearest tenth of a mile.

Solution   Since Randy drives 57 kilometers, and each kilometer is equivalent to 0.6215 mile, we multiply 57 by 0.6215 to compute the distance in miles:

57 × .6215 =

and the display reads 35.4255. Rounding to the nearest tenth, we have that Randy drives 35.4 miles to work.

Along the same lines as the comment made above about estimating an answer, if we keep in mind that 1 kilometer being equal to 0.6215 mile means that a kilometer is *shorter* than a mile, then we realize that 57 kilometers must be less than 57 miles. Thus if we inadvertently divide 57 by 0.6215 to get an answer of 91.7, we would immediately recognize that something must be wrong. ■

APPENDIX B EXERCISES

Use a calculator to compute each of the following. Round your answers to the nearest hundredth.

1. 23.57 + 832.18 + 5.8

2. 87.4 + 2.863 + 26 + 423.9

3. 95.75 − 27.324

4. 107.492 − 89.07

5. 36.5 + 58.7 − 29.7

6. 834.86 + 567.98 − 1,008.5

7. $5.8(53.2)$

8. $26.81(145.8)$

9. $2.09(14.8)(156.7)$

10. $75.4(1.03)(210.2)$

11. $\dfrac{235.24}{5.6}$

12. $\dfrac{27.76}{1.05}$

13. $\dfrac{5,642.89}{185.3}$

14. $\dfrac{135.6}{8,372.1}$

15. $9.43 + 6.2(2.85)$

16. $81.6 - 4.6(12.53)$

17. $\dfrac{0.0058}{0.00021}$

18. $\dfrac{0.000125}{0.0085}$

19. $\dfrac{(8.6)(24.5)}{(16.8)(86.3)}$

20. $\dfrac{8.6 + 24.5}{16.8 + 86.3}$

21. $\dfrac{35.8 - 17.62}{(2.3)(135.6)}$

22. $\dfrac{(48.63)(21.7)}{3.6 + 5.5}$

23. Sales tax in New York is computed at the rate of $8\frac{1}{4}\%$. What would be the total cost (cost of item plus sales tax) of an item priced $89.95?

24. The property tax in a certain town is computed as 0.675% of the assessed value. What would the property tax be on a house with an assessed value of $126,700?

25. Light travels at the approximate rate of 186,000 miles per second. How far does light travel in 1 minute?

26. One mile is approximately 1.609 kilometers. How many miles are there in 100 kilometers?

27. A wholesaler pays $12 for an item and marks up the price by 15%. A retailer buys the item from the wholesaler and marks up his cost by 12%. What is the final cost to a consumer who buys the item from the retailer?

28. In a certain town it is estimated that 6.3% of the population of 28,127 work at the local auto plant. Of the total number of locals who work at the plant, approximately 35.7% work on the assembly line. Approximately how many local people work on the assembly line? (Round off to the nearest whole number.)

Using a Graphing Calculator

For Section 1.2

In Section 1.2 we discussed the order of operations that we agree to use in algebra. Most graphing calculators follow this same order of operations.

The following display illustrates how a graphing calculator would compute the expressions appearing in Example 2, parts **(b)** and **(c).**

```
(8-3)(6-4)
              10
19-2*3+4
              17
```

Note that in part **(b)** the calculator computes the expressions within the parentheses first as required, and in part **(c)** the calculator carries out the multiplication first as required.

In Example 4 of Section 1.2 we evaluated the expression $\dfrac{20 - 4(3)}{10 - 2(3)}$. If we want to compute this expression on a calculator we must insert parentheses so that the calculator will recognize that the operations in the numerator and denominator must be completed before dividing, as illustrated in the following display.

```
(20-4(3))/(10-2(
3))
               2
```

For Section 1.3

When using a graphing calculator, it is important to distinguish between the key used for subtraction, $\boxed{-}$, and the key used for the negative (or opposite) of a number, $\boxed{(-)}$. The first display at the top of page 624 illustrates the expression $9 + (-4)$.

If we try to evaluate the same expression with the incorrect subtraction sign in place of the negative sign as it appears in the third line of the first display, we receive the error message appearing in the second display.

For Section 1.4

As noted above, we must be careful in using the negative key ⎡(−)⎤ and the subtraction key ⎡−⎤ appropriately. The following display illustrates the expression 5 − (−8).

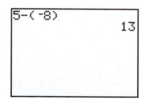

Note that the negative symbol is shorter and a bit higher than the subtraction symbol.

If you get a syntax error while doing a computation, two things to immediately check for are a minus sign error as described above and a parentheses error (such as having a closing parenthesis without a matching opening parenthesis).

Choosing the ⎡**2: Go to**⎤ option in an error message causes the calculator to place the cursor at the location at which the error was detected. This usually makes it easier to detect the error and fix it.

For Section 1.5

Note that the graphing calculator recognizes the implied multiplication in an expression such as 6(4 + 7) as illustrated in the following display.

```
6(4+7)
            66
```

In other words, the use of the multiplication symbol is optional on the graphing calculator in the same situations in which it would be optional when you write the expression.

For Section 2.2

In Example 8(g) of Section 2.2, we evaluated the expression $(x + y)^2 - (xy)^2$ for $x = 3$ and $y = -4$. We can evaluate such an algebraic expression using a graphing calculator in a variety of ways. One way is to replace the variables by their assigned values and have the calculator do the computation as illustrated in the following display.

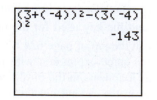

Another way to carry out the computation is to assign the given values to each of the variables and then enter the given expression. We assign the values by using the STO→ key. To assign the value 3 to x we use the keystroke sequence 3 STO→ X,T,θ,n. To assign a value to a variable other than x we must first use the ALPHA to access the letter designation of the keys. This is illustrated in the following display.

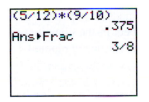

For Section 4.2

Normally when we do a fractional computation on a graphing calculator, the calculator will return a decimal answer. For instance, if we multiply $\frac{5}{12} \cdot \frac{9}{10}$ we get an answer of $\frac{3}{8}$. If we perform this computation on a graphing calculator we get an answer of .375 . Most graphing calculators can convert decimal answers into fractions. On the T1-83 Plus this is done using the MATH key and choosing the 1:▷ Frac as illustrated in the following display.

```
(5/12)*(9/10)
              .375
Ans▶Frac
               3/8
```

For Section 5.2

In Example 3 of Section 5.2, we sketched the graph of the equation $5y = 2x + 10$. In order to graph such an equation on a graphing calculator, it is necessary to first solve the equation for y. Solving for y, we get

$5y = 2x + 10$ *Divide both sides by 5.*

$y = \dfrac{2x + 10}{5} = \dfrac{2x}{5} + \dfrac{10}{5} = \dfrac{2x}{5} + 2$ *We can use any of these expressions for y.*

In addition, we must choose a window within which to view the graph. We can do this by specifying the window settings directly using the WINDOW key or by using one of the built-in window settings, which can be accessed from the ZOOM menu. The following displays illustrate how we would obtain the graph of $5y = 2x + 10$ and the window settings that match the graph drawn in Figure 5.9.

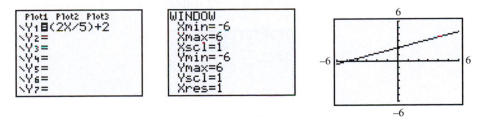

For Section 7.3

When we do computations that result in either very large or very small numbers, calculators will often automatically display these results in scientific notation. For example, if we compute $2,000^4$ we would see

```
2000^4
          1.6E13
```

Most graphing calculators allow you to choose the form for answers that appear on the display screen. In particular, on the T1-83 Plus the $\boxed{\text{MODE}}$ key allows you to choose to have answers appear in scientific notation by selecting $\boxed{\text{Sci}}$ in the $\boxed{\text{MODE}}$ menu, as illustrated in the first of the following displays.

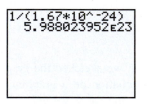

The second display above illustrates the computation done in Example 5 of Section 7.3. Recall that $\boxed{\text{1.2 E }^{-}\text{2}}$ means 1.2×10^{-2}. Note that when the calculator is in scientific notation mode, the answer will appear in scientific notation regardless of the form of the numbers in the computation.

In Example 7 of Section 7.3, we computed $1 \div (1.67 \times 10^{-24})$, which appears in the following display.

```
1/(1.67*10^-24)
    5.988023952E23
```

Rounding this answer to three decimal places gives us the same answer obtained in Example 7.

For Section 10.3

As we have noted, substituting a decimal approximation for a square root is not equivalent to simplifying it. However, we can use a calculator to check answers when working with radical expressions. For instance, in Example 3(a) of Section 10.3, we simplified $\sqrt{75} - \sqrt{12}$ and obtained $3\sqrt{3}$ as our final answer. We can compare the values of these two expressions to verify the correctness of our solution, as illustrated in the following display.

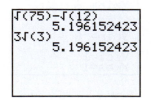

For Section 11.6

We can use a graphing calculator to investigate how changes to an equation affect the corresponding graph. For instance, the following display illustrates the graphs of $y = x^2$, $y = x^2 + 3$, and $y = x^2 - 4$ appearing in Example 1 of Section 11.6.

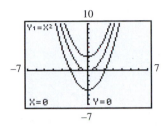

In addition, we can use a graphing calculator to check the graphs of all the parabolas in this section. For example, the following display illustrates the graph of $y = -x^2 - 2x + 4$ of Example 3 in Section 11.6.

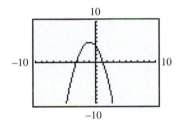

We can read off the vertex and intercepts fairly accurately to see that they agree with our algebraic results.

Introduction to Functions

The following advertisement appeared in the window of a travel agency advertising cellular phone rental.

TRAVEL TALK

The most convenient way to have phone access when you travel over-seas is to rent a cellular phone. For only $8 per week plus 0.45 per minute of airtime (up to 300 minutes) you can have the convenience of cellular service wherever you travel.

COME IN FOR DETAILS

Let's use the information in this advertisement to introduce the idea of a function. Clearly, the ad tells us that the amount you pay each week for the phone depends on the number of minutes of airtime you accumulate each week. In such a situation we say that the weekly cost of the phone is a *function* of the number of minutes of airtime each week. In particular, if we let m represent the number of minutes of airtime each week and C the cost for the phone service each week, we have

$$C = 8 + 0.45m$$

Let's examine a table of values for some values of m, the corresponding ordered pairs, and a graph of this equation.

Weekly airtime in minutes	Equation	Cost in dollars	Ordered pairs
m	$C = 8 + 0.45m$	C	(m, C)
0	$C = 8 + 0.45(0)$	8	$(0, 8)$
30	$C = 8 + 0.45(30)$	21.50	$(30, 21.5)$
60	$C = 8 + 0.45(60)$	35	$(60, 35)$
90	$C = 8 + 0.45(90)$	48.5	$(90, 48.5)$

Figure D.1

The graph of
$C = 8 + 0.45m$

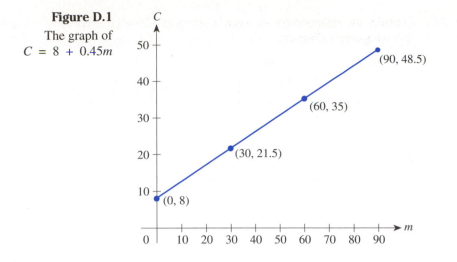

Before we proceed, a comment on how we labeled the horizontal and vertical axes. In our discussion of linear relationships, we noted that in some cases we had to decide which quantity to represent along the horizontal axis and which quantity to represent along the vertical axis. As we mentioned previously, when one of the quantities depends on the other, it is the accepted practice to use the horizontal axis for the "independent" variable and the vertical axis for the "dependent" variable. In this case, since we usually think of the cost depending on the number of minutes of airtime, we have labeled the horizontal axis with m and the vertical axis with C.

The key feature of the relationship between the number of minutes of airtime and the cost for the phone is that once we know the number of minutes of airtime, we can determine the exact cost of the phone rental. In other words, each value of m determines exactly one value of C, and we say that the cost C is a function of the number of minutes m. This is the distinguishing feature of a function, as described in the following definition.

DEFINITION

A *function* is a relationship between two sets so that to each element of the first set (called the *domain*) there corresponds *exactly* one element of the second set (called the *range*).

When we discuss a function we will think of the values from the domain as the *input* values and the values of the range as the *output* values. This illustrated in the following example.

EXAMPLE 1

Determine the domain and range of the cost function for the cellular phone rental described in the advertisement.

Solution

We have already determined that we can input the number of minutes m into the function equation $C = 8 + 0.45m$, which then outputs the cost C. Therefore the domain of this function is the set of possible values of m, which, according to the advertisement, is between 0 and 300 minutes.

By looking at the graph of the equation, we can see that the values of the cost vary between \$8 and \$143 and so these are the values in the range. Thus we have

$$\text{Domain} = \{m \mid 0 \le m \le 300\}$$
$$\text{Range} = \{C \mid 8 \le C \le 143\}$$ ∎

Note that the definition of a function does not require that the domain and range necessarily be sets of numbers. The following example illustrates a relationship between nonnumerical sets that is not a function.

EXAMPLE 2

Consider the relationships between mothers and biological children illustrated by the following arrow diagrams:

Mothers		Children		Children		Mothers
Martha	$\longrightarrow$	Steven		Steven	$\longrightarrow$	Martha
Fatima	$\longrightarrow$	Kisha		Kisha	$\longrightarrow$	Fatima
	$\searrow$	Prasheel		Prasheel	$\nearrow$	
Wai Man	$\longrightarrow$	Cheng		Cheng	$\longrightarrow$	Wai Man

The arrow means that we choose (input) the person on the left and associate with them (output) the person on the right. Explain whether each of these arrow diagrams illustrates a function.

Solution

Let's examine the first arrow diagram. The direction of the arrows tells us that the mothers make up the domain and the children make up the range. In other words, we choose a mother and the arrow associates her child(ren) with her. Note that if we choose Fatima, she has *two* children. For the relationship to be a function, each element in the domain (each mother) would have to have exactly one corresponding element in the range (exactly one child). Thus, the first arrow diagram does *not* represent a function. Even a single element in the domain that has more than one corresponding value in the range disqualifies the relationship from being a function.

Now let's examine the second arrow diagram. The direction of the arrows tells us that this time the children make up the domain and the mothers make up the range. In other words, we choose a child and the arrow associates his or her mother. The fact that both Kisha and Prasheel have the same mother is *not* a problem. Kisha is associated with only one range element: Fatima; and Prasheel is also associated with only one range element: Fatima. Each element in the domain (each child) has exactly one corresponding element in the range (his or her biological mother). Thus the second arrow diagram does represent a function. ∎

Frequently a function is defined by means of an equation. Unless otherwise specified, we will assume that the domain for the equation consists of all real numbers for which the equation makes sense. This is called the *natural domain* of the function. In other words, the natural domain consists of all real values of x for which the corresponding y value is also a real number.

Any equation describes a relationship between the variables that appear, but that relationship is not necessarily a function.

EXAMPLE 3

For each of the following, determine the domain and whether the equation defines y as a function of x.

(a) $y = 5x - 8$ **(b)** $y = x^2$ **(c)** $y^2 = x$ **(d)** $y = \dfrac{5}{x - 4}$

Solution

Let's analyze this question carefully.

THINKING OUT LOUD

What are we being asked?   To find the domain and whether the given equation defines y as a function of x

How do we find the domain?   Since the domain is the set of possible x values, we examine the equation and ask if there are any values of x that must be disqualified. We must disqualify any x value for which the corresponding y value is undefined or not real.

(continued)

THINKING OUT LOUD

(continued)

What does it mean for y to be a function of x?	In order for y to be a function of x, we must verify that to each value of x in the domain there corresponds exactly one value of y.
How do we determine that?	We consider a typical x value in the domain and see how many y values correspond to that x value.

(a) Examining the equation $y = 5x - 8$, we can see that there are no restrictions on x since we can take any number, multiply it by 5, and subtract 8 to obtain the y value. Therefore the domain is set of all real numbers.

We can also see that once we choose an x value, multiplying it by 5 and subtracting 8 gives a unique y value.

Therefore, the equation $y = 5x - 8$ does define y as a function x.

(b) A similar situation holds for the equation $y = x^2$. There are no restrictions on x (we can square any real number, and squaring a real number gives exactly one result). Therefore, the domain is the set of all real numbers and the equation does define y as a function of x.

(c) If we think about the equation $y^2 = x$, we notice that since x is equal to the square of a real number, x cannot be a negative number. Therefore, the domain is the set of nonnegative real numbers.

Additionally, we can see that if we choose x to be a positive number such as $x = 9$, then the equation becomes $y^2 = 9$, which makes two y values, $y = 3$ and $y = -3$, correspond to $x = 9$. Thus, the equation $y^2 = x$ does not define y as a function of x.

(d) Since the equation $y = \frac{5}{x-4}$ requires us to divide by $x - 4$, and we know that division by zero is undefined, we must disqualify any value of x that makes the denominator zero. Therefore, since $x = 4$ is the value that makes the denominator equal to zero, the domain consists of all real numbers except for $x = 4$. Alternatively, we can write that the domain is $\{x \mid x \neq 4\}$.

Notice that for all values except $x = 4$, each x value uniquely determines a corresponding y value; therefore, the equation does define y as a function of x. ■

Many everyday real-life situations can be described using the language of functions.

EXAMPLE 4   A fabric manufacturer has 100,000 square inches of fabric on hand. Express the number N of square (x-inch by x-inch) pieces of material that can be made from the fabric on hand as a function of x.

Solution   Let's begin with a few numerical examples (rounding to the nearest tenth where necessary).

If the pieces are 14 in. by 14 in., then the area of each piece is $14^2 = 196$ sq in., and so we could make $\frac{100,000}{196} = 510.2$ pieces (which means we could make 510 whole pieces and have some material left over).

If the pieces are 20 in. by 20 in., then the area of each piece is $20^2 = 400$ sq in., and so we could make $\frac{100,000}{400} = 250$ pieces.

If the pieces are x in. by x in., then the area of each piece is x^2 sq in., and so we could make $\frac{100,000}{x^2}$ pieces.

Thus the number of pieces that can be made is $N = \frac{100,000}{x^2}$. Clearly this is a function since each value of x gives only one value of N. ■

APPENDIX D EXERCISES

In Exercises 1–6, use the given arrow diagram to determine the domain and range and whether the diagram defines a function.

1. $A \longrightarrow 2$
    $B \longrightarrow 3$
    $C \longrightarrow 8$
    $D \longrightarrow 11$

2. $A \longrightarrow 2$
    $B \longrightarrow 3$
    $C \longrightarrow 8$
    $D \longrightarrow 11$

3. $F \longrightarrow L$
    $\qquad M$
    $G \longrightarrow 2$
    $3 \longrightarrow 8$

4. $P \longrightarrow 1$
    $Q \longrightarrow 5$
    $5 \longrightarrow T$
    $7 \longrightarrow M$

5. 1
    3
    $4 \qquad 5$
    9

6. $\qquad 1$
    $\qquad 3$
    $5 \qquad 4$
    $\qquad 9$

In Exercises 7–18, determine whether the given equation defines y as a function of x. If the equation does define a function, specify its domain.

7. $y = 4x - 7$

8. $y = \dfrac{x}{3} - \dfrac{6}{5}$

9. $y = x^2 + 1$

10. $y^2 = x + 1$

11. $y^2 = x^2 + 9$

12. $y = -x^2 + 7x - 2$

13. $y = \dfrac{4}{x - 3}$

14. $y = \dfrac{x - 3}{4}$

15. $y = \dfrac{x^2 - x - 12}{5}$

16. $y = \dfrac{5}{x^2 - x - 12}$

17. $y = 4x^3 - x^2 + 7x - 8$

18. $y^4 = 3x + 4$

In the following exercises, round answers to the nearest tenth where necessary.

19. If a ball is dropped from the top of CN Tower (which, at 1,821 feet tall, is the world's tallest self-supporting structure), then its height h above ground after t seconds is given by the equation $h = 1,821 - 16t^2$. Use this function to determine the height of the ball after 5 seconds and after 10 seconds.

20. If a ball is thrown vertically with an initial velocity of 60 feet per second, then its height h above the ground after t seconds is given by the equation $h = 60t - 16t^2$. Use this function to determine the height of the ball after 1.5 seconds and after 3 seconds.

21. A shoe manufacturer determines that the following function approximates the cost C (in thousands of dollars) to manufacture n pairs of shoes:

    $$C = 125 + 0.0213n$$

    Use this function to determine the cost of producing 2,500 pairs of shoes.

22. An economic consultant for a large computer manufacturer estimates that the cost C (in dollars) of manufacturing each computer is dependent on the number n of computers made according to the formula $C = \frac{34,800}{n} - 0.005n + 250$.

Use this function to determine the cost per computer if 1,000, 3,000, and 5,000 computers are manufactured.

23. A state Department of Highways determines that the cost C (in thousands of dollars) of building a particular stretch of highway is a function of the number m of miles that are paved according to the formula $C = 2,800 + 320m + 0.006m^2$. Use this function to compute the cost of paving 180 miles of this highway.

24. A consultant for a large industrial firm determines that the cost C (in dollars) to clean up g gallons of pollution can be approximated by the function $C = 450 + 52g + 0.0001g^3$. Use this function to determine the cost of cleaning up 5,000 gallons of pollution.

25. The area of a rectangle is to be 80 sq cm.

   (a) If the base of the rectangle is b, express the height h as a function of the base b.

   (b) Use the function you found in part (a) to determine the height of the rectangle if the base is 12 cm.

26. Selena drives 200 miles at a constant rate of speed r.

   (a) Express the time t (in hours) that it takes her to make the trip as a function of her rate r.

   (b) Use the function you found in part (a) to determine the time it takes to make the trip at a constant rate of 52 mph.

27. A printing company charges $0.06 per page plus a setup fee of $75 per job.

   (a) Express the cost of a job C as a function of the number p of pages printed.

   (b) Use the function you found in part (a) to determine the cost of a printing job containing 875 pages.

28. A building contractor charges a base fee of $750 plus construction costs of $25 per square foot of office space.

   (a) Express the costs of a job C as a function of the number s of square feet of construction.

   (b) Use the function you found in part (a) to determine the cost of building 1,650 sq ft of office space.

29. The length of a rectangle is 3 less than twice its width w.

   (a) Express the perimeter of the rectangle as a function of w.

   (b) Use the function you found in part (a) to determine the perimeter of the rectangle if the width is 9 in.

   (c) Express the area of the rectangle as a function of w.

   (d) Use the function you found in part (c) to determine the area of the rectangle if the width is 9 in.

30. The base of a rectangle is 4 more than three times its height h.

   (a) Express the perimeter of the rectangle as a function of h.

   (b) Use the function you found in part (a) to determine the perimeter of the rectangle if the height is 8 cm.

   (c) Express the area of the rectangle as a function of h.

   (d) Use the function you found in part (c) to determine the area of the rectangle if the height is 8 cm.

ANSWERS TO SELECTED EXERCISES

This answer section contains the answers to all odd-numbered exercises as well as the answers to all exercises in the Mini-Reviews, Chapter Reviews, Chapter Tests, Cumulative Reviews, and Cumulative Tests.

EXERCISES 1.1

1. True 3. True 5. True 7. True
9. False 11. {1, 2, 3, 4, 5, 6, 7}
13. {0, 2, 4, 6, 8, 10, 12, 14, 16, 18}
15. {0, 1, 2, 3, 4, 5, 6}
17. {0, 1, 2, 3, 4, 5}
19. {6, 7, 8, . . .}
21. {7, 8, 9, . . .} 23. {3, 4, 5}
25. {1, 2, 3, 4, 6, 8, 12, 24}
27. {1, 2, 3, 4, 5, 6, 12, 15}
29. {0, 4, 8, 12, . . .}
31. {0, 12, 24, 36, . . .} 33. ∅
35. {1, 3, 5, 7, 9, 11, 13, 15, 17, 19};
   {x | x is a natural number less than 20 that is not divisible by 2}
37. {5, 15, 25, 35, 45}; {t | t is a natural number less than 50 that is divisible by 5 but not by 10}
39. {100, 200, 300, 400, . . .}; {t | t is a natural number greater than 20 that is divisible by 100}
41. 4 > 2 43. 7 = 7 45. 19 > 14
47. 72 = 72 49. 0 = 0
51. <, ≤, ≠ 53. =, ≤, ≥
55. >, ≥, ≠ 57. <, ≤, ≠
59. >, ≥, ≠ 61. 2 · 7 63. 3 · 11
65. 2 · 3 · 5 67. Prime
69. 2 · 2 · 2 · 2 · 2 · 2
71. 2 · 2 · 2 · 2 · 2 · 3
73. 3 · 29
75. 2 · 2 · 2 · 3 · 3 · 5
77. 2 · 3 · 3 · 7
79. 2 · 3 · 11 · 13
81. 2 · 2 · 2 · 2 · 3 · 19
83. 2 · 2 · 2 · 2 · 7 · 17
85.
87.

EXERCISES 1.2

1. True, commutative property of addition
3. True, commutative property of addition
5. False 7. False
9. True, commutative property of multiplication
11. True, commutative property of multiplication
13. False

15. True, associative property of addition
17. True, associative property of addition
19. True, associative property of multiplication
21. True, associative property of multiplication
23. False 25. False
27. Associative property of addition
29. Associative property of multiplication
31. Commutative property of addition; associative property of addition
33. Commutative property of multiplication; associative property of multiplication
35. −4 37. 4 39. 4 41. 4
43. −4 45. −4 47. 5 49. 5
51. 2 53. 11 55. 5 57. 19
59. 18 61. 3 63. 0 65. 19 67. 6
69. 26 71. 20 73. 20 75. 50
77. 12 79. 3 81. 14 83. 3 85. 2
87. 1 89. 1 91. 29 93. 1
95. 57 97. 2

EXERCISES 1.3

1. −2 3. −12 5. 7 7. 5 9. 4
11. −10 13. −4 15. 10 17. −15
19. −4 21. 0 23. −7 25. −44
27. 6 29. −15 31. −1 33. −9
35. −5 37. −5 39. 1 41. −15
43. 6 45. −4 47. 6 49. −6
51. 5 53. −29 55. −67 57. −9
59. −215 61. 149 63. 268
65. −68 67. −1 69. 1 71. 12
73. 0 75. $13 77. 33°F 79. $33
81. 14-yard line 83. $14,196 profit
85. 9,973 feet

EXERCISES 1.4

1. −4 3. −11 5. 9 7. −6 9. 3
11. −3 13. 13 15. −13 17. 3
19. −3 21. −13 23. 13 25. 1
27. −11 29. −11 31. 3 33. −3
35. 3 37. −9 39. −12 41. 2
43. −2 45. −7 47. −13 49. 5
51. −1 53. 0 55. −16 57. 0
59. −17 61. −26 63. 15
65. −126 67. −52 69. 23

71. −31 73. −13 75. −1 77. −9
79. 0 81. 8 83. 0 85. 13°C
87. $726.41 89. 30,324 ft
91. 34,544 ft 93. −8, −2, −7
95. −3, −m, −n

EXERCISES 1.5

1. −28 3. −28 5. 28 7. −11
9. 11 11. −5 13. −5 15. 5 17. 0
19. Undefined 21. −22 23. −5
25. 9 27. −50 29. −45 31. 158
33. −240 35. 28 37. 32 39. −8
41. 8 43. −40 45. −231 47. 2
49. 10 51. 4 53. −19 55. 2
57. −4 59. −18 61. −12 63. 10
65. Undefined 67. 12 69. 56
71. −112 73. 32 75. 39 77. 75

EXERCISES 1.6

1. 10.1 3. 20.4 5. 4 7. 24.53
9. 70 11. −7.3 13. −10.6
15. 10.6 17. −7 19. −22
21. −0.209 23. −24.57 25. 1.1
27. 0.302 29. 2.26
31. Between 4 and 5
33. Between −2 and −3
35. Between 7 and 8
37. Between 0 and −1
39.
41.
43.
45.
47.
49. Members: 2.4, $\sqrt{5}$, $\frac{4}{7}$;
   Nonmembers: −3, 0, 6
51. No members; the set is empty
53. Members: $-\frac{4}{5}$, 5.1, $6\frac{2}{9}$;
   Nonmembers: −4, 6, 14

Chapter 1 Review Exercises

1. {2, 4, 6, 8, 10, 12, 14, 16, 18}
2. {1, 3, 5, 7, 9, 11, 13, 15, 17, 19}
3. {2, 3, 5, 7, 11, 13, 17, 19}
4. {4, 6, 8, 9, 10, 12, 14, 15, 16, 18}
5. {23, 29} 6. ∅ 7. 2 · 3 · 5

8. $2 \cdot 2 \cdot 7$ 9. Prime
10. $2 \cdot 2 \cdot 2 \cdot 3 \cdot 3$
11. $2 \cdot 2 \cdot 5 \cdot 5$ 12. $3 \cdot 19$
13. -4 14. -5 15. -12 16. -10
17. 4 18. -4 19. -15 20. -9
21. 0 22. -10 23. -3 24. -5
25. -3 26. -8 27. 9 28. -2
29. -8 30. 24 31. -1 32. -7
33. 26 34. 22 35. 11 36. -3
37. 144 38. 72 39. 25 40. 37
41. -10 42. -11 43. 4 44. 5
45. -20 46. -12 47. 1 48. -2
49. -1 50. 7 51. 4 52. -3 53. -1
54. -1 55. 54 56. 10 57. $-1,370$
58. 20 59. 38 60. 59 61. -25.39
62. 48.91 63. 0.97 64. -5

Chapter 1 Practice Test

1. (a) True   (b) False   (c) False
   (d) True   (e) $\varnothing$
2. -11 3. 0
4. -4 5. -24 6. 30 7. 17 8. 1
9. 7 10. -10 11. -12 12. 23
13. -19 14. 91 15. 29 16. -83
17. 729
18. (a) $2 \cdot 2 \cdot 3 \cdot 7$
    (b) $2 \cdot 2 \cdot 2 \cdot 2 \cdot 3 \cdot 3 \cdot 13$
    (c) Prime

EXERCISES 2.1

1. $xxxxxx$ 3. $(-x)(-x)(-x)(-x)$
5. $-xxxx$ 7. $xxyyy$
9. $xx + yyy$ 11. $xyyy$ 13. a^4
15. x^2y^3 17. $-r^2s^3$ 19. $-x^2(-y)^3$
21. x^3x^5 or x^8 23. 243 25. -8
27. 16 29. -45 31. -14 33. 17
35. 576 37. 243 39. x^8 41. a^7
43. $60x^3$ 45. $6r^5$ 47. $-15x^5$
49. $40c^5$ 51. $32x^4y^7$ 53. $6x^4y^4$
55. $9a^8$ 57. $-64n^6$ 59. x^{14}
61. 0.073 63. 14.758 65. 107.916
67. 48.337

EXERCISES 2.2

1. 33.07 3. 131.71 5. 16.6 ft
7. $3,141.20$ 9. 625 11. 24.28
13. 41.0 sq in.   15. 4.6 cu ft
17. $8,400$
19. Approx. 273 computers
21. 1993: 53.8%; 1994: 54.5%;
    1995: 55.2%; 1996: 55.7%;
    1997: 56.2%; 1998: 56.7%
23. 1990: 1.7%; 1991: 2.7%;
    1992: 3.2%; 1993: 3.2%;
    1994: 2.8%; 1995: 2.2%

25. 3 27. -3 29. -1 31. 5
33. 5 35. 5 37. -1 39. -5
41. 3 43. -2 45. 2 47. 1
49. 18 51. 11 53. 13 55. 25
57. -9 59. -16 61. -96 63. 2
65. -18 67. 100 69. 63 71. 20
73. 21 75. -113 77. 6 79. 0
81. 2 terms: $3x$, coeff. 3, literal part
    x; $-4y$, coeff. -4, literal part y
83. 1 term: coeff. -12, literal part xy
85. 1 term: coeff. 3, literal part
    $x(z - y)$
87. 3 terms: $4x^2$, coeff. 4, literal
    part x^2; $-3x$, coeff. -3, literal
    part x; 2
89. 3 terms: $-x^2$, coeff. -1, literal
    part x^2; y, coeff. 1, literal part y;
    -13
91. 2 terms: $6x^2$, coeff. 6, literal part
    x^2; $20y^2$, coeff. 20, literal part y^2
93. 2.031 95. -9.862 97. 0.046
99. 25.667
101. Rounding after: 2.44; rounding
     before: 2.47
103. Rounding after: -6.26; rounding
     before: -6.25

EXERCISES 2.3

1. Essential   3. Nonessential
5. Essential   7. Nonessential
9. Essential   11. Both essential
13. 1st nonessential, 2nd essential
15. Group 1: x, $2x$; coefficients: 1, 2
    Group 2: y, $3y$; coefficients: 1, 3
17. Group 1: $-x$, $-3x$; coefficients:
    -1, -3
    Group 2: $2x^2$, $-x^2$; coefficients:
    2, -1
    Group 3: $4x^3$; coefficient: 4
19. Group 1: 4, 5; coefficients: 4, 5
    Group 2: $4u$, $5u$; coefficients:
    4, 5
    Group 3: $4u^2$; coefficient: 4
21. No like terms; all coefficients
    are 5
23. Group 1: $-x^2y$, $-2x^2y$;
    coefficients: -1, -2
    Group 2: $2xy^2$, $3xy^2$;
    coefficients: 2, 3
    Group 3: x^2y^2; coefficient: 1
25. $3x + 12$ 27. $5y - 10$
29. $-2x - 14$ 31. $15x + 6$
33. $-12x - 4$ 35. $x^2 + 3x$
37. $x^3 + 3x^2$ 39. $10x^2 - 20x$
41. $16.8x - 0.42x$
43. $3.84x + 76.8$ 45. $2(x + 5)$
47. $5(y - 4)$ 49. $3(3x + y - 2)$

51. $x(x + y)$ 53. $7x$ 55. $7x^2$
57. $-3a$ 59. $-4y$ 61. $-6x$
63. $-5x + 2y$ 65. $3x + 5y + 2z$
67. $4x^2 + 10x$ 69. $2x^2 - 3x$
71. $6x^2y - 4x^2$ 73. $-2y$ 75. $3st$
77. $a^2b - ab^2$ 79. $-0.2x - 2.3$
81. $0.82x + 5.48$ 83. $7x + 2y$
85. $7x - y$ 87. $3x + 3y - 4xy$
89. $7x^3 + 21x$ 91. $11x^3 + 15x$
93. $0.15x + 450$
95. $-0.43x + 43.2$

EXERCISES 2.4

1. $8x + 5y$ 3. $8m + 7n$
5. $-7x + 11y$ 7. $3x - 1$
9. $-3x + 11$ 11. $2x - 4$ 13. $14m$
15. $10a - 8b$ 17. $-6x^2 + 8y$
19. $12 - 3x$ 21. $7y - 1$
23. $4x - 18y$ 25. $4x - 12y$
27. $5x - 15y + 3xy$ 29. $-12xy$
31. $-45x^2y^2$ 33. $5x^3 - 8x^2$
35. $13a^2 - 7a$ 37. $3x^2 + 21x$
39. $3a^3 + 9ab + 4a^2b^2 - 4b^3$
41. $-5x^2 - 8x + 1$
43. $-4x^3y^2 + 4x^3y$
45. $4u^3v - 5u^2v^2 - uv^3$
47. $8x + 15y + 12xy$
49. $12m - 14n - 12mn$
51. $9t^9 - 13t^5 - t^4$ 53. 0
55. $3a - 8$ 57. $x^2 - 16x$
59. $19x - 18$ 61. $8y - 22$
63. $-3t^2 - 12t$
65. $-5a^3 + 15a^2 - 16a$

EXERCISES 2.5

1. $n + 4$ 3. $n - 4$ 5. $n - 4$
7. $5n + 6$ 9. $2n - 9$
11. $n(n + 7)$ 13. $(n + 2)(n - 6)$
15. $2n - 8 = 14$
17. $5n + 4 = n - 2$
19. $r + s = rs$
21. $2(r + s) = rs - 3$
23. Let n = the first integer;
    $n + (n + 1)$
25. Let n = the first even integer;
    $n + (n + 2)$
27. Let n = the first odd integer;
    $n(n + 2)(n + 4)$
29. Let n = the even integer;
    $8n = 7(n + 2) - 4$
31. Let n = the first integer;
    $n^2 + (n + 1)^2 + (n + 2)^2 = 5$
33. $A = 8.65h$; 121.10
35. $C = 225 + 0.045p$; 441
37. $C = 42 + 9p$; 96
39. $C = 0.65 + 0.42(m - 1)$; 5.27

41. $T = 8d + 10k;$ 246 km
43. $T = 6L + 9D;$ 102 lawns
45. $0.30d$ dollars 47. $1.04d$ dollars
49. $0.80p$ dollars 51. $486.40
53. (a) $86.40 (b) No
55. $285 + 0.52x$ dollars
57. $6h + 14(h - 7)$ dollars
59. $48t + 54(t - 2)$ miles
61. (a) 8 (b) 40 cents (c) 12
    (d) 120 cents (e) 9
    (f) 225 cents (g) 29
    (h) 385 cents
63. (a) 200 (b) 15 (c) 3,000
    (d) 160 (e) 20 (f) 3,200
    (g) 6,200
65. (a) 100 m/min (b) 25 min
    (c) 2,500 m (d) 220 m/min
    (e) 35 min (f) 7,700 m
    (g) 10,200 m

Chapter 2 Review Exercises

1. $xyyy$ 2. $(xy)(xy)(xy)$ 3. $-xxxx$
4. $(-x)(-x)(-x)(-x)$ 5. $3xx$
6. $(3x)(3x)$ 7. $x^2 y^3$ 8. $x^2 + x^3$
9. $a^2 - b^3$ 10. $-a^2 b^3$ 11. -16
12. 16 13. -48 14. 144 15. 25
16. 29 17. 8 18. -2 19. 17
20. -23 21. x^8 22. r^9 23. $2a^9$
24. $5y^6$ 25. $4x^5 + 3x^6$
26. $5a^8 + 2a^7$ 27. $-2x^2 - 8x + 4$
28. $3m^3 - 4m^2 - 9$ 29. $6a^3 b^5$
30. $-10a^3 b^4$ 31. $6a^3 b + 2a^2 b^5$
32. $-10a^3 b - 5ab^4$
33. $11x^2 + 12x - 15$
34. $10z^2 - 10z + 4$ 35. $3y^3 - 3y^2$
36. $c^4 + 5c^3$
37. $3x^3 y - 6xy^2 + 4xy^3 - 4x^2 y^2$
38. $-3r^2 s^2 - 2r^3 s^2 - 5r^2 s^3$
39. $4x - 12y$ 40. $10a - 8b$
41. $-x^7 - 6x^4 y^2$ 42. $6x^3 y^5$ 43. 3
44. $3 - 3x + 3x^2 - x^3$
45. $0.16x + 87.5$ 46. $-0.69x + 70$
47. $5(x^2 + 2)$ 48. $2(a^5 + 8)$
49. $3(y - 2z + 3)$
50. $11(2x - 3y + 1)$
51. $n + 7 = 3n - 4$
52. $2n - 5 = 3n + 4$
53. $n + (n + 2) = n - 5$
54. $n + 4(n + 1) = 3(n + 2) + 8$
55. $0.65d$ dollars
56. $M = 4.50 + 0.02c$ dollars
57. (a) 12 (b) $2 (c) 9 (d) $5
    (e) $24 (f) $45 (g) $69
58. (a) n (b) $2 (c) $2n - 4$
    (d) $5 (e) $2n$ (f) $5(2n - 4)$
    (g) $2n + 5(2n - 4) =$
        $(12n - 20)$

Chapter 2 Practice Test

1. -81 2. 81 3. -1 4. -41
5. -1 6. -2 7. -24 8. 94
9. $3x^2 y - 8xy - y^2$ 10. $-24x^6 y^3$
11. xy 12. $3x^5$ 13. $3x - 12$
14. $0.18x + 60$
15. (a) $2n + 4$ (b) $5n - 20 = n$
16. (a) $3 (b) x (c) $2
    (d) $2x - 5$ (e) $3(2x - 5)$
    (f) $2x$
    (g) $2x + 3(2x - 5) =$
        $(8x - 15)$
17. $M = 99 + 0.15c$ dollars

EXERCISES 3.1

1. Identity 3. Contradiction
5. Identity 7. Contradiction
9. Contradiction 11. Identity
13. Contradiction 15. Identity
17. $x = -7$ 19. None 21. $y = \frac{3}{4}$
23. None 25. None 27. None
29. $x = 2$ 31. $a = -1, a = 4$
33. $y = 2$

Mini-Review

37. 2 38. -5 39. -3 40. 4
41. -16 42. 64 43. -3 44. -11
45. 23 46. -14 47. -33 48. 11

EXERCISES 3.2

1. $x = 5$ 3. $y = 11$ 5. $a = -2$
7. $a = -3$ 9. $x = 7$ 11. $x = \frac{15}{4}$
13. $x = 3$ 15. $x = -1$ 17. $x = -2$
19. $z = 2$ 21. No solutions
23. $x = 3$ 25. $x = \frac{13}{5}$ 27. $t = -5$
29. $a = -1$ 31. $w = 7$
33. $-3x + 20$ 35. $t = 3$
37. $14a + 20$
39. Let w = width; $6w = 40;$
    $\boxed{w = 6\frac{2}{3} \text{ in.}}$
41. Let L = length; $3.5L = 73.5;$
    $\boxed{L = 21 \text{ yd}}$
43. Let t = time needed to drive 234
    miles; $52t = 234;$ $\boxed{4.5 \text{ hours}}$
45. Let m = number of miles driven;
    $0.17m + 3(29.95) = 123.17;$
    $\boxed{196 \text{ miles}}$
47. (a) $0.375 (b) 46.7 mph
49. (a) 77.5°F (b) 200 times
51. (a) 19°F (b) Approx. 2,600 ft

53. 18.0 ft/sec
55. Let w = width; then $2w + 7 =$
    length; $2w + 2(2w + 7) = 50;$
    $\boxed{6 \text{ cm by } 19 \text{ cm}}$
57. Let L = length; then $5L - 10 =$
    width; $2L + 2(5L - 10) = 100;$
    $\boxed{10 \text{ yd by } 40 \text{ yd}}$

Mini-Review

64. -7 65. 1 66. -24 67. 22
68. $2 \cdot 2 \cdot 2 \cdot 3 \cdot 13$ 69. Prime

EXERCISES 3.3

1. $y = 5$ 3. $a = -\frac{7}{2}$ 5. $r = -6$
7. $x = -3$ 9. $u = -1$ 11. $x = 0$
13. Identity 15. No solutions
17. $x = 7$ 19. $t = \frac{9}{2}$ 21. $y = 4$
23. $y = 0$ 25. $a = 0$ 27. $z = -\frac{18}{5}$
29. $t = -\frac{1}{3}$ 31. $t = 5$ 33. $y = -1$
35. No solutions 37. $a = 0$
39. $x = 1$ 41. $x = -2$ 43. $z = 6$
45. $x = 6.5$ 47. $t = 0.03$
49. $t = -8.19$ 51. 1,750 items
53. Let x = length of the shorter
    piece; $x + 8$ = length of the
    longer piece; $x + (x + 8) = 30;$
    $\boxed{\text{shorter piece is 11 ft; longer piece is 19 ft}}$
55. The numbers are n and $3n + 4;$
    $n + 3n + 4 = 24;$ $\boxed{5, 19}$
57. The numbers are x and $4x - 5;$
    $4x - 5 = 11;$ $\boxed{\frac{16}{5}, \frac{39}{5}}$
59. The number is $y;$
    $y + 5y - 4 = 27;$ $\boxed{\frac{31}{6}}$
61. The number is $n;$
    $2n + 4 - n = 12;$ $\boxed{8}$
63. The smallest number is $x,$
    the largest is $2x + 10,$ the
    middle number is $2x - 5;$
    $x + (2x + 10) + (2x - 5) =$
    $80;$ $\boxed{15, 25, 40}$
65. Let n = first integer; $n + 1 =$
    second consecutive integer;
    $n + 2$ = third consecutive
    integer; $n + (n + 1) + (n + 2)$
    $= 45;$ $\boxed{14, 15, 16}$

67. Let n = first odd integer;
$n + 2$ = second consecutive
odd integer; $n + 4$ = third
consecutive odd integer;
$n + (n + 2) + (n + 4) =$
$2(n + 4) + 29;$ 31, 33, 35

69. Let w = width;
$2w + 1$ = length;
$2w + 2(2w + 1) = 26;$
4 cm by 9 cm

71. Let the three sides be represented
by $x, x + 1, x + 2;$
$x + (x + 1) + (x + 2) = 24;$
7 cm, 8 cm, 9 cm

73. Let w = width, $w + 6$ = length;
$2(w + 10) + 2 \cdot 3(w + 6) =$
$2w + 2(w + 6) + 56;$ 3 by 9

75. Let m = # of miles driven;
$2(45) + 0.40m = 170;$
200 miles

77. Let d = daily charge;
$125 + 5d = 275;$ \$30 per day

79. Let q = # of quarters; $20 - q =$
# of dimes; $25q + 10(20 - q) =$
$425;$ 15 quarters, 5 dimes

81. Let x = # of advanced-purchase
tickets sold; $150 - x$ = # of
tickets sold at the door;
$10x + 12(150 - x) = 1,580;$
110 advanced-purchase tickets

83. Let x = # of half-dollars;
$x + 11$ = # of quarters;
$45 - x - (x + 11) = 34 - 2x$
= # of dimes; $50x + 25(x + 11)$
$+ 10(34 - 2x) = 1,110;$
16 dimes, 20 quarters,
9 half-dollars

85. Let t = # of hours electrician
works; $45t + 24(t + 4) = 464;$
$5\frac{1}{3}$ hours

87. Let t = # of hours until they pass
each other;
$90t + 60t = 300;$ 12 noon

89. t = # of hours 55-kph person
travels; $55t + 45(t - 1) = 280;$
5:15 P.M.

91. t = # of hours to complete
running section;
$18t + 50(6 - t) = 172;$
4 hours, 72 km

93. Let r = slower rate;
$5r = 4(r + 15);$ 300 km

95. t = # of hours trainee works;
$7t + 15(t - 2) = 124;$ 4 P.M.

97. Let x = the number of shares of
stock bought at \$8.125 per share;
$8.125x + 9.375(200 - x) =$
$1,725;$ 120 shares at \$8.125,
80 shares at \$9.375

99. Let n = the number of heavier
boxes; $9.32n + 6.58(n + 89) =$
$1,974.28;$ 233 boxes

101. Let x = the number of deluxe
Web sites;
$1,850x + 675(17 - x) = 17,350;$
5 deluxe and 12 regular

EXERCISES 3.4

1. Yes 3. No 5. Yes 7. Yes
9. No 11. No 13. No 15. Yes
17. Yes 19. Yes
21. $x < 5$

23. $a > -3$

25. $w < -6$

27. $z > -3$

29. $x \le 4$

31. $y > -2$

33. $x > -2$

35. $x > 2$

37. $a > 0$

39. $a \le 0$

41. $x > -3$

43. $x > 3$

45. $-1 < a \le 6$

47. $-2 \le x \le 2$

49. $-2 < y < 1$

51. $0 \ge x > -1$

53. $5 > x > 1$

Mini-Review

59. $6x^5$ 60. $8x^6y^5$ 61. $10x - 14y$
62. $x^2 - xy + y^2$ 63. $-x^5$
64. $-27ab^2$

EXERCISES 3.5

1. $x < -2$ 3. $a > -1$ 5. $y < \frac{7}{2}$
7. $y > -\frac{7}{2}$ 9. $y > -\frac{7}{2}$ 11. $y < \frac{7}{2}$
13. $x > -4$ 15. $1 < y$ 17. $x \le 1$
19. $x \ge \frac{25}{2}$ 21. $z \ge -2$ 23. $x < 2$
25. $w \ge 4$ 27. $a > 6$ 29. $y \ge 2$
31. Contradiction 33. Identity
35. $x > -4$ 37. $t \ge \frac{13}{2}$
39. $a < -4$ 41. $w > -16$
43. $y \ge -10$ 45. $-5 < x < 3$
47. $-1 < a < 1$ 49. $-1 < y \le 2$
51. $3 < x \le 5$
53. $x > 5$

55. $a \le 4$

57. $-4 < x < -1$

59. $2 \le y \le 5$

61. $-2 \le t < 1$

63. $-3 \le x < 4$

65. $x \ge 3.2$ 67. $6.32 < t \le 11.51$
69. $7 > x$ 71. $t - 10$ 73. $x \ge 5$
75. $y \le 2$
77. Let n = the number;

  $3n - 4 < 17$; $\boxed{n < 7}$

79. Let n = the number;
  $6n + 12 > 3n$;

  $\boxed{\text{The number must be larger than } -4.}$

81. Let L = length; $2L + 2(8) \ge 80$;

  $\boxed{\text{The length must be at least 32 cm.}}$

83. Let W = width;
  $50 \le 2W + 2(18) \le 70$;

  $\boxed{\text{The length must be at least 7 in. and at most 17 in.}}$

85. Let d = price of a ticket at the door (in dollars); $d + 2$ = price of a reserved seat ticket;
  $150d + 300(d + 2) \ge 3{,}750$;

  $\boxed{\text{Reserved seat tickets must be at least \$9.}}$

87. Let s = annual sales (in dollars);
  $0.016(s - 82{,}000) \ge 1{,}800$;

  $\boxed{\text{At least \$194,500 in sales.}}$

Chapter 3 Review Exercises

1. Identity   2. Contradiction
3. Conditional   4. Identity
5. $x = -1$ 6. $x = 4$ 7. $y = \frac{5}{2}$

8. $w = \frac{4}{3}$ 9. $t = \frac{1}{2}$

10. Neither value   11. $x = -5$
12. None   13. $a = -2$ 14. None
15. $x = -5$ 16. $x = -4$
17. Identity   18. No solutions
19. $a = \frac{7}{3}$ 20. $t = 2$ 21. $x = 28$
22. $x = -1$ 23. $a = \frac{5}{2}$ 24. $w = -1$
25. $x > 3$ 26. $x \le 3$
27. $x \le -9$

28. $x > -5$

29. $-2 \le a < 4$

30. $-4 < t < 3$

31. Let n = one number, $2n - 3$ = the other number;

  $n + 2n - 3 = 18$; $\boxed{7, 11}$

32. Let s = smaller number, $3s - 7$ = larger number;

  $s + 3s - 7 = s + 8$; $\boxed{5, 8}$

33. Let w = width,
  $5w + 4$ = length;
  $2w + 2(5w + 4) = 80$;

  $\boxed{6 \text{ by } 34}$

34. Let L = length,
  $2L - 8$ = width;
  $2L + 2(2L - 8) =$
  $4(2L - 8) - 1$; $\boxed{8\frac{1}{2} \text{ by } 9}$

35. Let f = # of \$15 skirts,
  $150 - f$ = # of \$9 skirts;
  $15f + 9(150 - f) = 2{,}010$;

  $\boxed{110 \text{ \$15 skirts, } 40 \text{ \$9 skirts}}$

36. Let t = time going,
  $7 - t$ = time returning;

  $45t = 60(7 - t)$; $\boxed{360 \text{ km}}$

37. Let h = # of overtime hours worked; $12(40) + 18h \ge 570$;

  $\boxed{\text{At least 5 hours}}$

38. Let x = original price;
  $18.36 \le 0.80x \le 27.96$;

  $\boxed{\text{The original price range was \$22.95 to \$34.95.}}$

Chapter 3 Practice Test

1. (a) Contradiction   (b) Identity
   (c) Conditional
2. (a) Yes   (b) No   (c) No
3. (a) $x = \frac{8}{3}$ (b) $y = 0$
   (c) $a = -2$
   (d) $x \le 3$

   (e) $-2 \le x < 2$

4. Let w = width,
  $4w - 5$ = length;
  $2w + 2(4w - 5) = 7w + 11$;

  $\boxed{7 \text{ by } 23}$

5. Let n = # of new cassettes,
  $20 - n$ = # of old cassettes;
  $3n + 20 - n = 46$;

  $\boxed{13 \text{ new; } 7 \text{ used}}$

6. Let x = number;
  $4x - 8 = 2x + 1$; $\boxed{\frac{9}{2}}$

7. Let s = amount of sales;
  $0.0125(s - 10{,}000) \ge 15{,}000$;

  $\boxed{\text{She must have sales of at least \$130,000.}}$

CHAPTERS 1–3 CUMULATIVE REVIEW

1. -20 (§1.4)   2. 27 (§1.5)
3. 20 (§1.5)   4. 5 (§1.5)
5. -25 (§2.1)   6. 25 (§2.1)
7. x^6 (§2.1)   8. $2x^9$ (§2.1)
9. $-2x^2y - 3xy^2$ (§2.3)
10. $z^2 - 6z - 8$ (§2.3)
11. $6x^3 - 8xy$ (§2.3)
12. $-24x^3y$ (§2.1)   13. $15u^5v$ (§2.1)
14. $-3u^5 + 15u^2v$ (§2.3)
15. $10m - 15n$ (§2.4)
16. $-t^2 - 16r^3$ (§2.4)
17. $-2a^3b + 2a^2b^2$ (§2.4)
18. $3x^3yz^2 - 9x^2y^2z$ (§2.4)
19. $2x^2y - 2xy^2$ (§2.4)
20. $-7y$ (§2.4)   21. $10x - 60$ (§2.4)
22. $a - b + ab - a^2$ (§2.4)
23. $-6xy^2$ (§2.4)   24. 0 (§2.4)
25. 4 (§2.2)   26. -4 (§2.2)
27. -54 (§2.2)   28. 49 (§2.2)
29. 4 (§2.2)   30. 31 (§2.2)
31. -40 (§2.2)   32. 11 (§2.2)
33. $x = \frac{1}{3}$ (§3.3)   34. $t = -4$ (§3.3)
35. $a = -1$ (§3.3)
36. $w = \frac{15}{4}$ (§3.3)
37. Identity (§3.3)
38. $a \ge 9$ (§3.5)

39. $s > -7$ (§3.5)

40. No solutions (§3.5)
41. $3 < y \le 4$ (§3.5)

42. $-2 \le z \le 3$ (§3.5)

43. $d = 0$ (§3.3)   44. $x = 0$ (§3.3)
45. Let n = the number;

  $2n - 8 \ge n + 4$; $\boxed{n \ge 12}$ (§3.5)

46. Let w = width,
$3w + 8$ = length;
$2w + 2(3w + 8) = 44$;
$$\boxed{\text{Width} = 3\tfrac{1}{2}\text{ cm,} \\ \text{length} = 18\tfrac{1}{2}\text{ cm}}$$ (§3.3)

47. Let d = # of danishes;
$18 - n$ = # of pastries;
$40n + 55(18 - n) = 825$;
$$\boxed{11 \text{ danishes and } 7 \text{ pastries}}$$
(§3.3)

48. Let t = # of minutes old copier works, $t - 15$ = # of minutes new copier works;
$20t + 25(t - 15) = 885$;
$$\boxed{10\text{:}28 \text{ A.M.}}$$ (§3.3)

CHAPTERS 1–3 CUMULATIVE PRACTICE TEST

1. (a) -105 (§2.1) (b) 36 (§2.1)
2. (a) 13 (§2.2) (b) 5 (§2.2)
3. (a) $-2x^2 - 5x + 3$ (§2.3)
   (b) $-3x - y$ (§2.3)
   (c) $-2a^3 - ab$ (§2.4)
   (d) 0 (§2.4) (e) $4x^3y^3$ (§2.4)
   (f) $6 - 6a + 6a^2 - a^3$ (§2.4)
4. (a) $x = \frac{4}{5}$ (§3.3)
   (b) $w = -9$ (§3.3)
   (c) $x \geq 3$ (§3.5)
   (d) $a = -5$ (§3.3)
   (e) No solutions (§3.3)
   (f) Identity (§3.3)
5. (a) $a \leq -2$ (§3.5)

   [number line marked at -2 and 0]

   (b) $1 \leq x < 5$ (§3.5)

   [number line marked at 0, 1, 5]

6. (a) Let n = the number,
   $3n - 5$ = other number;
   $n + 3n - 5 = 27$;
   $\boxed{8, 19}$ (§3.3)

   (b) Let n = # of tires at \$32 each, $40 - n$ = # of tires at \$19 each; $32n + 19(40 - n)$
   $= 1{,}124$; $\boxed{28 \text{ new, } 12 \text{ old}}$
   (§3.3)

   (c) Let t = # of hours to complete the race at 15 kph, $t - 1$ = # of hours to complete the race at 25 kph, $15t = 25(t - 1)$;
   $\boxed{\text{distance} = 37.5 \text{ km}}$ (§3.3)

(d) Let x = least expensive original price;
$0.82x \geq \$36.49$; $\boxed{\$44.50}$
(§3.5)

EXERCISES 4.1

1. $\frac{3}{5}$ 3. $-\frac{3}{7}$ 5. $\frac{5}{2}$ 7. $\frac{1}{2}$ 9. $\frac{1}{3}$
11. x^2 13. $\frac{1}{x^2}$ 15. $\frac{5}{2x}$ 17. $-\frac{3z^4}{5}$
19. $\frac{2}{5t^5}$ 21. $-\frac{3b^3}{a^2}$ 23. 1 25. $-\frac{r^2}{2t^2}$
27. $-\frac{5b}{a}$ 29. $\frac{x^2}{2}$ 31. $-\frac{2}{x^6}$ 33. $\frac{1}{8x^3y^3}$
35. $\frac{1}{2}$ 37. $\frac{1}{12}$ 39. $\frac{4s - 3t}{8s - 9t}$ 41. a
43. 1

Mini-Review

49. Let w = width,
$2w + 5$ = length;
$2w + 2(2w + 5) = 34$;
$\boxed{4 \text{ by } 13}$

50. Let n = # of nickels,
$n - 5$ = # of dimes;
$5n + 10(n - 5) = 130$;
$\boxed{12 \text{ nickels, } 7 \text{ dimes}}$

EXERCISES 4.2

1. $\frac{8}{27}$ 3. -1 5. $\frac{2x}{15y}$ 7. $\frac{5x^3}{12y^2}$ 9. $\frac{16}{25}$
11. $\frac{12x^3}{y^3}$ 13. $\frac{w}{4}$ 15. $\frac{x}{3}$ 17. $\frac{48}{x}$ 19. $\frac{x}{48}$
21. $\frac{3}{2y}$ 23. $\frac{3m^2}{n}$ 25. $\frac{3v}{2u}$ 27. $4x^2$
29. $9y^2$ 31. $\frac{1}{9y^2}$ 33. $-\frac{2x^2}{3y^2}$ 35. 1
37. $\frac{9}{a^2}$ 39. $\frac{81}{a^4}$ 41. $\frac{81}{a^4}$ 43. $-\frac{175}{162}$
45. $-\frac{9}{343}$ 47. $\frac{3}{5}$ 49. $2x$ 51. $\frac{x^3}{y^3}$
53. $\frac{u^2}{2z^3}$ 55. $\frac{1}{y}$ 57. $\frac{3}{2y^2}$ 59. $-\frac{25}{9xy}$
61. $-4x^2$ 63. $62\frac{1}{2}$ miles 65. $\frac{3}{32}$
67. $39\frac{3}{8}$ 69. $8\frac{1}{3}$ 71. $181\frac{2}{3}$ miles
73. 6 75. 0.48 77. 0.45 79. 0.556

Mini-Review

84. Let m = # of miles driven;
$0.20m + 30 = 65$;
$\boxed{175 \text{ miles}}$

85. Let x = smallest angle,
$2x$ = largest angle,
$x + 20$ = third angle;
$x + 2x + (x + 20) = 180$;
$\boxed{40°, 60°, 80°}$

EXERCISES 4.3

1. 3 3. $-\frac{4}{5}$ 5. $-\frac{2}{3}$ 7. $\frac{22}{15}$ 9. $\frac{8}{15}$
11. $-\frac{1}{6}$ 13. $\frac{27}{8}$ 15. $\frac{7}{5}$ 17. $\frac{17}{24}$ 19. $\frac{11}{6}$
21. $\frac{4}{x}$ 23. $\frac{32}{9x^2}$ 25. $\frac{3}{2x}$ 27. $\frac{2y}{7x}$
29. 0 31. $-\frac{2a}{b}$ 33. $\frac{2x - 3}{3x}$ 35. $-\frac{y}{4}$
37. $\frac{2}{5}$ 39. -1 41. $\frac{2}{3w}$ 43. $\frac{-10t + 2}{5t^2}$
45. $\frac{3y + 2x}{xy}$ 47. $\frac{6}{xy}$ 49. $\frac{10 - 21x}{6x}$
51. $\frac{5y + 6x}{48xy}$ 53. $\frac{8 - 3x}{2x^2}$ 55. $\frac{6}{x^3}$ 57. $\frac{13}{6x^2}$
59. $\frac{49 - 9a}{84a^2}$ 61. $\frac{1 + 2x}{x}$ 63. $\frac{10y + x}{6xy^2}$
65. $\frac{14b^2 + 9a}{12a^2b^3}$ 67. $\frac{7}{8a^3b^4}$ 69. $\frac{35 + 2x}{120x^2}$
71. $\frac{45t - 14r}{144rt^3}$ 73. $\frac{15y^2 + 18x^3}{100x^2y}$
75. $\frac{14y - 9x}{6x^2}$ 77. $\frac{18n^2 - 20m + 3m^2n}{24m^2n^3}$
79. $\frac{5x^2 + 2y^2}{2xy}$ 81. $\frac{t^2 - 3}{t}$ 83. $\frac{3x^4 + x - 2}{x^2}$
85. $\frac{10x + 9}{72}$ 87. $\frac{x^2 + 4x + 6}{2x}$
89. $\frac{a^2 - 5a + 6}{2a}$

Mini-Review

96. Let d = # of dimes;
$28 - d$ = # of nickels;
$10d + 5(28 - d) = 200$;
$\boxed{12 \text{ dimes, } 16 \text{ nickels}}$

97. Let p = # of pounds;
$4p + 10 = 28$;
$\boxed{4\frac{1}{2} \text{ pounds}}$

EXERCISES 4.4

1. $x = 27$ 3. $a = -24$ 5. $y = \frac{15}{2}$
7. $a = -\frac{28}{3}$ 9. $x = -12$
11. $x = 20$ 13. $x = 8$ 15. $a = -1$
17. $u = 8$ 19. $y < \frac{9}{4}$ 21. $x = \frac{4}{7}$
23. $a \leq -30$ 25. $x = 5$
27. $x = 40$ 29. $m = \frac{34}{85} = \frac{2}{5} = 0.4$
31. $w = -7$ 33. $w = 5$
35. $x = \frac{11}{3}$ 37. $y < -6$ 39. $a = 1$
41. $y = 0$ 43. $x = -2$ 45. $t = -2$
47. $x = 4$ 49. $x = 6{,}000$
51. $x = 10{,}000$ 53. $y = 2$
55. $z = 0$ 57. $-39 \leq x \leq -21$
59. $\frac{31x}{30}$ 61. $x = 60$ 63. $x = -3$
65. $\frac{x + 11}{4}$

Mini-Review

68. Let L = length,
$3L - 20$ = width;
$2L + 2(3L - 20) = 32$;

> 7 ft by 9 ft

69. Let x = weight;
$7 + 4(x - 1) = 43$;

> 10 pounds

EXERCISES 4.5

1. $\frac{7}{5}$ 3. $\frac{5}{7}$ 5. $\frac{11}{5}$ 7. $\frac{1}{3}$ 9. $\frac{a}{b + c}$
11. $x = 20$ 13. $a = 10$
15. $y = 50$ 17. $y = 12$
19. $x = 0.12$ 21. $t = 11.59$
23. $y = -2.05$ 25. 294 27. 9
29. 8 cm 31. $\frac{2}{5}$ 33. $8\frac{1}{3}$ cm
35. 4.55 kg 37. 108.7 yd
39. 25.2 miles 41. $1,937.30
43. $6\frac{2}{3}$ tsp. of garlic;

   $16\frac{2}{3}$ Tbsp. of olive oil

45. $612.20
47. 44 oz of copper; 8 oz of tin
49. Samantha: $75,045.83;
Greg: $53,604.17
51. 366 losers 53. $88,471.43
55. 160 57. $x = 12, y = 12$
59. $x = 7.74, y = 6.45$ 61. 17.25 ft
63. 1,200 meters

EXERCISES 4.6

1. Let n = the number;
$\frac{2}{3}n + 5 = 9$; 6

3. Let n = the number;
$\frac{3}{4}n - 2 = \frac{1}{8}n - 7$; -8

5. Let L = length, $\frac{L}{2}$ = width;
$2L + 2\left(\frac{L}{2}\right) = 36$;

> length = 12 m, width = 6 m

7. Let L = length of longest side,
$\frac{3}{4}L$ = length of medium side,
$\frac{1}{2} \cdot \frac{3}{4}L = \frac{3}{8}L$ = length of

shortest side;
$L + \frac{3}{4}L + \frac{3}{8}L = 17$;

> short = 3 in., medium = 6 in.,
> long = 8 in.

9. Let c = # of combination tickets
sold, $350 - c$ = # of regular
tickets sold;
$22c + 15(350 - c) = 6,895$;

> 235 combination tickets,
> 115 regular tickets

11. Let q = # of quarters,
$2q + 3$ = # of dimes;
$25q + 10(2q + 3) = 255$;

> 5 quarters, 13 dimes

13. Let t = # of minutes new
machine works, $t + 15$ = # of
minutes old machine works;
$250t + 175(t + 15) = 13,675$;

> 10:41 A.M.

15. Let x = amount invested at
6.35%, $x + 4,000$ = amount
invested at 7.28%;
$0.0635x + 0.0728(x + 4,000) =$
972.70; $5,000 at 6.35%

17. Let x = amount invested at 9%,
$800 - x$ = amount invested at
6%; $0.09x + 0.06(800 - x) =$
67.50; $650 at 9%, $150 at 6%

19. Let x = amount invested at 8%,
$6,000 - x$ = amount invested
at 12%; $0.08x + 0.12(6,000 - x)$
$= 0.09(6,000)$;

> $4,500 at 8%; $1,500 at 12%

21. Let n = # of ml of 30% solution
needed; $0.30n + 0.50(30) =$
$0.45(n + 30)$; 10 ml

23. Let s = # of liters of 2.4% salt
solution, $90 - s$ = # of liters of
4.6% salt solution;
$0.024s + 0.046(90 - s) =$
$0.03(90)$;

> 65.5 liters of 2.4% solution,
> 24.5 liters of 4.6% solution

25. Let n = # of gallons of pure
antifreeze; $n + 0.30(10) =$
$0.50(10 + n)$; 4 gallons

27. Let n = # of pounds of $3.75
candy; $3.75n + 5(35) =$
$4.25(n + 35)$; 52.5 pounds

29. Let t = # of hours until they
meet; $4t + 8t = 9$; 8:45 A.M.

31. Let t = # of hours Susan jogs
until they meet;
$t + \frac{1}{4}$ = # of hours John walks
until they meet;
$4\left(t + \frac{1}{4}\right) + 8t = 9$; 8:40 A.M.

33. Let t = time jogging;
$2 - t$ = time walking;
$9t + 5(2 - t) = 16$; $1\frac{1}{2}$ hours

35. Let r = rate for the additional
$2,800; $0.072(3,200) + r(2,800)$
$= 390$; 5.7%

37. Let p = percent solution for
additional 60 ml; $60p + 0.20(30)$
$= 0.40(90)$; 50%

39. Let w = width;
$2w - 1$ = length;
$w = \frac{1}{7}[2w + 2(2w - 1)] + 1$;

> width = 5, length = 9

41. $25 \leq \frac{5}{9}(F - 32) \leq 40$;

> $77° < F° < 104°$

43. Let p = original price;
$12.60 \leq p - 0.20p \leq 20.76$;

> original price range:
> $15.75 to $25.95

Chapter 4 Review Exercises

1. $-\frac{3}{7}$ 2. $-\frac{5}{7}$ 3. $\frac{5x^4}{2}$ 4. $\frac{2}{7a^6}$
5. $-\frac{5x^2}{2y^5}$ 6. $-\frac{8x^4}{5y^4}$ 7. $\frac{1}{t}$ 8. $\frac{w}{z}$ 9. $\frac{a^2}{16}$
10. $\frac{a}{2}$ 11. $\frac{a}{3}$ 12. $\frac{7}{5}$ 13. $\frac{3x - 2}{6x}$
14. $\frac{5x + 1}{5x}$ 15. $\frac{y^2}{2}$ 16. 1 17. $\frac{3}{4y^3}$
18. $\frac{x^2y}{48}$ 19. $\frac{a^2}{8}$ 20. $\frac{3a}{4}$ 21. $\frac{2x^2}{3}$
22. $\frac{x^2}{4}$ 23. $\frac{3x + 8}{2x^2}$ 24. $\frac{9y + 10}{6y^2}$
25. $\frac{18b^2 - 20ab^2 + 21a^2}{24a^2b^3}$
26. $\frac{9t + 14t^2 - 40r}{24rt^3}$ 27. $x = 5$
28. $x = 7$ 29. $t < 0$ 30. $a = \frac{1}{2}$
31. $y = 2$ 32. $z < 8$ 33. $x = 1$
34. $x = 6$ 35. $x = 4$ 36. $x = 20$
37. Let x = # oz in 1 kg; $\frac{x}{1,000} = \frac{1}{28.4}$;

> 35.21 oz

38. Let n = the number;
$\frac{3}{4}n - 1 = n - 4$; 12

39. Let x = amount invested at 6%, $2x$ = amount invested at 7%, $7,000 - 3x$ = amount invested at 8%; $0.06x + 0.07(2x) + 0.08(700 - 3x) \geq 500$; $\boxed{\$1,500}$

40. Let n = number of nickels, $2n$ = number of dimes, $30 - 3n$ = number of quarters; $5n + 10(2n) + 25(30 - 3n) = 350$;

$\boxed{\text{8 nickels, 16 dimes, 6 quarters}}$

41. Let r = Bill's present speed; $5r = 3(r + 20)$; $\boxed{\text{30 mph}}$

Chapter 4 Practice Test

1. (a) $-\frac{5}{12}$ (b) x^8 (c) $2a^3$ (d) $-\frac{5t^2}{3r^2}$

2. (a) $\frac{3x}{2y}$ (b) $\frac{3x^2}{4y}$ (c) $\frac{3a^2}{4}$ (d) $\frac{2y^2}{81x^2}$

(e) $\frac{2a}{5}$ (f) $\frac{a^2}{25}$ (g) $\frac{3}{x}$ (h) $\frac{11a}{24}$

(i) $\frac{17}{12x}$ (j) $\frac{70xy + 9}{30x^2y}$ (k) $-\frac{1}{x}$

3. (a) $x = 15$ (b) $x \geq 15$
(c) $a = 2$ (d) $t = 20$

4. Let x = # of miles in 50 km; $\frac{x}{50} = \frac{1}{1.61}$; $\boxed{\text{31.06 miles}}$

5. Let x = the number; $x + \frac{2}{3}x = 2x - 5$; $\boxed{x = 15}$

6. Let x = # of tickets sold at the door; $15x + 18.50(400 - x) = 6,770$; $\boxed{\text{180 tickets}}$

7. Let d = amount invested at 13%; $7,000 - d$ = amount invested at 8%; $0.13d + 0.08(7,000 - d) = 750$; $\boxed{\$3,800 \text{ at } 13\%, \$3,200 \text{ at } 8\%}$

8. Let x = # of oz of 20% solution; $0.20x + 0.65(24) = 0.50(x + 24)$; $\boxed{\text{12 oz}}$

9. Let t = # of hours second person travels until they are 604 km apart, $t + 4$ = # of hours first person travels until they are 604 km apart; $48(t + 4) + 55t = 604$; $\boxed{\text{7:00 P.M.}}$

EXERCISES 5.1

1–20.

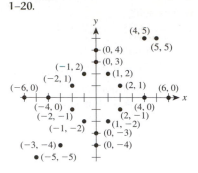

21. I **23.** III **25.** y-axis **27.** II
29. Origin **31.** IV **33.** $(1, 7)$
35. $(0, -1)$
37. All 3 points satisfy the equation.
39. $(-6, -10)$ **41.** $(1, -6)$
43. $\left(1, \frac{1}{3}\right)$ and $(-1, 1)$

45.

x	y
-1	1
-9	-5
$-\frac{7}{3}$	0
0	$\frac{7}{4}$

47.

x	y
8	$\frac{19}{3}$
$\frac{21}{2}$	8
$-\frac{15}{2}$	-4
-4	$-\frac{5}{3}$

Mini-Review

57. Let x = # of homework problems; $\frac{1}{3}x + 8 + \frac{2}{5}x = x$; $\boxed{\text{30 problems}}$

58. Let x = total # of questions; $0.46x + 5 + 0.44x = x$;
$\boxed{\text{23 multiple-choice questions}}$

EXERCISES 5.2

1. $(42, 15), (14, 3), \left(-2, -\frac{27}{7}\right)$
3. $(6, -9), (0, 18), (4, 0)$
5. $\left(-\frac{2}{3}, -\frac{12}{5}\right), \left(2, -\frac{4}{5}\right), (5, 1)$
7. $(-6, 8), (2, 4), \left(3, \frac{7}{2}\right)$
9. x-intercept: 7, y-intercept: -7
11. x-intercept: 4, y-intercept: -4
13. x-intercept: 6, y-intercept: 4
15. x-intercept: 0, y-intercept: 0
17. x-intercept: 0, y-intercept: 0
19. x-intercept: 5, no y-intercept
21. x-intercept: 4, y-intercept: $-\frac{4}{3}$

23. $x + y = -5$

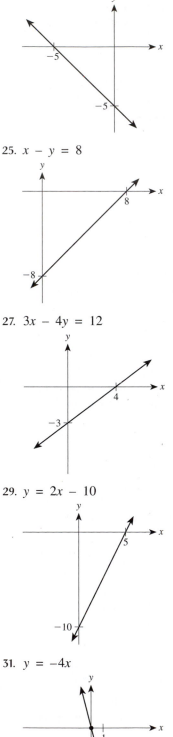

25. $x - y = 8$

27. $3x - 4y = 12$

29. $y = 2x - 10$

31. $y = -4x$

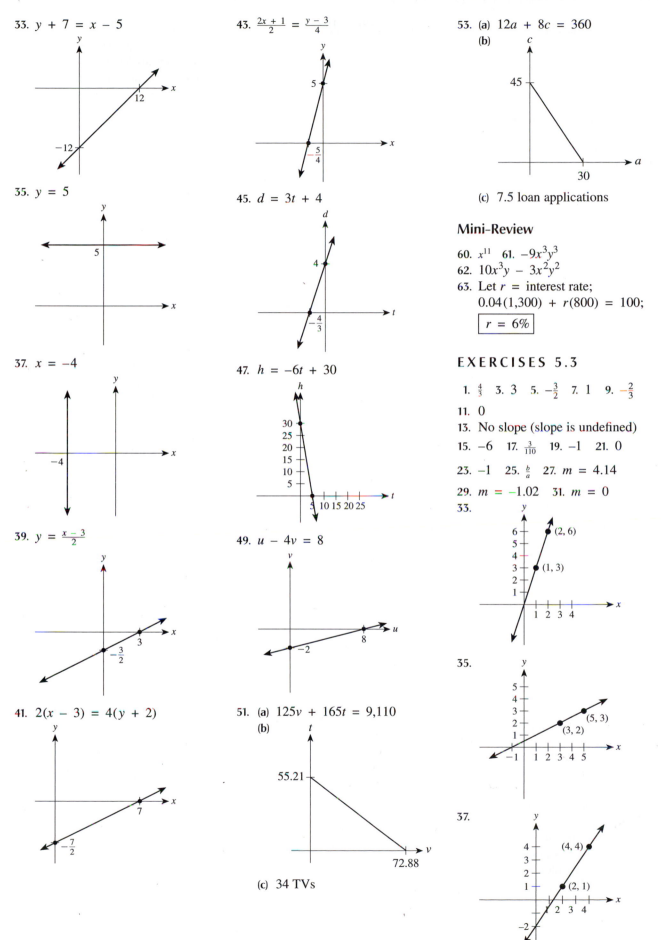

33. $y + 7 = x - 5$

35. $y = 5$

37. $x = -4$

39. $y = \frac{x - 3}{2}$

41. $2(x - 3) = 4(y + 2)$

43. $\frac{2x + 1}{2} = \frac{y - 3}{4}$

45. $d = 3t + 4$

47. $h = -6t + 30$

49. $u - 4v = 8$

51. (a) $125v + 165t = 9{,}110$
  (b)

  (c) 34 TVs

53. (a) $12a + 8c = 360$
  (b)

  (c) 7.5 loan applications

Mini-Review

60. x^{11} **61.** $-9x^3y^3$
62. $10x^3y - 3x^2y^2$
63. Let $r =$ interest rate;
  $0.04(1{,}300) + r(800) = 100$;
  $\boxed{r = 6\%}$

EXERCISES 5.3

1. $\frac{4}{3}$ **3.** 3   **5.** $-\frac{3}{2}$ **7.** 1   **9.** $-\frac{2}{3}$
11. 0
13. No slope (slope is undefined)
15. -6 **17.** $\frac{3}{110}$ **19.** -1 **21.** 0
23. -1 **25.** $\frac{b}{a}$ **27.** $m = 4.14$
29. $m = -1.02$ **31.** $m = 0$
33.

35.

37.

39.

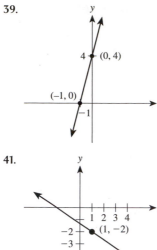

41.

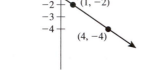

43.

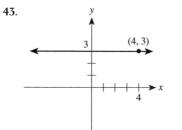

45.

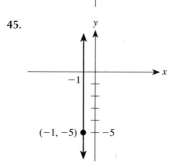

47. $m = -\frac{5}{4}$ 49. $m = 5$ 51. $m = \frac{7}{3}$
53. $m = 0$ 55. Positive slope
57. Zero slope 59. $m_3 < m_2 < m_1$
61. (a) 40 calories
    (b) $m = 5$
    (c) 5 calories per minute are burned
63. (a) 16 gallons
    (b) 180 miles
    (c) 22.5 mpg; the results are the same as they should be
    (d) Slope is 22.5
    (e) Slope of the line gives the car's gas consumption in miles per gallon

65. (a) $A = 80 + 15d$
    (b)

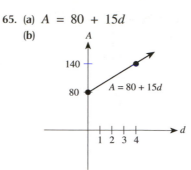

    (c) The slope of the line is 15, which is the rate Jane gets paid per delivery.

Mini-Review

68. $x = -\frac{3}{5}$ 69. $\frac{7x + 9}{12}$
70. Let t = the number of hours Tamika travels until they are 257.5 km apart;
    $60t + 70\left(t - \frac{1}{2}\right) = 257.5$;

    $\boxed{\text{12:15 P.M.}}$

EXERCISES 5.4

1. $y - 5 = 3(x - 1)$
3. $y - 4 = \frac{1}{2}(x + 1)$
5. $y + 5 = -\frac{2}{3}(x + 3)$
7. $y = 5x + 6$ 9. $y = \frac{1}{4}x - 2$
11. $y = -\frac{3}{4}(x + 2)$
13. $y + 2 = x - 4$
15. $y = 6$ 17. $y - 3 = 2(x - 2)$
19. $y - 4 = -2(x + 1)$
21. $y = -\frac{3}{5}x + 5$ 23. $y = -x - 1$
25. $y = 3$ 27. $x = 4$
29. $y = -\frac{2}{5}x + 2$ 31. $y = \frac{4}{3}x + 4$
33. $y = 7x + 61$; 96
35. $y = -\frac{3}{2}x + 17$; 11 min
37. $m = 5$ 39. $m = -3$
41. $m = -1$ 43. $m = -\frac{3}{2}$
45. $m = \frac{2}{5}$ 47. $y = \frac{2}{3}x$
49. $y = -\frac{3}{5}x + 3$ 51. $y = \frac{1}{3}x + 1$
53. $y = -4$ 55. $m = 3$
57. $y = -x - 3$
59. (a) $c = \frac{7}{20}f + 45$
    (b) $76.50
    (c) 130 franks
    (d) $45

61. (a) $N = \frac{2}{3}s + 70$
    (b) 83 beats per minute
    (c) 70 beats per minute
63. (a) $c = 1.5n + 0.20$ for $n \geq 1$
    (b) $5.45
    (c) 4.2 oz
    (d)

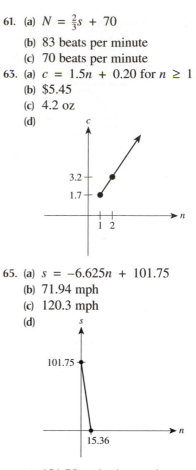

65. (a) $s = -6.625n + 101.75$
    (b) 71.94 mph
    (c) 120.3 mph
    (d)

    (e) 101.75 mph; the maximum speed on level ground
    (f) Approximately 15.4°; car can't climb a grade this steep
67. (a) $w = -0.09t + 3.66$
    (b) 1.14 cm
    (c) 40.7°C
    (d) 40.7°C = 105.3°F; seems possible but unlikely

Mini-Review

72. Let L = length, then $\frac{2}{3}L - 4$ = the width; $2L + 2\left(\frac{2}{3}L - 4\right) = $

    $3L;$ $\boxed{\text{length} = 24; \text{width} = 12}$

73. Let r = the alcohol percentage of the additional 30 ml.
    $30r + 0.60(4) = 0.45(70)$;

    $\boxed{25\%}$

Chapter 5 Review Exercises

1–6.

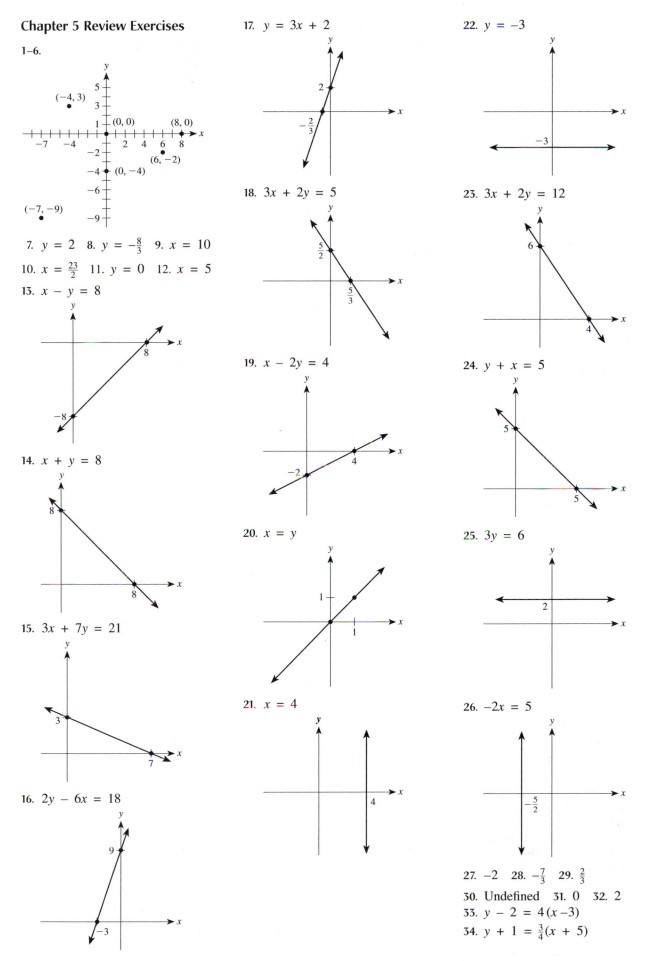

7. $y = 2$ 8. $y = -\frac{8}{3}$ 9. $x = 10$

10. $x = \frac{23}{2}$ 11. $y = 0$ 12. $x = 5$

13. $x - y = 8$

14. $x + y = 8$

15. $3x + 7y = 21$

16. $2y - 6x = 18$

17. $y = 3x + 2$

18. $3x + 2y = 5$

19. $x - 2y = 4$

20. $x = y$

21. $x = 4$

22. $y = -3$

23. $3x + 2y = 12$

24. $y + x = 5$

25. $3y = 6$

26. $-2x = 5$

27. -2 28. $-\frac{7}{3}$ 29. $\frac{2}{3}$

30. Undefined 31. 0 32. 2

33. $y - 2 = 4(x - 3)$

34. $y + 1 = \frac{3}{4}(x + 5)$

35. $y = -6(x - 1)$

36. $y = \frac{1}{7}x + 3$ 37. $y = 8$

38. $x = 3$ 39. $m = -\frac{3}{4}$ 40. $m = \frac{3}{8}$

41. $y - 5 = 5(x + 3)$

42. $y = -\frac{5}{2}(x - 3)$ 43. $y = \frac{4}{3}x$

44. $y = -\frac{2}{5}x + 2$

45. Let $h = $ # of overtime hours,
    $n = $ # of defective items;
    $n = \frac{5}{2}h - 8$;

    $\boxed{\text{42 defective items}}$

46. Let $h = $ # of hours studying,
    $g = $ grade on exam;
    $g = \frac{19}{3}h + \frac{178}{3}$; $\boxed{84.7}$

Chapter 5 Practice Test

1. No

2. (a) $y = 3$ (b) $x = 4$ (c) $y = 3$

3. (a) $3x - 5y = 15$

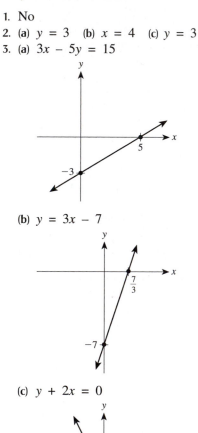

   (b) $y = 3x - 7$

   (c) $y + 2x = 0$

(d) $x = 4$

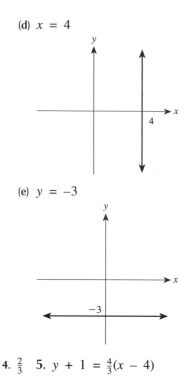

(e) $y = -3$

4. $\frac{2}{3}$ 5. $y + 1 = \frac{4}{3}(x - 4)$

6. $y = -\frac{5}{2}(x - 4)$

7. Horizontal: $y = 5$,
   vertical: $x = 3$

8. $m = \frac{5}{7}$ 9. $y = -3(x + 4)$

10. $y = -\frac{3}{7}x - 3$

11. Let $C = $ # of calls, $A = $ amount
    in pledges
    $A = \frac{35}{6}C - \frac{15}{3}$; $\boxed{\$167.50}$

EXERCISES 6.1

1. (a) B   (b) B   (c) A   (d) B

3. (a) 2,000   (b) 4:00 A.M.
   (c) 10,000
   (d) From 4:00 P.M. until 4:00 A.M.
   (e) From 10:00 A.M. until 4:00 P.M.
   (f) From 10:00 A.M. until 4:00 P.M.

5. (a) It contradicts this belief.
   (b) From sea level to 12 miles
       and from 50 to 80 miles
   (c) It increases from 45 to 50
       miles and decreases from
       50 to 55 miles.

7. (a) The group has successfully
       memorized all 20 words.
   (b) More after 4 hours
   (c) During the first two hours
   (d) 16 words; 2 words

9. (a) 150   (b) 1996–1997
   (c) 800

11. (a) A: \$10,000; B: \$7,500
    (b) A: 8 years; B: 10 years

(c) Yes; after 5 years

(d) For the first 5 years the value
    of machine A is greater than
    that of machine B. For the
    next 5-year period the
    opposite is true.

13. (a) Thursday; \$45/share
    (b) Tuesday
    (c) Thursday; \$25/share
    (d) \$20/share

15. The following is one possible
    graph that describes the situation;
    $m = $ # of minutes,
    $T = $ temperature

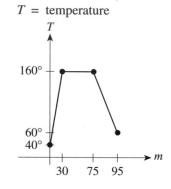

Mini-Review

17. $\frac{2x}{125y^2z^4}$ 18. $t \geq -18$

19. Let $n = $ the number of pairs
    of dress slacks;
    $40x + 25(36 - x) = 1{,}185$;

    $\boxed{\begin{array}{l}\text{19 pairs of dress slacks;} \\ \text{17 pairs of casual slacks}\end{array}}$

EXERCISES 6.2

1. (a)

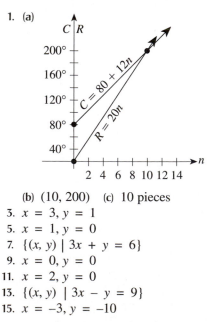

   (b) (10, 200)   (c) 10 pieces

3. $x = 3, y = 1$

5. $x = 1, y = 0$

7. $\{(x, y) \mid 3x + y = 6\}$

9. $x = 0, y = 0$

11. $x = 2, y = 0$

13. $\{(x, y) \mid 3x - y = 9\}$

15. $x = -3, y = -10$

17. $x = -5, y = -4$
19. No solutions; the lines are parallel
21. $x = 14, y = 36$
23. $x = -\frac{1}{2}, y = \frac{9}{2}$
25. $x = 8, y = 5$
27. $\{(x, y) \mid 2x + y = 10\}$
29. $x = 4, y = 4$
31. $x = 8, y = -2$
33. $x = -4, y = -10$
35. $x = -1, y = 4$

Mini-Review

40. $x \geq -5$

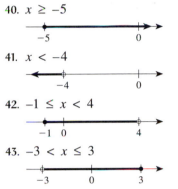

41. $x < -4$

42. $-1 \leq x < 4$

43. $-3 < x \leq 3$

44. Let x = price of tickets at the door, $x - 1.50$ = price of advanced-sale tickets; $200(x - 1.50) + 75x = 1,075$;
  $\boxed{\$5}$

45. Let $8x$ = number of members in favor, $7x$ = number of members against; $8x + 7x = 435$; $\boxed{232}$

EXERCISES 6.3

1. $x = 2, y = -1$
3. $x = 3, y = -1$ 5. $x = 1, y = 4$
7. $x = 6, y = 3$ 9. $x = -5, y = 2$
11. $s = \frac{3}{2}, t = \frac{2}{5}$ 13. $c = \frac{1}{3}, d = -\frac{3}{4}$
15. $x = 1, y = 4$ 17. $x = 4, y = 0$
19. $x = 2, y = -1$
21. $r = 2, s = -1$
23. $x = 1, y = 1$
25. $u = \frac{1}{4}, w = -\frac{1}{2}$
27. $w = \frac{1}{2}, t = -2$
29. $r = -8, t = 9$ 31. $x = \frac{1}{2}, y = 3$
33. No solution 35. $a = 2, b = -4$
37. $x = 0, y = \frac{3}{4}$
39. $x = 1, y = -2$
41. $a = \frac{9}{4}, b = \frac{5}{4}$
43. $w = 40, t = 45$
45. $r = 20, t = -16$

47. $x = 4, y = 0$
49. $x = 3, y = -4$
51. $x = 8, y = -9$ 53. No solutions
55. $x = 10, y = 20$
57. $x = 5, y = -4$
59. $x = 5.7, y = 1.1$
61. $x = 1.9, y = -0.6$

EXERCISES 6.4

The system of equations used to solve each odd-numbered exercise is included.

1. $x + y = 130, x - y = 28$;
  $\boxed{x = 79, y = 51}$
3. $n + q = 80, 5n + 25q = 1,360$;
  $\boxed{32 \text{ nickels, } 48 \text{ quarters}}$
5. $r_2 = r_1 + 15, 5r_1 + 5r_2 = 275$;
  $\boxed{r_1 = 20 \text{ kph}, r_2 = 35 \text{ kph}}$
7. $x + y = 1,700,$
  $0.07x + 0.06y = 110$;
  $\boxed{\$800 \text{ at } 7\%; \$900 \text{ at } 6\%}$
9. Let c = price of a cassette, d = price of a CD;
  $4c + 6d = 107.66,$
  $5c + 3d = 76.30$;
  $\boxed{c = \$7.49, d = \$12.95}$
11. $L = 2W, 2W + 2L = 28$;
  $\boxed{W = \frac{14}{3}, L = \frac{28}{3}}$
13. Let r = cost of a regular selection, s = cost of a special selection; $2r + 3s = 56.90,$
  $3r + 4s = 80.85$;
  $\boxed{r = \$14.95, s = \$9}$
15. $\frac{x}{y} = \frac{6}{5}, x - y = 8$;
  $\boxed{x = 48, y = 40}$
17. x = rate of slower plane, y = rate of faster plane;
  $y = 2x, 4x + 4y = 1,800$;
  $\boxed{x = 150 \text{ mph}, y = 300 \text{ mph}}$
19. r = price of receiver, t = price of turntable;
  $8r + 4t = 2,060,$
  $5r + 6t = 1,690$;
  $\boxed{r = \$200, t = \$115}$
21. p = cost of plain donut, f = cost of filled donut;
  $10p + 5f = 370,$
  $5p + 20f = 410$;
  $\boxed{p = 22 \text{ cents}}$

23. x = # of \$7 books, y = # of \$9 books;
  $x + y = 35, 7x + 9y = 271$;
  $\boxed{22 \text{ \$7 books, } 13 \text{ \$9 books}}$
25. x = speed of slower car, y = speed of faster car;
  $y = x + 40, 4x + 4y = 480$;
  $\boxed{x = 40 \text{ kph}, y = 80 \text{ kph}}$
27. p = speed of plane, w = speed of wind;
  $p + w = 150, p - w = 90$;
  $\boxed{p = 120 \text{ mph}, w = 30 \text{ mph}}$
29. x = # of pounds of \$3.35 candy, y = # of pounds of \$2.75 candy;
  $x + y = 60,$
  $3.40x + 2.75y = 60(3.10)$;
  $\boxed{x = 35, y = 25}$
31. Let d = # of desktop setups and C = total cost of the system;
  $C = 100,000 + 800d;$
  $C = 16,000 + 1,200d;$
  $\boxed{210 \text{ desktop setups}}$
33. Let c = cost of one computer and p = cost of one printer;
  $10c + 10p = 10,000,$
  $12c + 12p = 10,000$;
  $\boxed{\text{computer: \$800; printer: \$200}}$
35. Let W = width and L = length;
  $2W + 2L = 46; 2W = L + 1$;
  $\boxed{\text{width is 8 cm, length is 15 cm}}$
37. Let r = # of red marbles and b = # of blue marbles;
  $r + b = 36; 2r = b - 6$;
  $\boxed{10 \text{ red marbles and } 26 \text{ blue marbles}}$

Chapter 6 Review Exercises

1. $x = 2, y = 2$ 2. $x = 2, y = 0$
3. $x = 2, y = -3$
4. No solutions; the lines are parallel
5. $x = 6, y = 6$
6. $x = -12, y = -6$
7. $x = 5, y = -1$ 8. $x = 4, y = 6$
9. $x = \frac{11}{5}, y = \frac{7}{5}$
10. $x = \frac{14}{11}, y = \frac{-2}{11}$
11. $x = 2, y = -6$
12. $x = 6, y = -1$
13. $x = 1, y = 0$
14. $x = -2, y = 2$

15. $x = 1, y = -2$
16. $\{(x, y) \mid 2x - 4y = 3\}$
17. No solutions 18. $x = 4, y = 8$
19. $\{(x, y) \mid y - x = 4\}$
20. $x = 3, y = 0$
21. Let x = amount of pure water to be added, y = amount of 30% solution to be used;
$x + y = 30, 0.30y = 0.25(30)$;

| 5 gallons of water |

22. Let a = number of adult tickets sold, c = number of children's tickets sold; $a + c = 4{,}300$, $4.50a + 4c = 18{,}800$;

| $a = 3{,}200, c = 1{,}100$ |

23. Let w = rate walking, j = rate jogging;
$w + j = 17, 2w + \frac{1}{2}j = 16$;

| $w = 5$ kph, $j = 12$ kph |

24. $0.60x + 0.40y = 32$;
$0.50x + 0.30y = 25$

| $x = 20, y = 50$ |

25. (a) $R = 7$ (b) Decreasing
(c) $R = 0$ when $w = 9$
(d) $R = 9$ when $w = 7$
(e) The value of R when $w = 5$
26. (a) 65°F at 11:00
(b) 45°F at 7:00 (c) Yes; 55°F
(d) Decreasing; 5°

Chapter 6 Practice Test

1. (a) $32   (b) 10:00 to 11:00 and 12:00 to 1:00
(c) $30 (d) Yes; 11:00 to 12:00
(e) Rising
2. $x = 2, y = -1$
3. $x = 0, y = -3$
4. $x = 2, y = 0$
5. No solutions
6. $\{(x, y) \mid 4x - 6y = -8\}$
7. $x = -2, y = 4$
8. Let x = the price of an orchestra ticket, y = the price of a balcony ticket; $5x + 3y = 227$;
$6x + 2y = 238$

| $x = \$32.50, y = \$21.50$ |

9. Let x = amount invested at 4%, y = the amount invested at 5%;
$x + y = 5{,}550$;
$0.04x + 0.05y = 259$;

| $1,850 at 4%, $3,700 at 5% |

10. Let W = width and L = length;
$2W + 2L = 46$; $3W = 2L - 1$;

| width is 9 ft, length is 14 ft |

CHAPTERS 4–6 CUMULATIVE REVIEW

1. $-\frac{4}{7}$ (§4.1)   2. $\frac{3}{5x^4}$ (§4.1)
3. $\frac{9t}{5s}$ (§4.1)   4. $-\frac{1}{2a}$ (§4.1)
5. $\frac{12}{5}$ (§4.2)   6. $\frac{3x^2}{125}$ (§4.2)
7. $\frac{6x^2 + 250}{25x}$ (§4.3)   8. $\frac{2}{x}$ (§4.3)
9. $\frac{2}{t}$ (§4.3)   10. $\frac{1}{5u^2}$ (§4.3)
11. $\frac{6x^2y}{25z}$ (§4.2)   12. $\frac{2u}{3v^3w^4}$ (§4.2)
13. $-\frac{11}{6x}$ (§4.3)   14. $\frac{18z + 11y}{6yz}$ (§4.3)
15. $\frac{25y^2 - 27x}{30x^2y^3}$ (§4.3)   16. $\frac{3}{4x^3y^4}$ (§4.2)
17. $2x$ (§4.2)   18. $\frac{a}{3}$ (§4.3)
19. $\frac{2x^2 + 3x - 1}{x^2}$ (§4.3)
20. $\frac{120st - 4t^2 + 9s}{24s^2t^2}$ (§4.3)
21. $x = 8$ (§4.4)   22. $t = -3$ (§4.4)
23. Identity (§4.4)   24. $z = 0$ (§4.4)
25. $y = \frac{1}{2}$ (§4.4)   26. $x = -\frac{15}{4}$ (§4.4)
27. $x = 80$ (§4.4)
28. $x = 17$ (§4.4)
29. $y = 2x - 6$ (§5.2)

30. $3x - 5y = 15$ (§5.2)

31. $3y - 6x = 12$ (§5.2)

32. $4x + 3y = 10$ (§5.2)

33. $x - 2 = 0$ (§5.2)

34. $y + 3 = 0$ (§5.2)

35. $y = 5x$ (§5.2)

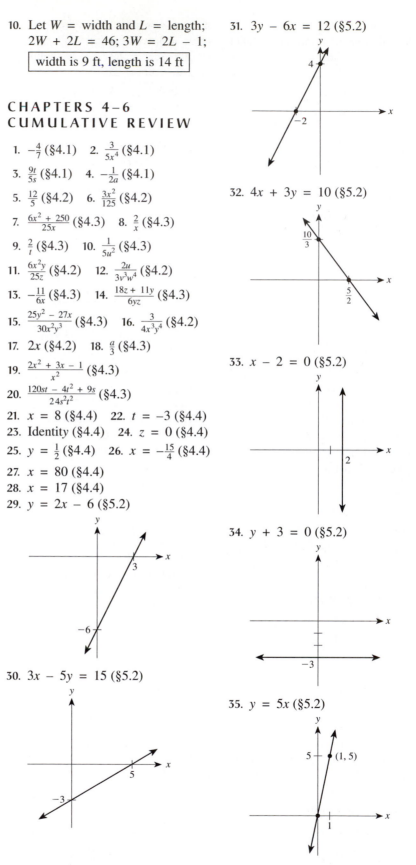

36. $3y = x$ (§5.2)

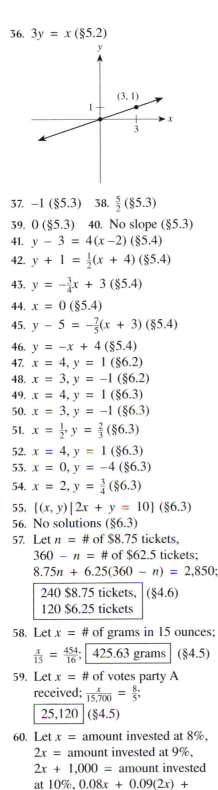

37. -1 (§5.3)   38. $\frac{5}{2}$ (§5.3)

39. 0 (§5.3)   40. No slope (§5.3)

41. $y - 3 = 4(x - 2)$ (§5.4)

42. $y + 1 = \frac{1}{2}(x + 4)$ (§5.4)

43. $y = -\frac{3}{4}x + 3$ (§5.4)

44. $x = 0$ (§5.4)

45. $y - 5 = -\frac{7}{5}(x + 3)$ (§5.4)

46. $y = -x + 4$ (§5.4)

47. $x = 4, y = 1$ (§6.2)

48. $x = 3, y = -1$ (§6.2)

49. $x = 4, y = 1$ (§6.3)

50. $x = 3, y = -1$ (§6.3)

51. $x = \frac{1}{2}, y = \frac{2}{3}$ (§6.3)

52. $x = 4, y = 1$ (§6.3)

53. $x = 0, y = -4$ (§6.3)

54. $x = 2, y = \frac{3}{4}$ (§6.3)

55. $\{(x, y) \mid 2x + y = 10\}$ (§6.3)

56. No solutions (§6.3)

57. Let n = # of \$8.75 tickets,
$360 - n$ = # of \$62.5 tickets;
$8.75n + 6.25(360 - n) = 2{,}850$;

$\boxed{\begin{array}{l} 240 \ \$8.75 \ \text{tickets,} \\ 120 \ \$6.25 \ \text{tickets} \end{array}}$ (§4.6)

58. Let x = # of grams in 15 ounces;
$\frac{x}{15} = \frac{454}{16}$; $\boxed{425.63 \ \text{grams}}$ (§4.5)

59. Let x = # of votes party A
received; $\frac{x}{15{,}700} = \frac{8}{5}$;
$\boxed{25{,}120}$ (§4.5)

60. Let x = amount invested at 8%,
$2x$ = amount invested at 9%,
$2x + 1{,}000$ = amount invested
at 10%, $0.08x + 0.09(2x) +$
$0.10(2x + 1000) = 3{,}090$;
$\boxed{\$33{,}500}$ (§4.6)

61. Let t = # of hours to overtake;
$80t = 65(t + \frac{1}{4})$;
$\boxed{1 \ \text{hour and 5 minutes}}$ (§4.6)

62. Let s = price of a shirt,
t = price of a tie;
$6s + 2t = 88.68$,
$4s + 3t = 68.27$;

$\boxed{\begin{array}{l} \text{shirt costs \$12.95;} \\ \text{tie costs \$5.49} \end{array}}$ (§6.4)

63. (a) 18
(b) No. It decreased between 1960 and 1980.
(c) 1900–1910; 6 people per square mile
(d) 1910–1920 (§6.1)

CHAPTERS 4–6 CUMULATIVE PRACTICE TEST

1. $\frac{4d^3s}{t}$ (§4.2)   2. $\frac{4x^2y^4}{81z^5}$ (§4.2)

3. $\frac{5a}{3x}$ (§4.3)   4. $\frac{-3x}{y}$ (§4.3)

5. $\frac{15 + 8ab}{18ab^2}$ (§4.3)   6. $\frac{9x}{10}$ (§4.2)

7. $a = 324$ (§4.4)

8. $t = -4$ (§4.4)

9. $x - 3y = 0$ (§5.2)

10. $x - 3y = 6$ (§5.2)

11. $x - 3 = 6$ (§5.2)

12. $\frac{4}{5}$ (§5.3)

13. $y + 4 = -\frac{7}{3}(x - 2)$ (§5.4)

14. $x = -1, y = -5$ (§6.3)

15. No solutions (§6.3)

16. Let L = length, $\frac{1}{3}L + 5$ =
width; $2L + 2\left(\frac{1}{3}L + 5\right) = 34$;
$\boxed{9 \ \text{by} \ 8}$ (§4.6)

17. Let n = # of 34-cent stamps,
$28 - n$ = # of 50-cent stamps;
$34n + 50(28 - n) = 1{,}096$;

$\boxed{\begin{array}{l} \text{nineteen 34-cent stamps,} \\ \text{nine 50-cent stamps} \end{array}}$ (§4.6)

18. Let t = # of hours they work
together, $t + 4$ = # of hours the
associate works; $40t + 24(t + 4)$
$= 480$; $\boxed{6 \ \text{hours}}$ (§4.6)

19. Let t = # of hours Tom walks
until they are 9 km apart,
$t + \frac{1}{3}$ = # of hours Terry walks
until they are 9 km apart;
$6\left(t + \frac{1}{3}\right) + 8t = 9$;

$\boxed{11{:}50 \ \text{A.M.}}$ (§4.6)

20. Let r = regular price,
s = special price,
$4r + 3s = 55.30$,
$2r + 5s = 50.40$;
$\boxed{r = \$8.95; \ s = \$6.50}$ (§6.4)

21. Graph C

EXERCISES 7.1

1. x^5 3. x^6 5. x^9 7. 10^9

9. 648   11. y^2 13. $3u^6v^4$ 15. a^7

17. x^8 19. x^4y^2 21. $x^{10}y^{15}$

23. $16r^{12}s^{20}$ 25. $-x^9y^3$ 27. $\frac{x^{12}}{y^8}$

29. $144x^{16}$ 31. x^7 33. xy^5

35. $\frac{x^8y^4}{256u^4}$ 37. $\frac{32x^{10}}{y^{10}}$ 39. $-\frac{27a^6b^9}{8c^3}$ 41. -1

43. -1 45. 1   47. $-x^2$ 49. -7

51. -2 53. 3   55. -1 57. $-\frac{1}{7}$

59. 196   61. $-\frac{12}{7}$

Mini-Review

67. $\frac{16}{xy}$ 68. $\frac{y}{4x}$ 69. $\frac{2y + 8x}{xy}$ 70. $\frac{3a - 8}{2a^2}$

71. $\frac{x}{12}$ 72. 36

73. Let m = # of miles driven,
$0.20m + 45 = 72$;

$\boxed{135 \ \text{miles}}$

74. Let q = # of quarters;
$40 - q$ = # of dimes;
$25q + 10(40 - q) = 670$;

$\boxed{\text{18 quarters, 22 dimes}}$

EXERCISES 7.2

1. (a) -6 (b) $x^{-1} = \frac{1}{x}$
   (c) $x^{-6} = \frac{1}{x^6}$ (d) $\frac{1}{8}$

3. (a) -12 (b) $x^{-1} = \frac{1}{x}$
   (c) $x^{-12} = \frac{1}{x^{12}}$ (d) $\frac{1}{81}$

5. (a) -1 (b) $x^{-1} = \frac{1}{x}$
   (c) -12 (d) $x^{-12} = \frac{1}{x^{12}}$
   (e) 7 (f) x^7 (g) $\frac{1}{81}$

7. 1 9. 5 11. x 13. $\frac{1}{25}$ 15. 25

17. 1 19. $\frac{1}{x^{10}}$ 21. $\frac{1}{a^8}$

23. 10^5 or $100,000$ 25. $\frac{1}{1,000}$ or 0.001

27. 1 29. x^4y^4 31. $\frac{2}{a^3}$ 33. $-\frac{3}{y^2}$

35. $-\frac{3}{y^2}$ 37. $\frac{x}{y}$ 39. $\frac{1}{a^8b^6}$ 41. $\frac{a^8}{b^6}$

43. $\frac{9y^6}{x^4z^8}$ 45. $\frac{4x^3}{y^3}$ 47. x^3 49. $-\frac{1}{3a^{12}}$

51. $\frac{1}{x^2} + \frac{1}{y}$ or $\frac{y + x^2}{x^2y}$ 53. x 55. $\frac{x^6}{y^5}$

57. 10^6 59. $3(10^4)$ or $30,000$

61. a^{15} 63. $\frac{y^5}{x^8}$ 65. $\frac{n^4}{4m^8}$ 67. $\frac{3}{xy}$

69. $\frac{n^{10}}{16m^4}$ 71. $\frac{1}{x^{20}y^4}$ 73. $\frac{1}{8}$ 75. 4

77. $\frac{1}{6}$

Mini-Review

81. Let t = # of hours Maria works;
$t - 2$ = # of hours Francis
works; $5t + 7(t - 2) = 70$;

$\boxed{\text{4:00 P.M.}}$

82. Let L = length of the pole;
$\frac{1}{4}L + 8 + \frac{2}{3}L = L$; $\boxed{96 \text{ ft}}$

EXERCISES 7.3

1. 4.53×10^3 3. 4.53×10^{-2}
5. 7×10^{-5} 7. 7×10^6
9. 8.537×10^4 11. 8.537×10^{-3}
13. 9×10 15. 9 17. 9×10^{-1}
19. 9×10^{-2} 21. 3×10^{-8}
23. 2.8×10 25. 4.75×10
27. 9.7273×10^3 29. 5.6×10^{-1}
31. 1.54×10^3 33. 2.84×10^7
35. $28,000$ 37. 0.00028

39. $42,900,000$ 41. 0.000000429
43. 0.00352 45. $352,860$
47. 0.000026 49. 1 51. 5×10^{-7}
53. 0.004 55. 0.03 57. (c)   59. (e)
61. (c)   63. (c)   65. (d)
67. 5.98×10^{24} kg
69. 7.44×10^{-18} g   71. 10^{-8}
73. 1.53×10^{-6} cm
75. 4.25×10^9 miles
77. 6.73×10^{21} tons
79. 5.87×10^{12} miles
81. 4.69×10^{22} km

EXERCISES 7.4

1. (a) 1 (b) 5 (c) 5
3. (a) 2 (b) $1, 0$ (c) 1
5. (a) 2 (b) $2, 3$ (c) 3
7. (a) 1 (b) 5 (c) 5
9. (a) 1 (b) 0 (c) 0
11. (a) 3 (b) $3, 2, 1$ (c) 3
13. (a) 2 (b) $3, 5$ (c) 5
15. (a) 1 (b) 8 (c) 8
17. (a) 4 (b) $5, 7, 5, 6$ (c) 7
19. (a) $2, 1, 0$ (b) 2 (c) $1, -5, 6$
21. (a) $2, 0$ (b) 2 (c) $1, 0, 4$
23. (a) $3, 0$ (b) 3 (c) $1, 0, 0, -1$
25. (a) $0, 5$ (b) 5 (c) $-1, 0, 0, 0, 0, 1$
27. $5x^2 - 10$
29. $4u^3 - u^2 + 5u + 7$
31. $2u^3 + u^2 - 9u + 7$
33. $3t^3 + 2t^2$
35. $2x^2y + 3xy - 6x^2y^2 - xy^2$
37. $3xy + 4x^2y^2 + xy^2$
39. $8y^2 - 11y - 1$
41. $-x^2 + 16$ 43. $3x^2 + 7x - 9$
45. $x^2 + 4x + 2$ 47. $a^2b + a^2 + b$
49. $-6x + 1$ 51. 33 53. 61 55. 8
57. -36 59. $-\frac{187}{60}$ 61. 22
63. $2x - 20$
65. $\$2,100; \$2,500; \$900$
67. 76 ft; 42 ft; 0 ft
69. 69 ft; 101 ft; 105 ft
71. 51 people; 84 people; 99 people

Mini-Review

76. -8 77. -3 78. 4 79. Identity
80. Let x = # of ml of 20% iodine
solution; $0.20x + 0.60(80) =$
$0.30(x + 80)$; $\boxed{240 \text{ ml}}$
81. Let r = rate for the $\$1,400$
investment; $1400r + 0.08(1800)$
$= 284$; $\boxed{10\%}$

EXERCISES 7.5

1. $60x^6$ 3. $15x^4 + 12x^3$
5. $-60x^2y^2z^2$
7. $12xy^2z - 20x^2yz$
9. $3x^3 + 13x^2y - 12xy^2$
11. $3x^2y^3 - 3xy^3$
13. $x^3 + x^2 + x + 6$
15. $y^3 - 3y^2 - 16y + 30$
17. $3x^3 + 7x^2 - 21x + 10$
19. $15z^3 + 16z^2 + 44z + 16$
21. $x^3 + y^3$
23. $x^4 + 2x^3 + x^2 - 1$
25. $x^6 - 2x^4y - 3x^2y^2 + 4xy^3 - y^4$
27. $x^2 + 8x + 15$
29. $x^2 - 8x + 15$
31. $x^2 - 2x - 15$
33. $x^2 + 2x - 15$
35. $x^2 + 5xy + 6y^2$
37. $a^2 + 3ab - 40b^2$
39. $12x^2 - 19x + 4$
41. $2r^2 + 5rs - 3s^2$
43. $x^4 + 5x^2 + 6$
45. $x^2 + 14x + 49$
47. $x^2 - 49$ 49. $x^2 - 8x + 16$
51. $2x^2 + 4x - 30$
53. $6a^2 - 36a - 96$
55. $x^3 + 6x^2 + 12x + 8$
57. $9x^2 - 30x + 25$
59. $4a^2 - 36ab + 81b^2$
61. $2x^4 - 8x^3 - 64x^2$
63. $45x^3 - 84x^2 + 36x$
65. $2x^2 - 7x$ 67. $-4a + 14$
69. $-24x$
71. $8x^3 - 36x^2 + 54x - 27$
73. $27a^3 + 108a^2b + 144ab^2 + 64b^3$
75. $y^3 - 5y^2 + 9y - 6$
77. $x^3 + 5x^2 + 7x + 4$
79. $P = 6w + 6, A = 2w^2 + 3w$
81. $9s + 20.25$ 83. $15a - 24$
85. $80 - x^2$ 87. $-4x + 24,000$

Mini-Review

92. $x \leq -6$ 93. $x > -9$
94. $-2 < x \leq 4$ 95. $8 < x \leq 12$
96. Let g = # of gallons in the tank;
$\frac{1}{5}g + 5 + \frac{3}{4}g = g$; $\boxed{75 \text{ gallons}}$
97. Let x = # of ounces of 55%
solution, $40 - x$ = # of ounces
of 35% solution;
$0.55x + 0.35(40 - x) = 0.51(40)$;

$\boxed{32 \text{ oz of 55\%; 8 oz of 35\%}}$

EXERCISES 7.6

1. $x^2 + 7x + 12$
3. $x^2 + 7x + 12$ 5. $x^2 + x - 12$
7. $5x - 12$ 9. $x^2 - x - 12$
11. $x^2 + 8x + 12$
13. $x^2 - 8x + 12$
15. $x^2 + 4x - 12$
17. $x^2 - 4x - 12$
19. $x^2 + 13x + 12$
21. $x^2 - 13x + 12$
23. $x^2 + 11x - 12$
25. $x^2 - 11x - 12$
27. $x^2 - 18x - 40$
29. $a^2 + 16a + 64$ (perf. sq.)
31. $a^2 - 16a + 64$ (perf. sq.)
33. $a^2 - 64$ (diff. of 2 sq.)
35. $c^2 - 8c + 16$ (perf. sq.)
37. $c^2 + 8c + 16$ (perf. sq.)
39. $c^2 - 16$ (diff. of 2 sq.)
41. $3x^2 + 25x + 28$
43. $3x^2 + 19x + 28$
45. $2y^2 - 13y + 15$
47. $3x^2 - 17x - 28$
49. $3x^2 + 17x - 28$
51. $15x^2 + 41x + 28$
53. $15x^2 + 47x + 28$
55. $15x^2 - x - 28$
57. $15x^2 + x - 28$
59. $4a^2 + 20a + 25$ (perf. sq.)
61. $4a^2 - 25$ (diff. of 2 sq.)
63. $4a^2 - 20a + 25$ (perf. sq.)
65. $x^3 + 2x^2 - 31x + 28$
67. $9x^2y^2$
69. $x^6 + 2x^3y^2 + y^4$ (perf. sq.)
71. $4a^2 + 20ay + 25y^2$ (perf. sq.)
73. $4a^2 - 25y^2$ (diff. of 2 sq.)

Mini-Review

78. $\frac{16x}{5z^2}$ 79. $\frac{10}{81z^2}$ 80. $\frac{31a}{30}$ 81. 0
82. Let $x =$ amount invested at 10%,
   $2x =$ amount invested at 12%;
   $0.10x + 0.12(2x) = 408$;

   $\boxed{\$1,200 \text{ at } 10\%, \$2,400 \text{ at } 12\%}$

83. Let $r =$ rate of \$1,100 investment;
   $1,100r + 0.09(1,400) = 203$;

   $\boxed{7\%}$

Chapter 7 Review Exercises

1. $\frac{1}{81}$ 2. 10 3. $\frac{49}{144}$ 4. $\frac{12}{7}$ 5. $\frac{y^2}{x^5}$
6. y^6 7. $\frac{9x^6}{y^4}$ 8. $\frac{1}{x^5}$ 9. x^6 10. $\frac{x^2}{y^2}$
11. $\frac{1}{4x^8}$ 12. $\frac{y}{x^{10}}$ 13. 5.87×10^7
14. 5.87×10^{-3}

15. 2×10^{-6} 16. 7×10^3
17. 0.00256 18. $879,000$
19. $577,300,000$ 20. 0.00000007447
21. $2,000$ 22. 1.8 23. 40
24. 5.7×10^5
25. (a) 3   (b) 2, 1, 0   (c) 2
26. (a) 4   (b) 3, 2, 1, 0   (c) 3
27. (a) 3   (b) 4, 2, 2   (c) 4
28. (a) 4   (b) 5, 6, 2, 1   (c) 6
29. (a) 2   (b) 1, 0   (c) 1
30. (a) 2   (b) 0, 1   (c) 1
31. (a) 1   (b) 0   (c) 0
32. (a) 1   (b) Undefined
   (c) Undefined
33. (a) 1   (b) 8   (c) 8
34. (a) 2   (b) 5, 3   (c) 5
35. $2x^3 - 7x^2 + 0x + 4$
36. $3t^5 + 0t^4 + 0t^3 - t^2 + 0t - 10$
37. $y^5 + 0y^4 + 0y^3 + y^2 - 2y - 1$
38. $-x^4 + 0x^3 + 0x^2 + 0x + 1$
39. $2x^2 + 2$ 40. $4y^4 + y^2 + 8y$
41. $4x^2 - 10x + 12$
42. $6y^4 - 3y^2 + 10y$
43. $xy^2 - 7x^2y^2 + 5x^2y$
44. $-8m^2 - 6m^2n$
45. $5x^2y - 5xy^2 - 8x^2y^2$
46. $3r^2s - 7rs^2 + 3r^2s^2$
47. $4a^3$
48. $7m^2n - 8m^2n^2 - n^3 + 2mn^2$
49. $8x$ 50. $5a^2 - 5b^2$ 51. $3x^2$
52. $-2a^3 + 2a^2 - 5a - 9$
53. $x^2 - 3x - 28$
54. $a^2 - 9a + 20$
55. $8x^2 - 22x + 15$
56. $30x^2 - 29x + 4$
57. $6a^2 + 7ab - 20b^2$
58. $28x^2 - 29xy + 6y^2$
59. $3a - 8ab - 20b^2$
60. $4x - 21xy + 6y^2$
61. $x^3 - 7x - 6$
62. $x^3 - 9x^2 + 26x - 24$
63. $x^2 + 12x + 35$
64. $a^2 + 9a + 18$
65. $x^2 - 12x + 35$
66. $a^2 - 9a + 18$
67. $x^2 - 2x - 35$
68. $a^2 - 3a - 18$
69. $x^2 + 2x - 35$
70. $a^2 + 3a - 18$
71. $x^2 - 10x + 25$
72. $a^2 + 12a + 36$
73. $x^2 - 25$ 74. $a^2 - 36$
75. $x^2 - 81y^2$ 76. $a^2 - 49b^2$
77. $2x^2 - 11x - 21$
78. $3a^2 + 14a - 24$
79. $5x - 21$ 80. $-a - 24$
81. $15x^2 + 14x - 8$

82. $28a^2 + 13ab - 6b^2$
83. $x^2 + 12x + 36$
84. $9x^2 - 12xy + 4y^2$
85. $x^3 - 15x^2 + 75x - 125$
86. $8x^3 - 12x^2 + 6x - 1$
87. $3x^4 - 6x^3 - 24x^2$
88. $6y^3 - 28y^2 - 10y$
89. $x^2 - 64$ 90. $9x^2 - 4y^2$
91. $x^3 - x^2 - 2x + 8$
92. $x^3 - 2x^2y - 7xy^2 + 12y^3$
93. $x^4 + 4x^3 + 4x^2 - 1$
94. $y^4 - 6y^3 + 9y^2 - 16$
95. $x^2 + 8x - 14$
96. $-2x + 5$ 97. $3w^2 - 5w$
98. $14x + 49$ 99. 0.3 meter
100. 2.99×10^{24}

Chapter 7 Practice Test

1. (a) $\frac{3}{2}$ (b) 6   2. $\frac{x^8}{y^2}$ 3. $\frac{1}{x^9}$
4. x^{20} 5. $\frac{4y^{19}}{x^6}$
6. (a) 4   (b) $-1, 0$ (c) 4
7. $24x^7y^2$ 8. $-12x^4y^2 + 6x^5y$
9. xy
10. $12x^3 - 23x^2 + 28x - 12$
11. $6x^3 - 4x^2y - xy^2$
12. $x^2 + x + 1$ 13. $-4a$
14. (a) 3.16×10^{-3}
   (b) 3.16×10^4 15. 2×10^5
16. $12s - 36$
17. 141.98 light years

EXERCISES 8.1

1. $4x$ 3. $5x$ 5. $6mnp$
7. $5(x + 4)$ 9. $6(x - 3)$
11. Not factorable   13. $14(2a - 3)$
15. $3(4a + 3)$ 17. Not factorable
19. $x(x + 3)$ 21. $2t(t + 4)$
23. $13y^2(2 - 3y)$
25. $2x(2x^4 + x - 4)$
27. $a(a + 1)$ 29. $x(x - 5 + y)$
31. $3c^3(c^3 - 2)$ 33. $xy(x - y)$
35. $3x(8x + 5)$ 37. Not factorable
39. $4c^2d^3(3cd^2 + 1)$
41. Not factorable
43. $2xyz^2(xz + 4 - 5xy)$
45. $6u^3v^2(1 + 3v - 2v^3)$
47. $(x - 5)(x + 4)$
49. $(y + 6)(y - 3)$
51. $(x + 8)(x + y)$
53. $(m + n)(m + 9)$
55. $(x - y)(x - 4)$
57. $(x + 2)(3xy - 5)$

Mini-Review

62. x^{12} 63. x^{12} 64. $\frac{27x^6}{2y^2}$ 65. $\frac{x^8}{y^3}$

66. Let x = amount of pure acid;
$x + 0.20(40) = 0.25(x + 40)$;

$\boxed{2\frac{2}{3} \text{ liters}}$

67. Let x = amount of pure water;
$0.18(36) = 0.12(x + 36)$;

$\boxed{18 \text{ liters}}$

EXERCISES 8.2

1. $x(x + 3)$ 3. $(x + 2)(x + 1)$
5. $(x - 2)(x - 1)$
7. Not factorable
9. $(x + 2)(x - 1)$
11. $(x - 2)(x + 1)$
13. $(a + 6)(a + 2)$
15. $(a - 4)(a + 3)$
17. Not factorable   19. $a(a - 12)$
21. $(a + 4)(a - 3)$
23. $(y + 7)(y + 4)$
25. $(x - 9)(x + 4)$
27. $(a + 10)(a - 4)$
29. $(z - 15)(z - 2)$ 31. $x(x - 9)$
33. $(x + 3)(x - 3)$
35. $(x - 10)(x + 1)$
37. $(x - y)(x - 2y)$
39. $(a + 6)(a + 4)$
41. $(y + 6)(y + 6)$
43. $(y + 6)(y - 6)$
45. $(x - 9)(x + 2)$
47. $(r - 5s)(r + 2s)$
49. $(c - 5)(c - 1)$
51. $4(x + 1)(x + 1)$
53. $(x + 6)(x - 5)$
55. $2(x + 5)(x - 5)$
57. $(x - 5)(x + 4)$
59. Not factorable
61. $(y - 7)(y - 4)$
63. $2(y + 7)(y - 6)$
65. $(7 + d)(7 - d)$
67. Not factorable
69. $10(x - 6y)(x + 2y)$
71. $(a + 13)(a + 1)$
73. $6(s - 4)(s + 3)$
75. $4(x + 4)(x - 4)$

Mini-Review

80. $25x^4y^2$ 81. $25x^4 + 10x^2y + y^2$
82. $3x^2 + 4x - 15$
83. $6x^2 - 13x - 28$
84. 4.3×10^{-3} 85. 2.87×10^4
86. $90,000$ 87. $90,000$

88. Let x = amount invested at 7%,
$4,000 - x$ = amount invested
at 6%; $0.07x + 0.06(4,000 - x)$
$= 253.60$;

$\boxed{\$1,360 \text{ at } 7\%, \$2,640 \text{ at } 6\%}$

EXERCISES 8.3

1. $x(x + 5)$ 3. $(x + 4)(x + 1)$
5. Not factorable
7. $(3x + 2)(x + 2)$
9. $(2x + 3)(x + 4)$
11. $2(x + 3)(x + 2)$
13. $(5x - 2)(x - 5)$
15. $5(x - 2)(x - 1)$
17. $(2y + 3)(y - 2)$
19. $(5a - 6)(a + 3)$
21. $(2t + 3)(t + 2)$
23. $2(t^2 + 3t + 3)$
25. $3(w^2 - 2w - 10)$
27. Not factorable
29. $(3x - 5y)(x - 3y)$
31. $(6a + 5)(a + 2)$
33. $(3a + 10)(2a - 1)$
35. $6(a - 4)(a + 1)$
37. $(x - 6y)(x + 6y)$
39. $4(x - 3y)(x + 3y)$
41. $x(x + 8)(x - 3)$
43. $6x(x - 4)$
45. $4x^2(x - 4)(x - 2)$
47. $2xy(3x - 4y + 6)$
49. $(3x - 16)(x + 3)$
51. $8x(x - 4)$
53. $-(x - 5)(x + 3)$
55. $-4xy(x + 7)(x - 3)$
57. $(5 - x)(5 + x)$

Mini-Review

60. Let t = time Sandy walks until
they are 12 km apart;
$10(t - 1) - 6t = 12$;

$\boxed{8:30 \text{ A.M.}}$

61. Let f = # of franks sold,
$f + 20$ = # of knishes sold;
$1.25f + 0.80(f + 20) = 169.75$;

$\boxed{75 \text{ franks, } 95 \text{ knishes}}$

EXERCISES 8.4

1. $x = 2, -3$ 3. $x = -4, 3$
5. $x = -12$ 7. $y = 0, 4$
9. $y = 6, -2$ 11. $x = 3, -2$
13. $x = 5, -2$ 15. $m = 4, -2$
17. $m = 8, 1$ 19. $p = 0$
21. $y = 0, 4$ 23. $w = 0, \frac{8}{5}$

25. $a = 4, \frac{3}{2}$ 27. $a = 0, -3$
29. $a = 2, -5$ 31. Not factorable
33. $a = 7, -2$ 35. $x = 3, -3$
37. Not factorable   39. $y = \frac{1}{2}$
41. $x = 3, -7$ 43. $x = 5$
45. $x = 2, -5$
47. Let L = length; $3L - 2 = $
width; $L(3L - 2) = 33$;

$\boxed{3\frac{2}{3} \text{ ft by } 9 \text{ ft}}$

49. Let x = width of frame;
$(x + 2)^2 - x^2 = 28$;

$\boxed{6 \text{ in. by } 6 \text{ in.}}$

51. Let x = width of path;
$(2x + 5)(2x + 7) = 63$; $\boxed{1 \text{ ft}}$

53. $y^2 - 65y = 1,200$; $\boxed{80 \text{ sq yd}}$

55. $240t - 16t^2 = 800$;

$\boxed{5 \text{ sec; } 10 \text{ sec}}$

The object is at a height of 800 ft
after 5 sec (on its way up) and
again after 10 sec (on its way
down).

57. $n^2 - n = 90$; $\boxed{10 \text{ teams}}$

EXERCISES 8.5

1. $\frac{x + 4}{2}$ 3. $\frac{t - 6}{6}$ 5. $x - 3y$

7. $\frac{x - 3y}{2xy}$ 9. $2ab - 3c - 4a^2c^2$

11. $x - 5, R = 12$ 13. $t + 2$
15. $w + 1, R = -24$
17. $2x - 1, R = 6$
19. $y^2 + 3y + 7$
21. $2a^2 - a - 2, R = 4$
23. $x^2 - 4x + 12$
25. $x^3 + 2x^2 + 4x + 8$
27. $x^2 + 4x - 2$
29. $2t^2 + 5t - 4, R = 4$

Mini-Review

32. 9 33. -9 34. -75 35. 225
36. $\frac{1}{5}$ 37. $\frac{1}{9}$ 38. $-\frac{3}{5}$ 39. $-\frac{1}{15}$
40. Let x = amount invested at 5%,
$x + 5,000$ = amount invested at
8%, $10,000 - (x + x + 5,000)$
$= 5,000 - 2x$ = amount invested
at 6%; $0.05x + 0.08(x + 5,000)$
$+ 0.06(5,000 - 2x) = 720$;

$\boxed{\$2,000 \text{ at } 5\%, \$7,000 \text{ at } 8\%, \\ \$1,000 \text{ at } 6\%}$

Chapter 8 Review Exercises

1. $(x + 4)(x + 3)$
2. $(x - 4)(x - 3)$ 3. $x(x + 7)$
4. Not factorable
5. $(x - 12)(x - 1)$
6. $(x - 4)(x + 3)$
7. $(x - 9y)(x + 3y)$
8. $(r - 6t)(r - 2t)$
9. $(x - 8)(x + 8)$
10. $16(x - 2)(x + 2)$
11. $(2x + 5)(x + 2)$
12. $2(x + 5)(x - 1)$
13. $3(x - 4)(x + 2)$
14. $(3x + 4)(x - 6)$
15. $6(a + 4)(a + 2)$
16. $(3a + 16)(2a + 3)$
17. $5xy(x - 4y)(x + 4y)$
18. $2(3m^2n - 4mr^3 + 4n^2r)$
19. $x(x + 9)$ 20. Not factorable
21. $(5t + 1)(5t - 1)$
22. $25(t + 2)(t - 2)$
23. $-(x - 6)(x + 5)$
24. Not factorable
25. $-3(x - 3)(x - 1)$
26. $-4(x - 5)(x + 1)$
27. $(x + 3)(x - 7)$
28. $(a + 5)(a + b)$
29. $(m + 2n)(m + 9)$
30. $(x - y)(x - 5)$
31. $x = 5, -4$ 32. $x = 6, -2$
33. $x = 8, -7$ 34. $x = 1, 10$
35. $x = \frac{7}{3}, -5$ 36. $x = 12, -2$
37. $x = 0, 4$ 38. $x = 0$
39. $x = 4, -4$ 40. $x = 4, -1$
41. Let $w = $ width; $2w - 3 = $ length; $w(2w - 3) = 14$; $\boxed{3\frac{1}{2} \text{ cm by } 4 \text{ cm}}$
42. Let $x = $ side of the square; $(x + 4)^2 - x^2 = 64$; $\boxed{6 \text{ ft by } 6 \text{ ft}}$
43. Let $x = $ side of the square; $(x + 4)^2 = 144$; $\boxed{8 \text{ ft by } 8 \text{ ft}}$
44. $94 + 25t - 16t^2 = 80$; $\boxed{2 \text{ sec}}$
45. $n^2 - n = 56$; $\boxed{8 \text{ teams}}$
46. $x - y$ 47. $\frac{3r - 2t^2 + 5rt^3}{t}$
48. $x - 3, R = -8$
49. $y^2 + y + 3, R = 5$
50. $2x^2 + 6x + 14, R = 38$
51. $2x^2 - x + 2$
52. $x^2 - 2x + 4$
53. $16x^3 + 32x^2 + 64x + 128$, $R = 192$

Chapter 8 Practice Test

1. $3x(2x^2 + 4x - 5)$
2. $2xy(2x - 4y - 1)$
3. $(x + 8)(x + 1)$
4. $(x - 10y)(x + y)$
5. $4x(x - 5)$ 6. $5x(x + 3)(x - 3)$
7. $6(x + 3)(x + 1)$
8. $(2x + 3)(x - 5)$
9. $(x + 2)(x + 2)$
10. $(3x - 2)(2x + 3)$
11. $(xy + 3)(xy - 3)$
12. $(w + 3)(2w + a)$
13. $\frac{6r - 9 + 10rt^2}{2rt}$
14. $2x^2 + 4x + 3, R = 12$
15. $x = 10, -2$ 16. $x = 5, 8$
17. $x = 5, 9$ 18. $x = \frac{5}{3}, 2$
19. Let $w = $ width of the path; $(9 + 2w)^2 - 81 = 40$; $\boxed{\text{path is 1 ft wide}}$
20. $300 - 16t^2 = 200$; $\boxed{2.5 \text{ sec}}$

EXERCISES 9.1

1. $\frac{4y^5}{5x^3}$ 3. $\frac{1}{3x}$ 5. $\frac{2(x + 4)^4}{3x}$ 7. $\frac{3}{2}$ 9. $\frac{3}{5}$
11. $\frac{1}{2}$ 13. $\frac{x - 2}{2(x + 2)}$ 15. $\frac{y}{2(y + 2)}$
17. $\frac{6}{x - 3}$ 19. Cannot be reduced
21. Cannot be reduced   23. $\frac{1}{2x}$
25. $\frac{y + 1}{y - 6}$ 27. $\frac{s + 3}{s - 1}$ 29. $\frac{x + 3}{x}$
31. $\frac{3a - 2}{a - 2}$ 33. $\frac{4x - 1}{x + 2}$ 35. $\frac{x - 2}{x}$
37. $\frac{2(x - 3)(x + 1)}{(x - 5)(x + 2)}$ 39. $\frac{x + 2}{x - 4}$
41. Cannot be reduced
43. $\frac{y(y + 1)}{6(y + 2)}$ 45. $\frac{1}{c}$

Mini-Review

48. $\frac{4xy}{3}$ 49. $\frac{27x}{5y^7}$ 50. $\frac{x}{2}$ 51. $\frac{1}{2x}$
52. $x = -18$ 53. $a = -10$

EXERCISES 9.2

1. $\frac{16ay^2}{3z^2}$ 3. $\frac{s^3}{25p}$ 5. $72x^3y^4$
7. $49ab^4z^4$ 9. $\frac{x + 4}{(x + 5)(x + 1)}$
11. $\frac{x(x + 4)}{(x + 2)(x - 2)}$ 13. $\frac{2r^2(r + 1)}{(r - 2)(r - 1)}$
15. $\frac{m + 6}{3(m + 3)}$ 17. $\frac{x + 1}{x^2 + 2}$

19. $\frac{(y + 1)(y + 2)}{(y - 4)^2}$ 21. $\frac{x + 2}{x + 1}$ 23. $\frac{6x + 1}{x - 3}$
25. $\frac{4}{5}$ 27. $\frac{x(x + 4)}{4}$ 29. $\frac{t^2}{(2t + 3)(t + 1)}$
31. $\frac{(x + 5)^2}{x - 5}$ 33. $\frac{(x + y)^2}{4(x - y)}$ 35. $\frac{x}{x - 3}$
37. $\frac{7}{c + 7}$ 39. $\frac{w^2 - 36}{8w}$

Mini-Review

43. $\frac{20x + 9}{24x^2}$ 44. $\frac{3y - 35x}{30x^2y}$ 45. $\frac{9x - 6}{x}$
46. $\frac{6x^2y + 4y - 3x}{6xy}$
47. Let $x = $ cost of a balcony seat, $x + 4 = $ cost of an orchestra seat; $250(x + 4) + 150x = 3{,}400$; $\boxed{\$10}$
48. Let $t = $ time running, $2 - t = $ time walking; $8t + 4(2 - t) = 13$; $\boxed{10 \text{ miles}}$

EXERCISES 9.3

1. $\frac{2x}{x + 2}$ 3. $\frac{x - 2}{2(x + 2)}$ 5. $\frac{3x + 8}{x + 2}$
7. 1 9. $\frac{y + 2}{4}$ 11. $\frac{13x + 10}{2x(x + 2)}$
13. $\frac{10}{x(x + 2)}$ 15. $\frac{5x + 12}{(x + 2)(x + 3)}$
17. $\frac{2a}{(a + 7)(a + 5)}$ 19. $\frac{-2x + 12}{3x^2(x + 3)}$
21. $\frac{2a^2 + 9a - 27}{12a^2(a - 3)}$ 23. $\frac{2x - 24}{x(x + 4)(x - 4)}$
25. $\frac{x - 2}{x - 1}$ 27. $\frac{2x}{x - 1}$ 29. $\frac{3}{2x(x - 4)}$
31. $\frac{8}{x(x + 2)}$ 33. $\frac{x^2 + 2x + 3}{(x + 1)(x + 3)^2}$
35. $\frac{3(a + 1)}{3a + 2}$ 37. $\frac{15x^2 - 31x + 9}{3x(x - 2)}$
39. $\frac{4}{x(x - 2)}$ 41. $\frac{x + 2}{2}$ 43. $\frac{1 - x}{2}$
45. a 47. $\frac{a^2b^2}{4(2a + b)}$ 49. $\frac{2}{x}$
51. $\frac{1}{(t + 1)^2}$ 53. 25.8

Mini-Review

57. $x = 3$ 58. $x = 6$ 59. $a = 15$
60. $m = 160$

EXERCISES 9.4

1. $x = 12$ 3. $y = -24$ 5. $a = 5$
7. $u = \frac{20}{3}$ 9. $\frac{3x}{12}$ 11. $\frac{x^3}{24}$ 13. $x = 6$
15. $x = \frac{1}{3}$ 17. $x = 5$ 19. $a = -7$
21. $x = 7$ 23. $r = 4$ 25. $x = \frac{3}{2}$
27. $t = \frac{10}{13}$ 29. $\frac{10 - 11x}{2x(x + 2)}$ 31. $x = 5$
33. No solution   35. $\frac{x + 18}{(x + 2)(x + 6)}$
37. $x = \frac{28}{3}$ 39. No solution
41. $m = 0$ 43. $x = 4$

45. No solution   47. $a = 1$
49. $\dfrac{10x + 2}{x(x + 1)(x - 1)}$ 51. $\dfrac{x - 3}{x + 1}$
53. $x = 8$ 55. $x = \dfrac{14}{3}$
57. Let w = width;
    $2w + 3$ = length;
    $\dfrac{w}{2w + 3} = \dfrac{2}{5}$; $\boxed{6 \text{ by } 15}$
59. $\dfrac{c - 3}{6} = \dfrac{-5}{3}$; $\boxed{c = -7}$
61. 2,000 CD players   63. 200 acres

Mini-Review

68. Let R = # of Republicans;
    $\dfrac{R}{2,520} = \dfrac{4}{7}$; $\boxed{1,440}$
69. Let x = # of students who take
    longer than 4 years to graduate;
    $\dfrac{2,600}{x} = \dfrac{13}{3}$; $\boxed{3,200}$

EXERCISES 9.5

1. $x = \dfrac{3y + 12}{4}$ 3. $y = \dfrac{6 - x}{2}$
5. $a = -2b - 5$ 7. $m = -\dfrac{2p}{7}$
9. $z = \dfrac{4x - y + 4}{3}$ 11. $r = \dfrac{16 - 5s}{4}$
13. $a = \dfrac{d}{b + c}$ 15. $x = \dfrac{-y - 2}{2}$
17. $x = \dfrac{5y + 6}{5}$ 19. $x = \dfrac{d - b}{a - 2}$
21. $y = \dfrac{6np - 9n - 3p}{3a + 1}$ 23. $u = \dfrac{y + 1}{y - 1}$
25. $x = \dfrac{c - ay - ab}{y + b}$ 27. $x = \dfrac{y - b}{m}$
29. $h = \dfrac{2A}{b_1 + b_2}$ 31. $P = \dfrac{A}{1 + rt}$
33. $F = \dfrac{9}{5}C + 32$
35. $v_0 = \dfrac{2S - 2S_0 - gt^2}{2t}$
37. $x < 2s + \mu$ 39. $f = \dfrac{f_1 f_2}{f_1 + f_2}$

Mini-Review

42. $\dfrac{1}{9x^{32}}$ 43. x^{19} 44. x^{17} 45. $\dfrac{17}{72}$

EXERCISES 9.6

We include the equation used to solve
each odd-numbered exercise.

1. Let $3x$ = width, $7x$ = length;
   $2(3x) + 2(7x) = 100$;
   $\boxed{\begin{array}{l}\text{width} = 15 \text{ meters,} \\ \text{length} = 35 \text{ meters}\end{array}}$

3. Let n = # of people who
   preferred brand X, $200 - n$ =
   # of people who did not prefer
   brand X; $\dfrac{n}{200 - n} = \dfrac{13}{12}$; $\boxed{104}$

5. Let x = # of additional hits.
   $\dfrac{120 + x}{450} = .400$; $x = 60$;
   $\boxed{\text{It cannot be done.}}$

7. Let x = # of hours to mow the
   lawn together. $\dfrac{x}{3} + \dfrac{x}{2} = 1$; $\boxed{1\frac{1}{5} \text{ hr}}$

9. Let x = # of hours electrician
   works, $2x$ = # of hours
   apprentice works. $\dfrac{6}{x} + \dfrac{6}{2x} = 1$;
   $\boxed{\begin{array}{l}\text{electrician works 9 hr;} \\ \text{apprentice works 18 hr}\end{array}}$

11. Let r = rate of train, $r + 100$ =
    rate of plane; $\dfrac{500}{r + 100} = \dfrac{300}{r}$;
    $\boxed{\begin{array}{l}\text{rate of train} = 150 \text{ kph,} \\ \text{rate of plane} = 250 \text{ kph}\end{array}}$

13. Let r = rate of current.
    $\dfrac{8}{4 - r} = \dfrac{24}{4 + r}$; $\boxed{r = 2 \text{ mph}}$

15. d = distance to the friend's
    house; $\dfrac{d}{6} + \dfrac{d}{14} = 3$; $\boxed{12.6 \text{ km}}$

17. Let the numbers be x and $3x$;
    $\dfrac{1}{x} + \dfrac{1}{3x} = \dfrac{5}{3}$;
    $\boxed{\text{the numbers are } \frac{4}{5} \text{ and } \frac{12}{5}}$

19. Let x = the number. $\dfrac{3 + x}{5 + x} = \dfrac{5}{6}$;
    $\boxed{x = 7}$

21. $\dfrac{x}{x + 2} = \dfrac{x + 2}{x}$; $\boxed{\text{fraction is } \frac{-1}{1}}$

23. $S = 252\pi$ 25. $R_T = 12$ ohms
27. $f = 2$ cm
29. Let h = # of hours for Taisha to
    finish; $\dfrac{2}{5} + \dfrac{h}{8} = 1$; $\boxed{6\frac{4}{5} \text{ hours}}$

31. Let h = # of hours Latifa works;
    $\dfrac{h + 1}{5} + \dfrac{h}{7} = 1$; $\boxed{11{:}20 \text{ A.M.}}$

33. Let h = # of hours it takes Sean
    to paint the apt. alone;
    $\dfrac{8}{h} + \dfrac{5}{2h} = 1$; $\boxed{10\frac{1}{2} \text{ hours}}$

Chapter 9 Review Exercises

1. Cannot be reduced   2. $\dfrac{x + 6}{3x}$
3. $\dfrac{x - 1}{x - 4}$ 4. $\dfrac{x + 5}{2x + 3}$ 5. $\dfrac{a + 4}{a + 2}$
6. $\dfrac{xy}{x + 2y}$ 7. $\dfrac{z - 2}{z + 1}$
8. Cannot be reduced   9. $\dfrac{x^2 + 4x}{2(x + 2)}$
10. $\dfrac{a - 4}{4a}$ 11. $\dfrac{3x + 12}{2x(x + 2)}$ 12. $\dfrac{x}{4}$
13. $\dfrac{4x^2}{x + 1}$ 14. $\dfrac{14y + 12}{y(y + 2)(y + 6)}$ 15. $\dfrac{2}{z(z - 2)}$

16. $\dfrac{1}{(x + y)(x + 2y)}$ 17. $\dfrac{2x^2 + 6x - 2}{x(x + 2)}$
18. $\dfrac{x + 3}{6}$ 19. $x = 2$ 20. $a = -1$
21. $y = -\dfrac{1}{2}$ 22. $x = 0$
23. No solution   24. $a = 10$
25. $x = \dfrac{3y + 4}{5}$ 26. $a = \dfrac{2cx}{c - 3x}$
27. Let b = # of black marbles,
    $b + 80$ = # of red marbles.
    $\dfrac{b + 80}{b} = \dfrac{7}{5}$;
    $\boxed{480 \text{ marbles altogether}}$

28. Let r = rate of current,
    $\dfrac{8}{4 - r} = 2 \cdot \dfrac{8}{4 + r}$; $\boxed{r = 1\frac{1}{3} \text{ mph}}$
29. Let x = # of hours for John to
    do the job alone. $\dfrac{4}{6} + \dfrac{4}{x} = 1$;
    $\boxed{12 \text{ hours}}$

30. Let x = the number.
    $\dfrac{8 + x}{13 + x} = \dfrac{4}{5}$; $\boxed{12}$
31. Let w = width; $2w - 5$ =
    length; $\dfrac{2w - 5}{w} = \dfrac{5}{3}$; $\boxed{15 \text{ by } 25}$
32. $c = 10$

Chapter 9 Practice Test

1. $\dfrac{4x}{x - 4}$ 2. $\dfrac{x + 3y}{3x}$ 3. $\dfrac{5x + 12}{(x + 3)(x + 2)}$
4. $\dfrac{x - 2}{4}$ 5. $\dfrac{5}{2(x + 4)}$ 6. $\dfrac{2(x + 4)}{x}$ 7. $\dfrac{5}{x}$
8. $x = 6$ 9. $t = \dfrac{14 - 2b}{2a - 3}$
10. No solution
11. Let d = distance from town A to
    town B; $\dfrac{d}{45} + \dfrac{d}{60} = 14$; $\boxed{360 \text{ km}}$
12. $b_2 = 60$ cm
13. Let x = # of hours Haleema
    works; $9 - x$ = # of hours
    Andrea works; $\dfrac{x}{8} + \dfrac{9 - x}{12} = 1$;
    $\boxed{\begin{array}{l}\text{Haleema works 6 hr;} \\ \text{Andrea works 3 hr}\end{array}}$
14. 90%

CHAPTERS 7–9 CUMULATIVE REVIEW

1. $x^2 + 3x - 40$ (§7.5)
2. $x^3 + 9x^2 + 3x - 40$ (§7.5)
3. $a^2 + 2ab + b^2 - c^2$ (§7.5)
4. $2z^3 + 6z^2 - 36z$ (§7.5)
5. $20a^2 + 3ac - 9c^2$ (§7.5)
6. $t^2 - t$ (§7.5)   7. $2x^2 + 3x$ (§7.5)
8. $-2x$ (§7.5)

9. $4a^3 - 12a^2 - 9a + 27$ (§7.5)

10. $x^4 - x^2 - x^2y^2 + y^2$ (§7.5)

11. $-4y^2 - xy$ (§7.4)

12. $-2a^3 + 3a - 9$ (§7.4)

13. (a) 4   (b) -3 (§7.4)

14. $3x^3 + 0x^2 - x + 4$ (§7.4)

15. $\frac{x+8}{2}$ (§8.5)

16. $\frac{2(2r - 3s^2 - r^2s)}{3rs}$ (§8.5)

17. $y - 1, R = 2$ (§8.5)

18. $2a^2 + a - 2$ (§8.5)

19. $6x^2 + 8x + 9, R = 8$ (§8.5)

20. $t^3 - 2t^2 + 3t - 6, R = 6$ (§8.5)

21. $\frac{19}{16}$ (§7.2)   22. 1 (§7.2)

23. x (§7.1)   24. $\frac{1}{a}$ (§7.2)

25. $\frac{4}{x^3}$ (§7.1)   26. $\frac{x^{13}}{y^{22}}$ (§7.2)

27. $\frac{a^{11}}{27t^2}$ (§7.2)   28. $\frac{x^3}{125y^{15}}$ (§7.2)

29. 4.39×10^{-4} (§7.3)

30. 5.78×10^5 (§7.3)

31. 10^5 (§7.3)   32. 0.002 (§7.3)

33. $(x + 1)(x + 5)$ (§8.3)

34. $x(x + 6)$ (§8.2)

35. $(x - 2)(x - 3)$ (§8.3)

36. $(x - 6)(x + 1)$ (§8.3)

37. $3xy(2x^2 - 4y - 3x)$ (§8.2)

38. $5m^2n^3(2mn^2 - 1)$ (§8.2)

39. $(u + 7)(u - 7)$ (§8.4)

40. $4(a - 3)(a - 3)$ (§8.3)

41. $(2r - 5)(r + 3)$ (§8.4)

42. $t^2(t + 6)(t - 6)$ (§8.4)

43. $5(t^2 + 2t + 3)$ (§8.3)

44. $xy(x + y)(x - y)$ (§8.4)

45. $(2x - 3y)(3x - 4y)$ (§8.4)

46. $-(x - 12)(x + 2)$ (§8.3)

47. $x(x + 16)$ (§6.2)

48. Not factorable (§8.3)

49. $(x + y)(x + a)$ (§8.2)

50. $(a - 3)(a + z)$ (§8.2)

51. $(x - a)(x - 4)$ (§8.2)

52. $(x^4 + y^4)(x^2 + y^2)(x + y)(x - y)$ (§8.4)

53. $\frac{x}{x+4}$ (§9.1)   54. $\frac{t-2}{t-3}$ (§9.1)

55. $\frac{23x + 12}{4x(x+4)}$ (§9.3)   56. $\frac{10x - 21y^3}{12x^2y^3}$ (§9.3)

57. $\frac{x^2}{10(x+5)}$ (§9.2)   58. $\frac{6r(r+t)}{t(r-t)}$ (§9.2)

59. $\frac{2x - 20}{x(x+2)(x-2)}$ (§9.3)

60. $\frac{31u - 24}{6u(u+3)}$ (§9.3)   61. $\frac{2(a+4)}{a(a-4)}$ (§9.3)

62. $\frac{-2t - 30}{t+2}$ (§9.2)   63. $\frac{1}{2z+6}$ (§9.4)

64. $x = 4$ (§9.4)   65. $a = -\frac{5}{3}$ (§9.4)

66. $t = -\frac{4x + 12}{3}$ (§9.5)

67. $y = \frac{30 - 20a}{17}$ (§9.5)

68. No solutions (§9.4)

69. $x = 0$ (§9.4)   70. $x = 2$ (§9.4)

71. $c = \frac{1}{6}$ (§9.4)

72. $x = 5, -7$ (§8.5)

73. $x = \frac{5}{2}, -3$ (§8.5)

74. $x = 2, 4$ (§8.5)

75. $x = -\frac{14}{5}$ (§8.5)

76. Let x = number of days to plow the field together; $\frac{x}{6} + \frac{x}{4} = 1$;

$\boxed{2\frac{2}{5} \text{ days}}$ (§9.6)

77. Let t = number of hours for Pat to overhaul the engine alone;

$\frac{5}{8} + \frac{5}{t} = 1;$ $\boxed{13\frac{1}{3} \text{ hours}}$ (§9.6)

78. Let the original fraction be $\frac{x-4}{x}$;

$\frac{x-1}{x+1} = \frac{1}{2};$ $\boxed{-\frac{1}{3}}$ (§9.6)

CHAPTERS 7–9 CUMULATIVE PRACTICE TEST

1. $x^3 - 5x^2y + 5xy^2 + 2y^3$ (§7.5)

2. $\frac{2y^8}{3x^4}$ (§7.2)

3. $7a^2 - 20a + 24$ (§7.5)

4. $-20x$ (§7.5)   5. $\frac{1}{16x^3}$ (§7.2)

6. $\frac{2(3t^3 - 2s)}{3st}$ (§8.5)

7. $4x^2 + 5x + 15, R = 10$ (§8.5)

8. $x^3 - 2x^2 + 3x - 6$ (§8.5)

9. (a) 9.16×10^{-4}

   (b) 9.16×10^5 (§7.3)

10. 222.22 (§7.3)

11. $-2x^2 + 7x + 1$ (§7.4)

12. $(x - 12)(x + 2)$ (§8.3)

13. $3ab^3(2ab^2 - 1)$ (§8.2)

14. $(2t - 3)(t + 4)$ (§8.4)

15. $6(x^2 - 6x + 12)$ (§8.2)

16. $3xy(x + 2y)(x - 2y)$ (§8.4)

17. $(a + 5)(a - 7)$ (§8.2)

18. $(x - 3)(x - y)$ (§8.2)

19. $2(u^2 + 4)(u + 2)(u - 2)$ (§8.4)

20. $\frac{3x}{x+3}$ (§9.1)

21. Cannot be reduced (§9.1)

22. $\frac{x}{(x+5)(x+4)}$ (§9.3)

23. $\frac{w}{4(w-5)}$ (§9.2)

24. $\frac{3a - 27}{a(a+3)(a-3)}$ (§9.3)

25. $\frac{1}{u(u+9)}$ (§9.2)   26. $\frac{x}{2x-1}$ (§9.3)

27. $t = 5$ (§9.4)

28. $u = \frac{15x + 35}{2 - 5a}$ (§9.5)

29. No solution (§9.4)

30. $x = 2, -5$ (§8.5)

31. Let r = Lamar's speed; $r - 5$ = Roni's speed;

$\frac{160}{r} = \frac{140}{r-5};$ $\boxed{4 \text{ hours}}$

32. Let x = the number of students in the larger group; $\frac{x}{228 - x} = \frac{12}{7}$;

$\boxed{\begin{array}{l} 144 \text{ in the larger group;} \\ 84 \text{ in the smaller group} \end{array}}$

EXERCISES 10.1

1. 2   3. -2 5. Undefined   7. 5

9. -10 11. 8   13. 11   15. 13

17. 15   19. 17   21. 19   23. 3

25. 29   27. 11   29. 33   31. x

33. 49   35. $49\sqrt{7}$ 37. 4   39. 2

41. 4   43. 16   45. 20.62   47. 25.24

49. 8.58   51. 1.45   53. 0.19

55. 4.1; 4.12; 4.123

57. 10.5; 10.49; 10.488

59. Between 8 and 9; 8.54

61. Between 14 and 15; 14.73

63. Between 25 and 26; 25.94

65. Between 16 and 17; 16.88

Mini-Review

71. $\frac{2}{x-6}$ 72. $\frac{y-4}{y}$ 73. $\frac{13x + 6}{2x(x+2)}$

74. $x = 0$

75. Let x = # of long-life bulbs, $40 - x$ = # of regular bulbs; $89x + 75(40 - x) = 3{,}168$;

$\boxed{\begin{array}{l} 12 \text{ at 89 cents each,} \\ 28 \text{ at 75 cents each} \end{array}}$

76. Let x = amount invested at 9.2%; $x - 1{,}000$ = amount invested at 7.6%; $2(x - 1{,}000)$ = amount invested at 6.5%; $0.092x + 0.076(x - 1{,}000) + 0.065(2x - 2{,}000) = 837$;

$\boxed{\$11{,}000}$

EXERCISES 10.2

1. 8   3. $3\sqrt{2}$ 5. $4\sqrt{2}$ 7. $5\sqrt{2}$

9. 20   11. x^3 13. $x^3\sqrt{x}$ 15. $4x^8$

17. $3x^4\sqrt{x}$ 19. $2x^4\sqrt{10}$

21. $5x^3\sqrt{x}$ 23. $2x^2\sqrt{3x}$ 25. ab^2

27. Cannot be simplified   29. $7a^4b^6$

31. $2x^4y^3\sqrt{7x}$ 33. $5m^3n^5\sqrt{2mn}$

35. Cannot be simplified

37. $4x^3y^4z^4\sqrt{3z}$ 39. $\frac{2}{3}$ 41. $\frac{\sqrt{7}}{5}$

43. $\frac{\sqrt{30}}{6}$ 45. $\frac{\sqrt{2}}{2}$ 47. $\frac{\sqrt{2}}{2}$ 49. $\frac{9\sqrt{10}}{5}$

51. $\frac{3\sqrt{x}}{x}$ 53. $\frac{15\sqrt{7}}{14}$ 55. $\frac{2\sqrt{6}}{5}$

57. $4\sqrt{2x}$ 59. $\frac{5\sqrt{x}}{2x}$ 61. $\frac{x\sqrt{xy}}{y}$

63. $\frac{2\sqrt{3}}{3}$ 65. 11.80% 67. 59 mph

69. 8.8 sec 71. 219 miles

73. 1.7 amp.

Mini-Review

79. $x = \frac{8-3y}{4}$ 80. $y = \frac{4}{5}x$

81. $a = 2 - 7b$ 82. $u = -\frac{8}{3}v$

83. Let t = # of hours 110-kph train
travels until they pass,
$t - 2$ = # of hours 140-kph train
travels until they pass;
$110t + 140(t - 2) = 595$;

$\boxed{2{:}30 \text{ P.M.}}$

84. Let $8x$ = number of members in
favor, $7x$ = number of members
against; $8x + 7x = 435$; $\boxed{232}$

EXERCISES 10.3

1. $6x$ 3. $6\sqrt{5}$ 5. $3x$ 7. $3\sqrt{6}$
9. Cannot be simplified
11. Cannot be simplified 13. x^2
15. 5 17. $2x$ 19. $2\sqrt{5}$
21. $-3\sqrt{5} - 2\sqrt{7}$ 23. $7\sqrt{3} - \sqrt{2}$
25. $-x + y$ 27. $-\sqrt{5} + \sqrt{3}$
29. $23\sqrt{6} + 6\sqrt{7}$
31. $3\sqrt{m} - 12\sqrt{n}$
33. $5\sqrt{2}$ 35. $5 + 2\sqrt{6}$
37. $\sqrt{5}$ 39. $3\sqrt{3}$
41. $2\sqrt{5} + 2\sqrt{10} + 2\sqrt{15}$
43. $-2\sqrt{2}$ 45. $30 + 4\sqrt{30}$
47. $11\sqrt{x}$ 49. $5\sqrt{3w}$ 51. $\sqrt{5x}$
53. $-y\sqrt{5y}$ 55. $5xy\sqrt{7xy}$ 57. 0
59. $\frac{3\sqrt{2}}{2}$ 61. $\frac{13\sqrt{3}}{3}$ 63. $\frac{53\sqrt{2}}{14}$
65. $\frac{9\sqrt{14}}{14}$

Mini-Review

69. $4x - 3y = 12$

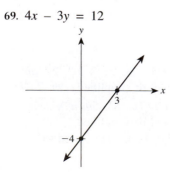

70. $5y - 2x = 10$

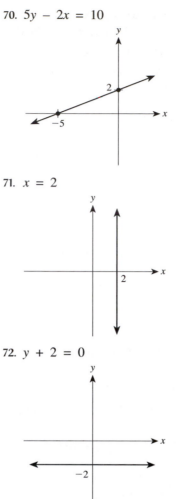

71. $x = 2$

72. $y + 2 = 0$

73. Let h = # of hours associate
works, $14 - h$ = # of hours
clerk works;
$120h + 40(14 - h) = 1,200$;

$\boxed{\begin{array}{l}8 \text{ hours for associate,}\\ 6 \text{ hours for clerk}\end{array}}$

74. Let t = # of hours biking;
$24t + 10\left(t + \frac{1}{2}\right) = 56$;

$\boxed{1\frac{1}{2} \text{ hours}}$

EXERCISES 10.4

1. $\sqrt{33}$ 3. $\sqrt{195}$ 5. $3\sqrt{6}$ 7. 12
9. $3\sqrt{10}$ 11. $\sqrt{15} + 3\sqrt{2}$ 13. 24
15. $6 - 3\sqrt{6}$ 17. $5x - 10\sqrt{5x}$
19. $4 - 7\sqrt{2}$ 21. $x - 2\sqrt{2x}$
23. $-7 - 3\sqrt{11}$
25. $x + 2\sqrt{3x} + 3$
27. $x - 3$ 29. $38 - 12\sqrt{10}$
31. $9x - 7$ 33. $6x - \sqrt{x} - 12$
35. $26 - 4\sqrt{42}$
37. $2t + 18\sqrt{t} + 90$

39. $4\sqrt{x} + 2$ 41. $\frac{10\sqrt{11}}{11}$ 43. $3\sqrt{2}$
45. $\frac{y}{x}$ 47. $\frac{\sqrt{ab}}{b^2}$ 49. $2(4 + \sqrt{11})$
51. $\frac{6(\sqrt{x} - \sqrt{3})}{x - 3}$ 53. $2\sqrt{3} - 3$
55. $4 + \sqrt{15}$ 57. $4\sqrt{5}$ 59. 9
61. $20 + 8\sqrt{5}$ 63. $\frac{2 + \sqrt{2}}{3}$
65. $\frac{6 - \sqrt{5}}{5}$

Mini-Review

70. $x(x - 6)$ 71. $(x - 7)(x + 1)$
72. $2(x + 5)(x - 5)$
73. Not factorable
74. $(a + 10)(a - 4)$
75. $w(w - 4)(w - 4)$
76. $(2y + 1)(y + 3)$
77. $(6u + 7)(u - 2)$
78. Let t = # of hours Marta drives
until they are 120 km apart,
$t - 1$ = # of hours Sarah drives
until they are 120 km apart;
$90t - 80(t - 1) = 120$;

$\boxed{1{:}00 \text{ P.M.}}$

79. Let x = price of ticket at the door,
$x - 1.50$ = price of advanced-
sale ticket; $200(x - 1.50) + 75x$
$= 1,075$; $\boxed{\$5}$

EXERCISES 10.5

1. $x = 25$ 3. $a = 81$ 5. $u = 1$
7. No solution 9. $x = 97$
11. $x = 49$ 13. $t = 13$ 15. $t = 18$
17. $x = -11$ 19. $u = -4$
21. No solution 23. No solution
25. $x = 16$ 27. $x = 1$ 29. $y = 9$
31. $y = 32$ 33. No solution
35. No solution 37. $c = 1$
39. $x = 12$ 41. $x = 7$ 43. $w = 64$
45. $w = 9$ 47. $x = \frac{13}{8}$ 49. $x = -\frac{1}{4}$
51. $x = 6$ 53. $x = 5$
55. 1,296 watts 57. Approx. 185 ft
59. Approx. 2.8 mi
61. Approx. \$4,450

Mini-Review

65. $4x^2 + 6 - 2x$ 66. $\frac{3b - 5 + 2a}{4}$
67. $3x - 2$
68. $4x^2 + 2x - 2, R = 1$

Chapter 10 Review Exercises

1. 7 2. -10 3. Undefined

4. $3\sqrt{10}$ 5. $4\sqrt{6}$ 6. $4x^8$

7. $3x^4\sqrt{x}$ 8. $2x^3y^5\sqrt{5x}$ 9. $\frac{2}{3}$

10. $\frac{\sqrt{3}}{2}$ 11. $\frac{\sqrt{15}}{5}$ 12. $\frac{2\sqrt{5}}{5}$ 13. $2\sqrt{7}$

14. $-4\sqrt{5} - \sqrt{3}$ 15. $\sqrt{5}$

16. $15\sqrt{2}$ 17. $7\sqrt{3x}$ 18. $x^2\sqrt{6x}$

19. $5\sqrt{3x}$ 20. $\frac{12\sqrt{35}}{35}$

21. $15 + \sqrt{10}$

22. $4x - 9\sqrt{3x} + 6$

23. $12 + 11\sqrt{21}$ 24. 4

25. $x - 6\sqrt{x} + 9$ 26. $a - 10$

27. $\frac{7\sqrt{3}}{3}$ 28. $\frac{5\sqrt{6}}{3}$ 29. $x\sqrt{x}$

30. $\frac{7\sqrt{6}}{18}$ 31. $3\sqrt{3}$ 32. $\frac{2m^2}{3}$

33. $2(3 + \sqrt{2})$ 34. $\frac{7\sqrt{2}}{3}$

35. $\frac{7 + 4\sqrt{5}}{31}$ 36. $3\sqrt{10}$ 37. $-2\sqrt{7x}$

39. $x = 44$ 40. $x = 4$ 41. $u = 16$

42. $a = 16$ 43. No solution

44. $z = -\frac{2}{3}$ 45. 55.9 cm

Chapter 10 Practice Test

1. $5x^8y^3$ 2. $10\sqrt{3}$ 3. $x\sqrt{2x}$

4. $x^2y^5\sqrt{6}$ 5. $5x^4y^4\sqrt{5y}$

6. $\frac{x\sqrt{6x}}{2}$ 7. $2x + 5\sqrt{5x} - 15$

8. 16 9. $-8\sqrt{x} + 20$

10. $\frac{5(\sqrt{7} + \sqrt{3})}{2}$ 11. It does satisfy.

12. $a = 12$ 13. 900 items

EXERCISES 11.1

1. $x = 4, -8$ 3. $x = 5, -13$

5. $x = \pm 5$ 7. $b = \pm 4$ 9. $b = \pm\frac{4}{3}$

11. No real solutions 13. $x = \pm\frac{2}{5}$

15. $x = \pm\frac{\sqrt{15}}{6}$ 17. $b = \pm\frac{\sqrt{33}}{3}$

19. $b = \pm 2$ 21. $a = \pm\frac{2\sqrt{5}}{3}$

23. $y = \pm\frac{4\sqrt{6}}{3}$ 25. $y = \pm\sqrt{5}$

27. $a = \pm\sqrt{3}$ 29. No real solutions

31. $y = \pm 2\sqrt{3}$ 33. $v = \pm\sqrt{7}$

35. $x = \pm 2\sqrt{6}$ 37. $t = 5, -1$

39. $a = -5 \pm \sqrt{7}$

41. $x = 6 \pm 2\sqrt{3}$

43. $x = -5 \pm \sqrt{10}$

45. $x = 3, 12$ 47. $x = 6, -2$

49. $m = 0, \frac{4}{3}$ 51. $x = \frac{-2 \pm \sqrt{3}}{5}$

53. $y = \frac{1}{2} \pm \frac{\sqrt{6}}{3}$ 55. $x = 1$

57. $x = 3$ 59. $x = 2, -1$

61. $a = 5$ 63. $x = 5$ 65. $x = -1$

67. $y = \pm 4$ 69. $y = \pm\sqrt{22}$

71. $x = 0, 3$ 73. $x = \frac{1}{2}, -\frac{5}{2}$

75. $x = \pm 2.65$ 77. $k = \pm 2.58$

79. $x = \pm 2.22$ 81. $a = 1.52, -1.22$

83. $y = 0.23, -1.19$

Mini-Review

87. $x = 5, y = -4$

88. $x = 4, y = 0$ 89. No solutions

90. $x = -1, y = -2$

91. Let $x = $ # of hours to prepare
the meal together; $\frac{x}{4} + \frac{x}{3} = 1$;

$\boxed{1\frac{5}{7} \text{ hr}}$

92. Let $x = $ # of votes A should
have received based on survey;

$\frac{x}{24,000} = \frac{7}{5}$; $\boxed{\text{less}}$

EXERCISES 11.2

1. $x = -4 \pm \sqrt{10}$ 3. $x = 2 \pm \sqrt{7}$

5. $x = 5 \pm 2\sqrt{10}$ 7. $a = 10, -2$

9. $x = -6 \pm \sqrt{30}$ 11. $z = 3 \pm \sqrt{7}$

13. $y = -2 \pm \sqrt{6}$

15. $u = -\frac{5}{2} \pm \frac{\sqrt{33}}{2}$

17. $z = \frac{7 \pm \sqrt{29}}{2}$ 19. $w = 2 \pm \sqrt{2}$

21. No real solutions

23. $x = 5 \pm \sqrt{30}$

25. $a = \frac{1 \pm \sqrt{33}}{2}$ 27. $x = \pm\sqrt{13}$

29. $x = \frac{3 \pm \sqrt{3}}{2}$ 31. $t = \frac{4 \pm \sqrt{11}}{5}$

33. $z = \frac{-5 \pm \sqrt{6}}{2}$ 35. $x = -3 \pm \sqrt{6}$

37. No real solution 39. $x = -9, 1$

41. $x = 2, \frac{2}{3}$

Mini-Review

45. $-\sqrt{3}$ 46. $5x\sqrt{5x}$ 47. $\frac{\sqrt{14}}{7}$

48. $\frac{5\sqrt{6}}{3}$ 49. $\frac{9 + 3\sqrt{2}}{7}$

50. $6\sqrt{5} + 6\sqrt{3}$ 51. $3 + 2\sqrt{3}$

52. $\frac{5 - \sqrt{5}}{3}$

53. Let $x = $ # of hours for Yvonne to
finish the puzzle; $\frac{3}{10} + \frac{x}{8} = 1$;

$\boxed{5\frac{3}{5} \text{ hr or 5 hr 36 min}}$

EXERCISES 11.3

1. $a = 1, b = 3, c = -5$

3. $a = 1, b = -7, c = -6$

5. $a = 2, b = -8, c = 0$

7. $a = 3, b = 0, c = -11$

9. $x = \frac{-3 \pm \sqrt{29}}{2}$

11. $y = -2 \pm \sqrt{10}$

13. No real solutions

15. $t = \frac{7 \pm \sqrt{73}}{2}$ 17. $x = \frac{3 \pm \sqrt{17}}{4}$

19. $x = \frac{1 \pm \sqrt{41}}{10}$ 21. $t = \frac{-7 \pm \sqrt{77}}{2}$

23. $w = \frac{3 \pm 2\sqrt{6}}{5}$ 25. $x = -\frac{1}{3}$

27. $x = 2 \pm \sqrt{2}$ 29. $u = \frac{3 \pm \sqrt{15}}{2}$

31. $x = 1, \frac{1}{2}$ 33. $x = 0, \frac{4}{9}$

35. $x = \pm\frac{2}{3}$ 37. $w = -1 \pm \sqrt{7}$

39. $y = 2, -\frac{1}{2}$ 41. $x = 1.62, -0.62$

43. $t = 8.22, -1.22$

45. $w = 1.72, -4.06$

47. $x = 3.06, -1.17$ 49. 380 sq ft

51. 0.7 sec and 1.8 sec

Mini-Review

56. $-8\sqrt{3} - 11\sqrt{5}$ 57. $8\sqrt{14} - 1$

58. $x - 6\sqrt{x} + 9$

59. $6x - 11\sqrt{5x} + 20$

60. Let $x = $ cost per lb for pistachios;
$y = $ cost per lb of cashews;
$2x + 3y = 23, 3x + 2y = 22$;

$\boxed{\begin{array}{l}\text{\$4 per lb for pistachios,} \\ \text{\$5 per lb for cashews}\end{array}}$

61. Let $p = $ cost of pastrami
sandwich; $t = $ cost of tuna
sandwich; $5p + 6t = 42$,
$4p + 9t = 47.25$;

$\boxed{\begin{array}{l}\text{\$4.50 for a pastrami sandwich,} \\ \text{\$3.25 for a tuna sandwich}\end{array}}$

EXERCISES 11.4

1. $x = -5, -1$ 3. $x = -3 \pm \sqrt{14}$

5. $r = \frac{1}{2}, 1$ 7. $w = 5, -1$

9. $x = -\frac{5}{2}$ 11. $a = 1$

13. No real solutions

15. $x = 1 \pm \sqrt{5}$ 17. $x = \frac{7 \pm 3\sqrt{5}}{2}$

19. $u = \pm 4$ 21. $y = 0, 9$

23. $t = 9, -5$ 25. $n = \frac{-3 \pm \sqrt{13}}{2}$

27. $z = 3$ 29. $z = 6, -\frac{2}{3}$

31. $x = 1$ 33. $x = 2, \frac{1}{2}$

35. $x = \pm 2$ 37. Contradiction

39. $x = -3, 2$

Mini-Review

43. (a) $-\frac{7}{4}$ (b) 0 (c) Undefined

44.

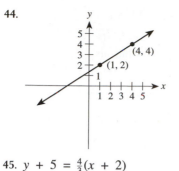

45. $y + 5 = \frac{4}{3}(x + 2)$

EXERCISES 11.5

The equation used to solve each odd-numbered exercise is given.

1. n = the number; $n + \frac{1}{n} = \frac{13}{6}$;

$$\boxed{n = \frac{2}{3} \text{ or } n = \frac{3}{2}}$$

3. x = one number, $20 - x$ = the other number; $x(20 - x) = 96$;

$\boxed{8 \text{ and } 12}$

5. W = width, $2W + 3$ = length; $W(2W + 3) = 90$; $\boxed{6 \text{ by } 15}$

7. x = length of the leg; $x^2 + 7^2 = 15^2$; $\boxed{4\sqrt{11} \text{ cm}}$

9. d = length of diagonal; $8^2 + 8^2 = d^2$; $\boxed{8\sqrt{2} \text{ in.}}$

11. h = height; $h^2 + 8^2 = 30^2$; $\boxed{2\sqrt{209} \text{ ft} \approx 28.9 \text{ ft}}$

13. Let the integers be $x, x + 1$, $x + 2$; $x^2 + (x + 1)^2 = (x + 2)^2$; $\boxed{3, 4, 5}$

15. Let the fraction be $\frac{x}{x + 1}$; $\frac{x + 3}{x + 1} = \frac{x}{x + 1} + 1$; $\boxed{\frac{2}{3}}$

17. x = # of seats per row, $x - 8$ = # of rows; $x(x - 8) = 768$;

$\boxed{32 \text{ seats per row}}$

19. The rates are r and $r - 20$; $\frac{120}{r} + \frac{30}{r - 20} = 2$; $\boxed{80 \text{ kph}}$

21. \$15,000; \$9,000   **23.** 9.9 meters

25. Let r = Yomar's speed; $r + 12$ = Neva's speed; $\frac{180}{r} - \frac{180}{r + 12} = \frac{1}{2}$;

$\boxed{72 \text{ mph}}$

27. (a) 804 ft
   (b) Approximately 4.3 seconds
   (c) Approximately 7.9 seconds

29. (a) \$119.92   (b) 440 TV sets
31. 127 ft
33. Let w = width of path; $(15 + 2w)(10 + w) - 150 = 100$;

$\boxed{2.5 \text{ ft}}$

35. Let x = # of days for father to paint alone; $x + 9$ = # of days for son to paint alone;

$\frac{6}{x} + \frac{6}{x + 9} = 1$;

$\boxed{\text{Father: 9 days; son: 18 days}}$

EXERCISES 11.6

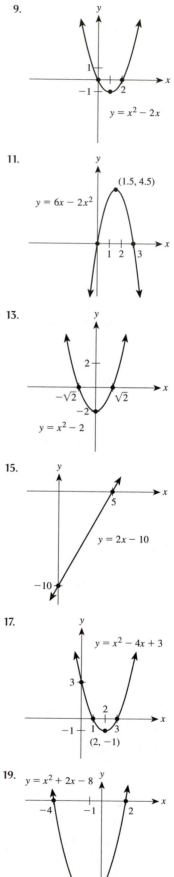

21.

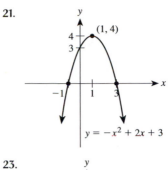

$y = -x^2 + 2x + 3$

23.

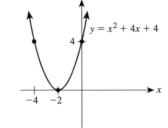

$y = x^2 + 4x + 4$

25.

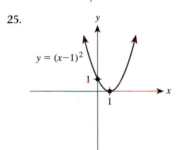

$y = (x-1)^2$

27.

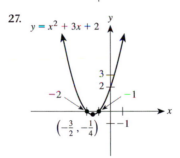

$y = x^2 + 3x + 2$

$\left(-\frac{3}{2}, -\frac{1}{4}\right)$

29.

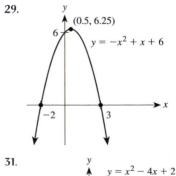

$(0.5, 6.25)$

$y = -x^2 + x + 6$

31.

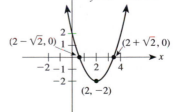

$y = x^2 - 4x + 2$

$(2 - \sqrt{2}, 0)$ $(2 + \sqrt{2}, 0)$

$(2, -2)$

33.

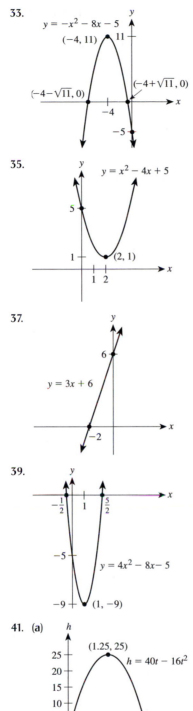

$y = -x^2 - 8x - 5$

$(-4, 11)$

$(-4 - \sqrt{11}, 0)$ $(-4 + \sqrt{11}, 0)$

35.

$y = x^2 - 4x + 5$

$(2, 1)$

37.

$y = 3x + 6$

39.

$y = 4x^2 - 8x - 5$

$(1, -9)$

41. (a)

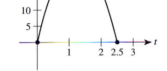

$(1.25, 25)$

$h = 40t - 16t^2$

(b) 25 feet   **(c)** 1.25 seconds

43. (a) $(3, 6)$ **(b)** $x = 3$
   (c) For $x \le 3$ **(d)** For $x \ge 3$
45. (a) $(-4, 0)$ **(b)** $x = -4$
   (c) For $x \ge -4$ **(d)** For $x \le -4$

Mini-Review

51. $x^2 - 10x + 25$
52. $x^2 + 16x + 64$
53. $x^2 - 25$ **54.** $x^2 - 64$
55. $x^2 + 2px + p^2$
56. $x^2 - 2hx + h^2$
57. Let $t = $ # of hours assistant works, $t - 1 = $ # of hours handyman works;
$11t + 19(t - 1) = 146$;

$\boxed{3{:}30 \text{ P.M.}}$

Chapter 11 Review Exercises

1. $x = \frac{7 \pm \sqrt{73}}{2}$ **2.** $x = 12, -2$
3. No real solutions   **4.** $x = -5, 1$
5. $u = 6 \pm \sqrt{13}$
6. No real solutions   **7.** $y = \frac{3}{2}, -5$
8. $y = \frac{3 \pm \sqrt{17}}{4}$ **9.** $x = \frac{1}{3}, 1$
10. $x = \frac{1 \pm \sqrt{21}}{5}$ **11.** $x = \frac{51}{10}$
12. $x = 5 \pm 2\sqrt{5}$ **13.** $u = \frac{2}{3}, \frac{3}{2}$
14. $z = 6, -3$ **15.** $x = \frac{\pm 3\sqrt{2}}{2}$
16. $w = \frac{2 \pm \sqrt{10}}{2}$ **17.** $x = 0$
18. $x = 2, \frac{9}{7}$ **19.** $x = \frac{7}{2} \pm \frac{\sqrt{37}}{2}$
20. $x = 2 \pm \sqrt{6}$

21.

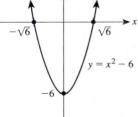

$y = x^2 - 6x$

$(3, -9)$

22.

$-\sqrt{6}$ $\sqrt{6}$

$y = x^2 - 6$

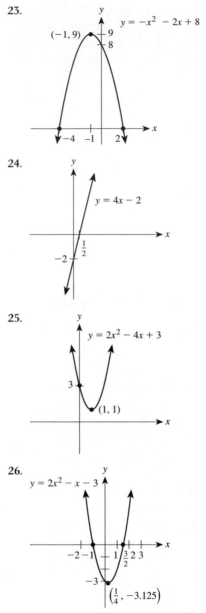

23. $y = -x^2 - 2x + 8$

24. $y = 4x - 2$

25. $y = 2x^2 - 4x + 3$

26. $y = 2x^2 - x - 3$; $\left(\frac{1}{4}, -3.125\right)$

Chapter 11 Practice Test

1. $x = -7, 4$ 2. $x = \frac{3 \pm \sqrt{33}}{4}$

3. $x = 7, -2$ 4. $x = -5 \pm \sqrt{10}$

5. No real solutions 6. $x = \frac{2}{3}, -5$

7. $x = 3 \pm \sqrt{6}$

8. $x = 4$ 9. $x = 2 \pm \frac{\sqrt{57}}{3}$

10. W = width, $2W + 7$ = length;
$W(2W + 7) = 30$;
$\boxed{\frac{5}{2} \text{ cm by } 12 \text{ cm}}$

CHAPTER 10–11 CUMULATIVE REVIEW

1. $6x^8 y^6$ (§10.2)
2. $3x^3 y^2 \sqrt{x}$ (§10.2)
3. $\frac{7\sqrt{6}}{6}$ (§10.2) 4. $\frac{9(\sqrt{6} + 2)}{2}$ (§10.4)
5. $5(3 + \sqrt{5})$ (§10.4)
6. $3\sqrt{5}$ (§10.2)
7. $2\sqrt{30}$ (§10.2)
8. $3xy^4 \sqrt{2x}$ (§10.2)
9. $\sqrt{5t}$ (§10.3)
10. $8\sqrt{6} - 16\sqrt{3}$ (§10.4)
11. $\frac{8\sqrt{21}}{7}$ (§10.3)
12. $2x + 7\sqrt{x} - 15$ (§10.4)
13. $x^3 y^2 \sqrt{3y}$ (§10.3)
14. $\frac{3\sqrt{5}}{2}$ (§10.3)
15. $64 - 22\sqrt{10}$ (§10.4)
16. $2x + 6\sqrt{x} + 12$ (§10.4)
17. $3\sqrt{7} - 2\sqrt{2}$ (§10.4)
18. $x = 0$ or $x = 3$ (§11.1)
19. $x = 5$ or $x = -2$ (§11.1)
20. $x = 3$ or $x = -2$ (§11.1)
21. $x = 1$ (§11.1)
22. $x = \frac{-3 \pm \sqrt{29}}{2}$ (§11.4)
23. $x = \frac{1}{2}$ or $x = 2$ (§11.5)
24. $x = -3$ or $x = 4$ (§11.1)
25. $x = -4$ or $x = 5$ (§11.1)
26. $x = \frac{-11 \pm \sqrt{97}}{4}$ (§11.4)
27. $x = \frac{5}{3}$ (§11.1)
28. $a = -8$ or $a = -2$ (§11.2)
29. $s = 2 \pm \sqrt{10}$ (§11.2)
30. $x = 19$ (§10.5)
31. $x = 49$ (§10.5)
32. No solutions (§10.5)
33. $x = \frac{28}{9}$ (§10.5)
34. $x = 3 \pm \sqrt{2}$ (§11.3)
35. $x = \frac{5}{2} \pm \frac{\sqrt{13}}{2}$ (§11.3)

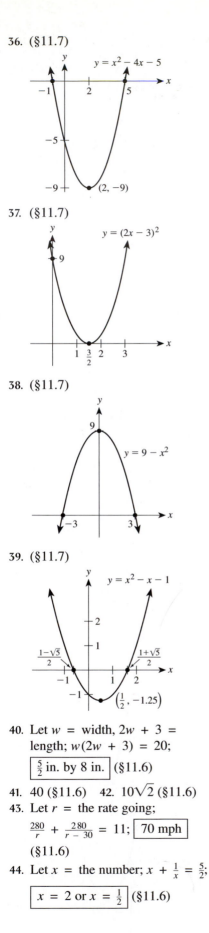

11. (3, 9); (0, 0); (6, 0); $y = 6x - x^2$

36. (§11.7) $y = x^2 - 4x - 5$; (2, −9)

37. (§11.7) $y = (2x - 3)^2$

38. (§11.7) $y = 9 - x^2$

39. (§11.7) $y = x^2 - x - 1$; $\frac{1 - \sqrt{5}}{2}$; $\frac{1 + \sqrt{5}}{2}$; $\left(\frac{1}{2}, -1.25\right)$

40. Let w = width, $2w + 3$ = length; $w(2w + 3) = 20$; $\boxed{\frac{5}{2} \text{ in. by } 8 \text{ in.}}$ (§11.6)

41. 40 (§11.6) 42. $10\sqrt{2}$ (§11.6)

43. Let r = the rate going;
$\frac{280}{r} + \frac{280}{r - 30} = 11$; $\boxed{70 \text{ mph}}$
(§11.6)

44. Let x = the number; $x + \frac{1}{x} = \frac{5}{2}$;
$\boxed{x = 2 \text{ or } x = \frac{1}{2}}$ (§11.6)

CHAPTERS 10–11 CUMULATIVE PRACTICE TEST

1. $6\sqrt{2}$ (§10.2)

2. $4xy^2\sqrt{3x}$ (§10.2)

3. $-6\sqrt{5}$ (§10.3)

4. $x^2y^3\sqrt{3x}$ (§10.3)

5. $\frac{8\sqrt{5}}{5}$ (§10.2)   6. $6y^{18}$ (§10.2)

7. $\frac{32 + 8\sqrt{7}}{3}$ (§10.4)

8. $37 - 20\sqrt{3}$ (§10.4)

9. $10t - 18\sqrt{3t} - 12$ (§10.4)

10. 9 (§10.4)

11. $x = 6$ or $x = -4$ (§11.1)

12. $x = 2 \pm \sqrt{7}$ (§11.2)

13. $x = \frac{20}{3}$ (§11.1)

14. $x = 7$ (§10.5)

15. $x = \frac{5 \pm \sqrt{17}}{4}$ (§11.4)

16. $t = \pm\frac{\sqrt{30}}{3}$ (§11.2)

17. $x = -\frac{25}{32}$ (§11.1)

18. $x = 5 \pm 2\sqrt{7}$ (§11.3)

19. $25\sqrt{3}$ sq in. (§11.6)

20. Let r = rate of the slower truck; $r + 10$ = rate of the faster truck; $\frac{400}{r} = \frac{400}{r + 10} + 2$;

   $\boxed{\text{40 mph and 50 mph}}$ (§11.6)

21. (§11.7)

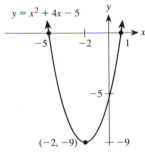

22. (§11.7)

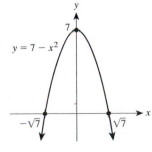

EXERCISE G.1

1. $\angle DEB$ and $\angle BEA$; $\angle EAD$ and $\angle DAB$

3. (a) 60° (b) 30° (c) 45°
   (d) 72° (e) 1°

5. 128° 7. 64° 9. $x = 20$

11. $x = 4$

13. $\angle 1 = 145°$, $\angle 2 = 35°$, $\angle 3 = 145°$

15. $\angle 1 = 65°$, $\angle 2 = 115°$, $\angle 3 = 65°$

17. $\angle 1 = \angle 3 = \angle 6 = \angle 8 = 55°$; $\angle 2 = \angle 5 = \angle 7 = 125°$

19. $x = 56$

21. $\angle 1 = \angle 4 = \angle 5 = \angle 7 = \angle 9 = \angle 11 = \angle 14 = \angle 16 = 140°$; $\angle 3 = \angle 6 = \angle 8 = \angle 10 = \angle 12 = \angle 13 = \angle 15 = 40°$

23. $\angle 1 = 50°$; $\angle 2 = 130°$; $\angle 3 = 50°$; $\angle 4 = 130°$

EXERCISES G.2

1. 123° 3. 70°, 70° 5. $x = 28$

7. $x = 28$ 9. $x = 67$ 11. $x = 26$

13. $|\overline{AB}| = 5$ 15. $|\overline{AC}| = \sqrt{65}$

17. $|\overline{AC}| = |\overline{BC}| = 4\sqrt{2}$

19. $|\overline{AC}| = 4\sqrt{5}$

21. $20\sqrt{2} \approx 28.28$ feet

23. 240.8 miles

25. 113.2 nautical miles

27. No 29. $x = 27$ 31. $x = 115$

33. $x = 70$ 35. $x = 40$

EXERCISES G.3

1. SSS 3. ASA 5. ASA 7. SAS

9. SAS 11. SAS 13. ASA

EXERCISES G.4

1. The corresponding pairs of angles are: A and D, B and E, C and F

3. The corresponding pairs of angles are: A and EDB, ACB and DEB, ABC and DBE

5. The corresponding pairs of angles are: A and D, B and E, C and F

7. $|\overline{DE}| = 24$, $|\overline{EF}| = 12$

9. $|\overline{DE}| = 7.5$, $|\overline{DF}| = 12.5$

11. $|\overline{AC}| = 20$ 13. $|\overline{EA}| = 10.8$

15. $|\overline{DE}| = 12$

17. $|\overline{AB}| = \sqrt{61}$, $|\overline{CD}| = 8\frac{1}{3}$,
    $|\overline{DE}| = \frac{5\sqrt{61}}{3}$

19. 27 ft 21. 10.2 cm 23. 2.9 ft

25. 0.3 cm; 0.1 cm

27. $|\overline{AC}| = 6$, $|\overline{BC}| = 6\sqrt{2}$

29. $|\overline{AB}| = 12\sqrt{3}$, $|\overline{BC}| = 24$

31. $|\overline{AB}| = |\overline{BC}| = \frac{20}{\sqrt{2}} = 10\sqrt{2}$

33. $|\overline{AB}| = 4\sqrt{3}$; $|\overline{BC}| = 8\sqrt{3}$

EXERCISES G.5

1. 80°

3. $B = 120°$, $C = 60°$, $D = 120°$

5. $x = 5$ 7. $x = 18$

9. $\sqrt{89}$ cm 11. $4\sqrt{2}$ in.

13. $\frac{5\sqrt{2}}{2}$ cm 21. $\sqrt{117} = 3\sqrt{13}$

23. $x = 5$ 25. $x = \sqrt{89}$

27. $x = \sqrt{85}$ 29. $x = \sqrt{37}$

EXERCISES G.6

1. $P = 26$ ft, $A = 40$ sq ft

3. $P = 36$ in., $A = 72$ sq in.

5. $P = 12$ in., $A = 9$ sq in.

7. $A = 48$ 9. $A = 60$

11. $A = 24$ 13. $A = 24$

15. $A = 12$ 17. $A = 51$

19. $P = 56$, $A = 180$

21. $P = 13 + \sqrt{89}$, $A = 20$

23. $P = 16 + 4\sqrt{2}$, $A = 14\sqrt{2}$

25. $P = 16 + 8\sqrt{5}$ cm, $A = 32\sqrt{5}$ sq cm

27. $P = 20\sqrt{2}$ mm, $A = 50$ sq mm

29. $P = 22$, $A = 26$

31. $P = 56$, $A = 108$

33. $A = 54\sqrt{2} - 18$ 35. $A = 37$

37. $A = 132$ 39. $A = 30$

41. $P = 147$ in.; $A = 945$ sq in.

43. $P = 128$ in.; $A = 5,120$ sq in.

45. 1,050 sq ft

EXERCISES G.7

1. 35° 3. 130° 5. 6π in.

7. 8π ft 9. $\frac{3\pi}{2}$ in.

11. $P = \frac{13\pi}{6} + 12$; $A = \frac{13\pi}{2}$

13. $P = \frac{5\pi}{18} + 10$; $A = \frac{25}{36}\pi$

15. 9π sq in. 17. 36π sq in.

19. $P = 20 + 10\pi$ in., $A = 50\pi$ sq in.

21. $\frac{50\pi}{9}$ sq in.

23. $P = 22 + 3\pi$, $A = 48 + \frac{9\pi}{2}$

25. $64 - 16\pi$ cm 27. 80π

29. $14 + 5\pi$

31. $P = 70 + 5\pi$ cm; $A = 300 + 12.5\pi$ sq cm

33. One 12-in. pie; they are both the same value.

662 ANSWERS TO SELECTED EXERCISES

EXERCISES G.8

1. $A = 396$ cm^2; $V = 504$ cm^3
3. $A = 40\pi$ ft^2; $V = 32\pi$ ft^3
5. $A = 24\pi$ m^2; $V = 12\pi$ m^3
7. $A = 275.6\pi$ ft^2; $V = 762.4\pi$ ft^3
9. $A = 113.1$ ft^2; $V = 113.1$ ft^3
11. $A = 226.2$ cm^2; $V = 251.3$ cm^3
13. $A = 134.6$ m^2; $V = 64.8$ m^3
15. $A = 502.7$ ft^2; $V = 599.5$ ft^3
17. $A = 4{,}712.4$ mm^2;
    $V = 29{,}452.4$ mm^3
19. $V = 144\pi$ m^3 21. $V = \frac{500}{3}\pi$ m^3
23. $A = 60\pi$ in^2 25. 15 cubic yards
27. $377.75   29. 859   31. $35.60
33. 790.3 cm^2 35. 11,196.6 ft^3
37. 360π cm^3

Chapter G Review Exercises

1. $\angle 1 = \angle 4 = 40°$;
   $\angle 2 = \angle 3 = 50°$
2. $x = 20, y = 110$ 3. $x = 65$
4. $x = 93$ 5. $x = 40$ 6. $x = 80$
7. $x = 115$ 8. $x = 20$
9. $|\overline{AC}| = 12$
10. $|\overline{AC}| = |\overline{BC}| = \frac{10}{\sqrt{2}} = 5\sqrt{2}$
11. $|\overline{AB}| = 5, |\overline{BD}| = 8, |\overline{BE}| = 10$
12. $|\overline{AB}| = 7.5, |\overline{EC}| = 6$
13. $x = 36$ 14. $x = 8, y = 4\sqrt{3}$
15. $a = 8, b = 8\sqrt{2}$
16. $a = \frac{10}{\sqrt{3}} = \frac{10\sqrt{3}}{3}, b = \frac{20}{\sqrt{3}} = \frac{20\sqrt{3}}{3}$
17. $x = 5$ 18. $x = 13$
19. $P = 24\sqrt{2}$ in.; $A = 72$ sq in.
20. $P = 28$ cm; $A = 48$ sq cm
21. $P = 14 + 2\sqrt{14}$ cm;
    $A = 5\sqrt{14}$ sq cm
22. $P = 30$ in.; $A = 36$ sq in.
23. $P = 32$; $A = 50$
24. $P = 56$; $A = 84$
25. $P = 44$; $A = 104$
26. $P = 20 + 10\sqrt{2}$; $A = 50$
27. $P = 36$; $A = 36\sqrt{3}$
28. 102   29. 160   30. 345   31. 132
32. (a) 40 (b) 30 (c) trapezoid
    (d) 110
33. (a) 15 (b) 49 (c) 72 (d) 56
34. (a) 22.5 (b) 24.5 (c) 24 (d) 150
    (e) 79
35. (a) 24 (b) 67.5 (c) 127.5 (d) 300
    (e) 81
36. (a) 64 (b) 48 (c) 112 (d) 96
37. (a) 48 (b) 108 (c) 12 (d) 120
38. $C = 20\pi$ in.; $A = 100\pi$ sq in.

39. $C = 32\pi$ in.; $A = 256\pi$ sq in.
40. 6π 41. 48π 42. $36 + 6\pi$
43. $16\pi - 32$ 44. $64 - 16\pi$
45. $68 + 12\pi$
46. $P = 60 + 10\pi$; $A = 400 - 50\pi$
47. $144 - 36\pi$ 48. $216 + 8\pi$
49. 10π 50. $216 + 18\pi$
51. $S = 158$ cm^2; $V = 120$ cm^3
52. $S = 256\pi$ in.^{2}; $V = \frac{2048}{3}\pi$ in.^{3}
53. $S = 78\pi$ ft^2; $V = 90\pi$ ft^3
54. $S = 24\pi$ m^2; $V = 12\pi$ m^3

Chapter G Practice Test

1. $x = 140$ 2. $x = 65$
3. $\angle 1 = \angle 2 = 53°$; $\angle 3 = 37°$;
   $|\overline{AC}| = 12$
4. $P = 60$; $A = 120$ 5. $9\sqrt{3}$
6. $24 + 4\pi$ 7. $216 + 9\pi$
8. 80   9. $\frac{25}{2}\pi - 24$ 10. $260

EXERCISES A.1

1. $\frac{2}{3}, \frac{12}{18}$ 3. $\frac{2}{5}, \frac{28}{70}$ 5. $\frac{3}{4}, \frac{6}{8}$ 7. $\frac{10}{2}, \frac{15}{3}$
9. $\frac{2}{2}, \frac{3}{3}$ 11. $\frac{4}{5}$ 13. $\frac{1}{7}$ 15. $\frac{1}{2}$
17. Can't be reduced
19. $\frac{7}{18}$ 21. 6   23. $\frac{11}{4}$
25. Can't be reduced
27. 20   29. 21   31. 64   33. 6
35. $\frac{5}{6}$ on the computer;
    $\frac{1}{6}$ on the software
37. $\frac{9}{10}$ hit the bull's-eye; $\frac{1}{10}$ missed
39. $\frac{6}{7}$ correct; $\frac{1}{7}$ incorrect
41. $\frac{9}{20}$ cheese; $\frac{3}{10}$ blueberry; $\frac{1}{4}$ cherry
43. $\frac{2}{15}$ LPs; $\frac{2}{5}$ CDs; $\frac{7}{15}$ cassettes

EXERCISES A.2

1. $\frac{10}{21}$ 3. $\frac{1}{36}$ 5. $\frac{2}{3}$ 7. $\frac{2}{9}$ 9. $\frac{25}{8}$
11. $\frac{4}{5}$ 13. $\frac{125}{144}$ 15. 1   17. $\frac{9}{25}$ 19. 27
21. 12   23. $\frac{1}{12}$ 25. $\frac{49}{50}$ 27. $\frac{8}{7}$ 29. 1
31. $\frac{9}{5}$ 33. $\frac{7}{4}$ 35. $18,000
37. 14 pieces   39. 45 hours
41. 1,320 tablets

EXERCISES A.3

1. $\frac{5}{7}$ 3. $\frac{3}{8}$ 5. $\frac{1}{2}$ 7. 1   9. $\frac{3}{4}$ 11. $\frac{11}{24}$
13. $\frac{5}{16}$ 15. $\frac{11}{12}$ 17. $\frac{13}{12}$ 19. $\frac{37}{60}$ 21. $\frac{23}{140}$

23. $\frac{28}{5}$ 25. $\frac{25}{3}$ 27. $\frac{15}{4}$ 29. $\frac{64}{5}$
31. $4\frac{3}{5}$ 33. $7\frac{1}{6}$ 35. $\frac{2}{3}$ 37. $\frac{5}{12}$
39. $10\frac{1}{6}$ cubic yards   41. $10\frac{2}{3}$ gallons
43. 63 acres   45. $6\frac{1}{9}$ mph

EXERCISES A.4

1. 29.9   3. 9.52   5. 82.956
7. 8.583   9. 1.42   11. 22.62
13. 5.97   15. 34.3215   17. 1.9092
19. 7.92   21. 420   23. 20.5
25. 0.0802   27. 0.0016   29. 7
31. 0.000006   33. 260   35. $99
37. 850 kilowatt-hours
39. No; owner gets 21.5 mpg
41. $660.21

EXERCISES A.5

1. 0.25   3. 78%   5. 5%   7. 0.09
9. 1.5   11. 0.28   13. 67%
15. 200%   17. 1.37   19. 0.7%
21. 0.624   23. 0.086   25. 21
27. 2.52   29. 0.04   31. 25%
33. 62.5%   35. 40%   37. 150
39. 82   41. $1.20   43. 13,250
45. 5%   47. In 1988

APPENDIX B EXERCISES

1. 861.55   3. 68.43   5. 65.5
7. 308.56   9. 4,847.04   11. 42.01
13. 30.45   15. 27.1   17. 27.62
19. 0.15   21. 0.06   23. $97.37
25. 11,160,000 miles   27. $15.46

APPENDIX D EXERCISES

1. Domain = $\{A, B, C, D\}$;
   range = $\{2, 3, 8, 11\}$, yes
3. Domain = $\{F, G, 3\}$;
   range = $\{L, M, 2, 8\}$, no
5. Domain = $\{1, 3, 4, 9\}$;
   range = $\{5\}$, yes
7. Yes; domain is all real numbers
9. Yes; domain is all real numbers
11. No   13. Domain = $\{x \mid x \neq 3\}$
15. Yes; domain is all real numbers
17. Yes; domain is all real numbers
19. 1,421 ft; 221 ft   21. $178,250
23. $60,594,400
25. (a) $h = \frac{80}{b}$ (b) $h = 6\frac{2}{3}$
27. (a) $C = 75 + 0.06p$
    (b) $C = $127.50
29. (a) $P = 6w - 6$ (b) $P = 48$ in.
    (c) $A = 2w^2 - 3w$
    (d) $A = 135$ in.^{2}

Index

Geometric Formulas

Rectangle

The length is L; the width is W.

The perimeter $P = 2L + 2W$.

The area $A = LW$.

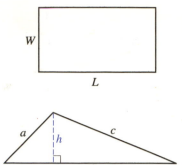

Triangle

The sides are a, b, and c, and the height (or altitude) to the base b is h.

The perimeter $P = a + b + c$.

The area $A = \frac{1}{2}bh$.

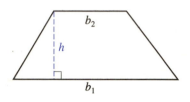

Trapezoid

The two parallel sides are b_1 and b_2, and the height is h.

The area $A = \frac{1}{2}h(b_1 + b_2)$.

Circle

The radius is r and the diameter is d. (π is approximately equal to 3.14.)

The circumference $C = 2\pi r$ or $C = \pi d$.

The area $A = \pi r^2$.

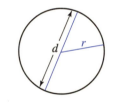

Other Formulas

Temperature

F = degrees Fahrenheit; C = degrees Celsius

$$F = \frac{9}{5}C + 32$$

$$C = \frac{5}{9}(F - 32)$$

Simple Interest

I is the interest earned on a principal (amount of money) P invested at $r\%$ per year (r is written as a decimal) for t years:

$$I = Prt$$